ASTRONOMICAL PAPYRI FROM OXYRHYNCHUS

(P. Oxy. 4133—4300a)

EDITED WITH TRANSLATIONS AND COMMENTARIES BY

ALEXANDER JONES

VOLUMES I AND II

American Philosophical Society
Independence Square • Philadelphia
1999

Memoirs
of the
American Philosophical Society
Held at Philadelphia
for Promoting Useful Knowledge
Volume 233

ISBN:0-87169-233-3
US ISSN:0065-9738

Library of Congress Cataloging-in-Publication Data

Jones, Alexander.
Astronomical papyri from Oxyrhynchus / by Alexander Jones.
p. cm. -- (Memoirs of the American Philosophical Society, ISSN 0065-9738 ; v. 233)
Includes bibliographical references.
ISBN 0-87169-233-3 (cloth)
1. Astronomy, Greek. 2. Manuscripts, Greek--Egypt--Bahnasā. 3. Manuscripts, Greek (Papyri) 4. Oxyrhynchus papyri. I. Title. II. Series.
Q11.P612 vol. 233
[QB21]
520'.938--dc21 99-22168
CIP

For Catherine

Reference to Papyri

To reflect the common provenance of the papyri published here and those published as *The Oxyrhynchus Papyri*, the editors of the Graeco-Roman Memoirs of the Egypt Exploration Society have assigned numbers 4133 through 4300a in the main *P. Oxy.* sequence to these papyri, and a list of their contents appears (without texts) in *The Oxyrhynchus Papyri* Part LXI. For purposes of citation, the papyri should be identified by that volume, e.g. as *P. Oxy.* LXI 4236, but with reference to the present publication. Within this work, heavy arabic numerals (e.g. **4236**) refer only to the papyri here printed. The abbreviations for other Greek papyri in published series, which in the main follow those in Oates *et al.* [1992], are explained in the bibliography. Bibliographical information concerning editions of astronomical papyri will be found in Appendix B.

TABLE OF CONTENTS

Volume I

TABLE OF PAPYRI vii

LIST OF PLATES xii

NUMBERS AND PLATES xii

INTRODUCTION
- 1. Preface 1
- 2. The Purpose and Provenance of the Papyri 4
- 3. Elementary Astronomical and Astrological Concepts 8
- 4. Calendars and Chronological Data 11
- 5. Concepts and Methods of Greek Predictive Astronomy 15
- 6. Varieties of Astronomical Papyri 35
- 7. Methods of Analysis and Dating 47
- 8. Documentary Aspects of the Texts 56
- 9. Editorial Principles and Notations 64

COMMENTARIES
- I. Theoretical and Instructional Texts 65
- II. Primary Tables 111
- III. Ephemerides and Almanacs 173
- IV. Miscellaneous Tables 229
- V. Horoscopes 247

APPENDICES
- A. Notations and Symbols 299
- B. Notes on Published Astronomical Papyri 301
- C. Published Horoscopes from Oxyrhynchus 308
- D. Regnal Canon 310
- E. Tables for Calendrical Conversions 312
- F. Conversion of Times 318
- G. Determination of Ascendant 320
- H. The Standard Lunar Scheme 321
- I. Conversion of Longitudes 343
- J. Zodiacal Signs of Saturn and Jupiter 344
- K. Solar Longitudes and Calendar Dates 349
- L. Astrological Houses, Terms, and Attributes 351

BIBLIOGRAPHICAL ABBREVIATIONS 353

SUBJECT INDEX 364

Volume II

TEXTS AND TRANSLATIONS
I. Theoretical and Instructional Texts 1
II. Primary Tables 47
III. Ephemerides and Almanacs 169
IV. Miscellaneous Tables 319
V. Horoscopes 371

INDEXES
I. Rulers and Regnal Years 451
II. Indictions 453
III. Eras 453
IV. Months and Days 454
V. Personal Names 455
VI. Geographical 455
VII. Authorities Cited 455
VIII. Astronomical and Astrological 455
IX. General Index of Words 460
X. Concordance of Publication and Inventory Numbers 467

PLATES I–XII

TABLE OF PAPYRI

		Vol. I	Vol. II
I. THEORETICAL AND INSTRUCTIONAL TEXTS			
4133–4147	General comment	67	
4133	Menelaus of Alexandria (?), Treatise on planetary theory	69	2
4134	Procedure text for planet?	80	6
4135	Procedure text for Venus	81	10
4136	Procedure text for the moon	84	14
4137	Predictions of lunar eclipses, A.D. 56 and 57	87	16
4138	On eclipse prediction	94	18
4138a	Treatise on eclipse prediction	95	20
4139	Treatise on lunar periods	97	22
4140	Tabulated lunar phenomena	100	24
4141	On latitudes of a planet	101	24
4142	Instructions for Ptolemy's *Handy Tables*	102	26
4143	Instructions for Ptolemy's *Handy Tables*	105	36
4144	Treatise on kinematic models	108	40
4145	Procedure text	109	42
4146	Procedure text	109	44
4147	Procedure text	110	45
II. PRIMARY TABLES			
4148–4174a	General comment	113	
4148	Epochs of the sun, A.D. 161–237	119	48
4149	Epochs of the moon, A.D. 96-166	121	54
4150	Epochs of the moon, A.D. 187–198, and template	121	58
4151	Epochs of the moon, A.D. 217–254	122	62
4152	Epochs of Mercury	123	66
4153	Epochs of Mercury, A.D. 206–215	125	68
4154	Epochs of Mercury	126	70
4155	Epochs of Mercury	128	72
4156a	Epochs of Mercury	133	76
4156b	Epochs of Mercury	134	78
4156c	Epochs of Mercury	134	78
4157	Epochs of Venus	135	80
4157a	Epochs of Venus, A.D. 106–121	137	80
4158	Epochs of Mars, A.D. 271–325	137	82
4159	Epochs of Mars, A.D. 225–235	140	84
4159a	Epochs of Mars, A.D. 101–133	141	86
4160	Epochs of Jupiter, A.D. 30–79	145	88

		Vol. I	Vol. II
4160a	Epochs of Jupiter, A.D. 6–13	148	90
4161	Epochs of Saturn, A.D. 23–92	148	92
4162	Template for the sun	150	94
4163	Template for the sun, and procedure text	151	100
4164	Template for the moon	153	104
4164a	Template for the moon	154	108
4165	Template for Mars	155	110
4165a	Template for Jupiter	157	112
4166	Template for Saturn with jottings	158	114
4167	Ptolemy, *Handy Tables*	160	118
4168	Ptolemy, *Handy Tables*	162	130
4169	Ptolemy, *Handy Tables*	164	142
4170	Ptolemy, *Handy Tables*	164	144
4171	Ptolemy, *Handy Tables*	165	148
4172	Handy Tables	165	150
4173	Handy Tables	166	152
4173a	Handy Tables	168	156
4174	Table of lunar mean motions	169	162
4174a	Table of solar mean motions	170	164

III. EPHEMERIDES AND ALMANACS

		Vol. I	Vol. II
4175–4215	General comment	175	
4175	Almanac–ephemeris, 24 B.C.	177	170
4176	Almanac–ephemeris, A.D. 111	179	174
4177	Almanac–ephemeris, A.D. 121-127	180	176
4177a	Ephemeris, A.D. 245	185	184
4178	Ephemeris, A.D. 261	186	186
4179	Ephemeris, A.D. 348	186	188
4180	Ephemeris, A.D. 465	190	192
4181	Ephemeris, A.D. 161	191	200
4182	Ephemeris	193	204
4183	Ephemeris	194	206
4184	Ephemeris	194	208
4184a	Ephemeris, A.D. 293	194	208
4185	Sign-entry almanac, A.D. 129–133	195	210
4186	Sign-entry almanac, A.D. 142–146	197	212
4187	Sign-entry almanac, A.D. 160/161	197	214
4188	Sign-entry almanac, A.D. 161–165	199	216
4188a	Sign-entry almanac, A.D. 201–208	202	220
4189	Sign-entry almanac, A.D. 136–144 (?)	202	224
4190	Sign-entry almanac, A.D. 241–243	204	228
4191	Sign-entry almanac, A.D. 236-243	206	230
4192	Sign-entry almanac, A.D. 276–280	206	236
4193	Sign-entry almanac, A.D. 195–203	207	238
4194	Sign-entry almanac, A.D. 282–286	207	246
4195	Sign-entry almanac, A.D. 300–304	213	250

		Vol. I	Vol. II
4196	Sign-entry almanac, A.D. 218–220	213	252
4196a	Sign-entry almanac	213	254
4197	Perpetual sign-entry almanac	214	256
4198	Perpetual sign-entry almanac	215	264
4199	Monthly almanac, Jupiter, 14 B.C.–6 B.C.	215	268
4200	Monthly almanac, Jupiter	216	274
4201	Monthly almanac, A.D. 126–128	216	276
4202	Monthly almanac, A.D. 152–155	217	278
4203	Monthly almanac, Saturn, A.D. 215/216	217	284
4203a	Monthly almanac, Mercury	218	286
4204	Almanac, A.D. 54	218	288
4205	5-day almanac, A.D. 257/258	219	294
4205a	5-day almanac	220	296
4206	5-day almanac, A.D. 272–274	220	298
4207	5-day almanac, A.D. 289–291	220	302
4208	5-day almanac, A.D. 295/296	222	304
4209	5-day almanac, A.D. 295/296	223	306
4210	5-day almanac, A.D. 300–302	224	308
4211	5-day almanac, A.D. 306/307	225	312
4212	5-day almanac, A.D. 217/218	226	314
4213	5-day almanac, A.D. 301/302	227	314
4214	Syzygies	227	316
4215	Mean motions of the moon, A.D. 287/288	228	316

IV. MISCELLANEOUS TABLES

		Vol. I	Vol. II
4216–4235	General comment	231	
4216	Arithmetical table: thirtieths of 40;33	231	320
4217	Arithmetical table: templates for Mercury?	232	322
4217a–c	Arithmetical table: templates for Mercury?	233	324
4218	Arithmetical table: fractions of 30	235	326
4219	Numbers (dates?) modulo 29 and 25	236	328
4220	Handy Tables	236	330
4221	Longitudes	238	334
4222	Table with zodiacal signs and latitudes (?)	238	336
4223	Sinusoidal function	239	338
4224	Sinusoidal function	239	342
4225	Numerical table	240	344
4226	Table with planets, zodiacal signs, and symbols	241	346
4227	Sexagesimals	241	350
4228	Sexagesimals	242	350
4229	Sexagesimals	243	352
4230	Tabular scraps	243	354
4231	Tabular scraps	244	356
4231a	Tabular Scraps	244	358
4232	Tabular scraps	244	360
4233	Tabular scraps	245	364

		Vol. I	Vol. II
4234	Tabular scraps	245	366
4235	Tabular scraps	245	368

V. HOROSCOPES

		Vol. I	Vol. II
4236–4300a	General comment	249	
4236	Horoscope, A.D. 63	250	372
4236a	Horoscope, A.D. 80	251	372
4237	Horoscope, A.D. 84	252	374
4238	Horoscope, A.D. 117	252	374
4239	Horoscope, A.D. 130	253	376
4240	Horoscopes, A.D.150, 183, and undated	254	376
4241	Horoscopes, A.D.162, 188, 191, and undated	254	378
4242	Horoscope, A.D. 212	256	378
4243	Horoscope, A.D. 217	257	380
4244	Horoscope, A.D. 217	257	380
4245	Deluxe horoscope, A.D. 218	258	382
4246	Horoscope, A.D. 223	259	384
4247	Horoscopes, A.D. 223	260	384
4248	Horoscope, A.D. 229	261	386
4249	Horoscope, A.D. 172 (?)	262	386
4250	Horoscopes, A.D. 237 and undated	263	388
4251	Horoscopes, A.D. 244 and undated	264	388
4252	Horoscopes, A.D. 247/248 and undated	264	390
4253	Horoscope, A.D. 249/250	265	392
4254	Horoscope, A.D. 254	265	392
4255	Horoscope, A.D. 256	266	394
4256	Horoscope, A.D. 269–275	266	394
4257	Horoscope, A.D. 279	267	394
4258	Horoscope, A.D. 283	267	396
4259	Horoscope, A.D. 284	268	396
4260	Horoscope, A.D. 296 (?)	268	398
4261	Horoscope, A.D. 289	269	398
4262	Horoscope, A.D. 292/293	269	400
4263	Horoscope, A.D. 299	270	400
4264	Horoscope, A.D. 300	270	402
4265	Horoscopes, A.D. 313 and 348	271	402
4266	Horoscopes, A.D. 312 and 314	272	404
4266a	Horoscope, A.D. 313	273	406
4267	Horoscope, A.D. 318	274	406
4268	Horoscope, A.D. 318	274	408
4269	Horoscope, A.D. 320	275	408
4269a	Horoscope, A.D. 325	276	410
4270	Horoscope, A.D. 350	276	410
4271	Horoscope, A.D. 354	277	412
4272	Horoscopes, A.D. 370 and 375	278	412
4273	Horoscopes, undated and A.D. 399	279	414

		Vol. I	Vol. II
4274	Horoscopes A.D. 480 and 503	279	414
4275	Horoscope, A.D. 508	281	416
4276	Deluxe horoscope	282	418
4277	Deluxe horoscope	284	420
4278	Deluxe horoscope	286	426
4279	Deluxe horoscope	287	428
4280	Deluxe horoscope	287	428
4281	Deluxe horoscope	288	430
4282	Deluxe horoscope	288	430
4283	Deluxe horoscope	289	432
4284	Deluxe horoscope	289	434
4285	Deluxe horoscope	290	434
4286	Tabular horoscope	290	436
4287	Horoscope, A.D. 221–234	290	438
4288	Horoscope	291	438
4289	Horoscope	291	438
4290	Horoscope, A.D. 285	291	440
4291	Horoscope, A.D. 285–293	292	440
4292	Horoscope	292	440
4293	Horoscope	292	442
4294	Horoscope, later than A.D. 284	293	442
4295	Horoscope	293	442
4296	Horoscope	293	444
4297	Horoscope (?)	294	444
4298	Horoscope	294	444
4299	Horoscope	294	446
4300	Horoscope	295	446
4300a	Horoscope	295	446

LIST OF PLATES

I. **4133, 4137**
II. **4136, 4139, 4144**
III. **4138a, 4152** front, **4154, 4155**
IV. **4157, 4158** fr. 1, **4161, 4163** verso
V. **4167** fr. 3, **4190** verso, **4205** verso
VI. **4165** fr. 1 1–9, **4173** front, **4174a**
VII. **4175, 4191** fr. 1 recto
VIII. **4140, 4217, 4226, 4228**
IX. **4160a, 4236, 4276**
X. **4203a, 4247**
XI. **4138, 4258, 4274** →
XII. **4261, 4275**

NUMBERS AND PLATES

4133	I
4136	II
4137	I
4138	XI
4138a	III
4139	II
4140	VIII
4144	II
4152 front	III
4154	III
4155	III
4157	IV
4158 fr. 1	IV
4160a	IX
4161	IV
4163 verso	IV
4165 fr. 1 1–9	VI
4167 fr. 3	V
4173 front	VI
4174a	VI
4175	VII
4190 verso	V
4191 fr. 1 recto	VII
4203a	X
4205 verso	V
4217	VIII
4226	VIII
4228	VIII
4236	IX
4247	X
4258	XI
4261	XII
4274 →	XI
4275	XII
4276	IX

I
Introduction and Commentaries

Introduction

§1. PREFACE.

This edition of astronomical papyri is both an adjunct of the series, *The Oxyrhynchus Papyri*, and a contribution to the goal of making accessible the astronomical and astrological documents of all kinds surviving from the time of the Roman Empire. My aim has been to include those papyri, among the vast collection recovered by Grenfell and Hunt at Bahnasa (ancient Oxyrhynchus) between 1896 and 1906 and now kept at the Ashmolean Museum, Oxford, that can be identified as directly reflecting the practice of technical astronomy, that is to say, astronomical treatises and instructional texts, tables, and horoscopes. I have not included works of a strictly literary character that allude to astronomical themes, or texts concerned with astrological interpretation and theory. Several horoscopes and a very few other astronomical papyri from this collection have been published before now in the *The Oxyrhynchus Papyri*; they are not reproduced here, but a list of them will be found in Appendices B and C. Otherwise I have consciously omitted only isolated fragments that were too small or illegible to be of value to scholarship.[1]

What we have known about the technical astronomy of the first centuries of our era was long limited almost exclusively to the works of Ptolemy and his subsequent commentators. Beginning in the 1940s, Neugebauer produced a series of studies that revealed certain papyri as pertaining to a wholly unsuspected astronomical tradition contemporary with Ptolemy, a tradition that was indebted to the Babylonian mathematical astronomy of the Seleucid period and that also had ties to later Indian methods. Astronomical papyri, supplemented occasionally by the astrological literature, are the key to rediscovering the nature of this non-Ptolemaic astronomy and its complex cohabitation with Ptolemy's tables.[2] Their importance is accordingly out of all proportion to the size of the published corpus. A recent inventory identifies 85 published Greek horoscopes on papyrus;[3] and Appendix B lists some eighty other published or identified astronomical papyri, of which perhaps fifteen have individually made a noteworthy addition to our

[1] See pp. 56–57 for details of the procedure of selection.

[2] In these general contexts 'papyri' should be understood to include ostraca and wooden tablets.

[3] Baccani [1992] 18–21. Inclusion of the Demotic horoscopes, horoscopes on ostraca, and a handful of Greek papyri published since 1992 would raise the count slightly.

understanding of Roman-period astronomy. The present volume more than doubles the number of known astronomical papyri; and the new material acts like a fresh set of puzzle pieces added to a first small selection: isolated pieces begin to form clumps, and larger patterns emerge. It therefore seems fitting to attempt in this introduction a general survey of the varieties of astronomical texts and tables so far known, and to discuss the underlying concepts and methods more broadly than would be required for the analysis of each single document. I hope that in so doing I will make it easier for others to identify unpublished papyri and situate them within the developing but still quite fragmentary picture.

Plan of the work

In deciding what to include in this book and how to arrange it, I have tried to keep in mind that it will be used not only by papyrologists but also by historians of science who are less familiar with the *arcana* of papyrological editions. The text editions and translations, on facing pages for ease of comparison, make up the bulk of the second volume, while the commentaries are in this volume. Brief introductions to the major groups of papyri precede the commentaries to the individual texts in each section; they are intended to orient the reader to the most important features of the texts and their broad significance for the historian. Most of the texts are of such a nature that the commentaries seemed to call more for a general interpretative discussion than for extended line-by-line annotation.

The papyri were grouped, ordered, and assigned *P. Oxy.* numbers in 1994, and a preliminary version of the list of numbers, titles, and inventory numbers has been published in vol. LXI of *The Oxyrhynchus Papyri* (pp. 142–145). The order reflects a classification of the varieties of astronomical documents into the following categories: (I) prose texts (other than horoscopes), (II) 'primary' tables that were used to predict the movements of the heavenly bodies and other phenomena, (III) ephemerides and almanacs that record the positions of the heavenly bodies and other information at regular intervals without directly showing the means of their computation, (IV) unclassifiable tables, and (V) horoscopes. Within parts II and III the tables are sorted by specific format (see §6). Additional texts (given 'lettered numbers' such as **4236a**) and fragments of existing ones turned up after the publication of the preliminary list, and I have revised some titles, but I felt obliged not otherwise to alter the published assignments. In particular the reader should note that some items in part IV were identified as belonging in the other categories too late to be renumbered, and among groups of papyri (such as the horoscopes) that were intended to appear in order of date, the sequence has occasionally been disturbed through revisions made since the numbers were assigned.

Acknowledgements
I must first of all thank the Egypt Exploration Society, the owners of the papyri published here, for granting me access, providing me with photographs, and for permission to publish them in the present form. The general editors of the Society's Graeco-Roman Memoirs have given the project every help and encouragement. Dr J. R. Rea and Dr R. A. Coles cheerfully endured three visitations upon the Papyrology Rooms of the Ashmolean Museum, which caused much interruption to their own work, and were always ready to answer the myriad technical questions that arise when a non-papyrologist undertakes to edit papyri. They are jointly responsible for the brief descriptions of non-astronomical contents and the datings of hands that appear in the commentaries below. On occasions too numerous to cite, Dr Coles identified joins and read near-illegible letters, and I have profited from his unequalled knowledge of the workings and history of the collection. Mr Gideon Nisbet made the digital photographs for the plates, and Dr C. Crowther and Dr D. Obbink converted them into a form that I could use. During my stays in Oxford in 1992 and 1994 I enjoyed the hospitality of Corpus Christi College and Christ Church College, for which I thank their Senior Common Rooms and in particular Prof. R. G. M. Nisbet and Prof. P. J. Parsons. I am indebted to the Warden and Fellows of All Souls College for granting me a visiting fellowship for part of my visit in 1996.

I am grateful to several institutions for their financial assistance. The first stages of this work were carried out during my tenure of a Canada Research Fellowship from the Social Sciences and Humanities Council of Canada; and in completing it I am aided by a research grant from the same source. In between, I have had the benefit of awards from the Connaught Fund of the University of Toronto, the Department of Classics of the University of Toronto (Gilbert Norwood Travelling Award), and the Dudley Observatory, Schenectady, New York. I also thank the American Philosophical Society for undertaking the publication of the book. It is no small honour to contribute to the same series in which Neugebauer and van Hoesen's *Greek Horoscopes* appeared forty years ago.

Colleagues and friends have patiently read or listened to my progressive attempts to make sense of these fragmentary remains. I am particularly indebted to John Britton, John Steele, Gerald Toomer, Asger Aaboe, Noel Swerdlow, and Benno van Dalen for comments on particular texts and help with technical matters, and to David Fowler for informing me of the existence of these papyri in the first place. Mr John Anderson spent many hours recalculating ephemerides and almanacs, and he was responsible for the preliminary datings of several of these tables. Dr Sarah Pothecary helped with bibliographical research, checking, and proofreading. Dr Kathryn Simonsen did much of the work of preparing the indexes, in the course of which she detected many mistakes. My editor, Carole

LeFaivre-Rochester, found more, in the course of guiding the unweildy manuscript to the press. To my colleagues and students at the University of Toronto I express my warmest thanks for years of encouragement.

Lastly, any scholar who undertakes the study of the astronomical papyri must acknowledge a profound obligation to Neugebauer. He was the first to recognize the historical significance of these documents; in more than a score of publications spanning five decades he almost single-handedly established the basic methods of dating and analysing them; and his *Astronomical Cuneiform Texts* supplies the key to the long-latent Babylonian elements in the papyri. Nor can I pass by the recollection of his tolerant encouragement of a beginner's efforts. I regret that he did not live to see these texts, which, however faulty the presentation, would I know have given him pleasure.

§2. THE PURPOSE AND PROVENANCE OF THE PAPYRI

Astronomical papyri constitute a minute fraction (perhaps less than one thousandth) of the rich trove recovered by B. P. Grenfell and A. S. Hunt from the rubbish heaps of ancient Oxyrhynchus.[1] Although records of where each lot of papyri was excavated were not kept (or at least have not survived), the inventory numbers occasionally provide a clue to the grouping and dating of fragments, for example when parts of several manuscripts were inventoried, and hence probably found, together (cf. pp. 56–60). Because the Oxyrhynchus papyri were *discarded* documents, we would anyway have profited little from precise knowledge of where any particular papyrus was found. The place and purpose of astronomy in Roman Oxyrhynchus have to be deduced primarily from the nature of the texts, supplemented by what little evidence of provenance exists from other Egyptian sites.[2]

The fragments tend to be of three kinds: horoscopes, numerical tables, and prose texts. Horoscopes record the positions of the sun, moon, planets, and ascendant point of the ecliptic computed for the birthdate of a person, who is often named. Positions of the heavenly bodies also appear in many of the tables ('almanacs'), computed now not for an arbitrary given date but for a succession of dates at regular intervals or determined by some other astronomical criterion. Other tables ('primary tables') were used in the intermediate steps of calculating positions, but do not necessarily contain positions pertaining to specific dates.

[1] See E. G. Turner's account of these excavations in James [1982] 161–178. It is impossible to give precise figures for the size of the Oxyrhynchus trove, which ranges from large fragments of rolls (comparatively rare) to many thousands of minute scraps.

[2] Aspects of this topic are dealt with more fully in Jones [1994b] and [1999b].

Despite the odd text concerned with theory and observation, such as **4133** (analysis of observations of Jupiter) or *P. Oslo* 73 (on the sun's apparent diameter), the prose texts are also by and large concerned with knowing the positions of the sun, moon, and planets and the disposition of the heavens at specific times, or the dates and circumstances of such conspicuous phenomena as eclipses and the alternating appearances and disappearances of the planets. In contrast to the modern conception of Greek astronomy as a theoretical enterprise, the papyri portray a science that was overwhelmingly directed towards prediction.[1]

One might not have been surprised to find the odd work on astronomical theory in the libraries of those cultured residents of Oxyrhynchus who left us so many fragments of classical literature.[2] The great majority, however, of the astronomical papyri were part of the equipment of practical astrology. Many astrologers were active in Egypt throughout the Roman period. They have left their traces in the horoscopes and in other strictly astrological papyri,[3] and the kinds of predicted data supplied by the astronomical texts and tables on papyrus coincide with the needs of Greek practical astrology. Astronomical and astrological ostraca have been discovered together at a site where they were actually used, the temple at Narmuthis (Medinet Madi)—the kind of collocation that we cannot expect to be repeated under the different archaeological conditions of Oxyrhynchus.[4] Lastly, in spite of the generally tidy distinction in contents between the astronomical and astrological papyri, astrological data sometimes find their way into astronomical documents, for example in the columns stating whether a day is auspicious or inauspicious in some ephemerides.

The papyrus horoscopes give an indication of the range of dates during which astrology flourished, from the earliest horoscope, a demotic ostracon cast for a native born in 38 B.C., to the latest (**4275**), for a native born in A.D. 508. The horoscopes may, of course, have been produced years or decades after the birthdates, so that we have to assume a lag of a generation or so in using the frequency of nativity dates to estimate the varying level of astrological activity in Egypt.[5] The graph (Fig. 1) reveals several pronounced trends. Demotic horoscopes

[1] In the context of ancient astronomy, 'prediction' should be understood in the sense of 'deduction from theory or rules' rather than 'forecasting'. Most astronomical predictions (e.g. the data in a horoscope) were made *after* the date to which they refer.

[2] The character of Roman Oxyrhynchus and the part of its population that owned manuscripts has been sketched by Turner [1952] and [1956].

[3] Inventory of published Greek astrological papyri: Baccani [1992] 32–34, updating Neugebauer & van Hoesen [1964]. The horoscopes are collected in Neugebauer & van Hoesen [1959] and Baccani [1992] (with a comprehensive list on pp. 18–21).

[4] Baccani [1992] 50–52, [1989], and [1995]. Few details of the astronomical ostraca are yet available.

[5] In Figs. 1 and 3, papyri in collections other than that of the Egypt Exploration Society are counted as 'other', although a few may be from Oxyrhynchus.

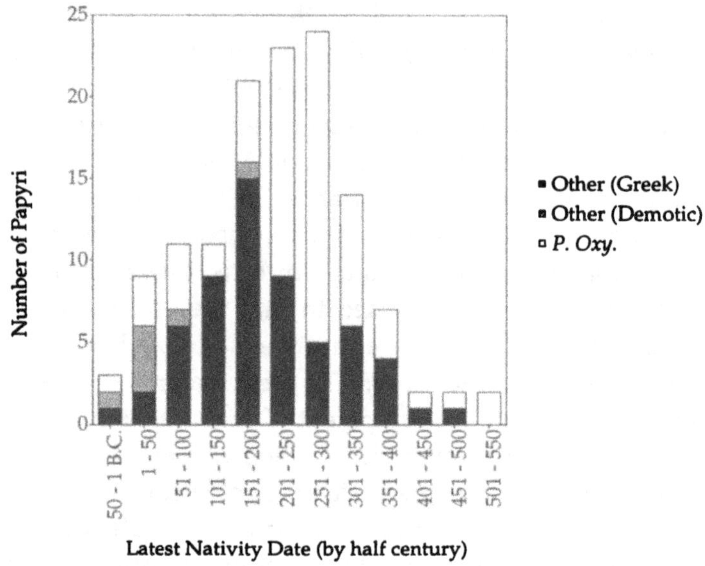

FIG. 1. Frequency of Documentary Horoscopes from Egypt
(by latest nativity date in each papyrus)

are few and early. The Greek ones increase in numbers from the beginning of Roman rule to the late second century, and decline again after the end of the third century. If we divide these into the *P. Oxy.* horoscopes and those in other collections, however, two very different patterns emerge. Those from Oxyrhynchus are fairly uniform in frequency (Fig. 2) until the end of the fourth century, except for a

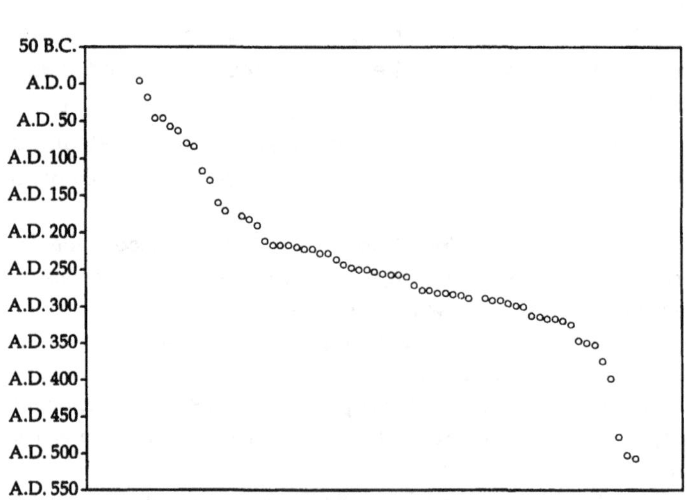

FIG. 2. Distribution of *P. Oxy.* horoscopes by last nativity date.

large surge in the third century. The papyri from other sources become gradually more and then less numerous, with the peak in the late second century. The dated astronomical papyri show much the same tendencies (Fig. 3; cf. Fig. 14).

It must be remembered that the number of published documents of a particular kind and date depends, not only on the number of such texts that were produced in antiquity, but also on the conditions of preservation at each site and on the biases of scholarship. Few of the exactly dated papyri published in *The Oxyrhynchus Papyri* belong to the first century B.C., and even the first half of the first century of our era is not nearly as well represented as subsequent intervals. This is because the damp level of the site was too high for deposits from earlier periods to survive, not because editors have neglected the oldest documents.

It is not so obvious that the fairly uniform frequency of dated non-astronomical papyri so far published for each half century from A.D. 50 to 300 is representative of the entire collection that Grenfell and Hunt excavated. The impression obtainable from the inventory of certain large categories of unpublished documents datable by contents or hand, for example private letters, is that the materials from the second century are roughly twice as numerous as those from the first, those from the third century roughly double those from the second, and the counts from the fourth century are about mid-way between those of the second and third. The number of texts from the fifth and sixth centuries is much smaller. It would appear that the distribution of the astronomical texts and horoscopes

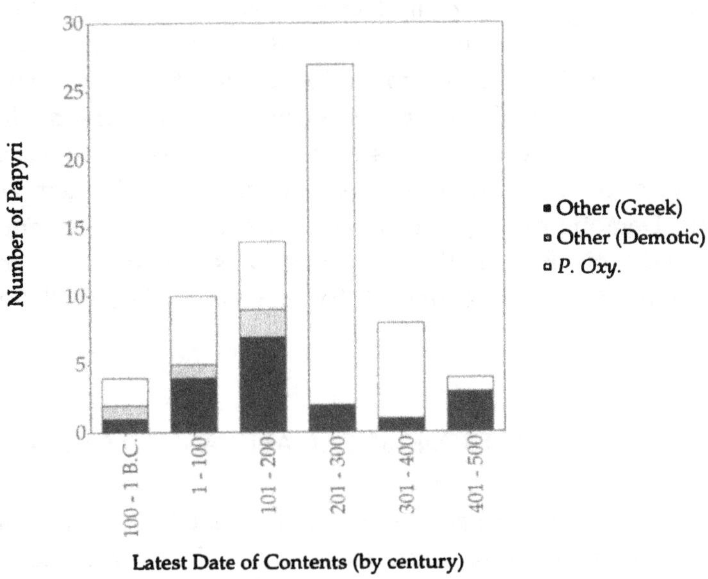

FIG. 3. Frequency of exactly dated astronomical papyri from Egypt (by latest date in text)

partly reflects this broader pattern up to the beginning of the fourth century, and that the popularity of astrology remained fairly steady in proportion to population throughout this period. Thereafter there appears to have been a marked decline.

What sort of people were the astrologers' clients? The documents give us few clues. Many horoscopes name the person whose nativity is recorded: for the most part they are the usual compote of Greek and Egyptian names that could pertain to almost any stratum of Oxyrhynchite society. The only recognizably Christian 'natives' among the horoscopes from Oxyrhynchus are the Anup of *P. Oxy.* 2060 and the Theodoros of **4275**, who were born in A.D. 478 and 508 respectively. Few women seem to have had their horoscopes cast: in fact among the horoscopes from Oxyrhynchus only **4258** and the second nativity in **4265** are identifiably women's. Many of the horoscopes are crudely written scraps, but such 'deluxe' productions as **4276** and **4277** must have been expensive. There are a fairly high number of fragments of deluxe horoscopes among the *P. Oxy.* (fifteen, including twelve in this volume), a circumstance that may reflect the relative prosperity or literacy of the residents of Oxyrhynchus. But in general the new papyri in this volume add little to the information that D. Baccani has recently assembled concerning the social position of the natives, which is not a great deal.[1]

We are no better informed about the astrologers themselves. None of the astronomical texts and tables gives an indication of its owner, and only two horoscopes are 'signed': the first horoscope in **4266** concludes with the words, 'Diapalos (is) the one who wrote it'; and the deluxe horoscope *P. Lond.* 130 (not from Oxyrhynchus), cast for one Hermon who was born in 81, closes, 'I, Titus Pitenius, computed [it] as it is set out.' Like Vettius Valens later in the second century, Titus Pitenius was presumably a Roman citizen. We do not know how or where the astrologers of Oxyrhynchus worked or were trained. The importance of the ostraca from Medinet Madi has been rightly stressed by Baccani: they reveal an astrological practice within a second-century Egyptian temple, conducted in Greek and Demotic by Egyptian priests. A similar arrangement seems to have existed in Tebtynis. But whether the temple was the normal setting for astrology in other towns of Roman Egypt (as it had been in Hellenistic Babylon) is far from clear.

§3. ELEMENTARY ASTRONOMICAL AND ASTROLOGICAL CONCEPTS

This section reviews some simple but fundamental concepts that are needed in order to follow the more technical treatment of the astronomical methods in the papyri, both in the following sections and in the commentaries to the individual

[1] Baccani [1992] 49–56.

texts. For a fuller orientation in the methods of ancient astronomy, see Neugebauer [1957b] and [1975] (in ascending order of technical detail and comprehensiveness); and for ancient astrology, Neugebauer & van Hoesen [1959] (introduction) and Bouché-Leclercq [1899].

Numbers. For whole numbers the astronomical papyri use the standard 'Milesian' alphabetic notation, which translates directly into our arabic numerals. The normal representation of fractions is, however, an adaptation of the Babylonian sexagesimal system that is found only in astronomical contexts. Fractional parts are expressed by sexagesimal (base 60) 'digits' or fractional places representing sixtieths, thirty-six-hundredths, and so forth. A special symbol ō is used for zero in any fractional place or in the units (see pp. 61–62); otherwise a fractional place can be filled by any whole number up to 59. Fractional places are separated by space or by tabular ruling. A digit comprising only tens is sometimes marked with an accent-like superscript to show that following numerals are not to be read as part of the same fractional place. In translation we use the convention of writing a semicolon to separate units from fractional places, and commas to separate fractional places. Thus τξε ιε λγ μς is translated as 365;15,33,46. When numerals appear in prose, the letters representing whole numbers and fractional places may be distinguished from words by horizontal strokes above the groups of letters.

In nontechnical contexts, and when greater precision is not required, fractions may be expressed in the Greek adaptation of the Egyptian system of unit fractions. In this notation, a fraction is written as the sum of one or more quantities from the series $^2/_3$, $^1/_2$, $^1/_3$, $^1/_4$, $^1/_5$, ... Of these, the first and second are represented respectively by the symbols ϖ and ∟, and the rest by the denominator of the fraction with an oblique or (occasionally) horizontal stroke above.

Positions of heavenly bodies. Locations on the celestial sphere are specified by two 'ecliptic' coordinates: the *longitude*, measured in degrees along the ecliptic (the great circle of the sun's apparent circuit), and the *latitude*, measured in degrees perpendicularly north or south from the ecliptic to the point in question (latitudes rarely appear in the papyri). The 360° of the ecliptic are divided into the twelve equal zodiacal signs (roughly corresponding in antiquity to their namesake constellations), so that a longitude will usually be expressed by the zodiacal sign, degrees counted from zero at the beginning of the sign, and fractions (minutes, seconds, etc.) of degrees. The zodiacal signs are usually identified by their names, in full or abbreviated; but many papyri represent them by their ordinal numbers counting from Virgo = 1 (reflecting the fact that the sun is in Virgo at the beginning of the Alexandrian calendar year). Alternatively, one can count the degrees continuously from a conventional starting point, Aries 0°; this is often more convenient for computation and analysis of data. Thus 65;30° is equivalent to

Gemini 5;30° or Gemini 5° 30' or sign 10, 5;30°. Appendix A, section 1 lists the signs with their degree equivalents and ordinal numbers for quick reference.

A special unit called 'step' (βαθμός), equivalent to 15°, was used to measure intervals in longitude starting from the northern limit of a heavenly body's (usually the moon's) orbit. The interval in steps from the moon's northern limit to the moon itself, called its 'argument of latitude', is featured in certain lunar tables; and a few elaborate horoscopes mention the number of the step occupied by a heavenly body.

Data in a horoscope. A Greek horoscope is a record of astronomical data computed for a given date, time, and place of birth of a 'native', compiled for the sake of astrological interpretation. The data reflect two relative configurations that change with time: the positions of the sun, moon, and planets along the ecliptic, and the instantaneous situation of the ecliptic in relation to the horizon and meridian at the place of birth. In documentary horoscopes the ecliptic is treated as the fixed frame of reference for both categories of information; that is, one reports the longitudes not only of the heavenly bodies but also of at least one of the four 'cardines' or intersections of the ecliptic with the horizon and meridian planes. These are, in order of decreasing longitude from the east horizon: the rising or 'ascendant' point (ὡροσκόπος), the culminating point or 'midheaven' (μεσουράνημα), the setting point (δύσις), and the 'lower midheaven' (ὑπὸ γῆν). Most horoscopes give only the ascendant's longitude.

Astrological associations of a zodiacal sign and its subdivisions. One of the means by which Greek astrology could extract multiple interpretations from the computed position of a single heavenly body was by dividing the ecliptic up in several different ways, and associating the divisions with certain qualities or with other heavenly bodies according to an established scheme. The terminology expressing these relations turns up frequently in deluxe horoscopes. In the present context it will be necessary only to review how these schemes worked, without considering the bearing they had on the interpretation of the horoscope.

To each zodiacal sign was attributed a list of qualities. Some follow established patterns. Alternate signs are male or female (cf. e.g. Ptolemy, *Tetrabiblos* I 12). Again, the signs associated with the sun's longitude at the equinoxes and solstices (Aries, Cancer, Libra, Capricorn) are 'tropical' (i.e. 'pertaining to the solstices'), the subsequent signs (Taurus, Leo, Scorpio, Aquarius) are 'solid', and the remaining four (Gemini, Virgo, Sagittarius, Pisces) are 'bicorporal' (*Tetrabiblos* I 11). The adjectives referring to this twofold and threefold scheme appear regularly in the deluxe horoscopes whenever a zodiacal sign is named. Other conventional qualifications, such as 'royal' for Leo or 'wet' for Aquarius, are less regularly included.

With each of the five planets is associated one sign as its 'exaltation', ὕψωμα, and another, diametrically opposite, as its 'depression', ταπείνωμα (*Tetrabiblos* I 19). Also, each sign is the 'house' (οἶκος) of one of the planets, the sun, or the moon (*Tetrabiblos* I 17). The former (when applicable to a particular sign) are usually, and the latter regularly, listed in the deluxe horoscopes. References to the 'triplicity' or 'triangle' to which a sign belongs are less rare; a triplicity comprises three signs one third of the ecliptic apart, and is 'ruled' by one of the seven heavenly bodies by day, another by night. The conventional assignments for all these schemes are tabulated in Appendix L.

Of the astrological subdivisions of the signs, the most important is the division into 'terms' (ὅρια) associated with the planets. Several schemes are known from the astrological handbooks, but the only one that turns up in the papyrus horoscopes is the so-called 'Egyptian' system, according to which every sign is broken up into five terms of irregular length, each ruled by one of the five planets (*Tetrabiblos* I 20). These are a regular feature of the deluxe horoscopes. The pattern is set out in Appendix L. A still finer partition used in a few horoscopes pairs each single degree or *monomoiria* in a sign with one of the planets.

Each sign is also divided into three equal 'decans' of 10° each. Although these had names derived from the names of the old Egyptian decan constellations, they are more commonly denoted in the papyrus horoscopes by their number (1, 2, or 3) within the relevant sign.

In addition to the concepts described above, the deluxe horoscope **4277** employs 'steps', 'winds', and 'zodiacal places'. These are discussed in the commentary to that text.

§4. CALENDARS AND CHRONOLOGICAL DATA

Full dates in the papyri are usually specified by emperor(s) and regnal year, calendar month, day number, and seasonal hour. In most instances the months and regnal years are reckoned according to either the Alexandrian (civil Egyptian) or the old Egyptian calendar (see Appendices D–E). It is sometimes convenient to denote the calendar months by roman numerals, counting from the first month of the year; numerals representing months are also sometimes found in the papyri. For modern equivalents I have used the Julian calendar, with the traditional designations A.D. and B.C. rather than the astronomical convention which uses 0 and negative numbers for years before A.D. 1, such that $-x$ is $(x+1)$ B.C. Most dates discussed here are, in fact, A.D., and I have omitted the abbreviation where context makes it obvious.

a. Egyptian calendar

Before the Roman period, the Egyptian civil calendar used an invariable year of 365 days comprising 12 months of thirty days followed by 5 epagomenal days. The use of a strictly 365-day year results in a shifting backwards of the beginning of the Egyptian year so that it falls one day earlier every four years with respect to the Julian calendar. Because of its convenience for calculating exact intervals of time, this calendar continued to be used in astronomical contexts long after it had ceased to be the civil calendar in Egypt; in horoscopes Egyptian calendar dates are sometimes specified as κατ' ἀρχαίουϲ, 'according to the ancients', or 'according to the Egyptians'. Julian calendar dates can be converted into Egyptian dates and *vice versa* with certainty for all periods (cf. Appendix E). The names of the months, following the orthography used in this work, are:[1]

I	Θώθ	Thoth	VII	Φαμενώθ	Phamenoth
II	Φαῶφι	Phaophi	VIII	Φαρμοῦθι	Pharmuthi
III	Ἁθύρ	Hathyr	IX	Παχών	Pachon
IV	Χοίακ	Choeac	X	Παῦνι	Payni
V	Τῦβι	Tybi	XI	Ἐπείφ	Epeiph
VI	Μεχείρ	Mecheir	XII	Μεσορή	Mesore
				Ἐπαγόμεναι	Epagomenals

b. Alexandrian calendar

The civil calendar of Roman Egypt used the same month names and lengths as the old Egyptian calendar, but inserted intercalary days. In the horoscopes, civil dates are sometimes designated as καθ' Ἕλληναϲ ('according to the Greeks'); the term 'Alexandrian', which will be used here, is also ancient. For the period following A.D. 8 the intercalations are known to have taken place every four years, at the end of the Egyptian year that immediately preceded the Julian intercalary year. For the earlier part of the reign of Augustus this scheme can be projected backwards hypothetically in two ways: (a) We can assume that intercalations in the Alexandrian calendar had, from its institution, always occurred at four year intervals. In this case the first four-year cycle, during which all years still coincided with the Old Egyptian calendar, began on Thoth 1 = Julian August 29, 26 B.C. (b) Alternatively, the intercalations could have been tied to the civil Roman calendar (see below), in such a way that intercalation took place in Egypt in the years preceding those in which the Roman calendar was intercalated, with all the attendant irregularities for the years before A.D. 8. On this hypothesis, the cycle in which the Alexandrian and old Egyptian calendars still coincided began Thoth 1 = Julian August 31 = Civil Roman August 29, 30 B.C. The tables for conversion of dates in Appendix E

[1] This is the traditional accentuation among papyrologists. See, however, Clarysse [1997] for argument that the names of the Egyptian months should all be oxytone.

assume case (a), i.e. intercalations at four-year intervals from the beginning.

c. Roman calendar

From March, A.D. 4 onwards, the civil Roman calendar's months exactly coincide with the conventional Julian calendar of modern chronography, in which all years have 365 days, with an intercalary day in February of every year A.D. $4i$ or $4i$–3 B.C. (where i is any whole number). However, for the 36 years following 45 B.C., when Julius Caesar's reform of the Roman calendar was instituted, Solinus (I 46-47) and Macrobius (I 13-15) report that the Roman pontifices incorrectly intercalated every third year instead of every fourth; thereafter Augustus decreed the omission of three consecutive intercalations in order to bring the calendar back into line with the course it should have taken if the intercalations had been made at the correct intervals. Hence for most dates between March of 42 B.C. and February of A.D. 4 a given date in the civil calendar should come later than the same date in the Julian calendar by as many as three days. Nevertheless the Roman calendar dates in the ephemeris **4175** for 24 B.C. are unexpectedly in agreement with the Julian calendar.

In ephemerides that have a column for the Roman calendar date, the days are counted by the usual 'backwards' reckoning from the Kalends, Nones, and Ides (conveniently tabulated by Bickerman [1980] 125). Some ephemerides of the fourth century and later also have a column numbering the days of the Roman months continuously forwards from the first through the last day in the modern fashion. In the '*hemerologia*' (tables for calendar conversions) incorporated in medieval copies of Ptolemy's *Handy Tables* this convention for counting the days is associated with a calendar there called 'of the Greeks' (and in modern works called the Antiochene or Common Macedonian), with months Audnaios, Peritios, etc. equivalent to the Roman months (cf. Kubitschek [1915] 101). We do not know what the papyrus ephemerides put in the heading of this column.

d. Other calendars

The eclipse canon **4137** specifies the date of a lunar eclipse by an Athenian month. This does not refer to the civil calendar of Athens, however, but to a schematic lunar calendar associated with the counting of years by Callippic periods (see below), using Athenian month names but following a fixed pattern of intercalated months to keep the beginnings of years aligned with the summer solstice. The first day of each month fell close to the conjunction of sun and moon.

Some ephemerides have a column counting days according to lunar months, with the first day coinciding with conjunction. We do not know whether in the column headings these months were named according to any calendar.

e. Regnal years and eras

Until the end of the fourth century, years in horoscopes, epoch tables, and almanacs are usually specified according to the reigning emperor and the regnal year. By convention the first regnal year of an emperor was the calendar year (of whichever calendar is being used) within which the accession occurred. Variants occur, e.g. in the naming of emperors and their colleagues, or in **4160**'s assignment of what is usually called Nero 14 to Galba. The most extensive surviving regnal canon is found in the medieval manuscripts of Ptolemy's *Handy Tables*, which extends from the reign of the Babylonian king Nabonassar (747 B.C.) to the Byzantine period. Appendix D excerpts from this version the portion from Augustus to Diocletian. With Diocletian's institution of co-emperors begins the practice of giving multiple regnal years; however, horoscopes from the fourth century and later are mostly dated simply according to years counted from the first year of Diocletian. In astronomical tables one sometimes finds years counted continuously from the first regnal year of Augustus (P. Berol. inv. 21226) or Titus (Bodl. MS Gr. Class. F 7 (P), P. Nelson, *P. Tebt.* 274).

The reckoning of years by Callippic Periods, which formerly was known only in observation reports from the third and second centuries B.C. quoted in Ptolemy's *Almagest*, can now be traced as late as the first century of our era in papyri. These were calendrical periods of 76 years, with the first year of the first period beginning about the summer solstice of 330 B.C. The new attestations are in the demotic eclipse canon P. dem. Berol. 13146+13147 (fourth Callippic Period, year 1 = 102/101 B.C.), the sign-entry almanac Tab. Amst. inv. 1 (fifth Callippic Period, year 1 = 26/25 B.C.), and the eclipse canon **4137** (sixth Callippic Period, year 1 = A.D. 51/52). The years of the sign-entry almanac are reckoned according to the Egyptian or Alexandrian calendar, while the eclipse canons both use schematic lunar calendar years beginning near the summer solstice.

f. Days, nights, and hours

Equinoctial hours, i.e. equal divisions of the whole day-and-night into 24 hours counted from noon or midnight, are used for computational purposes in Greek astronomy. In horoscopes and in certain varieties of almanac, however, times are given in *seasonal* hours, based on the division of the two variable intervals between sunrise and sunset and between sunset and sunrise into twelve equal parts; thus the first hour of day begins at sunrise, the sixth hour ends at noon, and the twelfth hour ends at sunset. To avoid ambiguity about whether a particular night 'belongs' to the preceding or the following day, for nocturnal times *both* day numbers are typically given, in the phrasing Παυνὶ η̅ εἰς θ̅, 'Payni 8th into the 9th', or simply Payni 8/9. Times specified by the seasonal hour could be understood as referring to the beginning, middle, or end of that hour. For approximate conversions of times in seasonal hours to equinoctial hours reckoned from local noon or midnight, see Appendix F; but in some situations the ancient practice was to ignore the seasonal variation and simply count equinoctial hours from 6 A.M. as hours of day, and from 6 P.M. as hours of night. Local mean time for the meridian of Alexandria is almost exactly two equinoctial hours later than for the Greenwich meridian (U.T.).

In epoch tables one encounters sexagesimal fractions following the dates of epoch events. These can sometimes (e.g. for the dates of solstices in **4148**) be a specification of the exact time of the epoch, as a fraction of a day counting from a conventional beginning of the day, say sunset or noon. For such epoch events as first appearances and disappearances of planets, which always happen at the same time of day, the fractions are merely a computational device required by the predictive method.

§5. CONCEPTS AND METHODS OF GREEK PREDICTIVE ASTRONOMY

The astronomy of the papyri had as its chief aim the production of horoscopes and almanacs, and this meant the continual performance of two tasks: to find the longitudes of the sun, moon, and planets for a given date, and to find the dates when each of these bodies reached given longitudes (specifically, the beginnings of the zodiacal signs). The papyri attest to two competing approaches to obtaining this information, the practical counterparts of two ways of describing the anomalistic motion of the heavenly bodies.

a. Kinematic models and tables.

One way is to ascribe to each body a definite path composed of uniform circular motions, and viewed in perspective from the earth. Such kinematic models are regarded, not without justice, as characteristically Greek. They are the models of Ptolemy's *Almagest* and *Planetary Hypotheses;* and through his and others' testimony they can be traced back to Hipparchus and perhaps to Apollonius of Perge, that is, to the second century B.C. Besides Ptolemy and the vast medieval tradition that derives from him, the kinematic approach passed in antiquity to India, where it became prevalent in the subsequent astronomical literature in Sanskrit.

Positions of heavenly bodies. The *Almagest* deals at length with the problem of computing apparent positions of the sun, moon, and planets (as well as certain other phenomena) in the best possible agreement with the models that Ptolemy there establishes; the tables of the *Handy Tables* are a revision and reorganization of their counterparts in the *Almagest.*[1] He reduces the problem to two steps: to tabulate the 'mean motions', that is, the uniformly revolving angular orientations

[1] The only complete edition of the *Handy Tables* is Halma [1822–1825]. For the numerical tables it is better to consult the more reliable Stahlman [1960], which also provides instructions and commentary. A more analytical treatment is Neugebauer [1975] 969–1028.

of each component of each model as a function of time, and to provide a second set of 'anomaly' tables that permit the conversion of the several instantaneous mean motions into the body's longitude and latitude. Ptolemy's mean motion tables are simple single-argument tables with calendrical units of time as their arguments: fixed 'Egyptian' years of 365 days, 30-day Egyptian months, days, and equinoctial or constant hours. The anomaly tables, however, are necessarily complex, since they must yield quantities that are functionally dependent on (usually) two independent arguments taken from the mean motion tables. To represent such anomaly functions with uncompromising accuracy would have demanded double-argument tables of enormous size. Ptolemy achieves a very close approximation to the ideal function in tables of remarkable economy. A single argument column is made to serve repeatedly for different kinds of argument (the several mean motions, and certain intermediate values), and the numbers read off from the other columns of the anomaly table are combined according to an arithmetical formula to obtain the longitude or latitude. The theoretical notion underlying these operations is that Ptolemy has computed the anomaly function according to the appropriate trigonometrical formula for a full range of values of its more important variable but only for extremes of the less important variable; a special interpolation function of the latter provides approximate intermediate values. The correct computation of a planetary longitude from Ptolemy's tables calls for several interpolations between tabulated values and multiplication as well as addition—all, of course, using sexagesimal fractions.

The Indian methods of computing planetary positions from kinematic models follows similar lines. Here too the apparent motion is computed as a trigonometrically derived correction to a mean motion. The mean motions are not tabulated, however, but obtained by direct calculation from period relations. One finds the anomaly by a formula combining separately tabulated functions, each dependent on just one mean motion.

There are strong arguments for believing that the indubitable Greek antecedents of the Indian kinematic models were untinctured with Ptolemy's influence, and hence perhaps also earlier than Ptolemy. A plausible speculation has it that certain 'Eternal Tables' (or 'Aeon-Tables') disparaged by Ptolemy and mentioned by two further second-century sources, Vettius Valens and a papyrus horoscope, were Greek counterparts of the Indian *Siddhāntas*. One would therefore expect to find in the documentary material from Roman Egypt kinematic tables other than Ptolemy's, employing the segregation of mean motions and anomaly. In fact the evidence for such tables among the papyri is sparse—a circumstance that, given the mass of texts and tables now available, is surely not accidental, although the reasons for it are far from clear. Kinematic tables were certainly not unknown in Egypt before Ptolemy, as we can verify from a procedure text of the first or second century (*PSI* XV 1490) that refers unambiguously to lunar mean motions in longitude, latitude, and 'depth' (βάθοc, i.e. anomaly). Nor can the difficulty of using such tables have daunted the astrologers; for their readiness to grapple with Ptolemy is apparent not only from the several fragments of copies of the *Handy Tables* that have come down to us on papyrus, but also from the steady stream of almanacs demonstrably computed using those tables from the middle of the third century on.

Spherical astronomy, eclipses, phases. In addition to the tables for computing planetary positions in longitude and latitude, the *Handy Tables* include other tables for astronomical computations of potential utility to the astrologer. For the conversion of times from seasonal to equinoctial hours and *vice versa* and the determinations of the ascendant and the other cardines in a horoscope, one would use Ptolemy's tables of oblique and right ascensions, which correlate simultaneously rising or culminating points along the ecliptic and the celestial equator. These tables are computed according to the appropriate formulas of spherical trigonometry.

Ptolemy also provides tables for the sake of predicting eclipses of the sun and moon and the planetary 'phases' (first visibilities, disappearances, stations). These are treated as secondary phenomena, the prediction of which involves testing for certain criteria in the course of calculating a more continuous run of positions of the heavenly bodies from the other tables. Although fragments from these and various other subsidiary tables that formed part of the *Handy Tables* are extant on papyrus, it is not clear that the owners of these copies made much use of them.

b. Arithmetical schemes.

The chief competition to Ptolemy's tables in the astronomy of the papyri was provided by methods of prediction involving only the arithmetical operations (addition, subtraction, multiplication, division) without recourse to trigonometry. The starting point of most of the arithmetical procedures for finding the longitudes of the heavenly bodies was the 'epoch table', comprising a sequence of computed dates and positions of a planet, sun, or moon at a specific stage or 'phase' of its anomalistic period. The existence of a range of epoch tables pertaining to each of the seven heavenly bodies has only now been revealed by the new Oxyrhynchus papyri (hitherto only a single table of this kind, *P. Lund* 35a, was known), and they are of the greatest interest for what they tell us about the nature of Roman period astronomy and its origins. In format these tables are an adaptation of the Babylonian 'ACT' tables of planetary phases.[1] The methods by which most of the planetary epoch tables on papyrus were computed were also substantially the same as

[1] 'ACT' (Astronomical Cuneiform Texts) is the acronym of Neugebauer [1955], and denotes the varieties of mathematical predictive methods comprised in that edition.

appear in the ACT tables; on the other hand the solar and lunar epoch tables have no direct Babylonian antecedents.

The sun. The sun's is the simplest of the epoch tables.[1] Since its period of anomaly was assumed to be constant, and identical to its period of longitude, the dates when it attains any particular longitude could serve as epochs. Because the longitude at the epoch dates is always supposed to be the same, the table need only list these dates, computed by repeatedly adding the assumed length of the year to some initial date. The natural stages to choose in its anomalistic cycle are the dates of slowest and fastest motion, corresponding to the apogee and perigee in a kinematic model. No example of a table of such epochs has yet turned up, but a table of dates of perigee can be inferred from the solar template **4162**, which uses this stage as its starting point. Alternatively, the equinoxes and solstices can be used as epoch dates. Thus the epoch table **4148** gives dates of summer solstices, extrapolated from an initial date for the solstice preceding the first regnal year of Augustus using a (sidereal) year of 365;15,33,46 days. A second template, **4163**, counts days from the summer solstice.

The moon. The epoch tables for the moon are a component of a method of computing lunar positions that I have called the 'Standard Lunar Scheme'.[2] Unlike Ptolemy's models, the Standard Scheme assumes only a single anomaly for the moon's motion, with the anomalistic month (roughly 27 $^1/_2$ days) for its constant period. In each anomalistic month, therefore, the moon progresses by a constant amount in longitude. In principle, one could use for epochs the consecutive occurrences of any stage of the anomalistic month, say the date of least motion; the epoch dates and the corresponding longitudes would be found by adding constants repeatedly to an initial date and position. In order to avoid the prolixity of a table that gives twelve or thirteen epochs in every year, the Standard Scheme uses as epochs only those dates of least motion that fall close to sunset (i.e. 6 p.m.), a device that also eliminates the need to keep track of fractions of days.

In the Standard Scheme it is assumed that 110 anomalistic months exactly equal 3031 days, and so epochs are tabulated at 3031-day intervals with constant increments in longitude of 337;31,19,7°. However, intervals of 3031 days, i.e. more than eight years between epochs, would be intolerably long. The scheme therefore adds ten more epoch dates at 248-day intervals following each of the epochs at 3031-day intervals, with a constant increment of 27;43,28,34° in longitude between each epoch and the next. In fact if one anomalistic month is $^{3031}/_{110}$ days, then nine anomalistic months amount to approximately 247.991 days. The tiny shortfall, which is taken up by a brief extra interval of lunar motion at close to minimum

[1] Detailed discussion in Jones [1997b].

[2] See Jones [1997a] for further details and bibliography.

speed, is reflected by a slight disproportionality in the increments used for 248-day and 3031-day intervals. After eleven epochs at 248-day intervals, an interval of 303 days and longitudinal increment of 32;33,44,51° makes up the remainder of the 3031 day period.

The procedure text *P. Ryl.* 27, which gives instructions for computing Standard Scheme epochs, also tells how to calculate the moon's argument of latitude, that is, its elongation from the northern limit of its orbit, for each epoch. The units are not degrees, but 'steps' (βαθμοί) of 15°, so that 24 steps amount to one revolution in latitude. Underlying the rules is the assumption that the northern limit shifts uniformly by –0;3,10,49,15° per day. Subtracting this motion from the increments in longitude, and converting from degrees to steps, one obtains the prescribed increments of 2;43,28,34,0 steps for each 248-day interval and 9;12,43,48,15 steps for each 3031-day interval.

The planets: System A. The points in the anomalistic cycles that serve as epochs form a sequence for the 'outer' planets Saturn, Jupiter, and Mars different from that for the 'inner' planets, Venus and Mercury, although in all cases they are coordinated with the planet's elongations from the sun, so that the cycle of variation in speed is referred to as a synodic period. The inner planets make alternate appearances as evening star (setting after the sun) and as morning star (rising before the sun), with intervening periods of invisibility. Venus's interval of invisibility between evening disappearance and first morning appearance always falls within its retrogradation, so that the two stationary points are also treated as distinct phases. Mercury's stations, on the other hand, are difficult or impossible to observe, and hence were assimilated in the schemes to the evening disappearance and morning appearance. The following lists the resultant sequence of phases during synodic period for these planets, together with Neugebauer's 'Greek-letter' symbols for them:

First morning visibility	Γ
Morning station	Φ
Disappearance (morning)	Σ
First evening visibility	Ξ
Evening station	Ψ
Disappearance (evening)	Ω

The relationship between these events and the continuous motion of the planet and the sun in longitude during the synodic cycle is displayed schematically in Fig. 3.

The outer planets have a single interval of visibility during each synodic period, with a retrogradation about half way through. The sequence is as follows:

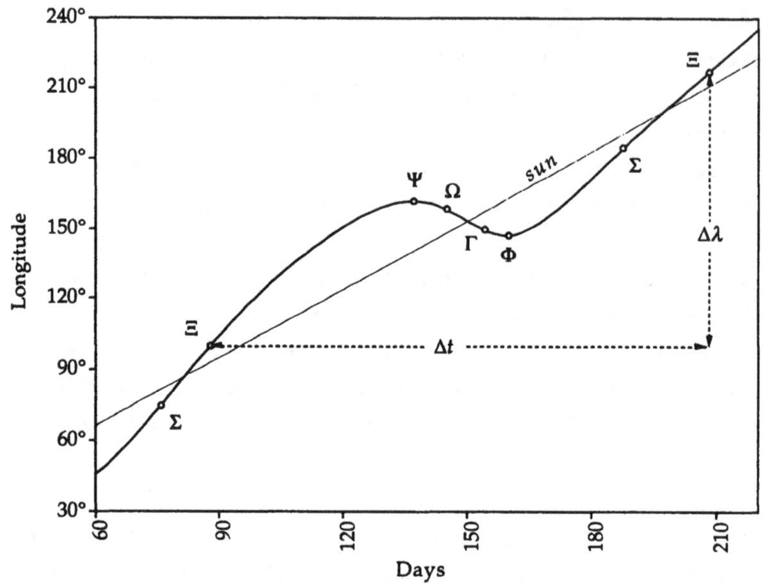

FIG. 3. Synodic cycle of an inner planet.

First visibility (morning)	Γ
First station (morning)	Φ
Acronychal (sunset) rising	Θ
Second station (evening)	Ψ
Disappearance (evening)	Ω

Fig. 4 shows how these phases are related to the motion of the planet and the sun.

The fact that each planet exhibits two anomalies is reflected in the arithmetical schemes by a variable spacing in both time and longitude between their epochs. This variation is assumed to depend only on the planet's longitude at the epoch beginning each synodic period. Babylonian astronomy had devised two general methods or models for quantifying the pattern of variation, known conventionally as System A and System B. The papyrus epoch tables provide direct and incontrovertible proof that planetary schemes of both System A and System B were known and used in Roman-period astronomy.[1]

I will here review the basic principles of these schemes, considered simply as computational devices for generating epoch tables, in order to provide a minimum conceptual background to the analysis of the individual texts. For a fuller treatment of the schemes and the theory underlying them, the reader should consult Neugebauer [1955] 279–315 and [1975] 347–473. The former work also

[1] See Jones [1998a] as well as the commentaries to the individual tables.

discusses computational shortcuts and analytical tricks, many of which I have found helpful in working with the papyri.

The System A model for computing longitudes of a particular phase of a particular planet divides the ecliptic into two or more zones of length a_i, with defined boundaries. Associated with each zone i is an 'ideal synodic arc', w_i, which is the amount by which the planet would progress between successive occurrences of the same phase so long as the planet is within the zone at *both* phases. If this much progress in longitude would put the planet past the boundary with the next zone, i+1, then the part of the synodic arc past the boundary must be scaled in the proportion w_{i+1}/w_i to reflect the change of zone. In certain situations the end of the synodic arc, after this adjustment, turns out to be past the *next* zone boundary, in which case the process must be repeated as if the synodic arc had originally begun in zone i+1 and surpassed its boundary with zone i+1.

The period P and mean synodic arc $\overline{\Delta\lambda}$ of any System A scheme are related to the chosen a_i and w_i values:

$$\sum \frac{a_i}{w_i} = P = \frac{360°}{\overline{\Delta\lambda}} \qquad (1)$$

Since P will not be a whole number for any nontrivial scheme, we define a whole number period Π as the least integer multiple of P. In other words, for each System A scheme there exists a whole number Π of iterations of the above

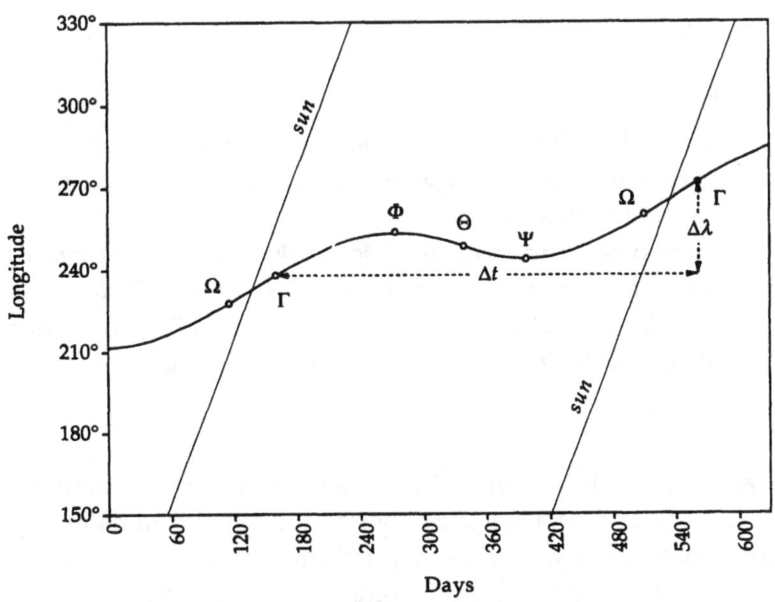

FIG. 4. Synodic cycle of an outer planet.

procedure that will bring the phase back to its initial longitude. Because each step represents a synodic cycle, the same period Π must contain a whole number of years Y and a whole number of revolutions of the phase in longitude R (we disregard any complete revolutions in longitude that the planet may have made during each synodic cycle). Thus we obtain the mean synodic arc ($\overline{\Delta\lambda}$), the mean synodic time ($\overline{\Delta t}$), and the difference between them (c):

$$\overline{\Delta\lambda} = \frac{R \times 360°}{\Pi} \tag{2}$$

$$\overline{\Delta t} = \frac{Y \text{ years}}{\Pi} \tag{3}$$

$$c = \overline{\Delta t} - \overline{\Delta\lambda} \bmod 360 \tag{4}$$

(Casting out 360s amounts to disregarding any complete lunar years of 12 months.) The value of c depends on the units in which $\overline{\Delta t}$ is reckoned. In the Babylonian texts time is counted in lunar months and thirtieths of months ('tithis'),[1] which are treated as approximately equivalent to calendar days within the lunar calendar. For the duration of the year one uses:

$$1 \text{ year} = 12;22,8 \text{ lunar months} = 371;4 \text{ tithis} \tag{5}$$

The planetary schemes assume that the actual synodic time (Δt) in tithis from one instance of a phase to the next is always greater than the synodic arc between the phases ($\Delta\lambda$) by c; thus:

$$\Delta t = \Delta\lambda + c \tag{6}$$

Dates of phases in the Babylonian lunar calendar are obtained by successively adding Δt to the previous date, if necessary also adding as many lunar years to the total as were cast out in determining c.

In the papyri this procedure is adapted to work with time intervals in true days instead of tithis, and with the Egyptian calendar year of 365 days instead of the Babylonian calendar year of 12 lunar months. Hence in the calculation of $\overline{\Delta t}$ (3) one uses a value for the length of the year in days; and instead of (4) one has:

$$c = \overline{\Delta t} - \overline{\Delta\lambda} \bmod 365 \tag{7}$$

The Babylonian and Greek versions of the rule are not exactly equivalent, although the discrepancies do not significantly effect the predicted dates of phases.

More than a dozen System A planetary schemes are known from the cuneiform sources. Fewer are as yet attested among the papyri, including schemes for

[1] This term is borrowed from Indian astronomy; Babylonian texts call these units simply 'days'.

Planet	*Scheme*	*Π*	*Z*	*c*	*i*	a_i	w_i	*boundary*	w_{i+1}/w_i
Mercury	A_1 for Γ	2673	848	1;40,8,...	1	165°	106°	286°	1;20
					2	134°	141;20°	60°	0;40
					3	61°	94;13,20°	121°	1;7,30
	A_1 for Ξ	1513	480	1;40,8,...	1	110°	160°	206°	0;40
					2	134°	106;40°	340°	0;54
					3	116°	96°	96°	1;40
	A_2 for Σ	1223	388	1;40,8,...	1	90°	107;46,40°	180°	1;12
					2	96°	129;20°	276°	0;45
					3	89°	97°	5°	1;20
					4	85°	129;20°	90°	0;50
	A_2 for Ω	684	217	1;40,8,...	1	180°	108;30°	270°	1;6,40
					2	60°	120;33,20°	330°	0;54
					3	60°	108;30°	30°	1;15
					4	60°	135;37,30°	90°	0;48
Mars	A for Γ, Φ, Ω	133	18	1;14,3,... + 2×365	1	60°	45°	90°	0;40
					2	60°	30°	150°	1;20
					3	60°	40°	210°	1;30
					4	60°	60°	270°	1;30
					5	60°	90°	330°	0;45
					6	60°	67;30°	30°	0;40
Jupiter	A (all phases)	391	36	0;44,43,... + 1×365	1	205°	36°	85°	0;50
					2	155°	30°	240°	1;12
	A' (all phases)	391	36	0;44,43,... + 1×365	1	120°	30°	219°	1;7,30
					2	53°	33;45°	272°	1;4
					3	135°	36°	47°	0;56,15
					4	52°	33;45°	99°	0;53,20
Saturn	A (all phases)	256	9	0;26,45,... + 1×365	1	200°	11;43,7,30°	330°	1;12
					2	160°	14;3,45°	130°	0;50

TABLE 1. Parameters of System A planetary schemes.

Mercury, Mars, and Jupiter. For ease of reference Table 1 sets out the defining parameters of the principal Babylonian System A schemes. The tabulated *c* are not the Babylonian parameters (based on the lunar calendar) but conjectural values for finding $\overline{\Delta t}$ in days, computed using a year length of 365;15,40 days.[1]

[1] Exceptionally, the two extant Babylonian System A_2 tables for Mercury do not use a constant *c* but vary its value according to a scheme that is not understood. In **4154**, however, the standard formula was used to compute dates, and this may reflect the normal Greek practice.

As an example of the use of this table, we compute a sequence of longitudes and dates of a phase (say, Φ) of Mars, starting with an initial phase at longitude Gemini 20;30° (i.e. 80;30°) on Tybi 10;55,20 (Egyptian calendar) of a year that we will call year 1. For *c* we will use the value 1;14,4 (plus two Egyptian years). First we need to find the zone of the System A scheme in which the initial phase falls. The column headed 'boundary' gives the *end* of each zone, from which it is apparent that 80;30° is in zone 1, with corresponding w_1 of 45°. Adding 45° to 80;30° gives 125;30°, which is past the zone boundary by:

$$\alpha = 125;30° - 90° = 35;30° \tag{8}$$

The next phase therefore falls in zone 2 or beyond. We calculate the corrected overrun:

$$\beta = \alpha \times {}^{w_{i+1}}\!/_{w_i} = 35;30° \times 0;40 = 23;40° \tag{9}$$

Adding β to the zone boundary yields 113;40° for the longitude of the second phase. Since this is not beyond the boundary ending zone 2, it requires no further correction. The synodic arc is:

$$\Delta\lambda = 113;40° - 80;30° = 33;10° \tag{10}$$

and the synodic time:

$$\Delta t = \Delta\lambda + c = 33;10 + 1;14,4 \text{ days} = 34;24,4 \text{ days} \tag{11}$$

which, added to the initial date of Year 1, Tybi 10;55,20, gives Year 3, Mecheir 15;19,24 for the date of the second phase.

Adding w_2 (30°) to the longitude of the second phase, we get 143;40°, which is still inside zone 2; this is therefore the longitude of the third phase. The synodic arc is of course 30°, the synodic time 31;14,4 days, and the resulting date Year 5, Phamenoth 16;33,28. However, the next addition of 30° brings us into zone 3, with an overrun (α) of 23;40°. This must be scaled up by a factor of 1;20, and the product (β = 31;33,20°) added to the zone boundary to obtain the longitude of the fourth phase, 181;33,20°. The synodic arc is 37;53,20°, hence the synodic time is 39;7,24°, and the date Year 7, Pharmuthi 25;40,52.

The computation of this series and its continuation is displayed more concisely in tabular format in Table 4158.1. In this and similar tables in the commentaries to the individual papyri, steps that take up two lines of the table are those where the subsequent phase falls *two* zones past its predecessor and therefore requires two successive corrections.

Figs. 6–12 show for each of the schemes in Table 1 how Δλ varies with initial longitude. The graphs provide a convenient tool for testing whether a particular scheme could have been used to generate a given sequence of longitudes or dates. At a more basic level, as fairly good imitations of the planet's actual patterns of synodic behaviour, they can help with the identification of a tabulated synodic event.[1] This is especially useful in the case of Mercury, since that planet's phases have the same mean period but very distinct distributions through the ecliptic.

The planets: simple longitudinal schemes. The System A rules for correcting the lengths of synodic arcs that straddle a zone boundary have the effect that the mean synodic arc and the period of each scheme are inherent in the relation (1). This is an important property, since control of the periodicity is necessary if prolonged sequences of computed phases are not to accumulate systematic errors. However, there existed certain situations in which arcs and times between phases were computed only singly rather than in a connected series, using a simple pattern of values assigned to various initial longitudes.

The extant Babylonian tables of phases of Mercury all use either System A_1 for the appearances or A_2 for the disappearances, but never both. To supply the disappearances in a System A table, one used a table of 'pushes', i.e. arcs and times for the interval of visibility assigned to the middle of each zodiacal sign. For initial longitudes between these points, one used linear interpolation to obtain the required push. A comparable scheme of pushes existed for the intervals of invisibility in System A_2.

The longitudes and times of Mars's Θ and Ψ were obtained by a rule of the same sort from the preceding Φ. There were several variants of this 'retrogradation scheme' associated with System A. Their purpose was to obtain long retrogradations (in both time and arc) in Cancer and Leo, and short ones in Capricorn and Aquarius, a pattern that independent computation of Θ and Ψ by System A would not produce. These special schemes for Mars and Mercury have not appeared in recognizable form in the papyri, although their presence in an early 6th century Sanskrit work, the *Pañcasiddhāntikā* of Varāhamihira, implies some means of transmission by way of a Greek intermediary.[2]

The planets: System B. Whereas in the System A schemes the synodic arc and time are computed from the initial longitude, the schemes referred to as System B derive each arc and time from its immediate predecessor; the correlation with the actual longitudes is maintained only approximately as a by-product of the chosen parameters. Consecutive synodic arcs and synodic times form 'linear zigzag functions', i.e. sequences of values that alternately increase and decrease linearly between fixed limits. The defining parameters of a zigzag function, for purposes

[1] Aaboe [1958] compares the patterns predicted by the System A and B planetary schemes with theoretical expectations derived from a simplified kinematic model, with conclusions generally affirming the quality of the Babylonian models.

[2] Neugebauer & Pingree [1971] v. 2, 119–120 and 121–123.

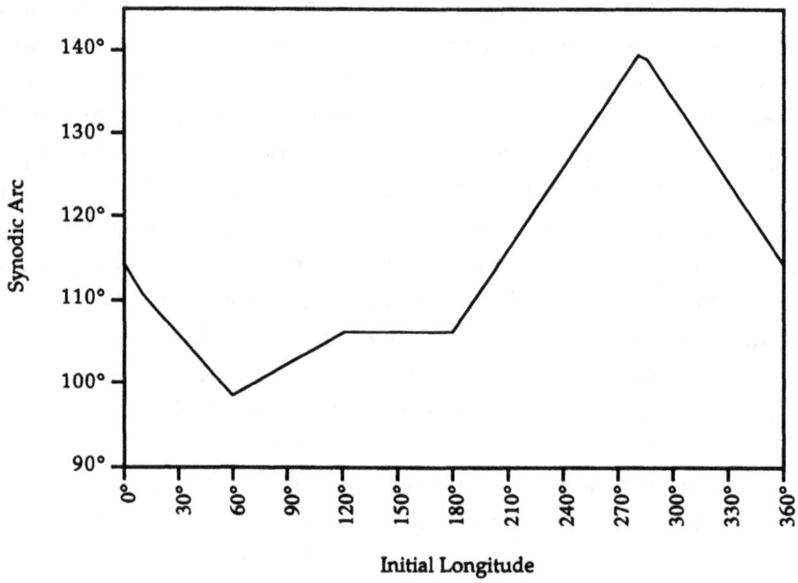

FIG. 6. Mercury, System A_1 for Γ.

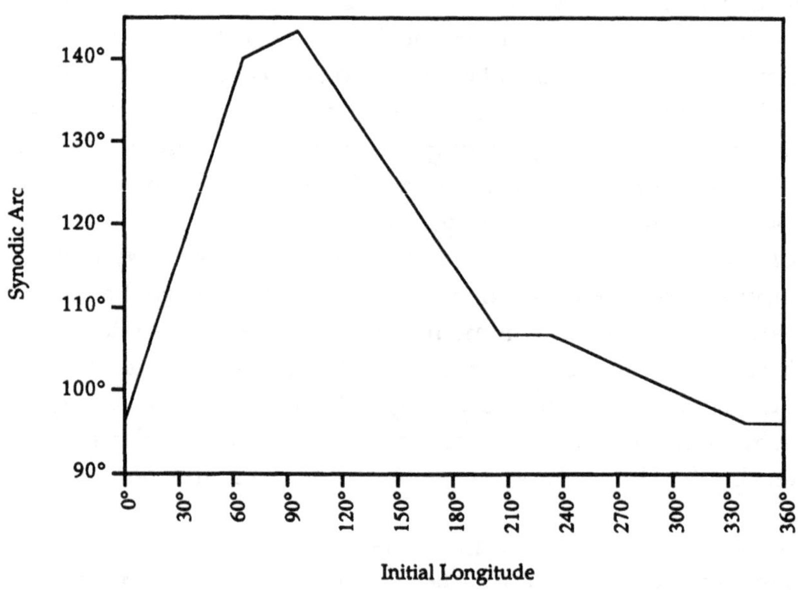

FIG. 7. Mercury, System A_1 for Ξ.

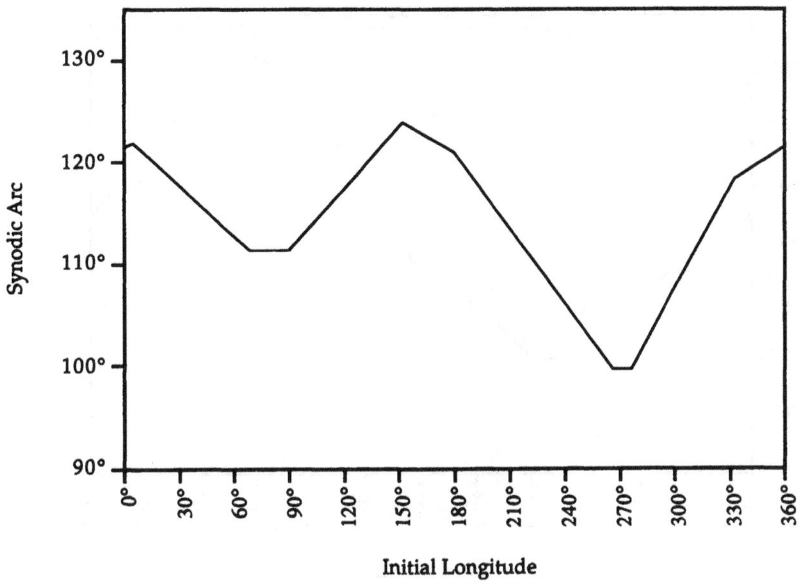

FIG. 8. Mercury, System A_2 for Σ.

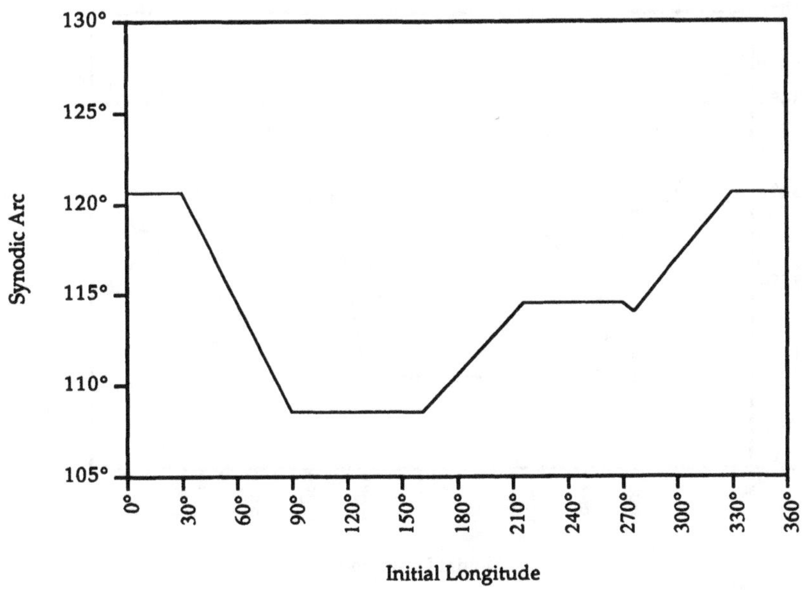

FIG. 9. Mercury, System A_2 for Ω.

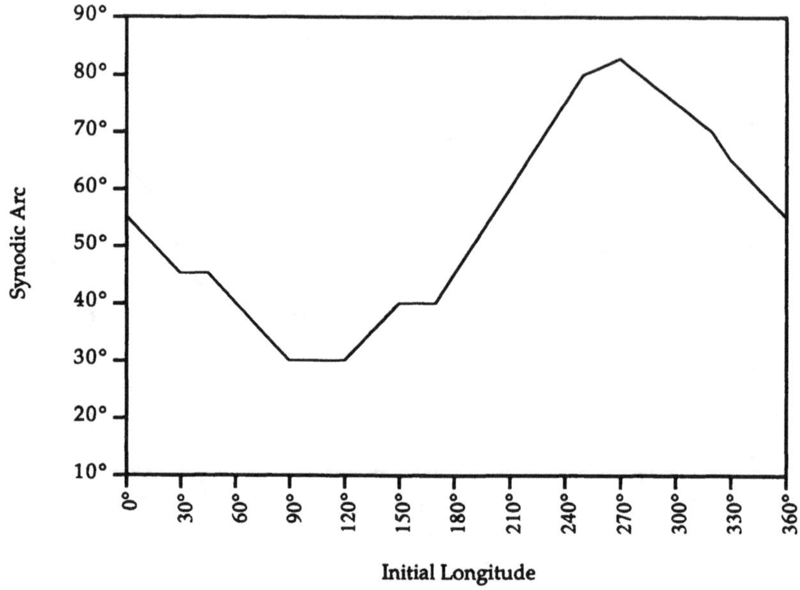

FIG. 10. Mars, System A for Γ, Φ, and Ω.

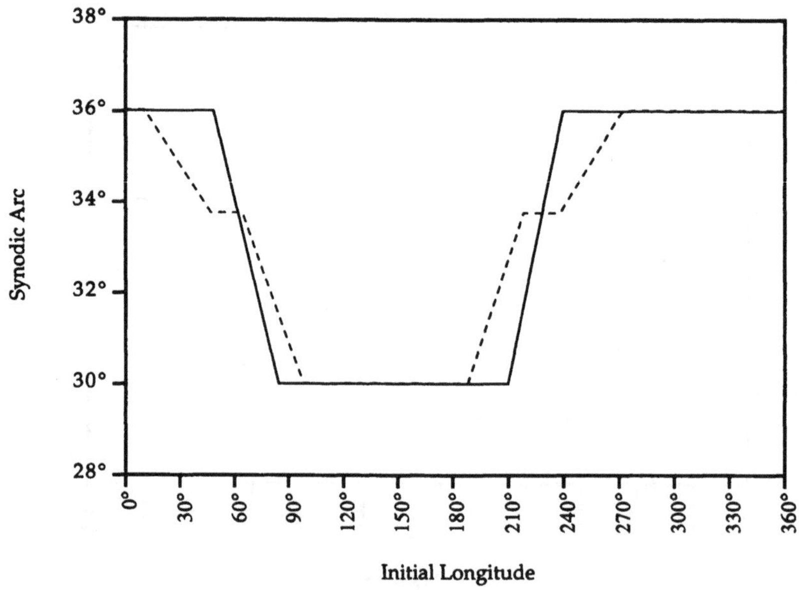

FIG. 11. Jupiter, System A (solid) and A' (broken) for all phases.

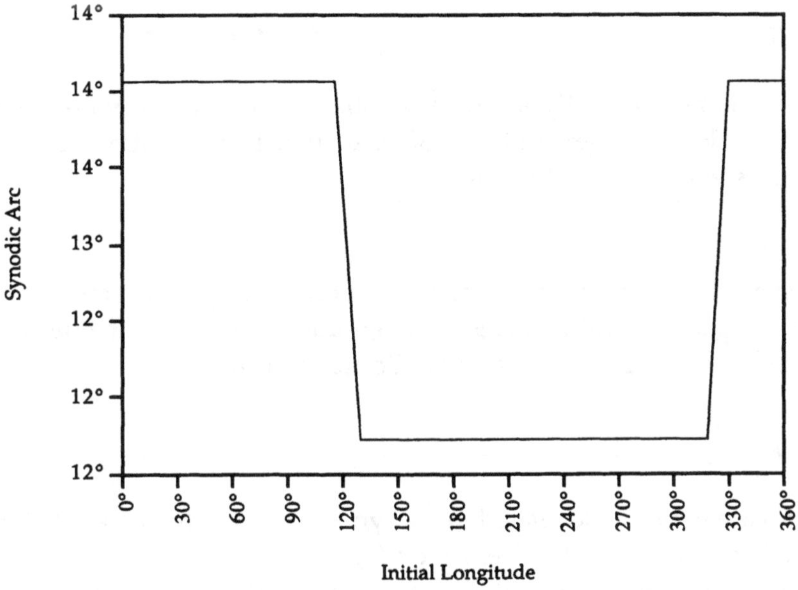

FIG. 12. Saturn, System A for all phases.

of computation, are the minimum limiting value m, the maximum M, and the increment/decrement d. Starting with some initial value z_0, each value of the function is said to belong to either an increasing branch or a decreasing branch. For any z_i, the next value z_{i+1} is determined as follows:

1) If z_i is on an increasing branch:
   a) If $z_i + d \leq M$, then $z_{i+1} = z_i + d$ on increasing branch
   b) If $z_i + d > M$, then $z_{i+1} = (2M - d) - z_i$ on decreasing branch

2) If z_i is on a decreasing branch:
   a) If $z_i - d \geq m$, then $z_{i+1} = z_i - d$ on decreasing branch
   b) If $z_i - d < m$, then $z_{i+1} = (2m + d) - z_i$ on increasing branch

The zigzag function is most easily (albeit anachronistically) visualized as a sequence of discrete and equally spaced values of a continuous function composed of equal stretches of linear increase and decrease reflected back and forth between m and M (Fig. 13). We may further define several other useful parameters: the amplitude Δ, the mean value μ, and the period P:

$$\Delta = M - m \tag{12}$$

$$\mu = \frac{M + m}{2} \tag{13}$$

$$P = \frac{2\Delta}{d} \tag{14}$$

Lastly, since P is not generally a whole number, we define the whole number period Π as the least integer multiple of P, so that in Π iterations the zigzag function returns to its initial value, that is,

$$z_{i+\Pi} = z_i \tag{15}$$

For the special case of a zigzag function representing synodic arcs of a planet (so that a running total of values gives the longitudes of the phase in question) we have the restriction relating the mean synodic arc to the period:

$$\overline{\Delta\lambda} = \mu = \frac{360°}{P}\text{, cf. (1)} \tag{16}$$

Synodic times are in principle related to the synodic arcs by the same rule (6) as in System A, where c is defined as before by (2)–(4). In practice one computes Δt using a second zigzag function with the same amplitude and period as the function for $\Delta\lambda$, but with m, M, and μ all increased by c. In the ACT tables the two functions are typically very slightly out of phase.

From the Babylonian texts we know of System B schemes for Mars, Jupiter, and Saturn. The parameters of these schemes are listed in Table 2. So far the schemes for Jupiter and Saturn have been found in the papyri (cf. **4160a**, **4161**).

Intermediate longitudes and daily motion. The problem of determining the motion of a heavenly body *between* epochs, and specifically to find its longitude on an

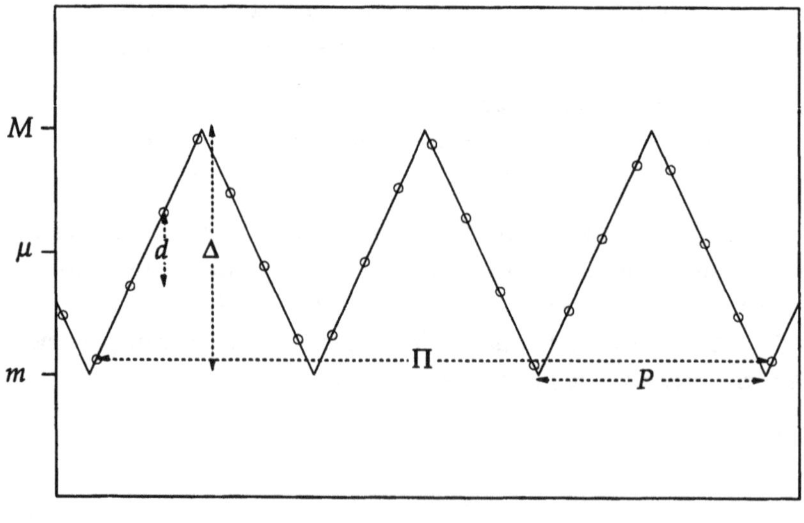

FIG. 13. Linear zigzag function.

	Mars	*Jupiter System B*	*Jupiter System B'*	*Saturn*
d	17°	1;48°	1;46,40°	0;12°
m	17;19,8,30°	28;15,30°	28;19,20°	11;14,2,30°
M	80;7,28,30°	38;2°	37;58,20°	14;4,42,30°
μ	48;43,18,30°	33;8,45°	33;8,45°	12;39,22,30°
P	7;23,20	10;51,40	10;51,33,45	28;26,40
Π	133	391	695	256

TABLE 2. Parameters of System B planetary schemes.

arbitrary date, was solved in various ways, which we need consider here only in outline. The methods can be grouped loosely into those that break the anomalistic period up into several intervals of linear (i.e. constant) motion, and those that use patterns involving continuously varying speed.

A pattern of motion that prescribes linear motion to bridge a series of dates tied to an epoch may be called a 'velocity scheme'. For example the procedure text **4134** sets out a velocity scheme for Venus that divides its synodic period into fixed intervals, mostly of 60 days, to each of which a constant daily motion is assigned. Because Venus's synodic cycles are nearly of constant length and show little variation in the pattern of longitudinal motion, this scheme could be used without adjustment to predict any number of consecutive cycles. For the other planets some account must be taken of the variation in synodic cycles through the ecliptic; hence from Babylonian astronomical texts we know of several velocity schemes for Saturn and Jupiter that had values for the constant speeds associated with each of the System A zones.[1] Examples of Babylonian velocity schemes are summarized in Table 3. Note that in some schemes so-called 'satellite phases' (Ω' and Γ') are introduced one month before Ω and one month after Γ in order to prolong the interval of fastest progress.

To generate a pattern of daily longitudes that avoids abrupt changes of speed at certain points, one could use arithmetical sequences of a higher order, i.e. 'second order' sequences of daily progress increasing or decreasing by a constant line-to-line difference, or 'third order' sequences such that the differences *of the differences* are constant. A second order sequence can be devised, for example, to bridge smoothly any interval from an outer planet's fastest motion (near Γ) to its stationary point Φ, given the dates and longitudes of the beginning and end of the interval.

[1] Babylonian methods of computing daily motion of planets are surveyed in Huber [1957].

Jupiter (System A'), scheme from procedure texts:

Interval	*Slow Arc*		*Medium Arcs*		*Fast Arc*	
	Daily	*Total*	*Daily*	*Total*	*Daily*	*Total*
Γ–Γ', 1 month	0;12,30°	6;15°	0;14,3,45°	7;1,52,30°	0;15°	7;30°
Γ'–Φ, 3 months	0;6,40°	10°	0;7,30°	11;15°	0;8°	12°
Φ–Ψ, 4 months	–0;4,10°	–8;20°	–0;4,41,15°	–9;22,30°	–0;5°	–10°
Ψ–Ω', 3 months	0;6,23,20°	9;35°	0;7,11,15°	10;46,52,30°	0;7,40°	11;30°
Ω'–Ω, 1 month	0;12,30°	6;15°	0;14,3,45°	7;1,52,30°	0;15°	7;30°
Ω–Γ, unspecified	0;12,30°		0;14,3,45°		0;15°	

Jupiter (System A), scheme from planetary tables:[1]

	Slow Arc	*Fast Arc*
Γ–Φ	16;15°	19;30°
Φ–Θ	–5°	–6°
Θ–Ψ	–5°	–6°
Ψ–Ω	17;45°	21;18°
Ω–Γ	6°	7;12°

Saturn (System A), scheme from procedure texts:

Interval	*Slow Arc*		*Fast Arc*	
	Daily	*Total*	*Daily*	*Total*
Γ–Γ', 1 month	0;5°	2;30°	0;6°	3°
Γ'–Φ, 3 months	0;3,20°	5°	0;4°	6°
Φ–Θ, 52;30 tithis	–0;4;13,40°?	–3;41,57,30°?	–0;5,4,24°	–4;26,21°
Θ–Ψ, 2 months	–0;3,20°	–3;20°	–0;4°	–4°
Ψ–Ω', 3 months	0;3,35,30°	5;23,15°	0;4,18,40°	6;28°
Ω'–Ω, 1 month	0;5°	2;30°	0;6°	3°
Ω–Γ, unspecified	0;5°	?	0;6°	?

Saturn (System A), scheme from planetary tables:[2]

Interval	*Slow Arc*	*Fast Arc*
Γ–Φ	7;30°	9°
Φ–Ψ	–6;40°	–8°
Ψ–Ω	7;33,7,30°	9;3,45°
Ω–Γ	3;20°	4°

TABLE 3. Representative Babylonian velocity schemes for Jupiter and Saturn.

[1] Aaboe & Sachs [1966] 4.

[2] Aaboe & Sachs [1966] 7.

Such patterns are used in the variety of table referred to as 'templates' (see §6c), which present a standardized pattern of longitudinal progress day by day starting from an epoch. For the planets, the templates use a division of the synodic period into intervals of linear and second order motion tailored to imitate the actual patterns of longitudinal progress as illustrated in Figs. 4–5. In some instances the parameters seem to have been adapted from the Babylonian velocity schemes.

The lunar template belonging to the Standard Scheme employs linear zigzag functions for the daily progress in longitude and argument of latitude, with the following parameters:[1]

	longitude	*argument of latitude*
d	0;12,50°	0;0,51,20 steps
M	14;38,59,7°	0;58,48,39,45 steps
m	11;42,10,37°	0;47,1,25,45 steps
μ	13;10,34,52°	0;52,55,2,45 steps
P	$\frac{3031}{110}$ days	$\frac{3031}{110}$ days
Π	3031 days	3031 days
Z	110	110

Consequently the actual longitudes and arguments of latitude in the lunar template form second-order sequences beginning and ending with each passage of a maximum or minimum in the zigzag functions.

Of the two extant templates for the sun, **4162** is computed from a kinematic model, not an arithmetical scheme, whereas **4163** uses linear interpolation between fixed stages of the annual solar motion.[2]

Ascensions. The remaining computational problem frequently encountered by the users of these arithmetical methods was to determine the longitudes of the ascendant and the other cardines for a given date and time in seasonal hours. Neugebauer [1975] 712–733 has assembled much evidence for the widespread use in the late Hellenistic and Roman periods of a set of tables of oblique ascensions based on simple arithmetical patterns rather than spherical trigonometry; these were adapted from Babylonian schemes that were originally part of the System A and B lunar theory. Although one suspects that this kind of table lies behind the computation of the ascendants in many papyrus horoscopes, it is seldom possible to check because the texts give only the zodiacal sign without degrees.

[1] Jones [1997a].

[2] Jones [1997b].

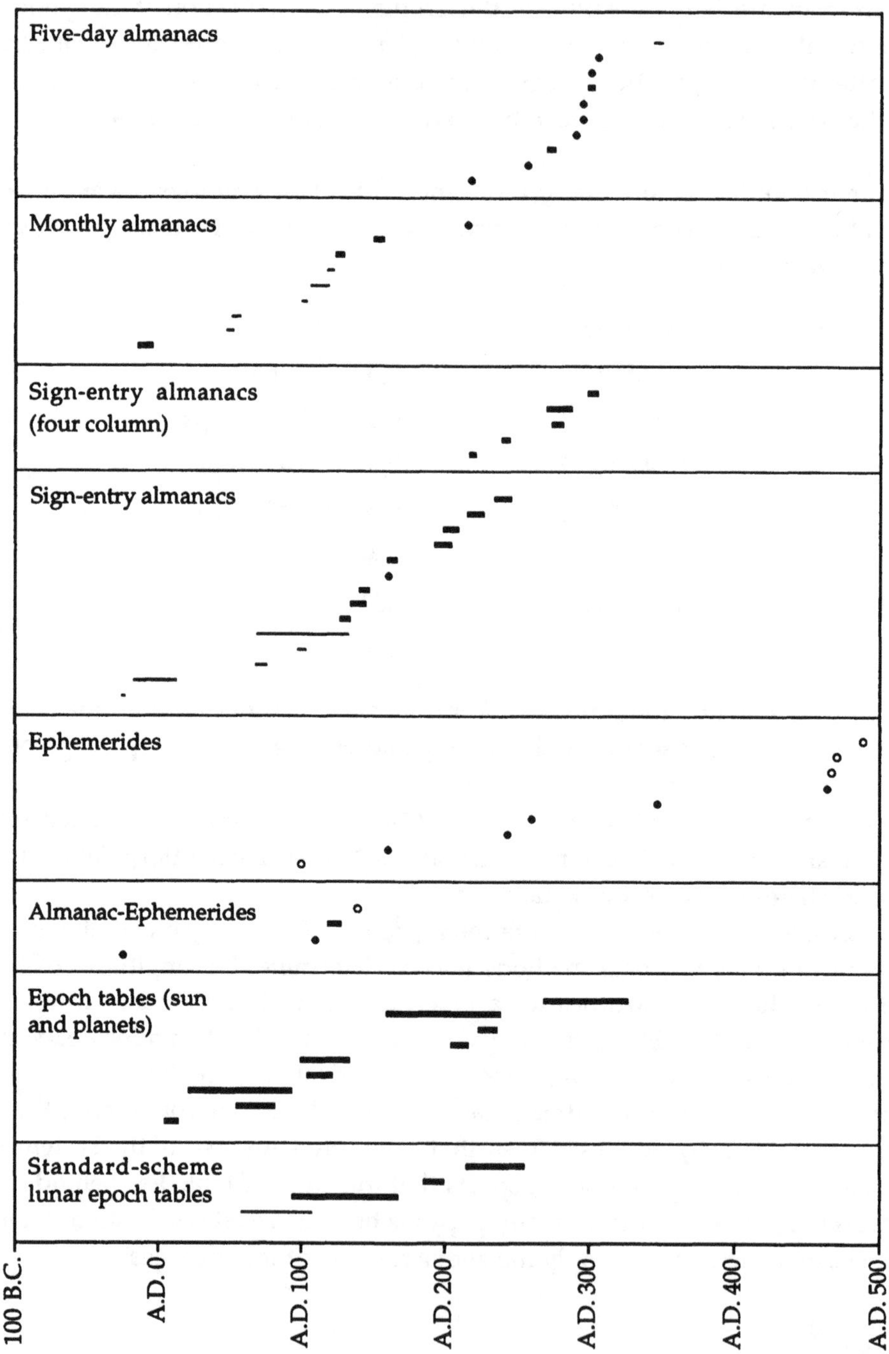

FIG. 14. Dated astronomical tables. (heavy = *P. Oxy.*, light = other)

§6. VARIETIES OF ASTRONOMICAL PAPYRI[1]

a. Texts

Unlike the tables and horoscopes that make up the bulk of the papyri edited in this work, the astronomical prose texts do not tend to break down into a few standard types or formats. Here I can only attempt to give some impression of the variety of the texts, leaving detailed treatment to the commentaries.

Theoretical texts, not directly concerned with prediction, are rare. Among the previously published Greek astronomical papyri, one can single out *P. Aberd.* 12 and 127, both discussing constellations; *P. Iand.* 84, from a work on spherical astronomy; *P. Oxy.* 303, on the motion of the moon; and *P. Oslo* 73, on measuring the apparent breadth of the sun's disk. Only the last of these is well enough preserved to make any connected sense. The new texts include a mutilated passage on an astronomical model illustrated by a diagram (**4144**), and three fragments of considerable historical interest. Two of these, **4138a** and **4139**, concern periodicities of the moon. The third, **4133**, is a substantial chunk of an astronomical treatise from the early second century, probably by Menelaus of Alexandria, in which observations of Jupiter are reported and subjected to analysis.

Instructions for the generation and use of tables and for other astronomical calculations are, not surprisingly, well represented. Of these 'procedure texts', the only one hitherto published that is at all well understood is the important *P. Ryl.* 27, the main part of which is on the computation of epochs for the Standard Lunar Scheme. Enough is preserved of three other procedure texts, *PSI* 1490 and 1491 and *P. Ant.* 141, to offer hope of eventual elucidation. In this publication, **4135** sets out a velocity scheme for Venus's progress through its synodic cycle that was previously known from a medieval Sanskrit source; **4136** concerns the computation of running totals in the Standard Scheme lunar template; and **4142** and **4143** are from manuals for Ptolemy's *Handy Tables*.

A third category, containing actual predictions of astronomical phenomena, is represented by the two eclipse canons P. dem. Berol. 13146+13147 and **4137**. The odd tabular text **4140** also seems to consist of predictions of some sort involving the moon.

b. Epoch tables

The purpose of an epoch table is to list successive occurrences of a specific kind of event in the motion of a heavenly body over an interval of years (cf. §5b). For a planet, the event in question is one of the phases, such as first evening visibility. Lunar epoch tables give dates of slowest motion (i.e., roughly, apogee). For the

[1] The classification of formats of table given here is explained in more detail in Jones [1999a].

Text	*Body*	*Years*	*Event*	*D*	*T*	*B*	*Other columns*
4148	Sun	161–237	Summer solstice	Yes	Yes	No	
P. Lund 35a	Moon	59–108	Slowest motion	Yes	Yes	Yes	
4149	Moon	96–166	Slowest motion	No	Yes*	Yes	
4150	Moon	187–198	Slowest motion	?	Yes	Yes	
4151	Moon	217–254	Slowest motion	Yes	Yes	Yes	*Ω* (vestigial)
4152	Mercury	?	Ξ, Σ	?	?	Yes	
4153	Mercury	206–215	Γ	Yes	Yes	?	
4154	Mercury	?	Ω	?	Yes	?	
4155	Mercury	?	Ω, Γ, [Σ]	?	?	Yes	$T_\Gamma - T_\Omega$, $T_\Sigma - T_\Gamma$
4156a	Mercury	?	Ω	?	Yes	Yes	
4156b	Mercury	?	Γ	?	Yes?	Yes	
4156c	Mercury	?	Ω?	?	?	Yes	
4157	Venus	?	?	?	Yes	?	ΔT
4157a	Venus	106–121	inf. conj.	?	Yes	?	ΔT?
4158	Mars	271–325	Φ, Θ?	?	Yes	?	ΔT
4159	Mars	225–235	Γ	?	Yes	?	
4159a	Mars	101–133	Ψ	?	Yes	Yes	
4160	Jupiter	30–79	Φ	?	Yes	?	
4160a	Jupiter	6–13	Γ	?	Yes	?	ΔT
4161	Saturn	23–92	Γ	No	Yes	Yes	

(*Alexandrian calendar)

TABLE 4. Epoch Tables.

sun, the only extant table uses summer solstices, but from the template **4162** we can infer that there were solar epoch tables listing dates of perigee. In each case, the event is tied to some stage in the periodic variations of the body's apparent speed. Although these events may have had some astrological significance, the main use of an epoch table was to provide a succession of fixed starting points or 'epochs' from which one could project the body's subsequent course.

The known epoch tables are listed in Table 4. Each table normally lists in parallel columns the accumulated discrepancy in days between the Egyptian and Alexandrian calendars (*D*), the regnal year and date of the event in the Egyptian calendar (*T*), and the body's corresponding longitude (*B*). In column *T* a sexagesimal fraction of a day sometimes follows the day number. Longitudes are given as zodiacal sign, degrees, and sexagesimal fractions. The longitude can be omitted from a solar epoch table because it is assumed to be constant for all epochs. We know from the procedure text *P. Ryl.* 27 that lunar epoch tables could have an extra column (Ω) for the moon's argument of latitude (i.e. its elongation from the northern limit of its orbit). The argument of latitude was measured in units called βαθμοί ('steps') such that one βαθμός is equivalent to 15°.

The format of the epoch tables resembles that of the most common type of Babylonian planetary tables, which similarly correlate the dates and longitudes of a planet's successive phases.[1] Since the computational methods underlying certain of the papyrus tables are demonstrably derived from the procedures of the Babylonian tables, it seems probable that the format is also an adaptation of the cuneiform models, adjusted to work with the Egyptian and Alexandrian calendars instead of the Babylonian lunar calendar. Counterparts of the solar and lunar epoch tables have not been found among the Babylonian texts.

Text	*Body*	*Epoch*	*Complete period?*	T'	B'	ΔB	*Other columns*
4162	Sun	Perigee	Yes	Yes	No	No	B
4163	Sun	Summer solstice	Yes	Yes	No	No	B
PSI 1493	Moon	Slowest motion	Yes	Yes	Yes	No	Ω'
4150	Moon	Slowest motion	Yes	Yes	Yes	?	Ω'
4164	Moon	Slowest motion	Yes	Yes	Yes	Yes	
4164a	Moon	Slowest motion	Yes	Yes	Yes	Yes	
P. Carlsberg 32 (demotic)	Mercury	Γ	?	Yes	Yes	Yes	
4217	Mercury	Γ	No	?	Yes	No	
4217a–c	Mercury	Ξ, Γ	No	?	Yes	No	
4165	Mars	Ψ	No	Yes	Yes	Yes	
P. Berol. inv. 21236	Mars	Γ?	?	Yes	Yes	No	
4165a	Jupiter	Γ	?	Yes	Yes	Yes	
PSI 1492	Saturn	Γ	Yes	Yes	Yes	Yes	
4166	Saturn	Ψ, Ω	No	Yes	Yes	Yes	

TABLE 5. Templates.

[1] Neugebauer [1955] v. 2, 280–281.

c. Templates

A template is a tabulation of the day-by-day progress of a heavenly body during part or all of an anomalistic period, with time and longitudinal progress measured from an arbitrary epoch. The known examples are listed in Table 5. The normal format comprises two parallel columns: whole numbers of days elapsed since epoch (T'), and progress in longitude relative to the longitude of the body at epoch (B'), measured in degrees and sexagesimal fractions of a degree. T' is normally tabulated only every five lines. A column for daily progress in longitude (ΔB), i.e. the line-to-line differences of B', sometimes also appears. In solar templates, instead of B' the longitude itself (B) is tabulated (zodiacal sign, degrees, sexagesimal fractions), because the epoch longitude is assumed never to vary. A lunar template may also have a column for progress in argument of latitude since epoch (Ω'), measured in βαθμοί of 15°.

d. Kinematic tables

The extant papyrus tables that reflect kinematic models by breaking apparent motion down into mean motions and equations can be subdivided into three groups: parts of Ptolemy's *Handy Tables*, in the form familiar from the later Byzantine copies; tables that are apparently derived from or influenced by the *Handy Tables*; and tables that have no obvious relation to Ptolemy's.

The medieval manuscripts of Ptolemy's *Handy Tables* contain a consistent kernel of tables that correspond closely to the list in Ptolemy's own *Introduction* as well as to the more detailed information in Theon's two commentaries. Papyrus fragments that substantially agree with their counterparts in the medieval tradition in both format and numerical contents will here be called 'normal' (Table 6). In Table 7, the papyri are listed again, this time according to the headings of Tihon's survey of the medieval tables.[1] One gets the impression that copies typi-

Text	*Codex?*	*Tables preserved (in part)*
4167	No	Oblique ascensions; anomaly table; planetary latitudes; stations; phases; greatest elongations
4168	Yes	Declinations & lunar latitude; solar & lunar mean motions; parallax; planetary anomaly; unidentified tables
4169	Yes	Planetary anomaly
4170	Yes	Anomaly table; planetary mean motions
4171	Yes	Oblique ascensions

TABLE 6. Manuscripts of the *Handy Tables*.

[1] Tihon [1992].

Table	*Description*	*Extant Papyri*
A1	Right ascensions	—
A2	Oblique ascensions	**4167, 4171**
A3	Mean motions, sun and moon	**4168**
A4	Anomaly, sun and moon	—
A5	Declinations and lunar latitude	**4168**
A6–12	Eclipses	—
A13	Parallax	**4168**
A14	Mean motions, planets	**4170**
A15	Anomaly, planets	**4168, 4169**
A16	Latitudes, planets	**4167**
A17	Stations, planets	**4167**
A18	Phases, planets	**4167**
A19	Greatest elongations, planets	**4167**
C1	Regnal canon	—
G1	Noteworthy cities	(*P. Ryl.* 523)

TABLE 7. *Handy Tables* MSS, classified by table types.

cally comprised the whole corpus of tables, even though such tables as those for parallax, eclipses, planetary latitudes, and planetary phases were probably of little use to their owners.

While the 'normal' *Handy Tables* papyri make a straightforward contribution to our understanding of the textual history of Ptolemy's work, other papyri add confusion by presenting authentically Ptolemaic data in a form not known from the *Almagest* or the medieval *Handy Tables*, or by using the structure of parts of the *Handy Tables* to present non-Ptolemaic data. These 'wild' papyri (Table 8) are open to interpretation as (1) adaptations of Ptolemy's published work by other hands, (2) authentic versions by Ptolemy that did not survive into the Middle Ages, or (3) compositions independent of Ptolemy, which resemble his tables only because he himself was following established norms. Although the first explanation seems in general the most economical, the relationship of these papyri to Ptolemy has to be considered on a case-by-case basis.

Only a few tables have so far turned up that make use of the analysis of apparent motion into mean motion and anomaly (also in Table 8). These too are difficult to situate in historical relation to the better-attested arithmetical and Ptolemaic traditions.

Text	*Codex?*	*Tables preserved (in part)*
P. Ryl. 522/523	Yes	Noteworthy cities; oblique ascensions
P. Lond. 1278	Yes	Geographical table; geog. latitudes and longest daylight; right asc. and eq'n of time; right asc. and decl.; planetary mean motions, with column of unknown meaning
4172	No	Planetary phases?
4173	Yes	Planetary mean motions
4173a	Yes	Anomaly? and others?
4174	Yes	Lunar mean motion (unrelated to Ptolemy)
4174a	No	Solar mean motions (unrelated to Ptolemy)
4220	Partly	Lunar mean motions, plan. anomaly? (unrelated to Ptolemy)

TABLE 8. 'Wild' *Handy Tables* manuscripts and other kinematic tables.

e. Ephemerides

An ephemeris is a table that gives day-by-day positions of some or all of the heavenly bodies for a range of dates organized according to calendar months.[1] The known ephemerides are listed in Table 9. Two varieties are known: 'pure' ephemerides in which the daily longitudes of the planets are tabulated in parallel columns along with the columns for calendar days, the moon's longitudes, etc.; and 'almanac-ephemerides' in which the dates and longitudes of critical stages in each planet's motion (e.g. sign-entries, phases, longitudes on the first day of the month) are recorded in a separate monthly almanac above the ephemeris (see below, §6g).[2] The almanac-ephemeris seems to be the older format. There seems also to be a transition from roll to codex, and (not simultaneously) from use of the Alexandrian calendar to the Roman as the calendar according to which the ephemerides were laid out, although the earliest ephemeris already contains a column giving Roman equivalents of the Alexandrian dates.

A text describing the arrangement of 'ephemerides' (περὶ τῆϲ τῶν ἐφημερίδων ἐκθέϲεωϲ) has come down to us in several manuscripts of Theon's shorter commentary on the *Handy Tables*.[3] The plan it prescribes is essentially that of our pure

[1] The term 'ephemeris' has often been applied loosely to tables of consecutive phenomena, e.g. Babylonian tables of planetary phases or syzygies. In the interests of clearer nomenclature it would be desirable to reserve this word for tables of daily positions, and to use more precise terms ('phase table', 'syzygy table', or more generally 'epoch table') for the other varieties of text.

[2] P. Mich. inv. 1454 was the first pure ephemeris to be analysed, by Curtis & Robbins [1935]. For the structure of almanac–ephemerides, see Jones [1994a].

[3] Text (corrupt) in Halma [1822–1825] v. 3, 38–42, and Curtis & Robbins [1935] 82–83, who also provide a translation. It is anonymous text no. 5 in the manuscript inventory of Tihon [1978] 359.

Text	*Years*	*Codex?*	*Calendar*	*Almanac?*	*Columns in ephemeris*
4175	24 B.C.	No	Alex.	Yes	Roman, Alex. calendars; moon
P. Dub. TCD F.7	100	No	Roman?	No	... planets; ...
4176	111	No	Alex.	Yes	... moon; moon's sign-entries
4177	121–127	No	Alex.	Yes	Alexandrian calendar; moon (only on dates of sign-entry)
P. Harr. 60	140	No	Roman	Yes	planetary weeks; Roman, Alex. calendars... moon?; ?
4181	161	No	Roman?	No	Roman, Alex., lunar cal.; moon; moon's sign-entries; ...planets
4177a	245	No	Alex.	?	moon; moon's sign-entries
4178	261	No	Roman	?	Roman, Alex., lunar calendars; moon; moon's sign-entries
4184a	293	No	Alex.	No	Alexandrian calendar; moon...
4179	348	No	Roman	No	Roman, 'Greek', Alex., lunar calendars; moon; moon's sign-entries; syzygies; sun; planets
4180	465	Yes	Alex.	No	...Roman, lunar calendars; evaluation of day; moon; sun; planets
P. Mich. inv. 1454	467	Yes	Roman	No	weeks; Roman, Alex., lunar calendars; moon; moon's sign-; entries; sun; planets; eval. of day
P. Vind. G. 29370b	471	Yes	Roman	?	weeks; Roman, Alex., lunar calendars; moon; ...
P. Vind. G. 29370	489	Yes	Roman	No	weeks; Roman, 'Greek', Alex., lunar calendars; moon; ...planets; ascend. node; evaluation of day
4182	?	Yes	Roman	No	Roman, Alex., lunar calendars; moon; ...planets; eval. of day?
4183	?	Yes	Roman	No	Roman, Alex., lunar calendars;...?
4184	?	No	?	?	...unknown calendar; moon...
P. Vind. G. 3231	?	Yes	?	No	...moon and lunar sign-entries; ...planets...
P. Berol. inv. 21240	?	No	?	No	...planets...

TABLE 9. Ephemerides.

ephemerides. For each Roman month, a table of thirty-one lines is to be laid out, with rulings at two-line intervals (excepting the last interval of three lines). From left to right, columns are to be assigned as follows:

1. prognostications (ἐπισημάσιαι) from the fixed stars
2. three or four columns for calendar dates, according to (a) the Roman, (b) the Alexandrian, (c) the local, and (d) the lunar calendar
3. four columns below the heading, 'motion of the moon': (a) the zodiacal sign, (b) degrees and minutes of longitude, (c) times when the moon enters a new sign, and (d) 'winds' (i.e. directions associated with the moon's latitudinal motion? cf. Vettius Valens I 18 Kroll = I 16 Pingree)
4. six columns containing the longitudes of the sun and the five planets
5. a calendrical column repeating Roman (or other) calendar dates
6. aspects of the moon and the other planets
7. general evaluations (καταρχαί) of the day (e.g. as auspicious or inauspicious)

The format of the ephemeris obviously allowed for much variation in the choice of information to be tabulated; and in fact no two preserved ephemerides have precisely the same columns. Not all the ephemerides contain explicitly astrological data. Those that do are directed especially towards catarchic astrology—in particular the determination of whether each day is auspicious or inauspicious for some activity—which only makes sense if the ephemeris was available *before* the dates it appraises.

f. Sign-entry almanacs

'Sign-entry almanacs' (Table 10) record the dates on which each planet makes its crossings from one zodiacal sign to a neighbouring sign. The usual format is year-by-year and, within each year, planet-by-planet in the standard Greek order: Saturn, Jupiter, Mars, Venus, Mercury. This sequence puts the planets in order of decreasing longitudinal and synodic periods, so that the number of sign-entries increases from one planet to the next. Sometimes the almanac is so arranged that each column starts a new year.

Each line of the almanac has three entries: the month (which may be omitted if it is the same as for the preceding line), the day of the month, and the sign entered. The month may be named or identified numerically by its sequence within the months of the Egyptian or Alexandrian calendar, starting with Thoth = 1. Similarly, the zodiacal signs are often represented as numbers in order of increasing longitude, starting with Virgo = 1. In addition to the sign-entries, other events such as phases or stations may be recorded in the almanac, in which case the sign specified is the one *within which* the event occurs. The sign occupied by each planet on the first day of the year is often recorded in this way.

Text	*Years*	*Calendar*	*Medium*	*Remarks*
Tab. Amst. inv. 1	26–24 B.C.	Egyptian or Alex.	boards	Callippic years
P. Berol. 8279 (demotic)	17 B.C.– A.D. 12	Egyptian	roll	
P. Laur. 144	69–75	Alexandrian	codex	
Bodl. MS Gr. Class. F 7 (P)	98–103	Egyptian	boards	years in era Titus
Stobart Tablets (demotic)	70–133	Alexandrian	boards	
4185	129–133	Alexandrian	roll	
4189	136–144?	Alexandrian	codex	
4186	142–146	Egyptian	roll	
4187	160/161	Alexandrian	roll	
4188	161–165	Alexandrian	codex	
4193	195–203	Alexandrian	codex	
4188a	201–208	Alexandrian	codex	
4196	218–220	Alexandrian	codex	four–column
P. Oxy. 3299	217–225	Alexandrian	codex	
4190	241–243	Alexandrian	codex	four–column
4191	236–245	Alexandrian	codex	
4192	276–280	Alexandrian	codex	four–column
4194	272–286	Alexandrian	codex	four–column
4195	300–304	Alexandrian	codex	four–column
4196a	?	Alexandrian	codex	four–column
P. Stras. inv. gr. 1097 (unpublished)	?	?	?	
P. Vind. G. 36041 + 26011	perpetual	Alexandrian	codex	
4197	perpetual	Alexandrian	codex	
4198	perpetual	Alexandrian	codex	

TABLE 10. Sign-entry almanacs.

Two variants of the sign-entry almanac should be noted. In some, the date of nocturnal sign-entries is given as a double-date (e.g. '16 17' signifies sign-entry during the night of the 16th leading into the 17th), and a fourth column of numerals follows the zodiacal sign. The numbers in this column represent the time of sign-entry in seasonal hours. A further peculiarity of the four-column almanacs is that the sign-entries that they contain were apparently always computed using the *Handy Tables*.

Secondly, we have examples of perpetual sign-entry almanacs that give entries for an entire recurrence period for each planet in turn. The same principle of arrangement is found in the medieval 'Almanacs'. To obtain the positions for a given planet and year, one would divide the year's number, counted from a specified era, by the number of years in the recurrence period, and use the remainder to locate the appropriate year's entries in the almanac. The recurrence periods are well known as the Goal-Year periods of Babylonian astronomy:[1]

Saturn	59 years	Venus	8 years
Jupiter	71 or 83 years	Mercury	46 years
Mars	47 or 79 years		

g. Monthly almanacs

A 'monthly almanac' (Table 11) is a tabular almanac in which columns correspond to the months of the year, and rows to the planets, or *vice versa*; hence there is space for recording at least one planetary position for each planet for each month. A single entry comprises a day number within the month, the planet's position on that day, and sometimes an indication of an event (e.g. sign-entry, station, or first appearance) to which the date and position apply. Whether a planet's progress through the year is to be read across a row or down a column is a matter of convenience of layout. When a monthly almanac is laid out above a lunar ephemeris for the same year (see §6e above), the planets are usually assigned rows so that the planetary positions for any month can be read directly above the ephemeris entries for the same month. On the other hand, most independent monthly almanacs assign the columns to the planets, to allow a wider space for the tabular entries. In addition, we have specimens of monthly almanacs for just one planet.

The monthly almanacs fall into two groups in another way as well. A simpler format records for a given planet and month, either (1) the day of the month on which the planet makes a sign-entry, with the newly occupied sign identified, or (2), if no sign-entry occurs during the month, the zodiacal sign occupied throughout the month. Such almanacs are close kin to sign-entry almanacs. A more

[1] Neugebauer [1975] 554–555.

elaborate kind reports the planet's phases and stations as well as sign-entries and gives the longitudes in degrees and even minutes; when nothing else happens to a planet during a month, the longitude on the first of the month is given.

h. 5-day almanacs

Whereas the monthly almanacs vacillate between noting planetary positions at fixed (monthly) intervals and at astronomically determined moments (phases, sign entries), the 'five-day' almanacs (Table 12) restrict themselves, with rare exceptions, to planetary longitudes at constant intervals of five days. In most instances the tabulated days begin with the first day of the Alexandrian year (and hence the first of each month), but two, **4205** and **4212**, begin with the fifth day for Venus's positions. The almanac is arranged year by year, with the entries for each planet in turn in the standard order. Some 5-day almanacs have carefully laid out columns of six months or five months, while others use the available space haphazardly. Zodiacal signs are usually denoted by their numbers, and may be omitted after the line marking the entry into the sign.

The 5-day almanacs have a remarkable uniformity in the underlying method of computation: the longitudes are derived from Ptolemy's *Handy Tables*, and converted to sidereal coordinates according to 'Theon's formula' (cf. Appendix I).[1]

Text	*Years*	*Calendar*	*Medium*	*Planets*	*Phases & Degrees?*	*Remarks*
4199	14-6 B.C.	Egy. or Alex.	codex	Jupiter	Yes	
PSI inv. 75D + E.E.S. 79/82(1)	48-55	Egyptian	roll	columns	Yes	
P. Berol. inv. 21226	52-56	Alexandrian	roll	Jupiter	Yes	Era Augustus
P. Nelson	101-103	Alexandrian	roll	columns	Yes	Era Titus
P. Tebt. 274	107-118	Alexandrian	roll	columns	No	Era Titus; new moons
P. Lund 35b	119-122	Alexandrian	roll	columns	No	
4201	126-128	Alexandrian	codex	columns	Yes	
4202	152-155	Alexandrian	roll	columns	Yes	
4203	215/216	Alexandrian	codex	columns	Yes	Saturn only?
4200	?	Alexandrian	codex	Jupiter	Yes	Unknown era
4203a	?	Alexandrian	roll	Mercury	Yes	

TABLE 11. Monthly almanacs.

[1] Burckhardt [1958] showed that P. Heid. inv. 34 used the *Handy Tables* and Theon's formula.

Text	*Years*	*Codex?*	*Remarks*
4212	217/218	No	
4205	257/258	Yes	
4206	272–274	Yes	
4207	290/291	Yes	
4208	295/296	No	
4209	295/296	No	Stations included
4210	300–302	?	
4213	301/302	No?	Perhaps same MS as 4210
4211	306/307	Yes	
P. Heid. inv. 34	345–349	Yes	Also daily lunar longitudes and syzygies
4205a	?	Yes	

TABLE 12. Five-day almanacs.

i. Miscellaneous almanacs

4204 is, so far, the only known specimen of a format that has some resemblance to the more detailed sort of monthly almanac. It covers the last six months of the Alexandrian year A.D. 54/55, recording, for each planet in turn (in the standard order), the dates and longitudes of its phases, stations, and sign-entries.

What seems to have been a standard format for a table of syzygys (consecutive lunar conjunctions or full moons) is represented by *P. Mich.* 150 and part of the five-day almanac P. Heid. inv. 34. The full moons or new moons for a year are each given a line in the table, with consecutive columns for the date (month, day, seasonal hour and fraction thereof) and longitude of the moon (zodiacal sign, degrees, minutes). P. Colker is a fragment of a table of syzygies of a different kind, related in structure and computation to the ACT System B syzygy tables.

j. Other tables

The classification of papyrus tables set out above accounts for most of the known documents; those left over tend to be puzzles. **4230–4235** are representative of the sort of tabular scraps that are often unidentifiable because too little is preserved. Among larger fragments, **4216**, **4218**, and **4219** have a recognizable arithmetical structure, although their application remains obscure. It is unusual to find a table like **4225** that exhibits no detectable pattern in the numbers it contains.

k. Horoscopes

Unlike the other varieties of astronomical papyri discussed here, the horoscopes have been the subject of excellent general discussions by Neugebauer & van Hoesen [1959] 161–175 and Baccani [1992] 39–79. It will only be necessary here to review the arrangement of the kinds of horoscope that I call 'standard' and 'deluxe'.

The standard horoscope is a terse document consisting of (1) optional opening formula (ἀγαθῇ τυχῇ, 'Good fortune!'), (2) optional name of the native, (3) regnal year, date, and seasonal hour of the geniture, (3) longitudes of the sun, moon, planets, and ascendant, and (4) optional closing formula (διευτυχεῖ, 'Good luck!'). The more common order for the data in section (3) follows the standard Greek sequence of the planets: Saturn, Jupiter, Mars, Venus, Mercury. The sun and moon may either follow or precede the planets; and the ascendant may appear either last or (especially in later texts) first. Most standard horoscopes give only the zodiacal sign occupied by each object without degrees, and when two or more objects occupy the same sign, they will usually be listed together, thus deviating from the standard order. The alternative arrangement is to sort the objects in ascending order of the zodiacal signs, beginning with the ascendant's sign.

Deluxe horoscopes provide more information of potential astrological significance, often set in a pretentious prose text with a paragraph devoted to each object (see e.g. **4277** or *P. Lond.* I 98 and 130). There is much variation in the selection of additional data—so that sometimes the distinction between a standard horoscope with supplementary data and a skimpy deluxe horoscope is subjective (and not terribly important)—but most commonly the longitudes are expressed precisely in degrees and fractions, and located in the astrological houses and terms.

§7. METHODS OF ANALYSIS AND DATING

The tables and horoscopes consist predominantly of quantitative data obtained by mathematical methods to describe astronomical phenomena. Depending on the genre of the document and its state of preservation, we may be interested in knowing (a) the astronomical meaning of the data, (b) the underlying mathematical methods and theory, and (c) the date or dates for which positions and other phenomena were computed. These problems can often be treated independently, but not always.

A complete ephemeris, for example, would probably have given the year and month in the top margin, calendar days in the leftmost columns, and headings identifying the purpose of each column. We would not expect to find any explicit mention of the methods by which the longitudes of the sun, moon, and planets for each day were computed. When we have to deal with fragments that consist of parts of a few contiguous columns without their headings, our first problem is often to identify the meaning of each preserved column, by recognizing a vertical sequence of numbers as characteristic of, say, Mars's day-to-day progress in longitude or the days of a Roman calendar month. Depending on the extent of preservation, we may then be in a position to combine information from more than one column to deduce the dates to which they must pertain, or to analyse the contents of a single column in search of evidence of how they were derived.

Broadly speaking, astronomical dating and analysis depend on different and even contrary aspects of the same text, the former counting on the recorded data to be correct, the latter seeking to exploit their inaccuracies. A date is established on the hypothesis that a selection of astronomical facts derivable from the text will, within acceptable margins of error, agree with calculations from modern astronomical theory for that date and for none other inside a historically plausible interval. The prerequisite for dating is thus a sufficiently wide range of independent but coordinated facts, which need not be stated to great precision; a horoscope giving only the zodiacal signs occupied by the sun, moon, and planets can normally be fixed within a margin of two days. On the other hand, a method of computation can be inferred when numbers exhibit symptomatic patterns that do *not* accurately mimic the phenomena they represent. For this one needs a reasonably large number of data of a single kind, stated with precision exceeding their objective accuracy.

Since the desiderata for dating and analysing are not the same, it is no surprise that we can sometimes know how a table of planetary positions was calculated without knowing the dates to which it refers, and conversely we can often establish the exact date of a table that is not otherwise amenable to analysis. However, in certain circumstances the date and means of computation can be determined simultaneously, namely when we can reproduce precise data using ancient predictive methods that are fully within our control. At present this control exists only for Ptolemy's tables and the Standard Lunar Scheme.

The fact that the astronomical papyri mostly belong to a few standard types and formats has the corollary that a few techniques of dating and analysis can be applied to many papyri. In order not to have to repeat the same basic principles throughout the commentaries to the individual texts, I will give a brief summary of them here. The reader should keep in mind, however, that these are idealized methods set out after the fact; the stages by which I originally tried to make sense of the papyri published here were often much less direct. The first way found to solve a problem seldom turns out to be the best way. Moreover, the work of transcription and joining of fragments is often made easier once one has a notion of the structure and meaning of the document, so that the editions presented here typically present more complete and accurate information than the preliminary transcriptions with which I began.

a. Recomputation by modern and ancient theory. Positions of the sun, moon, and the five planets in longitude and latitude computed according to modern theory for dates in classical antiquity are most conveniently tabulated (at five or ten day intervals) by Tuckerman [1962] and [1964]. In the case of Mars, for which Tuckerman's tables are vitiated by systematic errors, one should use the corrected tables in Houlden & Stephenson [1986]. Dates and longitudes of full moons and conjunctions may be found in Goldstine [1973]. These works are readily accessible, and, although they do not reflect the most up-to-date theory, they are more than accurate enough for most historical purposes. In the present work they will be commonly referred to by the author's name alone.

Valuable as these resources of modern astronomy are, it is very fortunate that we possess in Ptolemy's tables a complete *ancient* system for computing virtually all the phenomena recorded in the papyri. Wherever the *Handy Tables* were used to calculate data in an ancient document, we can confirm this fact directly by recomputation using the same tables. Even where dependence on Ptolemy is out of the question, his tables provide a coherent image of how the heavenly bodies were supposed to behave by a Greek astronomer of the mid second century—an excellent control for the interpretation of the fragmentarily preserved arithmetical systems of prediction.

To allow rapid and extensive recomputations, I transcribed into computer readable form the tables for mean motions and anomaly of the sun, moon, and planets from the *Handy Tables*, using the ninth-century manuscript *Vat. gr.* 1291 as copy text, and wrote software routines to carry out the calculations of longitudes as well as calendar conversions.

Modern tables such as Tuckerman's use a *tropical* frame of reference, counting longitudes from a zero point at the vernal equinoctial point. Ptolemy's longitudes are also professedly tropical, although his assumption of too long a value for the tropical year gives rise to a gradually increasing discrepancy between his zero point and the true vernal point. The longitudes in the papyri, on the other hand, are usually counted from a *sidereally* fixed zero point. A standard rule ('Theon's formula') was used in antiquity to convert Ptolemy's longitudes to this sidereal frame of reference; we can devise a similar formula for approximate conversion from modern tropical to ancient sidereal longitudes (cf. Appendix I). Roughly speaking, the sidereal longitudes in the papyri tend to be about 6° greater than tropical longitudes at the beginning of the first century, whereas they are nearly equal by the middle of the fourth century.

b. Characteristic behaviour of the heavenly bodies. The sun, moon, and five planets each have distinct patterns of longitudinal motion by which they can usually be identified from a sequence of recorded positions. In graphical terms, the longitude of a body plotted as a function of time forms a sinuous curve (e.g., for the planets,

Figs. 4–5 in §5), and a selection of several points along the curve will be enough to allow us to distinguish it from the curves for the other six bodies. In practice the parameters in Table 13 are especially useful. I have preferred to give coarse round values rather than precise ones derived from modern theory, because they are more easily remembered and fully adequate for working with ancient astronomical predictions. The only parameter calling for explanation is the 'rough recurrence': this is the smallest whole number of synodic periods that brings the planet or moon back to about the same longitude (within a zodiacal sign or so).

A further aid to the identification of planetary data in papyri is the fact that all planetary almanacs and ephemerides, and many horoscopes, employ the conventional Greek ordering of the planets: Saturn, Jupiter, Mars, Venus, Mercury.

c. Methods of dating. A complete dating formula in a horoscope or almanac would include the reigning emperor and regnal year, month, day, and perhaps the time of day. Normally such a date can be converted unambiguously into a Julian calendar date so long as we know which calendar (Egyptian or Alexandrian) was used. If the calendar is not expressly identified in the text, we will have just two possible dates, only one of which will usually be compatible with the astronomical information in the text.

Often part of the dating formula is missing, either because the papyrus is mutilated in a critical place, or because some information was omitted or written elsewhere in the original document. We may have the name of the emperor but not the regnal year, or conversely the regnal year but not the emperor's name. In the former case, we have a known interval of years within which the date must fall; in the latter, a sequence of unequally spaced single years. An extended series of regnal year numbers for consecutive years may be unique because of the irregular pattern of lengths of the emperor's reigns. In general the presence of partial dating information may either shorten the steps that would be required to date the text from its astronomical contents alone, or narrow down the possibilities when the astronomical contents do not lead to a unique date.

Whether we start with an incomplete dating formula or none at all, we may assume that we have an interval of time—it may be anything from several centuries to a few days—inside which we want to find one or more shorter subintervals such that any date falling outside these subintervals is incompatible with some of the astronomical information in the text. For this approach to be practicable, we need an astronomical datum, or a combination of data, that changes appreciably over time spans that are neither much longer nor very much shorter than the interval within which we are searching. If we obtain more than one subinterval, we try to rule out all but one using another datum with a comparable rate of change. We can then repeat the process, refining the interval as far as the data will allow.

	Saturn	*Jupiter*	*Mars*	*Venus*	*Mercury*	*Moon*	*Sun*
Synodic period (mean)	378d	398d	780d	584d	116d	29 1/2 d	
Synodic arc	11°–14°	28°–38°	30°–90°	205–225°	95–145°	28–30°	
Retrograde time	140d	120d	65d–80d	40d	20d		
Retrograde arc	7°	10°	10°–20°	15°	8°–15°		
Longitudinal period	30y	12y	2y – 60d	1y	1y	27 1/2 d	1y
Rough recurrence	29	11	7	5	3	12	
Max. daily motion	0;8°	0;14°	0;50°	1;15°	2°	15°	1;2°
Min. daily motion	–0;5°	–0;8°	–0;25	–0;40°	–1;20°	11°	0;56°

TABLE 13. Characteristic parameters for the heavenly bodies.

Dating of horoscopes. A particularly important application of this method is to the situation where the longitudes of the seven heavenly bodies are roughly known, for example in a dateless horoscope that specifies their positions only by the zodiacal signs.[1] The slowest planets, Saturn and Jupiter, have periods of longitude of about 30 and 12 years respectively; and a pair of simultaneous longitudes for Saturn and Jupiter will recur, with deviations of just a few degrees, after about 59 years. Since we know the longitudes only to within a tolerance of about ±15°, our subintervals will be on average about a year long (cf. Appendix I, where the applicable years are tabulated for each possible pairing of zodiacal signs). The sun's longitude, given to the same tolerance, reduces each subinterval to spans of about a month; and the moon's longitude further cuts these down to intervals of about two days each. There remain the three planets Mars, Venus, and Mercury to help eliminate multiple solutions.

As an illustration, in the undated horoscope **4263** we have Saturn in Leo, Jupiter in Scorpio, the sun in Gemini, and the moon in Aries. Using Appendix I to obtain years between 50 B.C. and A.D. 550 during which Saturn and Jupiter were in the correct signs, and then checking for the sun and moon in Tuckerman's tables, we find that the dates listed in Table 14 are compatible with the four positions used so far. Since the horoscope gives Mars in Aquarius, and Venus and Mercury with the sun in Gemini, the only possible date is A.D. 299, June 9 or 10.

Even if we know which zodiacal signs are occupied by the seven heavenly bodies, this information will not always lead to a uniquely determined date within as wide a span as six centuries. Fortunately we can often reduce the interval to be searched, exploiting (1) a partially preserved dating formula, (2)

[1] This procedure is due to Neugebauer; see for example Neugebauer & van Hoesen [1959] 1–2.

Date	*Mars*	*Venus*	*Mercury*
299 June 9–10	Aquarius	Gemini	Taurus/Gemini
358 June 16–17	Gemini	Gemini	Gemini
417 May 27–28	Leo	Taurus	Taurus
417 June 24–25	Virgo	Gemini	Gemini
477 May 24–25	Cancer	Cancer	Taurus
477 June 20–21	Leo	Leo	Cancer
536 May 30–31	Aquarius	Cancer	Taurus

TABLE 14. Example of dates matching positions in a horoscope.

other exactly dated or datable material on the papyrus, or (3) a paleographical estimate of the date of writing. Greater precision in the recorded longitudes, i.e. degrees within the signs, is also helpful up to a point. However, we must keep in mind that inaccuracies in the ancient predictive methods typically result in unpredictable deviations of several degrees from longitudes obtained by modern theory for the same dates. It is usually sufficient to assume a tolerance of ±5°, although larger discrepancies can occur. Hence in general, datings that depend on highly precise agreement of just one or two quantities with the predictions of highly refined modern tables are much less credible than those that use a wide range of crude data.

Dating of ephemerides and almanacs. The various formats of ephemerides and almanacs normally gave planetary longitudes, from which a set of approximate positions of the five planets for some date can be extracted. Thus in an ephemeris one can simply read off the longitudes of the planets in any row; and in a sign-entry or monthly almanac one can get at least the zodiacal signs occupied by the planets for, say, the first day of a tabulated year. The same procedure by which we date a horoscope can be used here, except that if the sun's longitude is not given, we may have to use Venus or Mercury instead.

On the other hand, these tables contain two kinds of additional information not provided in a horoscope. Firstly, we not only have the instantaneous positions, but also we can deduce their rates of motion, and hence roughly where they are in their cycles of anomaly. Secondly, calendrical information in the table can contribute to establishing the dates. For example an ephemeris may have columns giving the concordance of days according to the Roman, Alexandrian, and lunar calendars, and sometimes also indications of the days in the planetary week, all of which serve to restrict the possible dates that could be assigned to the astronomical columns. Or again, in an almanac that employs the old Egyptian calendar the relationship between the sun's longitude and the corresponding calendar date is a measure of the accumulated discrepancy between the Egyptian and Alexandrian calendars (cf. Appendix K).[1] When a papyrus table lacks some of the necessary planetary positions for 'horoscopic' dating, the problem has to be solved according to the particular combination of data that the fragment does preserve; there is no general procedure.

So far we have assumed that we do not know the method by which the astronomical information in the papyrus was computed, so that it can be compared with our recomputations only within fairly broad tolerances. Sometimes, however, we can show that precise numbers in a papyrus can be reproduced by applying a known ancient method or set of tables for a specific, unique series of dates. In some ephemerides, the lunar longitudes prove to be reproducible—and hence datable—by the Standard Scheme. On the other hand, certain ephemerides and sign-entry almanacs, and practically all the known five-day almanacs, were generated by Ptolemy's *Handy Tables*. Where the use of the *Handy Tables* is suspected, one can resort to the inelegant but effective 'brute force' stratagem of using a computer to calculate longitudes for all calendrically admissible dates within a plausible range of years, and then searching the list for patterns of numerals matching the readings of the papyrus.

d. Analysis of methods of computation. In seeking to determine how the astronomical data in the papyri were computed, we are impeded not only by the wretched condition of the texts, but also because we know so very little about how the predictive astronomy of the period worked, and because many papyrus tables by their very structure fail to supply us with the clues we want. It is hardly an exaggeration to state that without our much fuller understanding of Babylonian mathematical astronomy on the one hand, and on the other of Ptolemy's *Almagest* and *Handy Tables*, we would know nothing about the predictive methods underlying the papyri.

This introduction attempts no more than a cursory review of the small repertory of analytical strategies that have been successfully applied to astronomical papyri, leaving detailed discussion to the relevant commentaries. Much undoubtedly remains to be gleaned from the tables published here; it seemed neither necessary nor even prudent in an edition of the texts to aim to say the last word about them. The interesting questions of the theoretical and observational foundations that lie behind the methods of prediction of course fall entirely outside the scope of this work.

Primary tables. The most direct and cogent demonstrations of the use of arithmetical methods in the astronomy of the papyri come from the epoch tables and

[1] Though not explicitly recorded in an almanac, the sun's longitude can be deduced from the synodic cycle of one of the planets. Venus and Mercury are at approximately the same longitude as the sun at the midpoint of a retrogradation; the remaining planets are diametrically opposite the sun near the mid retrogradation (Θ).

templates. In the epoch tables for the sun and moon, the line-to-line differences in the columns for date, longitude, and argument of latitude are sexagesimal constants. In general, however, the synodic arcs and times in the planetary epoch tables vary in more complex ways.

If a column for longitudes (*B*) is extant, we can investigate how the synodic arcs (i.e. the line-to-line differences) vary as a function of the initial longitude. The Babylonian System A schemes lead to functions composed of stretches of constant or linearly varying synodic arc (see above, §5, figs. 6–12). The use of a known System A scheme can be shown by comparing the specific values of the synodic arc extracted from the papyrus with the expected function, or even more directly be recomputing the sequence of longitudes according to the scheme (e.g. **4152**). To test for the presence of a System B scheme, in which the synodic arcs are generated by a linear zigzag function, we can either check whether the *second* differences, i.e. the line-to-line differences in the synodic arcs, are usually constant, or whether plotting the synodic arcs as a function of initial longitude tends to yield a zigzag pattern (**4161**).[1]

If the longitudes are not preserved (e.g. **4153** and **4158**), we can nevertheless approximately reconstruct the sequence of synodic arcs by subtracting the constant *c* appropriate for the planet in question from the synodic times, i.e. the differences of the column for dates (*T*). The longitudes themselves, displaced by an unknown constant, can also be obtained as the running totals of the synodic arcs.

In analysing templates, the primary question is whether the numbers representing each day's progress since epoch were generated by an arithmetical sequence, and if so, what its defining parameters were. Hence one looks for constant values in the line-to-line differences, the second differences, or even the third differences. It is an easy matter to extrapolate a known arithmetical sequence in either direction indefinitely; the difficulty often lies in determining at what lines in a template a particular sequence began and ended. The first differences of the columns for progress in longitude and argument of latitude in the lunar Standard Scheme turn out to have been linear zigzag functions, which can be reconstructed exactly for the full extent of the table. The sinuous pattern of direct and retrograde motion of a planet, on the other hand, was typically stitched together out of several stretches of arithmetical sequences with different parameters, so that knowledge of the structure of one part of the template will not necessarily tell us anything about the remaining parts.

A difficulty frequently experienced in studying epoch tables and templates is

[1] The pattern of synodic arc as a function of initial longitude in System B is not really a zigzag function, but a composite of alternating parabolic arcs. However, for Saturn and Jupiter they are almost linear.

the loss of lower-order fractional places, either through damage to the papyrus or because the scribe chose to limit the precision of the numbers. Usually the two situations are equivalent because scribes tended simply to truncate fractional places even when they were quite close to the next higher number. When dealing with abbreviated numerals, it is often helpful to keep track of the tolerance, i.e. the range of possible values that could be represented. For example, the truncated numbers 19, 24, 30 may be treated as 19;30±0;30, 24;30±0;30, and 30;30±0;30; and their differences as 5;0±1;0 and 6;0±1;0. It is important to check that a hypothetical rule of computation inferred from truncated values really can be made to generate the numbers in the text. Thus in the example above, the difference between the first and second numbers could have been 4;30, and the difference between the second and third could have been 5;15, but no rule that produces *both* these differences can be made to generate the given sequence 19, 24, 30.

Almanacs. Some of the information in almanacs is of a similar character to the contents of primary tables, and hence in principle can be studied by the same techniques. Monthly almanacs contain, along with much else, computed dates and longitudes of planetary phases, which can be extracted and treated like an epoch table so long as there are enough of them to make the exercise worthwhile. Again, the daily longitudes of a heavenly body in a column of an ephemeris may be subjected to the same kind of investigation as a template.

The dates when planets are supposed to cross boundaries between zodiacal signs, recorded in sign-entry and monthly almanacs, are less amenable to analysis. Sign-entry dates have only an astrological, not an astronomical, significance, and are not a direct product of any of the known ancient methods of predicting planetary motion, whether arithmetical or kinematic. On the other hand, the papyri present us with a vast amount of data of this kind, large enough to motivate the search for ways to find out how they were obtained. The chief contributions in this line are van der Waerden's studies of the time intervals in the Stobart Tablets and P. dem. Berlin 8279, which revealed that they are in part based on arithmetical methods including the Babylonian System A scheme for Mars (van der Waerden [1972]). The problem of deducing how the almanacs were composed deserves to be readdressed with the enlarged corpus of available examples, and in the light of the new, more direct evidence for Babylonian planetary theory in the papyri; I have not attempted to deal with it in this edition.

Two varieties of tables, the four-column sign-entry almanacs and the five-day almanacs, proved unexpectedly easy to analyse. All the specimens that are well enough preserved to test have turned out to be computed using the *Handy Tables*, with longitudes converted to sidereal using 'Theon's formula'; it was therefore possible to date them, and reproduce their contents with insignificant discrepancies, using the computer simulation of the *Handy Tables*.

§8. DOCUMENTARY ASPECTS OF THE TEXTS

a. Provenance and method of selection.

Some understanding of the steps by which the papyri contained in these volumes came to be identified and selected for publication is necessary if one wants to appraise the causes, human and physical, that may have affected the distribution and representative character of the texts. The earliest stage was, of course, the discarding of the papyri in antiquity. We can do little more than hypothesize, on *a priori* grounds, under what conditions and after how long the possessor of a horoscope, a set of tables, or an astronomical treatise would have thrown it on the rubbish mounds outside the ancient town of Oxyrhynchus. Conditions of climate and soil of course influenced the survival of documents, most obviously by bringing about the almost total destruction of the papyrus record before the late first century B.C. Complete or near-complete manuscripts were gradually reduced to fragments, and surviving bits often became separated.

Over six seasons (1896/97 and 1902/03 through 1906/07) Grenfell and Hunt excavated the mounds, certainly hoping chiefly to find comparatively early papyri (at any rate earlier than the fourth century), although it is not clear how much a chronological bias might have contributed to the decisions of where to dig. For most of the seasons the excavation records, and therefore all correlation of the finds with the geography of the site, are lost. Grenfell and Hunt did not exhaust the site, as the subsequent Italian excavations demonstrated, but in the main the part that they recovered appears to have been representative of what was accessible at the time. Because of the great pace of discovery of papyri, there can have been little time for examination or sorting before they were placed in tin boxes for shipment. Later, but still in Grenfell and Hunt's time in the early decades of this century, fragments that appeared to be substantial and well enough preserved to merit further work were extracted from the original boxes, flattened, and filed in packets in a new series of boxes, each packet apparently corresponding to one of the former boxes, and each new box therefore containing the 'best' pieces from several. This partition of the fragments into two series of boxes, a main series containing larger pieces (but also numerous small bits, perhaps found intermingled with the others) and a series of 'scrap' boxes, survives still, and matters to us here (1) because most records of the correspondences between the new and old boxes are lost, and (2) because the main series is the more thoroughly inventoried and therefore the source of most of the papyri published here. It is therefore possible that small fragments of astronomical texts and tables, including bits that join papyri edited in these volumes, remain undetected and (for the time being) practically untraceable in the scrap boxes. For most folders of papyri almost the only information now available about provenance is the season of excavation.

However, the process of extraction does not seem to have resulted in unequal treatment of the astronomical and astrological documents, very few of which would, one suspects, have excited special interest at the time. Moreover, the lack of sorting means that fragments now filed together were probably found together, and hence in some instances perhaps discarded together in the first place.

The system of filing adopted by Grenfell and Hunt is partly reflected by the inventory numbers, which are indicated at the head of the commentary on each item and in Index I. For most papyri, the inventory number follows the pattern 21 3B.23/D(1-3)b. Here '21' is the number of the present cardboard box; '3B' refers to the third campaign at Oxyrhynchus (Bahnasa); '23' is the series number given within that year to the metal packing box; 'D' indicates a layer of papyri inside that box; and the remaining numbers and letters subdivide the layer. In some boxes the inventory numbers preserve no overt reference to the season or the original packing boxes; instead the box number is merely followed by a packet number, e.g. 100/124(a). Numbers preceded by a letter, e.g. A145/6H, refer to the series of 'scrap' boxes.

In pursuit of materials for publication, I examined all papyri that were inventoried as astronomical texts, tables, and horoscopes, as well as many items classified tentatively as mathematics, calculations, or astrology. When there was cause to suspect further unidentified fragments in nearby folders, an entire box would be browsed through. Fragments were rejected only for one of three reasons: (1) because their contents turned out not to be technical astronomy, (2) because the fragments were too small to be useful (often only one or two numerals or even just tabular ruling), or (3) because the condition of preservation was so poor that nothing worthwhile could be transcribed.

b. Date of contents and date of writing.

A large proportion of the tables and horoscopes contain dates or refer to astronomical conditions that applied, or were supposed to apply, to specific dates that we can deduce. How these dates relate to the dates when the texts were written is a question that has not yet been addressed systematically.

Horoscopes are of course inherently retrospective: the date on the horoscope always precedes that when the horoscope was cast. The interval must have varied considerably, for customers will have consulted an astrologer on behalf of their children as well as on their own behalf. Twenty or thirty years might be a plausible average age for a native at the time his horoscope was first written down, but we have to allow for great variation from case to case.

The astrologer who depended on almanacs for casting horoscopes would have needed at any given date to have at his disposal planetary positions extending from the present back more than half a century to allow for the likely range of ages of his customers, but certainly not more than a century. One might imagine an almanac covering years several decades before the present being copied out from time to time. It is more difficult to say how the continual need for fresh installments to cover the time currently passing was satisfied. Since the years in the almanacs are numbered according to the reigning emperor, they cannot be forecasts; so we have to imagine the astrologer periodically extending his almanacs to bring them up to date, but not usually prolonging them into the future years. On balance I would assume, given a small part of an almanac covering a single year *x* (or a few neighbouring years), that the copy was produced within the subsequent seventy years or so, the possible range diminishing with the span of years preserved. The same conditions ought to apply to epoch tables.

When we are in a position to check these hypothetical arguments on specific texts, by and large they seem to be valid. Thus of the two demotic sign-entry almanacs, P. dem. Berlin 8279 probably began when complete with the year Augustus 13 (18/17 B.C.) and was copied probably not long after A.D. 42, the date of a set of accounts on the front of the papyrus. This gives us a likely span of six decades. The Stobart Tablets (extant for A.D. 70 to 133) originally covered seven decades or more, from A.D. 63 through at least 133. Among the Greek sign-entry almanacs, the longest demonstrable continuous run of years is for *P. Oxy.* XLVI 3299, which is likely to have originally covered A.D. 201/202 through at least 224/225. Again, among the longer-running epoch tables, **4148** covers at least 76 years; **4149**, 70 years; *P. Lund* V 35a, 50 years; **4158**, 53 years.

On the other hand, there are rare instances where the hand in which a table is written appears on paleographical grounds to be much later than the contents. It is notoriously difficult to date tables of numerals by hand; but one can at least cite the extreme case of the monthly almanac for Jupiter **4199**, which covers dates in the late first century B.C. but was written not earlier than the middle of the second century of our era.

Were predictive texts ever written down before the dates to which they refer? In two contexts the answer must surely be yes, namely when the texts were directed towards either general ('omen') or catarchic astrology. Thus the eclipse canon **4137** was certainly composed, and our copy very likely written, before the eclipses of A.D. 56/57 that it forecasts. The later ephemerides, with their columns appraising each day as good or bad for activity, must also have been produced a little in advance. Before the fifth century we have no example of an ephemeris incorporating a column of this kind, but I suspect that this kind of table had always had an implicit catarchic application. (It is worth remarking on the other hand that the ephemeris is a bulky and inefficient way of storing the long runs of planetary positions required for horoscopic astrology.)

Clusters of manuscripts and archives; 'Group A'

Most of the papyri in this volume come to us as isolated fragments, and we have no way to prove that they were formerly owned by the same person. Pieces of a single manuscript are sometimes found in different folders. They may be close together, as in the case of **4149** and **4172**, where the inventory numbers practically imply discovery in consecutive spadefuls; but there are also instances where one fragment remained in the 'scrap box' while another was transferred to the main series of boxes (**4156a**, **4191**), or where fragments found their way into disparate boxes in the main series (**4199**). Where we lack the testimony of the inventory numbers, we need very strong internal grounds, ideally a direct join, for grouping fragments under one publication number as a single manuscript. Related contents and a similar hand might tempt one to unite **4173** and **4173a**, for example, but how can we be sure that they were not two separate productions of the same professional scribe?

Conversely, a single folder or nearby folders may contain bits of more than one manuscript of astronomical contents. Because of the comparative rarity of astronomical papyri, the probability of this occurring coincidentally is small, though not as small as if the allotment of papyri into their folders had been carried out without any consideration of their contents. We may presume that most such groupings are 'archives' in the papyrological sense, that is, manuscripts that were found in one place and that derive from a single ancient collection.

The archives may turn out to be trivial or uninformative. Thus **4160** shares its inventory number with a small scrap of a different table, about which little more can be said than that it *is* a different table. Again, a theoretical treatise, **4138a**, was discarded together with a set of tables, **4230**; but the latter are unfortunately too scrappy to be identified. Nor is it surprising that one folder has yielded a monthly almanac for Mercury (**4203a**) and a bit joining a monthly almanac for Jupiter, the rest of which was found in another box (**4199**).

On the other hand, there is one assemblage of fragments so large as to leave no doubt that it represents the remains of an astronomical library cast on the rubbish heaps in the early fourth century. This archive, which I will refer to as 'Group A', is dispersed in numerous folders from boxes excavated in Grenfell and Hunt's sixth season. Pieces from several of the constituent manuscripts are found in more than one folder, and in particular the *Handy Tables* fragments that make up **4167** are from no less than six folders. If we initially consider as parts of Group A only **4167** and those manuscripts of which at least one fragment shares a folder with another known part of Group A, then the list includes the following tables: **4153–4155**, **4159**, **4165**, **4167**, **4169**, **4173a**, **4184a**, **4194**, **4196**, **4196a**, **4209**, **4217a–c**, **4224–4226**, and **4231–4235**. Items in nearby folders that almost certainly also belong include the tables **4171**, **4182**, **4195**, **4205–4206**, **4208**, **4210–4213**, **4215**, **4220**, the texts **4138**, **4141**, **4145**, and perhaps also the horoscopes **4288–4289**.[1]

[1] Other folders close to those identified as Group A contain fragments of astrological texts.

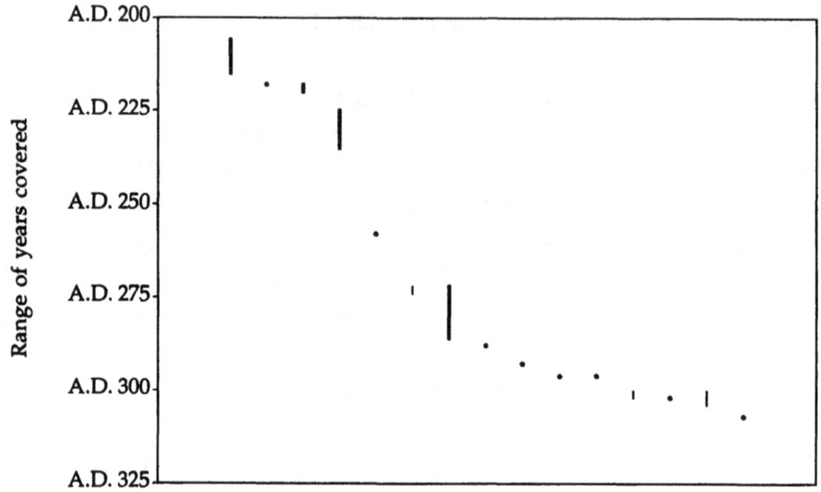

FIG. 15. Dated papyri certainly (heavy) and probably (light) belonging to Group A.

Fifteen of the papyri in Group A have datable contents, and these are distributed throughout an interval of almost exactly one century, from 206 to 307 (Fig. 15). All the major formats of table are represented except for monthly almanacs. In particular it is noteworthy that the archive includes arithmetically-based planetary epoch tables and templates as well as Handy Tables. There is some reason to suspect that the astrologer or astrologers who owned the manuscripts changed over from the arithmetical methodology to the kinematic about the middle of the third century, since the datable epoch tables are all earlier than 250 whereas most of the almanacs from after 250 were computed using Ptolemy's tables.

Codices

It is—or ought to be—well known that codices appear more frequently, and earlier, among astronomical texts and tables than in most other categories of literary and subliterary papyri. There are now enough examples to make it possible to draw more specific inferences about the use of codices for the various categories of astronomical papyri.

Most of the exactly dated astronomical codices are almanacs (cf. Tables 9–12). According to dates covered, the earliest example is the monthly almanac **4199**, which contains planetary positions for 14 through 6 B.C.; but the table is written in a much later hand, perhaps of the second century. The next codex, *P. Laur.* 144, is a sign-entry almanac for A.D. 69–75, and there is no reason to suppose that the date of writing was more than a few decades later. The majority of the later sign-entry almanacs, and all that are later than 161, are codices. The monthly and five-day almanacs are more mixed, while the ephemerides appear to have been published consistently in roll form until the fifth century. This perhaps reflects a difference in the application of the ephemerides from the other kinds of almanac: their connection with catarchic astrology suggests that they were primarily used close to the dates that they cover, whereas one would have continued to refer to random parts of a sign-entry almanac for many years.

The two solar templates **4162–4163** are codex pages, but most of the primary tables pertaining to arithmetical astronomy were either rolls in their own right or written on the backs of other documents (the lunar tables **4150** and **4151/4164** may have been loose sheets used on both sides rather than codices). The preference for the codex format for manuscripts of the *Handy Tables* and related sets of tables (cf. Tables 6 and 8) is no surprise; frequent reference to dispersed parts of a large compilation of tables would have rendered a layout in rolls very awkward.

Symbols

Astronomical papyri, like other genres of so-called subliterary texts, make free use of two methods of abbreviating often-repeated technical terms: truncation of the words (with or without letters or marks of abbreviation above the last conventionally written letter), and symbols. We are concerned here only with the latter.

The symbol for zero is a special case in that it probably does not stand in place of a specific word. It represents either an 'empty' place in a sexagesimal fraction or 'no whole units' (usually preceding a sexagesimal fraction). It was always closely tied to the sexagesimal notation, and only occurs in astronomical contexts. The most common form of zero in the papyri is a small circle with a horizontal stroke above it, but there are many variations (Fig. 16).[1] The stroke is not the horizontal bar sometimes written above alphabetic numerals; the latter is never drawn above a zero, and is generally written further above the base line. I share Neugebauer's doubts of the supposition often repeated in paleographical handbooks that the symbol was an abbreviation of οὐδέν (Neugebauer [1957b] 14). It was the counterpart of the cuneiform punctuation mark 𒑱, which was used in the same way in sexagesimal numerals, and the direct ancestor of the zero of Arabic astronomical manuscripts (Irani [1955]). A Demotic counterpart is known only from P. Carlsberg 32, where the occurrences are unfortunately damaged so that it is not clear whether the horizontal stroke that can be seen was the entire symbol or only part of it.

Symbols for the common units of longitude (degrees = μοῖραι, minutes = λεπτά) and time (hours = ὧραι) were derived from their initial letters. The monogram for ὧραι, a tall rho drawn through the middle of omega, was standardized from an early period. The other abbrevations, however, vary in form, μοῖραι being represented by a broad mu with either omicron or omicron and iota written

[1] In its more common forms the zero is indistinguishable from the conventional symbol for *artaba* (unit of volume), on the development of which see Blanchard [1974] 34–38.

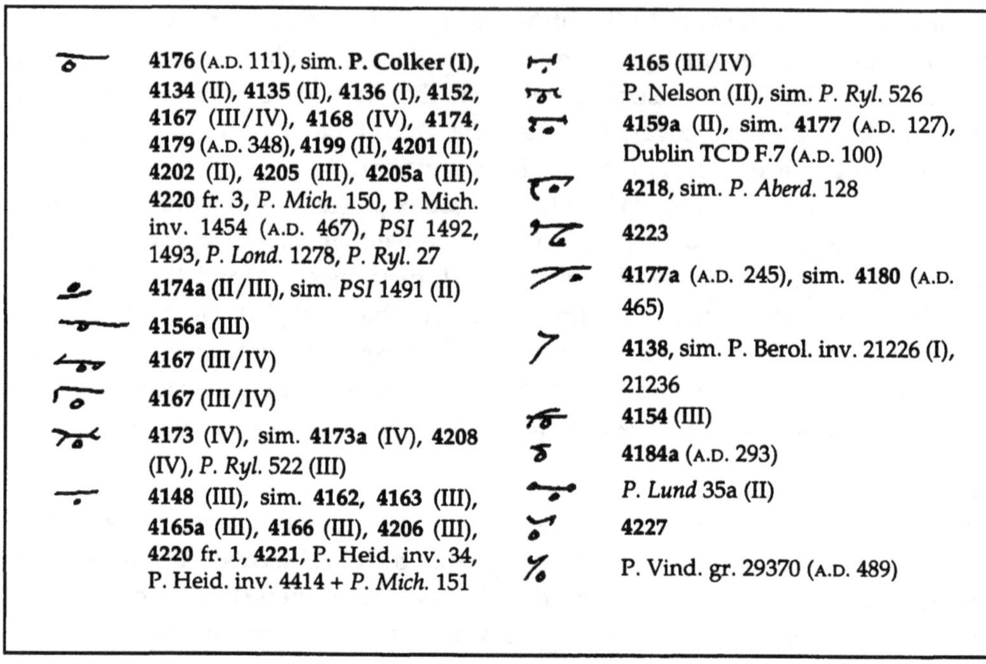

FIG. 16. Forms of the zero symbol in astronomical papyri.

small and directly above, and λεπτά by a lambda with some mark of abbreviation either above and to the right or occasionally centred below the letter.

Of the symbols for the heavenly bodies as they appear frequently in Byzantine manuscripts, only ☉ for the sun and ☾ for the moon have hitherto turned up in papyri (as early as the third century). It is now possible to trace the medieval symbols for at least four of the five planets to forms that occur in some of the latest papyrus horoscopes (**4272, 4274, 4275**, cf. Fig. 17). That for Jupiter is an obvious monogram derived from the initial letter of the Greek name; Saturn's has a similar derivation, as is clear from the form in **4272**, but underwent simplification. The ideal form of Mars' symbol is uncertain, and perhaps not related to the later circle with an arrow through it. Mercury's is a stylized caduceus. The monogram for the ascendant (ὡροσκόπος) is indistinguishable from that for ὧραι, which perhaps explains some of the instances where the word ὥρα, spelled out, means the ascendant.

No precedents for the medieval symbols representing the zodiacal signs have yet turned up in a Greek papyrus. Aside from the common practice of curtailing the names, the only symbolic forms seen so far are the monograms in **4175** and P.S.I. inv. 75D, and what seems to be a Demotic sign for Aquarius in **4184**. Among the abbreviations for the planetary phenomena in monthly almanacs (**4199–4203a**) and almanac-ephemerides (**4177**), the stations and acronychal risings are truncations of the Greek words, but the first and last appearances are marked by small loops (filled or hollow) with tails pointing to the upper right. These notations seem to have died out in late antiquity, perhaps with the diminishing importance of the phenomena they signified.

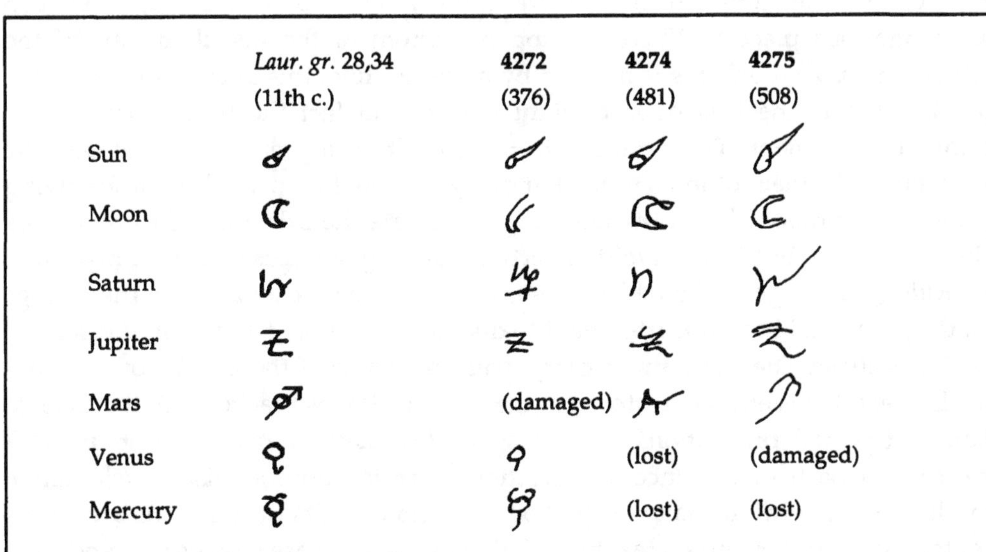

FIG. 17. Symbols for the heavenly bodies in late horoscopes compared with early medieval forms.

§9. EDITORIAL PRINCIPLES AND NOTATIONS

In most respects the transcriptions of texts in this edition follow standard papyrological practices. Word division, puctuation, capitalization, and accents and breathings are editorial supplements. The apparatus notes abbreviations, lectional signs, peculiar and dubious scripts, standard orthographies, and occasional errors of the scribe. The following editorial signs are used in the texts:

( )	Resolution of an abbreviation or symbol
[ ]	Text lost through abrasion or broken off the papyrus
⟦ ⟧	Text deleted by the scribe
< >	Text omitted in error by the scribe
{ }	Text included in error by the scribe
` ´	Text inserted, usually above the line, by the scribe
α̣β̣γ̣ *etc.*	Letters that cannot be read with conviction
. . .	Letters that cannot be plausibly identified

Note on use of brackets. A closing bracket, ']', at or towards the beginning of a line indicates loss of all text to the left, and conversely at the end of a line an opening bracket, '[', indicates loss of all text to the right. In many texts, however, especially tables with several columns or ruling where it would otherwise be difficult to see at a glance whether a particular bit of text is present on the papyrus or restored, additional brackets have been included for the sake of legibility.

Line numbering. Lines are numbered continuously from the top visible line of each column, including lines that are entirely restored (e.g. in gaps between noncontiguous pieces). Where the top or bottom of the inscribed part of the papyrus is broken off, this is marked by a dashed line; this does not exclude the possibility that the first or last extant line was in fact the first or last of the complete document. The symbols ⇐ ⇒ in the heading of a table indicate the existence of further columns to the left or right, edited on preceding or following pages. The symbols ⇑ ⇓ in the left margin indicate the existence of further rows above or below; but I have avoided vertical splitting of tables wherever possible.

Rulings. Rulings in the tables attempt to reproduce schematically the rulings on the papyrus. I have not attempted to distinguish visible and restored rules.

Translations. The translations carry over only some of these notations. In particular, text that is entirely restored in the transcription is bracketed in the translation, but partial restorations are not specially noted. A raised question mark indicates a particularly uncertain reading or restoration; a raised exclamation mark, a noteworthy mistake in the text. Uncertain digits of numerals are represented by 'x', and uncertain text by '...'. Parentheses enclose glosses not rendering specific words in the Greek but added to the translation to clarify the meaning.

Part I

Theoretical and Instructional Texts

4133–4147. Theoretical and Instructional Texts

The prose texts presented below are as varied in historical interest as in content. A handful (**4133, 4138a, 4139, 4144**) are evidently parts of scientific treatises that dealt with the analysis of observations, the periodicity of phenomena, and kinematic modelling at a theoretical level. Two, **4137** and **4140**, record series of predicted or—perhaps—observed phenomena of a kind that demanded verbal descriptions instead of the more usual numerical tables. The rest are procedure texts, a category that embraces short independent sets of instructions as well as larger works capable of being circulated as books in their own right. The text on lunar syzygies edited as part of **4163**, which is written in the vacant space surrounding a table, illustrates the former possibility, whereas **4142–4143** are obviously fragments of long manuals. In other cases the loss of context makes it difficult to appraise the scale and genre of the whole.

We may also group the prose texts, for the purposes of the following summary, by subject rather than manner of treatment. Five have to do with one or more of the planets, and seven (including **4163**) concern the moon. One text addresses fundamental topics of astronomical time-reckoning. The scope of four less well preserved texts remains uncertain.

4133 is an especially precious discovery. It represents a kind of text—a theoretical treatise that embodies original research, including the analysis of observations—such as we might not expect to find among the papyri from a provincial centre even of Oxyrhynchus' intellectual stature. Moreover, the surviving portion of this book, though not extensive, by good luck coincides with a passage rich in information. The author's account of his own observation of Jupiter on December 31, A.D. 104, is the first scientific observation report in Greek from the time before Ptolemy that has come down to us independent of Ptolemy's *Almagest*. On the basis of the date and manner of the observation we can identify the author with some likelihood as Menelaus of Alexandria, a prominent mathematician of the generation before Ptolemy whose astronomical work is otherwise known to us only through two observations cited in the *Almagest*. The author of **4133** compares his sighting of Jupiter with a much older one, from the middle of the third century B.C., and this too can be attributed to one of Ptolemy's sources. The juxtaposition of the two observations points to the kinds of problem that were current in planetary theory at the beginning of the second century.

The description of planetary motion by arithmetical sequences, so abundantly documented in practice in the epoch tables and templates **4148–4166**, is also the concern of two procedure texts. **4135** sets out a standard pattern of daily motion in longitude for Venus, by which the planet's apparent velocity is assumed to take on a sequence of constant values during prescribed numbers of days starting with its first appearance as evening star. For most of this 'velocity scheme' the time intervals and associated speeds exactly match those given in an early sixth-century Sanskrit treatise, illustrating in the most concrete way the historical ties between the kind of mathematical astronomy found in the papyri and certain strands of Indian astronomy. The papyrus also contains an enigmatic reference to the early third-century B.C. Alexandrian astronomer Timocharis. **4134**, by contrast, has yet to reveal most of its secrets: its extensive but much damaged manipulations of sexagesimal numerals seem to have something to do with the computation of a planetary template, but even the identity of the planet is not clear.

The predictions of planetary positions in the tables and horoscopes on papyrus are almost exclusively restricted to longitudes. Nevertheless some interest in planetary latitudes is reflected in two texts. A small fragment, **4141**, describes in broad terms the stages of latitudinal motion of a planet, probably Mars. **4143** explains the use of Ptolemy's tables for computing planetary latitudes, with a worked example that allows us to date the text approximately. This clearly belonged to a general manual to the whole of the *Handy Tables*.

4136 adds a further document, and so far the earliest one, to the substantial body of texts and tables relating to the procedure for calculating lunar positions called the Standard Scheme. The problem with which this text deals originates in an astronomical table, but its greatest interest is in the competent handling of the mathematical properties of a summed linear zigzag function, a subject that we would have expected to find in a Seleucid cuneiform text sooner than in a Greek papyrus.

4137 is a part of a canon of predicted lunar eclipses pertaining to the middle of the first century; the only known parallel to this remarkable text is a demotic eclipse canon from the early first century B.C. This appears to be the kind of canon that Ptolemy had in mind in his treatment of eclipse prediction in *Almagest* VI, directed towards a branch of general astrology that descended ultimately from the Mesopotamian eclipse omens. Among the noteworthy details of the text are the predicted προϲνεύϲειϲ or 'inclinations' of the eclipse, a stellar nomenclature with ties to Hipparchus, and the reappearance of dating by Callippic periods and Athenian months centuries after the latest recognized attestations of this convention.

Eclipses are the focus of two further texts. The procedure text **4138**, perhaps from a treatise, is part of a worked example of the prediction of an eclipse, not using Ptolemy's tables but perhaps based on kinematic modelling. **4138a** is a fragment of a theoretical work, perhaps not at a very advanced level, and discusses the relationship between eclipse periods (time intervals that bring about the approximate repetition of lunar eclipses) and eclipse intervals (the number of synodic months between consecutive eclipse possibilities). The debt of Greek lunar theory to Babylonian astronomy, already evident in this text, is still more pronounced in **4139**; for here we find, for the first time in a Greek document, one of the fundamental numerical parameters of the Babylonian 'System A' lunar theory, together with a probable reference to the astronomers of Uruk. **4140**, a list in tabular form of some sort of event connected with the moon's motion in longitude and latitude, remains an enigma.

4142 is, like **4143**, part of a manual of instructions for the use of Ptolemy's *Handy Tables*. Here the preserved sections must be from near the beginning of the work, dealing with the fundamental problems of conversion of time reckoning to and from the 'Alexandrian mean time' employed in the tables. Such manuals are most valuable as evidence of the popularization of Ptolemy's tables and the level of explication—far more elementary and detailed than Ptolemy's own—that the user of the tables expected in late antiquity. In **4144** we have an example of a theoretical discussion—again at an elementary level?—of kinematic models, illustrated by a geometrical diagram but unfortunately much damaged.

The nature of the remaining texts, **4145–4147**, is not clear to me; I have included them, with necessarily minimal commentary, for the sake of completeness and in the hope that further work may shed some light on their contents.

4133. Menelaus of Alexandria (?), Treatise on Planetary Theory

100/124(a) 11.5×14.5 cm + debris Second century

Plate I

This unusual fragment comes from a treatise on planetary theory nearer in genre to Ptolemy's *Almagest* than any hitherto known. By happy chance the one well-preserved column of text dates the work and allows a probable identification of its author. The papyrus is darkened and badly shattered towards the bottom. The back is blank, and it can be presumed that the fragment derives from a roll, although no sheet joins survive. The text was written in small informal rounded capitals, in columns of 7 cm width separated by 2 cm; the same hand may have written **4134**. The breadth of one column and part of the next (written with slightly wider line spacing) are preserved, as is the top margin to a height of 4 cm. In this margin a larger, cruder hand has written—as a pen-trial?—καιcαρ καιcαρ |² λιβια |³ αυτοκρατω[ρ] |⁴ δρουcοc. A large numeral ιδ is also written above the centre of the first column, perhaps indicating the fourteenth column of the roll.

Lines 3–22 of the first column report the writer's observation, on the night of December 31, A.D. 104/January 1, A.D. 105, of a heavenly body when it was close to two fixed stars in the constellation Cancer. Recomputation of the appearance of the heavens on that date shows that the object in question was Jupiter. The planet's location is fixed by two observed distances measured in 'lunar diameters': perpendicularly from an imaginary 'straight line' drawn through the stars δ Cnc (the southern of the two 'Aselli', cf. note to fr. 1 line 8 below) and θ Cnc, and radially from δ Cnc. In lines 18–22 the author deduces from these data that the planet was 'east' of δ Cnc (apparently reckoning parallel to the ecliptic) by some fraction less than one lunar diameter, and slightly north of the ecliptic.

The remains of the continuation of column i and the first half of column ii are too fragmentary to reveal the nature of their contents. This may be where conclusions were drawn from the comparison of the two observations. Lines 15–19 of column ii gave part of the discussion of an observation report similar in character to the preserved one in column i. Two distances, again expressed in 'lunar diameters', were given, of which the second was in relation to a 'southern' star, not necessarily again the southern Asellus. Line 20, with the surprising word δαψιλῶϲ, must be part of a comment on either what preceded or what followed. From line 22 to perhaps the end of the preserved text we have bits of another reported observation, apparently not by the author. Jupiter is here named by its Hellenistic alternate name Phaethon.

Fig. 4133.1 shows the pertinent part of the sky above the eastern horizon, as it would have appeared in the early evening of December 31. (Since Jupiter was travelling only about 8' per day retrograde, a different time of observation would have had a negligible effect on the relative positions—our author was justified in not specifying the hour.) Although the reported displacements of Jupiter with respect to the two stars are correct in terms of directions and approximate relative proportions, the author depreciated the actual distances. Taking one 'lunar diameter' as approximately 30', we obtain 15' for Jupiter's reported distance from the line between the two stars, and between 30' and 36' for its reported distance from δ Cnc. The measurements may have been performed without any instrument; and the moon rose only towards sunrise as a mere sickle above the opposite horizon, quite useless for reference.

So far as we can tell, the author does not explain how he derived the distance by which Jupiter lagged eastwards (ὑπελείπετο) of δ Cnc or the fact that the planet was slightly north of the ecliptic from its observed situation relative to the stars. His silence suggests that the analysis was a straightforward reduction using plane triangles to approximate the small spherical angles involved (cf. Fig. 4133.2a). Using the 'theorem of Pythagoras' he would find that Jupiter was between 0.87 and 1.1 lunar diameters eastwards of δ Cnc along the line through the two stars

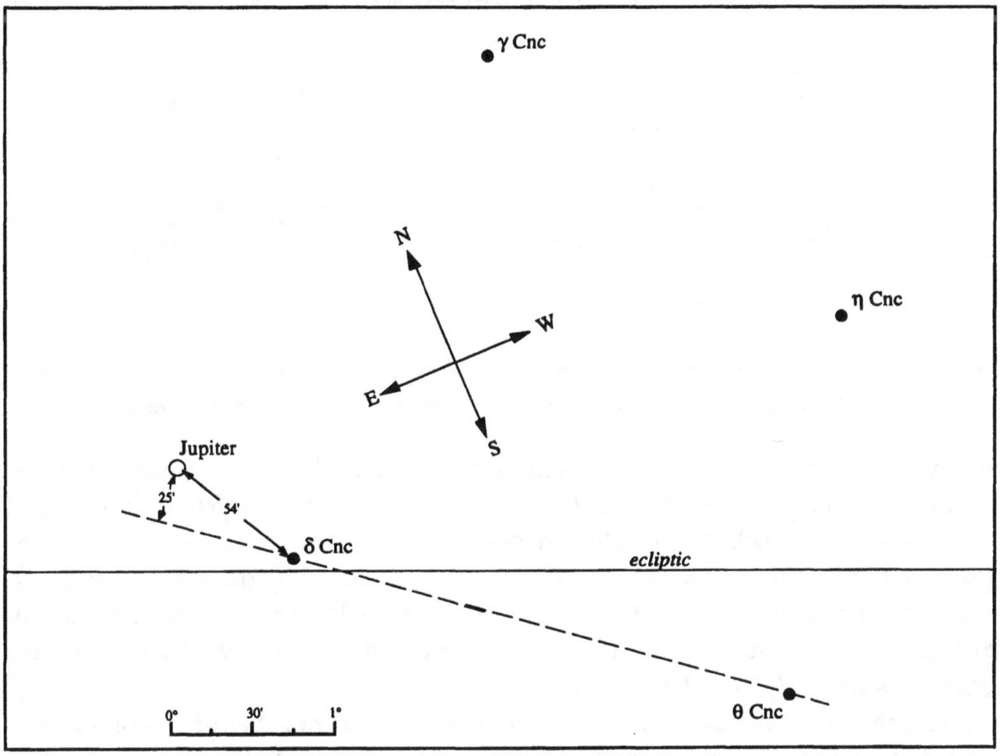

FIG. 4133.1. Jupiter and vicinity, A.D. 104 December 31, 18:00 U.T.

(*c*). To reduce this to components parallel and perpendicular to the ecliptic, he must then have made some assumptions about how the two stars are placed with respect to the ecliptic. In Ptolemy's catalogue (*Almagest* VII 5, ed. Heiberg v. 2, 94–96, trans. Toomer 366) their ecliptic coordinates are given for the epoch year (A.D. 137) as:

	Longitude	Latitude
δ Cnc	♋ 11 1/3°	–0 1/6°
θ Cnc	♋ 8°	–1 1/6°

Hipparchus (*in Arat.*, ed. Manitius 104) sets δ Cnc right on the northern tropic circle, and θ Cnc 'not less than' 1° south of it; this would give δ Cnc a latitude of a small fraction of a degree north, and θ Cnc a latitude of just under 1° south, if one takes into account the small deviation of the ecliptic south of the tropic circle in Cancer. The correct tropical coordinates of δ Cnc according to modern theory were approximately longitude 102;14°, latitude 0;5° at the date of observation, and those of θ Cnc were approximately 99;15° and –0;45°.

Assuming, for the sake of argument, that the stars are supposed to be 3° apart and that δ Cnc is 0;50° north of θ Cnc, our author could have broken down the

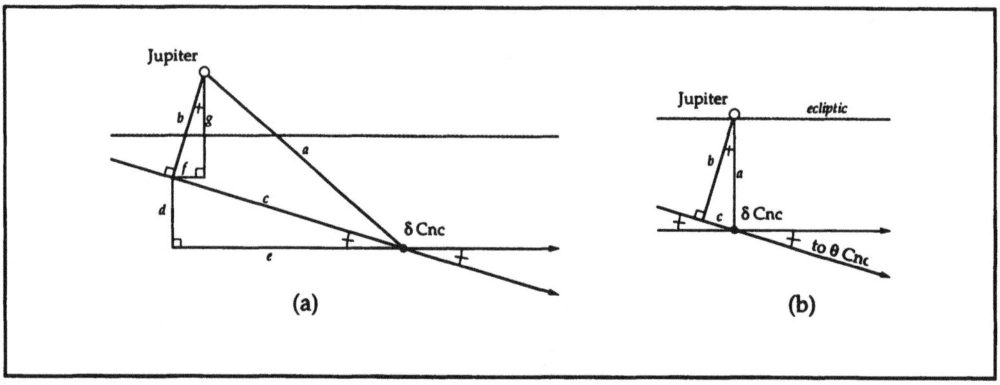

FIG. 4133.2. Hypothetical reduction of observed positions of Jupiter to the ecliptic.

known distances *b* and *c* into their ecliptic components *d*, *e*, *f*, and *g*, and estimated that Jupiter was between 0.7 and 0.8 lunar diameters north of δ Cnc ($d + g$), and east of it along the ecliptic by between about 0.7 and 0.9 lunar diameters ($e - f$). The text's statements at this point are unfortunately obscured by damage to the papyrus; the expression 'slightly' (βραχύ) north of the ecliptic might suggest that our author placed δ Cnc even further south than Ptolemy (Fig. 4133.2a shows the ecliptic as lying $^1/_4$ ° north of δ Cnc).

The observation is said to have been made a stated number of Egyptian years and days later than some occurrence, which was undoubtedly another observation of Jupiter discussed immediately before. The interval is given in the papyrus as 341 Egyptian years and 87 days, but the first figure must be a scribal error for 344; for the corrected interval amounts to almost exactly 344 sidereal years, an accurate recurrence period of Jupiter:

$$\begin{aligned} \text{344 sidereal years} &= \text{29 sidereal revolutions of Jupiter} \\ &= \text{29 latitudinal periods} \\ &= \text{315 synodic periods} \end{aligned} \quad (1)$$

Because of this relation, the previous observation, made on the night of December 30/31, 241 B.C., would have found Jupiter in almost the identical situation with respect to δ and θ Cnc. The actual configuration for that date, computed by modern theory, is shown in Fig. 4133.3. Except for the end of the word cελήνηc ('moon', here a metrological unit, cf. note to fr. 1 line 1) we have none of the original observation report, but only the reduction to the ecliptic. This differs from the later observation in that Jupiter is here said to have been 'in line with the star', which probably means that it had the same longitude as δ Cnc; and the planet was computed to be far enough south that there was doubt whether it lay north of the ecliptic or on it. Again this would make the best sense if the author assumed that δ Cnc was, say, $^1/_4$ ° south of the ecliptic (cf. Fig. 4133.2b). Perhaps the lost report simply stated that Jupiter was seen, say, half a 'moon' to the north of δ Cnc.

Who, then, were the two astronomers who observed Jupiter's position 344 years apart? Our only other source of scientific observational reports in Greek up to the beginning of the second century is the *Almagest*. Therein, Ptolemy cites observations by three people who might be candidates for the authorship of our text: Agrippa, Menelaus, and Theon. Agrippa (Ἀγρίππας) is known to us only from a single lunar observation of his from the night of November 29/30, A.D. 92, used by Ptolemy in his investigation of precession:

> Agrippa, who observed in Bithynia, records that in the 12th year of Domitian, in the local calendar Metroos 7, at the beginning of the 3rd hour of night, the moon covered with its southern horn the trailing and southern part of the Pleiades. (*Almagest* VII 3, ed. Heiberg v. 2, 27)

Theon too (if we disregard the improbable identification of him with Theon of Smyrna) is known only from the *Almagest*, where Ptolemy cites four planetary observations from the years 127–132 that he 'received from Theon'. Ptolemy's direct acquaintance with Theon, and the fact that he incorporates no correction for difference in terrestrial longitude in analysing the observations, imply that Theon worked at or near Alexandria:

> In the 14th year of Hadrian, in the Egyptian calendar Mesore 18 (July 4, 130) in the evening, as we find among the observations received from Theon, (Mercury) was, he says, at its greatest elongation from the sun, being 3 1/2 1/3 ° behind the heart of Leo. (IX 9, ed.

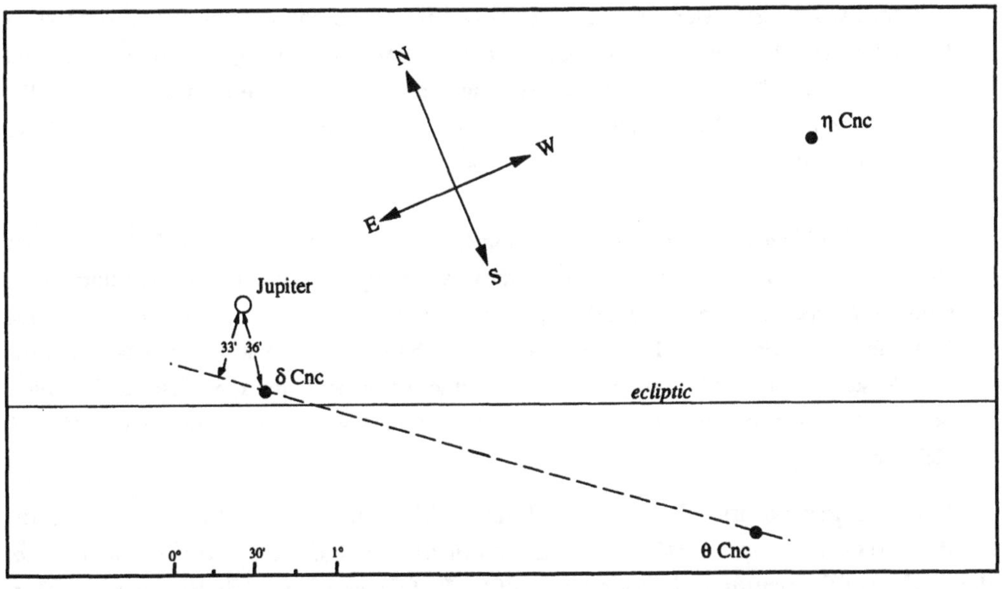

FIG. 4133.3. Jupiter and vicinity, 241 B.C. December 30, 16:00 U.T.

Heiberg v. 2, 275)

Among those given us by Theon the mathematician, we find recorded an observation in the 16th year of Hadrian, in the Egyptian calendar Pharmuthi 21/22 (March 8/9, 132), in which he says that Venus was at its greatest elongation from the sun as evening star, being ahead of the middle of the Pleiades by the breadth of the Pleiades; and it seemed also to be passing a little to the south of (the Pleiades). (X 1, ed. Heiberg v. 2, 296)

Likewise among the (observations) from Theon we find that in the 12th year of Hadrian, in the Egyptian calendar Hathyr 21/22 (October 11/12, 127), Venus was at its greatest elongation from the sun as morning star, being behind the star on the southern wing of Virgo by the breadth of the Pleiades or less (than that) by (Venus's) own size; and it seemed to be passing the star to the north by one moon. (X 1, ed. Heiberg v. 2, 297–298)

Among the (observations) given us by Theon we find that in the 13th year of Hadrian, in the Egyptian calendar Epeiph 2/3 (May 19/20, 129), Venus was at its greatest elongation from the sun as morning star, being ahead of the straight line through the leading one of the three stars in the head of Aries and the star on its hind leg by 1 2/5 °, and it was making its distance from the leading one of the stars in the head approximately twice its distance from the star on the leg. (X 2, ed. Heiberg v. 2, 299)

Menelaus made two lunar observations in A.D. 98 that Ptolemy uses in the same chapter as the observation by Agrippa:

Menelaus the geometer says that he observed at Rome in the 1st year of Trajan, Mecheir 15/16 (January 10/11, 98), at the end of the 10th hour, Spica obscured by the moon: for it could not be seen. But when the 11th hour was coming to an end (Spica) was seen in advance of the centre of the moon by less than its diameter, being equidistant from (both) horns. (VII 3, ed. Heiberg v. 2, 30–31)

Likewise Menelaus, who observed at Rome, says that in the 1st year of Trajan, Mecheir 18/19 (January 13/14, 98), when the 11th hour was coming to an end, the southern horn of the moon appeared to be in a straight line with the middle and southern of the stars in the forehead of Scorpius, while its centre was behind the straight line and was as far from the middle star as the middle star is from the southern star, and (the moon) seemed to have occulted the northern one of the stars in the forehead; for it was nowhere apparent. (VII 3, ed. Heiberg v. 2, 33)

Unlike Agrippa and Theon, Menelaus of Alexandria is well known from ancient sources as a mathematician, and two of his works, the *Sphaerica* and the *On the Method of Determining the Magnitude of Each of Several Mixed Bodies*, are extant in

Arabic translations. The dedication of the latter work to Domitian—the name is corrupt in the unique manuscript, but confirmed by secondary Arabic sources—shows that it must have been written before 96. More problematic is the mute presence of 'Menelaus the mathematician' in the fictitious dialogue of Plutarch's *De Facie* (930A), which is *probably* supposed to be set in Rome and refers to a solar eclipse that was *probably* the eclipse of January 5, 75 as 'recent' (ἔναγχος, 931D–E).[1]

The year of the observation made by the author of our papyrus, 104/105, is near enough to the attested dates for both Agrippa and Menelaus that it could have fallen within the working life of either man. In the case of Theon, the span of 27 years between 105 and his latest known observation is just barely credible for an observer's career. The papyrus does not mention a place of observation, but it is surely significant that the author specifies the date, not only in the astronomically useful Egyptian calendar, but also in the Roman calendar. This evidently represents the civil calendar in local use, just as Agrippa, living in Bithynia, gave the Bithynian calendar date in the observation quoted by Ptolemy. Of the three astronomers in question, one would expect only Menelaus to have used the Roman calendar.[2]

Some details of the observation report in the papyrus are also reminiscent of the observations by Menelaus in the *Almagest*. Apparent distances between heavenly bodies are often measured in 'moons' (cελῆναι: thus the 3rd century B.C. observers who used the 'calendar of Dionysius'), or 'full moons' (cελῆναι διχόμηνοι: Ptolemy himself, ed. Heiberg v. 2, 297–298); but the expression 'lunar diameter' used for this measure in the papyrus is only paralleled in the first of Menelaus's observations from 98. In the second observation from 98 we also find the same method of specifying a position in relation both to single fixed stars and 'straight lines' through pairs of stars as in the papyrus (also paralleled, however, in Theon's observation of 129 and the 'Dionysian calendar' observations).

All in all, the report in the papyrus is compatible with what we know or would expect of Menelaus's date and observational practice, and is unlikely to be the work of the other observers who we know could have been active about A.D. 104/105. Of course one must admit the possibility that there was another astronomer using similar conventions about the same time, and that Ptolemy either did not have access to his observations or did not find occasion to incorporate any of them in his own treatise.

[1] Bulmer–Thomas [1974]; Sezgin [1974]. Arabic references to Menelaus's astronomical observations are entirely dependent on the *Almagest*.

[2] There are no Roman calendar dates in the observation reports in the *Almagest*; but Ptolemy is inconsistent elsewhere in preserving or altering the dating formulas of his sources. The given dates have in any case been altered, perhaps by Menelaus himself, since the death of Nerva did not take place until January 25, A.D. 98.

Date		Observed	ed. Heiberg (v. 2)
13 Aigon 25	Jan. 17/18, 272 B.C.	Mars	352
21 Scorpion 22	Nov. 14/15 & 18/19, 265 B.C.	Mercury	288
23 Hydron 21	Feb. 11/12, 262 B.C.	Mercury	264
23 Tauron 4	Apr. 25/26, 262 B.C.	Mercury	265
24 Leonton 28	Aug. 23/24, 262 B.C.	Mercury	267
28 Didymon 7	May 28/29, 257 B.C.	Mercury	265
45 Parthenon 10	Sept. 3/4, 241 B.C.	Jupiter	386

TABLE 4133.1. 'Dionysian Calendar' observations in the *Almagest*.

The *Almagest* offers us two possible origins for the observation of Jupiter from 241 B.C. that was cited in the lines preceding the preserved beginning of our papyrus: the anonymous observer or observers who used the 'calendar of Dionysius', or a translated Babylonian observation[1]—to which list we must again add the imponderable 'unknown others'. But here there is no difficult choice, for among Ptolemy's 'Dionysian calendar' observations is one of Jupiter that was made on September 3/4, 241 B.C., just four months before the one in the papyrus:

> Next for the sake of the periodic motions (of Jupiter) we again took one of the unambiguously recorded ancient observations, according to which it is declared that in the 45th year according to Dionysius, Parthenon 10, Jupiter in the east covered the southern Asellus. (ed. Heiberg v. 2, 386)

There can be little doubt that Ptolemy and our author were both selecting observations of Jupiter from a series conducted over an unknown span including 241 B.C.

We do not know the name of the observer. In all Ptolemy cites seven observations of planets in relation to the fixed stars that are dated by year number, month, and day κατὰ Διονύcιον, 'according to Dionysius', (Table 4133.1), and it is often presumed that they were made by a single astronomer named Dionysius. Ptolemy, however, consistently uses an impersonal, passive phrasing for these reports, and it is likely that he did not know who made them. There may even have been more than one person involved, for the spread of 31 years from the earliest to the latest is wide, and the sighting of Jupiter is separated from the immediately preceding observation in the series by 16 years. Unfortunately our papyrus adds nothing new to the evidence from the *Almagest* for the calendar of Dionysius. What we do know is that the year count began with the summer solstice of 285 B.C., which immediately preceded the beginning of the year that Ptolemy Philadelphus retrospectively called his first regnal year (actually the first year of his coregency with

[1] An extant Diary tablet covering the latter half of 241/240 B.C. (S.E. 71) reveals that no observations were made at Babylon between December 29 and January 4 because of bad weather: Sachs & Hunger [1988–] v. 2, text –240 rev. line 2.

his father)—hence suggesting Egypt as the site of observation—and that the months were named for the signs of the zodiac.[1] It may be significant, in the light of column ii line 23, that in the *Almagest* instances of the alternative names of the planets ('Phaethon' = 'the brightly shining one' instead of the usual 'the star of Zeus', etc.) are restricted to two of the 'Dionysian calendar' sightings of 'Stilbon', i.e. Mercury.[2]

The author of the papyrus had a clear theoretical purpose in juxtaposing the observation from 241 B.C. with his own. We have seen that Jupiter was in very nearly the same position with respect to the fixed stars at both dates: at the second date it was about half a degree (i.e. one lunar diameter) further east than at the earlier, in fairly good agreement with what the author himself deduces, and its latitude was very slightly more northerly at the later date than at the earlier (Table 4133.2). Moreover, the sun was also at nearly the same longitude, a consequence of the fact that the interval of 344 Egyptian years and 87 days is only a day and a half less than 344 sidereal years.[3] Both these returns follow from the 344-year recurrence period for Jupiter (1).[4] In fact, both observations were at or near the date of Jupiter's acronychal (sunset) rising, since true opposition took place a little over a day after the first observation, and three days after the second.

Date	Sun Longitude	Jupiter Longitude	Latitude	Elongation
241 B.C., Dec. 30, 16:00 U.T. (evening, Egypt)	276° 22'	97° 45'	+0° 35'	181° 23'
A.D. 104, Dec. 31, 18:00 U.T. (evening, Rome)	275° 7'	98° 13'	+0° 37'	183° 4'
A.D. 105, Jan. 1, 18:00 U.T.	276° 8'	98° 5'	+0° 37'	181° 57'

TABLE 4133.2. Sidereal coordinates of sun and Jupiter according to modern theory (longitudes normed to equal tropical longitudes in 241 B.C.).

[1] The derivation of the calendar's epoch is due to Boeckh [1863] 286–340, who further argues that the calendar was modelled on the Egyptian (with 30-day months and 5 extra days at the end), but was intercalated with an extra day every four years. For differing appraisals of Boeckh's (and others') reconstruction of the calendar's details see Samuel [1972] 50–51, Neugebauer [1975] 1066–1067, Toomer [1984] 13–14, van der Waerden [1984b] (a persuasive restatement of Boeckh's thesis).

[2] Toomer [1984] 450 note 59. For the nomenclature, see Cumont [1935].

[3] Taken as an exact number of sidereal years, 344 Egyptian years + 87 days would imply a sidereal year of approximately 365.253 days. I do not know why the author observed one day early.

[4] The 344-year period relation for Jupiter is indeed mentioned in one 'atypical' cuneiform text, Sp. II 985, for which see Kugler [1907–1935] v. 1, 48–53.

One possible reason, then, for comparing the two observations was to determine or confirm an accurate value for the period of anomaly of Jupiter. Assuming the period relation (1), Jupiter should have been at the same sidereal longitude after 344 sidereal years. However, it is not necessary to assume that the author began with the presumption that (1) was exact. He might, for example, have started with such a period relation as the one underlying the Babylonian ACT schemes:

$$427 \text{ sidereal years} = 36 \text{ sidereal revolutions} = 391 \text{ synodic periods} \qquad (2)$$

To check (2), he would have needed suitable observations of Jupiter at least 427 years old, i.e. before the 320s B.C., and it is likely that such were not available.[1] Instead, one could derive a shorter period relation by deducting from (2) the rougher 'Goal-Year' period:

$$83 \text{ sidereal years} = 7 \text{ sidereal revolutions} = 76 \text{ synodic periods}$$

which leads directly to (1). Any inaccuracy in (1) should be expressible by introducing a correction as follows:

$$\begin{aligned} 344 \text{ sidereal years} + \frac{365\frac{1}{4}}{360}x \text{ days} &= 29 \text{ sidereal revolutions} + x^\circ \\ &= 315 \text{ synodic periods} \qquad (3) \end{aligned}$$

Simply taking $^{315}\!/_{391}$ of the numbers in (2) leads to an expected value of 0.92 for x. On the other hand, our author could estimate x from the observations along the following lines. Jupiter has been observed close to opposition at an interval of approximately 1.4 days less than 344 sidereal years, and its longitude was about 0;24° greater at the second observation. Supposing a retrograde motion of, say, –0;8° per day for Jupiter at opposition, we can extrapolate its longitude forward in time from the second observation until its advance in longitude with respect to the first observation is approximately equal to the number of days in excess of 344 sidereal years. In this way we arrive at a smaller value for x, about 0.5.[2] This correction can be introduced into either (3) or, proportionally adjusted, into (2), to obtain an improved estimate of the anomalistic period of Jupiter.

If the foregoing reconstruction is near the truth and the author was primarily interested in Jupiter's period of anomaly, then the criterion for selecting observations was, firstly, to have them separated by an interval known to be close to an exact restitution of anomaly, and secondly, to have the difference between Jupiter's true and mean velocity as great as possible; this second consideration would

[1] Ptolemy cites no planetary observation, Greek or Babylonian, earlier than 272 B.C.

[2] The above analysis treats the fraction in (3) as equal to 1, and neglects the small effect of the zodiacal anomaly.

lead to the selection of dates near either opposition—as here—or first or last visibility. On the other hand there would be no significance in the specific location of the planet in Cancer or in its latitude, and it is not obvious what would be gained by adducing more observations in the continuation of the text. The approach could instructively be contrasted to the one Ptolemy prescribes for establishing an accurate period of anomaly for a planet. Ptolemy's overt procedure is to measure the two anomalistic components of each planet's motion (in terms of his models, the eccentricity, apsidal line, and epicycle's radius) *before* using a pair of widely-spaced observations to correct the initially assumed anomalistic period. In this way, Ptolemy does not need to select observations separated by a whole number of synodic cycles (e.g. oppositions) because he is able to deduce the planet's mean position from its true position at any location on the epicycle and eccentre.[1] Our author seems to be much closer in methodology to Hipparchus, whose investigation of the periods of lunar anomaly (as described by Ptolemy, *Almagest* IV 2) similarly employed observations containing whole numbers of anomalistic periods in order to minimize the influence of the moon's anomaly.

Alternatively, the author may have wished to establish the location of Jupiter's nodal line. Suppose, for example, that he assumed an epicyclic model for the planet, with the essentially correct provision that the epicycle was parallel to the plane of the ecliptic and the deferent inclined with respect to it. Then since at opposition Jupiter is situated approximately in line longitudinally with the centre of its epicycle, the planet's observable position north or south of the ecliptic at opposition indicates whether the epicycle's centre itself is north or south of the ecliptic, and at the nodes the planet's latitude will be zero. One could therefore locate one of the nodes between a pair of observed positions of Jupiter at opposition where the planet was respectively a little north and south of the ecliptic. Another pair of observations at the same longitudes and separated from the first by a long interval would serve to demonstrate that the nodal line remains fixed. The longitude of Jupiter's ascending node about A.D. 100 was approximately 81;25°. It is quite plausible that the two observations in our text were each the northerly member of a pair of oppositions supposed to be flanking this node; indeed, since the author evidently believed that Jupiter was nearer the ecliptic than it really was, he would likely have situated the node at a greater longitude. A generation later we find Ptolemy in *Almagest* XIII 1 assuming nodal lines for the planets without citing empirical evidence (which suggests some dependence on his predecessors' work), and he turns out to place Jupiter's ascending node at the beginning of Cancer, i.e. 90°.

[1] *Almagest* IX 10, X 4 and 9, XI 3 and 7. Ptolemy's actual planetary mean motions are slightly discrepant from those that would have resulted from the analyses in these chapters. For details and proposed explanations, see Newton [1977] 320–327, Toomer [1984] 669–672, Rawlins [1987].

Notes:

Col. 1

1: (cε)λήνηc: This word is probably all that remains of the actual report of the observation in 241 B.C. The moon, if visible at all, was near the opposite horizon to Jupiter at the time, so that the reference here is surely to the *measure* of a moon's diameter somehow involved in the determination of Jupiter's location with respect to the nearby stars. Three of the 'Dionysian Calendar' observations of Venus cited by Ptolemy (*Almagest* IX 7 and 10, ed. Heiberg v. 2, 265 and 288) use 'moons' (cελῆναι) in this way. Here the most probable assumption is that Jupiter was reported as being a fraction (half or less) of a 'moon' north of the line through δ and θ Cnc, much as in the preserved observation of A.D. 104/105.

2: ἐξ ἴcου τῷ ἀcτέρι, 'in line with the star', probably means that Jupiter had the same longitude as δ Cnc (cf. *Almagest* ed. Heiberg v. 2, 382 for Ptolemy's use of the expression with this sense). Like the statement about Jupiter's latitude, this is the author's deduction, not part of the original observation report.

7: Mecheir 7 (Egyptian) = December 31 in A.D. 104.

8: For the nomenclature of the stars cf. Ptolemy's catalogue (*Almagest* VII 5, ed. Heiberg v. 2, 94–96, trans. Toomer 366): τῆc ἐν τῷ cτήθει νεφελοειδοῦc cυcτροφῆc καλουμένηc Φάτνηc τὸ μέcον (=CGal 2632)... τοῦ περὶ τὸ νεφέλιον τετραπλεύρου τῶν προηγουμένων β̅ ὁ βορειότεροc (= η Cnc)... ὁ νοτιώτεροc τῶν προηγουμένων β̅ (= θ Cnc)... τῶν ἑπομένων τοῦ τετραπλεύρου β̅ καλουμένων δὲ Ὄνων ὁ βόρειοc (= γ Cnc)... ὁ νότιοc τῶν προειρημένων β̅ (= δ Cnc). The identifications of the stars are not in doubt.

4134. Procedure Text for a Planet?

16 2B.47/B(c) Fr. 1 6×19 cm Fr. 2 3.5×4 cm Second century

These fragments appear to be written in the same skilled informal hand as **4133**, and were found during the same season (the second) of excavation; but I see no reason to assume that they derive from the same treatise or manuscript. The text is on one side only. The larger fragment preserves the lower margin to 4 cm, whereas top 3 cm has been stripped of its surface fibres. Several lines between 6 and 14 almost certainly reach the left margin of a column, but all lines are incomplete on the right; and abrasion has left several lines illegible in whole or part. The smaller fragment is broken on all sides and abraded. The backs of both pieces bear a vertical line from top to bottom. If it is the same line, then fr. 2 must have been above fr. 1, with at least eight intervening lines lost, which would make the whole column comprise more than fifty lines. It seems more likely that fr. 2 comes from another column.

The text describes manipulations of numbers with three fractional sexagesimal places, referred to as 'degrees' in fr. 1 line 28, and probably representing the longitudinal motion of a planet. I would guess that we are dealing with the construction of a template based on arithmetical sequences. Beyond this few details are clear, and I am unsure even of the identity of the heavenly body concerned.

One apparent element of structure is the parallelism between lines 3–11 and 19–27 of fr. 1. In both passages a quantity is subtracted from 11 plus some unknown fraction (lines 3, 19). The difference obtained in line 4 is 1 plus a lost fractional part, which as we shall see was very close to 0;3. In line 20 the difference obtained is 1;57,5x,20 (the damaged digit seems to have been either 51 or 54). Next a division occurs. In lines 5–6 the remainder of approximately 1;3 is divided by 101 (perhaps plus a fraction) to yield 0;0,37,25. In the corresponding lines 20–21 1;57,5x,20 must have been divided by a number between 78 and 89 to yield 0;1,2x. The subsequent steps are obscure, although one has the impression that a sequence is being generated, and a total increment is set out in lines 11 and 26–27.

Notes:

Fr. 1

6: –ται: presumably γίνονται.

12: I know of no other instance of ἀντιϲτροφή in an astronomical or computational sense.

16: Probably either μεγίϲτη or ἐλαχίϲτη πρόϲθεϲιϲ.

20: The trace from the bottom right of the numeral before $\overline{\kappa}$ resembles either δ or α.

23: The sense of ἀπαρτίζειν here is obscure; cf. **4145**, back, line 6.

30: I can make no sense of the first seven letters of this line.

Fr. 2

5: 0;0,4,10 is exactly half the 0;0,8,20 mentioned in fr. 1 line 25.

8: The terminology is suggestive of kinematic modelling, although 'uniformity' might here mean no more than a constant rate of progress.

4135. Procedure Text for Venus

8×22.5 cm Second century

The middle and part of the right edge of one column, written in a second-century semi-documentary hand, are preserved on the back of what appear to be a register from the second half of the first century. No margin is visible at top or bottom. Three small and rather rubbed fragments belong to an inter-columnar space with some illegible traces of writing on their left edges.

Lines 1–35 set out a velocity scheme for Venus, that is, a description of the planet's longitudinal motion during one synodic period approximated as a sequence of stretches of motion at constant speeds. Such schemes are an important feature of both Babylonian and early Indian planetary theory, and it has long been hypothesized that Greek intermediaries also existed. The present text is the first direct evidence for such a scheme in Greek before the Byzantine period; and it is the more remarkable in that it is manifestly the same scheme for Venus that we have in corrupted form in the sixth-century *Pañcasiddhāntikā* of Varāhamihira (XVII 1–5), in a passage expressly attributed to the earlier writer Vasiṣṭha.[1] Such a close correspondence between Sanskrit and Greek texts has hitherto only been demonstrated for Vasiṣṭha's lunar theory.[2]

The papyrus specifies, for each stretch of constant speed, the number of days, the total number of degrees travelled, and the number of degrees per day; and running totals are given at the principal phases as well as at the end of the whole cycle. The redundancy makes it possible, in spite of the mutilated condition of the fragment, to restore all details of the scheme with complete certainty except for the retrogradation in lines 12–20. The Indian text has no such checks, and where its numbers diverge from the papyrus they seem to be garbled (Table 4135.1).

In the complete synodic cycle of the papyrus scheme, Venus travels 215;30° (beyond one longitudinal revolution) in 584 days. A mean synodic arc of 215;30°, which is also explicitly attested in a Babylonian ACT text (no. 400), is equivalent to the period relation:

1151 sidereal years = 720 synodic periods

This periodicity underlies other ACT schemes for Venus, and reappears in *Pañcasiddhāntikā* XVII 75. *Pañcasiddhāntikā* XVII 1, however, gives 215;50°. For 420 days Venus travels faster than the sun, advancing a total of 513°, so that the greatest elongations at the beginning and end of this interval are approximately 49;30°. This value is about 3;30° larger than the true value (greatest elongations of 46°, 48°, and 50° are attested in various classical sources).[3] The essentially accurate maximum speed between Σ and Ξ, $1;15^{\circ/d}$, is again a Babylonian parameter, and a recently published Babylonian velocity scheme from the fourth century B.C., B.M. 33552, actually prescribes a 60 day interval between Σ and Ξ during which this speed applies.[4] The gradations of speed between this sector and the greatest elongations are not closely paralleled in any known Babylonian scheme. Van der Waerden has suggested that they were derived from a simple epicyclic model; but

[1] Neugebauer & Pingree [1970–1971] v. 1, 156–157.
[2] Neugebauer & Pingree [1970–1971] v. 2, 16–18.
[3] Neugebauer [1975] 804.
[4] Britton & Walker [1991].

then one would expect the scheme to yield the same greatest elongation, 48°, that van der Waerden hypothesizes was the empirical datum determining the model.[1]

According to the papyrus, Venus slows in two stages as it approaches station, with a small adjustment of 0;18° in 3 days at the end. The Indian version here is obviously corrupt. An attempt was made by P. Wirth to produce an astronomically plausible scheme by retaining the first stage, 20° in 27;30 days, and deriving a second stage by *adding* the numbers in the text to obtain 21;15° in 30;30 days.[2] This reconstruction finds no support in the papyrus; and in any case intervals containing fractions of days would be highly unusual, and highly inconvenient, in a velocity scheme. One might be tempted to rescue 1;15° as an error for 0;15°, representing a variant of the papyrus scheme in which the retrogradation was exactly an integer number of degrees.

From the known totals for the direct motion and the complete cycle, we can deduce that the retrogradation was 18;6° in 38 days; the latter number can be read in line 19. Unfortunately the details of the pattern of retrograde motion are poorly preserved in the papyrus, and some of the legible numerals, e.g. the 20° in line 14 and the 60 days in line 15, make no sense. The deletion at the beginning of line 14 suggests that this section gave the copyist trouble. The Indian version is also

	Papyrus	*Vasiṣṭha*
Ξ to Ψ	74° in 60^d	74° in 60^d
	73° in 60^d	73° in 60^d
	72° in 60^d	72° in 60^d
	30° in 40^d	20° in $27;30^d$
	10° in 20^d	
	0;18° in 3^d	1;15° in 3^d
(Total of Ξ to Ψ)	*(259;18° in 243^d)*	
Ψ to Φ	?	–2° in 15^d
		–2°? in 5^d (to Ω)
		–4°? in 10^d (to Γ)
		–4° in 20^d
(Total retrogradation)	*(–18;6° in 38^d)*	
Φ to Σ	reverse of Ξ to Φ	reverse of Ξ to Φ?
Σ to Ξ	75° in 60^d	75° in 60^d
(Total)	*(360° + 215;30° in 584^d)*	

TABLE 4135.1. Velocity Schemes for Venus.

[1] Van der Waerden [1988].
[2] Wirth [1974].

incoherent at this stage and offers no help.

Lines 35–39 seem to have set out an epoch for the scheme. The reference to Timocharis is enigmatic. In the *Almagest* Ptolemy cites four lunar occultations and a stellar passage of Venus observed by Timocharis at Alexandria between 295 and 272 B.C., as well as several undated measurements of stellar declinations. Did our text extrapolate an epoch for the Venus scheme from an observation by Timocharis? I do not think, at any rate, that the appearance of his name at this stage in the text justifies the inference that he had a hand in devising the scheme.

The short line 40, with a horizontal stroke above it, seems to mark the title of a new section of the text. What kind of calculation was discussed is probably beyond recovery, given the very uncertain readings of these lines.

4136. Procedure Text for the Moon

12 1B.133/C(d) 12×13.5 cm First century

Plate II

Much of the upper part of a column of text in an elegant first-century semi-documentary hand is preserved, along with traces of the initial letters of some lines of a second column (not transcribed). The copyist made a few mistakes in transcribing numerals; these are reported in the apparatus and tacitly corrected in the translation. On the back are the horoscopes for dates in A.D. 162, 191, and 201 edited below as **4241**, as well as other jottings described under that number.

The text concerns the computation of the template component of the 'Standard Scheme' for the moon (see Appendix H and **4164**). This template gives the daily and cumulative progress of the moon in longitude and argument of latitude, starting from an epoch coinciding with a date of least motion; the daily progress of each quantity is computed as a linear zigzag function with the anomalistic month as its true period P, and 3031 days (= 110 anomalistic months) as its whole number period, Π. Procedure texts must have existed for the line-by-line calculation of the tabulated values of these two zigzag functions and their running totals. What we have here, however, are rules for finding the total progress in longitude in the template during each anomalistic month, counted from the last day before a minimum to the last day before the next minimum, an interval always of either 27 or 28 days. The text assumes that the template will be laid out with a new set of columns starting after each minimum. We may describe its purpose as to show how one can find the running total of the zigzag function (i.e. progress in longitude since epoch) for the last line of every column without having to proceed through all intervening lines. This would be useful as a way of checking the sums.

The present commentary aims only to restate in a clearer form the operations by which the author finds the desired totals, leaving for discussion elsewhere the interesting question of how he derived his rules (Jones [1997a] 11–18). The first, shorter, part of the text finds a total progress in longitude that can be applied every time that the interval under consideration is 28 days.

We take the mean value of the zigzag function, which is the mean daily motion of the moon in (sidereal) longitude,

$$\mu = 13;10,34,52°, \tag{1}$$

and multiply it by 28 (for the 28 days being totalled), getting 368;56,16,16°. Then we encounter a new number, 0;5,43, which is described as 'the amount added for 28-line (columns)'. This alludes to the fact that whenever the interval covered by one 'column' of the template is 28 days, the value of the zigzag function for daily progress in longitude at the top of the next column is 0;5,43,0° greater than at the top of the preceding column. We multiply this by 7 (described as $^1/_4$ of the 28 days) to get 0;40,1°. Finally, this is subtracted from the previous result, yielding 368;16,15,16°, which is the constant progress in longitude in a 28-line column.

Whereas all 28-line columns have the same total progress, the totals for each of the 27-line columns is different. The text accordingly shows how to find the total for the first 27-line column (the third column of the whole template) as a correction to the total for 28-line columns, and then how to find each subsequent 27-line total as a correction of the previous one.

From 368;16,15,16°, obtained above, the author subtracts the minimum value of the zigzag function,

$$m = 11;42,10,37°, \tag{2}$$

leaving 356;34,4,39°. The number next mentioned, 0;1,24°, is described as 'the excess of the amount subtracted for 27-line (columns) over the amount added for 28-line (columns).' In fact from the first line of any 27-line column to the first line of the next column, the value of the zigzag function decreases by 0;7,7,0°, and this is greater than the increase in the zigzag function for 28-day columns (0;5,43,0°) by 0;1,24,0°. Adding 0;1,24° to 356;34,4,39°, we get the progress in longitude over the third column, 356;35,28,39°, or equivalently a shortfall of 3;24,31,21° from one full revolution in longitude.

For the next 27-line column (the fifth of the template) we subtract the same 0;1,24,0° from the shortfall for the third column, which leaves a smaller shortfall for this column: 3,23,7,21°. (Equivalently, the author could have *added* 0;1,24,0° to the total *progress* for column 3).

For the seventh column, we get a new shortfall of 3;23,7,21° by adding 0;1,17,0° to the shortfall for the fifth column (or equivalently, *subtracting* 0;1,17,0° from the

Column	Day	Difference	Longitude	Difference
	0		0; 0, 0, 0°	
1	28	28	8;16,15,16°	8;16,15,16°
2	56	28	16;32,30,32°	8;16,15,16°
3	83	27	13; 7,59,11°	-3;24,31,21°
4	111	28	21;24,14,27°	8;16,15,16°
5	138	27	18; 1, 7, 6°	-3;23, 7,21°
6	166	28	26;17,22,22°	8;16,15,16°
7	193	27	22;52,58, 1°	-3;24,24,21°
8	221	28	31; 9,13,17°	8;16,15,16°
9	248	27	27;43,24,56°	-3;25,48,21°
10	276	28	35;59,40,12°	8;16,15,16°
11	303	27	32;33,44,51°	-3;25,55,21°

TABLE 4136.1. Progress in longitude at intervals of 1 anomalistic month in the Standard Scheme template.

total progress). This 0;1,17,0° was obtained by dividing the line-to-line increment of the zigzag function,

$$d = 0;12,50,0°, \tag{3}$$

by 110 (the number of anomalistic months in the zigzag function's whole number period) and subtracting the quotient (0;0,7,0°) from the 0;1,24,0° that we have used twice already. The shortfall for the ninth column, 3;25,48,21°, is again obtained by simply adding 0;1,24,0° to the shortfall for the seventh column. An addition of something to something, possibly of 0;1,24° to the shortfall for the third column, yields the shortfall for the eleventh (and last) column of the template.

Table 4136.1 extracts from the Standard Scheme template (Appendix H, Table 1) the total progress in longitude tabulated in the last line preceding each minimum. As the differences show, our text derives the correct advance or shortfall in every instance that we can check.

Notes:

5: One expects ō μ α (0;40,1 as in the translation, rather than 0;40,0,1). It is conceivable that the extra ō is not an error, but rather acts as a separator to prevent reading μα. Such a use of the 'zero' sign is well documented in Babylonian astronomical tablets; see Neugebauer [1941].

8: A paragraphus marks the new section.

4137. Predictions of Lunar Eclipses, A.D. 56 and 57

38 3B.85/H(1)c | 8×16 cm | First half of first century
Plate I

The fragment is written on the front only in an uneven documentary hand with frequent abbreviations. The bottom margin is preserved to 2 cm, and what was almost certainly the left edge of the column of text is just extant in many lines. On the remaining two sides the text is broken off. Lines 1–17 contain the predicted circumstances of a lunar eclipse, the date of which was presumably given in the lost preceding text. Lines 18–22 belong to the prediction of another lunar eclipse or eclipse possibility six months later, of which little more than the date is preserved. Each prediction would have included the following data:

(a) the number of lunar months since the preceding eclipse,
(b) a year number (regnal?),
(c) a year number according to the current Callippic period,
(d) Athenian month without specific day number,
(e) Egyptian month and day,
(f) the time of opposition (i.e. mid-eclipse)? and perhaps other information,
(g) the direction and magnitude of obscuration (as a fraction of the disk),
(h) the duration of the eclipse in hours,
(i) the inclination (πρόcνευcιc) of the eclipse at certain critical stages,
(j) the location of the moon with respect to fixed stars near the ecliptic.

Data (g) through (j) are preserved for the first eclipse, and (a) through (e) and part of (f) for the second; there is no overlap.

It would be natural to suppose that this is a fragment of a canon of lunar eclipse predictions, which, when complete, listed the circumstances for a series of consecutive eclipses over an interval of perhaps several years. We can confirm this hypothesis by comparing it with a recently published demotic papyrus (P. dem. Berlin 13146 + 13147) listing on its 'recto' lunar eclipses during the interval 85–74 B.C.[1] The demotic canon gives some or all of the following information for each eclipse or eclipse possibility:

(Da) a year number according to an era such that year 1 = 102/101 B.C.,
(Db) the number of lunar months since the preceding eclipse or eclipse possibility,

[1] Neugebauer, Parker, & Zauzich [1981].

(Dc) the location of the moon with respect to a zodiacal sign (or constellation),

(Dd) Egyptian month and day,

and for actual eclipses:

(De) time of the (mid?-)eclipse in seasonal (?) hours,

(Df) the general direction of obscuration, and

(Dg) past-tense remarks about planets visible in the sky, apparently derived from observations.

There is an obvious correspondence between elements (Db) and (a), (Dc) and (j), (Dd) and (e), (Df) and (g), and probably (De) and (f). Moreover the original editors of the demotic canon did not recognize that the year numbers (Da) pertain to the—at that time unattested—fourth Callippic period, so that this datum too has its counterpart in (c). There are no remarks on the planets in the preserved part of the Greek text, which gives, however, more detail about the appearance of the eclipse and alternative specifications of the date. Overall the structural resemblance between the two texts is remarkably close; these must be widely-spaced instances of a standard variety of text.

In order to establish the dates of the eclipses or eclipse possibilities in our text, we have the following facts to take into account. (1) We know that the first date was of an actual partial eclipse expected to be visible at night, whereas we can *assume* only that the second date was of an eclipse possibility six months later. (2) The second date fell within the sixth Callippic period, i.e. between A.D. 51/52 and 126/127, and (3) during the month Epeiph in either the Egyptian or Alexandrian calendar. (4) The eclipse at the first date was partial but not small, with an expected duration of about 3 hours and expected magnitude less than 8 digits. (5) The moon's disk was eclipsed from the south. (6) The moon was situated in the constellation Gemini, in the neighbourhood of the star δ Gem ('the star in the navel of the trailing twin', cf. note to l. 16).

Now the longitude of δ Gem during the sixth Callippic period was about 82°, so that for the moon to be eclipsed within, say, 15° of this star, the sun would have to be in the range 247° to 277°, and this corresponds approximately with the month of December. We isolate, therefore, all lunar eclipses that took place during December between A.D. 51 and 127 (Table 4137.1).[1] We can eliminate from the list the total eclipses 4, 5, and 10, and eclipses 2, 3, 7 which occurred during the daylight hours for Egypt. There remain eclipses 1, 6, 8, and 9. Of these, however, eclipses 6, 8, and 9 all had the obscuration on the north side of the moon's disk, because its latitude was south of the ecliptic. Hence we can identify the eclipse of lines 1–17 as the only remaining eclipse, that of December 10/11, A.D. 56.

[1] I am indebted to Dr John Steele for calculating the data in this table.

	Date	Time (U.T.)	Local time (Alexandria)	Magnitude	Duration (hours)
1)	56 Dec. 10(/11)	23.34	1:20 A.M.	0.43 (south)	2.51
2)	65 Dec. 31	12.26	2:16 P.M.	0.46 (south)	2.55
3)	74 Dec. 22	7.53	9:32 A.M.	0.42 (south)	2.48
4)	75 Dec. 11	6.77	8:46 A.M.	1.64 (total)	3.91
5)	93 Dec. 21	14.84	4:50 P.M.	1.63 (total)	3.90
6)	94 Dec. 10	16.39	6:23 P.M.	0.76 (north)	3.08
7)	103 Dec. 1	12:76	2:46 P.M.	0.21 (north)	1.75
8)	112 Dec. 21	0.77	2:46 A.M.	0.77 (north)	3.08
9)	121 Dec. 11	21.28	11:17 P.M.	0.22 (north)	1.76
10)	122 Dec. 1	8.26	10:16 A.M.	1.57 (total)	3.53

TABLE 4137.1. December lunar eclipses within the 6th Callippic Period (local times are not corrected for the equation of time).

Consequently, the date six months later, discussed in lines 18–22, was of the total lunar eclipse of June 5/6, A.D. 57. For an observer in Egypt, the eclipse began more than an hour before sunset, with totality for about an hour starting half an hour after moonrise. The predictions no longer survive in the papyrus; but with knowledge of the actual date in question, we can now look more closely at the text's redundant dating formula. In the first place, we note that June 5, 57 was Epeiph 1 in the old Egyptian calendar but Payni 11 in the Alexandrian. Hence there is, surprisingly, no trace in the papyrus of the civil date.

Still more unexpected is the use of the Callippic periods to specify the year. This method of dating has hitherto been found only in twelve observation reports in Ptolemy's *Almagest*, with none later than year 51 of the third period, corresponding to 128 B.C.[1] It is now clear that the Callippic periods continued in use in astronomical texts well into the first century of our era. With this knowledge, we can also recognize Callippic year counts (without explicit references to the period number or Callippus's name) in two earlier papyri: the demotic eclipse canon discussed above (period 4, years 17–28), and *Tab. Amst.* inv. 1, a sign-entry almanac on a wooden tablet, covering the years 26/25 and 25/24 B.C. (period 5, years 1 and 2).[2]

The use of Athenian months must be taken as part of the 'Callippic' dating convention. The earliest attested Callippic dates are in Ptolemy's citations (*Almagest* VII 3) of four observations of lunar occultations made by Timocharis at Alexan-

[1] Listed in Samuel [1972] 47.

[2] Jones [1993].

dria in the years 295–283 B.C. (period 1, years 36–48). These all give dates according to Athenian months, with Egyptian calendar equivalents. It has long been assumed that these are not references to the civil Athenian calendar—in any case irrelevant in Alexandria—but to a schematic 'astronomical' calendar with a fixed pattern of intercalated months, coordinated with the Callippic periods.[1] Moreover, since the seven observation reports from the second and third periods (period 2 years 54 and 55 = 201–200 B.C.; period 3 years 32–51 = 146–128 B.C.) that Ptolemy derives from Hipparchus's works have only Egyptian calendar dates, the suggestion has been made that the schematic Athenian calendar had dropped out of use by Hipparchus's time.[2] Our papyrus emphatically confirms the former hypothesis, and refutes the latter.[3]

How were the predictions in the eclipse canon generated? One possibility is that the author possessed an advanced mathematical theory of eclipses, such as Ptolemy's (*Almagest* VI) or the Babylonian 'ACT' eclipse schemes. Methods of this kind yield precise numerical values for, e.g., the magnitude of the eclipse, which the author would have had to convert into the rougher estimates in the papyrus; for in general, the description of the predicted eclipse is much closer to the observation reports of eclipses in the *Almagest* than to the results of using Ptolemy's predictive tables. Another hypothesis that would explain the 'look' of the prediction is that the author made use of an eclipse period to project the observed characteristics of a past eclipse forward by one or more periods. The most suitable eclipse period would have been the ἐξελιγμός or 'revolution', an interval of 669 synodic months that was supposed to bring about a repetition of eclipses of the same magnitude in exactly 19756 days (or 54 Egyptian years and 46 days), and with an advance in longitude of 32°.[4] The eclipse preceding that of December 10/11, A.D. 56, by 19756 days was that of November 8/9, A.D. 2. It took place only about 40 minutes earlier in the night than the eclipse of A.D. 56, and the magnitude and duration of the two eclipses were almost identical. The author could have reproduced without alteration from an observation report the duration, the fraction of the disk obscured, and the directions of the shadow; and if the situation of the moon with respect to the stars was reported, he could have extrapolated its situation at the later eclipse using a list of longitudes of stars in the zodiacal belt

[1] The range of opinion on this question is exemplified by van der Waerden [1984a] (developing a detailed reconstruction of the Callippic calendar by Fotheringham [1924]); Neugebauer [1975] 617 (arguing for civil dates); Toomer [1984] 12–13 (arguing for an astronomical calendar, but denying that it can be reconstructed from the available evidence); Goldstein & Bowen [1989] (arguing that the 'Callippic' calendar was in fact a version of the Egyptian lunar calendar).

[2] Goldstein & Bowen [1989] 287.

[3] I will discuss elsewhere the bearing of **4137** on the reconstruction of the Callippic calendar.

[4] The period relation, obtained by tripling the 18-year 'Saros' period, is described in similar terms by Ptolemy (*Almagest* IV 2, ed. Heiberg v. 1, 270) and Geminus (XVIII, ed. Manitius 200–202).

and the 32° surplus of motion prescribed by the period relation. It goes without saying that, to compile an eclipse canon in this way, one would require a continuous record of eclipse observations over a span of years, and we have no direct evidence for the existence of this during the late Hellenistic or early Roman periods.

The purpose of canons such as this one and P. dem. Berlin 13146 + 13147 is also a matter for conjecture. Predictions of eclipses might be made for the sake of comparison with subsequent observations, in order to test or demonstrate the accuracy of theoretical assumptions underlying the methods of prediction. Such a 'scientific' aim seems unlikely here, given the text's provenance and the lack of precision in its forecasting. The attention paid to the 'inclinations' of the eclipse invites comparison with Ptolemy's chapter on the computation of these directions of obscuration (*Almagest* VI 11), where he refers to the very same critical moments of an eclipse that appear in the papyrus as 'those possessing a certain "sign"' (ἐπὶ μόνων τῶν ἐπιϲημαϲίαν τινὰ ἐχουϲῶν). Ptolemy applies this term ἐπιϲημαϲία in the *Almagest* not only to eclipse inclinations (ed. Heiberg v. 1, 536 and 537) but also to eclipse magnitudes (v. 1, 512) and to the appearances and disappearances of fixed stars (v. 2, 204). In this last context he has in mind specifically 'signs... concerning the states of the airs' (περὶ τὰ καταϲτήματα τῶν ἀέρων), that is, a sort of weather prediction, and this is borne out by the frequent occurrence of the term in the weather calendar of his *Phaseis*.[1] But in the *Tetrabiblos*, where the word is applied to characteristics of eclipses and planetary phases and stations (e.g. II 5 and 7 Boll–Boer = 4 and 6 Robbins), it has a broader meaning of 'ominous moment', foretelling the full range of beneficial and calamitous outcomes for peoples and places that falls within the scope of 'general' astrology. It is true that the *Tetrabiblos* does not include inclinations among those things that one is to take into account in interpreting an eclipse; on the other hand the Babylonian eclipse omen texts, with their extensive use of the quadrants of the moon obscured at the various stages of an eclipse, provide a precedent for such a practice,[2] and it is possible that a later Greco-Egyptian stage of this tradition has simply failed to reveal itself in the documents now available.

Notes:

1–2: For the specification of magnitude by comparison with a fraction of the moon's diameter cf. *Almagest* IV 6, ed. Heiberg v. 1, 303 (Babylonian report of the eclipse of September 1/2, 720 B.C.): ἐξέλειπεν ἀπ' ἄρκτων πλεῖον τοῦ ἡμίϲουϲ. Similarly ed. Heiberg v. 1, 418–419, 332, 206, 315 (Babylonian reports, April 21/22, 621 B.C. and July 16/17, 523 B.C.; Nov. 19/20, 502 B.C.; anonymous

[1] Neugebauer [1975] 141–142 therefore argues that the inclinations too were weather signs as distinct from astrological data.

[2] Rochberg [1988] 36–63, esp. 51–55.

Alexandrian report, April 5/6, 125 B.C.; Ptolemy's own observations, October 20/21, A.D. 134 and March 5/6, A.D. 136). Rochberg [1988] 49–51 reviews the use of magnitudes (expressed in 'digits' or in fractions of the disk) in Babylonian omen commentaries and related texts. In the present instance the prediction is correct for the eclipse of December 10/11, A.D. 56, but imprecise; modern theory yields 5.3 digits from the south, i.e. slightly less than half the disk, while the *Almagest* tables yield 6 digits.

3–4: None of the eclipse reports in the *Almagest* gives the number of hours of obscuration, although it can sometimes be directly inferred from reported times of beginning, middle, or end. In *Tetrabiblos* II 7 Boll–Boer (= II 6 Robbins) Ptolemy derives the duration of the omen's terrestrial outcome from the duration of the eclipse. In Babylonian omen texts, account is taken of the 'watches' in which the eclipse begins and ends rather than the time interval itself (Rochberg [1988] 44–47). The actual duration of the eclipse according to modern theory was 144 minutes. The 'about 3 hours' of our text is rather too long, whether or not it takes account of the longer seasonal hours of night in December; a duration of 3 equinoctial hours would correspond to an eclipse of about 9 digits.

4–9: The meaning of the term πρόcνευcιc is made clear in *Almagest* VI 11: the direction of the horizon in which the shaded part of the sun or moon points at critical stages of an eclipse, or equivalently the direction of the line through the centres of moon and shadow (or sun and moon for a solar eclipse). For a partial lunar eclipse, *Almagest* VI 11 (ed. Heiberg v. 1, 537) recommends inspecting the inclination at three critical stages: (1) that at first obscuration (ἡ τοῦ πρώτου ἐκλείποντοc), when the shadow first touches the moon's disk; (2) that at maximum obscuration (ἡ τοῦ πλείcτου ἐκλείποντοc), coinciding with mid-eclipse; and (3) that of final emersion (ἡ τοῦ ἐcχάτου ἀναπληρουμένου), when the last part of the moon's disk leaves the shadow (cf. Fig. 4137.1, showing the approximate situations predicted in our text for the eclipse of December 10/11, A.D. 56). These are precisely the three stages named in our text. This incidentally confirms that Heiberg's reading at ll. 12–13, rather than the reading of MS D preferred by Toomer ([1984] 314), is correct: 'maximum obscuration' pertained to partial, not total, eclipses, since it is only in a partial eclipse that an inclination can be observed at mid-eclipse.

In Ptolemy's treatment the determination of the inclinations is a complex problem in spherical geometry involving not only the situation of the moon and shadow with respect to the ecliptic but also the simultaneously shifting situation of the ecliptic with respect to the cardinal directions of the horizon; and he points out that in practice it would be much easier to observe the directions within the rather coarse tolerance expected for this kind of inquiry. Our text gives only rough directions, which could have been estimated without difficulty from the expected magnitude and direction of the obscuration.

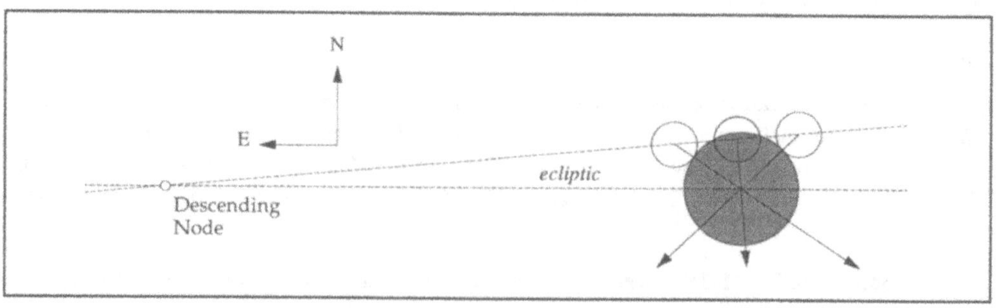

FIG. 4137.1. Inclinations of a partial lunar eclipse.

14–17: The second star named here, 'the star in the navel of the trailing twin', presents little difficulty. Ptolemy's star catalogue in *Almagest* VII 5 (ed. Heiberg v. 2, 92–94) lists no such star; but in the list of stars near the ecliptic in the *Handy Tables* (ed. Halma v. 3, 44–58; *Vat. gr.* 1291 f. 94ᵛ) the star δ Gem is described as 'the (star) over the knees, which Hipparchus calls navel' (ὁ ὑπὲρ τὰ γόνατα ὃν Ἵππαρχος ὄμφαλον καλεῖ).[1] In the *Almagest* catalogue it is called 'the star in the left groin of the trailing twin'. The tropical longitude of δ Gem at this date was 81;31°, about three degrees to the east of the moon during the eclipse (true opposition at 78;23° according to modern theory; cf. Fig. 4137.2).

The identity of the other star is uncertain. Among the names of stars in the *Almagest* catalogue, the only obvious candidates are the stars 'on the right shoulder' (ἐπὶ τοῦ δεξιοῦ ὤμου, υ Gem), 'on the right side' (ἐπὶ τοῦ δεξιοῦ πλευροῦ, 57(A) Gem), and 'on the right foot' (ἐπὶ τοῦ δεξιοῦ ἀκρόποδος, ν Gem) of the leading twin. Of these, υ Gem is ruled out because its longitude is almost three degrees *higher* than that of δ Gem, so that the moon was not between these stars during the eclipse. 5 (A) Gem is amost directly north of δ Gem, and although no stage of the eclipse fell exactly between them, the

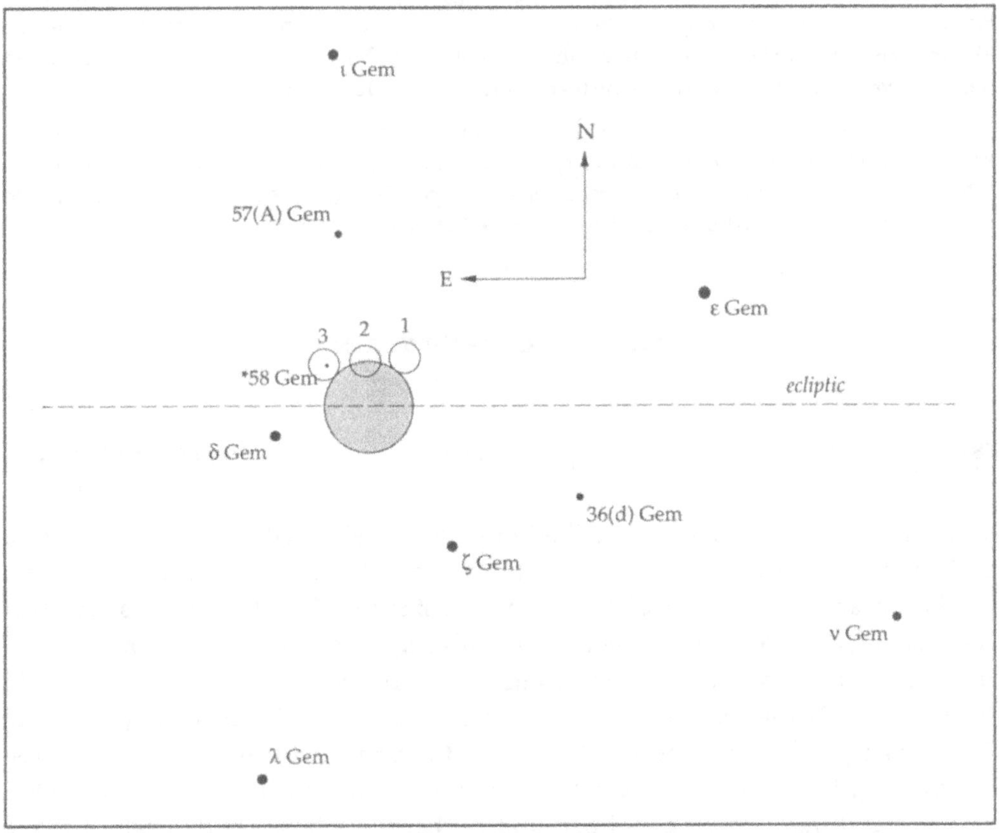

FIG. 4137.2. The eclipse of December 10/11, A.D. 56.

[1] The star catalogue of the *Handy Tables* has been neglected in spite of its significance for the text of the *Almagest* catalogue and the history of stellar nomenclature. The only discussion of its relation to and variants from the *Almagest* catalogue is Neugebauer [1975] 1026–1028.

moon was to the east of 57(A) at the end of the eclipse. The remaining star, ν Gem, was about 11° east of δ Gem and 8° east of the moon. This would give a needlessly lax specification of the moon's location, since there are two comparably bright stars close to the ecliptic in the catalogue that would have situated the moon in tighter bounds: 'the star on the left knee of the leading twin' (ε Gem, at 72;55°) and 'the star under the left knee of the trailing twin' (ζ Gem, at 77;59°). The *Handy Tables* as well as Hipparchus (*In Arat.* ed. Manitius 206, 210, and 280) call these respectively 'the leading' and 'the middle' of the 'three stars in the knees', without specifying which knee of which twin each star belongs to. In the *Almagest* the leading twin has no right knee; perhaps the list used by the author of the eclipse canon assigned the 'knees' differently, either giving the leading twin two knees (so that ζ Gem would have been his 'right knee') or making his only stellated knee the right one (ε Gem).

18: The reading and interpretation of the end of this line are very uncertain. After the conventional sign for ἔτοc we expect a year number. The following letter superficially resembles an uncharacteristically straight-sided π, but on closer inspection the right vertical stroke seems to cross the horizontal stroke rather than continuing it, suggesting a crossed γ like the common abbreviation for γίνεται. After this is, apparently, κ written in two not-quite-touching strokes (similar instance in line 9). June 5/6, A.D. 57 in fact fell during the third regnal year of Nero. If this was somehow indicated here—of which I am far from sure—the text would have had to be written after 54.

20: If the restoration of the end of the preceding line is correct (and I see no alternative), no day number within the Athenian month was specified. Curiously, this parallels the three Babylonian eclipse observations from 383–382 B.C. reported in *Almagest* IV 11 (ed. Heiberg v. 1, 340–343), which are dated by Athenian archon and Athenian month but no day number.

4138. On Eclipse Prediction

68 6B.25/H(1–2)a   7×7 cm   Fourth or fifth century

Plate XI

The fragment, unusually, is inscribed only on the side with vertical fibres, in a serifed informal round hand, and is broken on all sides. The text is part of a worked example of the prediction of circumstances of eclipses of—as we shall see—the moon. In lines 4–13 the time of mid-eclipse (or true opposition) is stated to be later than midnight by a sexagesimal fraction, 0;50,25, of a day, which would be 8:10 P.M. Of the date of the eclipse only the month, Mecheir, is preserved. Immediately before this specification of the time, the text was evidently discussing the longitude of the eclipse and its situation relative to one of the lunar nodes. The zodiacal sign Gemini could pertain to either the eclipsed body itself or to the node, which of course must have been nearby.

Taking the specified longitude and month together, we can deduce both that the eclipse in question was of the moon and that the calendar was the Egyptian. For if the eclipse was solar, then the sun would have had to be in or near Gemini, and the corresponding dates (late May to early July) are impossible for Mecheir in either the Alexandrian or the Egyptian calendar throughout antiquity. In a lunar eclipse the sun would have been in or near Sagittarius, which would put the date between late November and early January. Only in the Egyptian calendar did Mecheir overlap this interval, from roughly the middle of the second century to the middle of the fourth. Obviously the Egyptian month and approximate time of day are not enough information to establish which eclipse during this long span is the one of our text.

The tables used to find the longitude and time of the opposition must have used midnight epoch. This shows that they were neither Ptolemy's tables, which use noon epoch, nor the 'Standard Scheme', which was based on evening epoch. It seems as if the moment was determined in a crude fashion from the sun's and moon's longitudes for the midnights before and after. If we call their respective longitudes at the previous midnight s_0 and m_0, and those at the following midnight s_1 and m_1, then

$$t = \frac{s_0 - m_0 - 180°}{(s_0 - m_0 - 180°) + (s_1 - m_1 - 180°)}$$

yields the time of opposition as a fraction of a day after the former midnight, on the assumption that a constant rate of progress for both luminaries. By a similar interpolation one would have arrived at the longitude. Perhaps in lines 3–4 the elongation of the moon from the node was used in a rough-and-ready way to obtain the eclipse's magnitude or duration.

4138a. Treatises on Eclipse Prediction

35 4B.101/G(1–3)a 10.5×11 cm Second century
Plate III

The text, apparently from a theoretical work on lunar theory and eclipses, is written in an elegant rounded bookhand, rather plain but with some serifs, on one side only of the papyrus. Parts of the tops of two columns are preserved with part of the upper margin (3 cm). Restorations of the text of column ii, as well as evident crowding of the last letters of line 11, suggest that the right margin is just reached in the extant part of line 10, making a column width of 6 cm or approximately 22 letters.

The topic is the distribution of 'eclipse possibilities', that is, those oppositions or conjunctions when the moon is closest to either of its nodes, so that its latitude may be small enough for an eclipse to occur. In the better-preserved column ii the text mentions eclipse periods, i.e. intervals that bring about a repetition of eclipses of similar character, which exist only for lunar eclipses; hence the eclipse possibilities in question are probably oppositions. Since the mean longitudinal progress of the sun and moon in one synodic month is close to 29 $^1/_6$ ° (over one complete revolution in the moon's case), while the nodes recede by about 1 $^1/_2$ ° in the same time, the interval from an eclipse possibility near one node to the next possibility near the opposite node will usually be six synodic months, with a five-month interval inserted in the series after every six or seven six-month intervals.

In determining the correct distribution of these intervals, it is more convenient to consider, not the progress in longitude of the place of opposition, but instead the progress relative to the nodes, i.e. motion in 'argument of latitude'. The time from the sun's passage of one node to its passage of the opposite node will correspond to exactly a semicircle of progress in argument of latitude. Six synodic months' progress will exceed a semicircle by about 4°, while five months' progress will fall short by about 26 $^2/_3$ °. One method of forecasting the pattern of five-month and six-month intervals between eclipse possibilities, and incidentally to estimate whether an eclipse will occur at each and of what magnitude, is to continue adding 4° to the argument of latitude for some initial eclipse possibility, until a threshold of about 15° from the node is reached, at which stage one assumes a five-month interval and subtracts the corresponding 26 $^2/_3$ °. Procedures analogous to this are attested in Babylonian astronomy and in the demotic *P. Carlsberg* 31; see Aaboe [1972].

As well as comprising a whole number of synodic months, an eclipse period must be close to an integer multiple of the moon's period of latitude, and preferably will also contain nearly a whole number of periods of lunar anomaly and of years. One consequence is that the period ought to bring about a return of the nodes to their initial longitudes. This may be what the text stated in lines 1–3.

Lines 6–12 apparently say that the moon 'deducts' one semicircle for every five-month or six-month eclipse interval. This must refer to the relationship between the synodic and draconitic months (revolutions of the moon relative to the node) in an eclipse period, which can be expressed as:

$$^1/_2 p = d - s,$$

where p is the number of eclipse intervals, s of synodic months, and d of draconitic months. Unfortunately the preceding assertion, from which this relation is supposed to follow, is damaged in the papyrus, and I have not succeeded in restoring a text concerning the five-month intervals that is both pertinent and valid.

The remains of column i are clearly part of the same discussion, although too little survives to make any continuous sense. In line 1 we can probably restore τοῦ πλάτουϲ ('latitude'), in line 3 ἐν τῇ περιόδῳ ('in the period'), while the 'complete circles' of line 5 are presumably the disregarded complete revolutions of the moon in some number of synodic months, beyond which the moon 'was taking up' some additional arc (ἐπελάμβανε, same verb as col. ii line 5, but why the past tense?).

4139. Treatise on Lunar Periods

82/87(b) 3×3.5 cm Second century

Plate II

This small fragment preserves the top right corner of a column of text in a tiny round serifed hand. Probably more than half the column width is missing to the left (cf. note to lines 4–5), and a continuous text cannot be restored with any certainty. As a discussion of periods of the moon's motion, it can be compared with *Almagest* IV 2 and with passages from an anonymous third-century commentary on Ptolemy's tables.[1] The larger context was probably a work on lunar theory or eclipses.

In the second line the number 6695 certainly refers to the number of anomalistic months in the fundamental period relation for lunar anomaly of the Babylonian System A:[2]

6247 synodic months = 6695 anomalistic months (1)

This is of considerable interest in itself as the first appearance of this period relation, or indeed of any explicit parameter of the System A lunar theory, in a Greek text.[3] In the papyrus, an incompletely preserved interval of days and hours is associated with (1), and it is probable that a total progress in longitude was also given (cf. note to line 2).

The relationship between the original Babylonian relation (1) and the augmented relation of the papyrus would have been analogous to the relationship between the System B period relation:[4]

251 synodic months = 269 anomalistic months (2)

[1] Jones [1990b] 20–23, 28–51.

[2] Neugebauer [1975] 478.

[3] System A's arithmetical pattern for the rising times of zodiacal signs is widely attested in classical sources (Neugebauer [1975] 715–721). Jones [1991b] and Bowen & Goldstein [1988] 69 hypothesize that Hipparchus might have known and used the synodic arcs associated with Column B of System A.

[4] Neugebauer [1975] 478.

and the relation that Ptolemy ascribes to Hipparchus (*Almagest* IV 2, ed. Heiberg v. 1, 270–272):

$$\begin{aligned} 4267 \text{ synodic months} &= 4573 \text{ anomalistic months} \\ &= 4612 \text{ longitudinal rotations} - 7\,{}^{1}/{}_{2}{}^{\circ} \\ &= 126007 \text{ days} + 1 \text{ hour} \end{aligned} \tag{3}$$

For if we divide all the numbers in (3) by 17, a multiple evidently chosen to produce an eclipse period, we obtain

$$\begin{aligned} 251 \text{ synodic months} &= 269 \text{ anomalistic months} \\ &= 271 \text{ longitudinal rotations} + 105;26,\ldots{}^{\circ} \\ &= 7412 \text{ days} + 4;17,\ldots \text{ hours} \end{aligned} \tag{4}$$

Aaboe[1] has shown that Hipparchus derived (4) by combining (2) with the System B mean synodic month (attested in the same passage of *Almagest* IV 2):

$$1 \text{ synodic month} = 29;31,50,8,20 \text{ days} \tag{5}$$

and a value for the length of the sidereal year close or identical to the Babylonian

$$1 \text{ sidereal year} = 12;22,8 \text{ synodic months.} \tag{6}$$

or, equivalently,

$$1 \text{ synodic month} = \frac{360^{\circ}}{12;22,8} = 1 \text{ longitudinal rotation} + 29;6,19,0,55,\ldots{}^{\circ} \tag{7}$$

If, instead, we combine (1) with (5) and (7), we obtain

$$\begin{aligned} 6247 \text{ synodic months} &= 6695 \text{ anomalistic months} \\ &= 6752 \text{ longitudinal rotations} + 20;41,\ldots{}^{\circ} \\ &= 184477 \text{ days} + 14;55,\ldots \text{ hours} \end{aligned} \tag{8}$$

Since the preserved text confirms the number of days but is missing the number of additional hours, we can deduce the underlying value for the synodic month only within bounds:

$$29;31,49,46 \text{ days} < 1 \text{ synodic month} < 29;31,50,22 \text{ days}$$

This range is consistent with either the System B value (5), which is attested in Greek sources, or the System A mean synodic month[2]

$$1 \text{ synodic month} = 29;31,50,19,11,\ldots \text{ days}$$

[1] Aaboe [1955].

[2] Neugebauer [1975] 501.

Lines 4–5 of the papyrus apparently assert that eclipses at intervals of 6247 synodic months were supposed to be always separated by a constant time interval in days and hours. In fact (8) is not an eclipse period at all: it corresponds to a little over 6779 $^1/_5$ draconitic months, which is not near enough to a whole number to make a repetition of eclipses possible. The statement makes more sense if it is to be interpreted as referring to eclipses at intervals of a *multiple* of 6247 synodic months, although it could not have been confirmed from observations since one period of 6247 synodic months is already more than 505 years.

The remaining lines are enigmatic, with an apparent allusion to a different, later period relation and to the Babylonian city of Uruk, one of the two sites (with Babylon) where 'ACT' texts have been excavated in modern times. It is tempting to guess that the text was here comparing relation (1) with the Hipparchian relation (3) founded on System B. Strabo (XVI 1.6) lists the Orchenoi among the 'many tribes of astronomical Chaldeans'; and, whether or not by chance, about two thirds of the surviving lunar System B tablets come from Uruk, whereas nearly all System A tablets are from Babylon.[1] Leaving speculation aside, our text is an important addition to the evidence, not merely for Greek knowledge of elements from Babylonian astronomy, but also for Greek awareness that these elements were in fact Babylonian in origin.

Notes:

1: εὑρι̣[is also consistent with the traces, but a past tense seems most likely here. The traces of bottoms of letters towards the right end of the line are mostly indeterminate.

2: I can find no good alternative to the restoration κατὰ μ]ῆκος, 'in longitude'. This is awkward, because the number that follows is without question the number of anomalistic months in the recurrence period, and this would normally be expressed as revolutions κατὰ βάθος.

3: The total of days would have been 18 myriads and 4477 days. This might have been expressed using a compendium for μυρίαδες or by overdotting the pertinent numerals.

At the ends of lines 6, 7, 8 (see below), and 9 the scribe has employed line-fillers in the form of a *diple* with the lower stroke prolonged down and, in most cases, back towards the right.

7–8: Perhaps τού/[του].

8: At the end of the line is ʅ, which could either be read as a vertically elongated version of the end-of-line mark (see note to line 3) or as a curvaceous ι. I can find no plausible restoration assuming the latter reading, nor indeed any except an indeterminate case of 'Ορχηνοί, 'people of Orchoe', i.e. Uruk.

9: ἀποκατάστασις is the technical astronomical term for a recurrence period of a heavenly body; a form of the related verb is also possible here.

10: Perhaps through lack of foresight, the scribe allowed this line and the next to extend one letter beyond the margin set in lines 1–9.

[1] Neugebauer [1955] v. 1, 10, 42; [1975] 610. Pliny (*NH* VI 30.123) refers more vaguely to '*Orcheni,... tertia Chaldaeorum doctrina*' without specific mention of astronomy.

4140. Tabulated Lunar Phenomena

18 2B.64/F(3–4)b 8×5.5 cm Second century
Plate VIII

The fragment is broken on all sides; on the back are ends and beginnings of lines from two columns of text of unknown nature in a second-century hand. The contents of the front are arranged in a tabular format with red ruling. A column of cells appears to have existed to the left of column i, but no writing is visible. The cells of column i are approximately square. It is not certain how wide column ii was, although a minimal restoration of line 4 would double the extant breadth; the space between the horizontal rulings is just enough for three lines of text.

Each row of cells seems to correspond to a time or event in a sequence that is somehow numbered in column i. In the second extant row (lines 3–5) the moon is first said to be in something, presumably a zodiacal sign which, according to the traces, must be Pisces. Then there may or may not have been other information about its position in the lost continuation of line 3. By the beginning of line 4 the text has gone on to specify where the moon is relative to its nodes, using one of the conventional formulas 'ascending the north' (i.e. in the quadrant following the ascending node) or 'descending the north' (i.e. in the quadrant preceding the descending node). The following three words seem to state that the moon may even be crossing the node (ϲύνδεϲμοϲ) at the time in question; if so, then we can probably restore καταβαίνουϲα τὰ βόρια, 'descending the north', since only this quadrant leads *into* a node. I have no suggestion as to what line 5 is about.

The third row (lines 6–8) states that the moon reached something named in the lost part of line 6. The verb ϲυνάπτειν might be applied to proximity to another heavenly body, as in the astrological sense of ϲυναφή, but more likely refers here to the crossing of a boundary of a zodiacal sign (cf. the use of ϲυνάψει for these points in the planetary data of the ephemeris **4175**). Line 7 again gives the moon's position relative to the nodes, this time in the quadrant following the ascending node. Line 8 said something about the 'rest of the stars', which probably means the planets rather than the fixed stars. The fourth row (line 9) begins like line 3. Returning to the first row, the traces in line 1 (i.e. the second line of the row) are compatible with the restoration καταβαίνουϲα, 'descending', so that the moon would have been in either the quadrant before or the quadrant after the descending node. The bearing of line 2, with its reference to 'geniture' or 'generation', is again obscure.

What are the dates or events described in the rows of this table? And what is being counted in column i? I do not think it can be consecutive days or consecutive new or full moons; for at these intervals the moon progresses with respect to the nodes, whereas from the second to the third row the moon seems to have regressed from the second quadrant to the first following the ascending node. Could these be eclipses or eclipse possibilities, with column i giving either the year reckoned from some era, or the ordinal number within a long eclipse period? Again, since the events in question lay in the past (cf. line 7), we may ask whether any of the information was derived from observations. The moon's position in longitude and argument of latitude is more likely to have been computed; but it is hard to say what the third line of any of the rows is describing. A better understanding of this baffling fragment will probably have to await the discovery of a parallel text.

Notes:

2: One might restore either βαθμῷ or ἀνεμῷ, i.e. either one of the 'steps' or 15° divisions of the moon's motion in argument of latitude or one of the wind-directions associated with these steps (for which cf. Vettius Valens I 18 Kroll). But the qualification τῆϲ γενέϲεωϲ is unfamiliar.

4141. On Latitudes of a Planet

68 6B.19/G(3)e 1.5×3 cm

A tiny fragment, blank on the back and broken on all sides. The text lists the longitudes from which a certain heavenly body begins each quadrant of its latitudinal motion, i.e. the two nodes and the northern and southern limit. Since the longitudes are fixed, the body must be one of the planets rather than the moon, and indeed one of the superior planets since only they have fairly well defined geocentric nodes. Mars, with its ascending node located at approximately Taurus 5° about A.D. 100, alone fits the zodiacal signs partly legible in lines 4 and 6. Hence lines 2–3 specify Mars' ascending node, beginning the quadrant of 'ascending the north'; lines 4–5 specify the northern limit, beginning the quadrant of 'descending the north'; lines 6–7 specify the descending node, beginning the quadrant of 'descending the south'; and the southern limit was probably given in the lost continuation of the text.

Notes:

5: The contents of this line are shifted about one letter to the right of their counterparts in line 3; line 7 is a little further to the right again but not by a letter's breadth. The displacement could be explained by the restoration of ἀναβαίνει in line 3 and καταβαίνει in 5 and 7, assuming that the rest of the wording was identical.

4142. Instructions for Ptolemy's *Handy Tables*

100/3–5 | 10×14.5 cm (fr. 1) 27.5×34 cm (fr. 2) 24×16.5 cm (fr. 3) 6×10.5 cm (fr. 4, not transcribed) | Fourth century

Parts of at least five columns of text written in a sloping informal hand with extensive use of abbreviations, on the back of a register (third century) with receipts in money. The surface of the papyrus has been damaged in many parts through salt encrustation. Fr. 1 is from the bottom of a column, with 1.5 cm lower margin, and the full width (10 cm) of the column of text in some lines with negligible margin on the right. Fr. 2 preserves the full height of the roll, with 2 cm top and bottom margin. Col. i (in poor condition) has lost its left edge; between cols. i and ii is a 1.5–2 cm intercolumnium; col. ii is 13 cm wide; to the right of col. ii, as much as 1 cm margin. Fr. 3 is from the top of the roll, with 1.5 cm upper margin. It has 1 cm margin on the left edge, the full width (11 cm) of col. i, 2 cm intercolumnium between cols. i and ii, and more than half the breadth of col. ii (badly damaged). Fr. 4, in extremely poor condition, appears to come from the middle of a column. On the basis of the contents, it seems likely that frs. 1 and 2 represent three consecutive columns of text, and that at least one column is lost between fr. 2 col. ii and fr. 3 col. i.

The work is a handbook of instructions to Ptolemy's *Handy Tables*, at a more elementary level than Ptolemy's own introduction, and with worked examples. Similar ground is covered by Theon of Alexandria's 'shorter commentary' to the *Handy Tables*, but in spite of the fact that both Theon and our text use the geographical location of Rome to illustrate the methods of conversion of local times, I have detected no significant verbal parallels. The order of topics was more or less dictated by the arrangement of the tables. Our fragments come from towards the beginning, and deal with parts of the following basic problem: to convert a given time reckoned in seasonal hours relative to sunrise or sunset in a particular location to the mean equinoctial time employed in the *Handy Tables*, or *vice versa*. Before one can carry out these conversions, one would have to be acquainted with (1) the geographical table ('Noteworthy cities') of the *Handy Tables*, in order to be able to situate oneself relative to the equator and the meridian of Alexandria, (2) the regnal canon, in order to express a given date in the chronological system of the tables, and (3) the tables of solar mean motion and anomaly, in order to compute the sun's longitude for noon (Alexandria) on the day in question. These topics are not treated in any of the extant fragments of **4142**.

The first stage of time conversion is to translate a given time in seasonal hours past local sunrise or sunset to the equivalent time in equinoctial hours past the preceding local noon. Fr. 1 contains the end of the general instructions for this (lines 1–12), followed by the beginning of an example (13–20). In fr. 2 col. i we resume the text in the middle of its general treatment of the converse problem, to convert equinoctial hours past noon to seasonal hours (lines 7–10), followed again by an example (11–21).

In the continuation of fr. 2 col. i, the text must have turned to the slightly more complex problem of converting a given local time in seasonal hours to the equivalent time in equinoctial hours counted from the preceding noon at Alexandria. An example of the procedure started a little before the beginning of col. ii, and takes up the whole of the column.

Fr. 3 col. i has part of the instructions for converting equinoctial hours reckoned from true noon, Alexandria, to hours reckoned from mean noon and *vice versa*, i.e. for the 'equation of time'. Col. ii presumably belongs to an example of the full conversion from local seasonal hours to equinoctial hours past mean noon at Alexandria; the preserved part comes from the first step, i.e. the conversion to equinoctial hours past local noon.

Notes:

Fr. 1

4–12: Conversion of a given local time in seasonal hours to local time in equinoctial hours, cf. Neugebauer [1975] 980. (The instructions consider the case of a time during the day.) The given time S_S will originally have been expressed in seasonal hours after sunrise. This has been converted to $S_N = |6 - S_S|$, the number of seasonal hours before or after noon. Also for the given date one will have computed the sun's longitude $\lambda_{\odot}$, and found the corresponding duration $h(\lambda_{\odot})$ of 1 seasonal hour in time-degrees from the appropriate column in Ptolemy's table of oblique ascensions for the *klima* (latitudinal zone) of the observer.

In lines 4–7, $h(\lambda_{\odot})$ (which was named in the lost preceding text) is multiplied by S_N; the product is the number of time-degrees between the given time and noon. In 7–9 this number is then divided by 15°, the number of time-degrees in 1 equinoctial hour. The quotient, as stated in 9–12, is E_N, the number of equinoctial hours between the given time and noon.

14–20: The foregoing procedure is illustrated by an example for a specific date. This date had evidently been stated already in a previous example, surely the one in which it was shown how to compute a solar longitude. There it was found that $\lambda_{\odot}$ = ♎ 6;26°. The rest of the present example is lost, but almost certainly used the same data in reverse as fr. 2 i 11–19.

Since the handbook is only now explaining how to convert a given time into the form needed for entry into Ptolemy's tables, it is probable that $\lambda_{\odot}$ was calculated (as a first approximation) for noon of the day in question. Dates on which the *Handy Tables* yield a solar longitude of ♎ 6;26° at noon occur at 73-year intervals. The only possibilities falling between Ptolemy's time and the latest admissible date of writing of **4142** are 207 October 2, 280 October 1, and 353 October 1.

Fr. 2, col. i

7–10: The context is obscure. The 'diametrically opposite time-degrees' might refer to the duration of 1 seasonal hour of night, which is equal to $h(\lambda_\odot + 180°)$.

11–21: An example of the reverse process, conversion of equinoctial hours to seasonal hours. Again we have $\lambda_\odot$ = ♎ 6;26°, and we are told (line 17) that $h(\lambda_\odot)$ = 14;44°. This shows that the author was using Ptolemy's table of oblique ascensions for the *klima* through Lower Egypt (appropriate, e.g., for Alexandria). Also we are given E_N = 4;54 e.h. In the broken lines 13–14 E_N was multiplied by 15° to obtain the number of time-degrees between the given time and noon. The product should, of course, be 73;30°, but the author actually uses 73;40°. This, when divided by $h(\lambda_\odot)$, yields 5 seasonal hours *exactly*. Obviously the author has simply reversed the steps of his previous example for the conversion from seasonal to equinoctial hours, forgetting that the 4;54 e.h. that he had found there was a truncated answer (from 4;54,40).

20: The number of equinoctial hours, expressed as a sum of unit fractions, is exactly equivalent to 4;54.

Fr. 2, col. ii

1–40: An example of the conversion of a given local time, S_S, to the number of equinoctial hours from noon for the meridian of Alexandria. It is assumed that one is at Rome, which according to the coordinates in the Table of Noteworthy Cities has latitude 41;20°, longitude 36;20° east of the prime meridian through the Islands of the Blest. The appropriate table of oblique ascensions would be the one for the fifth *klima* through the Hellespont (computed for latitude 40;56°). Since $h(\lambda_\odot)$ = 18;40° (line 1), $\lambda_\odot$ was approximately ♊ 21°, corresponding to a date close to June 15 of an undetermined year. The given time is the end of the second hour, i.e. S_S = 2 s.h. past sunrise.

1–9: The beginning of the example, which stated the foregoing data, is lost. The first preserved lines recapitulate the procedure for converting seasonal to equinoctial hours (cf. fr. 1). We find $S_N = 6 - S_S$ = 4 s.h. before noon, hence $S_N \times h(\lambda_\odot)$ = 74;40°, and E_N = 74;40° ÷ 15° = 4;58,40 e.h. ≈ 5 e.h. before noon = 19 e.h. after noon.

10–20: Determination of the difference in local time between Rome and Alexandria. The longitudes of the meridians (inadvertently called parallels in line 11) are taken from the Table of Noteworthy cities. Since 15° of longitude correspond to 1 e.h. of time difference, the 24;10° between the meridians through Rome and Alexandria correspond to 1;36,40 e.h., which the author rounds to 1;36.

21–37: The time difference is added to E_N to obtain the number of equinoctial hours from noon at Alexandria, 20;36 e.h. past noon (= 3;24 e.h. before noon). The general remarks about the relation between longitude and time are confusingly expressed, but correct if 'earlier' in line 24 is understood as applying to the local time corresponding to the same absolute moment.

40–45: These lines seem to set out an alternative method of carrying out the same conversion as in the foregoing example. We find E_N as before, but now we change this into E_S, the number of equinoctial hours reckoned from 6 A.M. local time, i.e. $|E_N - 6|$. In the present example E_S = 1 e.h. We again add the time difference between the meridians, here 1;36 e.h., and finally add the 18 e.h. between noon and 6 A.M. The result is of course the same as before.

Fr. 3, col. i

1–15: Observed local time, and hence also the equivalent times in equinoctial hours for the observer's meridian and the meridian of Alexandria, are all relative to true noon, that is, the time of the sun's meridian crossing. Ptolemy's tables, on the other hand, assume that time is reckoned from *mean noon*, which is supposed to occur at exactly constant intervals of time starting with the epoch date of the tables, 324 B.C. November 12 at (true) noon. The difference between the two quantities is the 'equation of time'.

The present lines belong to the general instructions for converting a given number of equinoctial hours past true noon to the equivalent number of equinoctial hours past mean noon, and *vice versa*. The equation of time, expressed as a sexagesimal fraction of 1 e.h., is tabulated in Ptolemy's table of right ascensions (*'sphaera recta'*) as a function of $\lambda_{\odot}$. Mean noon always precedes true noon, so that the conversion from time past true noon ('non-uniform days-and-nights') to time past mean noon ('uniform days-and-nights') is effected by adding the equation of time to the given number of equinoctial hours. The reverse conversion requires subtracting the equation of time from the number of hours past mean noon. This is the rule set out in the text.

Fr. 3, col. ii

5–8: One would expect at this point an example of the procedure for correcting a given time for the equation of time. What can be read here is evidently a conversion of given seasonal hours to local equinoctial hours. Probably this was merely the beginning of a cumulative example that illustrated the several stages involved in proceeding from seasonal hours to equinoctial hours after mean noon.

4143. Instructions for Ptolemy's *Handy Tables*

96/69(b) 9×17 cm After A.D. 329

The upper inside portion of a codex page (vertical fibres on recto), with 1 cm top and 2 cm inner margin, written in a sloping 'severe' hand of the fourth century. Like **4142**, this was part of a set of instructions for the use of the *Handy Tables* with worked examples. The preserved parts belong to the general instructions for computing a planet's latitude on a given date (recto), and an example demonstrating the calculation of Venus' latitude on September 23/24, A.D. 329 (verso; for this date see below). This makes the composition of **4143** a generation older than Theon of Alexandria's 'shorter commentary' on the *Handy Tables*, which incorporates worked examples for A.D. 360, and approximately contemporary with Pappus's *Almagest* commentary. There do not seem to be any nontrivial verbal parallels between our text and the equivalent section of Theon's instructions.

In the following description of the method of calculating a planet's latitude according to the *Handy Tables* I follow the notations of Neugebauer [1975] 1002–1011, which the reader should consult for details including the relation to Ptolemy's planetary models. The lengthy procedure may be divided into several stages, only the last two of which are represented by passages in the extant papyrus:

(1) For the given date we find the elongation of the centre of the planet's epicycle from the apogee of the eccentre ($\bar{\kappa}$), and the position of the planet on its epicycle relative to the mean apogee of the epicycle ($\bar{\alpha}$). For Venus on Sept. 23/24, A.D. 329 = Era Philip 653, Egyptian Choeac 24, 13 e.h. past noon,

$\bar{\kappa} = 124;9°$ $\bar{\alpha} = 348;46°$

(2) Most of the computation of the planet's longitude has to be carried out, using the planet's anomaly table. Specifically, we have to find the 'corrected' positions of the epicycle's centre (κ_0) and of the planet on the epicycle (α), which take account of the fact that the observer is on the earth rather than at the centre of uniform motion of the epicycle, and the planet's elongation from the apogee of the eccentre (κ). In our example,

$\kappa_0 = 122;9°$ $\alpha = 350;46°$ $\kappa = 118;15°$

(3) The latitude of the epicycle's centre (β_0) is found as a function of κ, using the planet's table of latitudes. In our example,

$\beta_0 = 0;4°$ south

(4) The latitude of the planet relative to the epicycle's centre (β_1) is found as a function of κ_0 and α, using the latitude table again. The verso of our papyrus contains part of these steps as carried out for the example. First we find in the table the quantities c_4, c_5, c_6 corresponding to α in the columns headed 'Difference for greatest' (*scil.* distance of the epicycle from the earth), 'Mean latitude', and 'Difference for the least' (*scil.* distance), which in the present instance are:

$c_4 = 0;1°$ $c_5 = 1;29°$ $c_6 = 0;1°$

Secondly we find the interpolation constant c_3 corresponding to κ_0 in the first column headed 'sixtieths', in this case:

$c_3 = 0;32$ additive

Since c_3 turns out to be additive, we correct the 'mean latitude' c_5 by adding the proportionate fraction of the 'correction for least distance' c_6, to obtain the relative latitude of the planet on the hypothesis that it is at the northernmost point of the epicycle:

$$\beta' = c_5 + c_3 \times c_6 \approx 1;29° + 0;0° = 1;29°$$

Now we determine the elongation Ω_P of the planet on its epicycle from the northernmost point. This angle incorporates a constant Ω_A which is different for each planet; for Venus, $\Omega_A = 270°$:

$$\Omega_P = \Omega_A + \kappa_0 + \alpha = 270° + 122;9° + 350;46° = 742;55° = 22;55°$$

Corresponding to this in the last column, headed 'Common sixtieths', in the latitude table is the coefficient necessary to take account of the planet's actual position on the epicycle:

$c_7 = 0;56$ north

Hence

$\beta_1 = 0;56 \times \beta' = 1;23$ north

(5) Finally, we form the total latitude:

$\beta = \beta_0 + \beta_1 = 1;17°$ north

The general instructions on the recto of the papyrus are mostly concerned with this step, and the rules for handling the possible combinations of northerly or southerly components.

At this point we may justify the identification of the planet and the date used in the example on the verso. The number 742 in line 20 can only represent Ω_P, since nowhere else in the computation of a latitude can such a large number arise. The preceding lines must accordingly set out the addition that leads to this sum, and in lines 16–17 we are given the elongation of the epicycle, κ_0, namely 122;10°. Supposing the maximum possible α, Ω_A has to be greater than 260°, and this is true only for Venus. Thus we know that $\alpha \approx 350°$. It is not difficult to establish that this pair of values for κ_0 and α is generated by Ptolemy's tables only for Era Philip 653 Choeac 24/25, at approximately 13 e.h. past noon.

This reconstruction is based on the assumption that the author calculated Ω_P as $\Omega_A + \kappa_0 + \alpha$. Since, however, it is always the case that $\kappa_0 + \alpha = \bar{\kappa} + \bar{\alpha}$, we also have to consider the possibility that the number 122;10° in line 17 is $\bar{\kappa}$. This hypothesis would lead by the same argument to the date A.D. 345 Sept. 21/22 = Era Philip 669 Choeac 26/27, 15 e.h. past noon. The difference in the initial data has almost no effect on the later stages of the computation. Nevertheless I think it is unlikely that 122;10° represents an uncorrected mean motion, because it is impossible to get this precise number from Ptolemy's mean motion table for Venus unless the given time included a fraction of an hour.

Notes:

Recto

1: The line juts slightly (about one letter's breadth) into the right margin.

2–4: These lines probably belong to the determination of whether the component β_1 of the latitude is northerly or southerly. If Ω_P, the 'number entered' in the table of latitude to obtain c_7, is between 90° and 270°, β_1 is southerly, otherwise northerly. The illegible number at the end of line 3 would be the number of the relevant column in the table.

4–9: The two components β_0 ('the mean latitude derived from the longitude', i.e. from κ) and β_1 are compared and combined.

10: ἄνεμοϲ, literally 'wind', here means direction north or south of the ecliptic. For this usage, which appears in neither Ptolemy nor Theon, one can compare a scholion at the bottom of the table of

Saturn's latitudes in the *Handy Tables* manuscript *Vat. gr.* 1291, f. 78v: τινὲc ἔφηcαν ὅτι ἐὰν οἱ β̅ ἄνεμοι βό(ρειοι) ὦcιν, βό(ρειοc) ἀνα(βαίνει?)· ἐὰν δὲ οἱ β̅ νό(τιοι), νό(τιοc) κατέρχεται· ἐὰν δὲ ὁ μὲν βό(ρειοc), ὁ δὲ νό(τιοc), ὑπερέχῃ δὲ ὁ βό(ρειοc), βό(ρειοc)· εἰ δὲ ὑπερέχῃ νό(τιοc), νό(τιοc).

Recto

1–2: This refers to c_3, which was found in the part of the table of latitude where its value is additive.

3–4: When c_3 is additive, it is supposed to be multiplied by c_5, the difference between the maximum latitudinal component at least and at mean distance of the epicycle. Since c_5 never reaches 1°, it is appropriately called 'sixtieths'.

6–9: It appears that the sexagesimal multiplication of 0;1 and 0;31 was carried out digit by digit.

12: The number is probably β' rather than c_5.

13–15: Apparently telling the reader that the result obtained so far is the latitude of the planet relative to the centre of the epicycle.

16–17: 122;10° is either the corrected elongation of the epicycle's centre, κ_0, or the mean elongation, $\bar{\kappa}$. See the general comment above.

18: The other variable component of Ω_p, either α or $\bar{\alpha}$ (depending on whether κ_0 or $\bar{\kappa}$ was given in the preceding lines), must have been named here. I do not understand the traces at the beginning of the line.

19: The προcθήκη is the constant angle between the apogee of the epicycle and its northernmost point, for Venus 270°.

20: This is the integer part of Ω_p.

21: Two whole circles of 360° would have been subtracted from 742°, leaving 22° (plus some fraction).

24–25: I cannot identify the numbers in these lines.

4144. Treatise on Kinematic Models

100/103(d) 5.5×10.5 cm Second or third century

Plate II

Written in a sloping 'severe' hand on the front of the papyrus (the back is blank), and is broken on all sides. Below line 3 a space of 3.5 cm was left for the figure, of which only the lower right portion survives. The circle was drawn crudely freehand. Corrections have been made above the line in lines 6 and 13, perhaps in the same hand as the text.

The presence of a geometrical figure with lettered points, as well as the allusion in line 2 to dropping a perpendicular, identify the text as part of a geometrical demonstration. This is not pure geometry, however, since the sequel speaks repeatedly of motions. Possible contexts for such a discussion include optics and mechanics as well as astronomy, but the emphasis on 'nonuniform' motion points to the behaviour of kinematic models for the sun, moon, or a planet compounded from circular revolutions. The 'coming together' of more than one motion in line 8 could, for example, refer to the stage in the operation of an epicyclic model when the body borne by the epicycle is travelling parallel to the motion of the epicycle's centre; and the resulting maximum velocity of the body (both real and apparent as seen from the centre of the system) could be described in line 10 as the 'peak and relaxation of the motion', since thereafter it slows. A discussion of this kind would be expected in a fairly elementary treatise on astronomy, comparable to the pertinent sections in Theon of Smyrna.

4145. Procedure Text

68 6B.24/D(1–3)c 6.5×11.5 cm Fourth century

Probably from a codex, written in a large informal hand. The side with vertical fibres appears to preserve line beginnings; the text is broken on all other edges.

The text consists of a series of arithmetical operations on whole numbers, in the manner of such crude procedure texts as **4163** (prediction of syzygies) or the astronomical chapters of Vettius Valens book I. I am unable to identify the meaning of the numbers, or to connect the disjointed operations into a coherent procedure. On the front, lines 2–3, 4 is multiplied by 13, and the product, 52, is added to 77, yielding 129. The next lines (4–5) are obscure. In 6–7, 13 and 77 are added together, and the sum, 90, is divided by 45, yielding 2. The back mentions 13 again (line 5), something is 'completed' (line 6), then perhaps 7 is multiplied by 11 to obtain 77. I can make no sense of the rest.

4146. Procedure Text

72/29(b) 4×10.5 cm (fr. 1) 2.5×7 cm (fr. 1) Second century

Written on the front of the papyrus only in a rounded, informal but serifed hand; the back is blank. Fr. 1 has part of the top margin (3 cm), with the numeral $\overline{\epsilon}$ about 2 cm above the text, evidently identifying the number of the column; the other sides are broken off. Fr. 2 is broken on all sides. I suspect that fr. 2 belongs directly below fr. 1, with their left edges lining up.

I have made little sense of this text. It is to some extent calendrical: the first day of Thoth is referred to (tautologously) as τὴν Θωὺθ $\overline{\alpha}$ νεομηνίαν in line 9, and line

3 probably mentions the rising of Sirius, κυνὸϲ [ἐπιτολή, cf. line 10. Something is periodic (line 1), and it seems that lunar months as well as solar and calendar years are involved (line 11). These elements are also all present in two other fragmentary procedure texts, the astronomical ostracon *O. Bodl. Tait* II 2176 (with an example for A.D. 257) and *P. Oxy.* LXV 4473 lines 22–30 (following a text concerning astrological predictions of the height of the Nile flood). These texts seem to concern the determination of the conjunction preceding the rising of Sirius in a given year; but I am not certain whether that was also the case in **4146**.

4147. Procedure Text

103/40(c) 10×8 cm (fr. 1) 5×6.5 cm (fr. 2) Second century
2.5×3 cm (fr. 3) 10×8 cm (fr. 4, not transcribed)
1.5×1.5 cm (fr. 5, not transcribed)

Fragments from a roll, with at least two columns of text in a broad 'severe' hand, perforated and abraded in parts. Frs. 1, 2, and 4 are from the bottom, with 2 cm margin; the other fragments are broken on all sides. There is a narrow (0.5 cm) intercolumnium on fr. 1.

The text seems to have been divided into sections, a new one being marked by a *paragraphos* in fr. 1 col. i, below line 8. The subject in col. ii is some sort of computation involving the moon in zodiacal signs; the only detail that is vaguely intelligible is a division by some quantity (the number of hours?) between the rising and the setting of some heavenly body (the sun or moon?). The purpose of the text may be astrological rather than astronomical.

Part II

Primary Tables

The so-called primary tables were the fundamental tables used in the prediction of astronomical phenomena, especially the dated positions of the heavenly bodies that played such an important role in later Greek astronomy and its astrological applications. We discover here the most perspicuous demonstration of the dichotomy of predictive astronomy in this period: on the one hand, arithmetical methods expressed in the epoch tables and templates, on the other, the kinematic modelling underlying the *Handy Tables*. Basic as this distinction of methods is to our conception of the historical development of ancient astronomy, it seems to have mattered little to the users of the tables; for we find both varieties in simultaneous use, even mingled in the same 'archive' (cf. **4153–4155, 4165, 4167, 4169**).

Epoch tables

Nowhere is the significance of the new astronomical papyri from Oxyrhynchus more pronounced than in the epoch tables; for only one specimen of this variety, the Standard Scheme lunar epoch table *P. Lund.* V 35a, had hitherto come to light. The epochs are periodic events related to each heavenly body's anomalistic motion, rather than the arbitrary fixed intervals in days and months or the crossings into zodiacal signs that characterize the several types of almanac. Hence they reveal directly and precisely certain assumed properties and parameters of the motion of the sun, moon, and planets that we can extract only with difficulty and uncertainty from the almanacs.

The three new lunar epoch tables, **4149–4151**, belong to the same Standard Scheme as *P. Lund.* V 35a. The epoch dates in any of these tables would, if extrapolated, coincide exactly with the others, and the divergences in the associated longitudes are astronomically insignificant. As early as the first century this one method for computing the moon's daily positions had established itself to the practical exclusion, apparently, of all rivals until the advent of Ptolemy's tables.

The appearance of a table of solar epochs, containing dates of epoch at intervals of one solar (anomalistic) year, is not in itself unexpected. The simple periodicity of the sun's anomalistic motion, and its near coincidence with the period of the sun's revolution in longitude, make the sun even more suitable than the moon for the combination of epoch table and template. However, one may be surprised to find that the epochs in **4148**, although manifestly intended for summer solstices, are separated by an interval that is effectively a sidereal year rather than a tropical one, a circumstance that tends to subvert the presumption that the theory of precession was widely accepted following Ptolemy.

The epoch tables for the planets are of the utmost importance, because they settle once and for all the question of how much of the Babylonian mathematical theories of planetary motion was transmitted into Greek astronomy. Of the fifteen planetary epoch tables currently known, all but four were demonstrably computed using known Babylonian schemes or new variants of known schemes. All four major System A schemes for Mercury's four synodic events are attested on papyrus, as well as the System A scheme for Mars, a version of System A' for Jupiter, and System B for both Jupiter and Saturn. This accounts for the majority of the schemes known to us from cuneiform texts, even though the number of papyri is much smaller than the extant corpus of Babylonian planetary tables.

Estimates of the extent of Greek knowledge of Babylonian planetary theory have varied considerably. Neugebauer was sceptical of claims that went beyond the transmission of a few period relations, and in general judged the technical level of the astronomy in the papyri to be much lower than its Mesopotamian counterpart (Neugebauer [1975] 604–607, 789–790). Yet as long ago as 1947 van der Waerden had conjectured that Babylonian, or Babylonian-style, methods lay behind part of the computations of planetary sign-entries in the *Stobart Tablets* and P. dem. Berlin 8279 (van der Waerden [1947], [1960], [1972]). At least for the use in the *Stobart Tablets* of the System A scheme for Mars, according to which the boundaries of the zones conveniently coincide with the beginnings of alternate zodiacal signs, his arguments had the force of proof. Indirect evidence of a different kind is furnished by certain Indian texts, in particular Varāhamihira's *Pañcasiddhāntikā*, in which planetary schemes of the System A variety—albeit seldom agreeing with the parameters of the Babylonian schemes—seem to lead us to a line of transmission that had passed through Greek intermediaries. Taking account of the evidence then known to me, I suggested a few years ago that a substantial part of the Babylonian planetary theory could have been accessible to Greek astronomers (Jones [1993b] 82–83, 87–88); but I anticipated that by the Roman period the Babylonian predictive methods would be more likely to appear in heavily modified forms, as in the later Indian versions (Jones [1991c]).

The fidelity with which the Greco-Egyptian tradition kept these half-a-millenium-old technical procedures in working order is therefore almost as remarkable as the fact of transmission. Still there are some divergences in practice, small but significant. The most general difference is that in the papyri the computation of dates of planetary phenomena is adapted to the Egyptian calendar and reckoning in true days, in place of the Babylonian lunar calendar. This transition, which Neugebauer thought would have posed a 'major obstacle' (Neugebauer [1975] 790) turns out to have been effected in an economical way that demonstrates understanding of the theory behind the rules on the part of the adaptors.

There are also some interesting differences between the cuneiform tablets and the papyri in the handling of the System A schemes for Mercury. Leaving aside a peculiar and apparently archaic scheme (System A_3) attested in only two tablets, we know of a pair of schemes, called System A_1, for predicting Mercury's first visibilities as evening and morning star, and a second pair, called System A_2, for predicting the last visibilities. In the Babylonian tables these schemes are never used together. Rather, the tables that contain first visibilities computed by System A_1 derive the corresponding last visibilities by a special pattern of intervals in time and longitude, known as 'pushes', that are functionally dependent on the planet's longitude at first visibility; and conversely the tables in which the last visibilities are according to System A_2, the subsequent first visibilities are obtained by a scheme of pushes. The known System A_2 tables, which are in fact two copies of the same text, are exceptionally early (308–289 B.C., more than sixty years before the next surviving planetary tables), and they do not use the standard rule for deriving the intervals between dates of the phenomena by adding a constant to the synodic arcs. One might have guessed that System A_2 represented an early attempt to establish a theory for Mercury, which was superseded by System A_1. But now the papyri show Systems A_1 and A_2 side by side in the same table (**4152** and **4155**), and the dates associated with last visibilities calculated by System A_2 obey the standard rule (**4154**, **4155**, and **4156a**). It is not yet clear whether these divergences reveal an innovation or a survival of alternative Babylonian practices that by accident are not attested in the extant cuneiform texts.

Templates

The Babylonians had adopted two contrasting approaches to the problem of finding day-to-day positions of the heavenly bodies. In the case of the sun and moon, daily longitudes were computed as unbroken arithmetical sequences, each line derived from the previous one by adding (for the sun, Neugebauer [1955] texts 185–187) a constant increment or (for the moon, texts 191–196) consecutive values of the linear zigzag function; thus a single epoch initiated a run of positions that could be prolonged indefinitely. A planet's daily positions, on the other hand, had to be fitted to a scaffolding of dates and positions of the synodic phenomena previously found by means of the System A or B schemes. Since the interval from one phenomenon to the next in both longitude and time is variable, the series of intermediate positions would never follow exactly the same pattern. We do not know a great deal about this aspect of Babylonian planetary theory, but we do possess tables in which the daily motion is assumed to be constant from one stage of the synodic period to the next, so that the change in velocity is abrupt at these points, and other tables in which the acceleration and deceleration of the planet before and after the stations is smoothed by allowing either the daily increments or even the rate of change of the increments (i.e. the second differences of the longitudes) to vary linearly (Huber [1957]). The numerical parameters of the required bridging sequence seem to have been derived afresh in each application from the predetermined endpoints.

The template tables found on papyrus are an advance on the Babylonian methods of predicting daily positions with respect to consistency, convenience, and precision. One format of table suffices for the sun, moon, and planets, and each is used in the same way: one finds the epoch preceding a given date in the appropriate epoch table, counts the days elapsed since epoch, and finds the corresponding progress relative to the epoch position in the template. All the difficult mathematical work has been subsumed in the composition of the template. With a single addition one is able to find the longitude for an arbitrary date without having to calculate all the preceding longitudes. And, at least in the case of the sun and moon, the fact that the template has been prepared beforehand means that fewer compromises need be made for ease of computation, so that the templates can present a more faithful representation of anomalistic motion with more accurate parameters than we find in the Babylonian tables of daily positions.

The best understood of the templates at present is the one associated with the lunar Standard Scheme, which in its most complete form gives the moon's relative position since epoch in both longitude and argument of latitude, both being computed as the running total of a linear zigzag function. One and the same pattern of increasing and diminishing speed was imposed on every period of lunar anomaly, reflecting the ignorance of the moon's second anomaly that prevailed in ancient astronomy before Ptolemy. The first copy to have come to light of the Standard Scheme template, as yet only partially published, is *P.S.I.* XV 1493. It contains both the columns for longitude and argument of latitude, but with all values truncated to one fractional sexagesimal place. With this incomplete information it was possible to recover approximate values for the defining parameters of the two zigzag functions with sufficient accuracy to be able to recompute the preserved values, with occasional deviations of 0;1 (Jones [1983] 27–30). Parts of three new copies of the template appear here as **4164**, **4164a**, and **4150**. In all these copies, the relative longitudes are given to three fractional places, so that it is now possible to recover the exact parameters of the zigzag function for daily progress in longitude, and to reconstruct the entire template to its full precision of three places (Appendix H, Table 2). This in turn makes sense for the first time of certain details of the rules for computing the lunar epoch tables of the Standard Scheme, and is the key to interpreting the procedure text **4136** (Jones [1997a] sections 1–3).

When we refer to the lunar templates and epoch tables as components of the 'Standard Scheme', this name reflects the remarkable uniformity that they exhibit. Each template is, barring truncation or the possible omission of the arguments of latitude, essentially the same table, and each epoch table is part of an unbroken series of cycles of dates, with only negligible deviations in the epoch positions.

Hence we could use any of the epoch tables together with any of the templates and obtain equivalent and valid results.

But the consistency of the Standard Scheme tables is not typical of the arithmetical astronomy as a whole. Consider, for example, the two solar templates **4162** and **4163**, the first such tables to come to light. In **4162** the longitudes are derived (exceptionally for a template) by trigonometry from a kinematic solar model, the epoch is the sun's annual passage of its perigee, and the summer solstitial point is assumed to be at Cancer 8°. In **4163** the sun's rate of progress is treated as constant through whole zodiacal signs, the epoch is the summer solstice, and the solstitial point is placed at Cancer 0°. One might at least hope that **4163** is the kind of template intended to be used with the summer solstice epochs in **4148**; but even this is doubtful, because **4148** operates with sidereal years whereas the convention that the equinoctial and solstitial points are at 0° in their respective signs is associated with a tropical frame of reference.

Any attempt to devise template schemes for the planets would have had to deal somehow with the variability of their synodic cycles in both time and longitudinal progress. Our understanding of how the planetary templates worked, though still extremely defective, has advanced considerably with the new examples from Oxyrhynchus. One point that is now obvious is that there was no single uniform set of planetary templates.

In the case of Saturn, for example, we already had part of a template, *P.S.I.* XV 1492, which we can now compare with **4166**. The former papyrus contains a stylized version of Saturn's mean synodic cycle, starting with an epoch at Γ, and representing the accelerations and decelerations of the planet by alternating second-order sequences, that is, the daily motion increases or decreases linearly, reaching 0° per day at the stations. This template would have been applied to an epoch table containing dates and positions of Γ, such as **4161**, although it is not clear how one allowed for the variations from the mean case of the synodic cycles. In **4166**, a separate template was provided for the intervals Ψ–Ω and Ω–Γ, and presumably there were once tables for the remaining intervals between the synodic phenomena. This suggests that the template was meant to be used with an epoch table giving the dates and positions of more than one of the phenomena, such as we are familiar with from the Babylonian planetary tables. The motion of Saturn in the two extant sections of the template is described respectively by an accelerating second-order sequence and linear motion, as in *P.S.I.* 1492; but the defining parameters are different, and in fact turn out to be derived from a velocity scheme associated specifically with the slow zone of the Babylonian System A scheme for Saturn. This means that the complete scheme comprised an epoch table of Saturn's phenomena based on System A as well as templates associated with both the fast and slow zones of the System A model. One might have had to switch from one template to the other whenever the longitudes crossed a zone boundary.

Similar schemes with multiple templates associated with the System A zones existed for the other superior planets. **4165a** describes Jupiter's motion from Γ to Φ as a decelerating second-order sequence, attaining a total progress that seems to have been chosen to fit the fast zone of either System A or System A'. For Mars too, we now have a template, **4165**, that describes the planet's motion after an epoch at Ψ as an accelerating second-order sequence fitted to the slowest zone of the System A scheme. This may be compared with the tiny template fragment P. Berl. 21236, which gives a different value for Mars' fastest motion in the same part of the ecliptic.

Mercury presented the ancient table-maker with an extreme case. Each of Mercury's four phases has its own characteristic pattern of distribution over the ecliptic, which was reproduced by a distinct System A zone scheme in the epoch tables. As a result, the intervals between consecutive phases varied considerably, and in an erratic way that could not easily be accommodated by one or even a small number of templates; the one fixed point was that for each of the four kinds of interval the number of days separating the phases differs by a constant from the number of degrees travelled between them. The importance of these time intervals is made evident in the epoch table **4155**, which devotes a column to them. To bridge the intervals of invisibility, one probably was satisfied to use linear interpolation, that is, to derive a constant daily motion by dividing the total longitudinal interval by the number of days. But during the intervals of visibility Mercury shifts from being almost stationary to near its maximum velocity or *vice versa*, and this called for accelerating or decelerating second-order sequences. It was almost certainly to this end that tables with an unusual triangular layout were composed, such as **4217** and **4217a–c**, in which a series of columns either increasing or decreasing in length by steps of 1 row present second order sequences with the line-to-line differences increasing from about zero to about two or decreasing from two to zero.

Handy Tables and other kinematic tables

If the arithmetical astronomy of the epoch tables and templates is a thing of which we would have been utterly ignorant were it not for archeologically recovered documents, the medieval tradition of Greek astronomical texts has prepared us much better for the kinematic tables among the papyri. Even so, we stand to learn much from these manuscripts, which average about half a millenium older than the oldest Byzantine codices of Ptolemy's *Almagest* and *Handy Tables*. Of the *Almagest*, it is true, the papyri have so far yielded us no ancient copy: whatever was the readership that Ptolemy imagined for this work, it was not to be found among the practising astrologers. But an impressive number of manuscripts of the

Handy Tables, in substantial accord with the medieval copies, have come to light (**4167–4171**), and it will be necessary to take account of these when the still murky textual history of Ptolemy's *Handy Tables* comes to be elucidated.

But there are also other kinematic tables that do not correspond to any known medieval copy of Ptolemy's tables. The example of *P. Lond.* III 1278 had raised the possibility of tables that to some extent mimic the format of Ptolemy's *Handy Tables* while going back to the *Almagest* for their numerical content. **4173**, and possibly also **4173a**, are tables of this kind. Neugebauer, influenced by the poorly founded but tenacious modern dogma that the medieval *Handy Tables* were heavily edited by Theon of Alexandria in the late fourth century, suspected that variant formats in the papyri pointed to the existence of intermediate stages in the evolution of the tables from their *Almagest* originals to the form in which we have them. I would conjecture that they are really offshoots of the tradition, influenced by both the *Almagest* and the *Handy Tables* but incorporating innovations from elsewhere. The format of the planetary tables in **4173** seems to imply different underlying planetary models, basically eccentric instead of epicyclic.

In a third category of table in this section are the mean motion and anomaly tables unrelated to Ptolemy's parameters and uninfluenced by his formats; these include **4220** as well as **4174** and **4174a**. The lunar tables in **4220** are interesting because they appear to be based again on a model in which the moon's principal anomaly was produced by an eccentre rather than an epicycle as in Ptolemy's system. By way of contrast, the lunar mean motions in longitude in **4174** are rather crude and uninformative about the underlying theory. **4174a** is a solar mean motion table that, unexpectedly, has different periodicities for longitude, anomaly, and latitude according to a scheme reported by Theon of Smyrna.

4148. Epochs of the Sun, A.D. 161–237

103/49(a) 15×25.5 and 4.5×4.5 cm After A.D. 237

The table is written with ruling in black, on the back of a sheet of papyrus that bears on its front (inverted with respect to the table) accounts in talents and denarii concerning commodities (wheat, wine, etc.) connected with the *annona*, badly preserved but mentioning dates of A.D. 305/306, 306/307, and 307/308. Since the astronomical table is unlikely to have been written much later than the middle of the third century, we have to suppose that the better side was originally left substantially blank. The larger of the two disconnected pieces making up **4148** contains parts of cols. i–viii, while the smaller has parts of cols. x–xii. The original width was more than 25cm.

Remains of two tables are preserved. Column i belonged to a table of unknown nature in which the lowest place digits increased by 0 or 1 from line to line, at least in the few extant lines. The epoch table comprises three sets of columns: ii–v, vi–ix, and x–xii (with restored xiii). In each set, the second, third, and fourth columns tabulate epoch dates (T = iii–v, vii–ix, xi–xii) in successive regnal years, beginning with Marcus Aurelius and Lucius Verus 1 (A.D. 160/161). Dates are expressed as day number and three-place sexagesimal fraction within Egyptian months. The first column of each set (ii, vi, x) gives the number of days that must be subtracted from the Egyptian date to obtain the date in the Alexandrian calendar (D = ii, vi, x).

With two exceptions that are certainly errors (col. v line 11 and vi–ix line 32), the interval of time from line to line is invariably 1 year plus 0;15,33,46 days. This shows, first, that the phenomenon in question is strictly annual, and secondly that the years are of uniform length, i.e., the calendar is the Egyptian. The tabulated dates are toward the end of June, and the phenomenon in question is certainly the summer solstice. Actual solstices fell several days earlier; for example, in A.D. 161 the solstice occurred on June 23, in the late afternoon Egyptian time, five days before the tabulated Mesore 20;16,55,40 (June 28). This discrepancy reflects, not bad observations of the second or third century, but extrapolation from much earlier dates using an increment that is significantly too long for the tropical year, although it is in fact a very good value for the sidereal year. If we extrapolate the table backward using the constant difference of 365;15,33,46 days, we find that the solstice for 30 B.C., immediately preceding the beginning of Augustus 1, was assumed to be Epeiph 1;0,0,0 (June 27) exactly. This was therefore the primary epoch date for the scheme; but even in that year the true solstice took place two days earlier. Perhaps the scheme's inventor calculated the date of solstice for 30 B.C. from still older observations, and rounded the result to the nearest integer day. It is likely that the fractions of a day are supposed to be counted from 'sunset' (6 P.M.), on analogy with the Standard Lunar Scheme.

The purpose of a table of predicted solstices was to provide epoch dates at which the sun was assumed to be at the same longitude, presumably either Cancer 0° or Cancer 8°, and at the same stage of its anomalistic period. To obtain the sun's longitude at other dates, one would also need a template covering one year's progress, starting with the solstice as day 0. **4163** is part of just such a template, with Cancer 0° as its initial longitude.

Notes:

col. v:

11: The date should be 22 36 59 34 (an isolated error).

cols. vi–ix:

32: Here, where the epoch dates cross from the end to the beginning of the Egyptian years, the computer of the table apparently neglected to skip over the first year of Severus Alexander, which ought to have contained no epoch. The result was to have two dates separated by just a quarter day and presumably a consequent displacement of all subsequent lines.

4149. Epochs of the Moon, a.d. 96–166

93/Dec. 11/C.3 + 93/Dec. 11/E.7 — 5×8 cm (fr. 1), 2.5×4.5 cm (fr. 2) — After a.d. 166

A Standard Scheme epoch table, sloppily written without tabular framework, from a codex or loose sheet. Eight lines, or about 5 cm, are missing between the bottom of fr. 1 and the top of fr. 2. The table has preserved columns for regnal year and Alexandrian calendar date (*T* = fr. 1, front i–ii, back i–iii) and for epoch longitude (*B* = fr. 1, front iv; fr. 2, front i–ii) to at least one fractional sexagesimal place. The position of the emperors' names in fr. 1, line 1 shows that there was room for another column to the left, which might have been the discrepancy between Egyptian and Alexandrian calendar dates (*D*) or the Egyptian calendar date itself. The use of the Alexandrian calendar is unusual for an epoch table.

The years covered are (fr. 1, front) Nerva 1 through Trajan 6 = Augustus 126–132; (fr. 2, front) Trajan 13–18 = Augustus 139–144; (fr. 1, back) Marcus Aurelius and Lucius Verus 1–6 = Augustus 190–195. When complete, the table cannot have had more than about eight more epochs (or five years) at the end, since there are no traces on the back of fr. 2. All dates and longitudes, so far as they are preserved, are in exact agreement with the corresponding parts of Appendix H, Table 1, which was computed according to Rule B of *P. Ryl.* I 27.

4150. Epochs of the Moon, a.d. 187–198, and Template

A 387/5D — 4.5×8 cm — After a.d. 198

An unusual feature of this papyrus is that, to go from front to back, one has to flip along a horizontal axis; this suggests that the original sheet was detached rather than part of a codex. The Standard Scheme epoch table on what I (arbitrarily) call the 'front' covers the years Marcus Aurelius and Lucius Verus 28 through Septimius Severus 6 = Augustus 217–227. Parts of columns for the Egyptian calendar date (*T* = i) and the longitude (*B* = ii–iv) to one fractional place are preserved. Doubtful traces of a subsequent column (col. v) seem to be written slightly higher than the corresponding lines of the preceding columns; too little is preserved for a secure identification of the contents. The epoch longitudes are higher by approximately 0;3,30° than their counterparts in Appendix H, Table 1.

On the other side is a less well preserved fragment of a Standard Scheme template, giving parts of the lines corresponding to days 231 through 248 counted from epoch. The preserved columns are respectively the lowest order digits of the progress in longitude since epoch, originally given to the full three fractional places (B' = i); the number of days since epoch (T' = ii), tabulated at five-day intervals; and the progress in argument of latitude measured in 'steps', to at least one fractional place (Ω' = iii). All legible numerals are in agreement with Appendix H, Table 2.

4151. Epochs of the Moon, A.D. 217–254

64 6B.60/G(1)a 8.5×13.5 cm After A.D. 254

Like **4150**, this papyrus has a Standard Scheme epoch table on the 'front' (along the fibres) and a template on the 'back' (**4164**). Both are in tabular frameworks. The ruling for the epoch table was clearly intended to have each set of columns contain three 3031-day epoch cycles of twelve epochs each, i.e. roughly 25 years; the preserved part extends from the top margin down nearly half way through the second cycle of each of two consecutive trios of cycles. (Grouping the lunar epoch dates in 25-year triple cycles is also a feature of Rules A and B in the procedure text *P. Ryl.* I 27.) In writing out the table, the copyist used the twelfth line of the first preserved cycle (line 15) for an emperor's name, forcing him to use the spare line between cycles (line 16) for the cycle's twelfth epoch. By the second set of columns, he has learned to write emperors' names in vacant space of col. vii, but a mistake in line 13, crossed out, again pushed the twelfth epoch into the spare line.

The table covered at least Elegabalus 1 through Gallienus 1 = Augustus 247–283. The sequence of columns included the accumulated difference between Egyptian and Alexandrian calendar dates (D = vii); the regnal year and Egyptian calendar date (T = viii–x); and the longitude (B = i–v) to either one fractional place (as far as line 16 of cols. i–v) or, apparently, three places (lines 17–21). The reason for the change is that the first epoch cycle (only!) uses cols. iv and v for numbers that can be recognized from their line-to-line differences as arguments of latitude (Ω) given to one fractional place. Perhaps the copyist put together his table from two different examplars without noticing that they had different contents; in any event the template on the other side omits the column for progress in argument of latitude without which the epoch positions would be useless.

All preserved epoch dates are in agreement with those in Appendix H, Table 1. The epoch longitudes in the first cycle of the papyrus are consistently about 0;1,40° higher than their recomputed counterparts. In the second cycle the difference drops to approximately 0;1,5°, which supports the assumption of a separate source for this part of the table. The arguments of latitude for the first cycle are inaccurate and indeed increase steadily in inaccuracy, from 0;29 steps less than the value in Appendix H, Table 1 in line 8 to 0;32 steps less in line 14. The discrepancy grows because the computer of the table has used exactly 2;43 steps for the progress in 248 days instead of the correct value of 2;43,28,34 steps required by the Standard Scheme. Prolonged use of truncated epoch increments from an initial epoch five or six cycles back could account for the whole discrepancy.

Notes:

ccol. iii:

10 and 17: Apparently a symbol to indicate lack of units was written after the numerals, but in both cases it is too damaged to identify.

col. iv:

20: A stroke at bottom level precedes γ.

col. vii:

6: Indistinct traces of numeral and of hooks above and below (cf. 12).

12: Hooks (rather like c with the top stroke prolonged) above and below the numeral.

13: Messily written and blurred. The scribe began Decius's reign one year too early, and crossed out the entire line on recognizing the error.

4152. Epochs of Mercury

51 4B.21/D(1–3)b 2×5.5 cm Second or third century

Plate III

The fragment, broken on all sides, is written in a small, neat tabular hand, with black ruling. The back has legible mirror-reversed offset traces transferred from another part of the same tables.

Both sides give parts of columns for the longitude of epoch phenomena (*B*), readily identifiable as belonging to Mercury from the mean line-to-line increment of about 115°. The better preserved front gives a sequence of longitudes of Mercury at first evening appearance (Ξ), computed according to the Babylonian System A_1 rules (cf. Table 4152.1). The longitudes on the back, however, are of morning disappearance (Σ); System A_2 for this phase can generate a sequence apparently matching the traces (Table 4152.2).

λ	*i*	λ + *w*	α	β
♓ 12;36°	3	78;36°		
♊ 18;36°	3	174;36°	78;36°	131°
	(1)	227°	21°	14°
♏ 10°	2	326;40°		
♓ 26;40°	2	73;20°	93;20°	84°
♊ 4°	3	160°	64°	106;40°
♎ 22;40°	1	2;40°	156;40°	104;26,40°
♒ 10;26,40°	2	57;6,40°	77;6,40°	69;24°
♉ 19;24°	3	145;24°	49;24°	82;20°
♍ 28;20°	1	338;20°	132;20°	88;13,20°
♑ 24;13,20°	2	40;53,20°	60;53,20°	54;48°
♉ 4;48°				

TABLE 4152.1. Recomputation of Ξ according to System A_1.

λ	*i*	λ + *w*	α	β	*Text*
♊ 23;20°	4	212;40°	122;40°	102;13,20°	♊ 2x°
	(1)	192;13,20°	12;13,20°	14;40	
♎ 14;40°	2	324°	48°	36°	♎ 14°
♒ 12;0°	3	49°	44°	58;40°	♒ 12°
♊ 3;40°	4	193°	103°	85;50°	♊ 3°
♍ 25;50°	1	283;36,40°	103;36,40°	124;20°	♍ 25°
	(2)	304;20°	28;20°	21;15°	
♑ 27;15°	3	34;15°	29;15°	39	♑ 2x°
♉ 14°	4	173;20°	83;20°	69;26,40°	♉ 14°
♍ 9;26,40°	1	267;26,40°	87;26,40°	104;40°	♍ 9°
	(2)	284;40°	8;40°	6;30°	
♑ 12;30°	3	19;30°	14;30°	19;20°	♑
♈ 24;20°	4	153;40°	63;40°	53;3,20°	♈
♌ 23;3,20°					♌

TABLE 4152.2. Recomputation of Σ according to System A_2.

4153. Epochs of Mercury, a.d. 206–215

68 6B.20/L(8)a, part 5×14 cm After a.d. 215

A fragment of an epoch table in a red tabular framework, from the bottom of the sheet or roll, and broken on the top and both sides. The inventory number to which it belongs is part of Group A.

The preserved columns include the deviation between Egyptian and Alexandrian calendar dates (D = i); and regnal year, month, day, and fraction of day (T = ii–iv). The numerals in col. i identify the years as Augustus 236–244 = Septimius Severus and Caracalla 15–23 (cf. Appendix E, Table 2 and Appendix D).

The frequency of the phenomena, e.g. mean interval of 117 days between lines 15 and 18, is characteristic of Mercury. In order to determine the specific phenomenon tabulated, we may compare representative dates in the table, say the epochs on Thoth (I) 8, Choeac (IV) 22, Pharmuthi (VIII) 15, and Mesore (XII) 29 of Severus 20 (a.d. 211/212) with the motion of Mercury and the sun computed by modern theory for that year (Fig. 4153.1). If the dates in the table are in the Egyptian calendar (circles in the graph), the phenomenon clearly is first morning visibility (Γ); whereas if the dates are Alexandrian, i.e. 59 days later than the same dates in the Egyptian calendar (crosses in the graph), the phenomenon would have to be first evening visibility (Ξ). The time intervals between the four chosen epoch dates are respectively 104, 113, and 134 days, which would correspond to longitudinal intervals of approximately 100°, 109°, and 130°. Assuming that the dates are of Γ in the Egyptian calendar, this pattern would give Mercury a small synodic arc of about 100° starting near the beginning of Cancer, and a large synodic arc of about 130° starting near the beginning of Aquarius, in excellent agreement with the behaviour of System A_1 for Γ. If the dates are of Ξ in the Alexandrian calendar, the 100° synodic arc would apply about the beginning of Libra, where System A_1 for Ξ

λ	i	$\lambda + w$	α	β	$\Delta\lambda$	Δt	T
♒ 26;30°	2	107;50°	47;50°	31;53,20°			19 IX 6;38,40
♋ 1;53,20°	3	186;6,40°	65;6,40°	73;16°	125;23,20°	127;3,20	20 I 8;42
♎ 14;15°	1	300;15°	14;15°	19°	102;21,40°	104;1,40	20 IV 22;43,40
♒ 5°	2	86;21,20°	26;20°	17;33,20°	110;45°	112;25	20 VIII 15;8,40
♊ 17;33,20°	3	171;46,40°	50;46,40°	57;7,30°	132;33,20°	134;13,20	20 XII 29;22
♍ 28;7,30°	1	284;7,30°			100;34,10°	102;14,10	21 IV 6;36,10
♉ 14;7,30°					106	107;40	21 VII 24;16,10

Table 4153.1. Recomputation of Γ according to System A_1.

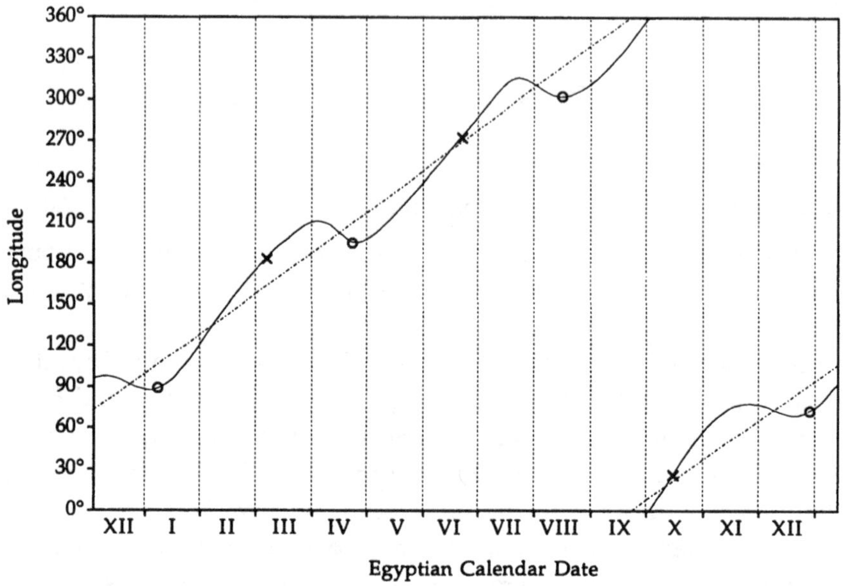

FIG. 4153.1. Mercury and the sun in A.D. 211/212.

yields 115°; and the 130° arc would apply near the beginning of Taurus, where System A_1 yields only about 117°. The identification of both the phase and the method of computation is finally secured by the possibility of generating an actual sequence of epoch longitudes and dates of Γ according to System A_1 fully compatible with the readings of the papyrus (Table 4153.1).

Notes:

14: λ̣ ̣ : κ̣ ̣ (2x) is also possible.

15: μβ̣: μ β̣ (40 2) is also possible.

19: λ[̣ : The second stroke of the λ is sloped as if to join a following letter.

4154. EPOCHS OF MERCURY

68 6B.20/L(8)a, part — 2.5×7.5 cm — Late second or early third century

Plate III

The table is inscribed in a red tabular framework; the surviving part is from the bottom of the sheet or roll, and broken on the top and both sides. Its inventory number is part of Group A. **4155** is in the same hand, and may be a different part of the same manuscript. The preserved columns give the month, day, and fraction of day of an epoch event (T), the frequency of which identifies the table as pertaining to Mercury. The time intervals between epochs are always either 110;10 days (for intervals beginning in Mesore and Thoth), 116;10 days (for intervals beginning in Hathyr, Choeac, and Tybi), or 122;10 days (for intervals beginning in Pharmuthi and Pachon). Now the System A_2 for last evening visibility (Ω) is notable for containing three longitudinal intervals of constant synodic arc, with the Δλ values 108;30°, 114;30°, and 120;30° (see p. 27, fig. 8). These intervals are so situated that consecutive occurrences of Ω can cycle through each of them three times or more before falling into the intervening sections where synodic arc varies continuously with longitude. Using c = 1;40 to convert synodic arcs to synodic times, we obtain precisely the three synodic times of the papyrus.

A sequence of longitudes and dates of Ω, computed by System A_2, that matches the contents of the papyrus (except for two isolated errors) is shown in Table 4154.1. The longitude for line 3 is the greatest that can keep all longitudes within the intervals of constant $\Delta\lambda$ and Δt, whereas a longitude less than ♍ 6;30° will push some longitudes later in the series out of the intervals. Assuming that Mercury's last evening visibility was predicted for Thoth 28 of some year about longitude ♍ 10°, the sun must have been at a longitude about 20° lower, i.e. roughly ♌ 20°, on that day. This is possible only if the dates are in the Egyptian calendar, and implies a date for the table close to A.D. 200 (cf. Appendix K). It is not possible to prove whether the longitudes of Ω in **4155** fr. 1 belonged to a continuous series with those of **4154**. If they did, the closest possible relative placement of

line	λ	$\Delta\lambda$	Δt	*T*	*text*
1	♑ 18;20°	115;25°	117;5	VI 9;12	VI
2	♉ 13;45°	117;45°	119;25	X 6;17	X
3	♍ 11;30°	108;30°	110;10	I 28;42	I 28;42
4	♑ 0°	114;30°	116;10	V 18;52	V 18;52
5	♈ 24;30°	120;30°	122;10	IX 15;2	IX 15;2
6	♌ 25°	108;30°	110;10	I 12;12	I 12;12
7	♐ 13;30°	114;30°	116;10	V 2;22	V 2;22
8	♈ 8°	120;30°	122;10	VIII 28;32	VIII 28;32
9	♌ 8;30°	108;30°	110;10	XIII 0;42	XII! 0;42
10	♏ 27°	114;30°	116;10	IV 15;52	IV 15;52
11	♓ 21;30°	120;30°	122;10	VIII 12;2	VIII 12;x
12	♊ 22°	108;30°	110;10	XII 14;12	XII 11! 12
13	♏ 10;30°	114;30°	116;10	III 29;22	III 29;22
14	♓ 5°			VII 25;32	VII 25;32

TABLE 4154.1. Recomputation of Ω according to System A_2.

the two sequences would have **4154** line 1 either 19 lines (about 6 years) earlier than **4155** fr. 1 line 3, or 22 lines (about 7 years) later. In the former case there would have been no lost lines between the bottom line of **4155** fr. 1 and the top preserved line of **4154**, which is unlikely if they were a single manuscript.

4154 and **4156a** illustrate an interesting feature of the use of Mercury's System A_2 in the papyri. The cuneiform texts pertaining to System A_2 are from the beginning of the third century B.C., about eighty years before the remainder of the ACT planetary tables, and they are unique among the System A schemes in having a variable difference between synodic time and arc instead of the constant *c*. When System A_2 reappears four centuries later in these papyri, the dates are obtained by the usual rule of adding a constant.

Notes:

4: At the right edge of the fragment in this line and lines 5, 11, 12, and 13 there is a small loop that could be part of ō (though no upper bar appears), α (though it appears rather high for that), or ϕ.

9: ō: The expected date is Mesore 30 (i.e. Epagomenae 0). The horizontal bar of the ō is crossed by a curved diagonal, which may be meant as a correction.

10: At the right edge, a stroke near the lower level resembles the beginning of an μ.

11: The fractional part of the day number is an unrecognizable script. β (2) is the expected reading.

12: 14;12 is required for numerical consistency. 11;12 appears to be a copying error.

4155. Epochs of Mercury

68 6B.25/H(3)b, part	aggregate 5×10 cm (fr. 1)	Late second or early third century
+ 68 6B.25/H(4)b, part	aggregate 5.5×6.5 cm (fr. 2)	Plate III
+ 66 6B.29/L(1–4)d		

The remains of handsomely written tables in a red tabular framework, written in the same hand as **4154**, which may be from the same manuscript The back is blank. Fr. 1 and fr. 2 seem to come from different parts of the original set of tables; each consists of two pieces that do not all touch but can be placed in relation to each other. The lower piece of fr. 1 preserves some bottom ruling; all other sides are broken. The papyrus belongs to Group A. My previous discussion of this table in Jones [1998a], section 2d preceded the discovery of two of the bits; the most important conclusions drawn there remain valid, but some details require correction and the analysis can now be more straightforward.

In both fragments there are parts of four columns. Col. i, which is very poorly preserved, contains sexagesimal fractions of uncertain purpose. Col. ii, marked off on both sides by double ruling, contains whole numbers ranging from 12 to 31 in fr. 1, and from 22 to 47 in fr. 2. Cols. iii–iv contain the longitudes (B = iii–iv) of an epoch event, identifiable from its frequency as pertaining to Mercury in both fragments. Only a few traces of the numbers of degrees survive in fr. 1, with no fractions, but in fr. 2 are several lines with complete longitudes to two fractional places. These prove to be longitudes of Mercury's first evening appearance (Γ), computed by the rules of System A_1. A restoration of the sequence for the entire fragment is given in Table 4155.1.

The almost complete loss of the degree values in fr. 1 is partly compensated for by the extended sequence of zodiacal signs. It is not possible to reproduce the sequence by means of the System A_1 schemes for first appearances (Γ or Ξ), or the System A_2 scheme for last evening appearance (Σ). On the other hand, the System A_2 scheme for last morning appearance (Ω) can produce the desired pattern, as shown in Table 4155.2. There is small margin (less than 1° either way) in the choice of the initial longitude.

The numbers in fr. 1, col. ii, which we will refer to as d_Ω, are correlated with the epoch longitudes (cf. Fig. 4155.1, small circles), with a maximum value of 35 about the beginning of Gemini, a minimum of 12 near the beginning of Aquarius, and a mean of about 20. The pattern of d_Ω has a close enough resemblance to that of the 'pushes' in the Babylonian System A_2 prescribing the time intervals in *tithis*—thirtieths of a synodic month—from a calculated date of Ω to the subsequent Γ (solid line in Fig. 4155.1, from Neugebauer [1955] 298–299) to suggest identifying d_Ω as the difference in days between the date of the occurrence of Ω tabulated in

λ	i	$\lambda + w$	α	β
♓ 1°	2	112;20°	52;20°	34;53,20°
♋ 4;53,20°	3	189;6,40°	68;6,40°	76;37,30°
♎ 17;37,30°	1	303;37,30°	17;37,30°	23;30°
♒ 9;30°	2	90;50°	30;50°	20;33,20°
♊ 20;33,20°	3	174;46,40°	53;46,40°	60;30°
♎ 1;30°	1	287;30°	1;30°	2°
♑ 18°	2	69;20°	9;20°	6;13,20°
♉ 6;13,20°	3	160;26,40°	39;26,40°	44;22,30°
♍ 15;22,30°	1	271;22,30°		
♑ 1;22,30°	1	377;22,30°	91;22,30°	121;50°
♉ 17;50°	2	189;10°	129;10°	86;6,40°
	(3)	146;6,40°	25;6,40°	28;15°
♌ 29;15°	1	255;15°		
♐ 15;15°	1	361;15°	75;15°	100;20°
♈ 26;20°				

TABLE 4155.1. Recomputation of Γ according to System A_1.

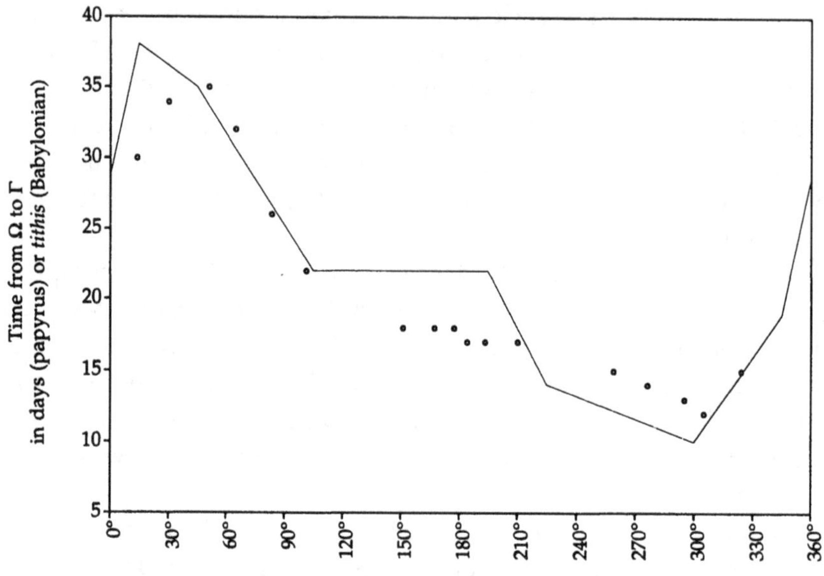

FIG. 4155.1. d_Ω (circles) compared with System A_2 pushes for Ω–Γ.

the same line and the date of the immediately subsequent Γ.

There is an important difference between the pushes in the Babylonian tables for Mercury and d_Ω in the papyrus. The pushes are a preordained scheme, through which one found the positions and dates of Γ from the computed positions and dates of Ω. In **4155**, however, the series of Ω and Γ have been obtained independently by means of their respective System A schemes, so that the tabulated d_Ω must be derived from the longitudes rather than the other way around. This can be proved. No dates survive on either fragment of the papyrus; but it is enough for our purposes that if the dates of Ω and Γ were calculated according to the standard rule relating synodic time to synodic arc ($\Delta t = \Delta\lambda + c$, with c constant), then a similar relationship will hold between the intervals of time and longitude separating each pair of Ω and Γ. That is, the number of days from Ω to Γ, i.e. d_Ω, can be obtained by adding some constant to the difference in longitude between the two phases, neglecting fractions.

Now between Ω and Γ Mercury travels a small number of degrees, usually retrograde, but occasionally direct. The only way of aligning the sequences of phases in fragments 1 and 2 that makes astronomical sense is that according to which the Ω of fr. 1 line 16 immediately preceded the Γ of fr. 2 line 1. If we subtract the longitudes in the sequence of fr. 2, which we know exactly, from the corresponding longitudes in the sequence of fr. 1, which we have reconstructed within a tolerance of less than 1°, and then add the corresponding d_Ω from the papyrus,

λ	*i*	λ + *w*	α	β
♊ 12°	4	207;37,30°	117;37,30°	94;6°
♎ 4;6°	1	292;36°	22;36°	25;6,40°
♑ 25;6,40°	2	55;40°	85;40°	77;6°
	(3)	47;6°	17;6°	21;22,30°
♉ 21;22,30°	4	187	97	77;36°
♍ 17;36°	1	276;6°	6;6°	6;46,40°
♑ 6;46,40°	2	37;20°	67;20°	60;36°
	(3)	30;36°	0;36°	0;45°
♉ 0;45°	4	166;22,30°	76;22,30°	61;6°
♍ 1;6°	1	259;36°		
♐ 19;36°	1	8;6°	98;6°	109°
	(2)	19°	49°	44;6°
♈ 14;6°	3	122;36°	92;36°	115;45°
	(4)	145;45°	55;45°	44;36°
♌ 14;36°	1	243;6°		
♐ 3;6°	1	351;36°	81;36°	90;40°
	(2)	0;40°	30;40°	27;36°
♓ 27;36°	3	106;6°	76;6°	95;7,30
	(4)	125;7,30°	35;7,30°	28;6°
♋ 28;6°	1	226;36°		
♏ 16;36°	1	335;6°	65;6°	72;20°
	(2)	342;20°	12;20°	11;6°
♓ 11;6°	3	119;36°	89;36°	112°
	(4)	142°	52°	41;36°
♋ 11;36°	1	240;6°		
♏ 0;6°	1	318;36°	48;36°	54°
♒ 24°	2	84;33,20°	114;33,20°	103;6°
	(3)	73;6°	43;6°	53;52,30°
♊ 23;52,30°	4	219;30°	129;30°	103;36°
♎ 13;36°	1	302;6°	32;6°	35;40°
♒ 5;40°	2	66;13,20°	96;13,20°	86;36°
	(3)	56;36°	26;36°	33;15°
♊ 3;15°	4	198;52,30°	108;52,30°	87;6°
♍ 27;6°				

TABLE. 4155.2. Recomputation of Ω according to System A_2.

the sums are consistently within 1 of a constant 31 (Table 4155.3). This successful result simultaneously confirms (i) the restoration of fr. 1 according to System A_2 for Ω, (ii) the hypothesis that dates of both Ω and Γ were obtained from the longitudes by the standard rule, (iii) the identification of d_Ω as the time interval from Ω to Γ, and (iv) the alignment of the phases in the two fragments.

The corresponding column ii in fr. 2 (d_Γ) therefore may be expected to represent the time intervals in days between Γ and the subsequent Σ; and there is again a close resemblance between the pattern of attested values and the Babylonian scheme of pushes between these phases (Fig. 4155.2, cf. Neugebauer [1955] 294–295). Between Γ and Σ Mercury's average progress is close to 1° per day, because

d_{Ω}	λ_{Ω}	λ_{Γ}	$\lambda_{\Omega}-\lambda_{\Gamma}$	$d_{\Omega}+\lambda_{\Omega}-\lambda_{\Gamma}$
(fr. 1 lines 2–10)				
17	185;36°	172;15°	13;21°	30;21
13	296;46,40°	278;15°	18;31,40°	31;31,40
35	53;15°	57°	–3;45°	31;15
18	169;6°	156;7,30°	12;58,30°	30;58,30
14	278;26,40°	262;7,30°	16;19,10°	30;19,10
34	32;37,30°	35;30°	–2;52,30°	31;7,30
18	152;36°	140°	12;36°	30;36
15	261;6°	246°	15;6°	30;6
30	15;36°	14°	1;36°	31;36
(fr. 1 lines 17–24)				
22	103;6°	94;53,20°	8;12,40°	30;12,40
17	211;36°	197;37,30°	13;58,30°	30;58,30
15	325;40°	309;30°	16;10°	31;10
26	85;45°	80;33,20°	5;11,40°	31;11,40
17	195;6°	181;30°	14;36°	31;36
12	307;20°	288°	19;20°	31;20
32	65;7,30°	66;13,20°	–1;5,50°	30;54,10
18	178;36°	165;22,30°	13;13,30°	31;13,30

TABLE 4155.3. Fr. 1 col. ii (d_{Ω}) interpreted as days from Ω to Γ.

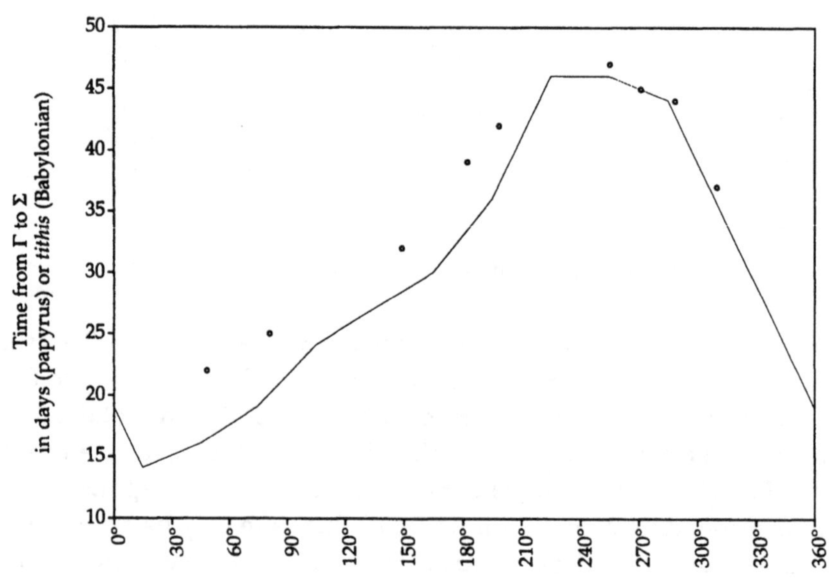

FIG. 4155.2. d_{Γ} (circles) compared with System A_1 pushes for Γ–Σ.

d_Γ	λ_Γ	$\lambda_\Sigma = \lambda_\Gamma + d_\Gamma$	λ_Σ (System A_2)
42	197;37,30°	239;37,30°	240°
37	309;30°	346;30°	346°
25	80;33,20°	105;33,20°	105;50°
39	181;30°	220;30°	220;20°
44	288°	332°	331;15°
?	66;13,20°	?	89;20°
?	165;22,30°	?	200;40°
45	271;22,30°	316;22,30°	316;30°
22	47;50°	69;50°	69;40°
32	149;15°	181;15°	181°
47	255;15°	302;15°	301;45°
2x	26;20°	?	50°

TABLE 4155.4. Longitudes of Σ reconstructed from fr. 2, and according to System A_2.

its elongation from the sun is nearly the same at the beginning and and of the period of visibility. We therefore hypothesize that the missing longitudes of Σ can be approximately obtained by adding d_Γ to the longitudes of Γ. In fact the sums fit (within a tolerance of 1° because d_Γ is truncated) a hypothetical sequence of longitudes of Σ computed by System A_2 as shown in Table 4155.4. Thus three of the four Babylonian schemes for Mercury's phases are present in this single papyrus.

4156a. EPOCHS OF MERCURY

19 2B.78/B(1–3)c + A188/3F 6×10 cm Third century

Part of an epoch table for Mercury ruled in red, broken on the top, left, and right sides, and almost entirely stripped of its upper layer of fibres below line 4.[1] No traces of ink are visible below the ruling under line 13. The back is blank.

Cols. i–iii contain the month, day, and sexagesimal fraction of a day of the epoch date (*T*, Egyptian calendar); cols. iv–vi, the corresponding epoch longitude (*B*). The horizontal rulings correspond to the ends of years. The event in question is last evening visibility (Ω), computed according to System A_2 using the constant value $c = 1;40$ for the dates. The correspondence of Egyptian calendar dates with longitudes would be acceptable for the first half of the third century; for example the year of lines 2–4 could have been A.D. 218/219.

[1] Additional fragments were joined to this papyrus subsequent to the inconclusive discussion in Jones [1998a] 26.

4156b. Epochs of Mercury

20 3B.30/C(4–7)b 6×8 cm Second or third century

From the bottom of an epoch table for Mercury, in a red tabular framework, with lower margin preserved but broken on the top and both sides. Columns iv–v contain the epoch longitudes (*B*). The preceding columns might contain the epoch date (*T*), in the form (i) month, (ii) day, (iii) sexagesimal fraction.

From line 3 to line 4 the longitude has advanced between 120° and 150° from an initial longitude in Capricorn. This is possible only if the tabulated phase is Γ. System A_1 for Γ can generate a similar sequence of longitudes and, perhaps, dates (Table 4156b.1).

4156c. Epochs of Mercury

38 3B.81/B(1–2)c, part 2.5×2.5 cm Second or third century

A small fragment, broken on all sides, from an epoch table for Mercury ruled in red. The back is blank.

Col. i is very poorly preserved, and its contents are uncertain. Col. ii holds the zodiacal signs of the epoch longitude (*B*). If the longitudes were computed using one of the four System A schemes for Mercury's phases, the preserved sequence is compatible only with System A_2 for last evening visibility (Ω).

Longitude (λ)	$\Delta\lambda$	Δt (= $\Delta\lambda$ + 1;40)	Date (t), assuming Egyptian years
♍ 26°	106°	107;40	II 11;56
♉ 12°	139;20°	141	V 29;36
♊ 1;20°	98;32,30°	100;12,30	X 20;6
♍ 9;52,30°			I 25;48,30

Table 4156b.1. Recomputation of Γ according to System A_1. (Months are hypothetical)

4157. Epochs of Venus

93/Jan. 1/E.2 3×9.5 cm Second or third century
Plate IV

A small scrap from the bottom of an epoch table ruled in red, with column iii giving the synodic time leading up to an epoch event (ΔT) and column iv giving the month of the epoch date (T); the format resembles **4158**. The synodic time of 610;7 days can only pertain to Venus. Since Venus's mean synodic period is about 584 days, this is a fairly high value, though not much longer than the maximum intervals of 20;28 synodic months (i.e. approximately 604;24 days) that are prescribed for some phases of Venus in the Babylonian so-called System A_1 and A_2 schemes, reflecting the extremely variable periods of invisibility of this planet in different parts of the ecliptic (Neugebauer [1955] 301–302). Col. ii may have been vacant. It is not clear what col i contained.

A passage in the second-century astrologer Vettius Valens may have some relation to this fragment. Valens devotes a chapter (I 20 Kroll = I 18 Pingree) to easy methods of obtaining a rough longitude of each of the five planets for a given date. While the procedures for the other planets use only coarse period relations and neglect anomaly entirely, that for Venus is more complex:

> Venus (is to be computed) as follows. Take the (years) from Augustus (1) to the year you seek, and divide by 8. See if the number remaining inside 8 has an epoch of station; and using this, take the number of days until the day in question, if possible. If not, use the one above, just like for the moon. For if the epoch was found preceding the geniture, this is to be used, but if (it was found) following the geniture, (use) the one above. Add up the sum (of days from the beginning of the year) and subtract[1] 120 units, which are for the station, and count off the remainder from the sign corresponding to the epoch, giving 25 to each sign. And wherever the number falls, there will be Venus. The epochs of the stations will be clear from the remainder of years. If the excess (over multiples of 8 years) is say 1, 3, 4, 6, or 7, it has a station (in that year), but if it is 2 or 5, it is in direct motion. (Remainder) 1 has an epoch on Phamenoth 10 in Taurus; 3 on Phaophi 10 in Sagittarius; 4 on Payni 22 in Leo; 6 on Tybi 8 in Pisces; and 7 on Mesore 14 in Libra, with the station extending into the eighth year.

The principle behind these instructions has been elucidated by Tihon [1968] and Neugebauer [1975] 796–800. Valens assumes a repeating cycle of five epochs in

[1] I delete the words τοῦ ζῳδίου τὴν ἐποχὴν ἀπόλυε ἑκάστου ζῳδίου, which are a garbled anticipation of a phrase occurring two lines later.

Year	*T (Alexandrian)*	*ΔT*	*B*
1 (+ 8*i*)	Phamenoth 10	566	♉
3 (+ 8*i*)	Phaophi 10	576	♐
4 (+ 8*i*)	Payni 22	617	♌
6 (+ 8*i*)	Tybi 8	581	♓
7 (+ 8*i*)	Mesore 14	580	♎

TABLE 4157.1. Epochs of Venus from Vettius Valens I 18.

eight (Alexandrian) years. After each epoch it is assumed, for the purposes of rough estimation, that Venus remains stationary for 120 days, and thereafter progresses at a constant rate of 30° in 25 days until the next epoch. Neugebauer realized that the epochs must have been originally part of a more refined scheme, perhaps making use of a template to bridge the intervals between epochs, and moreover argued that the dates are garbled in our text since the intervals between them are unequal. In fact Valens uses expressions that suggest that the dates and zodiacal signs were originally found in an epoch table, which would have allowed for variable synodic time; there is even a recognizable allusion to the Standard Scheme lunar epoch table, with which Valens evidently expected his reader to be familiar. Valens' epochs of Venus tend to precede the planet's actual evening station (Ψ) by about a month or less. As far as it can be reconstructed (without emending Valens' dates), the epoch table would have been as shown in Table 4157.1. One immediately notices that an unusually large synodic time of 617 days corresponds to an epoch in Payni, very much like the surviving entry in **4157**.

The Babylonian tables for Venus ACT 410–430 seem to be based on a rule that synodic arcs and times are constant for all initial longitudes within each zodiacal sign. Table 4157.2 compares the variation according to initial longitude exhibited by the synodic times in Valens' epoch table and by the Babylonian schemes for each of the phases (expressed in days). It is curious that the closest match is with Φ, which terminates the retrogradations whereas Valens' epochs precede them.

	Valens	*Γ*	*Φ*	*Σ*	*Ξ*	*Ψ*	*Ω*
♉	580	578	579	604	590?	591	588
♌	566	582	561	575	553	589	594
♎	576	589	587	561	575	576	591
♐	617	585	604	587	601?	579	570
♓	581	582	585	589	597	581	573

TABLE 4157.2. Synodic times (ΔT) of Valens' epochs compared with Babylonian parameters.

4157a. EPOCHS OF VENUS, A.D. 106–121

38 3B.81/B(3–4)c, part 5.5×10.5 cm After A.D. 121

A fragment of an epoch table ruled in red, broken on the left, right, and bottom, and with the fibres disintegrating towards the bottom. The top margin is present (3.5 cm including the heading to the table); in it another hand has written four lines of text, none of which I am able to read. On the back, across the fibres, are parts of 14 mostly illegible lines from the top of a document.

Four columns are preserved. Col. i contains numerals of uncertain significance, and cols. ii–iv contain the date (*T*) of a phenomenon of Venus, expressed as regnal year (of Trajan and Hadrian), Egyptian calendar month, and day with one place of sexagesimal fractions. The event in question is Venus' inferior conjunction, i.e. the conjunction falling between last evening visibility (Ω) and first morning visibility (Γ). The heading describes this as 'morning conjunction', evidently meaning the conjunction preceding morning rising.

The Babylonian schemes for Venus are only imperfectly understood (Neugebauer [1975] 460–466), but it is at least clear that they predicted only the planet's observable synodic events, i.e. the first and last visibilities and the stations. The fact that our papyrus is a table of dates of conjunction demonstrates that for once we are not dealing with a Babylonian scheme. Unfortunately the dates are too fragmentary to allow us to reconstruct the rules for their computation.

4158. EPOCHS OF MARS, A.D. 271–325

62 6B.82/H(2–4)b 8×16 cm (fr. 1) After A.D. 325
3×7.5 cm (fr. 2) 2×4 cm (fr. 3) debris Plate IV

Shattered remains of a handsomely written epoch table in red ruling; the back is blank. The preserved columns of fr. 1 are, from the left, the time since previous epoch in days and sexagesimal fractions (ΔT = i); date of epoch in the Egyptian calendar, with sexagesimal fractions (T = ii–iii); and a narrow column (col. iv) vacant in most preserved lines, and of uncertain purpose. Months repeat at intervals of between seven and eight lines (e.g. in the piece preserving fr. 1, ii 1–9), with the progress from line to line varying between one month and three months (above whole years); this periodicity identifies the tabulated phases as belonging

Longitude (λ)	Δλ	Δ*t* (= Δλ + 1;14,4)	Date (*t*), assuming Egyptian years
♊ 20;30°			0 Tybi 10;55,20
♋ 23;40°	33;10°	34;24,4^{d}	2 Mecheir 15;19,24
♌ 23;40°	30°	31;14,4^{d}	4 Phamenoth 16;33,28
♎ 1;33,20°	37;53,20°	39;7,24^{d}	6 Pharmuthi 25;40,52
♏ 17;20°	45;46,40°	47;0,44^{d}	8 Payni 12;41,36
♑ 26°	68;40°	69;54,4^{d}	10 Mesore 22;35,40
♈ 12°	76°	77;14,4^{d}	13 Hathyr 4;49,44
♊ 3°	51°	52;14,4^{d}	15 Choeac 27;3,48
♋ 12°	39°	40;14,4^{d}	17 Mecheir 7;17,52
♌ 12°	30°	31;14,4^{d}	19 Phamenoth 8;31,56
♍ 16°	34°	35;14,4^{d}	21 Pharmuthi 13;46,0
♎ 26°	40°	41;14,4^{d}	23 Pachon 25;0,4
♐ 24°	58°	59;14,4^{d}	25 Epeiph 24;14,8
♓ 15;45°	81;45°	82;59,4^{d}	28 Phaophi 12;13,12
♉ 15;30°	59;45°	60;59,4^{d}	30 Choeac 13;12,16
♋ 0;20°	44;50°	46;4,4^{d}	32 Tybi 29;16,20
♌ 0;20°	30°	31;14,4^{d}	34 Mecheir 30;30,24
♍ 0;26,40°	30;6,40°	31;20,44^{d}	36 Pharmuthi 1;51,8
♎ 10;26,40°	40°	41;14,4^{d}	38 Pachon 13;5,12
♐ 0;40°	50;13,20°	51;27,24^{d}	40 Epeiph 4;32,36
♒ 16°	75;20°	76;34,4^{d}	43 Thoth 16;6,40
♈ 27°	71°	72;14,4^{d}	45 Hathyr 28;20,44
♊ 13°	46°	47;14,4^{d}	47 Tybi 15;34,48
♋ 18;40°	35;40°	36;54,4^{d}	49 Mecheir 22;28,52
♌ 18;40°	30°	31;14,4^{d}	51 Phamenoth 23;42,56
♍ 24;53,20°	36;13,20°	37;27,24^{d}	53 Pachon 1;10,20

TABLE 4158.1. Fragment 1: Phases of Mars computed by System A scheme. (Year numbers are relative to first line)

to Mars. In fact the dates in fr. 1 were computed according to the Babylonian System A zone scheme (see Introduction, §5b), using the value c = 1;14,4 for the constant difference between synodic arcs and synodic times in days. Because the dates and their differences were given (and are partly preserved) to the full two fractional places required by the scheme, the exact sequence of corresponding longitudes and synodic arcs can be restored (Table 4158.1).

The System A Mars scheme was used in ACT texts to compute only three of the planet's synodic phenomena: first visibility (Γ), first station (Φ), and last visibility (Ω), whereas special retrogradation schemes determined the acronychal rising (Θ) and second station (Ψ) directly from the preceding **Φ**. Assuming that this was still true in the Greek version of the scheme, we can narrow the possible phases for the present table down to just **Φ**. We choose any of the computed longitudes and Egyptian calendar dates from Table 4158.1, say the second line, which gives a phase at longitude Cancer 23;40° on Mecheir 15. During a span of years plausible for the papyrus, say from 50 B.C. to A.D. 500, Mecheir 15 in the Egyptian calendar corresponds to Julian calendar dates shifting backwards from mid February to late September, so that the sun's longitude was somewhere in the part of the ecliptic between Libra (assuming a late date) and Aquarius (assuming an early one), whereas the phases Γ and Ω require that the sun be well within a zodiacal sign of the planet. (Choosing a date and longitude from any other line would lead, of course, to the same result.) Moreover, we know (e.g. from inspection of Tuckerman's tables) that Φ occurs in the middle of Cancer only when the sun's longitude is about 130° higher, i.e. in this instance about Sagittarius 4°. Allowing a liberal tolerance, this limits consideration to those years in which Egyptian Mecheir 15 fell between November 15 and December 5, i.e. between A.D. 240 and 323. Within this range there are only six years when Φ occurred reasonably close to Mecheir 15 (Table 4158.2). Of these, A.D. 273 yields the best agreement in longitude, and this dating alone leads to acceptably small deviations in both longitude and date when the whole series of epochs in the papyrus is compared with the actual occurrences of Φ as computed by modern theory. Hence we can say with some confidence that 'year 0' in Table 4158.1 was the Egyptian year A.D. 271/272 (Era Augustus 301), and the last year in the table was A.D. 324/325.

Fr. 2 preserves the bottom lines of another part of the table, which does not match the extrapolation of the sequence of dates in fr. 1 either forwards or backwards. If set next to the corresponding lines of fr. 1, the dates of fr. 2 appear to be consistently just over a month later than fr. 1's dates of first station (Table 4158.3). Since Mars's mean retrogradation is about 70^d, intervals of between 34 and 36 days would be suitable for the half of the retrograde motion from first station (Φ) to acronychal rising (Θ). Unfortunately the day numbers in the fragment are too badly preserved to permit reconstruction of the method of computation; the evidence from ACT texts for this aspect of the System A retrogradation schemes is too fragmentary to be of much help.[1]

Mecheir 15 (Egyptian)	*Date of Φ*	*Mars*	*Sun*
226 December 9	Nov 27	116.11	245.34
258 December 1	Dec 4	123.93	253.28
273 November 27	Nov 21	111.05	240.17
305 November 19	Nov 29	118.93	248.05
320 November 15	Nov 15	105.88	234.85
367 November 4	Nov 11	100.56	229.56

TABLE 4158.2. Mars, dates and sidereal longitudes of Φ (modern theory).

[1] Neugebauer [1955] 302.

Fragment 1	Fragment 2	Difference
Epeiph 4;32,36	Mesore 9 [	34;27 < Δt < 35;28
Thoth 16;6,40	Phaophi 21 [	34;53 < Δt < 35;54
Hathyr 28;20,44	Tybi 4 [	35;39 < Δt < 36;40
Tybi 25;34,48	Mecheir 30 [	34;25 < Δt < 35;26
Mecheir 22;28,52	Phamenoth 28 [	35;31 < Δt < 36;32
Phamenoth 23;42,56	Pharmuthi 2x [	Δt < 36;18
Pachon 1;10,20	Pachon 6 [	34;49 < Δt < 35;50

TABLE 4158.3. Time intervals between corresponding lines of frs. 1 and 2.

Fr. 3 contains part of a column for dates, probably computed according to the same System A rules as fr. 1, but apparently not connectible with that sequence. The month names are lost, so that we cannot identify which phase was tabulated.

Notes:

fr. 1:

22: The time interval in col. i should be 72;34,4. The error is isolated, having no effect on subsequent dates.

fr. 2:

3: possibly ε̣ (5).
4: possibly κδ̣ (24).

4159. EPOCHS OF MARS, A.D. 225–235

68 6B.20/D(4)a, part 2.5×4.5 cm After A.D. 235

A small scrap of an epoch table in red ruling from Group A; the other side (the front) has parts of seven lines of a document in a third-century hand. Only the dates of the epochs (T) are preserved. Like **4158**, this table was computed using the System A scheme for Mars, probably with the same value c = 1;14,4 for the difference between synodic arcs and synodic times. The sequence of longitudes can be reconstructed only approximately because of the loss of most fractional places in the dates (Table 4159.1).

The epoch in line 6 was the first to have taken place in the reign of Maximinus, and therefore was in either his first or his second regnal year (234/235 or 235/236). In the Egyptian calendar Hathyr 2 corresponded to August 26 in both years, while in the Alexandrian calendar it corresponded to October 29/30. In neither 234 nor 235 did any phase of Mars occur close to the end of October. Second station (Ψ) was passed a few days before August 26, 235; but we would not expect

λ	*i*	λ + *w*	α	β	Δλ	Δ*t*	*T*
♒ 10°	5	40°	70°	52;30°			5 VIII 25;39
♈ 22;30°	6	90°	60°	40°	72;30°	73;44,4	7 XI 9;23
♊ 10°	1	115°	25°	16;40°	47;30°	48;44,4	9 XII 28;7
♋ 16;40°	2	136;40°			36;40°	37;54,4	12 II 1;1
♌ 16;40°					30°	31;14,4	1 III 2;15

TABLE 4159.1. Mars, recomputation of Γ according to System A.

this phase to be computed by the System A zone scheme. There remains only August 26, 234; on this date the longitude of Mars was about 18° less than the sun's, which is plausible for first visibility (Γ).

4159a. EPOCHS OF MARS, A.D. 101–133

19 2B.83/E(c), part 7.5×9 cm After A.D. 133

The papyrus consists of four pieces that do not quite touch but can be relatively placed. The top, left, and right sides are broken off, but a little of the bottom margin (1.5 cm) is preserved. The table is ruled in red, and the hand is that of the ephemeris **4177**, which is securely dated to the A.D. 120s. The back is blank.

Cols. i–ii are part of the columns for the date (*T*) of a phenomenon of Mars, containing respectively the day of the Egyptian calendar month and one sexagesimal fractional place. Columns lost to the left must have contained the year and month. The longitude of the phenomenon (*B*, zodiacal sign, degrees, and one fractional place) occupies cols. iii–v. Both the longitudes and the dates show symptoms characteristic of the Babylonian System A scheme for Mars. In particular, the longitudes in lines 8–9 give a synodic arc of exactly 30° for initial longitude Gemini 30° (i.e. Cancer 0°). Again, if we subtract c = 1;14 from the differences between the dates, we get whole numbers or numbers with simple fractions. Neither series, however, can be generated by the basic System A rules, nor is the difference between the synodic times and the corresponding synodic arcs constant. This suggests that the phenomenon in question was either acronychal rising (Θ) or second station (Ψ), since in the Babylonian texts these events were derived secondarily from the preceding first station (Φ) by adding intervals of longitude and time dependent on the longitude of Φ. We know that there existed several such 'retrogradation schemes', but our information about how they worked is incomplete except for the patterns of intervals in longitude from Φ to Θ (Neugebauer [1955] 305–306). The effect of all these schemes was to produce a

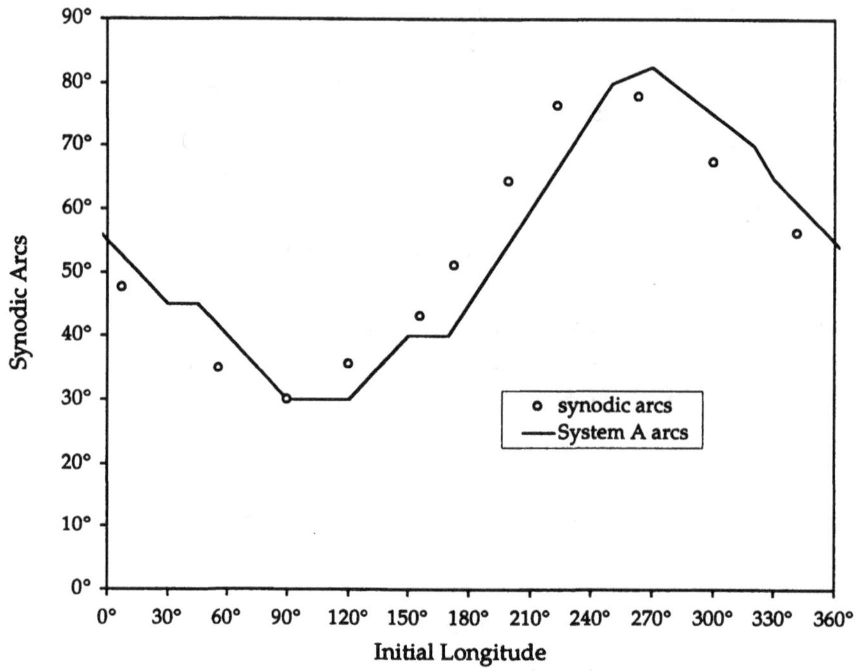

FIG. 4159a.1. Synodic arcs in **4159a** compared with System A model.

variation in the retrograde arcs and times such that they will be largest about Cancer and Leo, and smallest about Capricorn and Aquarius, as is true in nature. Independent use of the System A model for Θ and Ψ would put the maxima and minima diametrically opposite where they ought to be.

As a first step in analysing **4159a**, we may attempt to establish whether the tabulated event is Θ or Ψ. In fig. 4159a.1 the synodic arcs from the papyrus are plotted as a function of initial longitude, together with the synodic arc function generated by the System A model (cf. p. 28 fig. 10). The arcs in the papyrus follow a pattern closely resembling the model, but leading the model by a mean shift of about 19°. This behaviour fits Ψ as computed by the Babylonian retrograde schemes. Θ, on the other hand, would lead the model by only about 8°.

Within the first century and the first half of the second, there were only two sequences of occurrences of Mars' Φ that fit the longitudes in the papyrus. In the earlier sequence, line 3 of the papyrus would represent the station that actually occurred on A.D. 26 April 29 (Egyptian Pachon 16), whereas in the second sequence, line 3 would represent the station of A.D. 105 May 1 (Egyptian Payni 8). The day numbers in the papyrus fit the two possible datings about equally well, with the earlier dating giving the closest matches where the longitude of the station is in the region around 0°, and the later one working best around 180°. The

Date of Ψ (Era Augustus)	*Longitude of Ψ*	*Longitude of Φ (System A)*	*Date of Φ (System A)*	*Δλ*	*Δt*
134 X 8;9	172;xx°	189;33,40°	VII 22;2	17°	76;6
136 XI 27;23	223;xx°	239;20°	IX 13;3	16°	74;20
139 I 9;37	300°	314°	XI 28;57	14°	75;40
141 IV 22;6	7;30°	25;30°	II 6;41	17°	75;25
143 VI 13;35	55°	72°	III 24;25	17°	79;10
145 VII 20;39	90°	108°	V 1;39	18°	79
147 VIII 21;53	120°	138°	VI 2;53	18°	79
149 IX 26;17	155;4x°	174°	VII 10;7	18;15°	76;10
151 XI 8;41	198;53°	216°	VIII 23;21	17;6°	75;20
154 I 8;15	263;20°	279°	X 27;35	15;40°	75;40
156 III 29;49	341;15°	359;15°	I 14;4	18°	75;45
158 V 28;43	37;30°	54;30°	III 10;33	17°	78;10
160 VII 12;47	—	96;20°	IV 23;37	—	79;10

TABLE 4159a.1. Dates and longitudes in papyrus, with reconstructed intervals from Φ to Ψ.

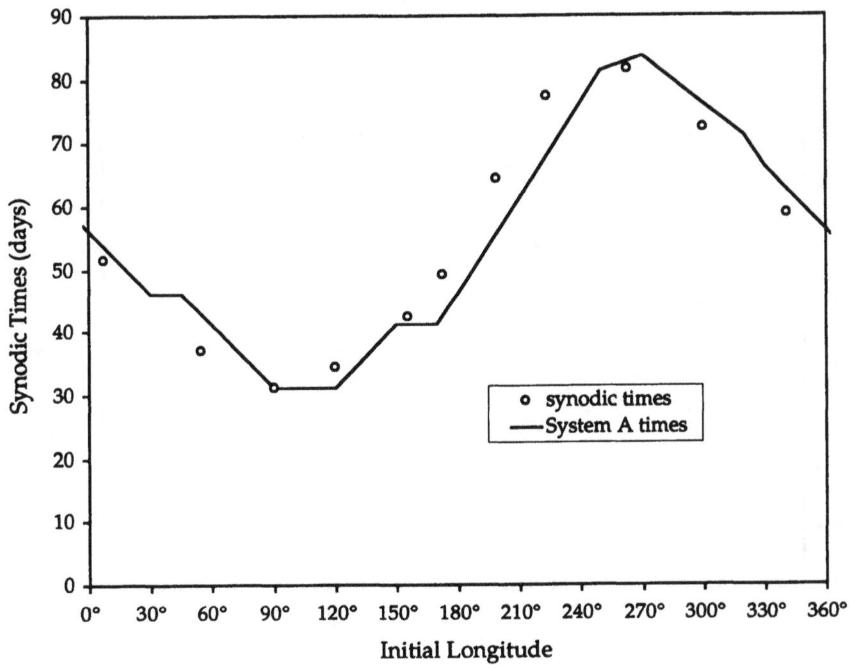

FIG. 4159a.2. Synodic times in **4159a** compared with System A model.

later dating is almost certainly the true one, because according to it the epoch table covers the years A.D. 100/101 through 132/133, thus overlapping the ephemeris **4177** which was written by the same scribe; a range of dates eighty years eariler seems implausibly remote. The rightmost column of Table 4159a.1 gives the restored dates, and Fig. 4159a.2 shows how the synodic times separating the dates lead the System A model by a mean of about 14° when plotted as a function of initial longitude.

The third and fourth columns of Table 4159a.1 present a hypothetical sequence of longitudes and dates of Φ calculated by System A, aligned so that the retrograde arcs in Cancer and Leo are 18° (as in the Babylonian retrogradation scheme S) and the retrograde times are 79 days. The retrogradations, tabulated in the last two columns, exhibit more or less the expected variation when plotted against the longitude of Φ (Figs. 4159a.3–4), that is, they are greatest in the neighbourhood of Cancer and Leo where the retrograde arcs are smallest, and they are least in the neighbourhood of Capricorn and Aquarius. The patterns seem to be irregular, however, and do not conform to any of the known Babylonian retrogradation schemes. Varying the initial longitude of the sequence of Φs changes the shape of the graphs, but without bringing about any clearer pattern. The means by which the longitudes and dates in the papyrus were calculated remains unknown despite its evident connection with System A.

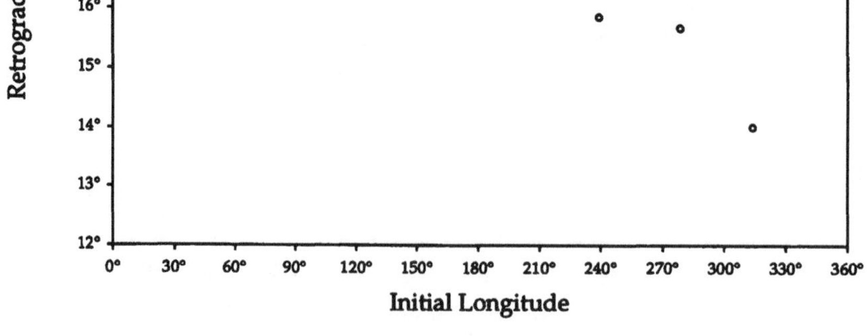

FIG. 4159a.3. Reconstructed retrograde arcs as a function of longitude of Φ.

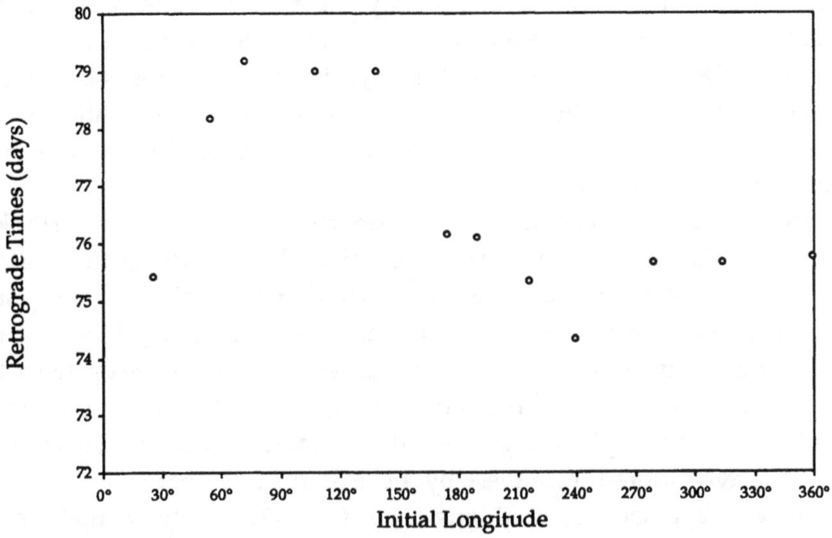

FIG. 4159a.4. Reconstructed retrograde times as a function of longitude of Φ.

4160. EPOCHS OF JUPITER, A.D. 30–79

P. Berol. inv. 16511	9.5×9.5 cm	After A.D. 79
+ 30 4B.39/D(1–3)a	4×11.5 cm	

The table is written in a frame ruled in black, on the back of a document (court proceedings?) in a late first-century hand.[1] (30 4B.39/D(1–3)a also includes a small scrap from a different table, ruled in red, not edited here.) The Berlin fragment (lines 1–13) preserves a top margin of 2.5 cm, with headings of uncertain meaning; the remaining sides, and all sides of the Ashmolean fragment (lines 26–47), are broken off. The combined height, incorporating the thirteen lost intervening lines of the table, would have been about 28 cm. Columns for the date of epoch (*T*) are preserved in cols. iii–v. In the light of the vacant col. ii, col. i seems to another series of epochs, either a different phenomenon of the same planet or the preceding occurrences of the same phenomenon; the sexagesimals are probably longitudes.

[1] I am grateful to Dr W. Brashear and the Ägyptisches Museum und Papyrussammlung (Berlin) for making the Berlin fragment of this manuscript available to me. A separate edition and discussion of P. Berol. 16511 are presented in Brashear & Jones [1999], and a deduction of the underlying model will appear in Britton & Jones [2000].

The synodic times of between 30 and 37 days (in excess of one year) identify the planet as Jupiter. To determine which phase is tabulated, we take any epoch date, e.g. from line 28, Nero 5, Tybi 14. In the Egyptian calendar this is equivalent to A.D. 58 December 20, on which date Jupiter was close to first station (F) at approximate longitude 151°; on the other hand, no phase of Jupiter was near Alexandrian Tybi 14 (January 9).

The fractional parts of the epoch dates are imperfectly preserved; nevertheless approximations of the synodic times can be obtained by tabulating their differences (Table 4160.1). To investigate the possibility that the table was computed using a scheme of the System A variety, we then use an estimate of 0;44 for the constant difference between Jupiter's synodic times and arcs to reconstruct the synodic arcs themselves. The corresponding longitudes of Φ were calculated independently using Ptolemy's *Handy Tables* (the actual longitudes underlying the table would not have differed from these by more than a few degrees).

Plotting the arcs against the initial longitudes (Fig. 4160.1) shows that we are not dealing with any of the known System A, A', or B schemes for Jupiter. The synodic arcs obviously do not follow the almost-linear zigzag pattern characteristic of System B. Use of System A would have resulted in two broad plateaus with constant synodic arcs of 30° and 36°, while System A' (or any of its variants) would have narrowed these plateaus and added two intervening ones at 33;45°. Our graph indicates plateaus at all these values; but they are narrow compared

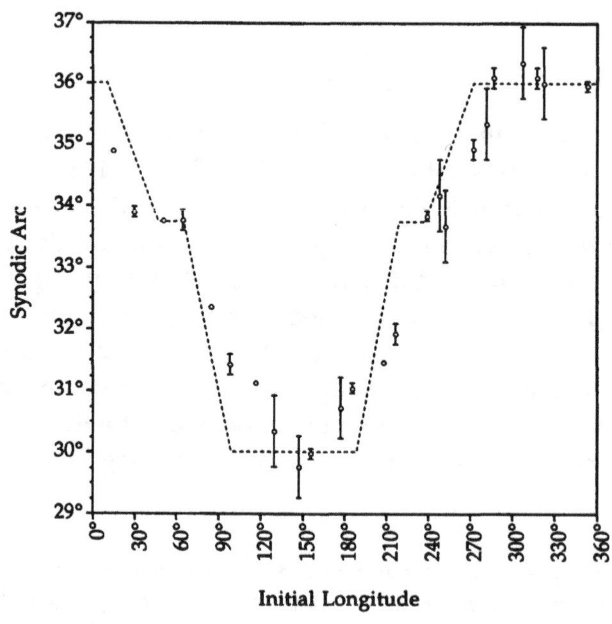

FIG. 4160.1. Synodic arcs from the papyrus compared with System A' (broken line).

Line	*Date*	*Synodic Time*	*Synodic Arc*	*Longitude*
2	XII 19;55	35;37	34;53	14;55
3	I 20;32	34;30	33;46	50;37
4	II 25;2	33;6	32;22	84;39
5	III 28;8	31;52	31;8	116;49
6	V 0;0	30;30±30	29;46±30	147;31
7	VI 0;xx	31;28±30	30;44±30	177;34
8	VII 1;58	32;12	31;28	207;48
9	VIII 4;10	34;35±5	33;51±5	239;8
10	IX 8;4x	35;40±10	34;56±10	272;9
11	X 14;2x	37;5±35	36;21±35	307;4
12	XI 21;xx			
28	V 14;2x	30;43 ±5	29;59 ±5	155;50
29	VI 15;8	31;47 ±5	31;3 ±5	185;52
30	VII 16;5x	32;40 ±10	31;56 ±10	216;19
31	VIII 19;3x	34;55 ±35	34;11 ±35	242;59
32	IX 24;xx	36;5 ±35	35;21 ±35	281;38
33	XI 0;3x	36;50 ±10	36;6 ±10	312;31
34	XII 7;2x	36;42 ±5	35;58 ±5	348;37
35	I 9;7	34;38 ±5	33;54 ±5	24;35
36	II 13;4x	34;30 ±10	33;46 ±10	58;29
37	III 18;1x	32;10 ±10	31;26 ±10	92;15
38	IV 20;2x	31;5 ±35	30;21 ±35	124;11
39	V 21;xx			
42	VIII 27;xx	34;25 ±35	33;41 ±35	247;50
43	X 1;5x	36;50 ±10	36;6 ±10	281;31
44	XI 8;4x	36;45 ±35	36;1 ±35	317;37
45	XII 15;xx			

TABLE 4160.1. Reconstruction of synodic arcs and longitudes.

with the plateaus of System A'. In order to account for the distribution of synodic arcs between 30° and 33;45°, one would have to hypothesize further intermediate steps (say at 31;15°). Hence the scheme underlying the papyrus seems to be a new System A variety model involving (at least) six zones.

Notes:

32: The spacing suggests loss of only one numeral at the end of the line.

47: ι̣[would also be possible.

4160a. Epochs of Jupiter, A.D. 6–13

19 2B.81/C(s), part | 5×5.5 cm | After A.D. 13

Plate IX

A fragment broken on the left, right, and bottom, but with 1.5 cm top margin, from an epoch table ruled in black. The back is blank. Col. i contains the synodic time in days (ΔT) between epochs, with two sexagesimal fractional places; cols. ii–iii hold the regnal year and Egyptian month of the epoch date (T). Regnal years above 40 can only belong to Augustus, and the time intervals of just over one year would be possible for either Jupiter or Saturn. In fact we will presently see that the planet in question was Jupiter. Jupiter's only synodic event in Augustus 41 Thoth (Egyptian) was first visibility (Γ), about Thoth 14, so this is a table for Γ.

The synodic times in the papyrus decrease from line to line by 1;48 days, which is recognizable as the increment/decrement d of the zigzag functions for synodic arc and time in the Babylonian System B scheme for Jupiter (Introduction, §5, Table 2). On the evidence of the fractional places, the zigzag function for synodic time was derived from the function for synodic arc by adding the constant c = 0;44,37 to the maximum and minimum. This value of c implies a sidereal year of approximately 365;15,33,44,... days.

4161. Epochs of Saturn, A.D. 23–92

A 145/6H | 5×12 cm | After A.D. 92

Plate IV

The fragment, which is broken on all sides (except perhaps the bottom) and rather rubbed, comes from an epoch table written neatly in a tabular frame ruled in black. The first three columns of each set (i–iii and vii ff.) give the regnal year, month number, and day of the epochs (*T*); the next two (iv–v) give the longitude (*B*) as zodiacal sign number (counted from Virgo = 1) and integer degrees. Synodic arcs of about 14° and synodic times of about 14 days over one Egyptian year are characteristic of Saturn.

The chronology of the papyrus and the identity of the phenomenon it records can be established from the data in line 15, where the phenomenon occurs on Phaophi 4 of a first regnal year at a longitude about Virgo 5° (extrapolated from earlier lines). In the Alexandrian calendar, Phaophi 4 is equivalent to October 1/2; this possibility can be ruled out because the sun's longitude on October 1 is about Libra 10°, and no phase of Saturn would be expected at an elongation of 30°. Using the Egyptian calendar, one finds that Saturn was about Virgo 7° on Phaophi 4 of year 1 of Gaius (36 September 16); for no other first regnal year does its longitude fall within 10° of Virgo 4°. This result is consistent with the traces of the emperor's name in line 14. Since the sun's longitude on the same day was Virgo 22°, the tabulated phenomenon is Saturn's first appearance (Γ). Assuming that columns vii–viii are part of a continuation of the same table, extrapolation shows that they must refer to years of the reign of Domitian. In lines 1–11 the corresponding events in columns i–v and vii–viii are 60 years or 58 synodic periods apart, so that the complete table likely had sixty lines, some of them occupied by emperors' names. (I have no guess concerning what was written in line 3 of the otherwise vacant column vi.)

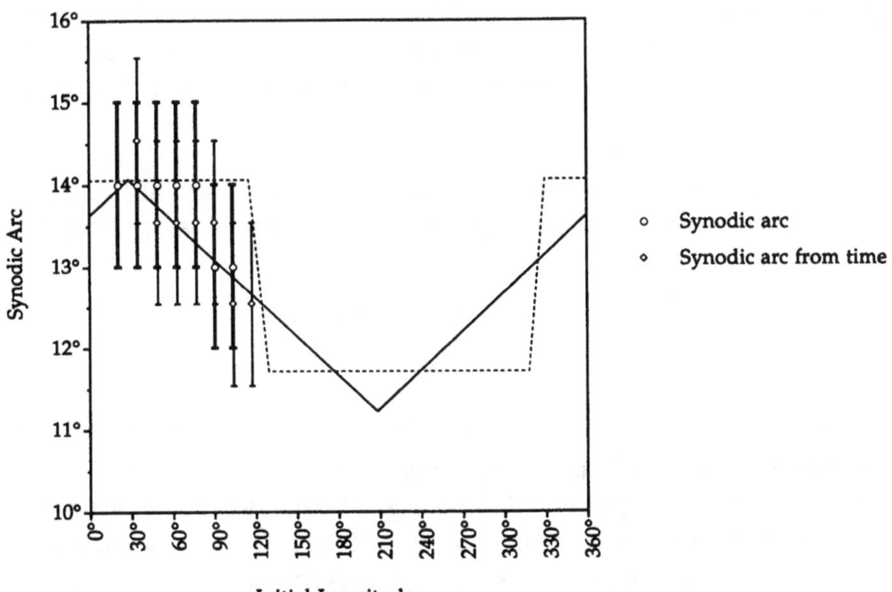

Fig. 4161.1. Synodic arcs from the papyrus compared with System A (broken line) and System B (solid line).

Δλ	λ	*Papyrus*	Δt	t	*Papyrus*
13;2	341;40		13;28	VIII 12;52	
13;14	354;54		13;40	VIII 26;32	
13;26	8;20		13;52	IX 10;24	
13;38	21;58	♈ 21°	14;4	IX 24;28	
13;50	35;48	♉ 5°	14;16	X 8;44	X 8
14;2	49;50	♉ 19°	14;28	X 23;12	X 23
13;55,25	63;45,25	♊ 3°	14;21,25	XI 7;33,25	XI 7
13;43,25	77;28,50	♊ 17°	14;9,25	XI 21;42,50	XI 21
13;31,25	91;0,15	♋ 1°	13;57,25	XII 5;40,15	XI 5
13;19,25	104;19,40	♋ 14°	13;45,25	XII 19;25,40	XII 19
13;7,25	117;27,5	♋ 27°	13;33,25	XIII 2;59,5	XIII 2
12;55,25	130;22,30		13;21,25	I 11;20,30	I 11
12;43,25	143;5,55		13;9,25	I 24;29,55	
12;31,25	155;37,20		12;57,25	II 7;27,20	II 4!
12;19,25	167;56,45		12;45,25	II 20;12,45	
12;7,25	180;4,10		12;33,25	III 2;46,10	III 2

TABLE 4161.1. Recomputation of longitudes and dates of Saturn's Γ by System B.

Synodic arcs can be obtained directly as differences between consecutive longitudes in the table, and also by subtracting a constant c of approximately 0;26 from the differences between consecutive dates. The tolerance in either case is ±1°. In Fig. 4161.1 the synodic arcs are plotted as a function of the initial longitude. The pattern clearly rules out System A as the method of computation, and is strongly suggestive of System B.[1] In Table 4161.1 the data in the first set of columns of the papyrus are compared with a hypothetical sequence of longitudes and dates computed using the System B zigzag function for synodic arcs and $c = 0;26$.

Notes:

15: δ: Probably an error. Together with line 17 it would require implausibly high synodic times of 14 days in Virgo, and the recomputation by System B confirms that line 17 is approximately correct.

4162. TEMPLATE FOR THE SUN

32 4B.1/E(1–2)a 14×20 cm Second or third century

This is a substantial part of a page from a codex, broken on both sides but with the top and bottom margins preserved. The original page was roughly square, with one more set of three columns at the right side of the front and one more at the left side of the back. The fragment is extensively perforated, and the back is rubbed and darkened. Red ink was used for the day numbers and the names of zodiacal

[1] In the cuneiform texts System B is in fact much better attested for Saturn than System A.

signs (indicated by heavy type in the edition), as well as for the tabular ruling.

The template has columns for the number of days since epoch (*T′*) and the sun's longitude (*B*). The longitude is tabulated (to one fractional place) for 365 days of motion, with a final entry for the completion of the progress after 365 $^1/_4$ days, replicating the longitude for day 0. This epoch longitude, Sagittarius 13;30°, is Hipparchus's and Ptolemy's solar perigee (Sagittarius 5;30°) adjusted to the convention that the vernal equinoctial point is at Aries 8° rather than Aries 0°. When complete, the table had thirteen sets of columns, of which the first five were presumably on the back of the lost preceding codex page. Two further sets have been lost at the outer edge of the preserved page, so that we have close to half the template, although many single entries have been lost due to the wretched condition of the fragment. The template was presumably accompanied by a table of epoch dates, i.e. dates of solar perigee.

The theoretical basis and method of computation lying behind the longitudes in **4162** are discussed in Jones [1997b], section 2. Among the templates that have so far come to light, this one is unique in representing the anomalistic motion of a heavenly body by a sequence of longitudes computed trigonometrically from a kinematic model instead of Babylonian-style arithmetical sequences. The assumed model was apparently the simple eccentric model of Hipparchus and Ptolemy, in which the sun travels uniformly in one year around a circle, the centre of which is displaced from the earth by an eccentricity $^1/_{24}$ of its radius in the direction of an apogee 65;30° greater in longitude than the vernal equinoctial point. In principle, therefore, the numbers in the table should be reproducible using Ptolemy's solar tables, so long as one adds 8° to the computed longitudes because of the difference in the assumed location of the equinoctial and solsticial points. In fact there are small systematic discrepancies which probably reflect a different and less accurate method of obtaining the longitudes from the assumed model rather than any difference in the model itself.

4163. Template for the Sun and procedure text

85/2(d) 5×13 cm After A.D. 259

Plate IV

A shattered part of a leaf from a codex, with 1cm top and bottom margins (the top margin on the front containing the first line) and, on the front only, a 0.5cm right margin. The preserved part was the outer edge of the page, since there is no gap between the preserved sequence of longitudes on front and back. Red rulings delimit the top and bottom margins and, on the front, the right margin. On the back the end of the table does not take up the full height, and is bordered to the right by a double ruling in black, and below by a single black ruling which was originally prolonged to the left of the table. The continuation of this ruling appears to have been rubbed out. The procedure text on the back is written surrounding the table by the same hand in more closely spaced lines, beginning just below the upper ruling to the right of the table and continuing (lines 19ff) below the table itself and down to the bottom margin.

Like **4162**, this fragment gives the last part of a template containing daily solar longitudes through the course of one year. Here, however, the longitude corresponding to the beginning (day 0) and end ($^1/_4$ day past day 365) of the year is Cancer 0°, that is, the summer solstitial point according to the convention of Hipparchus and Ptolemy. It is worth remarking that the only known solar epoch table, **4148**, is a list of dates of summer solstices, although it does not necessarily follow that the specific rule for calculating the epoch dates in **4148** was intended to be used with the template of **4163**.

For a detailed discussion of the method of computation of the longitudes in **4163**, see Jones [1997b], section 1. They seem to have been obtained by assuming a constant rate of progress of exactly 0;57° per day in Taurus and 0;56,50° per day in Gemini. It is not clear how these numbers were derived, although 0;56,50° is close to the mean daily motion of the sun in Gemini according to the Hipparchus–Ptolemy eccentre model.

The text written in the vacant space around the end of the table is unfortunately quite broken, and in many lines little connected sense can be recovered. It appears to set out a rough procedure for estimating the dates of syzygies in a given Alexandrian year, with an example that seems to be calculated for the year 7 of Gallienus (A.D. 259/260). The rule for conjunctions, as I reconstruct it, is based on the following assumptions: (a) that 19 Alexandrian years correspond to a whole number of synodic months; (b) that a conjunction occurred on the day preceding the first day of the year Augustus 10, i.e. August 28, 21 B.C. (it actually took place about 8:30 A.M. on August 29); (c) that the first conjunction of each year occurs either 19 calendar days later or 11 calendar days earlier than in the preceding year; and (d) that the interval between consecutive conjunctions is 29 days plus 12 $^3/_4$ hours (29;31,52,30 days). Obviously no account is taken of solar and lunar anomaly, which can affect the date of a syzygy by several hours.

In the example (lines 16–22), the given year Valerianus (i.e. Gallienus) 7 is represented as 288 completed years since the beginning of Augustus' reign (i.e. the year is equivalent to Augustus 289). 10 years are added to this number to obtain a number of years since an epoch year in which Thoth 1 was supposed to be the day following conjunction. The residue of this total, modulo 19, yields the number of years since the *most recent* such epoch year. One then multiplies this number by 19 days, takes the residue modulo 30, and treats the result as the day in Thoth of the first conjunction of the year. In the example, the conjunction would have been found as Gallienus 7, Thoth 7, i.e. 259 September 5. The true conjunction actually occurred about 6 P.M. on the preceding day.

Notes:

Front:

iii 8: the longitude for day 334 was inadvertently omitted.

Back (table):

ii 13–14: one expects simply λ ♈. Why this has been repeated twice is not clear.

Back (text):

12: Below this line is a horizontal stroke, perhaps marking the beginning of a new section.

4164. Template for the Moon

64 6B.60/G(1)a) 8.5×13.5 cm After A.D. 254

This Standard Scheme template occupies the the 'back' of the Standard Scheme epoch table **4151**. The top parts of two sets of columns are partially preserved. Each set comprised columns for the increment in longitude leading up to the day's position (ΔB = vi–ix and the restored –IV to –I) and for the total progress in longitude since epoch (B' = i–iv) are given. The number of days since epoch (T'), serving as an index number, must have occupied a column between ΔB and B' (restored [0] and [X]). There were no columns for the argument of latitude.

The template is laid out in a tabular frame allowing for three groups of 14 lines in each column. It seems that each 14-line group was intended to hold a single branch of the Standard Scheme zigzag function, either increasing or decreasing. Some branches would of course require only 13 lines. The extant fragment has parts of the first and second sets of columns (cf. Appendix H, Table 2). A template covering the maximum 303 day interval between epochs would have taken up eight sets.

The previously known Standard Scheme template *P.S.I.* XV 1493 gives B' truncated to only one fractional place (and no column for ΔB), so that it was formerly not possible to establish the exact parameters of the Standard Scheme zigzag function for ΔB (Jones [1983] 17–23). By providing values of both ΔB and B' to the full three fractional places, **4164** has been the key to reconstructing the exact function (Jones [1997a], section 1).

4164a. Template for the Moon

15 1B.201/F(b),part 7.5×23 cm Second or third century

The only Standard Scheme template so far known to preserve entries for the days beyond day 248 after epoch, written across the fibres on the back of a coarse, tattered, and abraded piece of papyrus broken on the left and right sides (on the front there are only one or two specks of ink and, at the very bottom, slight traces of writing). The table was laid out so that each set of columns represented one anomalistic month, i.e. either 27 or 28 days, with a vacant line in the middle separating the accelerating from the decelerating half of the month. The extant columns, repectively for daily progress in longitude, index number (at five line intervals), and total progress in longitude since epoch, belong to the last anomalistic month, i.e. days 277 through 303. The expected line for day 277 is not visible at the top of the fragment, but may have been rubbed beyond recognition where there is now 1 cm of apparently vacant margin; at the bottom, a margin of 1.5 cm is present.

The interest of this papyrus is that it shows us the derivation of the part of the Standard Scheme template corresponding to the last 55 days preceding the beginning of a new epoch cycle. The template for days 1 through 248 after epoch was, as we know, computed as the running totals of the first 248 values of a linear zigzag function having the whole number period 3031 days. This period is the same as that of the Standard Scheme epoch cycles, i.e. every group of eleven consecutive epochs amounts to eleven intervals of 248 days plus one of 303 days for a total of 3031 days. An ideal version of the Standard Scheme would have used only epochs at these 3031-day intervals together with an enormous template of 3031 lines, containing the full period of the underlying zigzag function. The economy of the attested scheme is to use the first 248 lines of this template repeatedly in place of lines 249–496, 497–744, etc. which have a similar, though not quite identical, sequence of daily increments. The resulting discrepancies in the running totals attain a maximum of a few minutes of longitude mid-way through each anomalistic month, and vanish at the 248-day intervals, so that the epoch positions recorded for the intermediate epochs in each cycle are exactly the positions that would result from use of a complete 3031-line template.

The template must have had 303 lines if it was to be usable for the longer interval between the eleventh epoch of a cycle and the first epoch of the next cycle, and the obvious method of extending it would be to continue the zigzag function for daily motion until the 303rd value in the 3031-line period. This procedure would, however, bring with it the slight infelicity that the total obtained for day 303 would be 32;32,27,51°, whereas the shortfall between eleven intervals of 248 days' motion and one full cycle of 3031 days' motion according to the Standard Scheme should be 32;33,44,51°. The discrepancy, although astronomically insignificant, is just the kind of numerical misfit that the inventor of the Standard Scheme seems to have tried to avoid. His expedient, as can be seen from the present fragment, was to use, not values 249–303 of the zigzag function, but the final 55 values (i.e. 2977–3031) which would have been the daily motions if the ideal 3031-line template were employed. The result is that the precise desired total of 32;33,44,51° is obtained.

By chance we possess an ephemeris, **4177a**, in which the lunar longitudes were computed using the sequence of template entries for days 235 through 255. This is the more interesting because the date treated as day 249 after epoch, A.D. 245 (= Era Augustus 27), Alexandrian Pachon 24, was in fact a Standard Scheme epoch, and hence the subsequent lines in the ephemeris really ought to have been obtained by adding the first lines of the template to the position recorded for this epoch in the epoch table. The computer of the ephemeris presumably found it more convenient to continue using the old epoch at least until the end of the month.

4165. Template for Mars

68 6B.20/D(4)a, part	aggregate approx. 22×9.5 (fr. 1)	Late third/early fourth century
+ 68 6B.25/H(3)b, part	3×5.5 cm (fr. 2)	Plate VI
+ 65 6B.34/G(1–2)f		

The table is in a red tabular framework, on the back of a document in a third-century hand. Fr. 1 comprises three noncontiguous pieces, the largest of which (6.5×6.5 cm) preserves 1 cm top margin, whereas the others (1.5×4 and 1.5×2.5 cm) are broken on all sides. Fr. 3, from elsewhere in the manuscript, has 1.5 cm top margin. The full height, allowing for 54 rows of numerals, was about 30 cm.

Columns ii and v of fr. 1 contain a sequence of numbers increasing uniformly by 0;0,28,0 per line, so that, if i is the index number tabulated in columns i and iv, then the corresponding value in columns ii and v is

$$d = 0;0,4,30 + 0;0,28,0\ i$$

Column iii is the summation of column ii, starting with an initial value of 0;0,0,0 in the untabulated line 0, so that

$$s = 0;0,4,30\ i\ +\ 0;0,14,0\ i\ (i + 1)$$

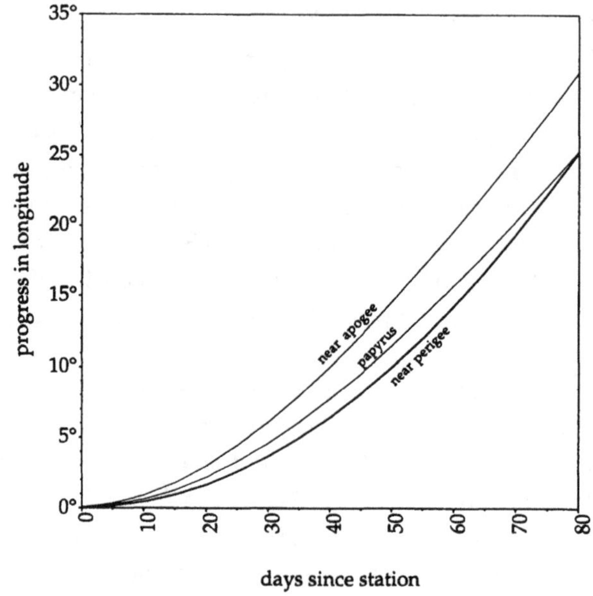

FIG. 4165.1. Longitudinal progress of Mars after second station.

This sequence was sustained to at least line 62, where the running total in the lost column to the right of column v would have been 15;16,3,0. From the preserved parts of fr. 1 it cannot be seen whether columns iv and v continued below line 9.

Columns i and iv are in fact days elapsed since epoch (T'); columns ii and v, the daily progress of the planet (ΔB); and column iii, the accumulated progress in longitude since epoch (B'). Since ΔB is increasing from zero and B' is also increasing, the epoch in question is second station (Ψ). The planet advances approximately 15° in the 62 days following the station, and its velocity after 62 days is about 0;30°/ᵈ. This is too fast for Jupiter or Saturn, while the rate of change is too gradual for Mercury or Venus. Thus Mars is the only planet to which our template could belong.

Fig. 4165.1 compares the motion set out in fr. 1 with typical intervals of Mars's actual motion following second station where the station occurred near perigee (towards the end of Capricorn) and near apogee (towards the end of Cancer). The template gives a good approximation of motion near apogee. The further circumstance that the sign Leo is named at the top of the table confirms that this template was designed expressly to describe Mars's motion following second station only in a zone of the ecliptic near apogee. This zone was probably the familiar Babylonian System A zone of 60° coinciding with the signs Cancer and Leo.

Fr. 2 comes from a different part of the template, although the heading shows that we are again dealing with motion in the same zodiacal region Cancer and (presumably) Leo. The index numbers start again from 1 (col. i), but now the daily progress (col. ii) is diminishing from an initial value of about 0;29°. This was probably the decelerating counterpart of the section of Mars' synodic cycle tabulated in fr. 1, i.e. the days leading up to first station (Φ). Assuming that the sequence was approximately the mirror image of that in fr. 1, it would have had to begin just over 60 days before Φ; and conversely we can infer that the accelerating sequence in fr. 1 was of about this duration, in other words scarcely extending beyond the last preserved line. For the remainder of the intervals of direct motion before last visibility (Ω) and after first visibility (Γ) a constant velocity of about 0;30° per day would have been acceptable in this slow part of the ecliptic.

4165 casts light on how templates could be applied to the motion of a planet such as Mars, with its widely fluctuating patterns of synodic motion. Instead of having a single template describing a 'mean' synodic cycle, one had several templates for different zones of the ecliptic, apparently associated with the System A zones. Moreover, separate templates were made for the parts of the synodic cycle in which velocity changes most rapidly, in the present instance before first station and following second station; this would allow greater freedom in assigning longitudes and dates to the various phases of the planet.

4165a. Template for Jupiter

115/81 8×9 cm (fr. 1) 2×7 cm (fr. 2) debris Third century

The table is on the back of an astrological (?) text (quoting *Iliad* IV 75) written in biblical uncials (first half of the third century). In the larger fragment, the text on the front is broken on the top, bottom, and left side; on the right there is a 1.5 cm intercolumnium followed by traces of initial letters of a second column. The table on the back seems to be broken on all sides. The smaller fragment has 4.5 cm top margin on the front, 1 cm top margin on the back, the other sides being broken.

The template was laid out in sets of three columns: a column for daily progress in longitude, an index column counting days since epoch, and a column for total progress in longitude since epoch, i.e. the running totals of the first column. In fr. 1, col. i is an index column, tabulated as usual at intervals of five lines. Col. ii is the running total expressed as sexagesimal numbers with three fractional places, increasing by differences that are themselves decreasing by a constant 0;0,8,48° per line. A velocity of 0;11°/d at 50 days past epoch and decelerating would be appropriate for the motion of Jupiter after its first visibility (Γ). Col. iii is the daily progress belonging to the next set of columns, still decreasing by constant 0;0,8,48°. Since the decrease from the difference of 0;11,0,0° in col. ii lines 5–6 to the entry 0;5,43,12° in col. iii line 5 is 0;5,16,48° = 36×0;0,8,48°, the number of lines in a column was approximately 36.

If the daily progress had been decreasing by 0;0,8,48° per line since the beginning of the table, the initial velocity associated with longitude 0° on day 0 would have been 0;18,20°, and the total progress on day l would be:

$$s = 0;18,20^\circ\, l \;-\; 0;0,4,24^\circ\, l\,(l-1)$$

This formula does in fact reproduce the totals in fr. 1 col. ii; and it allows us to identify fr. 2 as the top lines (roughly days 73 and following) of the total progress column associated with fr. 1 col. iii, such that the last line of fr. 2 may have lined up with line 1 of fr. 1. The daily progress diminishes to zero, i.e. first station (Φ) on day 125, with total progress 19;15°. This number is rather large to represent Jupiter's mean progress from Γ to Φ, but compares well with the parameter 19;30° used in Babylonian texts for Jupiter's fast arc according to the velocity scheme associated with System A' (cf. Introduction, §5, table 3).

The remainder of the template cannot be reconstructed with certainty, but we may make plausible guesses. For the fast zone of the System A or A' scheme for Jupiter, we expect a total progress of 36° in approximately 400 days. We have accounted already for 19;15° in 125 days from Γ to Φ, and can expect the section to be mirrored from Ψ to Ω. The interval of invisibility, Ω to Γ, should be about 30 days, and the System A' velocity scheme prescribes a constant rate of progress of 0;15°/ᵈ, for a total of 7;30°. Thus we are left with an estimated retrogradation (Φ to Ψ) of 10° in 120 days, in excellent agreement with the velocity scheme's 10° in 4 months. This interval would likely have been effected by making Jupiter's retrograde velocity increase linearly from zero at Φ to a maximum of –0;10°/ᵈ at Θ, and then decrease again to zero at Ψ.

4166. Template for Saturn with Jottings

100/52(c) 10×20 cm, and debris Third century

The table is on the back of a fragmentary register in a third-century hand. The section of it to which columns i and ii belong ends with line 21. Since the index number (column iii) of the next section has reached 15 in line 9, we can deduce the loss of six lines, i.e. at least 5 cm, from the top. In the wide space between the two preserved columns of the register on the front are inserted a few lines in the same hand as the table, apparently setting out the defining parameters from which part of the table was computed.

We may begin with the first section of the table. Column ii gives running totals of the numbers in column i, which themselves increase by a constant increment of

0;0,4,0,0. If we call the number in line l of column i d, and the number in the same line of column ii s, then:

$$d = 0;6,19,27,16 + 0;0,4,0,0\ l$$

$$s = 5;1,47,26,44\ +\ 0;6,19,27,16\ l\ +\ 0;0,2,0,0\ l\ (l + 1)$$

Hence for $l = -89$, $s = 0;0,0,0,0$ exactly, and $d = 0;0,23,27,16$. Interpreting d as daily progress of some planet (ΔB) and s as its accumulated progress in longitude since epoch (B'), we can say that the planet is assumed to progress almost exactly 7;30° in 110 days following second station. These round figures were undoubtedly the theoretical parameters that the template function was designed to reproduce. The non-zero but tiny initial velocity for $l = -89$ was necessary to allow the choice of a single-digit increment in velocity for the sake of easier calculation of the table. The shortfall of 0;0,0,0,20 in the final total is the accumulated error resulting from limiting all arithmetic to four fractional places.

The planet in question is Saturn, and the 110 days are evidently intended to represent the interval from Ψ to Ω. An advance of 7;30° in the 110 days following second station is a little too small for Saturn's mean synodic cycle, so that the template would be more appropriate as a description of Saturn's motion specifically near its apogee. The longitudinal interval was originally a Babylonian parameter. According to a Babylonian velocity scheme for Saturn found in certain planetary tables based on the two-zone System A model the progress from Ψ to Ω in the slow zone is 7;33,7,30°, and the corresponding progress from Γ to Φ is 7;30° exactly.[1]

The association of **4166** with System A is confirmed by the second table (columns iii–iv), which consists of multiples of 0;6,20 up to 30 × 0;6,40 = 3;20,0. For the same Babylonian planetary tables prescribe precisely 3;20° for Saturn's motion in the slow zone during its interval of invisibility, which lasts about one month. The two tables are therefore partial templates covering Saturn's daily motion in the slow zone from second station to last visibility and from last visibility to first visibility. Presumably **4166** once had templates for motion from first visibility to first station, mirroring the 7;30° motion from second station to last visibility, and for the retrogradation. The retrogradation would have had to be about −6;40° (the parameter in the Babylonian scheme) in about 126 days in order to obtain a total synodic arc of about 11;40° for Saturn's slow zone. There must of course also have existed templates for the fast zone.

Below the second table are the beginnings of six lines of writing, some of which seem to be parts of the names of zodiacal signs. I know of no relationship between this text and the tables, and I have no conjecture about its meaning. On

[1] See Introduction, §5, Table 3, and Neugebauer [1975] 439–440.

the other hand the text on the 'front' of the papyrus is clearly manipulating the numerical parameters of the first table, and may in fact be part of a procedure text setting out the table's derivation. Certain of the numbers are at once recognizable: 0;0,4 (line 6 and perhaps line 4) is the constant increment of ΔB; 0;7,43,27,16 (line 9) is the final value of ΔB for day 110 past station; 0;0,27,27,16 (line 8) is the value of both ΔB and B' for day 1 past station. The double of this number appears in line 7; and this in turn was apparently obtained by adding 0;0,4 to the number in line 5. However, without more of the context it is difficult to explain the reason for these particular arithmetical operations.

Notes:

Front:

4: I do not know what the symbol at the beginning of this line represents.

6: The traces do not appear to fit the expected reading ⊽ ⊽ κζ̣ κζ ι̣ς̣ (0;0,27,27,16).

4167. Ptolemy, *Handy Tables*

65 6B.30/M(1)a, part	22×22 cm (fr. 1)	Third to fourth century
+ 66 6B.27/G(5–6)b, part	5.5×5 cm (fr. 2a) 3×2 cm (fr. 2b)	Plate V
+ 67 6B.16/K(2–4)b, part	15×27 cm (fr. 3) 9×11 cm (fr. 4)	
+ 68 6B.20/L(8)a, part	1.5×2.5 cm (fr. 5) 2×2.5 cm (fr. 6a)	
+ 68 6B.25/H(3)b, part	2×1.5 cm (fr. 6b) 5×1 cm (fr. 7)	
+ 68 6B.25/H(4)b, part	4.5×3 cm (fr. 8) 2×3 cm (fr. 9)	

Among the numerous bits of astronomical tables inventoried under the folders that we here call 'Group A', the manuscript represented by the most fragments is the present copy of Ptolemy's *Handy Tables*. Rather than the usual codex format, it was a roll—indeed several rolls in all likelihood, since the 30 metres or so required for the complete *Handy Tables* would be utterly unwieldy as a single roll. The tables are laid out neatly in red tabular frames. The presence of the tables of planetary visibility and greatest elongations is noteworthy, since there is little other evidence that these parts of Ptolemy's collection obtained much use.

For this and the subsequent papyri of the *Handy Tables* the main burden of the commentary will be to identify the specific tables from which each fragment comes. In the absence of a critical edition, the most precise means of reference is by the folio numbers of the ninth century codex *Vat. gr.* 1291 (V), which are also cited by Neugebauer [1975] 969–1028, Stahlman [1960], and Tihon [1992]. Attention will be drawn to major deviations from the format and contents of V and the other Byzantine codices of the *Handy Tables*, but not to minor textual and numerical variants.

Fragment 1 belongs to the tables of oblique ascensions for the sixth *klima* (latitude of the middle of the Pontus). As in the medieval copies, the table gives, for each degree of each zodiacal sign, the simultaneously rising degree of the equator counted from the vernal equinoctial point (ἀναφοραί), and the number of equatorial degrees that rise in one equinoctial hour when the sun is at that longitude (ὡριαῖοι χρόνοι). Both quantities are given to one fractional sexagesimal place. In the codices each page contains an index column (running from 1° to 30°) followed by the columns for three consecutive zodiacal signs, so that in *V* the ascensions for Aries through Virgo take up two pages, ff. 34–34ᵛ. There is no vestige of such page breaks in **4167**, so that one index column, not extant, seems to have served for the entire table for each *klima*.

Fragment 2a is part of the oblique ascensions for the fourth *klima* (latitude of Rhodes), zodiacal sign Aquarius (*V* f. 31ᵛ), lines corresponding to 17° through 25°. 2b belongs to the corresponding table for the seventh *klima* (latitude of Borysthenes), zodiacal sign Leo (*V* f. 36ᵛ), lines corresponding to 11° through 14°.

Fragment 3 belongs to the tables of planetary latitudes for Venus (cols. i–ii) and Mercury (iii–xi). Its left edge has parts of the last two columns of numbers from the table for Venus (*V* f. 81ᵛ). Then, after the intercolumniar space, come columns corresponding to the leftmost eight columns of the table for Mercury (*V* f. 82). However, the quantity represented as a unit and sexagesimal fraction in columns v–vi of the papyrus fragment is expressed in *V* as a single-place sexagesimal fraction only, and is allotted just one column. In fact the numbers in this column reach a maximum of exactly 1 (i.e. 60 minutes) in each of the latitudinal tables.

Fragments 4–6 come from the tables for first and last visibilities of Venus and Mercury. Fragment 4 has parts of the table for Venus's first evening visibility (Ξ), which in *V* is the lower table on f. 87 (there mislabelled 'morning setting'). The small scrap fragment 5 apparently belongs to the nearby table for the same planet's first morning visibility (Γ, *V* f. 87ᵛ, upper table). Fragment 6a is from the table for Mercury's Σ (*V* f. 88, upper table, mislabelled as 'evening rising'), and 6b is from the table for the same planet's Ξ (*V* f. 88, lower table, mislabelled as 'morning setting').

Fragment 7 is a bit from the table for Mercury's greatest elongations as evening star (*V* f. 90, upper right hand table). Fragment 8 is from the top of two consecutive 'pages' from one of the anomaly tables, either for the sun and moon (*V* ff. 43–43ᵛ) or for one of the planets (*V* ff. 65–65ᵛ, 68–68ᵛ, 71–71ᵛ, 74–74ᵛ, 77–77ᵛ). Fragment 9 belongs to the table of stations of Jupiter (*V* f. 83ᵛ).

Notes:

Fragment 1:

ix–x 2: χρ(όνοι): ⁎.

Fragment 8:

i 1: διαφορ(ά): apparently the final letter was simply omitted.

4168. Ptolemy, *Handy Tables*

48 5B.107/D(1) | 6×27 cm (fr. 1) 8.5×24 cm (fr. 2) 3×2.5 cm (fr. 3) 2×4 cm (fr. 4) 1.5×4 cm (fr. 5) 7.5×3.5 cm (fr. 6) 2.5×2.5 cm (fr. 7) debris | Fourth century

These fragments derive from several pages of a codex of the *Handy Tables*. The tables are ruled in red. Fragment 1 preserves the full height of the original pages, with 2 cm margin at the top and 3 cm at the bottom.

The front side of fragment 1 is part of the table for solar declination and lunar latitude (*V* f. 44). In the medieval copies this table is ruled in six pairs of columns, with intercolumnial spaces between each pair and the next. Thus the first and second columns and the seventh and eighth columns contain the arguments ('common numbers'); the third and fourth and the ninth and tenth contain the declination in degrees and minutes; and the fifth and sixth and the eleventh and twelfth contain the lunar latitude in degrees and minutes. In the papyrus, columns i–vi correspond to the fourth through ninth of the complete table, so that the fragment evidently represents the middle of the original page. However, there is only single ruling between the declinations and the lunar latitudes (cols. i and ii–iii), no ruling between the degrees and minutes of the latitudes, and again only single ruling separating the latitudes from the following arguments (iv–v) and the arguments from the declinations (vi). The column headings are abbreviated and use the conventional symbols for the sun and moon. Below the table are the beginnings of two lines of a marginal note of unknown import.

The back of this fragment is part of the first page of the tables of parallax for the first *klima*, latitude of Meroe (*V* ff. 50–50ᵛ). Hence in this ancient copy there were no tables between the table for declination and lunar latitude and the parallax tables, a position that in medieval copies is occupied by a miscellany of tables including the eclipse tables, which were certainly an original part of Ptolemy's compilation. As in the medieval copies, the parallax tables were laid out with six zodiacal signs on a page, three above three. The medieval arrangement has the six consecutive signs from Cancer through Sagittarius on the first of the pair of pages belonging to each *klima*, and the rest of the signs on the second page, so that the middle strip of the first page would contain the parallaxes for Leo above those for Scorpio. In the papyrus manuscript the sequence of signs evidently ran continuously across the two facing pages, and hence we have the table for Aquarius below that for Leo.

Fragment 2 is the part nearest the binding of a page from the table for Saturn's anomaly (*V* ff. 64–64ᵛ). The arrangement is essentially the same as in the medieval copies, including the numbering of the columns. The ruling is simpler and less clear in the papyrus, in the same way as for the front of fragment 1: where the medieval copies put an intercolumniar space between columns of distinct meaning, the papyrus has only a single ruling, and where they separate units and fractional places by a single ruling, the papyrus has no ruling at all. The column headings are again abbreviated.

Fragments 3 and 4 are disconnected bits of the table for Mercury's anomaly (*V* ff. 75–75ᵛ). Fragment 3 is from near the centre of the page, with parts of the lines corresponding to arguments 12 through 16 (front) and 41 through 44 (back). Fragment 4 preserves the page's outer edge near the bottom corner, corresponding to arguments 26 through 30 (front) and 53 through 59 (back). The side with vertical fibres of fragment 5 is from the right edge of the table of mean motions of sun and moon for single Egyptian years (*V* f. 39), corresponding either to years 14 through 18 or to 20 through 24. The other side (not transcribed) has ruling from what appears to be the bottom left corner of a table, perhaps the table of solar and lunar mean motions for Egyptian months (*V* f. 39ᵛ). The only writing preserved on this side is a speck from the bottom left of a single letter, which might be λ, μ, or ν.

The tables on the two sides of fragment 6 are not part of any of the *Handy Tables* as we know them. The numerals on the side with horizontal fibres might perhaps be from a table for calendrical conversions, with parallel columns correlating the days of different local calendars. On the other side is part of an enigmatic geographical table. This may have had a similar function to the Table of Noteworthy Cities in the *Handy Tables*, which is a catalogue of cities with their positions in longitude and latitude extracted from Ptolemy's *Geography* and organized province by province following a general sequence from the northwest to the southeast of the known world. Here we seem to have certain cities and provinces of Asia Minor (Apamea in Phrygia, Cilicia) and Syria (Damascus); but Gallia is unexpected in this context, and may be a copyist's misunderstanding (of Galatia?). I cannot explain the numerals—if that is what they are—in line 4, the small N's along the left edge of the fragment, or for that matter the arrangement of the localities.

Fragment 7 has traces of two lines of unidentified text on one side, perhaps from the bottom margin of a page.

Notes:

Fragment 1, front:

iv 3–24: From the top of this column the initial ρ of each numeral is followed by a space, as if it were a separate numeral from the following tens and units. The space gradually diminishes until, from line 25, each number is clearly written as a single numeral.

v 3–12: As in column iv, the hundreds place of the numerals in the top lines of this column are separated from the tens and units, with the space gradually diminishing. In lines 29–30 there is no superfluous space between the digits.

Fragment 1, back:

i 7: ☉: scil. ἡλίου (ἀνατολή). This use of the solar symbol to represent the times of sunrise and sunset seems to be standard in the medieval copies of the parallax table.

i 21: ☉: scil. ἡλίου (δύcιc).

4169. PTOLEMY, *HANDY TABLES*

66 6B.27/G(5–6)b, part 6×8 cm Third century

A fragment from a neatly executed codex of the *Handy Tables*, with red ruling. The piece is part of the bottom (including 3 cm margin) of a page of the table of anomaly for Mercury (*V* ff. 75v–76). There are small but systematic deviations from the readings of *V* in column iv of the front and column v of the back.

The same inventory number includes fragment 2 of **4167** as well as three small tabular scraps (not transcribed). **4169** therefore belongs to Group A, although no further fragments of the same manuscript have been identified under the other related inventory numbers.

4170. PTOLEMY, *HANDY TABLES*

93/Dec. 8/B1 13.5×18.5 cm Fifth century?

The fact that both sides of this fragment contain tables suggests that it comes from a codex, although the preserved margins are exceptionally large (5 cm right margin on the 'front', and 8.5 cm left margin on the 'back'). Moreover the tables are not such as one would expect to find consecutively in a complete manuscript of the *Handy Tables*. Perhaps we have here the end of a papyrus roll used on both sides. The tables are ruled in red.

The side here designated as the front contains the upper right portion of the table of mean motions for Mars, Venus, and Mercury at 25-year intervals (*V* f. 57v).

In the medieval codices this table has been prolonged far beyond Ptolemy's time, and takes up more than one page at thirty epochs to the page. The last line of the first page in this layout has the entries for the year Era Philip 726 (A.D. 402/403). In the papyrus there are traces of one additional line for the year Era Philip 751 (A.D. 427/428). It is not likely that this line would have been added unless the owner of the table anticipated a need for computing positions later than that year.

On the back little is preserved beyond a pair of columns for arguments. Since these are at intervals of one unit, the table was presumably the last page of one of the anomaly tables, either for the sun and moon or for the planets; the uncertain traces of column iii appear to fit the table for Jupiter (*V* f. 68[v]).

4171. Ptolemy, *Handy Tables*

68 6B.22/P(1–2)a 3.5×4.5 cm

A small scrap from the top of one of the tables for oblique ascensions, from a codex, ruled in red. The side designated as the back must be from the table for the third *klima* (latitude of Lower Egypt), since only in that table is one equinoctial hour equivalent to 17 and a fraction degrees of ascension for solar longitude Cancer 1°. The ruling implies that this is the top right corner of the page. In the medieval copies, the ascension tables are laid out with three signs to a page, and Cancer is always the leftmost of a set of three signs; in the present manuscript it may have been the rightmost of two signs, on the back of a page containing the tables for just Aries and Taurus.

Notes:

Front:

4: ☉: scil. ἡλίου.

Back:

2–3: l. ὡριαῖοι χρόνοι, 'Time-units for one hour'.

4172. Handy Tables

12 1B.142/D(b) 4×10.5 cm (fr. 1) Late fourth to early fifth century
+ 12 1B.142/E(b), part 5.5×10.5 cm (fr. 2)

Fragments of a table ruled in black, written on the back (across the fibres) of a document in a large late cursive hand. Fr. 1 is from the bottom (with 5.5 cm bottom margin), fr. 2 from the top (with 3 cm top margin including heading). It seems likely that the original manuscript had two sets of tables, one (including fr. 2) above the other (including fr. 1), but there are no clues to the correct relative placement of the fragments.

The identity of the tables is obscure. A table or set of tables was associated with a number of hours; each row is associated with a zodiacal sign, in order starting from Cancer. The entries in the table are numbers (units and one fractional sexagesimal place) that appear to increase from a minimum less than ten for Cancer to a maximum about twice as large for Capricorn and back again. The numbers match no table known to me. The format and the general trend of the numbers are reminiscent of Ptolemy's tables for planetary visibility in the *Handy Tables* (*V* ff. 85^v–88^v). These tables present the minimum elongation of a specific planet from the sun necessary for the planet to be visible, either before or after conjunction, calculated as a function of the planet's longitude and tabulated for 0° in each zodiacal sign and for each of the seven standard terrestrial latitudes, identified in part by the length of longest daylight in equinoctial hours. But none of the standard latitudes is far enough north to have a longest day of 20 hours or more, and Ptolemy's tables list the signs in order from Aries, not Cancer.

4173. Handy Tables

72/94(a) 12×8 cm Fourth century

Plate VI

The fragment is from the bottom of a page of an elegantly written codex (2 cm margin on the front, 4 cm on the back), with ruling in red. Both sides belong to a set of tables for planetary mean motions, computed from Ptolemy's parameters but presented in a different format from either the *Almagest* or the *Handy Tables*. The correspondences may be summarized as follows:

Front, column i: seconds place of mean motion of Regulus for 30-day intervals (Egyptian months). Indistinguishable from *Handy Tables* (*V* f. 60). There is no such table in the *Almagest*.

Front, columns ii–iv: Saturn, mean motion for 30-day intervals of the centre of the epicycle, to two fractional places. This is a trunction of the table in *Almagest* IX 4, which gives the mean motion to six fractional places. In the *Handy Tables* (*V* f. 60) the values are rounded (not truncated) to just one fractional place.

Front, columns v–vii: Jupiter, mean motion for 30-day intervals of the centre of the epicycle. Columns viii-x: Mars, mean motion for 30-day intervals of the centre of the epicycle. Columns xi–xiii: Venus, mean motion for 30-day intervals of the planet on its epicycle. Columns xiv–xv: Mercury, mean motion for 30-day intervals of the planet on its epicycle. These are again truncations to two fractional places of the values tabulated in *Almagest* IX 4, whereas the *Handy Tables* counterparts (*V* ff. 60–60^{v}) are always rounded to the first fractional place.

Back, column i: index number for day of Egyptian month.

Back, columns ii–iv: daily mean motion of Regulus, truncated to two fractional places. In the *Handy Tables* (*V* f. 61) the values are rounded to the second fractional place, so that the values there are sometimes greater than in the papyrus by one second. No counterpart in the *Almagest*.

Back, columns v–vii: daily mean motion of Saturn's epicycle. Columns viii–x: daily mean motion of Jupiter's epicycle. Columns xi–xiii: daily mean motion of Mars' epicycle. Columns xiv–xv: daily mean motion of Venus on its epicycle. These are truncations to two fractional places of the values in *Almagest* IX 4, while the *Handy Tables* again round to the first place. The format, however, is that of the *Handy Tables*, that is, the index numbers are actual calendar days rather than the elapsed days of the *Almagest* mean motion tables.

The set of tables to which **4173** belonged obviously cannot have been derived from the *Handy Tables*, since the mean motions are presented to a greater precision than in that work. Instead they must have been computed directly from the *Almagest*. Nonetheless, the tables share certain features with the *Handy Tables*: the inclusion of Regulus, and the measurement of time in true calendrical units rather than the more awkward intervals of time elapsed since epoch employed in the *Almagest*.

In another respect the papyrus diverges from both Ptolemy's works: for each planet it gives only a single mean motion, omitting the mean anomaly (i.e. the motion of the planet on its epicycle) for the superior planets and the mean longitude (i.e. the mean motion of the epicycle itself) for the inferior planets. In the latter case the economy is an obvious one, since Venus and Mercury share the same mean motion in longitude as the sun, and this would have been tabulated elsewhere. The mean motion in anomaly of the remaining planets can always be obtained by subtracting their mean motion in longitude from that of the sun. For the sake of consistency with respect to Ptolemy's planetary models one might just as well have tabulated only the mean anomaly for all five planets. The arrangement followed here would indeed make sense if one substituted for Ptolemy's epicyclic models for the superior planets the equivalent eccentre models (a possibility that Ptolemy himself alludes to in *Almagest* XII 1). Precisely the same expedient of independently computing only the mean longitude for the superior planets and only the mean anomaly for the inferior ones is found regularly in medieval Indian astronomy.

4173a. Handy Tables

65 6B.30/M(1)a, part	aggregate 2.5×7.5 cm (fr. 1)	Fourth century
+ 66 6B.26/D(1)c, part	aggregate 4×14 cm (fr. 2)	
+ 68 6B.25/H(3)b, part	1.5×3.5 (fr. 3) 4.5×5 (fr. 4)	
+ 68 6B.25/H(4)b, part	1.5×1 (fr. 5)	

Fragments of more than one page of a codex, belonging to 'Group A', in a hand resembling that of **4173** and ruled in red. The original organization of the tables is obscure; one clue may be that one of the pieces that constitute fr. 1 and one of the pieces of fr. 2 were found adhered togeter by the faces with vertical fibres. Fr. 3 preserves part of the headings of a table, and fr. 4 is from one of the bottom corners of a page (2.5 cm bottom margin); otherwise the fragments are all broken on all sides. All have tables on both sides except the tiny fr. 5, which is blank on the side with vertical fibres.

Frs. 1–3 all come from tables that were arranged like the anomaly tables of the *Handy Tables*. The leftmost pair of columns of each page gave the arguments ('common numbers'), counted up in the first column from 1° to 180° by steps of 1°, and down in the second column from 359° to 180° by steps of 1°, and to the right there were several columns containing the various functions involved in computing the positions of the sun, the moon, or one of the planets. Each page had thirty lines, making six pages in all. In fr. 1, the side with horizontal fibres is part of a page with index numbers 31–60 (and 329–300); the other side was therefore either the first or the third page of the same table. The side of fr. 2 that has vertical fibres was likely the next page (either forwards or backwards); it belonged to a page with index numbers 151–180. The simplest reconstruction of the sequence—though not necessary the correct one—would have fr. 2 belonging to the last two pages of one anomaly table (recto with horizontal fibres), and fr. 1 belonging to the first two pages of the next table (recto with vertical fibres). Fr. 3, side with horizontal fibres, has index numbers that could belong to any page of an anomaly table except the first two, though the spacing of the ruled lines is too wide for this to be part of the same page as fr. 2. I have not been able to identify the specific contents of any of these tables from the remnants of the other columns.

Fr. 4 is from a different kind of table. The side with vertical fibres has only a single column of index numbers, reaching 60 (by steps of 1?) at the bottom line of the page. Again I do not recognize the purpose of the table.

4174. Table of Lunar Mean Motions

72/29(d) 3×7 cm

A scrap from a codex, cut to an irregular curved outline along one side (perhaps, Dr Coles suggests, to make part of a shoe). The tables on both sides are ruled in red, except that the middle line of the triple rulings is black. There is a narrow margin or intercolumniar space along the right edge of both sides.

The longitudes in the table on the side here designated as the 'front' are clearly mean lunar longitudes tabulated at intervals of civil Alexandrian years. In 365 days the moon's mean progress in longitude is approximately 129;22° beyond complete revolutions, and in 366 days it is approximately 142;33°. In the papyrus the line-to-line differences follow the uniform sequence 142, 129, 129, 129, except that the computer of the table seems to have misread line 7 as Aries 3°, thereafter adding an extra 3° to the longitudes for two subsequent lines. We may conclude therefore that the table gives the moon's mean longitude for a particular date (say Thoth 1) of consecutive calendar years, and that the years of lines 2, 6, and 10 were intercalary.

The table on the 'back' presents the moon's mean longitudinal progress through the days of a calendar year. Column i contains the number of the day within the month, so that lines 1–9 represent Hathyr 22–30 and lines 11–15 represent Choeac 1–5. Column ii gives the whole number of degrees of motion in longitude. The progress from line to line is either 13° or 14°, with a mean of approximately 13;10° per day. Extrapolating backward, we find that the longitude assigned to Thoth 1 must have been roughly 15°. One would have preferred to find either 0° (for 0 days' progress since the beginning of the year) or 13° (for 1 day's progress).

I would conjecture that the two tables constituted a coarse but handy method of estimating lunar longitudes, without taking any account of the lunar anomaly. One had merely to add together the longitude corresponding to the year in question (representing the moon's position on the last day of the preceding year?) and the progress corresponding to the month and day. The results would have been subject to errors of as much as 7° even disregarding the effects of using truncated values in computing the annual positions.

Notes:

Front:

4a: Perhaps this is part of an indication of a change of reign.

8–10: The expected sequence is Leo 9°, Sagittarius 21°, Aries 27°.

4174a. Table of Solar Mean Motions

9 1B.173/J(c),part 14.5×8.5 cm Late second/early third century

Plate VI

The table, ruled in black, is written on the back of a register (late second or early third century). Part of the top margin (1.5 cm) is present, with no headings visible; the other sides are broken. There are also two small detached bits, one with ruling and , the other with ruling and trace of a single letter. A peculiarity of the hand is that the zeros are written upside down compared to the usual form, i.e. with a small circle *above* a horizontal stroke.

The table is effectively divided into two parts by a narrow vacant column between cols. iii and iv. In the left portion, cols. ii and iii contain multiples by 0, 1, 2, ... of 0;59,5,48 and 0;59,9,30. Col. i appears to preserve the last fractional place of a similar series of multiples of a number ending with 16. In the right portion, col. iv is an index column starting with 1 in the first row; col. v contains the corresponding multiples of 0;2,27,50,40,0; and col. vi contains the corresponding multiples of another base between 0;2,20 and 0;2,30.

This is a table of mean motions, arranged in a format roughly similar to that of the *Handy Tables*. Cols. i–iii give the mean motions for days 1, 2, ... of an arbitrary Egyptian month, to be added to the mean motion for the beginning of the first day of the month in question; hence the entry in the first line is always zero. The remaining columns give the mean motions for 1, 2, ... equinoctial hours counted from the beginning of the day in question. Multiplying the first entry in col. v by 24, we obtain 0;59,8,16 as the equivalent daily mean motion, and this is likely to be the same as the base number of col. i. Again, dividing the base of col. ii by 24, we obtain 0;2,27,44,30, which could be the base of col. vi. In other words, the table seems to have presented three uniform motions, reducible to daily motions respectively of 0;59,8,16, 0;59,5,48, and 0;59,9,30. The corresponding periods of revolution are 365;14,55,32,... days, 365;30,10,15,... days, and 365;7,18,39,... days, or approximately $^1/_4$, $^1/_2$, and $^1/_8$ day above 365 days.

These numbers find their elucidation in an unexpected place: Theon of Smyrna reports a version of solar theory according to which the sun returns to the same solstice or equinox in 365 $^1/_4$ days, to the same position in anomaly in 365 $^1/_2$ days, and to the same position in latitude in 365 $^1/_8$ days (ch. 27, ed. Hiller p. 172). Theon's solar parameters, which he neglects to ascribe to any authority, are peculiar in more than one respect. They are unique in antiquity in attributing to the sun a period of anomaly sidely different from its period in longitude (whether thought of as tropical or sidereal). The notion that the sun has a latitudinal deviation from the ecliptic recurs in various contexts in Greek astronomy from Eudoxus on (Neugebauer [1975] 629–631), but it is only in Theon that we find a specific periodicity for this alleged phenomenon. And as Neugebauer remarked, the three periods with their neat progression of unit fractions looks suspiciously like the product of a numerological speculation. Yet it is clear that Theon is referring to the same model that underlies the mean motion tables in **4174a**, and these were undoubtedly meant to be used with a lost table of equations for the practical end of calculating solar longitudes and positions in latitude. The sexagesimal fractions in the papyrus were not simply obtained by dividing Theon's periods into 360°, which would have led to daily motions differing from those of the papyrus by 1 or 2 in the third fractional place.

Part III

Ephemerides and Almanacs

4175–4215. EPHEMERIDES AND ALMANACS

The tables in this section are for the most part records of the longitudes of the sun, moon, and planets, computed at frequent intervals over a range of months or years in order to provide a convenient reference for the astrologer. This is the class of table that has been best represented among the astronomical papyri published before now. Nevertheless the examples from Oxyrhynchus deepen our understanding of these intermediaries between the basic methods of predicting positions of the heavenly bodies and their practical applications.

Ephemerides

Formerly the only recognized examples of ephemerides on papyrus were P. Mich. inv. 1454, ably studied by Curtis & Robbins [1935] and dated by them to A.D. 467, and P. Vind. G. 29370 and 29370b, published by Gerstinger & Neugebauer [1962] and dated by Neugebauer to 348 and a nearby year. Helped largely by the twelve new ephemerides presented here (**4175–4184a**), I was able to identify three further published papyri as fragments of ephemerides: P. Dublin TCD F.7, for A.D. 100 (Jones [1995]), *P. Harris* I 60, for 140 (Jones [1994a]), and P. Vind. G. 3231; as well as to redate P. Vind. G. 29370 and 29370b to 489 and 471 respectively (Jones [1994c] 121–126). We now possess a spread of ephemerides, most of them exactly dated, from the late first century B.C. to the late fifth century of our era—the longest span attested for any variety of table.

All the ephemerides are organized in sets of columns corresponding to civil calendar months, either Alexandrian or Roman, and tabulate lunar longitudes through the course of each month. We now see, however, that they fall into two formats on the basis of the way they record the motions of the planets. The almanac-ephemerides, which are on the whole the older group (**4175–4177** and *P. Harris* 60, ranging from 24 B.C. to A.D. 140), set this information out in the form of a monthly almanac above the ephemeris proper, whereas the pure ephemerides (**4179–4181** and the remaining published ephemerides, ranging from 100 to 489) devote columns in the ephemeris to the daily longitudes of each of the heavely bodies.

Another transition, from arithmetical methods to the *Handy Tables* as the basis for the longitudes, is less surprising than the fact that it occurs so late. The lunar longitudes in all the ephemerides from the second, third, and even fourth century (**4179**) are consistently derived from the Standard Scheme. When we come to the cluster of fifth-century ephemerides, however, all the recorded data agree with recomputation using the Handy Tables. The longitudes in these late tables are tropical, that is, 'Theon's formula' was not applied to them.

Among the ephemerides, **4175** stands out by its early date, more than a century before the next dated example, making it also the earliest dated Greek astronomical table. It has several unusual features: abbreviations unattested elsewhere, pink pigment for certain tabular entries, vacant columns. Special interest attaches to the calendrical information that can be derived from this table, since it falls within the years during which the pattern of intercalations in the Roman and Alexandrian calendars is in dispute.

4179 is noteworthy for the large number of planetary longitudes that can be read. Like the longitudes of the moon in this papyrus, they were evidently obtained by arithmetical schemes, a surprising thing to find in the middle of the fourth century. **4180** has in common with the other fifth-century ephemerides the presence of a column appraising each day as good or bad for (unspecified) activities. These καταρχαὶ were certainly based in part on the relative configurations of the moon and the other heavenly bodies, but the details of the rules for obtaining them await investigation.

Sign-entry Almanacs

While the papyri from Oxyrhynchus have notably added to the known ephemerides in the earlier centuries, they have had the opposite effect on the corpus of sign-entry almanacs. The latest instance of this format from elsewhere, the demotic Stobart Tablets, ands with A.D. 133, just overlapping the earliest of the sign-entry almanacs from Oxyrhynchus (**4185**). The new texts extend the series into the early fourth century. This is the variety of almanac that must have been most useful to the astrologer whose trade was primarily in plain horoscopes, and we find a striking continuity in the details of the format from the earliest to the latest examples, with a tendency in the later ones to settle exclusively on the Alexandrian calendar, the representation of months and zodiacal signs by their ordinal numbers (counted respectively from Thoth and Virgo), and the medium of the codex.

Two special varieties of sign-entry almanac deserve notice. Almanacs with a fourth column for the time of sign-entry have not been seen hitherto. The specimens included here (**4190, 4192, 4194–4196a**) turn out, remarkably, to have been computed using the *Handy Tables* by what must have been an exceedingly laborious process of calculating longitudes day by day and even hour by hour. **4197–4198** are perpetual almanacs incorporating the Babylonian Goal-Year periods; a third example of this type can now be recognized in P. Vind. G. 36041+26011 fr. n. (Jones [1994c] 111–121). In the cyclic structure of their planetary positions these tables are manifestly the ancestors of the medieval almanacs.

Monthly almanacs

The monthly almanac is a composite format, merging the contents of a sign-entry almanac with planetary phases (presumably taken over directly from an epoch table) and computed longitudes at the beginnings of months. The potential richness of these almanacs as sources of information about ancient planetary theory has unfortunately not been fully realized because the examples that have so far come to light are in poor condition and sometimes marred by errors. Nevertheless the larger ones, **4199** and **4202**, will likely repay further study. An unexpected feature recurring in the monthly almanacs is the use of symbols to represent the planetary phases: see especially **4199, 4202, 4203a**, and the almanac portion of **4177**.

5-Day almanacs
The 5-day almanacs are the most homogeneous in arrangement of the major categories. Like the four-column sign-entry almanacs, they were all computed using the *Handy Tables*, and consequently they are a useful index of Ptolemy's growing influence among the astrologers of the third and fourth centuries. At Oxyrhynchus the range of dates covered by these almanacs is comparatively brief, less than a century. Nevertheless their influence can be felt in the much later medieval almanacs, which also tabulate planetary positions at intervals of five (or ten) days.

4175. ALMANAC–EPHEMERIS, 24 B.C.

93/Dec. 28/F.1 14×13.5 cm About 24 B.C.
Plate VII

The ephemeris from which **4175** derives was presumably in the form of a roll; the back has parts of two columns of a document in a hand of the late first century B.C. or early first century A.D. The original height was at least 22 cm, and a 3 cm upper margin is preserved. The table is ruled in black, but the entries for the Roman calendar dates and the names of the Egyptian months along the top are written with a pink (not red) pigment. These are represented by heavy type in this edition. The hand is small and careful, and there are many abbreviations and compendia.

The lower part of the table, lines 8–22, is a lunar ephemeris giving the moon's daily longitude truncated to the degree. Each set of columns, bounded by double or (at the beginning and end of the year) triple rulings and separated from its neighbours by intercolumniar spaces (columns iv and xii), corresponds to an Egyptian month. The day number is written on each line in the third column (vii and xv). The first column (v and xiii) indicates the cardinal days of the Roman calendar, while the second column (vi and xiv) is vacant. To the right of these calendrical columns are those for the moon's zodiacal sign and degree (i, viii–ix, and xvi). The remaining two columns (ii–iii and x–xi), which are also empty, may have been intended to contain minutes and seconds of lunar longitude.

The five larger cells above the lunar ephemeris give dates and longitudes of planetary phases and sign entries, in the conventional order Saturn to Mercury. Entry into the first, or retrograde entry into the thirtieth, degree of a sign is indicated as ϲύναψιϲ, 'boundary', a term apparently not otherwise attested in this sense. The terminology for the other phases is unfortunately obscured by damage to the papyrus. As in the lunar ephemeris, the zodiacal signs are represented by unusual biliteral and triliteral compendia.

The date of the table was established from the planetary and lunar longitudes and the dates of minimum lunar daily motion. Day 1 (line 8) of the month Mesore in columns v–xi coincides with Julian July 25, 24 B.C., which shows incidentally that the Roman calendar indications in column v are in agreement with the Julian calendar dates. Again, Thoth 1 (line 8 of columns xiii–xvi) coincides with Julian August 29. These equations are correct for the old Egyptian calendar, but also for the Alexandrian calendar if that calendar was instituted in 26 B.C. and intercalated at four-year intervals from the start. In all other known ephemerides the civil Alexandrian calendar, not the Egyptian, is employed, and the same is surely the case here. The number 7 in xiii 1 turns out to indicate the beginning of the seventh regnal year of Augustus. This is also in accord with the use of the month name *Sextilis*, which was superseded by *Augustus* in 8 B.C. The inclusion of a concordance with the Roman calendar is noteworthy in a table of this early date (it may in fact be the oldest known Greek astronomical table).

In general, agreement with computation according to modern theory is good (Table 4175.1), with a systematic excess of about 5° over modern tropical longitudes. For Venus, however, the excess is consistently about 12°; and Mercury's longitudes fall short by 6°, declining to 1° near station. The lunar longitudes exhibit a variation in their line-to-line differences greater than that produced by the Standard Scheme; they may have been generated using the Babylonian System B zigzag function for daily motion (F*), or a daily motion pattern with nearly the same amplitude.[1] For example, taking the longitude on Mesore 3 (July 27) as 2 19;36°, and the increment from this to the next day as the first value (11;7,10°) in the F* table ACT 190, we obtain exact agreement by truncation for all preserved longitudes in Mesore and for the two preserved longitudes in the preceding month, Epeiph (1 5° and 18°). Extrapolating forward, we find that the moon should have been predicted as entering Scorpio on Thoth 8 (September 5), not on the next day. Perhaps a new epoch was taken for the start of the new year; or the computer may have accidentally added a low value of the daily motion function twice in a row, resulting in a systematic error in subsequent longitudes.

[1] Jones [1983] 2–6.

Text	Recomputation
Saturn	
July 25–August 28: Ω in ♍	149°–153°, elongation 30°–1°
Jupiter	
August 18 (mid-month): ♎ 6°	182°
September 21: Ω at ♎ 14°	189°, elongation 13°
Mars	
August 23: ♏ 1°	205°
September 14: ♏ (mid-sign?)	223°
Venus	
June 25–July 24: ♊ 1°	45°–76°
July 26: ♋ 1°	79°
August 20: ♌ 1°	108°
September 14: ♍ 1°	138°
Mercury	
August 5: ♍ 1°	157°
August 21: Ω at ♍ 11°	162°, elongation 17°
Moon	
July 27: ♉ 19°, near least speed	42°, mean anomaly 353° from apogee (*Handy Tables*, 6 P.M.)

TABLE 4175.1. Concordance of astronomical data.

4176. ALMANAC–EPHEMERIS, A.D. 111

50 4B.34/A(1–3)a 8×26 cm About A.D. 111

The table is written entirely in black on the back of a list of accounts in a late first-century hand, and is much rubbed. A top margin of 3 cm is preserved. On the front and along the fibres, accounts. The format of the table is similar to **4175**, with five rows for planetary data above a lunar ephemeris. The calendrical columns of the latter are all lost to the left of the extant columns for lunar sign and degrees (columns i–ii) and the time (in seasonal hours) of the moon's entry into each sign (iii). Allowing for lost lines of the lunar ephemeris and bottom margin, the full height was about 40 cm.

Examination of the line-to-line differences of the lunar longitudes reveals that they were obtained from the Standard Scheme template, such that a Standard

Text	Recomputation
Saturn	
November 28: ♓ ? 12;2x	337°
Jupiter	
November 28: ♑ 27°	295°
December 7–16: ♒ 0°	297–299°
Mars	
November 28: 1x° (sign lost)	7°
Venus	
Nov. 28?: ♏ 0°	205°
December 22: ♐ 0°	235°
Mercury	
Nov. 28 or later: ♐ 0°	223°+
December 18: Σ at ♐ 2x°	252°, elongation 14°

TABLE 4176.1. Concordance of planetary data.

Scheme epoch (i.e. passage of the lunar apogee) occurred on the eighth day of the month at longitude Cancer 13;3°. This date can be identified as the epoch of Era Augustus 141, Alexandrian Choeac 8 (A.D. 111, December 5), for which the rules of *P. Ryl.* I 27 yield longitude Cancer 13;21,31,1° (cf. Appendix H). All longitudes in **4176** are therefore approximately 0;18° less than if computed using the tables in Appendix H. The poorly legible traces of the planetary data in lines 1–5 are also in agreement with the equation of the month in question with Alexandrian Choeac in A.D. 111/112 (Table 4176.1). From the times of sign entry in column iii it is evident that the lunar longitudes are meant to apply to sunset, i.e. effectively 6 P.M., on the day in question, which is consistent with the alignment of epochs that I have inferred for *P. Ryl.* I 27 (Jones [1997a] section 4).

4177. ALMANAC–EPHEMERIS, A.D. 121–122 AND 127

118/11(b) 13×12 cm (fr. 1) About A.D. 127
5×12 cm (fr. 2) 3.5×4 cm (fr. 3)

The fragments come from a set of tables, perhaps originally consisting of separate sheets covering single calendar years rather than one long roll. Fragments 1 and 2 have been spliced together, with 2 overlapping 1 by 2.5 cm on the back, which was used (after the splicing) for a grammatical text. Fragment 3's two sides have the same orientation, whereas the text on the back of frs. 1 and 2 is inverted with respect to the tables on the front. The tables are ruled in red, and a bottom margin of 2.5 cm is preserved in fragment 1. Since the upper table of planetary positions had one line for each Egyptian month, the full height was at least 16 cm plus the top margin.

The plan of **4177** resembles other almanac-ephemerides in placing an almanac of planetary longitudes, sign-entries and phases above a lunar ephemeris. It is unique, however, in two respects. The planetary table exchanges the usual role of the rows and columns: it has five columns for the five planets, in the conventional order Saturn to Mercury, and a row for each month's entries. Consequently the planetary data for a given month are not directly above the contemporaneous lunar data. Secondly, the lunar ephemeris records the longitude only on the first day of the month and subsequent days before which the moon has entered a new zodiacal sign, so that this part of the table requires only fifteen lines instead of the usual thirty. The columns in the ephemeris are, respectively, the day of the month (fr. 1 i, iv, vii, x; fr. 2 ii; fr. 3 i, iv) and the sign and whole number of degrees of the moon's longitude (fr. 1 ii–iii, v–vi, viii–ix, xi; fr. 2, i, iii–iv; fr. 3, ii–iii).

Because of the atypical format of **4177**, the method of establishing their date will be described in more detail than for the other ephemerides. It was first necessary to establish the relative dates of the three fragments. Because the longitudes are given without fractions, the variation in the moon's daily progress cannot be accurately determined from single line-to-line differences. However, when the lunar longitudes in fragment 1 were graphed as a function of time, it became apparent that the lunar anomalies were consistent with those generated by the Standard Scheme template. By extrapolation using the template, it was possible to show that if fragment 2 covered the month Mesore of year *x*, then fragment 1 covered Phamenoth through Payni of year *x*–1, and fragment 3 covered Thoth of year *x*+6. Because 32 epagomenal days must be assumed between fragments 2 and 3, the calendar must have been the Alexandrian. (The relation between fragment 3 and the others is obscured by an isolated error on the papyrus: the longitude for Thoth 1 should have been ♏ 19°, not 17°.) Moreover, Mesore 1 on fragment 1 proved to fall within one day of minimum lunar daily motion.

Lunar longitudes and the numbers of days since minimum motion were recomputed using the Standard Scheme for Mesore 1 (Alexandrian) of all years between 30 B.C. and A.D. 400, yielding five years in which the longitude was within 15° of ♍ 24 and the minimum daily motion occurred within two days of Mesore 1. Further constraints were imposed by the planetary data in the upper tables of fragments 1 and 3. In fragment 1, cols. i–ii, Venus enters Aries on Pachon 15 of year *x*–1, and the subsequent four signs at 25-day intervals. In cols. iii–iv, Mercury was in Gemini on Epeiph 3, and in Virgo on Mesore 4. In fragment 3, Saturn was in Libra during Epeiph and Mesore of year $x+6$. Among the possible datings according to the lunar data, the only one that even nearly fitted the planetary positions was x = A.D. 121/122. For this dating, the agreement is excellent (Table 4177.1). The lunar longitudes turn out to match longitudes computed using the epochs in Appendix H, rounded to the nearest degree; the few exceptions may be due to errors of calculation or erratic rounding (Table 4177.2). The beginning of the zodiacal sign is consistently taken to be 1° rather than 0°.

The planetary data preserved in the fragments may usefully be supplemented by comparison with modern recomputation. We begin with the data for Mercury in fragment 1, columns iii–iv of the upper table. Mercury's last evening appearance (Ω) took place in 120/121 towards the beginning of Pachon (late April) in Gemini, and its first morning appearance (Γ) occurred in Taurus about the end of the same month. The events recorded for Pachon (line 1) therefore originally would have included last appearance, retrograde sign entry into Taurus, and first appearance in Taurus; only the record of the first of these events is partly extant. The reentry into Gemini in Payni was probably the only recorded event for that month (line 2). During Epeiph (line 3) Mercury's last morning appearance (Σ) was followed by three other events: its entries into Cancer and Leo as well as its first evening appearance (Ξ) in Leo. Again only the first of these has survived. In Mesore (line 4) Mercury entered Virgo, and then Ω occurred again about the end of the month near the end of Virgo. It is not certain whether a sign entry into Libra would have been recorded for that month. For the following year 121/122 (fr. 2, line 1) only a tiny remnant of the line for Mecheir is preserved; it probably was the record of Mercury's last morning appearance in Aquarius.

The recorded events for Venus in 120/121 (fragment 1, columns i–ii of the upper table) are sign-entries belonging to the planet's slowly accelerating motion as morning star. The longitudes and dates in Pachon through Mesore (lines 1–4) appear to have been computed using the 'Vasiṣṭha' velocity scheme (see the commentary to **4135**). From Pachon 15 to Payni 10, and from thence to Epeiph 5 the motion is 30° in 25 days, a velocity of exactly 1;12° per day. Between Epeiph 30 and Mesore 25 the progress appears to be 30;25° (the readings for these longitudes are not certain), hence a velocity of exactly 1;13° per day. The progress of 30;18° between Epeiph 5 and Epeiph 30 would correspond to 7 days at 1;12° per day followed by 18 days at 1;13° per day, so that the 60 day interval of motion at 1;12° per day should have begun 3 days before Pachon 15, i.e. on Pachon 12. Extrapolating according to the scheme, conjunction should be predicted 210 days after Pachon 12, i.e. on Choeac 7 = 121 December 3, at longitude 9 13;12°. Tuckerman's tables for 16:00 U.T., December 3, give the longitude of Venus as 249.57°, and that of the sun as 251.14°, with the actual conjunction occurring about six days later.

Text	Recomputation
Saturn (A.D. 127) Epiphi 1? = June 25: in ♎ Mesore 1? = July 25: in ♎	 180 181°
Venus (A.D. 121) Pachon 15 = May 10: ♈ 0;18° Payni 10 = June 4: ♉ 0;18° Epeiph 5 = June 29: ♊ 0;18° Epeiph 30 = July 23: ♋ 0;36° Mesore 25 = August 18: ♌ 1;1°	 2° 28° 56° 84° 116°
Mercury (A.D. 121) Pachon 1 = April 26: Ω in ♊ Payni 15 = June 10: entry into ♊ Epeiph 3 = June 27: Σ in ♊ Mesore 4 = July 28: entry into ♍ (A.D. 122) Mecheir = Jan 26–Feb 24: (at least two recorded events)	 54°, elongation 19° 57° 86°, elongation –8° 144° Feb 9±: entry into ♒ Feb 14±: Σ about ♒ 9°

TABLE 4177.1. Concordance of planetary data.

Fragment 3's poorly preserved positions of Saturn for 127 call for little comment. Saturn's second station (Ψ) took place about June 5 = Payni 11 at a longitude of 179°, so that it would fall in Libra according to ancient predictions. This station was probably the only recorded event for Payni. Because of Saturn's slow progress, the subsequent two months would have had neither sign-entries nor phases, so that the table recorded merely the planet's longitude on the first day of the month.

Notes:

Fragment 1, upper table:

iii 1 and 3: The symbols appear also in **4203a**, a monthly almanac for Mercury. It appears from that text that a pair of empty circles with diagonals ascending to the right signifies a morning phase, either appearance (Γ) or disappearance (Σ); filled circles with diagonals signify an evening phase, either appearance (Ξ) or disappearance (Ω). The phase indicated here are Ω (line 1) and Σ (line 3).

Fragment 1, lower table:

i 11: ιβ: error for ια (11).

iii 12: γ: error for ιγ (13).

vi 17: α: error for γ (3).

Julian	Alexandrian	Text	Recomputation
121 Feb 25	Pham 1	♏ 12°	221;55°
121 Feb 27	Pham 3	♐ 9°	249; 6°
121 March 1	Pham 5	♑ 7°	277; 9°
121 March 3	Pham 7	♒ 6°	306; 2°
121 March 5	Pham 9	♓ 5°	334;53°
121 March 7	Pham 11	♈ 3°	2;53°
121 March 10	Pham 14	♉ 3°	43;16°
121 March 12	Pham 16	♊ 9°	69; 7°
121 March 14	Pham 18	♋ 4°	94; 7°
121 March 17	Pham 21	♌ 10°	130; 0°
121 March 19	Pham 23	♍ 4°	153;57°
121 March 22	Pham 26	♎ 12°!	191;29°
121 March 24	Pham 28	♏ 7°!	217;35°
121 March 26	Pham 30	♐ 5°	244;32°
121 March 27	Pharm 1	♐ 18°	258;20°
121 March 28	Pharm 2	♑	272;20°
121 March 30	Pharm 4	♒	301; 0°
121 April 2	Pharm 7	♓	344;18°
121 April 4	Pharm 9	♈	12; 6°
121 April 6	Pharm 11	♉ 9°	39; 3°
121 April 8	Pharm 13	♊ 5°	65; 8°
121 April 11	Pharm 16	♋ 13°	102;40°
121 April 13	Pharm 18	♌ 7°	126;37°
121 April 16	Pharm 21	♍ 13°	162;31°
121 April 18	Pharm 23	♎ 7°!	187;31°
121 April 20	Pharm 25	♏ 1°!	213;22°
121 April 23	Pharm 28	♐ 14°	253;45°
121 April 25	Pharm 30	♑ 12°	281;45°
121 April 26	Pachon 1	♑ 26°	296; 4°
121 April 27	Pachon 2	♒ 11°	310;36°
121 April 29	Pachon 4	♓ 10°	339;30°
121 May 1	Pachon 6	♈ 8°	7;32°
121 May 3	Pachon 8	♉	34;43°
121 May 5	Pachon 10	♊ 1°	61; 2°
121 May 8	Pachon 13	♋ 9°	98;55°
121 May 10	Pachon 15	♌ 3°	123; 7°
121 May 13	Pachon 18	♍ 9°	158;48°
121 May 15	Pachon 20	♎ 4°	183;33°
121 May 18	Pachon 23	♏ 12°	222;18°
121 May 20	Pachon 25	♐ 9°	249;12°
121 May 22	Pachon 27	♑ 7°	276;58°
121 May 24	Pachon 29	♒ 6°	305;35°
121 May 26	Payni 1	♓	334;40°
121 May 28	Payni 3	♈	2;56°
121 May 31	Payni 6	♉	43;45°
121 June 2	Payni 8	♊	69;53°
121 June 4	Payni 10	♋	95;10°
121 June 7	Payni 13	♌	131;29°
121 June 9	Payni 15	♍	155; 9°
121 June 12	Payni 18	♎	192;16°
121 June 14	Payni 20	♏	218; 4°
121 June 16	Payni 22	♐	244;44°
121 June 18	Payni 24	♑	272;15°
121 June 20	Payni 26	♒	300;38°
121 June 23	Payni 29	♓	344; 5°
122 June 25	Epeiph 1	♌ 21°	141; 0°
122 June 26	Epeiph 2	♍ 3°	153;16°
122 June 29	Epeiph 5	♎	188;57°
122 July 1	Epeiph 7	♏ 3°	213; 9°
122 July 2	Epeiph 8	♐ 11°	225;34°
122 July 4	Epeiph 10	♑	251; 2°
122 July 6	Epeiph 12	♒ 5°	277;22°
122 July 8	Epeiph 14	♓ 3°	304;33°
122 July 10	Epeiph 16	♈	332;35°
122 July 12	Epeiph 18	♉	1;29°
122 July 15	Epeiph 21	♊	44;26°
122 July 17	Epeiph 23	♋	72; 0°
122 July 19	Epeiph 25	♌	98;42°
122 July 21	Epeiph 27	♍	124;34°
122 July 24	Epeiph 30	♎	161;44°
122 July 25	Mesore 1	♏ 24°	173;42°
122 July 26	Mesore 2	♐ 5°	185;27°
122 July 29	Mesore 5	♑ 12°	221;42°
122 July 31	Mesore 7	♒ 7°	246;56°
122 August 2	Mesore 9	♓ 3°	273; 2°
122 August 5	Mesore 12	♈ 14°	313;46°
122 August 7	Mesore 14	♉ 12°	342; 0°
122 August 9	Mesore 16	♊ 11°	11; 5°
122 August 11	Mesore 18	♋ 10°	39;45°
122 August 13	Mesore 20	♌ 8°	67;33°
122 August 15	Mesore 22	♍ 4°!	94;30°
122 August 17	Mesore 24	♎ 1°	120;35°
122 August 20	Mesore 27	♏ 8°	158; 7°
122 August 22	Mesore 29	♐ 2°	182; 4°
122 August 25	Epagom 2	♑ 8°	217;57°
122 August 27	Epagom 4	♒ 3°	242;58°
127 August 30	Thoth 1	♏ 17°!	229; 1°
127 August 31	Thoth 2	♐ 3°	243;16°
127 Sept 2	Thoth 4	♑ 1°	271; 6°
127 Sept 5	Thoth 7	♒ 11°	311;16°
127 Sept 7	Thoth 9	♓ 7°	336;59°
127 Sept 29	Phaophi 1	♐	266;31°
127 Sept 30	Phaophi 2	♑	280;14°
127 Oct 2	Phaophi 4	♒	307; 2°
127 Oct 4	Phaophi 6	♓	332;59°
127 Oct 7	Phaophi 9	♈	10;18°

TABLE 4177.2. Concordance of lunar longitudes.
(! indicates disagreement with recomputation using the Standard Scheme.)

Fragment 2, lower table:

ii: 22–23 are squeezed into the space of a single line. Line 24, which could not fit inside the ruled frame, is written immediately below the ruling.

Fragment 3, lower table:

ii 5: ιζ: error for ιθ (19).

4177a. Ephemeris, A.D. 245

15 1B.201/F(b), part 5×12 cm About A.D. 245

The table, which is ruled in red, is broken on the top and both sides. The bottom is identifiable from the double ruling, although only about 0.2 cm of margin is present. Columns i–iii contain the zodiacal sign, degrees, and minutes of the moon's longitude; the poorly preserved column iv contained the times of the moon's passage into each new sign. Nothing survives of the calendrical columns or the planetary motions.

The longitudes in lines 1–14 exactly match those generated by the Standard Scheme for Alexandrian Pachon 10–23 of A.D. 244/245 (hence the last line corresponds to Pachon 30, and the calendar of the ephemeris is the Alexandrian). Pachon 24 was a Standard Scheme epoch, with epoch longitude Leo 11;26°, and one would therefore expect the subsequent longitudes to have been computed from this new epoch. In fact there are small systematic discrepancies (increasing to 0;8° in line 20), which reveal that the computer of the table chose to continue using the entries beyond day 248 in the Standard Scheme template together with the previous epoch. (See the commentary to **4164a** for discussion.)

Notes:

11: The sign-entry should have been noted in column iv as taking place in the second or third hour of day.

15: Expected: Leo 11;25°. (Isolated error.)

18: Expected: Virgo 29;36°. (Isolated error.)

4178. Ephemeris, A.D. 261

25 3B.58/L(a) 7.5×11 cm About A.D. 261

A neatly written table, with red ruling. The bottom margin (2 cm) is preserved. The complete table may have been an almanac-ephemeris, with planetary data in the lost upper part; alternatively, columns for longitudes of the other heavenly bodies besides the moon may have been lost to the right of the extant columns.

Columns i–iii are calendrical. The first, column i, counts the days up to 28 in line 14, and below this line is a vacant space between rulings slightly wider than the two-line intervals higher up. These are obviously the days of the Roman month February, counted ordinally, and the ephemeris itself was therefore laid out according to the Roman calendar. We may assume that a column giving the Roman reckoning of the days has been lost to the left of column i. Column ii gives the Alexandrian calendar equivalents, in the months Mecheir and Phamenoth; and column iii counts days from new moon (hence February 17 was a new moon, being also the thirtieth day of the preceding lunar month).

The longitudes of the moon in columns iv–vi (zodiacal sign, degrees, and minutes) were computed by the Standard Scheme. Line 1 of the papyrus, i.e. February 15 = Alexandrian Mecheir 21, was the 185th day past epoch. The epoch was that of A.D. 260 August 14, for which Appendix H gives lunar longitude Taurus 1;1,50,11°. The longitudes in the papyrus are always identical to those obtained from Appendix H or, in marginal cases, one minute greater, implying epoch longitudes less than 0;1° higher than according to the rules of *P. Ryl.* I 27.

As in **4176**, the times of the moon's entry into each zodiacal sign are recorded in column vii. These once more show that the longitudes apply to sunset. The conjunction of sun and moon actually occurred about 1 A.M. on the night of February 16/17. It is probably this event, rather than the visibility of the new moon crescent, that determined the beginning of the lunar month in column iii.

4179. Ephemeris, A.D. 348

67 6B.10/L(1)a 26×25 cm About A.D. 348

The table is from a roll, which presumably covered at least a year when complete. The ruling is in red, and the back is blank.

The ephemeris is organized according to Roman months. The extant sections are for May (columns i–xiv) and June (xv–xvi). The year was identified as A.D. 348

by the alignment of the planetary week (e.g. June 25, 348 was a Saturday), the Standard Scheme lunar longitudes, and the planets' longitudes and velocities.

Hooks attached to the double ruling along the left edge of each month's first column indicate the Saturdays. These marks are extant for June 11, 18, and 25; the others shown in the transcription are restored. The leftmost columns are calendrical, giving the days of the Roman calendar (May in i, June in xv), the ordinal days of the Roman month (ii, xvi), those of the Alexandrian calendar (Pachon and Payni in iii), and the date within the lunar month beginning with conjunction (iv).

The next four columns, marked off as a distinct section by double rulings, concern the moon. First we have the moon's longitude: zodiacal sign (v), degrees and minutes (vi). The lunar longitudes, which apply to 'sunset', were computed by the Standard Scheme, using epoch and template values truncated to minutes. The epoch for the first nine days of the ephemeris was Era Augustus 337, Alexandrian Thoth 7 at 332;31°, and a new epoch occurred on Pachon 15 at 0;14°; both longitudes are approximately 0;4° lower than their counterparts in Appendix H, Table 1. Column vii gives the time of sign-entry in seasonal hours, and column viii the longitude and time of the syzygies. The contents of both columns were probably derived from the lunar and solar longitudes in columns v–vi and ix.

The remaining six columns contain the longitudes (degrees and minutes) of the sun and the five planets in the conventional order. Since the Standard Scheme lunar longitudes are sidereal and refer to sunset, it may be safely assumed that the positions of the other heavenly bodies are to be understood in the same way. As a substantial collection of planetary positions computed by methods different from Ptolemy's tables, these columns are worth discussing in greater detail.

During the second half of May the sun (ix) is near its slowest motion, reflected in the table by line-to-line differences of 0;57° or occasionally 0;56°. If we extrapolate forward, the longitude Cancer 8°, marking the sidereal 'solstice', is passed about one quarter day after the position tabulated for June 30. This is approximately half a day later than the epoch predicted by the scheme of **4148**.

Throughout the month Saturn (x) slowly advanced through the beginning of Aries, with decreasing velocity. The preserved entries in the papyrus are in good agreement with modern theory, but are insufficient to cast light on the methods of computation. Jupiter (xi) passed its first station about May 9 near Capricorn 21° (tropical). Fig. 4179.1 contrasts the line-to-line differences in the papyrus (circles) with Jupiter's actual daily progress calculated by modern theory and truncated to minutes (crosses). The most obvious difference is that in the papyrus Jupiter remains effectively stationary much longer, and the beginning of its retrogradation is abrupt. Mars (xii) crossed from Aquarius to Pisces, gradually slowing in its direct motion following conjunction. The line-to-line differences show a pattern similar to that produced by modern theory (Fig. 4179.2). Venus (xiii) passed from

Pisces into Aries, moving slower than the sun at the beginning of the month but faster at the end; the trend of line-to-line differences is roughly correct, but there are unexpected intervals of several days during which the difference remains constant (Fig. 4179.3). Mercury (xiv) was in the course of its fast motion around superior conjunction—noted in the papyrus—and progressed from the end of Aries through Taurus and Gemini. In the case of this planet the behaviour of the line-to-line differences in the papyrus is conspicuously different from modern theory (Fig. 4179.4): instead of the expected acceleration leading up to conjunction and deceleration afterwards, the daily progress stays nearly constant. This strongly suggests that Mercury's daily positions were interpolated between a few calculated positions, perhaps in this instance last morning visibility, conjunction, and first evening visibility.

Notes:

vi 30: The longitude should be 27;56°.

vii 24: 6 $^1/_2$ would be more accurate.

viii 26: 8 would be more accurate.

xiii 5: The name of the new sign entered (Aries) ought to appear between this line and the next. It may have disappeared through abrasion.

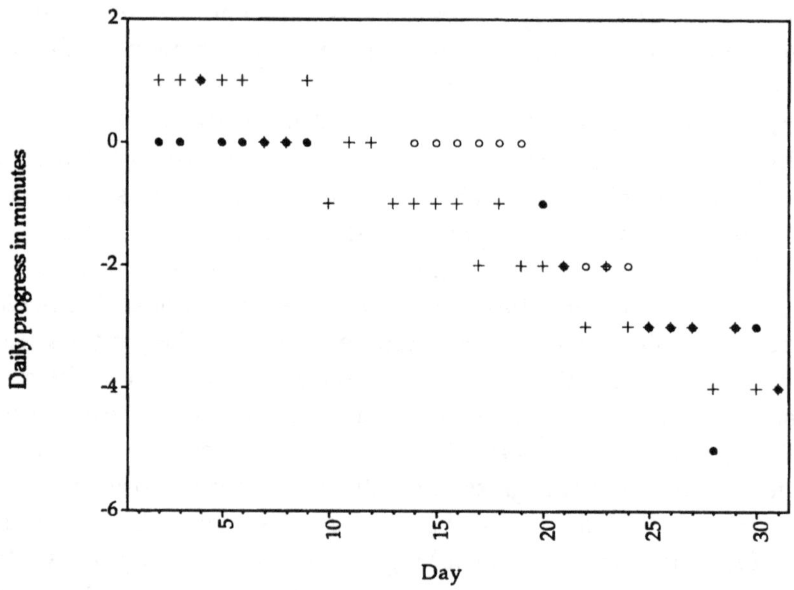

FIG. 4179.1. Daily motion of Jupiter in May, A.D. 348.
Solid circles: papyrus. Empty circles: restored. Crosses: modern theory.

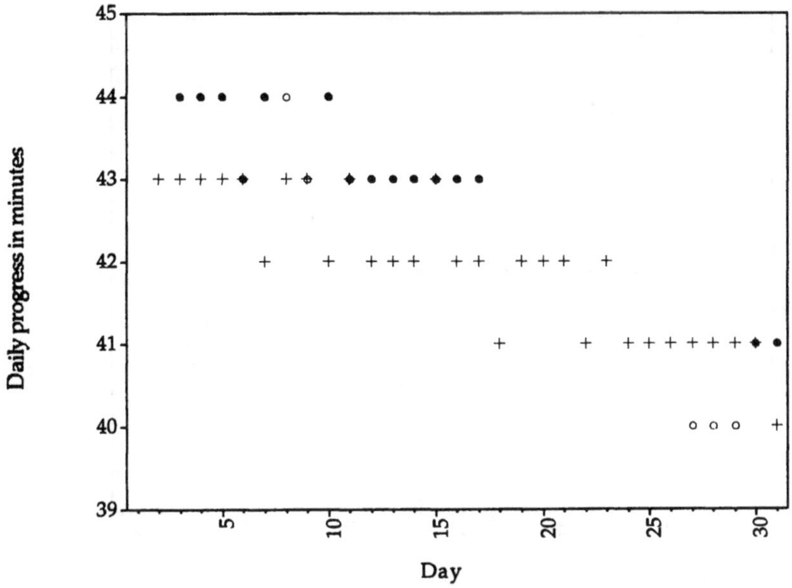

FIG. 4179.2. Daily motion of Mars in May, A.D. 348.

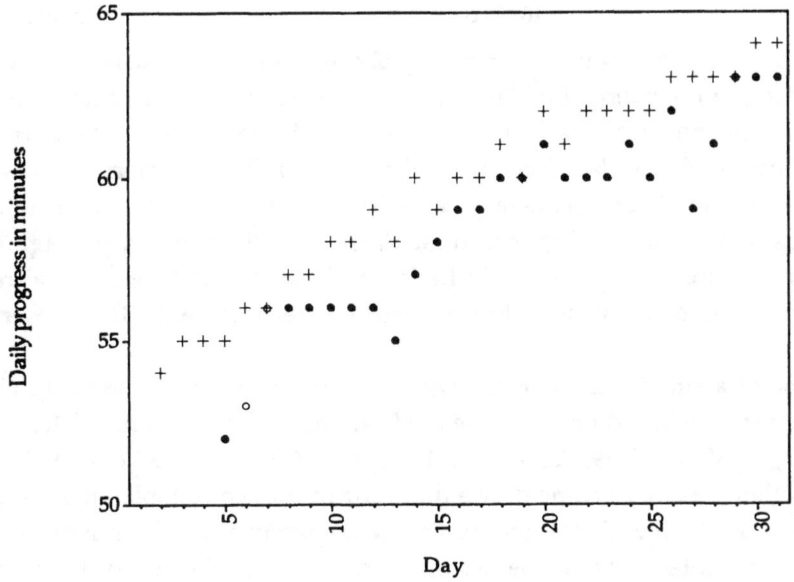

FIG. 4179.3. Daily motion of Venus in May, A.D. 348.

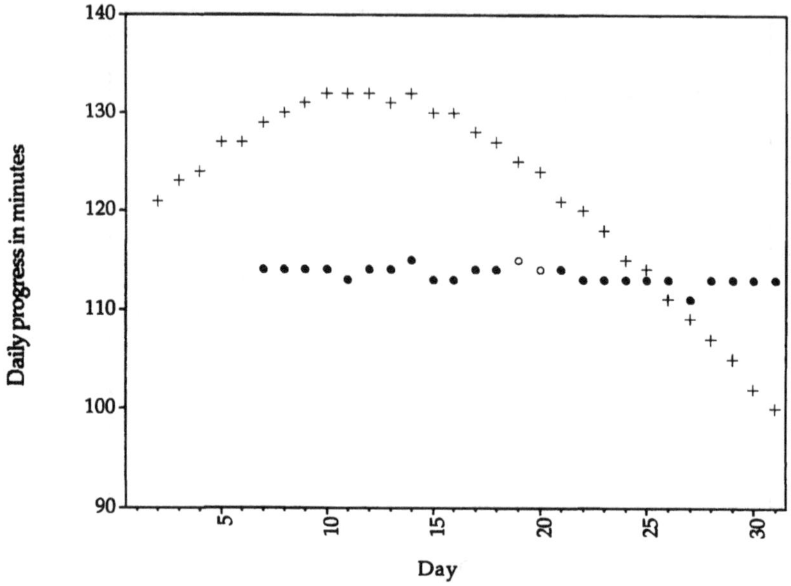

FIG. 4179.4. Daily motion of Mercury in May, A.D. 348.

4180. EPHEMERIS, A.D. 465

96/69(a) 16×30 cm + debris About A.D. 465

The remains of a large codex page (originally about 30×30 cm), consisting of eight detached pieces, with ruling in black ink. The recto gives the ephemeris for Hathyr of the Alexandrian year A.D. 465/466 and the verso the ephemeris for Choeac, i.e. late October to late December. The date of the ephemeris was established from the calendrical, planetary, and lunar data (e.g., on the recto, the zodiacal signs of Saturn and Jupiter, lunar dates, and the moon's passage of a node). Recomputation using the *Handy Tables* for 6 P.M., *without* converting tropical longitudes to sidereal, yields almost perfect agreement with the preserved positions.

4180 is one of a small cluster of known ephemerides from the second half of the fourth century, all based on the *Handy Tables*. The other three are P. Mich. inv. 1454 (A.D. 467), P. Vind. G. 29370b (A.D. 471), and P. Vind. G. 29370 (A.D. 489). It is not clear whether it is mere coincidence that four of the ten datable ephemerides known so far should fall within a twenty-five year span in a century not otherwise abundantly represented among the astronomical papyri. No two of them have exactly the same format. The arrangement of **4180** may be summarized as follows:

recto	*verso*	*contents*
[0]	[–II]	Alexandrian calendar date (not preserved)
i	[–I]	Roman calendar date
ii	[0]	Lunar calendar date
iii	i	Evaluation of day. Usually the first term indicates the auspiciousness of the day, and the second is the conventional 'nature' of the moon's zodiacal sign. In recto, lines 29–30, the latter is omitted to make room for the indication that the moon is close to its ascending node. Of the three terms for bad days, ἀποφρὰϲ corresponds to *dies nefas*, a day that is inauspicious on calendrical rather than astrological grounds. I do not know what the distinction between κακὴ and φαύλη was.
iv	ii	Zodiacal sign occupied by the moon, and time of its entry into that sign in seasonal hours.
v	iii	Longitude of the moon in degrees and minutes
vi	iv	Longitude of the sun in degrees and minutes
vii–viii	v–ix	Longitudes of the five planets (in the conventional order) in degrees and minutes. The column headings state the zodiacal sign occupied at the beginning of the month, and the stage of the synodic cycle.
bottom margin		Date and longitude of syzygies.

4181. Ephemeris, a.d. 161

5 1B.41/C(c) 7.5×22 cm (fr. 1) About a.d. 161
6×11 cm (fr. 2) 2.5×3 cm (not transcribed)

Each of the two transcribed fragments is composed of disjointed pieces, those of fr. 1 being (from top to bottom) 3×5.5, 2.5×3.5, 5×5, 2×3, and 1×2.5 cm, and those of fr. 2 being 3.5×9.5 and 3.5×2.5 cm. The small remaining bit is illegible. The table is written on the side with horizontal fibres, and is ruled in red, except that the rightmost ruling of fr. 1 and the leftmost of fr. 2 are black. Red ink was also used for alternate lines of the column headings (fr. 1, lines 1 and 3), the cardinal days of the Roman calendar (fr. 2, i line 9), and the planetary phases (fr. 1, iv line 18a). On the back are abraded literary remains in a sloping bookhand.

Fr. 1 preserves parts of the four last columns of an ephemeris for a month of thirty days, specifically the columns for the longitudes of the planets Jupiter through Mercury. We may note (a) that Jupiter began the month in Aries, and was progressing at a rate of about 0;11° per day; (b) that Mars began the month in Leo, and was in its direct motion; (c) that Venus's daily motion was approximately 1;14°; and (d) that Mercury's morning setting (Σ) took place after the middle of the month, so that its conjunction with the sun would fall later in the month or early in the following month. Fr. 2 is part of the first columns for a month, with columns i–iii giving respectively the date in the Roman, Alexandrian, and lunar calendar, columns iv–v giving the moon's longitude as sign, degrees, and minutes, and column vi giving the times of the moon's sign-entries. A conjunction occurred between the 16th and 17th of the Alexandrian month at a longitude near the end of Cancer or the beginning of Leo. Together with the calendrical data, this establishes that fr. 2 pertains to Alexandrian Epeiph, i.e. Roman July. If the ephemeris was laid out according to the Roman calendar, the only likely hypothesis would be that fr. 1 belonged to the month (June) preceding fr. 2; but if it was laid out according to the Alexandrian calendar, fr. 1 could give planetary positions for either Payni or Epeiph.

In order to determine the relation between the two fragments and their date, years were searched for in the span from 30 B.C. to A.D. 570 such that a conjunction occurred within or just outside the interval between 6 P.M., July 10 and July 11 (i.e. Epeiph 16 and 17), and moreover such that Jupiter and Mars were in or very near the required zodiacal signs at the beginning of either June or July. Only two years satisfied these criteria: 30 B.C. and A.D. 161. The former year is ruled out because Mercury's conjunction had occurred already about June 9. For A.D. 161 we obtain good agreement with modern theory, assigning fr. 1 to the month before fr. 2. The match of planetary longitudes is most satisfactory if we assume that the ephemeris was organized according to the Roman calendar, so that for example line 15 corresponds to June 12.

There is little to be said about the computational methods underlying most of the data in the ephemeris. The scanty lunar longitudes could have been computed by the Standard Scheme, assuming epoch longitudes approximately higher than those of Appendix H, Table 1. The apparently constant daily motion of Venus need not be indicative of arithmetical methods, since this planet's velocity changes very slowly in any case.

The longitudes of Mercury, however, are clearly derived from an arithmetical scheme. Most of the interval after Σ is traversed at an exactly constant rate of 2;5° per day; and if we make the obvious correction to 3;3° in line 15, the line-to-line differences leading up to Σ form the linearly increasing sequence 1;30°, 1;37°, 1;43°, 1;50°. These differences would arise if one attempted to bridge an interval of 18° in 18 days between Γ and Σ using a second order sequence of the kind tabulated in the columns of **4217**. The longitude Gemini 10;7° in line 19, which does not exactly fit either the preceding or the following sequence, is probably the independently computed position at Σ.

Notes:

Entries written in red are represented by heavy type.

fr. 1:

ii 3: This entry is very unclear. If correctly read, it may mean that Mars is in an interval (the second?) of its synodic period when its daily motion is assumed to change rapidly.

fr. 2:

v 20: If the lunar longitudes were computed by the Standard Scheme, one would expect here either Aquarius 23;58° or 23;59°. The last digit might be an η written differently from elsewhere in the preserved fragments.

4182. Ephemeris

68 6B.24/D(1–3)d 7.5×5 cm

The fragment is from the upper edge of an ephemeris in codex form, and preserves the top margin (2.5 cm) and part of the column headings. The ink is brown, with rulings in the same colour.

On the side designated the verso (horizontal fibres) are headings for the first columns of a month's ephemeris. These were for (i) the date in the Roman calendar, (ii) the Alexandrian calendar date, (iii) probably the lunar date, and (iv) the moon's zodiacal sign. Since the beginning of the month is a date in both March and Pharmuthi, the calendar must have been laid out according to the Alexandrian calendar, because Pharmuthi 1 = March 27 but March 1 = Phamenoth 5.

On the other side the line above the double ruling probably set out the dates and positions of the syzygies. Below this are the headings of the last columns of an ephemeris: those for the longitudes of Mars, Venus, and Mercury and, probably, an indication of the character of the day. The headings tell us that at the beginning of the month Venus and Mercury were both at longitudes less than the sun's, in Capricorn and Pisces respectively. This would not be possible for the month following Pharmuthi (Pachon), but could occur in the preceding month, Phamenoth. Obviously there is not sufficient information to identify the year.

4183. Ephemeris

93/Dec. 16/E.1, part 5×4.5 cm (fr. 1) 3×6 cm (fr. 2)

The fragments are from a codex, though there is no writing visible on the back of fr. 2. The ruling is black.

Both fragments are from the column headings, which, to judge by the heading for Jupiter's column in fr. 2, were quite elaborate. I can make no sense of the 'verso' (along the fibres) of fr. 1. The 'recto' has the column headings for the first, calendrical, columns of a month: the date in the Roman, Alexandrian, and lunar calendars. Since the month's beginning was in both March and Phamenoth, the ephemeris must have been laid out according to Roman months (cf. **4182**).

4184. Ephemeris

2 1B.100/K(c) 5×6 cm

This is a piece from the bottom of an ephemeris, including a 4 cm bottom margin. The ruling is red, and the back blank.

Column i is calendrical, probably giving the lunar calendar date. Columns iii–iv are the lunar longitudes, in degrees and minutes, with daily motions of about 14;10° and 14;0°. Hence column ii ought to indicate the moon's zodiacal sign. The entry in line 4 is clearly a symbol, consisting of three horizontal strokes resembling a Ξ. I have no notion of what it signifies.

4184a. Ephemeris, A.D. 293

66 6B.26/D(1)c, part 3.5×5.5 cm About A.D. 293
+ 68 6B.20/L(8)a, part

A small piece from the lower left corner of an ephemeris, ruled in red. 2 cm margin is preserved at the bottom, and 0.5 cm on the left; the other sides are broken off. The back is blank. The folders in which the pieces of the papyrus were found belong to 'Group A.'

The ephemeris is unusual in that it has only one calendrical column, and the rulings are at intervals of five days instead of the usual two days. Col. i gives the day number, either in an Egyptian month or in a Roman month comprising thirty days. Cols. ii–iii give the sign and degrees of the moon's longitude; sexagesimal fractions of a degree were presumably given, but have broken off.

To date the ephemeris, it was assumed that the lunar longitudes were computed by the Standard Scheme, and a search was made by computer for a sequence of longitudes matching those in the papyrus, and such that the corresponding days fell at the end of either an Alexandrian or a Roman month. The only dates that fitted were A.D. 293 March 17–26 = Phamenoth 21–30. The single discrepancy is for Phamenoth 30 (line 10), where recomputation gives Aries 22° 2' instead of the papyrus' 21°. On the other hand the recomputed longitude for Phamenoth 26 (line 6), Pisces 0° 8', is in agreement with the papyrus. This suggests that the epoch longitudes were lower than those given in Appendix H by between 0' and 8'.

4185. SIGN-ENTRY ALMANAC, A.D. 129–133

5 1B.43/B(b) — 9×4.5 cm (fr. 1) 10×11 cm (fr. 2) 2×4.5 cm (fr. 3, illegible) debris — After A.D. 133

Fragments of thick and badly abraded papyrus, with writing on the front only. Fat red ruling for the verticals and the horizontal in fr. 1, but black rules separate the entries for the planets. Fr. 1 column iii seems to give a regnal year (Hadrian 14), and probably was situated some way above fr. 2 column i. Fr. 1 column ii seems to be empty, but there are traces in the preceding column.

Months and zodiacal signs are represented by numerals in the standard fashion, and there are no headings to identify the planets. Whenever two consecutive entries belong to the same month, the month's number is omitted in the second instance. The sign occupied by a planet on Thoth 1 is sometimes entered at the beginning of its section, but not consistently. The format seems to be one column per year. The surviving entries for Mars and Venus agree with recomputation (*Handy Tables*, sidereal longitudes) for the Alexandrian calendar years A.D. 129/130 (fr. 2, col. i) through 131/132 (fr. 2, col. iii). The month numbers in col. iv probably belong to Venus, and the entries in fr. 1 to Saturn and Jupiter.

A.D. 129/130

Saturn

Handy Tables			4185		
1	1	2	*(damaged)*		
1	17	3			

Jupiter

Handy Tables			4185		
1	1	12	*(damaged)*		
3	28	1			
5	9	12			
11	16	1			

Mars

Handy Tables			4185		
1	1	1	*(lost)*		
1	13	2			
2	26	3			
4	7	4			
5	17	5			
6	25	6			
8	4	7			
9	13	8			
10	25	9			
12	11	10			

Venus

Handy Tables			4185		
1	1	12	*(lost)*		
1	16	1			
2	10	2			
3	4	3	3	[ ]	3
3	27	4	3	28	4
4	21	5	4	24	5
5	15	6	5	20	6
6	9	7	6	[ ]	7
7	3	8	7	[ ]	8
7	28	9	7	29	9
8	23	10	8	24	10
9	18	11	9	19	11
10	15	12	10	[ ]	12
11	13	1	*(lost)*		
12	16	2			

Mercury

Handy Tables			4185		
1	1	12	1	1	12
1	10	1	*(lost)*		
1	30	2			
2	18	3			
3	7	4			
5	15	5			
6	1	6			
6	17	7			
7	5	8			
9	12	9			
9	28	10			
10	14	11			
11	2	12			

A.D. 130/131

Saturn

Handy Tables			4185		
1	1	3	*(lost)*		

Jupiter

Handy Tables			4185		
1	1	1	*(lost0*		
4	28	2			
6	10	1			
12	15	2			

Mars

Handy Tables			4185		
1	1	10	*(lost)*		
2	8	11			
4	7	10			
7	25	11			
9	24	12			
11	13	1			
12	29	2			

Venus

Handy Tables			4185		
1	1	2	*(lost)*		
2	3	1			
3	4	2			
4	10	3			
5	7	4			
6	3	5			
6	28	6			
7	23	7	7	[ ]5	7
8	18	8	8	1[ ]	8
9	13	9	*(lost)*		
10	8	10			
11	2	11	11	1	11
11	27	12	11	27	12
12	21	1			

Mercury

Handy Tables			4185		
1	1	12			
1	5	1	1	3	1
1	23	2	1	22	2
2	11	3	2	12	3
3	7	4	4	16	4
3	15	3			
4	18	4			
5	7	5	5	2	5
5	23	6	5	21	6
6	10	7	6	10	7
8	17	8	8	22	8
9	4	9	9	[ ]	9
9	20	10	*(lost)*		
10	7	11			
12	14	12			
13	3	1			

A.D. 131/132

Saturn

Handy Tables			4185		
1	1	3	*(lost)*		
4	14	4			
11	8	3			

Jupiter

Handy Tables			4185		
1	1	2	*(lost)*		
5	24	3			
7	15	2			

Mars

Handy Tables			4185		
1	1	2	*(lost)*		
2	6	3			
3	17	4			
4	26	5			
6	4	6			
7	12	7			
8	21	8	8	20	8
10	2	9	10	2	9
11	16	10	11	25	10
13	4	11			

Venus

Handy Tables			4185		
1	1	1	1	1	1
1	9	2	1	8	2
2	3	3	2	2	3
2	27	4	3	5	4
3	21	5	3	29	5
4	15	6	4	2[ ]	6
5	10	7	5	1[ ]	7
6	7	8	6	13	8
7	6	9	7	16	9
11	13	10	11	1[ ]	10
12	12	11	12	[ ]	11

Mercury

Handy Tables			4185		
1	1	1	*(lost)*		
1	14	2			
2	4	3			
4	11	4			
4	28	5			
5	14	6			
6	3	7			
7	2	6			
7	18	7			
8	9	8			
8	26	9			
9	12	10			
9	30	11			
10	30	10			
11	16	11			
12	7	12			
12	24	1			

TABLE 4185.1. Sign-entries, *Handy Tables* (left) and **4185** (right).

4186. SIGN-ENTRY ALMANAC, A.D. 142–146

A 20/1H 4×13 cm After A.D. 146

The fragment, badly stripped towards the top and rubbed at the bottom, is written in a tiny hand on the front only, with ruling in black (less dark than the writing). At the top, 1 cm of margin is extant; the other sides are broken.

This almanac takes the numerical representation of the data one step further than we have seen elsewhere: not only are the months and zodiacal signs represented by numerals, but the planets themselves are numbered from 1 = Saturn through 5 = Mercury. When two consecutive entries belong to the same month, the month's number is omitted in the second instance. The sign occupied by a planet on Thoth 1 is sometimes entered at the beginning of its section, but not consistently. The format is one column per year, with regnal years of Antoninus Pius recorded at the top. (I do not know what the letters following the symbol for ἔτος mean in line 1). Table 4186.1 compares the preserved entries with recomputation (*Handy Tables*, sidereal longitudes) for the Egyptian calendar years A.D. 142/143 (col. i) through 145/146 (col. iv).

4187. SIGN-ENTRY ALMANAC, A.D. 160/161

19 2B.78/D(11–13) 2×7.5 cm After A.D. 161

A small fragment, written on the front only with black ruling. 1 cm top margin preserved, but broken on the other sides. The zodiacal signs are all lost; presumably they were represented as numbers like the months. The only first regnal year for which the dates of sign entry approximately fit the text is Marcus Aurelius (or Commodus) 1 = A.D. 160/161, Alexandrian calendar (Table 4187.1, with recomputation by the *Handy Tables*, sidereal longitudes).

A.D. 142/143

Saturn

1	1	8	(*lost*)		

Jupiter

1	1	1	(*lost*)		
5	6	2			
9	5	1			

Mars

1	1	12	(*damaged*)		
2	5	1			
3	21	2			
5	5	3			
6	18	4	6	[ ]	4
7	29	5	[ ]	[ ]4	5
9	9	6	9	10	6
10	20	7	10	[ ]8	7
12	3	8	12	6	8

Venus

1	1	11			
1	17	12	1	[ ]	12
2	11	1	2	13	1
3	5	2	(*lost*)		
3	29	3			
4	23	4			
5	17	5			
6	10	6			
7	5	7			
7	29	8			
8	25	9			
9	22	10			
10	21	11			

A.D. 143/144

Saturn

1	1	8	1	1	8
11	6	9	[ ]	[ ]	9

Jupiter

1	1	2	(*lost*)		
6	5	3			
10	10	2			

A.D. 143/144, continued

Mars

1	1	8			
1	24	9	1	[ ]	9
3	21	8	3	27	8
6	21	9	7	4	9
8	23	10	8	23	10
10	16	11	10	15	11
12	5	12	12	1	12

Venus

1	1	11	1	1	11
2	22	12	2	16	12
3	21	1	3	24	1
4	18	2	4	19	2
5	13	3	5	14	3
6	7	4	6	6	4
7	1	5	7	1	5
7	25	6	7	28	6
8	19	7	8	23	7
9	14	8	9	17	8
10	8	9	10	11	9
11	3	10	11	5	10
11	28	11	11	29	11
12	22	12	12	23	12

A.D. 144/145

Saturn

1	1	9	1	1	9
5	11	8			
7	20	9			

Jupiter

1	1	2	1	1	2
1	29	3	(*damaged*)		
6	28	4			
11	26	3			

Mars

1	1	12			
1	17	1	1	[ ]	1
3	2	2	3	4	2
4	16	3	4	16	3
5	27	4	5	28	4
7	7	5	7	10	5
8	16	6	8	18	6
9	25	7	9	26	7
11	4	8	11	7	8
12	17	9	12	18	9

A.D. 144/145, continued

Venus

1	1	12			
1	12	1	1	13	1
2	6	2	2	6	2
3	1	3	3	5	3
3	26	4	3	27	4
4	21	5	4	22	5
5	18	6	5	18	6
6	19	7	6	28	7
8	15	6	8	11	6
9	13	7	9	16	7
10	21	8	10	28	8
11	20	9	11	24	9
12	17	10	12	2[ ]	10

A.D. 145/146

Saturn

1	1	9	(*lost*)		
12	20	10			

Jupiter

1	1	3	1	1	3
2	19	4	3	4	4
7	16	5	7	11	5

Mars

1	1	9	1	1	9
2	1	10	2	2	10
9	21	11	9	15	11
11	15	12	11	9	12
13	3	1			

Venus

1	1	10			
1	7	11	1	1[ ]	11
2	3	12	2	2	12
2	27	1	2	2[ ]	1
3	21	2	3	[ ]	2
4	15	3	4	[ ]	3
5	8	4	5	9	4
6	2	5	6	2	5
6	26	6	6	2[ ]	6
7	20	7	7	[ ]	7
8	14	8	8	[ ]	8
9	9	9	9	[ ]	9
10	4	10	10	4	10
10	30	11	10	2[ ]	11
11	26	12	(*damaged*)		
12	25	1			

TABLE 4186.1. Sign-entries, *Handy Tables* (left) and **4186** (right).

A.D. 160/161

Planet	Handy Tables			4187		
Saturn	1	1	3	1	1	3
	6	6	4			
	8	20	3			
Jupiter	1	1	7			
	6	20	8	6	28	8
	11	9	9	10	25	9
Mars	1	1	9			
	1	12	10	1	24	10
	3	5	9	3	15	9
	6	19	10	6	25	10
	8	20	11	8	13	11
	10	11	12	10	5	12
	11	28	1	11	27	1
Venus	1	1	2			
	1	18	3	1	18	3
	2	13	4	*(remainder lost)*		

TABLE 4187.1. Sign-entries, *Handy Tables* (left) and **4187** (right).

4188. SIGN-ENTRY ALMANAC, A.D. 161–165

11 1B.151/C(d) aggregate approx. 18×17 cm After A.D. 165

The fragment comprises two disconnected pieces, the larger one (18×7.5 cm) from the top of a roll, with slight upper margin extant, the smaller one (5×5.5 cm) from about the middle height and broken on all sides. The table is on the back (some traces of writing can be seen on the front), and is ruled in red except for the black horizontals separating the sections for the planets. Line 1 has part of a heading giving a regnal formula for Marcus Aurelius and Lucius Verus—fairly formal for this genre of text. Months and zodiacal signs are represented as numerals, with omission of repeated month numbers. Positions for Thoth 1 seem to be included only when the planet makes no sign-entry during the year. Retrograde entry into a sign is indicated, at least sometimes, by a diagonal stroke after the sign's number (col. iv line 8, v line 5, vi line 8). Each year's entries take up nearly two columns, in a continuous flow (but the scribe began a new column for Mercury's entries in Year 5, making col. v line 23 prematurely the end of the column).

Tables 4188.1–2 compare the preserved entries with recomputation (*Handy Tables*, sidereal longitudes) for the Alexandrian calendar years A.D. 161/162 (col. i) through 164/165 (cols v–vi). The regnal years in question are Marcus Aurelius and Lucius Verus 2–5.

A.D. 161/162

Mercury					
1	1	1			
2	7	2	2	10	2
2	25	3	2	29	3
3	13	4	3	18	4
4	5	5			
4	22	4			
5	20	5	5	[ ]	5
6	9	6	6	[ ]	6
6	25	7	6	[ ]	7
7	11	8		(lost)	
8	3	9			
8	17	8			
9	17	9			
10	6	10			
10	22	11			
11	8	12			

A.D. 162/163

Saturn					
1	1	4		(lost)	

Jupiter					
1	1	10		(lost)	
4	27	9			
6	27	10			
11	30	11			

Mars					
1	1	10		(lost)	
1	26	11			
5	5	10			
7	13	11			
9	18	12			
11	9	1	11	[ ]	1
12	25	2	12	1[ ]	2

Venus					
1	1	1			
3	9	2	3	8	2
4	10	3	4	14	3
5	7	4	5	9	4
6	2	5	6	[ ]	6
6	27	6	6	2[ ]	7
7	22	7			
8	17	8			
9	11	9			
10	6	10			
11	1	11			
11	25	12			
12	19	1			

Mercury					
1	1	12		(lost)	
1	10	1			
1	30	2			
2	18	3			
3	7	4			
5	14	5			
6	1	6			
6	17	7			
7	5	8	7	[ ]	8
9	11	9	9	[ ]	9
9	28	10	9	22	10
10	14	11	10	12	11
11	2	12	11	[ ]	12

TABLE 4188.1. Sign-entries, *Handy Tables* (left) and **4188** cols. i–iii.6 (right).

A.D. 163/164

Saturn

1	1	4			
5	25	5	[ ]	[ ]	5
11	21	4	[ ]	[ ]	4

Jupiter

1	1	11	*(damaged)*		
12	24	12			

Mars

1	1	2	*(lost)*		
2	2	3			
3	12	4			
4	22	5			
5	30	6			
7	8	7			
8	17	8			
9	28	9			
11	11	10			
12	28	11			

Venus

1	1	1	*(lost)*		
1	8	2			
2	2	3			
2	26	4	2	26	4
3	20	5	3	20	5
4	14	6	4	15	6
5	9	7	5	10	7
6	6	8	6	7	8
7	7	9	7	7	9
9	13	8	8	9	9
10	2	9	10	9	9
11	13	10	*(damaged)*		
12	11	11			

Mercury

1	1	12	*(damaged)*		
1	4	1			
1	21	2			
2	9	3			
4	17	4			
5	6	5			
5	22	6			
6	8	7			
8	16	8			
9	3	9			
9	19	10			
10	5	11			
12	13	12			
13	2	1			

A.D. 164/165

Saturn

1	1	4			
2	21	5	2	18	5

Jupiter

1	1	12	1	1	12

Mars

1	1	11			
2	19	12	2	30	12
6	18	11	6	17	11
8	7	12	8	17	12
10	14	1	10	1[ ]	1
12	3	2	12	[ ]	2

Venus

1	1	11			
1	3	12	1	[ ]	12
1	28	1	1	28	1
2	22	2	*(lost)*		
3	16	3			
4	10	4			
5	4	5			
5	28	6			
6	22	7	6	22	7
7	16	8	7	17	8
8	10	9	8	10	9
9	5	10	9	5	10
9	30	11	9	29	11
10	25	12	10	21	12
11	21	1	11	19	1
12	16	2	12	14	2

Mercury

1	1	1	1	1	1
1	14	2	1	22	2
2	4	3	2	10	3
4	11	4	4	5	4
4	28	5	5	10	5
5	14	6	5	27	6
6	3	7	6	2	7
6	28	6	7	6	6
7	19	7	*(damaged)*		
8	9	8			
8	25	9			
9	11	10			
9	30	11			
10	26	10			
11	16	11			
12	7	12			
12	24	1			

TABLE 4188.2. Sign-entries, *Handy Tables* (left) and **4188** cols. iii.7–vi (right).

4188a. Sign-Entry Almanac, a.d. 201–208

40 5B.95/J(1–4)c 4.5×7 cm After a.d. 208

A small piece from the top centre of a codex bifolium. On the side with horizontal fibres, along the left side, is a rather faint column (col. i) of sign-entries, with no ruling visible. Along the right is another column (col. ii) and the edge of a third (col. iii). This part has ruling: a red rule along the left edge of col. ii and double along the top, double black ruling between cols. ii and iii, and black horizontals to mark off sections belonging to the years and planets. The 1 cm space between cols. i and ii is almost certainly where the sheet was bound, so that col. i is part of a verso, and cols. ii–iii of a recto. On the other side of the papyrus, the left side, i.e. verso of cols. ii–iii, has two more columns of sign-entries (cols. iv–v) ruled in black except for the red couble rule along the top; the remainder of this side is blank.

The almanac uses the standard numerical representation for months and zodiacal signs, omits repeated month numbers, and sometimes gives the planet's position for Thoth 1. A regnal year 12 in column ii together with the positions of Saturn, Jupiter, and Mars identify the year as Septimius Severus (or Caracalla) 12 = a.d. 202/203, Alexandrian calendar. Col. v, being on the verso, must be later, and the next year to which its entries (Venus and Mercury) could pertain is 207/208. Col. iii, from the slight traces, appears to contain the entries for Venus and Mercury of 202/203, and col. iv the end of Mercury for 206/207 and Saturn, Jupiter, and Mars for 207/208. We can estimate that one side of a codex page had four or possibly five columns covering about two and a half years. The entries in col. i are for Venus in a year such as 199/200 (plus or minus any small multiple of eight years).

4189. Sign-Entry Almanac, a.d. 136–144 (?)

105/147(c) 6×9.5 cm After a.d. 144?

From the part towards the binding of a codex page, with part of the inner margin extant but broken on the other sides. The side with vertical fibres is the recto. Vertical rulings are red; the horizontals separating the sections for the planets, black. There were originally four columns on each page, with each year's entries assigned a column. The usual conventions apply: numerical representation of months and zodiacal signs, month numbers omitted when they repeat, but no definite instance of an initial line for a planet's location on Thoth 1.

The page is broken in such a way that only entries for Venus and Mercury survive for each represented year. If we call the year of recto column i *y*, then the data for Venus would be compatible with *y* = A.D. 136/137 or any year a small multiple of eight years before or after. The data for Mercury best fit 136/137 (cf. Table 4189.1–2 for recomputation using the *Handy Tables*, sidereal longitudes). The hand gives the impression of a rather later date, however (some letter forms seem most appropriate for the fourth century).

A.D. 136/137

Venus					
1	1	2		(lost)	
1	19	3			
2	14	4			
3	9	5			
4	6	6			
5	7	7	5	[ ]	7
7	11	6	6	27	6
7	27	7	8	5	7
9	9	8	9	16	8
10	9	9	10	12	9
11	5	10	11	7	10
12	1	11	12	1	11
12	26	12	12	26	12
Mercury					
1	1	12	1	3	12
1	8	1	1	6	1
1	27	2	1	25	2
2	15	3	2	15	3
3	5	4			
4	10	3			
4	16	4	4	20	4
5	11	5	5	9	5
5	28	6	5	29	6
6	14	7	6	12	7
7	3	8			
7	30	7			
8	18	8	8	14	8
9	8	9	9	1	9
9	25	10	9	[ ]7	10
10	11	11	[ ]	[ ]	11
10	30	12	[ ]	[ ]	12
11	25	11	[ ]	[ ]	11
12	15	12		(lost)	

A.D. 137/138

Venus					
1	1	12		(lost)	
1	15	1			
2	9	2			
3	3	3			
3	27	4			
4	20	5			
5	14	6			
6	8	7			
7	2	8	7	5	8
7	27	9	7	29	9
8	22	10	8	24	10
9	18	11	9	18	11
10	15	12	10	16	12
11	13	1	11	8	1
12	18	2	12	16	2
Mercury					
1	1	12	1	[ ]	12
1	2	1		(damaged)	
1	19	2			
2	8	3			
4	15	4	4	15	4
5	3	5	4	30	5
5	20	6	5	18	6
6	7	7	6	4	7
			6	21	6
			7	[ ]	7
8	14	8		(damaged)	
9	1	9			
9	17	10			
10	4	11			
12	12	12			
12	29	1			

TABLE 4189.1. Sign-entries, *Handy Tables* (left) and **4189** recto cols. i–ii (right).

A.D. 142/143

Venus

1	1	1	(*lost*)		
1	24	2			
2	18	3			
3	12	4			
4	6	5			
4	29	6			
5	24	7	5	26	7
6	18	8	6	22	8
7	14	9	7	14	9
8	11	10	8	12	10
9	10	11	9	16	11
			10	25	12
			[ ]	[ ]	11

Mercury

1	1	12			
1	11	1	1	[ ]1	1
2	2	2	(*lost0*)		
2	19	3			
3	8	4			
			[ ]	[ ]	[ ]
			[ ]	[ ]	4
5	16	5	5	14	5
6	3	6	6	1	6
6	19	7	6	17	7
7	6	8	(*lost*)		
9	13	9			
9	30	10			
10	16	11			
11	3	12			

A.D. 143/144

Venus

1	1	11	(*lost*)			
1	10	12				
2	9	1				
3	6	2				
4	1	3				
4	25	4				
5	19	5				
6	13	6				
7	7	7	[ ]	[ ]	7	
8	2	8	8	[ ]	8	
8	26	9	8	29	9	
9	21	10	10	8	10	(!)
10	16	11	10	17	11	
11	10	12	11	12	12	
12	5	1	12	6	1	
12	29	2	13	1	2	

Mercury

1	1	12			
1	5	1			
1	23	2	1	7	2
2	11	3	2	13	3
3	4	4			
3	22	3			
4	18	4	4	18	4
5	8	5	5	15	5
5	24	6	5	24	6
6	10	7	6	14	7
7	4	8	[ ]	[ ]	8
7	13	7			
8	17	8			
9	5	9	8	28	9
9	21	10	9	15	10
10	7	11	10	2	11
10	30	12	(*lost*)		
11	10	11			
12	14	12			
13	3	1			

TABLE 4189.2. Sign-entries, *Handy Tables* (left) and **4189** verso cols. ii–iii (right).

4190. SIGN-ENTRY ALMANAC, A.D. 241–243

103/14(b) 5.5×6 cm After A.D. 243

Plate V

The top outer corner of a codex page, with 1 cm upper margin and 2 cm outer margin preserved. The side with vertical fibres is the recto. The table is ruled in red, and in a careful hand. This is one of the series of almanacs in four-column format (cf. also **4192**, **4194–4196a**), which all share the following features. Each Alexandrian year starts a new set of columns on the page, which likely contained three years on each side. A sign-entry is expressed by the usual three numerals, i.e. the number of the month, the day within the month, and the number of the zodiacal sign entered; in the fourth column is the hour of day or night when the sign-entry occurred. When the sign-entry takes place at night, this is indicated by a double date in the column for days. Thus the entry for recto line 3 may be read, 'Saturn enters Virgo (sign 1) on Phaophi 2, hour 5 of day,' and line 4 means 'Saturn enters Leo (sign 12) on Phamenoth 7/8, hour 5 of night.' The first entry for each planet is regularly its position on Thoth 1, with nothing in the hours column. The month number is omitted when it would repeat the line above. In **4190** we have entries for Saturn, Jupiter, and Mars for two consecutive years named in the column headings as Gordian 5 and 6 (A.D. 241/242 and 242/243).

The four-column almanacs were all computed using the *Handy Tables* with longitudes converted to sidereal. Consequently our recomputation using the same tables leads to exactly the same sequence of dates; we may therefore dispense with thorough side-by-side comparisons for these texts. For the slowest planets it would be difficult to check the derivation of the times of sign-entry, because different users of the *Handy Tables* can obtain planetary longitudes varying by one or two minutes, and this is enough to introduce many hours' uncertainty into the times when Saturn or Jupiter crosses a sign boundary. Even in the case of Mars fluctuations of one or two hours can be expected. The times recorded for Mercury in **4194** (see the commentary to that text) lead us to believe that the times were found by checking the longitudes at every equinoctial hour past noon close to the anticipated sign-entry, and that the 'seasonal' time in the almanac actually expresses equinoctial hours past 6 A.M. or 6 P.M., that is, the seasonal variability of the times of sunrise and sunset was simply ignored. Applying this procedure to the sign-entries of Mars in 242/243, we find times either in agreement with **4190** verso or not more than two hours earlier (Table 4190.1).

Text				*Recomputation*			
Month	*Day*	*Sign*	*Hour*	*Month*	*Day*	*Sign*	*Hour*
[2]	[2]	3	3rd of day	2	2	3	1st of day
[3]	12/13	4	6th of night	3	12/13	4	6th of night
[4]	[21/22]	5	11th of night	4	21/22	5	10th of night
[5]	[30]	6	5th of day	5	30	6	4th of day
[7]	[8]	[7]	11th of day	7	8	7	11th of day

TABLE 4190.1. Sign-entries of Mars, A.D. 242/243.

4191. Sign-Entry Almanac, a.d. 236–245

118/9(a) + A22/C251 6×15.5 cm (fr. 1) 1.5×4 cm (fr. 2) After a.d. 245

Plate VII

Fr. 1 is from the top outer corner of a codex page (horizontal fibres on recto), fr. 2 from the bottom edge. The table is ruled in red except for black horizontals used to separate the entries for the planets. The almanac allots a column to each year. The arrangement is the standard one, with numbers representing months and zodiacal signs, month number omitted when it would repeat, and an initial entry for position on Thoth 1 when no other entry occurs in that month. The larger fragment has entries for all the planets except Mercury, and is easily dated to the run of consecutive Alexandrian years 236/237–239/240 (recto) and 240/241–242/243 (verso). The dates for 241/242 and 242/243 may instructively be compared with those in **4190**, which were computed using the *Handy Tables*. The dates of Mercury's sign-entries on the presumed verso of fr. 2 would fit the years 243/244 and 244/245 quite well; this would situate the fragment on the same page as fr. 1, but closer to the inner margin, so that the recto would have the last lines for 235/236. The correspondence of dates for that year with the contents of the fragment are unfortunately not close enough to inspire confidence.

4192. Sign-Entry Almanac, a.d. 276–280

103/20(a) 4×6 cm After a.d. 280

From the top of a codex page, but without surviving headings or upper margin. The side with horizontal fibres is the recto. The table, which is ruled in red, might have formed part of the same manuscript as **4190**. This is a sign-entry almanac in the four-column format as described in the commentary to **4190**. The preserved entries are for Saturn, Jupiter, and Mars, with the first lines for Venus on recto cols. iv–v line 11 and verso cols. iv–vi line 12.

The recorded dates are in exact agreement with those recomputed using the *Handy Tables* (sidereal longitudes) for the four consecutive years a.d. 276/277 (recto i–iii), 277/278 (recto iv–v), 278/279 (verso i–iii), and 279/280 (verso iv–vi).

4193. Sign-Entry Almanac, a.d. 195–203

51 4B.19/C(1)c approx. 19×9 cm After a.d. 203

Shattered and disconnected remains of a codex bifolium, the middle one of a gathering since the run of years in the almanac is unbroken. The horizontal fibres were on the outside, i.e. folio A recto and B verso. The top and bottom margins are entirely lost, but there remains a little of the outer margins as well as a gap at the binding of about 2 cm. Only on folio A recto are there vertical rulings (in red) to separate the columns. The horizontal rules separating the entries for each planet are in black ink.

In the almanac, months and zodiacal signs are represented by numerals in the standard fashion, and there are no headings to identify the planets. When two consecutive entries belong to the same month, the month's number is omitted in the second instance. For the sake of clarity the day numbers have been fully indented in this edition when there is no month number; the scribe tended to use only a slight indentation in such situations. The sign occupied by a planet on Thoth 1 is entered at the beginning of its section whenever there is no further entry for Thoth. Each page has four columns of entries, two for each year, with the entries for a new year always starting a new column even if space remains at the bottom of the previous column. The dimensions of a page must have been approximately 11×8 cm exclusive of margins.

The surviving entries agree very well with recomputation (*Handy Tables*, sidereal longitudes) for the Alexandrian calendar years a.d. 195/196 (A recto, cols. i–ii) through 202/203 (B verso, cols. iii–iv).

4194. Sign-Entry Almanac, a.d. 272–286

68 6B.20/L(8)a, part 4×7 cm After a.d. 286

A fragment of a codex page, broken on all sides, but from near the bottom. The side with horizontal fibres is the recto. The table is ruled in red, and written in a hand close in character to that of **4190** and **4192**. This is a sign-entry almanac in the four-column format as described in the commentary to **4190**. The preserved entries are for Mercury, with a few remains of those for Venus in the top lines.

Recomputation using the *Handy Tables* (sidereal longitudes) for the years a.d. 282/283 (recto i–iv) and 285/286 (verso ii–v) leads to exactly the dates in the

A.D. 195/196

Saturn					
1	1	5			
6	28	6		(*damaged*)	
12	16	5			

Jupiter					
1	1	6	1	8	6
4	13	7	4	14	7
8	30	8	9	[ ]	8

Mars					
1	1	2	1	1	2
1	28	3		27	3
3	8	4	3	7	4
4	17	5	4	17	5
5	26	6	5	26	6
7	4	7	7	8	7
8	13	8	8	9	8
9	24	9	9	22	9
11	7	10	11	3	10
12	23	11	12	[ ]	11

Venus					
1	1	1			
1	6	2			
1	30	3			
2	24	4			
3	18	5			
4	13	6		(*lost*)	
5	8	7			
6	6	8			
7	11	9			
8	19	8			
10	9	9			
11	13	10			
12	10	11	12	14	11

Mercury					
1	1	12			
1	9	1	1	7	1
1	29	2	1	27	2
2	16	3	2	13	3
3	6	4	4	20	4
5	13	5	5	10	5
5	29	6	5	24	6
6	15	7	6	14	7
7	4	8			
8	8	7			
8	17	8	8	16	8
9	10	9	9	3	9
9	26	10	9	[ ]	10
10	12	11	10	[ ]	11
11	1	12			
12	3	11			
12	15	12	12	[ ]	12

A.D. 196/197

Saturn					
1	1	5		(*lost*)	
3	23	6			

Jupiter					
1	1	8		(*lost*)	
9	10	9			

Mars					
1	1	11		(*lost*)	
2	12	12			
10	8	1			
11	28	2			

Venus					
1	1	11		(*lost*)	
1	2	12			
1	26	1			
2	21	2			
3	15	3			
4	8	4			
5	2	5			
5	26	6	5	27	6
6	20	7	6	[ ]3	7
7	14	8		(*lost*)	
8	9	9			
9	3	10			
9	28	11			
10	24	12			
11	19	1			
12	16	2			

Mercury					
1	1	12		(*lost*)	
1	3	1			
1	21	2			
2	9	3			
4	17	4			
5	5	5			
5	21	6	5	25	6
6	8	7	6	13	7
8	15	8	8	16	8
9	3	9	9	2	9
9	18	10	9	18	10
10	5	11	10	6	11
12	13	12	12	12	12
13	1	1			

TABLE 4193.1. Sign-entries, *Handy Tables* (left) and **4193** fol. A recto (right).

A.D. 197/198

Saturn					
1	1	6	(*lost*)		

Jupiter					
1	1	9	(*lost*)		
9	21	10			

Mars					
1	1	2	(*lost*)		
1	7	3			
2	18	4			
3	27	5			
5	6	6			
6	14	7			
7	20	8	7	26	8
9	4	9	9	20	9
10	17	10	10	5	10
12	2	11	11	1	11

Venus					
1	1	2			
1	9	3	1	[ ]	3
2	13	4	[ ]	[ ]	4
3	24	3			
5	4	4			
6	9	5	6	8	5
7	7	6	7	14	6
8	3	7	(*lost*)		
8	29	8			
9	24	9			
10	19	10			
11	14	11			
12	8	12			
13	2	1			

Mercury					
1	1	1	(*lost*)		
1	14	2			
2	4	3			
4	10	4			
4	27	5			
5	14	6			
6	4	7			
6	24	6			
7	20	7			
8	9	8			
8	25	9			
9	11	10			
9	30	11			
10	23	10			
11	17	11	11	1[ ]	11
12	6	12	12	16	12
12	23	1	13	[ ]	1

A.D. 198/199

Saturn					
1	1	6	(*lost*)		
6	7	7			

Jupiter					
1	1	10	(*lost*)		
10	14	11			

Mars					
1	1	11	(*lost*)		
1	16	12			
3	10	1			
10	29	2			
12	19	3			

Venus					
1	1	1	(*lost*)		
1	21	2			
2	15	3	2	[ ]	3
3	9	4	3	14	4
4	3	5	4	8	5
4	27	6	5	3	6
5	21	7	5	28	7
6	16	8	6	22	8
7	12	9	7	17	9
8	10	10	8	[ ]	10
9	13	11	9	[ ]	11
11	4	10	12	[ ]	10
12	11	11	12	[ ]	11

Mercury					
1	1	1	(*lost*)		
1	6	2			
3	13	3			
4	3	4	4	[ ]	4
4	19	5	4	24	5
5	7	6	5	12	6
7	15	7	7	5	7
8	1	8	8	[ ]	8
8	17	9	8	[ ]	9
9	4	10	9	6	10
11	12	11	11	16	11
11	29	12	12	[ ]	12
12	16	1	12	[ ]	1

TABLE 4193.2. Sign-entries, *Handy Tables* (left) and **4193** fol. A verso (right).

A.D. 199/200

Saturn					
1	1	7	(lost)		

Jupiter					
1	1	11	(lost)		
11	8	12			

Mars					
1	1	3			
1	25	4	1	[ ]	4
3	5	5	3	9	5
4	14	6	4	1[ ]	6
5	23	7	5	2[ ]	7
7	3	8	7	4	8
8	14	9	9	17	9
9	28	10	9	26	10
11	13	11	11	9	11
12	30	12	13	2	12

Venus					
1	1	11			
1	11	12	1	1	12
2	8	1	2	12	1
3	4	2	3	7	2
3	28	3	4	2	3
4	22	4	4	27	4
5	16	5	5	22	5
6	10	6	6	[ ]	6
7	5	7	7	[ ]	7
7	29	8	(lost)		
8	23	9			
9	18	10			
10	13	11			
11	7	12			
12	2	1			
12	27	2			

Mercury					
1	1	2	1	1	2
2	1	1	2	10	1
2	10	2	2	17	2
3	6	3	3	13	3
3	24	4	3	27	4
4	11	5	5	17	5
6	19	6	6	20	6
7	6	7	7	8	7
7	22	8	7	26	8
8	9	9	8	12	9
9	4	10	9	[ ]	10
9	7	9			
10	16	10			
11	4	11	11	7	11
11	20	12	(lost)		
12	7	1			

A.D. 200/201

Saturn					
1	1	7	(lost)		
8	1	8			

Jupiter					
1	1	12	(lost)		
12	6	1			

Mars					
1	1	12	(lost)		
2	14	1	[ ]	[ ]4	1
4	6	2	[ ]	[ ]	2
6	8	3	[ ]	[ ]8	3
9	10	2			
11	4	3			
13	2	4			

Venus					
1	1	2			
1	17	3	1	[ ]	3
2	12	4	2	[ ]	4
3	7	5	3	[ ]	5
4	6	6	4	1	6
8	8	7	5	7	7
			6	28	6
			8	5	7
9	9	8	9	16	8
10	6	9	10	[ ]	9
11	3	10	11	[ ]	10
11	28	11	(lost)		
12	23	12			

Mercury					
1	1	1	(lost)		
2	10	2			
2	29	3			
3	16	4			
4	5	5			
5	8	4	5	[ ]	4
5	21	5	5	22	5
6	12	6	6	15	6
6	29	7	7	1	7
7	14	8	7	18	8
8	3	9	8	8	9
9	5	8			
9	17	9			
10	9	10	10	12	10
10	26	11	10	26	11
11	12	12	11	11	12
12	2	1	12	9	1
12	26	12	12	29	12

TABLE 4193.3. Sign-entries, *Handy Tables* (left) and **4193** fol. B recto (right).

A.D. 201/202

Saturn

1	1	8	(*lost*)		

Jupiter

1	1	1	(*lost*)		

Mars

1	1	4	(*lost*)		
2	10	5			
3	20	6			
5	1	7	5	[ ]	7
6	12	8	6	11	8
7	25	9	7	2[ ]	9
9	9	10	9	[ ]	10
10	20	11	10	21	11
12	12	12	12	[ ]	12

Venus

1	1	12			
1	12	1	1	16	1
2	6	2	2	[ ]	2
2	30	3	3	[ ]	3
3	23	4	3	[ ]	4
4	17	5	4	21	5
5	11	6	5	16	6
6	5	7	(*lost*)		
6	29	8			
7	24	9			
8	19	10			
9	16	11			
10	13	12			
11	14	1			

Mercury

1	1	12	(*lost*)		
1	11	1			
2	3	2			
2	21	3			
3	9	4			
5	17	5	5	[ ]	5
6	5	6	6	8	6
6	21	7	6	28	7
7	7	8	7	1[ ]	8
9	15	9	9	20	9
10	2	10	10	1	10
10	18	11	10	15	11
11	5	12	11	1	12

A.D. 202/203

Saturn

1	1	8	(*lost*)		
9	18	9			

Jupiter

1	1	1	(*lost*)		
1	3	2			

Mars

1	1	12	(*lost*)		
1	24	1			
3	12	2			
5	1	3			
6	20	4			
8	15	5	[ ]	[ ]	5

Venus

1	1	1	1	1	1
3	11	2	3	11	2
4	10	3	4	21	3
5	5	4	5	[ ]1	4
5	30	5	(*damaged*)		
6	25	6			
7	20	7			
8	14	8			
9	9	9			
10	4	10			
10	29	11			
11	23	12			
12	17	1	[ ]	[ ]	1
			[ ]	[ ]	2

Mercury

1	1	12	(*damaged*)		
1	8	1			
1	26	2			
2	14	3			
3	5	4			
3	30	3			
4	19	4			
5	10	5	5	6	5
5	27	6	6	1	6
6	13	7	6	19	7
7	4	8			
7	22	7			
8	19	8	8	[ ]	8
9	8	9	9	7	9
9	24	10	9	[ ]	10
10	10	11	[ ]	[ ]	11
10	30	12	1[ ]	[ ]	12
11	18	11			
12	16	12			

TABLE 4193.4. Sign-entries, *Handy Tables* (left) and **4193** fol. B verso (right).

Text				*Recomputation*			
Month	*Day*	*Sign*	*Hour*	*Month*	*Day*	*Sign*	*Hour*
1	18	2	1st of day	1	18	2	1st of day
2	7	3	2nd of day	2	7	3	1st of day
4	14	4	12th of day	4	14	4	12th of day
5	1/2	5	12th of night	5	1/2	5	12th of night
5	18	6	5th of day	5	18	6	5th of day
6	6	7	2nd of day	6	6	7	2nd of day
8	13	8	1st of day	8	13	8	1st of day
8	29	9	12th of day	8	29	9	12th of day
9	15	10	8th of day	9	15	10	8th of day
10	2/3	11	10th of day	10	2/3	11	9th of day
12	10/11	12	7th of night	12	10/11	12	7th of night
12	28	1	2nd of day	12	28	1	2nd of day

TABLE 4194.1. Sign-entries of Mercury, A.D. 312/313 (**4194** recto).

papyrus; hence recto col. v can be identified as 283/284, and verso col. i as 285/286. The times of sign-entry are effectively equinoctial hours counted from 6 A.M. or 6 P.M. With the sole exception of a four-hour discrepancy in verso line 15, they are consistently either identical to the times obtained by our recomputation or at most one hour later (Table 4194.1–2). The agreement is so good, even near retrogradations, that one is forced to infer that the computer of the almanac calculated Mercury's longitude at one-hour intervals when approaching a sign-entry.

Text				*Recomputation*			
Month	*Day*	*Sign*	*Hour*	*Month*	*Day*	*Sign*	*Hour*
2	12/13	2	5th of night	2	12/13	2	4th of night
3	3	3	7th of day	3	3	3	7th of day
3	20/21	4	4th of night	3	20/21	4	4th of night
4	8/9	5	2nd of night	4	8/9	5	2nd of night
6	16/17	6	3rd of night	6	16/17	6	3rd of night
7	3	7	7th of day	7	3	7	6th of day
7	19	8	2nd of day	7	19	8	2nd of day
8	6	9	6th of day	8	6	9	6th of day
10	13/14	10	1st of night	10	13/14	10	1st of night
10	30	11	7th of day	10	30	11	6th of day
11	16/17	12	2nd of night	11	16	12	10th of day (!)
12	5	1	3rd of day	12	5	1	3rd of day

TABLE 4194.2. Sign-entries of Mercury, A.D. 315/316 (**4194** verso).

4195. Sign-Entry Almanac, a.d. 300–304

65 6B.36/E(1–3)d 3×3.5 cm After a.d. 304

A fragment of a codex page, broken on all sides. The side with horizontal fibres is the recto. The table is ruled in black. This is a sign-entry almanac in the four-column format as described in the commentary to **4190**. The preserved entries are for Saturn, Jupiter, and Mars. All dates and sign numbers are in agreement with recomputation using the *Handy Tables* (sidereal longitudes) for the four consecutive Alexandrian years a.d. 300/301 (recto i–iii), 301/302 (recto iv–v), 302/303 (verso i-ii), and 303/304 (verso iii–iv).

4196. Sign-Entry Almanac, a.d. 218–220

68 6B.20/L(8)a, part 4.5×5.5 cm After a.d. 220

A fragment of a codex page, with 3 cm outer margin, broken on the other sides. The side with vertical fibres is the recto. The table is ruled in black, and on the recto a superfluous series of vertical rulings has been drawn in the margin, as the scribe originally contemplated squeezing one more year on the page. This is a sign-entry almanac in the four-column format as described in the commentary to **4190**. The preserved entries are for Venus. The dates of sign-entry on the verso are consistent with recomputation using the *Handy Tables* (sidereal longitudes) for only two Alexandrian years: a.d. 211/212 or 219/220. The times of sign-entry on the recto are in satisfactory agreement with recomputation for 218/219 (text consistently one hour later than recomputation), confirming the later date for the verso.

4196a. Sign-Entry Almanac

65 6B.30/M(1)a, part 1.5×2 cm

A tiny scrap from a codex page, preserving a little of either the inner or outer margin, broken on the other sides, and ruled in red. This is a sign-entry almanac in the four-column format as described in the commentary to **4190**. The 'front' looks as if it is from the end of a column, i.e. the last entries of Mercury for a year; if so, then the other side also pertains to Mercury. However, I did not succeed in finding years even roughly contemporary with the other four-column almanacs such that, according to the *Handy Tables*, Mercury entered the signs Cancer, Gemini, Cancer, and Leo at the times indicated in the papyrus.

4197. PERPETUAL SIGN-ENTRY ALMANAC

28 4B.62/E(1–5)a 11×14.5 cm Third century

A codex page (the recto having vertical fibres), shattered but preserving parts of the top, bottom, and outer margins. The table is ruled in black, and written in a coarse third-century hand.

The plan of a perpetual almanac is to present for each planet *Y* consecutive Alexandrian years' positions starting from an epoch year, where *Y* is a Goal-Year period associated with the planet (e.g. 79 years for Mars, 8 for Venus). The user of the tables wishing to find the positions applicable to a given year would count the years since epoch, take the remainder *r* after division by *Y*, and read off the entries in the almanac for year *r*. **4197** applies this principle to the sign-entry almanac, that is, the positions recorded for each planet are sign-entries (other examples include **4198** and P. Vind. G.. 36041 + 26011, for which see Jones [1994c]). The sign-entries themselves are presented in the usual format, i.e. numerals represent months and zodiacal signs, repeated month numbers are omitted, but the position on Thoth 1 of each year is not normally given since it would merely repeat the position in the last line of the year immediately above. The recto contains sign-entries for Mars, years 53–74; the verso continues with years 75–79, followed by all eight years of Venus' period.

Although the contents of a perpetual almanac are intended to be valid for a series of years at intervals of *Y*, they were presumably computed in the first instance for a specific range of years. Moreover one would expect that the epoch year would be the same for all the planets, perhaps the first regnal year of an emperor. These assumptions might lead to clues concerning the date of composition of the almanac. For **4197**, however, the results are inconclusive. The entries for Venus seem to have been computed for a cycle of 8 years such as that beginning A.D. 203/204, i.e. the epoch year was 202/203 plus or minus some small multiple of 8 years. Assuming that the cycles for Mars were counted from the same epoch year, the only epoch year tolerably close to 202/203 that works at all well for Mars is 179/180, a year of no obvious significance.

4198. Perpetual Sign-Entry Almanac

106/80(a) 9×10.5 cm (fr. 1) 2.5×3 cm (fr. 2) debris

Shattered remains of a codex page, including parts of the outer margin: 2.5 cm on the 'front', 4 cm on the 'back'. The table is ruled in red. The table consists entirely of sign-entries of Mercury for a succession of Alexandrian calendar years, evidently part of a perpetual almanac like **4197**, laid out in sets of four columns respectively for the year number since epoch (only surviving example: verso col. v line 3), month number, day, and number of zodiacal sign. Month numbers are omitted when they would repeat; there are no special entries for the position occupied on Thoth 1 of each year. Because of the poor condition of the papyrus, and the uncertainties that typically adhere to ancient computations of Mercury's motions, I have not succeeded in securely establishing the sequence of the preserved parts of the almanac.

4199. Monthly Almanac, Jupiter, 14 B.C.–6 B.C.

81 2B.85/17(a) + 15 2B.31/A(b), part 8×22.5 cm Second century

From a codex page (recto has horizontal fibres), written in a hand that can scarcely be earlier than the second quarter of the second century. The top margin is partly preserved (1 cm including heading), as well as 0.5 cm vacant to the left of the table on the recto and to the right of the table on the verso. Bottom and outer edge are broken. The original page would have been about 27 cm tall, and at least 10 cm wide—roughly twice that broad if further sets of columns have been lost on the side towards the binding. One fragment (now joined to the rest) was found in the same folder as **4203a**.

Each side bears positions of a planet for the twelve months of four consecutive years, numbered (in the space above each year's entries) as 17–20 on the recto and 21–24 on the verso. In each line representing a month, col. i gives the date of the recorded event, i.e. sign-entry, phase, or longitude on the first of the month, and col. ii gives the longitude (sign, degrees, minutes). If the event is a phase, it is identified by an abbreviation or, for first and last visibility, by a small circle with ascending diagonal tail. The contents show that we are dealing with Jupiter on both sides, and the years turn out to be regnal years of Augustus, equivalent to

14/13 through 7/6 B.C. It is not certain whether the calendar is Alexandrian or Egyptian, since the deviation between them was no more than five days in the years in question, that is, within the acceptable margin of error for dating phenomena of Jupiter. The interval of well over a century between the dates of the contents and of the writing is exceptional. P. Berlin inv. 21226 (Brashear & Neugebauer [1974]) is a very similar almanac for Jupiter in codex form, with data for years 82–85 counted from the first year of Augustus, i.e. A.D. 52/53–55/56. It is probable that we are here dealing with two copies of a standard set of tables for Jupiter covering a much longer span than would have been practically necessary for a user in the middle of the second century.

4200. Monthly Almanac, Jupiter

100/65(f) 9.5×7 cm Fourth century

From a codex page (recto has horizontal fibres). Bottom margin is present (1.5 cm), and the outer edge is effectively the original edge of the page. The page was tall enough to have room for two and a quarter years' entries per column, and there were at least two columns on each side.

Positions of a planet are recorded for the twelve months of a series of Alexandrian years, identified by a year number above the first line. The entries on the recto belong to year 57, those on the verso to years 59–60 (col. i) and 62 (col. ii). It is not clear what epoch these are counted from (it is not year 1 of Augustus); they may situate the years in a recurring cycle. Each entry consists of the number of the month, the day, abbreviated name of the phase if it is one, number of the zodiacal sign, degrees, and minutes.

4201. Monthly Almanac, A.D. 126–128

19 2B.75/D(d) 6×4 cm After A.D. 128

The fragment consists of two brittle fragments adhering together on the side with horizontal fibres. A text of uncertain nature can be seen on the inside; our table is on both sides of the outside, although only one face is legible enough to transcribe. It is possible that both faces are from a single sheet that has become folded and stuck, or that we have a deliberate attempt to form a codex page by gluing together two sheets of used papyrus with the blank sides outermost. On both faces are parts of two columns of planetary positions ruled in red. All the edges are broken.

The complete table probably had parallel columns for the five planets; cols. i and ii are those for Saturn and Jupiter. Each line, representing an Alexandrian calendar month, has the month number, the day, the number of the zodiacal sign, degrees and minutes, and identification of the phase if it is one (Jupiter's Ω? by circle with ascending tail, col. ii 6). The years in question were A.D. 126/127 (lines 1–8) and 127/128 (lines 10–12).

4202. MONTHLY ALMANAC, A.D. 152–155

8 1B.197/F(5) 15×17 cm After A.D. 155

The table is written, with red ruling, on the back of accounts from the late first century (parts of two columns extant). The surface is badly abraded. Part of the lower margin (2 cm) is present, but the other sides are broken off.

Col. i, probably the original leftmost column, lists the months of each consecutive Alexandrian calendar year, identifiable as A.D. 152/153 (lines 1–6), 153/154 (lines 8–19), and 154/155 (lines 21–32). To the right of this were five pairs of columns for the five planets in the standard order, of which those for Saturn (cols. ii–iii), Jupiter (iv–v), and Mars (vi–vii). When a single event is noted for a particular month, i.e. a phase, sign entry, or longitude on the first of the month, the date is recorded in the narrower left column of the pair, and in the other column, the zodiacal sign, degrees and minutes, and an indication of the phase if it is one. First and last visibility are marked by circles with ascending tails, the other phases by abbreviations. If there is an additional event within the same month, the date of this event (with a bar over it) and the other information are written in the remaining space on the same line of the table.

4203. MONTHLY ALMANAC, SATURN, A.D. 215/216

104/153(b) 4.5×9 cm After A.D. 216

A fragment from the top of a codex page, with 2 cm upper margin, broken on the other sides. The side with vertical fibres, here designated the verso, bears positions of Saturn for a year 24, including an occurrence of first visibility (Γ) in

Phaophi, marked by an empty circle with rising tail (verso line 4); this year is securely identifiable as Septimius Severus 24 (Alexandrian), i.e. A.D. 215/216. The other side is more problematic. The sequence of longitudes makes no sense for any planet, but if one assumes that the recurring sign Virgo is correct but that errors have been made in copying the longitudes, the planet would again be Saturn, with positions probably for A.D. 213/214.

4203a. Monthly Almanac, Mercury

15 2B.31/A(b), part 3×17.5 cm Second or third century
Plate X

A narrow vertical strip, with 4 cm bottom margin, but broken on the other sides. The table, ruled in red, is on the front only; the back is blank. There is also a small scrap (1×3 cm, not transcribed) from elsewhere in the table; and a part of **4199** was found in the same folder.

The planet in question is easily identified as Mercury, but there is no possibility of establishing the (Alexandrian) years in question. Each month's entry seems to have given the date of the recorded event, the longitude (sign, degrees, minutes), and the identity of the phase if it is one. Morning phases (both Γ and Σ) are marked by a pair of small empty circles with ascending diagonal tails; evening phases (Ξ and Ω) by a pair of filled circles with tails. If there is a second event in the same month, as often happens for Mercury, it is written in the continuation of the line, with a bar over the date (bold type in the translation).

4204. Almanac, A.D. 54

103/142(a) 22×21 cm After A.D. 54

A shattered, brittle, and distorted fragment. The tables, ruled in black (but the middle line of the triple rulings is faint grey) and written in a documentary hand, are on the front of the papyrus; on the back, faint remains of a document. Margins are present on the top (3.5 cm) and bottom (2.5 cm). There is also 1 cm vacant space to the right of col. iii. 3.5cm top, 2.5 cm bottom, and (apparently) 1.5cm right margin preserved; about 2cm of column i is lost to the left. Shattered and brittle. Written on front (along the fibres) only in a documentary hand.

This almanac has a unique format, although its contents most nearly resemble the monthly almanacs. Half an Egyptian year is covered, from Phamenoth to the end of the year. Col. i, presumably the first column of the almanac (or of this section of it at least) begins with an introduction, lines 1–5, very poorly preserved, but evidently describing the contents (e.g. sign-entries mentioned at the end of line 2) and giving the date (year [14] of Tiberius [Claudius], line 3). Then each planet is allotted a section headed by its name: Saturn (col. i 6–9), Jupiter (i 10–15), Mars (ii 1–6), Venus (ii 7–15), and Mercury (iii). The events listed are sign-entries and phases, described by month and day, zodiacal sign, sometimes degrees, and the name of the phase if it is one.

4205. Five-Day Almanac, a.d. 257/258

65 6B.30/C(1)b   4.5×4 cm   After a.d. 258
Plate V

From a codex page (recto has vertical fibres), broken on top, bottom, and inner side, but the remaining edge (1cm blank on recto but not on verso) appears to be the outer margin.

The preserved longitudes on the recto and column i of the back are of Venus, while column ii pertains to Mercury. Each column of the almanac covered eight or nine months' motion, with one planets' entries followed immediately by the next. Venus's longitudes are tabulated for the 5th, 10th, etc. of each month, but Mercury's follow the more usual norm starting with the 1st, 6th, etc. For comparison with longitudes recomputed using the *Handy Tables*, see Table 4205.1.

Julian	Alexandrian	Text	Recomputation	Julian	Alexandrian	Text	Recomputation
Venus				258 July 19	Epeiph 25	♍ 0;11°	♍ 0;21°
257 Nov 1	Hathyr 5	♏ 11;15°	♏ 11;17°	258 July 24	Epeiph 30	♍ 1; 5°	♍ 1; 7°
257 Nov 6	Hathyr 10	♏ 17;40°	♏ 17;38°	258 July 29	Mesore 5	♍ 1;10°	♍ 1;10°
257 Nov 11	Hathyr 15	♏ 24; 0°	♏ 23;58°	258 Aug 3	Mesore 10	♍ 0;15°	♍ 0;13°
257 Nov 16	Hathyr 20	♐ 0;19°	♐ 0;19°				
257 Nov 21	Hathyr 25	♐ 6;37°	♐ 6;40°	*Mercury*			
257 Nov 26	Hathyr 30	♐ 12;58°	♐ 13; 1°	258 March 27	Pharm 1	♓ 27°	♓ 27;57°
257 Dec 1	Choeac 5	♐ 19;20°	♐ 19;22°	258 April 1	Pharm 6	♈ 7°	♈ 7;23°
				258 April 6	Pharm 11	♈	♈ 16;54°
258 June 29	Epeiph 5	♌ 19;35°	♌ 19;47°	258 April 11	Pharm 16	♈ 2x°	♈ 26;22°
258 July 4	Epeiph 10	♌ 23; 4°	♌ 23;14°	258 April 16	Pharm 21	♉ 5°	♉ 5;44°
258 July 9	Epeiph 15	♌ 25;55°	♌ 26;12°	258 April 21	Pharm 26	♉ 14°	♉ 14;50°
258 July 14	Epeiph 20	♌ 28;23°	♌ 28;36°				

Table 4205.1. Concordance of planetary longitudes
(recomputation: *Handy Tables*, sidereal, 6 p.m.).

4205a. Five-Day Almanac

66 6B.26/K(1–2)g 2×3 cm

From a codex page, broken on all sides, written in the same (?) hand as **4205**. On the 'front', the format of the table seems to be nonstandard, with some positions tabulated at ten-day intervals starting with the 10th of the month. I have not succeeded in finding a date for the longitudes (Mercury in Choeac?) on the 'back'.

4206. Five-Day Almanac, A.D. 272–274

68 6B.25/H(3)d, part 5.5×7 cm (fr. 1) 3×5 cm (fr. 2) After A.D. 274

Two fragments from a single codex page (recto has horizontal fibres). Fr. 1 has part of the top of the page, albeit with no margin; otherwise all sides of both fragments are broken. Each page of this almanac covered one year, with two columns for each planet in the standard order. The four numbers in each line are, respectively, the date, the number of the zodiacal sign, degrees, and minutes. The recto of fr. 1 bears the beginning of the second half of the year A.D. 272/273 for Mars, and the beginning of the first half of the same year for Venus. On the verso are the beginnings of the second half of A.D. 273/274 for Jupiter and of the first half of the same year for Mars. Fr. 2 is from towards the bottom of the page, with longitudes of Venus from both halves of 272/273 on the recto, and longitudes of Saturn (second half of 273/274) and Jupiter (first half) on the verso. The recorded positions are compared with recomputation using the *Handy Tables* in Table 4206.1.

4207. Five-Day Almanac, A.D. 289–291

103/70(b) 2×4 cm After A.D. 291

An abraded fragment of a codex page (recto with vertical fibres), broken on all sides, ruled in red. The arrangement was like that of **4206**, with each page bearing two columns (Thoth–Mecheir, Phamenoth–Epagomenae) for each planet in the standard order. Each entry consisted of four numbers: the day of the month, the number of the zodiacal sign, and the longitude in degrees ande minutes. The longitudes on the verso are of Saturn for Pharmuthi and Pachon, A.D. 290/291, i.e. the second column on the page; see Table 4207.1 for comparison with recomputed

Julian	Alexandrian	Text	Recomputation
Saturn			
274 June 20	Payni 26	♍ 2x°	♍ 29; 2°
274 June 25	Epeiph 1	♍ xx;x1°	♍ 29;12°
274 June 30	Epeiph 6	♍ xx;12°	♍ 29;25°
274 July 5	Epeiph 11	♍ xx;26°	♍ 29;41°
274 July 10	Epeiph 16	♍ xx;x3°	♍ 29;59°
274 July 15	Epeiph 21	♎ 0; 3°	♎ 0;19°
274 July 20	Epeiph 26	♎ x;xx°	♎ 0;27°
274 July 25	Mesore 1	♎ x;47°	♎ 0;51°
274 July 30	Mesore 6	♎ x;x4°	♎ 1;16°
274 Aug 4	Mesore 11	♎ 2;xx°	♎ 1;44°
Jupiter			
273 Dec 27	Tybi 1	♎	♎ 18;10°
274 Jan 1	Tybi 6	♎	♎ 18;41°
274 Jan 6	Tybi 11	♎	♎ 19; 8°
274 Jan 11	Tybi 16		♎ 19;31°
274 Jan 16	Tybi 21	♎	♎ 19;51°
274 Jan 21	Tybi 26	♎	♎ 20; 6°
274 Jan 26	Mecheir 1	♎ 20°	♎ 20;16°
274 Jan 31	Mecheir 6	♎	♎ 20;22°
274 Feb 25	Pham 1	♎ 19;43°	♎ 19;44°
274 March 2	Pham 6	♎ 19;22°	♎ 19;23°
274 March 7	Pham 11	♎ 18;57°	♎ 18;58°
274 March 12	Pham 16	♎ 18;29°	♎ 18;29°
274 March 17	Pham 21	♎ 17;59°	♎ 17;58°
274 March 22	Pham 26	♎ 17;23°	♎ 17;23°
274 March 27	Pharm 1	♎ xx;46°	♎ 16;46°
274 April 1	Pharm 6	♎ xx; 8°	♎ 16; 8°
274 April 6	Pharm 11	♎ xx;31°	♎ 15;31°
274 April 11	Pharm 16	♎ xx;52°	♎ 14;53°
274 April 16	Pharm 21		♎ 14;15°
274 April 21	Pharm 26	♎ xx;36°	♎ 13;37°
Mars			
273 Feb 25	Pham 1	♒ 11;49°	♒ 11;51°
273 March 2	Pham 6	♒ 15;46°	♒ 15;46°
273 March 7	Pham 11	♒ 19;41°	♒ 19;43°
273 March 12	Pham 16	♒ 23;35°	♒ 23;36°
273 March 17	Pham 21	♒ 27;27°	♒ 27;29°
273 March 22	Pham 26	♓ 1;20°	♓ 1;22°
273 March 27	Pharm 1	♓ 5;x4°	♓ 5;15°
273 April 1	Pharm 6	♓ 9; 6°	♓ 9; 7°
273 April 6	Pharm 11	♓ 12;57°	♓ 13; 0°
273 April 11	Pharm 16	♓ 16;47°	♓ 16;49°
273 April 16	Pharm 21	♓ 20;36°	♓ 20;38°
273 April 21	Pharm 26	♓ 24;19°	♓ 24;26°
273 April 26	Pachon 1	♓ 2x°	♓ 28;13°
273 May 1	Pachon 6	♈ 1°	♈ 1;57°
273 Aug 29	Thoth 1	♊ 22;2x°	♊ 22;24°
273 Sept 3	Thoth 6	♊ 25; 7°	♊ 25; 9°
273 Sept 8	Thoth 11	♊ 27;44°	♊ 27;45°
273 Sept 13	Thoth 16	♋ 0;17°	♋ 0;20°
273 Sept 18	Thoth 21	♋ 2;48°	♋ 2;49°
273 Sept 23	Thoth 26	♋ 5;13°	♋ 5;13°
273 Sept 28	Phaophi 1	♋ 7;31°	♋ 7;30°
273 Oct 3	Phaophi 6	♋ 9°	♋ 9;40°
273 Oct 8	Phaophi 11	♋ 11;x1°	♋ 11;43°
273 Oct 13	Phaophi 16	♋ 13;41°	♋ 13;42°
273 Oct 18	Phaophi 21	♋ 15;30°	♋ 15;30°
273 Oct 23	Phaophi 26	♋ 17; 2°	♋ 17; 2°
273 Oct 28	Hathyr 1	♋ 18;26°	♋ 18;26°
Venus			
272 Aug 29	Thoth 1	♎ 15°	♎ 15;23°
272 Sept 3	Thoth 6	♎ 21;20°	♎ 21;20°
272 Sept 8	Thoth 11	♎ 27;1x°	♎ 27;15°
272 Sept 13	Thoth 16	♏ 3;1x°	♏ 3;12°
272 Sept 18	Thoth 21	♏ 9°	♏ 9; 3°
272 Sept 23	Thoth 26	♏	♏ 14;53°
272 Sept 28	Phaophi 1	♏ 20°	♏ 20;42°
272 Oct 3	Phaophi 6	♏ 2x°	♏ 26;28°
272 Oct 8	Phaophi 11	♐ 2°	♐ 2; 8°
272 Oct 13	Phaophi 16	♐ 7°	♐ 7;43°
272 Oct 18	Phaophi 21	♐	♐ 13;13°
272 Oct 23	Phaophi 26	♐	♐ 18;42°
272 Dec 27	Tybi 1	♒ x;49°	♒ 1;47°
273 Jan 1	Tybi 6	♒ x;30°	♒ 0;27°
273 Jan 6	Tybi 11	♑ x8;x3°	♑ 28;12°
273 Jan 11	Tybi 16	♑ x5;37°	♑ 25;33°
273 Jan 16	Tybi 21	♑ x2;44°	♑ 22;41°
273 Jan 21	Tybi 26	♑ 20;19°	♑ 20;15°
273 Jan 26	Mecheir 1	♑ 18;32°	♑ 18;32°
273 June 20	Payni 26	♉ 26°	♉ 26;56°
273 June 25	Epeiph 1	♊ 2;3x°	♊ 2;51°
273 June 30	Epeiph 6	♊ 8;3x°	♊ 8;47°
273 July 5	Epeiph 11	♊ 14°	♊ 14;46°
273 July 10	Epeiph 16	♊ 20;2x°	♊ 20;44°
273 July 15	Epeiph 21	♊	♊ 26;44°
273 July 20	Epeiph 26	♋	♋ 2;30°
273 July 25	Mesore 1		♋ 8;34°

TABLE 4206.1. Concordance of planetary longitudes (recomputation: *Handy Tables*, sidereal, noon).

Julian	Alexandrian	Text	Recomputation	Julian	Alexandrian	Text	Recomputation
Saturn							
291 April 1	Pharm 6	♈ xx;49°	♈ 19;43°	291 April 26	Pachon 1	♈ 23; 7°	♈ 23; 6°
291 April 6	Pharm 11	♈ 20;29°	♈ 20;24°	291 May 1	Pachon 6	♈ 23;4x°	♈ 23;46°
291 April 11	Pharm 16	♈ 21; 9°	♈ 21; 5°	291 May 6	Pachon 11	♈ 24;2x°	♈ 24;26°
291 April 16	Pharm 21	♈ 21;45°	♈ 21;45°	291 May 11	Pachon 16	♈ 25;xx°	♈ 25; 3°
291 April 21	Pharm 26	♈ 22;26°	♈ 22;26°				

TABLE 4207.1. Concordance of planetary longitudes
(recomputation: *Handy Tables*, sidereal, noon).

longitudes according to the *Handy Tables*. The recto is the second from last column, belonging to Mercury for Phaophi and Hathyr of 289/290, but preserving only the numbers for the signs (Libra throughout, which is valid for 289/290 but not 290/291 or 291/292).

4208. FIVE-DAY ALMANAC, A.D. 295/296

68 6B.25/H(3)d, part 3.5×6 cm After A.D. 296

The fragment is written on the front only, and broken on all sides. The traces of column i are too slight for identification. Column ii belongs to Jupiter, covering the end of the year 295/296. Since this is not an intercalary year in the Alexandrian calendar, the second line for the Epagomenae presumably gives the longitude on the last day of the year even though this falls out of the regular five-day sequence. Each line consists of the date, the number of the sign (usually omitted unless different from the preceding entry, and marked by a diagonal stroke), the degrees and minutes.

Julian	Alexandrian	Text	Recomputation	Julian	Alexandrian	Text	Recomputation
Jupiter				296 Aug 4	Mesore 11	♍ 3; 9°	♍ 3;11°
296 July 10	Epeiph 16	♌	♌ 28; 9°	296 Aug 9	Mesore 16	♍ 4;13°	♍ 4;14°
296 July 15	Epeiph 21	♌ 29; 5°	♌ 29; 7°	296 Aug 14	Mesore 21	♍ 5;17°	♍ 5;18°
296 July 20	Epeiph 26	♍ 0; 5°	♍ 0; 7°	296 Aug 19	Mesore 26	♍ 6;22°	♍ 6;22°
296 July 25	Mesore 1	♍ 1; 6°	♍ 1; 8°	296 Aug 24	Epagom 1	♍ 7;26°	♍ 7;28°
296 July 30	Mesore 6	♍ 2; 7°	♍ 2; 9°	296 Aug 28	Epagom 5	♍ 8;18°	♍ 8;20°

TABLE 4208.1. Concordance of planetary longitudes
(recomputation: *Handy Tables*, sidereal, noon).

4209. Five-Day Almanac, a.d. 295/296

68 6B.20/L(8) a, part 7×7 cm After a.d. 296

The table is written in a black tabular framework on the front only of the papyrus, along the fibres (the back is blank). The fragment is broken on the bottom and both sides. There is a 2 cm margin at the top including the headings above the ruling. The tops of three consecutive columns are partly preserved, giving longitudes respectively of Saturn for Phamenoth and Pharmuthi of 295/296, Saturn for Epeiph, and Jupiter for Choeac. Thus each column contained five months' entries, and just one column for the first five months of Saturn is missing on the left. The complete table for the year would have required twelve columns, making it about 30×17cm, not including margins. In addition to the longitudes at five-day intervals, the almanac reports Saturn's second station on Mecheir 24 and Jupiter's first station on Choeac 17. Each entry consists of the date, the number of the zodiacal sign (usually omitted, and marked with a diagonal stroke when included), the degrees and minutes, and, where relevant, a notation to identify planetary phases. Table 4209.1 compares the contents of the papyrus with recomputation according to the *Handy Tables*.

Julian	Alexandrian	Text	Recomputation
Saturn			
296 Jan 27	Mecheir 1	♊ 21;53°	♊ 21;54°
296 Feb 1	Mecheir 6	♊ 21°	♊ 21;40°
296 Feb 6	Mecheir 11	♊ 21°	♊ 21;30°
296 Feb 11	Mecheir 16	♊ 21°	♊ 21;22°
296 Feb 16	Mecheir 21	♊ 21°	♊ 21;18°
296 Feb 19	Mecheir 24	♊ 21°	♊ 21;15°
296 Feb 21	Mecheir 26	♊ 21;16°	♊ 21;15°
296 Feb 26	Pham 1	♊ 21;18°	♊ 21;16°
296 March 2	Pham 6	♊ 21;20°	♊ 21;19°
296 March 7	Pham 11	♊ xx;x7°	♊ 21;26°
296 June 25	Epeiph 1		♋ 2;31°
296 June 30	Epeiph 6		♋ 3;12°

Julian	Alexandrian	Text	Recomputation
296 July 5	Epeiph 11	♋ 3;47°	♋ 3;52°
296 July 10	Epeiph 16	♋ 4;17°	♋ 4;18°
296 July 15	Epeiph 11	♋ 4°	♋ 4;57°
296 July 20	Epeiph 26	♋ 5°	♋ 5;35°
Jupiter			
295 Nov 28	Choeac 1		♌ 28; 7°
295 Dec 3	Choeac 6	♌ 28;1x°	♌ 28;19°
295 Dec 8	Choeac 11	♌ 28;24°	♌ 28;25°
295 Dec 13	Choeac 16	♌ 28;27°	♌ 28;28°
295 Dec 14	Choeac 17	♌ 28;27°	♌ 28;28°
295 Dec 18	Choeac 21	♌ 28;25°	♌ 28;27°
295 Dec 23	Choeac 26	♌ xx;20°	♌ 28;21°

Table 4209.1. Concordance of planetary longitudes (recomputation: *Handy Tables*, sidereal, noon).

4210. Five-Day Almanac, a.d. 300–302

68 6B.20/L(14–16)a+b 4.5×5 cm (fr. 1) 4.5×6 cm (fr. 2) After a.d. 302

Fragment 1, comprising two almost adjacent pieces, is written on both sides, that with horizontal fibres being the recto. Along the left of the recto and the right of the verso, 1 cm vacant space; the other sides are broken. Fragment 2, although it belongs to the same page of tables, has the almanac on the verso only. On the recto are parts of six lines of text, apparently subscriptions to a contract, written the other way up, and below this a margin of 2.5 cm. Hence the corresponding *top* edge of the verso, containing the regnal year, is probably a margin reserved for headings. This is therefore probably not a true codex page, but rather a reused part of a roll in which part of the front as well as the back was vacant before the table was put on it. Cf. also **4213**, which may be a further fragment of the same manuscript.

The entries give the date and the longitude in degrees and minutes; the zodiacal sign was specified by inserting its number after the date when a new sign was entered (the only surviving instance is fr. 2, line 2). In chronological order, we

Julian	Alexandrian	Text	Recomputation	Julian	Alexandrian	Text	Recomputation
Saturn				302 April 1	Pharm 6	♍ 12;24°	♍ 12;25°
300 Dec 28	Tybi 1	♍ 5;21°	♍ 5;22°	302 April 6	Pharm 11	♍ 12; 6°	♍ 12; 6°
301 Jan 2	Tybi 6	♍ 5;13°	♍ 5;13°	302 April 11	Pharm 16	♍ 11;4x°	♍ 11;48°
301 Jan 7	Tybi 11	♍ 5; 3°	♍ 5; 4°	302 April 16	Pharm 21	♍ xx;x3°	♍ 11;33°
301 Jan 12	Tybi 16	♍ 4;49°	♍ 4;50°				
301 Jan 17	Tybi 21	♍ 4;34°	♍ 4;35°	*Jupiter*			
301 Jan 22	Tybi 26	♍ 4;17°	♍ 4;17°	302 April 11	Pharm 16	♓ 4;58°	♓ 4;59°
301 Jan 27	Mecheir 1	♍ 4; 0°	♍ 3;58°	302 April 16	Pharm 21	♓ 5;57°	♓ 5;58°
301 Feb 1	Mecheir 6	♍ 3;37°	♍ 3;38°	302 April 21	Pharm 26	♓ 6;56°	♓ 6;57°
301 Feb 6	Mecheir 11	♍ 3;13°	♍ 3;14°	302 April 26	Pachon 1	♓ 7;50°	♓ 7;52°
301 Feb 11	Mecheir 16	♍ 2;50°	♍ 2;51°	302 May 1	Pachon 6	♓ 8;44°	♓ 8;45°
				302 May 6	Pachon 11	♓ 9;33°	♓ 9;34°
302 Feb 25	Pham 1	♍ 1x; 8°	♍ 15; 9°	302 May 11	Pachon 16	♓ xx;20°	♓ 10;21°
302 March 2	Pham 6	♍ 14;46°	♍ 14;46°	302 May 16	Pachon 21		♓ 11; 5°
302 March 7	Pham 11	♍ 14;19°	♍ 14;20°	302 May 21	Pachon 26		♓ 11;46°
302 March 12	Pham 16	♍ 13;53°	♍ 13;54°	302 May 26	Payni 1	♓ 12;21°	♓ 12;22°
302 March 17	Pham 21	♍ 13;30°	♍ 13;31°	302 May 31	Payni 6	♓ 12;53°	♓ 12;54°
302 March 22	Pham 26	♍ 13; 7°	♍ 13; 8°	302 June 5	Payni 11	♓ 13;22°	♓ 13;24°
302 March 27	Pharm 1	♍ 12;43°	♍ 12;45°	302 June 10	Payni 16	♓ xx; 1°	♓ 14; 2°

Table 4210.1. Concordance of planetary longitudes (recomputation: *Handy Tables*, sidereal, noon).

have (fr. 1, recto) Saturn, A.D. 300/301; (fr. 2, verso) heading for year 18 of Diocletian, i.e. A.D. 301/302, and Saturn's longitudes for the second half of that year; and (fr. 1, verso) Jupiter again for the second half of 301/302. To the left of these longitudes of Jupiter are traces of two lines of text or table with a double ruling below the second line. I was unable to identify these lines as part of the almanac. There is obviously much uncertainty about the original layout of the tables.

4211. Five-Day Almanac, A.D. 306/307

68 6B.25/H(3)d, part 3.5×6 cm After A.D. 307

From a codex page (vertical fibres on front). The outer margin is partly preserved (2 cm on right of recto, 1.5 cm on left of verso); broken on other sides, although line 1 of the back was probably the top line. Entries normally consist of the date and the degrees and minutes of longitude, with the number of the zodiacal sign included only when clarity required it (e.g. at the top of the column on the verso). The longitudes on both sides pertain to Saturn in A.D. 306/307. We have the beginning of the first half of the year on the recto, and the beginning of the second half on the verso; hence the format was six months per column, but not a year to a page.

Julian	Alexandrian	Text	Recomputation	Julian	Alexandrian	Text	Recomputation
Saturn				307 Feb 25	Pham 1	♏ 16;20°	♏ 16;22°
306 Aug 29	Thoth 1	♏ 0;37°	♏ 0;38°	307 March 2	Pham 6	♏ 16;18°	♏ 16;20°
306 Sept 3	Thoth 6	♏ 1; 3°	♏ 1; 4°	307 March 7	Pham 11	♏ 16;14°	♏ 16;14°
306 Sept 8	Thoth 11	♏ 1;32°	♏ 1;33°	307 March 12	Pham 16	♏ 16; 6°	♏ 16; 7°
306 Sept 13	Thoth 16	♏ 2; 2°	♏ 2; 3°	307 March 17	Pham 21	♏ 15;57°	♏ 15;58°
306 Sept 18	Thoth 21	♏ 2;33°	♏ 2;34°	307 March 22	Pham 26	♏ 15;4x°	♏ 15;46°
306 Sept 23	Thoth 26	♏ 3; 5°	♏ 3; 6°	307 March 27	Pharm 1	♏ 15;31°	♏ 15;32°
306 Sept 28	Phaophi 1	♏ 3;36°	♏ 3;38°	307 April 1	Pharm 6	♏ 15;10°	♏ 15;15°
306 Oct 3	Phaophi 6	♏ 4;10°	♏ 4;12°	307 April 6	Pharm 11	♏ 14;56°	♏ 14;57°
306 Oct 8	Phaophi 11	♏ 4;44°	♏ 4;46°	307 April 11	Pharm 16	♏ 14;37°	♏ 14;38°
306 Oct 13	Phaophi 16	♏ 5;20°	♏ 5;21°	307 April 16	Pharm 21	♏ 14;15°	♏ 14;18°
306 Oct 18	Phaophi 21	♏ 5;54°	♏ 5;55°	307 April 21	Pharm 26	♏ 13;53°	♏ 13;55°
306 Oct 23	Phaophi 26	♏ 6;33°	♏ 6;32°				

Table 4211.1. Concordance of planetary longitudes (recomputation: *Handy Tables*, sidereal, noon).

4212. Five-Day Almanac, a.d. 217/218

68 6B.25/H(1–2)i, part + 68 6B.25/H(3)d, part	6.5×5.5 cm	After a.d. 218

The almanac is written on the back only, and is rather abraded. On the front, in another, larger hand, are parts of lines from two columns of an astrological text concerning decans. Another more or less illegible fragment (1.5×4.5 cm, not transcribed) cannot be placed in relation to the main piece.

Of the five-day almanacs this one covers the earliest range of dates, and it also differs from the others in giving longitudes for the slower planets only at ten-day intervals (a practice revived in medieval almanacs). The entries, as usual consist of the date and the degrees and minutes of longitude, with the number of the zodiacal sign inserted when a new sign is entered (e.g. col. iii, lines 6 and 12). Col. i belongs to Jupiter, col. ii (down to line 10) to Mars, and the rest to Venus, all for a.d. 217/218. The longitudes were evidently computed using the *Handy Tables*, but fairly carelessly (Table 4212.1).

Julian	Alexandrian	Text	Recomputation	Julian	Alexandrian	Text	Recomputation
Jupiter				*Venus*			
218 March 26	Pham 30	3;27°	– 3;51°	218 March 26	Pham 30	♉ 8?°	♉ 8;55°
218 April 5	Pharm 10	5; 9°	– 5;25°	218 March 31	Pharm 5	♉ 14;4x°	♉ 14;49°
218 April 15	Pharm 20	6;40°	– 6;48°	218 April 5	Pharm 10	♉ 20;50°	♉ 20;42°
218 April 25	Pharm 30	7;49°	– 7;55°	218 April 10	Pharm 15	♉ 26;35°	♉ 26;32°
218 May 5	Pachon 10	8;43°	– 8;47°	218 April 15	Pharm 20	♊ 2;17°	♊ 2;21°
				218 April 20	Pharm 25	♊ 7;59°	♊ 8; 6°
				218 April 25	Pharm 30	♊ 1x;57°	♊ 13;50°
Mars				218 April 30	Pachon 5	♊ 1x;54°!	♊ 19;31°
218 July 24	Epeiph 30	6;46°	♐ 5;45°	218 May 5	Pachon 10	♊ 25; 8°	♊ 25; 9°
218 August 3	Mesore 10	9; 7°	– 8;52°	218 May 10	Pachon 15	♋ 0;39°	♋ 0;47°
218 August 13	Mesore 20	13;27°	– 12;58°	218 May 15	Pachon 20	♋ x;2x°	♋ 6;19°
218 August 23	Mesore 30	18;34°	– 17;52°	218 May 20	Pachon 25	♋ x2?°	♋ 11;44°

Table 4212.1. Concordance of planetary longitudes (recomputation: *Handy Tables*, sidereal, 6 p.m.).

4213. Five-Day Almanac, a.d. 301/302

68 6B.20/D(5–8)f 1.5×3 cm After a.d. 302

A small scrap, broken on all sides. The table is written only on the back; the front has, the other way up, parts of 6 lines of documentary text of uncertain character. This may be from the same manuscript as **4210**, although **4210** fr. 1, which has the table on both sides, would have to be situated *between* fr. 2 and **4213** on the page.

Each entry consists of the date and the degrees and minutes of longitude; no example of a specification of the zodiacal sign survives. The longitudes are in good agreement with Venus' positions as computed with the *Handy Tables* for Hathyr and Choeac, a.d. 301/302; no other year is possible.

4214. Syzygies

101/106(c) 3.5×3 cm

Broken on all sides, and written on the front only, with red ruling. The back bears parts of five lines of a document. The longitudes increase from line to line by about 28;7° (with little room for variation). Increments of about this amount in the vicinity of Sagittarius would be appropriate for the moon's longitude at successive full moons. In fact the Babylonian System A lunar theory prescribes constant increments of 28;7,30° for initial solar longitudes between ♓ 27° and ♌ 14;52,30°, i.e. for lunar longitudes between ♍ 27° and ♒ 14;52,30°. However, without better preserved longitudes we can only regard the use of System A in this papyrus as an interesting possibility.

Julian	*Alexandrian*	*Text*	*Recomputation*
301 Nov 12	Hathyr 16	19;xx°	♎ 19;11°
301 Nov 17	Hathyr 21	20;xx°	♎ 20;38°
301 Nov 22	Hathyr 26	22;xx°	♎ 22;42°
301 Nov 27	Choeac 1	25;2x°	♎ 25;28°
301 Dec 2	Choeac 6	28;xx°	♎ 28;48°

Table 4213.1. Concordance of longitudes of Venus (recomputation: *Handy Tables*, sidereal, noon).

4215. Mean Motions of the Moon, a.d. 287/288

68 6B.20/L(3)d 5.5×5 cm After a.d. 288

The table is written in a black ruled framework on the back of a document, and is broken on the bottom, right, and left sides but apparently not the top. The traces of ink in line 1, mere specks, were probably column headings. The numbers, extracted from the table of mean motions of the sun and moon in the *Handy Tables*, are the components of two of the four mean motions involved in computing the moon's position: the motion of the epicycle's centre relative to the apogee of the moon's deferent circle (col. i), and the motion of the planet on the epicycle (col. ii). These are respectively the third and fourth quantities tabulated in Ptolemy's table, and it is probable that the other three were added up in lost columns to the left and right of the extant columns in our papyrus. The four rows of numerals in the papyrus correspond respectively to the year Era Philip 601 (in the table of mean motions at 25-year intervals), 10 complete Egyptian years, month Tybi, 19th day. Hence the date in question is Era Philip 611, Egyptian Tybi 19, which is equivalent to a.d. 287, October 29. Since there is no trace of a further line for elapsed equinoctial hours, the time seems to have been noon. The sums of each column would have been given further down.

It is not clear whether this fragment of a computation of the moon's position was a practical one by the scribe or a 'textbook' example to demonstrate the method. The date is roughly contemporary with the hands.

Part IV

Miscellaneous Tables

4216–4235. MISCELLANEOUS TABLES

This section contains those tables that did not find a place in the scheme of classification according to which the foregoing parts of this edition are organized.[1] In practice this has generally turned out to mean that the purpose of the table is unknown. In some cases the difficulty of identification arises because too little can be read. Although tiny or illegible scraps have generally not been incorporated in the present selection of texts, it seemed worthwhile to include some collections of smaller bits that were presumably excavated together, especially the remnants of the folders that contributed to the 'archive' here called Group A after the extraction of the more substantial fragments (**4231–4235**).

But for many of the tables that follow, the absence of an adequate commentary can be attributed to ignorance: either our collective lack of the information about ancient astronomy needed to provide the right context, or my own failure to recognize the obvious. **4223–4228** stand out in my mind as texts of which enough survives to offer hope of an explication. I hope that some of the *lacunae* will come to be filled with the aid of fresh eyes or fresh texts.

4216. ARITHMETICAL TABLE: THIRTIETHS OF 40;33

31 4B.9/H(1–4)b 6×11 cm

The fragment is from a set of tables, ruled in red and written on the front side only; the back is blank. Two columns of one table are substantially preserved, except for the top lines. It is not clear whether there were more columns or further tables to the right; but on the left edge are parts of the ruling of another table, apparently of similar structure and separated from the extant table by a 1 cm intercolumniar space.

Column ii, unusually, acts as an index column, with numerals ascending by ones to 30. Corresponding to 30 in column i is 40;33, and the other entries in this column are all in approximate proportion to this number; that is, if c_1 and c_2 are the entries in any line of columns i and ii respectively, then:

$$c_1 = 40;33\frac{c_2}{30}$$

[1] **4217** and **4217a–c** were identified as templates, and **4220** as 'Handy Tables', too late to be assigned numbers in the appropriate sections. They are discussed on pp. 118–119.

All entries match the numbers generated by this formula, truncated or rounded to one fractional place. If we adopt a slightly higher constant, say 40;33,20 (40 5/9), we can obtain nearly perfect matches with consistently truncated values, but it is impossible to reconcile the entries corresponding to 17 and 26 without assuming some rounding up.

The structure of **4216** is very similar to that of P. Heid. inv. 4144 + *P. Mich.* III.151. In the latter papyrus we again have a set of tables comprising pairs of columns: in the right column, index numbers beginning with 0;20, 0;40, and then 1 through 30, and in the left column, proportional numbers given to one fractional sexagesimal place. The defining constants of the five preserved tables, i.e. the entries that correspond to c_2 = 30, range between 40;45 and 57. Following a conjecture of Neugebauer, I attempted in Jones [1990] to explain the tables as components of a modification of the Babylonian System A scheme for predicting Mars' synodic phenomena. The table in **4216** does not exactly match any of the tables in my reconstruction, although its defining constant, approximately 40;33, is close to the smallest of the attested values, 40;45. Thus while my interpretation of P. Heid. inv. 4144 + *P. Mich.* III.151 is not refuted by the new table, it may now appear less probable. Nevertheless I can offer no better suggestion of the purpose of these tables.

4217. Arithmetical Table: Templates for Mercury?

39 5B.121/B(1–2)a — 7.5×19 cm — Second or third century

Plate VIII

The table is written on the front of the papyrus in a neat hand with ruling and column headings in red ink; the back is blank. A top margin of 3 cm (including column headings) and the staggered bottom margin are partially preserved.

The table comprises a series of paired columns (units and one sexagesimal fractional place), with headings decreasing by ones from at least 31 to 27 or less, and as many lines in each pair of columns as the numeral in the heading. If the complete table extended rightwards all the way down to 1, it would have appeared more or less triangular. The numerical structure of the table is easy to recover. If we designate the number in a column heading as *c*, and count the rows *r* from 1 at the top (i.e. line 2 of the transcription), then the entry for any column and row is:

$$l = \frac{r^2}{c}$$

In other words, any column contains a second-order arithmetical sequence progressing from 0 to *c* in exactly *c* steps, and with the line-to-line differences ascending from just above 0 to just below 2. The values are consistently truncated to one fractional place.

The only purpose that I can suggest for such a table is to represent the daily longitudinal progress of Mercury in degrees between first morning visibility (Γ) and last morning visibility (Σ). We know, from the example of **4155**, that there existed epoch tables for Mercury that gave explicitly the whole number of days from Γ to Σ, which was contrived to be equal (disregarding fractions) to the corresponding progress in longitude. If we call this interval *c*, then for each date *r* (*c*) days past Γ, the present table may be said to yield the number of degrees that must be added to the computed longitude at Γ to obtain the longitude on that date. In terms of velocities, this would amount to assuming a steady acceleration from station at Γ to approximately 2° per day at Σ. We do not know the precise range of values of *c* that would have been considered possible; the attested numbers in **4155** (fr. 2) range between 22 and 45, while the scheme of 'pushes' that serves an analogous function in Babylonian tables allows for longitudinal intervals from 12° to 44° (Neugebauer [1955] 293).

4217a–c. Arithmetical Tables: Templates for Mercury?

a: 68 6B.20/D(4)a, part	3.5×5 cm	Third century
b: 65 6B.30/M(1)a, part	approx. 9×12 cm (aggregate)	
+ 66 6B.26/D(1)c, part		
+ 68 6B.20/L(8)a, part		
c: 68 6B.20/L(8)a, part	1.5×2.5 cm	

These scraps from Group A all derive from similarly laid out tables, with red ruling at one-line intervals and double red ruling separating the columns (cf. also **4217**). All are written on the front side, with blank backs. **4217b** and **c** appear to be in the same hand, which may also be that of the epoch table for Mercury, **4155**.

4127a. The numbers (units and one fractional place) in column i increase by steadily decreasing differences, such that the entry 33;0 in line 11 appears to terminate the sequence. Let us number the rows of this column so that line 11 represents $r = 34$. Then we find that all preserved entries are exactly reproduced by taking truncated values of the formula:

$$l = 33 - \frac{(34 - r)^2}{35}$$

Column i thus presents a second-order sequence progressing from close to 0 when $r = 0$ to exactly 33 in 34 steps, with the increments decreasing linearly from a little less than 2 down to 0. This makes sense as a template describing Mercury's decelerating progress from first evening visibility (Ξ) to last evening visibility (Ω), supposing that the interval to be covered is 33° in 34 days.

Column ii is less well preserved, but the legible numbers fit the following formula designed to reach exactly 32 in 33 steps (i.e. 32° in 33 days):

$$l = 32 - \frac{(33-r)^2}{34}$$

It may be presumed that similar columns existed for the full range of possible intervals (in whole numbers of days) between Ξ and Ω. The table as a whole would have had a similar triangular appearance to **4217**.

4217b. As in the foregoing fragment, the columns of this table are decelerating second-order sequences that probably were meant to serve as templates for Mercury's progress from Ξ to Ω. From left to right each sequence was one line shorter than its predecessor, but also began one line lower, so that all the columns ended on the same line whereas the top of the table sloped downwards to the right.

The kind of arithmetical sequence making up the columns of **4217b** is also different from those of **4217** and **4217a**. In **4217** the entry for any line is proportional to the square of the line number, with the result that the sequence of line-to-line differences takes the form $^1/_2\,d$, $^3/_2\,d$, $^5/_2\,d$, ... $^{(2n-1)}/_2\,d$, where n is the number of lines in the column and d is a constant determined by the desired total at the end of the sequence. The differences in **4217a** are of the same kind but in reverse order, i.e. ending ...$^5/_2\,d$, $^3/_2\,d$, $^1/_2\,d$. In **4217b**, however, the decreasing differences follow the more straightforward pattern nd ... $3d$, $2d$, d. Consequently the expression for the entry in row r becomes:

$$l = r(2n+1-r)d$$

The extant columns i–v of the papyrus correspond to $n = 23$ through 19. In each case the constant d was chosen such that the final total $l(n)$ is close to n + 0;40, equivalent to assuming that Mercury's longitudinal progress from Ξ to Ω is always about 0;40° in excess of 1° per day; but the precise value of this excess varies irregularly from column to column in a manner that I am unable to explain (Table 4217.1).

4217c. This small fragment seems to come from a table in which the columns began on progressively lower lines as one read from left to right, i.e. the same plan as **4217b**. The partially preserved numeral above column i is presumably the number of steps in the sequence tabulated below. Since the first entry in column ii

column	*n*	*d*	*l(n)*
i	23	0;2,34,1,40?	23;37,3?
ii	22	0;2,41,14,35	22;39,49
iii	21	0;2,49	21;41,18
iv	20	0;2,57,11,43	20;40,22

TABLE 4217.1. Parameters of the arithmetical sequences in **4217b**.
(N.B. except for column iii, the last place of *d* and *l*(*n*) is uncertain)

was less than 1, the table presumably comprised templates for Mercury's accelerating motion from Γ to Σ. Too little remains to allow reconstruction of the sequences.

Notes:

4217b:

ii 5: error for 7;20,45.

4218. ARITHMETICAL TABLE: FRACTIONS OF 30

69/10(h) 4.5×12 cm

The upper right-hand corner of a set of tables, with 2.5 cm top margin (including heading) and 1 cm vacant space on the right. The ruling is in red ink, except for the extra rulings marking the rows between the tables (lines 12 and 17), which are black. The heading in line 17 is also in red. There are several layers of papyrus stuck together, perhaps to form a book cover. No traces of writing are visible on the back or between the layers.

Three tables are preserved, giving exact sexagesimal fractions of 30. Thus the top table has $^1/_6 \times 30$, $^2/_6 \times 30$,... $^6/_6 \times 30$; the second has $^1/_8 \times 30$, $^2/_8 \times 30$,... $^8/_8 \times 30$; and the third has $^1/_{12} \times 30$, $^2/_{12} \times 30$,... $^{12}/_{12} \times 30$. An index column giving the numerator of these fractions may have been lost to the left. I do not know what the purpose of these tables was, or what further tables of fractions may have completed the set.

Notes:

1: perhaps μοῖ]ραι.

4219. Numerical Table

18 2B.70/C(1–3)c 2.5×5 cm

From the left side of a table (there is a narrow vacant space left of the double ruling), broken on the remaining sides. The ruling is in black ink, but the second numeral in each tabular cell is red.

In general both the black and the red numerals increase by 1 as one reads across, and by 4 as one reads down, suggesting that the complete table was four cells wide, with contents as follows:

4	**5**	5	**6**	6	**7**	7	**8**
8	**9**	9	**10**	10	**11**	11	**12**
12	**13**	13	**14**	14	**15**	15	**16**
16	**17**	17	**18**	18	**19**	19	**20**
20	**21**	21	**22**	22	**23**	23	**24**
24	**25**	25	**1**	26	**2**	27	**3**
28	**4**	29	**5**	1	**6**	2	**7**
3	**8**	4	**9**	5	**10**	6	**11**
7	**12**	8	*(remainder lost)*				

The black numerals return to 1 after passing 29, and the red numerals return to 1 after passing 25. The meaning of the numbers is unfortunately obscure. They might have some chronological significance, counting days or years according to some unidentified convention, but I am unable to suggest any interpretation that gives meaning to a cycle of 25, a cycle of 29, and a layout in four columns.

4220. Handy Tables

68 6B.25/E(1–2) 3.5×12 cm (fr. 1) 2×6.5 cm (fr. 2)
2×7 cm (fr. 3) 2×3 cm (fr. 4, not transcribed)

Fragments 1 and 2 are from the same manuscript, written rather sloppily on the front only, with ruling in black. The sexagesimal numbers in fragment 1 increase uniformly by approximately 40;38,35,30. Extrapolating backwards we obtain 0;0,0,0 four lines above line 1. In other words line 22 is actually the 25th line of the complete table, and since this is the only line between the bottom two rulings, it must have been the end of the table.

In fact 40;38,35,30° represents the mean motion in longitude of the moon's apogee in 365 days, so that the fragment belongs to a table of lunar mean motions for 1 through 25 Egyptian years. The format using 25-year cycles is the same as in Ptolemy's *Handy Tables* (as distinct from the *Almagest*'s 18-year cycles), but the papyrus gives two fractional places where Ptolemy is content in the *Handy Tables* with just one. Moreover Ptolemy does not tabulate the longitudinal mean motion of the apogee directly, although it can be deduced from the mean motions that he does give. This is because, in terms of Ptolemy's epicycle-with-eccentre model for the moon, the apogee as we understand it is not an obvious element (the apogee of Ptolemy's quickly revolving eccentre is an entirely different thing). Our table seems to presume a model in which the moon's principal anomaly is effected by a slowly advancing eccentre, such as the model that Ptolemy ascribes to Hipparchus in *Almagest* IV 11.

If we subtract Ptolemy's value for the moon's mean motion in anomaly in one Egyptian year from the simultaneous mean motion in longitude (*Almagest* IV 4), we obtain 40;39,38,45,...° as the progress of the apogee. The excess of this number over the parameter in the papyrus, approximately 0;1,3,15°, is too large to be due solely to a difference in the assumed period of lunar anomaly, and must mean that the longitudes in the papyrus are sidereal.

Fragment 2 is more immediately recognizable as the index and units columns from a table of the moon's mean motion in elongation (i.e. the difference between the lunar and solar mean longitudinal motion) for single days. The preserved numbers match those in Ptolemy's equivalent table in *Almagest* IV 4, but because of the loss of all fractions we cannot tell how close the underlying daily motion in elongation was to Ptolemy's 12;11,26,41,...° per day. The counting of days is as in Ptolemy's *Handy Tables*, namely with 0° corresponding to day 1 rather than to day 0 as in the *Almagest*. The presence of a table of mean motion in elongation is interesting. It probably means that these tables were based on a theory incorporating Ptolemy's discovery of the moon's second anomaly, which is dependent on the elongation, even though the model was, as we have seen, probably not Ptolemy's.

Fragment 3 is from a codex, in a different hand but also ruled in black. On the side with horizontal fibres is part of a column of numbers with one fractional place, attaining a maximum of 47;0 in line 2 and dropping with increasing differences in subsequent lines. This is probably the column for synodic anomaly from a table of equations for Venus, with the arguments tabulated at 1° intervals. It was not a version of Ptolemy's anomaly table, which is based on a maximum equation of 46°. The other side probably has other columns from the preceding or following page of the same table, but without more understanding of the table's structure it is not possible to identify the purpose of the columns.

The small Fragment 4 is a badly abraded bit from another codex, with red ruling and only a few isolated numerals legible.

Notes:

Fr. 3, along the fibres:

13: 44 or 45 is expected here, but the abraded traces are incompatible with μ.

4221. Longitudes

68 6B.25/H(3)b, part 4×6 cm (fr. 1)
+ 68 6B.25/H(4)b, part 2×2.5 cm (fr. 2) 1.5×4 cm (fr. 3)

Three fragments, broken on all sides, from a neatly executed table with red ruling; the back is blank. Columns ii and iii of Fragment 1 obviously represent a series of longitudes, with constant differences of 0;56,40°. The format resembles an ephemeris. None of the heavenly bodies, however, could exhibit a series of daily positions like this. When the sun is near the beginning of Scorpio, its progress is close to maximum speed, which exceeds 1° per day; and the only planets that can progress 0;56,40° per day, namely Venus and Mercury, change velocity too swiftly when travelling about this rate. Another possibility is that the table has something to do with ascensions, i.e. correlating the rising of the ecliptic and the equator; but I have not found a plausible interpretation of the numbers along these lines either.

4222. Table with Zodiacal Signs and Latitudes (?)

50 4B.30/J(1–2)a, part 4.5×7.5 cm (fr. 1) 2×2.5 cm (fr. 2) debris

The larger fragment preserves the leftmost two columns and the edge of a third from a table ruled in red. The top lines are lost; below the apparent bottom of the table is a broad space with what appears to be the beginning of a title, and below that (along the extreme edge of the fragment) a further red ruling, suggesting that there was at least one more table below. Column i lists the zodiacal signs in order of increasing longitude, from Aries (two lines above the top preserved line) to Pisces. The last six signs are designated in column ii as 'southerly', and the first six were probably all 'northerly'. The traces of column iii appear to be numerals, which could be taken together with column ii as a measure of distance to the north or south. The northerly and southerly signs in the table coincide with those that are respectively north or south of the equator, which might suggest that this is a table of declinations; but then the numerals in column iii are only about one third of the expected numbers of degrees. On the other hand the quantities are too large and not suitably aligned for any of the planets' latitudes.

The identification of the table is not made easier by the second fragment, which contains the heading and top two lines of another column of zodiacal signs. This is not another part of column i of the main fragment, since it is delimited on the left by only a double ruling. A single trace is visible from a preceding column, perhaps of numbers. The signs this time begin with Sagittarius and, initially at least, work backwards.

Notes:

Fr. 2:

i 2: a high horozontal stroke, perhaps from a ς.

4223. Sinusoidal Function

20 3B.31/F(1–3)b approx. 6×27 cm

Parts of two sets of columns from a table, ruled in red. On the back are traces of a document. The format of the table, with an index column increasing by ones but tabulated only every five lines, resembles a template, but the preserved numbers make no sense as daily positions of a heavenly body. The pattern, reaching a maximum of 7;30 for index number 96 and descending to zero (or nearly) at index number 180, suggests the sinusoidal behaviour of an equation table or latitudinal motion. The heading above columns i and ii *might* have been βορεῖα, 'northerly', but the maximum value seems implausibly high for any planet.

4224. Sinusuidal Function

68 6B.20/D(4)a, part 6×12 cm (fr. 1)
2×3.5 cm (fr. 2) 2×4.5 cm (fr. 3)

The table is written on the back of a document, and is ruled in red. The largest fragment preserves 6 cm of bottom margin.

Column ii of Fragment 1, presumably tabulated every five lines, must be the index for the sexagesimals in columns iii–v. These descend, approximately following a second-order difference sequence (which incidentally reveals a mistake in line 8), as can be seen from the first and second differences:

index	*cols. iii–v*	*differences*	*second difference*
98	[19;24],1[9]	0	
99	[19;2]3,58	–0;0,21	–0;0,21
100	[1]9;23,17	–0;0,41	–0;0,20
101	[1]9;22,15	–1;0,2	–0;0,21
102	[1]9;20,53	–1;0,22	–0;0,20
103	19;19,10	–1;0,43	–0;0,21
104	19;17,6	–2;0,4	–0;0,21
105	19;*14*,4[1]	–2;0,25	–0;0,21
106	19;11,56	–2;0,45	–0;0,20

If we extrapolate the sequence forward using a constant second difference of approximately 0;0,20,30, we obtain a value at or very close to zero for index 180. This seems to be a similar type of table to **4223**, but with a larger maximum value. Again I am unable to suggest what theoretical significance the numbers had; nor can I relate the contents of Fragments 2 and 3 to the main fragment.

Notes:

Fr. 1:

iv 8: expected value: 14.

4225. Numerical Table

68 6B.25/H(3)b, part   5.5×15 cm

The fragment is from the bottom of a table (with 1.5 cm margin including the decorative rulings), and preserves parts of the leftmost four columns; a red ruling along the left edge may have been the right edge of another table. The ruling is red except for the middle lines of the triple rulings, which are black.

Column i is clearly an index number. The numbers in the remaining columns are baffling. They range from 1 to 24, and there is no visible pattern to them, whether one reads down the columns or across the rows. Smaller numbers predominate, so that there are 31 legible numerals less than 10, 15 between 10 and 19, and only one above this range.

4226. Table with Planets, Zodiacal Signs, and Symbols

65 6B.30/M(1)a, part	3.5×12 cm (fr. 1) 6×5 cm (fr. 2)	Third century
+ 66 6B.26/D(1)c, part	2.5×2.5 cm (fr. 3) 2.5×2 cm (fr. 4)	Plate VIII
+ 68 6B.20/L(8)a, part	1.5×3 cm (fr. 5) 1×1.5 cm (fr. 6)	
+ 68 6B.25/H(3)b, part		
+ 68 6B.25/H(4)b, part		

The remains of what may have been a fairly extensive set of tables, the purpose of which eludes me. The rulings as well as some of the symbols are in red ink. Fragment 1 has 3.5 cm of bottom margin, and Fragments 5 and 6 have part (2 cm and 1 cm respectively) of the top margin. The backs are all blank.

The structure of an individual 12-line table seems to have been a column listing the zodiacal signs in order of increasing longitude, but not all beginning with the same sign, and a second column containing symbols in red or black ink (represented here as outlines and filled symbols respectively). Each zodiacal sign has one or two symbols beside it, and these are variously red and black dots and red and black eight-pointed stars. Above each table is the name of one of the planets. Reading a set of tables from left to right, the planets may have been in the standard order Saturn–Mercury, but there was more than one table for each planet. I have been unable to find pattern or meaning in either the symbols or the starting points chosen for the lists of zodiacal signs.

4227. Sexagesimals

22/C287 3×3.5 cm

The fragment, broken on all sides, contains parts of 6 lines of a table of sexagesimal numerals ruled in black. On the other side are traces of three lines of a document, written at 90° to the table, so that both sides are along the fibres.

The alignment of the numerals suggests that the leftmost are units. Each line is nearly two less than its predecessor, but there is no obvious arithmetical pattern in the differences, and I have failed to identify the significance of these numbers:

	difference	*second difference*
33;35,...		
31;41,...	−1;54	
29;47,29,15,...	−1;54	0;0
27;51,35,34,...	−1;55,54,41	0;1
25;54,29,43,...	−1;57,5,51	0;1,11,10
23;55,0,5,3x,...	−1;59,29,38	0;2;23,47

4228. Sexagesimals

9 1B.172/D(3) 6×9 cm Second or third century

Plate VIII

The table is written across the fibres on the back of a document, to which at least one layer of a (different?) document adheres. The ruling is in black ink. 3 cm of margin is preserved at the bottom, but the remaining sides are broken.

Because of its arithmetical structure the complete table can be reconstructed with some plausibility; its purpose is unfortunately less clear. As well as the extant columns i–iv, there probably existed an index column (or columns) and another column to the left of column i, which we may refer to as column [0]. Each column had thirty-one lines. Column [0] contained a sequence beginning 0;0,0 and ending 103;45,0, with the line-to-line differences decreasing from 3;31,38 to 3;23,38 by constant steps of 0;0,16. Column i, of which the ends of the last lines are preserved, contained a sequence beginning 180;0,0 and ending 103;45, with line-to-line (subtractive) differences increasing from 2;28,22 to 2;36,22. Column ii began 180;0,0 and ended 256;15,0, with the same line-to-line differences—additive now—as column i. Column iii began 360;0,0 and anded 256;15,0, with the same differences—subtractive—as column [0]. Column iv actually contains the differences for columns [0] and iii, and we may perhaps conjecture a further column giving the differences for columns i and ii.

Hence, reading first down column [0], then up column i, then down ii, then up iii, one had an unbroken series of numbers from zero to 360 by steps averaging 3. These are surely to be understood as degrees, and the steps represented by consecutive lines of the table are probably of 3°. In other words, the table correlates two angles or arcs that have some fixed geometrical relation to each other in such a way that they are equal at 0° and 180° and reach a maximum difference of 13;45° when one of the arcs is 90° or 270°. This behaviour is suggestive of mean and true longitude in a kinematic model, or longitude and oblique ascension; but the specific parameter of 13;45° maximum difference seems to be unattested elsewhere, and I have failed to identify it.

Whatever the two correlated quantities are supposed to be, the mathematical pattern of the table is most peculiar. We have seen that rather than representing their relationship by a smooth trigonometrical function, the table breaks the cycle up into four parts, each of which is a second-order arithmetical sequence. What is particularly odd is that in each sequence the first differences vary only by 0;8° from smallest to greatest, while there is a large jump of 0;47,16° between the greatest difference in a 'slow' sequence and the adjacent least difference in the preceding or following 'fast' sequence.

4229. Sexagesimals

22 3B.21/B(c) 4×7 cm

A fragment broken on all sides from a table ruled in red; the back is blank. Of the three partially preserved columns, we have only the ends of the numbers in column i and the units (always 1) of those in column iii. In column ii the scribe originally wrote a series of numbers with two fractional sexagesimal places, declining by steps of 0;0,16 or 0;0,17 . Later he replaced these numbers (crossing out only the changed digits and writing their replacements above) with a new sequence, close to the original values, but now decreasing by steps of 0;0,48. It is not clear what these numbers represent; perhaps they are daily longitudes of a planet in slow retrograde motion.

4230. Tabular Scraps

35 4B.101/G(1–3)c

The scraps of tables in this folder seem to be from at least two manuscripts; none has been identified.

(1) 3×3.5 cm, broken on all sides, red ruling. Tables on both sides, but not a conventional codex since the axis of flipping is horizontal.

(2) 2×3 cm, broken on all sides, red ruling. Table on front, back blank.

(3) 1.5×1 cm, broken on all sides, red ruling. Table on front, back blank. Same manuscript as (2).

Not transcribed:

(4) 1×2.5 cm, broken on all sides, red ruling. Table on front, back blank. Five lines from a column, all ō, double ruling on the left edge.

(5) 1×2 cm, broken on all sides, red ruling. Table on front, back blank. Three lines from a column, numerals ending]η ι^{2}]ζ ι^{3}]ς̣.

(6) 1×1 cm, broken on all sides, red ruling. Table on front, back blank. Two lines from a column, both δ.

(7) 2.5×2 cm, broken on all sides, red ruling. Tables on both sides. Same manuscript as (1)? On side with horizontal fibres, ruling and one number,] κη λ. On the other side, ruling and indeterminate trace from one letter.

4231. Tabular Scraps

65 6B.30/M(1)a, parts

This folder belongs to Group A. Several manuscripts are represented.

(1) 2.5×5.5 cm, left edge of table intact but broken on remaining sides, red ruling. On back of a document. Preserved is part of the index column and the units (all zero) from a template, obviously not identifiable. A second fragment, 1.5×5.5 cm, has only ruling.

(2) 4.5×6.5 cm, 1 cm bottom margin (not counting vacant lines of tabular frame), broken on remaining sides, red ruling. On back of a document. Column i is an index column running up to 30 (days? years?). Columns ii–iv are sexagesimals increasing by a constant close to 45;6,45. The entry for index number 1 was 0;0,0. This appears to be a mean motion table; but I am unable to identify a mean motion that increases in steps of just over 45° per day or year.

Not transcribed: approximately 14 bits with numerals and red ruling, some on both sides; approximately 20 with ruling only—some of these do not appear to be meant for tables—or with negligible or no traces of writing.

4231a. Tabular Scraps

66 6B.26/D(1)c, parts

This folder belongs to Group A. Several manuscripts are represented.

(1) 3×4 cm, broken on all sides, red ruling (exceedingly faint). On back of a document.

(2) 2×2.5 cm, broken on all sides, red ruling. From a codex.

Not transcribed: 2 bits with numerals and red ruling, 2 with ruling only.

4232. Tabular Scraps

68 6B.20/D(4)a, parts
+ 68 6B.20/L(8)a, parts in (1) and (2)

This folder belongs to Group A. Several manuscripts are represented.

(1) 3.5×5.5 cm (fr. 1), 1×3 cm (fr. 2), some top margin extant, broken on remaining sides, black ruling. On back of a document.

(2) 2.5×3.5 cm (fr. 1), 2×2 cm (fr. 2), 1×2.5 cm (fr. 3). Fr. 1 has the right edge of a table; all remaining sides are broken. Ruling in red and (for middle of triple ruling) black. Backs blank.

(3) 4×5 cm, broken on all sides, red ruling. Back blank. Probably longitudes of consecutive conjunctions.

(4) 3×3.5 cm, right edge of table extant, broken on remaining sides, red ruling. Traces of beginnings of lines of a document on the back.

Not transcribed: 3 bits with numerals and red or black ruling (one has heading ιχ]θυες above table); approximately 12 with ruling only or traces of writing.

4233. Tabular Scraps

68 6B.20/L(8)a, parts

This folder belongs to Group A. Several manuscripts are represented.

(1) 3.5×3.5 cm, left edge of table extant(?), broken on remaining sides, red ruling (faint). Traces of writing on back.

(2) 1×3 cm, broken on all sides, red ruling. Back blank.

(3) 1×3.5 cm, broken on all sides, red ruling. Back blank.

Not transcribed: approximately 17 bits, most from tables with red or black ruling.

4234. Tabular Scraps

68 6B.25/H(3)b, parts
+ 68 6B.25/H(4)b, parts in (1) and (2)
+ 65 6B.30/M(1)a, part in (1)

This folder belongs to Group A. Several manuscripts are represented.

(1) 5×8.5 cm, left edge of table extant, broken on remaining sides, red ruling. Back blank.

(2) 1×5.5 cm (fr. 1) and 1×2 cm (fr. 2), broken on all sides, red ruling. Back blank.

Not transcribed: approximately 9 bits from tables with red or black ruling.

4235. Tabular Scraps

68 6B.25/H(4)b, parts

This folder belongs to Group A. Several manuscripts are represented.

(1) 1×2 cm, broken on all sides, red ruling. Back blank.

(2) 1×1.5 cm, broken on all sides, red ruling. Back blank.

Not transcribed: approximately 10 bits from tables with red ruling.

Part V

Horoscopes

4236–4300a. HOROSCOPES

Much of the interest of the horoscopes is collective: they are the written record of the most common practical application of astronomy in Roman Egypt, and the majority of them that can be exactly dated, if gauged against the chronological distribution of all kinds of papyri, provide a measure of the popularity of horoscopic astrology through five centuries. The apparent monotony of the briefer texts taken singly is, however, often broken by unexpected details.

Collections of more than one horoscope on a single document are surprisingly common. Some undoubtedly are the astrologer's own transcripts: for example **4241**, which is written on the back of **4136** (an astronomical text that was more than a century old at the time), or **4274**, which was found among fragments of an astrological text in the same hand. The multiple horoscope **4247** is an unusual case where the same native seems to have consulted two astrologers, who produced quite differing versions of the horoscope for the same date. Many of the individual horoscopes were written on a ticket, often cut from a sheet or roll previously used on the other side. Almost any available surface, however, might be used, even the inside of a codex cover (**4254**).

The impressive number of fragments of deluxe horoscopes (**4245**, **4276–4286**) perhaps reflects the quality of the local clientèle at Oxyrhynchus. Most of these pieces cannot be dated except on paleographical grounds, but they are useful to us in other ways. They show the astrologer using more resources and making more precise computations than in the other horoscopes. A basic horoscope like **4242** could have been cast in a few minutes by an astrologer equipped with a sign-entry almanac and a set of tables for the sun and moon; but **4277** was the work of several hours of reckoning and careful writing.

For horoscopes that bear an explicit date, or that can be dated from the astronomical positions that they contain, the commentaries below give the approximate longitudes of the heavenly bodies and the ascendant calculated according to modern theory for the same date.[1] Where a horoscope specifies the *n*th hour of day or night, the ascendant has been recomputed for the *end* of that seasonal hour, so that the zodiacal sign according to modern theory may sometimes be one sign later than that of the text through no fault of the ancient astrologer. For consistency, all tropical longitudes from modern theory have been converted to sidereal longitudes using formula (3) in Appendix I, and rounded to the nearest degree. A single exclamation mark (!) following a recomputed longitude indicates that the

[1] For these calculations I used the *Mathematica* software package *Scientific Astronomer*.

zodiacal sign is different from but adjacent to the one in the papyrus; a double mark (!!) indicates a larger discrepancy.

The grammar, and particularly the orthography, of the horoscopes very often deviates from convention. I have not thought it necessary to signal each instance in the commentaries, but refer the reader to Gignac [1976–1981] for discussion and rationalizations of most of the phonological and morphological phenomena. The preserved dates in the horoscopes published here are all in the Alexandrian calendar (there are very few indications of an equivalent in the old Egyptian calendar, none fully preserved). Up to A.D. 284, specified years are regnal years of the current Roman emperor according to the Alexandrian calendar. After that date, the horoscopes are dated either by the Alexandrian regnal years of the current Augusti and Caesars (which are tabulated conveniently by Bagnall & Worp [1978] 67–94) or by the extrapolation of Diocletian's Alexandrian regnal years, the so-called 'Era Diocletian.' In the commentaries, brief mention is made of the two conventional orderings of the data in the papyrus horoscopes: (a) the 'standard planetary order' Saturn, Jupiter, Mars, Venus, Mercury (with the sun, moon, and ascendant either before or after the planets), and (b) ordering by zodiacal signs, in which the heavenly bodies are listed by the signs that they occupy, in order counting eastwards from the sign of the ascendant.

4236. Horoscope, A.D. 63

21 3B.23/D(1–3)b 8×10.5 cm After A.D. 63

Plate IX

The top half of the horoscope is preserved, with 1.5 cm top margin, but the text is broken on both sides as well as the bottom; the back is blank. The inclusion of degrees, houses, and terms, and the quality of the writing (an informal book hand) mark this as a deluxe horoscope, although it lacks the usual elaborate prose text. The nearest parallel seems to be the duplicate horoscope P. Par. 19 and *P. Lond.* I 110, dating from A.D. 137 (Neugebauer & van Hoesen [1959] nos. 137a and 137b).

The regnal year and the names of the celestial bodies involved in the horoscope are lost. Nevertheless assumption of the standard order with sun, moon, Saturn through Mercury leads to a satisfactory dating. The first body, which ought to be the sun, is in Sagittarius, and in the terms of Jupiter (0° to 12°). In the Alexandrian calendar this makes sense, since Hathyr 28 is November 25/26, with solar longitudes about Sagittarius 2°. On the other hand, during the reign of Nero (A.D. 54 to 67) Hathyr 28 in the Egyptian calendar was in the range November 2–5, with corresponding solar longitudes between Scorpio 8° and 12°.

We expect lines 6–7 to pertain to the moon. Its longitude was therefore in Gemini and in the terms of Venus (12° to 17°). The only year between 54 and 67 in which the moon was in Gemini on Alexandrian Hathyr 28 was 63; and Saturn (lines 8–9) was in Leo, the house of the sun, only in 62 through 64. The lines for Jupiter (10–11) are unfortunately too illegible to serve as further verification of our dating. The statement that Saturn was rising in the east does not necessarily mean that the date is close to the planet's first visibility, but only that it is between first visibility and first station; in fact the first station occurred only a few days later.

	Text	63 November 25, 6th hour of day
☉	♐ 0°–12°	♐ 6°
☾	♊ 12°–17°	♊ 16°
♄	♌ 11°–18°	♌ 19°
♃	♉ ? 14°–22°?	♉ 14°
♂	lost	♏ 4°
♀	lost	♎ 24°
☿	lost	♐ 25°

4236a. Horoscope, A.D. 80

103/124(c) 5.5×8 cm After A.D. 80

An unusually terse horoscope, apparently complete (1 cm margin at top and 3.5 cm at bottom), written in a semicursive hand across the fibres. On the other side, the other way up, are traces of cursive. Neither the date nor the name of the native is given. The hand belongs to the late first or early second century. No date leads to complete agreement with the papyrus. The only plausible match is for August 4, A.D. 80; Jupiter and Venus were outside but acceptably close to the signs in the papyrus, but Mercury's longitude was seriously discrepant. If 'the rest' includes the ascendant as well as the unnamed planets, the time was close to sunrise.

	Text	80 August 4, 6th hour of day
☉	♌	♌ 14°
☾	♌	♌ 16°
♄	♒	♒ 25°
♃	♊	♋ 4° !
♂	♌	♌ 26°
♀	♌	♍ 5° !
☿	♍	♌ 19° !!

4237. HOROSCOPE, A.D. 84

75/82(b) 11×8 cm After A.D. 84

The top part of a horoscope, written in a sloping hand on the back of a rectangular piece cut from a register. At least ten lines of the horoscope are missing. Two further pieces with blank backs, 6×7.5 and 4.5×7.5 cm, are from the bottom of the sheet. As in **4236** the degrees of longitude (for the moon, with a fraction instead of minutes), houses, and terms are included. The calendar is the Alexandrian.

	Text	84 September 6, 6th hour of day
☉	♍ 11°	♍ 16°
☾	♈ 23;30°	♈ 28°
♄	♈ 24°	♈ 19°
♃	lost	♎ 17°
♂	lost	♎ 14°
♀	lost	♌ 5°
☿	lost	♍ 12°

4238. HOROSCOPE, A.D. 117

B3.6/C320 5×8 cm After A.D. 117

The bottom left corner of a horoscope in an informal book hand, with 1.5 cm space on the left and 3.5 cm lower margin; the back is blank. The date, time, and position of the sun are missing at the top. The calendar is the Alexandrian. Except for the inclusion of the midheaven, this is a standard horoscope notwithstanding the good hand. Because the date is lost and the lunar sign illegible, we can establish the date of the horoscope only within a margin of several days.

	Text	117 June 25, 9th hour of day
☉	lost	♋ 5°
☾	lost	♎ 20°
♄	♊	♊ 5°
♃	♋	♋ 30°
♂	♉	♉ 26°
♀	♌	♌ 6°

☿	♌	♌ 2°
H	♏	♏
M	lost	♌

4239. Horoscope, a.d. 130

47 5B.46/H(3–6)d	17×5 cm	After a.d. 130

The text is written in a crude hand across the fibres, although the other side is blank. The top and right edge are preserved (with no appreciable margins), but the left side is broken off. One line is probably lost from the end, since the ascendant is not mentioned in the extant lines. The writing in line 7 is compressed as if space was running out.

The horoscope deviates from the standard format in that precise longitudes are given to the degree and minute (these are, however, not accurate). The order in which the heavenly bodies are listed is fundamentally the standard sequence, but modified in order to keep together those that occupy the same zodiacal sign: thus Mars and Mercury come immediately after the sun, and Jupiter immediately after the moon. This procedure is unusual in a horoscope that gives the positions to greater precision than the zodiacal sign.

The name of the native is mutilated beyond recognition, but seems to be in the nominative instead of the genitive expected after γένεϲιϲ. For parallels, see Baccani [1992] 177.

	Text	130 January 14, 5th hour of day
☉	♑ 20°	♑ 27°
☾	♌ 9;25°	♌ 20°
♄	♏	♏13°
♃	♌ 9°	♌ 29°
♂	♑ 10°	♑ 2°
♀	♒	♒ 6°
☿	♑ 3;3°	♑ 9°

4240. HOROSCOPES, A.D. 150, 183, AND UNDATED

12 1B.133/C(d), part 6×14 cm After A.D. 183

A series of horoscopes written in a practised sloping hand on the back of a document (other way up). The left edges of the horoscopes are preserved, but the remaining sides are broken off. There is nothing in the writing to suggest that the horoscopes were not all written down at approximately the same time. The first horoscope is nearly complete, and lists the heavenly bodies in the standard order except for grouping together those that occupy the same sign. It is dated Phaophi 24 (Alexandrian), late in the night, but since the text unusually fails to give a double date for this nocturnal nativity, it is not clear whether the night preceding or following the day Phaophi 24 is intended, i.e. whether the equivalent date was October 21 or 22, A.D. 150. The recorded positions of Mars and the moon are seriously in error.

	Text	150 October 21/22, 11th hour of night
☉	[♏] same as ♀, ☿	♏ 1°
☾	♊	♈ 8° !!
♄	♋	♋ 30°
♃	♉	♊ 2° !
♂	♉	♐ 8° !
♀	[♏]	♏ 10°
☿	[♏]	♏ 23°
H	♎	♎

Following this is the dating formula for a second horoscope, for a date between May 26 and June 24, A.D. 183 (assuming the use of the Alexandrian calendar). No astronomical data are given (cf. **4248**). It is impossible to tell whether the third horoscope was complete or restricted to the date.

4241. HOROSCOPES, A.D. 162, 188, 191, AND UNDATED

12 1B.133/C(d), part 12×13.5 cm After A.D. 191

The back of the procedure text **4136** bears the following texts: (i) close to the right edge, sideways (i.e. along the fibres), Διογένηϲ ὁ καὶ Ἀπόλλω; (ii) about the upper middle, oriented like the foregoing, Θέων |² Θέωνοϲ; and (iii) occupying the left part, across the fibres and in a tiny semidocumentary hand, the series of horoscopes edited under the present number. The text of the horoscopes is broken on the left, apparently with the loss of at least half the line length, so that it is not certain whether lines 4 and 12, which are vacant in the extant fragment, originally had text further to the left. The ends of lines 13 are also broken off, and the remaining writing is abraded and in part illegible. It is impossible to tell whether the persons named in (i) and (ii) had anything to do with the horoscopes. The circumstance that the horoscopes are written on the back of a (much older) astronomical text suggests the possibility that these were an astrologer's own notes rather than, say, a family's records.

The first of the horoscopes takes up lines 1–3, and may have continued on line 4. Line 1 preserves the regnal year 28, which cannot belong to any emperor contemporary with the hand except Commodus. One would expect the month, day, and hour to have followed, but the traces preceding the numeral 6 in line 2 make no sense as either a month name or an indication of day or night. Of the planetary positions we know only (line 3) that Mercury was in Leo and Venus, probably, in another sign. In A.D. 187/188 (= Commodus 28) this would have been valid only towards the end of the year, i.e. in July and August of 188.

The second horoscope begins in line 5 with some illegible text—part of the name of the native?—followed by the date: the third regnal year (emperor's name lost), Thoth 5, sixth hour of either day or night. The statement in line 7 that Jupiter was in Gemini fits (within a half century either way of A.D. 188) only the third regnal year of Marcus Aurelius, i.e. A.D. 162/163. Since the ascendant was in Scorpio while the sun would be in Virgo on Thoth 5, the time must have been the sixth hour of day. The computed planetary positions allow us to identify the symbol at the beginning of line 7 (see apparatus) as representing Mars, and the damaged abbreviation at the beginning of line 8 as Mercury.

	Text	162 September 2, 5th hour of day
☉	lost	♍ 11°
☾	lost	♏ 24°
♄	lost	♐ 7°
♃	♊	♊ 8°
♂	♊ ?	♊ 20°
♀	lost	♍ 30°
☿	♌	♌ 23°
H	♏	♏

The first line (line 9) of the third horoscope is entirely illegible, but the following two preserve the positions of Saturn, Jupiter, Mercury, and the moon. If these are correctly given, the only acceptable dates would be A.D. 191 September 10–11.

	Text	191 September 10/11, 5th hour of night
☉	lost	♍ 19°
☾	♏	♏ 10°
♄	♐	♐ 2°
♃	♎	♎ 23°
♂	lost	♍ 23°
♀	lost	♌ 3°
☿	♎	♎ 9°
H	♊ ?	♊

A single line (line 13) of a fourth horoscope preserves only positions of Mercury and Jupiter, which are insufficient for reconstructing the date.

4242. Horoscope, A.D. 212

36 4B.110/H(1–3)a 18.5×9 cm After A.D. 212

The horoscope is written in an informal hand on a rectangular sheet; all edges are preserved, with 1 cm margin at the top and 2 cm on the other sides. The back is blank. The order in which the celestial bodies are given does not follow the usual patterns. Perhaps the writer had a different ideal sequence (disturbed when more than one body falls in the same sign) in mind: ascendant, planets from Mercury to Saturn (reversing the conventional order), sun, moon.

For 212 May 17, the equivalent of Alexandrian Pachon 22, all positions agree with computation except for the two fastest bodies, Mercury and the moon. Mercury's longitude is at least 20° too great on May 17, while the lunar longitude is entirely off. Good agreement is obtained ten days earlier, suggesting that Pachon 22 is an error for Pachon 12.

	Text	212 May 17, 3rd h of day	212 May 7, 3rd h of day
☉	♉	♉ 27°	♉ 18°
☾	♐	♉ 12° !	♐ 30°
♄	♌	♌ 22°	♌ 22°
♃	♋	♋ 25°	♋ 23°
♂	♌	♌ 16°	♌ 11°
♀	♈	♈ 17°	♈ 14°
☿	♉	♊ 19° !	♊ 11° !
H	♋	♋	♋

Notes:

3: The symbol for the ascendant is written in the left margin, directly beside the spelled-out word.

4243. Horoscope, a.d. 217

71/84(e) 7×8.5 cm After a.d. 217

The horoscope, which is rather abraded, is written along the fibres on the back of a cursive text of uncertain nature (also along the fibres, oriented perpendicular to the horoscope). At least one line is lost at the beginning, but there appear to be extant margins on the other sides. The traces of writing below line 7 may be extraneous to the horoscope.

The recorded positions of the heavenly bodies are correct for a.d. 217, May 1, or for a date one day earlier or later.

	Text	217 May 1/2, 3rd hour of night
☉	♉	♉ 12°
☾	♌	♌ 15°
♄	♎	♎ 24°
♃	♐	♐ 4°
♂	♉	♉ 24°
♀	♓	♓ 28°

4244. Horoscope, a.d. 217

A188/4E 7×7.5 cm After a.d. 217

The fragment is broken on the top, bottom, and right edges; on the left a 1 cm margin is preserved. The hand is an unhandsome cursive. The traces reported as line 1 may have belonged to another horoscope or an unrelated text. The back is blank.

Neither the date nor the planetary positions are complete. We know that the date fell within the brief reign of Elagabalus, designated ὁ κόρυφος, 'the catamite', as in the horoscopes *P. Oxy.* XLVI 3298 line 2 (see introduction and note *ad loc.* for further discussion) and *P. Warren* 21 line 51. Within the four possible years a.d. 217/218 through 220/221 Jupiter was in Sagittarius only towards the beginning of 217/218, i.e. Elagabalus 1. The further requirement that Mars and Venus should be in the same sign is satisfied in (Alexandrian) Thoth of that year only for the first ten days or so of the month.

	Text	217 August 29 (= Thoth 1), 4th hour of day
☉	lost	♍ 7°
☾	lost	♉ 3°
♄	lost	♎ 25°
♃	♐	♐ 25°
♂	same as ♀	♌ 11°
♀	same as ♂	♌ 19°
☿	lost	♍ 16°
H	lost	♏

4245. Deluxe Horoscope, A.D. 218

22 3B.20/E(1–3)b 14.5×23 cm After A.D. 218

The horoscope is written in an upright hand on the back of a lease of a house dated Gordian (I & II) 1, Payni 22 = A.D. 238 June 16. Parts of the top margin (1 cm, marked off by a ruled line) and the bottom (4 cm) are present; some line beginnings and endings also survive, with slight margin on the right. At the bottom, the text seems to end abruptly, with barely room to give the sign occupied by the ascendant, let alone the remaining information supplied for the other elements of the horoscope; yet there can have been no further columns to the right.

The date is almost entirely missing at the top. The planetary positions are as follows, with sidereal positions recomputed by modern theory and also by Ptolemy's *Handy Tables*[1] for the night of A.D. 218, November 27/28:

	Text	218 Nov. 27/28 (= Choeac 1/2), 6th h of night	
		Modern	*H.T.*
☉	♐ 7;55°	♐ 8°	♐ 7° 44'
☾	♍ 6;51°	♍ 8°	♍ 7° 42'
♄	♏ 11;3x°	♏ 15°	♏ 14° 40'
♃	♒ 4;xx°	♒ 5°	♒ 6° 2'
♂	♒ 20°–30°	♒ 20°	♒ 20° 56'
♀	♎ xx;45°	♎ 22°	♎ 21° 47'
☿	♏ 25;50°	♐ 3° !	♐ 2° 36' !
H	lost	♍	

[1] The conversion to sidereal longitudes is by 'Theon's formula', for which see Appendix I (1).

For the sun, moon, and planets, positions are specified by sign, degrees, and minutes, as well as the number of the decan with the sign. Signs are characterized with the usual adjectives, and the lords of the sign and terms are given (cf. Appendix L, 1 and 2). Mercury's proximity to Saturn is mentioned.

Notes:

3: The reason for the horizontal stroke above the tau is not clear. One does not expect a numeral in the hundreds in a horoscope.

7: ἡ δὲ θεὰ καὶ cελαcφόροc cελήνη occurs in the deluxe horoscope P. Lond. I 130 (= No. 81 in Neugebauer & van Hoesen [1959]) lines 73–74. It was evidently a stock phrase in the composition of this kind of horoscope.

8: There does not seem to have been room at the beginning of the line for the word cελήνη.

10: The words ἀφαιροῦcα – ἀριθμοῖc are evidently a statement describing the moon's current speed and acceleration. The moon was in fact accelerating, having just passed its stage of least speed, and hence was trailing behind its mean longitude.

11. The moon's motion in latitude is characterized by its elongation from the northern limit of its orbit, measured in 'steps' of 15°. On the date in question, the moon was actually close to the beginning of the fourth step.

14. 'Rising in the east' signifies merely that the planet has already had its first appearance (which actually occurred about the middle of November), but not yet its morning station.

17. Jupiter was not in fact 'rising in the east,' but between its evening station and evening disappearance.

27. A statement of proximity of two planets is unusual in a horoscope. The interval in question, 14° according to the longitudes of the papyrus, is not small.

28. Mercury was rising slightly ahead of the sun, but probably not visible before sunrise. Conjunction occurred a day later.

4246. Horoscope, A.D. 222

C 319 = 10 B 4/4 9.5×5.5 cm After A.D. 222

The horoscope is written in a cursive hand across the fibres, although the other side is blank. All edges are preserved, with a 1.5 cm margin at the bottom. The heavenly bodies are all accounted for in the extant text, so that it is not clear whether any text has been lost at the beginning of line 6.

The positions are given in zodiacal order starting from the ascendant. Since Mars is near the end of Aries, the crossed out position in Pisces is more likely the effect of misreading an almanac than an erroneous computed longitude.

	Text	222 Dec. 31/223 Jan. 1, 7th h of night
☉	♑	♑ 12°
☾	♉	♉ 21°
♄	♐	♐ 29°
♃	♊	♊ 24°
♂	♈ (♓ deleted)	♈ 29°
♀	♒	♒ 13°
☿	♐	♐ 27°
H	♏	♏

4247. Horoscope, A.D. 223

31 4B.13/H(6–8)a, part 11×20 cm After A.D. 223

Plate X

This unusual composite horoscope is written on a rectangular ticket of papyrus, of which the top has broken off with at least one line of text; the back is blank. Lines 1–7, written in crude large capitals and with erratic orthography, present the horoscope of someone born in A.D. 223, December 14 (= Severus Alexander 3, Choeac 16/17). Below this, after a gap of 3 cm, a second hand writing an experienced cursive has copied out the identical horoscope, correcting the phonetic misspellings, but not otherwise altering the data. Lastly, a third, upright, hand writing with a paler ink has marked certain of the elements of this copy with tick marks, apparently to register agreement with them, but has written out different zodiacal signs for the moon, Mars, and Venus. It would appear that the owner of the horoscope had obtained a 'second opinion'—and indeed a better one, as can be seen from the recomputation below.

	Text (m. 1 & 2)	(m. 3)	223 Dec. 14, 5th h of night
☉	♐	♐	♐ 24°
☾	♓	♒	♒ 13°
♄	♑	♑	♑ 7°
♃	♌	♌	♌ 2°
♂	♏	♐	♐ 2°
♀	♐	♏	♏ 25°
☿	♐	♐	♐ 10°
H	♌ (corr. from ♎?)	♌	♍ !

4248. Horoscope, a.d. 229

20 3B.37/J(5)a 4×17 cm After a.d. 229

The horoscope is written in an upright hand across the fibres, although the other side has no writing. In spite of the seeming omission of the emperor's name, line 1 looks as if it was the top. The bottom 10 cm of the fragment is blank. Both edges are broken off.

The dating formula can be restored on the basis of the astronomical data as Severus Alexander 8, Pachon 22, end of hour 4 of day (= a.d. 229 May 17). From the pairing of the sun and Mercury in line 3 and Mars and the moon in line 8, it is clear that the positions were given only to the sign; nevertheless the ascendant seems to be specified in line 10 to the degree. The order is the standard planetary sequence except for Venus, which ought to follow Mars. I cannot explain the interlinear insertion (line 7) of the preceding day and the day of the nativity.

	Text	229 May 17, 4th hour of day
☉	♉	♉ 27°
☾	same as ♂	♌ 25°
♄	♓	♓ 18°
♃	♑	♑ 8°
♂	same as ☾	♌ 27°
♀	♊	♊ 26°
☿	♉	♉ 22°
H	♋ 14°?	♌ !

Notes:

10: Degrees are not expected after the sign of the ascendant in a horoscope that otherwise specifies only the signs occupied by the heavenly bodies. The reading is doubtful.

4249. Horoscope, A.D. 172 (?)

19 2B.78/B(4–7)c 6.5×21.5 cm After A.D. 172?

Written in a rounded serifed semiliterary hand on what was probably once a rectangular piece of papyrus. 1 cm margins at the top and left, 9 cm vacant at the bottom, broken on the right. The back is blank.

This horoscope is difficult to date securely. The dating formula in lines 1–4 is damaged, leaving the regnal year, the identity of the emperor, the calendar in use, and the day of the month all in doubt (see notes). A complete list of positions of the heavenly bodies survives, but the assumption that these were all correct or nearly correct leads to no acceptable date in the first four centuries. The date diffidently proposed here was arrived at on the assumption that the date of the nativity was Payni 8 in either the Alexandrian or the Egyptian calendar, and that of the zodiacal signs given for Saturn, Jupiter, and Mars, at least two were approximately correct. The range of years to be searched may be estimated in the first instance by the character of the hand, which appears to belong to the second or early third century (although the late first century cannot be excluded). Allowing for the possible interval between the nativity and the casting of the horoscope, consideration may therefore safely be limited to the years between A.D. 20 and 250. The statement that the sun was in Gemini would be compatible with any year in the Alexandrian calendar, but in the Egyptian calendar it would be unacceptable later than the middle of the first century. Only one date meeting these criteria could also be reconciled with the legible traces of the dating formula: A.D. 172 June 2 = 'Commodus' 12, Alexandrian Payni 8. On this day all the heavenly bodies were in or within one or two degrees of the expected signs except for Mars and Venus, the locations of which may have been inadvertently exchanged.

	Text	172 June 2, 3rd hour of night
☉	♊	♊ 13°
☾	♓	♈ 3° !
♄	♈	♈ 13°
♃	♓	♓ 29°
♂	♉	♊ 30° !!
♀	♊	♉ 2° !!
☿	♊	♋ 1° !
H	♑	♑

Notes:

1: The remains of the year number, following the symbol for ἔτουc, consist of a fairly clear ι and the bottom serif from another letter (most likely β, γ, or η); and above these, a long, slightly curved horizontal stroke marking the numeral. After this, were we expect the emperor's name, is a faint but unambiguous κ and, cut off by the edge of the fragment, an arc that could be the left half of ο (definitely not α). Κọ[μμόδου seems to be the only plausible reading, although it should be noted that because Commodus reckoned his regnal years from the accession of Marcus Aurelius, the earliest year properly attributed to him alone was the 20th. A parallel instance of this usage is the horoscope *P. Oxy.* XLVII 3353 dated Commodus 19 (= A.D. 178/179); cf. also a retrospective dating to Commodus 18 in a document, *P. Oxy.* 988. The attribution of regnal years of Marcus Aurelius to Commodus in astronomical contexts could have originated in the use of regnal canons in which only the name of Commodus was set beside the whole block of 32 years, cf. *P. Oxy.* I 35 (line 13), *P. Oxy.* XXXI 2551 (verso line 16), and *P. Ryl.* I 27 (line 76).

2: There is a gap of about 0.5 cm between Παῦνι and the numeral (similarly before γ̄ in line 4). Only the lower parts of the numeral survive; these are just like the bottom of η elsewhere in the papyrus, but one could conceivably read ιϛ. Since this is a nocturnal nativity, one would expect the line to have continued [εἰc τὴν θ̄, i.e. Payni 8/9.

3: The formula signifying the use of the old Egyptian calendar. This could pertain to the month and day in the preceding line, or to an equivalent date lost to the right. The latter is the case if the date proposed above is correct, and we might restore the equivalent Egyptian date Epeiph 27(/28).

4250. Horoscopes, A.D. 237 and undated

15 2B.39/A(e), part — 7.5×6 cm — After A.D. 237

Two horoscopes (the last line of one, and the first half of a second) in an upright hand, all sides broken. The back is blank. The recorded positions in the second horoscope compare as follows with recomputation:

	Text	237 July 7, 5th h of day = Maximinus 3, Epeiph 13
☉	♋	♋ 16°
☾	lost	♊ 16°
♄	lost (not ♋)	♋ 4° !
♃	lost	♍ 5°
♂	lost	♈ 1°
♀	♌	♌ 26°
☿	♋	♋ 5°
H	♎	♍ !

4251. Horoscopes, a.d. 244 and undated

100/127(b) 7×20 cm After A.D. 244

The horoscopes are written in a practised sloping hand on a tall narrow rectangular ticket, of which parts of all edges survive; the back is blank. The positions are recorded in order of zodiacal signs, making the identification of the heavenly bodies in the second horoscope impossible. For the first horoscope, the only uncertainty is concerning the objects in the signs named in lines 6 and 10 (neither corresponds to the moon's location). We have the following comparison with modern theory:

	Text	244 February 12, 3rd hour of night
☉	♒	♒ 25°
☾	lost	♍ 10°
♄	♍	♍ 29°
♃	♓	♓ 28°
♂	♋	♌ 3° !
♀	♈	♈ 9°
☿	♒	♒ 29°
H	♍	♎ !

4252. Horoscopes, a.d. 248/249 and undated

49 5B.98/K(1–4)a 8×11 cm After A.D. 249

Broken on top, bottom, and right, but with 2 cm left margin intact; the back is blank. The hand is upright, semiliterary but rather unsteady, with some attempt at serifs. A second hand has added something unreadable about Venus after the first horoscope. At the bottom edge are traces of writing in, apparently, a third hand, perhaps not relevant to the horoscopes.

The date of the second horoscope is incomplete, and we also lack the positions of the faster heavenly bodies. Within the sixth regnal year of the Philips (A.D. 248/249), Mars was in Leo only during the approximate interval between December 31 and February 9. Within a few decades of this year, there are several ranges of dates that fit the preserved positions of Saturn and Mars in the first horoscope, including September–October, A.D. 215; August–September, A.D. 217; September–November, A.D. 245; May–June, A.D. 274; and October, A.D. 275 – June, A.D. 276. The information that Venus and Mercury were in the same sign, and the moon in Cancer, does not suffice to determine which date is correct.

4253. Horoscope, A.D. 249/250

22 3B.19/F(a) 4×8.5 cm After A.D. 250

The left part of a rectangular ticket cut from a document (traces of writing on the front). The horoscope is written in a rather coarse upright hand across the fibres.

Within the Alexandrian calendar year Decius 1 (A.D. 249/250), Mars was in the same sign as Saturn (Sagittarius) approximately from the end of February to the beginning of September.

4254. Horoscope, A.D. 254

16 2B.45/D(b), part 30×11.5 cm After A.D. 254

This was the inside lining of a codex cover, a large rectangular slab composed of several layers of reused papyrus, with holes for the binding along one of the long edges and remains of leather (no writing) on one side. Traces of documentary writing, including an agricultural document of the third century, can be seen where the layers have come unstuck at the edges. The horoscope is written in the middle, parallel to the binding edge, and along the fibres. At the right end of the same face as the horoscope, but sideways and between two horizontal ruled lines, are written the first line of *Iliad* V followed by what appears to be an argument summarizing the book; perhaps this identifies the contents of the lost codex.

The horoscope is to be read downwards through the second and third columns, and follows neither of the conventional orders. It is not clear why Taurus is named by itself in the first column. The moon's position is missing (it cannot have been in Taurus if the other positions are correct). The planetary positions fit the last ten days of October, A.D. 254.

	Text	254 October 25, 11th hour of day
☉	♏	♏ 4°
☾	lost	♍ 26°
♄	♑	♑ 19°
♃	♒	♒ 15°
♂	♉	♉ 11°
♀	♏	♏ 22°
☿	♏	♏ 17°
H	♈	♈

4255. Horoscope, a.d. 256

85/72(a) 10.5×7.5 cm After a.d. 256

Written in browned ink on the front of a rectangular ticket; all edges are intact, but the text of the horoscope is damaged by a large hole. The back is blank. No name or date was given, and the standard order is disturbed by having the sun and Mercury precede Venus. The positions fit either August 28 or 29, a.d. 256.

	Text	256 August 28, 7th hour of day
☉	♍	♍ 6°
☾	♉	♉ 10°
♄	♒	♒ 14°
♃	♉	♉ 7°
♂	♊	♊ 11°
♀	♎	♎ 15°
☿	♍	♍ 8°
H	♐	♐

4256. Horoscope, a.d. 269–275

114/68(b) 9.5×10 cm After a.d. 269

Written in a neat, slightly serifed upright hand. Part of the top and left margins survive; the other sides are broken. The back is blank. A date in the old Egyptian calendar followed the Alexandrian date, but both are lost. We cannot deduce which regnal year of Aurelian was intended, because the moon always progresses through all the signs during the thirty days or so that the sun is in Aries.

4257. Horoscope, a.d. 279

101/155(b) 24.5×12 cm After a.d. 279

Written in clumsy capitals along the fibres in the top half of a large rectangular ticket of coarse papyrus, shattered but substantially complete; the back is blank. Planetary positions are separated by blots of ink, sometimes with tails but apparently lacking meaning. The sun's position is left out (inadvertently?); the ascendant, on the other hand, appears twice, with orthographical variations. The order of presentation is confused, even the year appearing unexpectedly at the end.

	Text	279 June 7, 3rd hour of night
☉	lost	♊ 17°
☾	♎	♎ 20°
♄	♏	♏ 29°
♃	♈	♈ 7°
♂	♊	♊ 11°
♀	♊	♊ 5°
☿	♉	♉ 25°
H	♒	♒
M	♉	♉

4258. Horoscope, a.d. 283

101/168(a) 18×5 cm After a.d. 283

Plate XI

Note for a horoscope, written in clumsy capitals along the fibres on a rectangular ticket; the back is blank. Only the name of the native and the date and time of birth are given, without any astronomical data (for this genre of document, see Baccani [1989]). Assuming that the calendar is the Alexandrian, the date is equivalent to a.d. 283 June 19/20.

4259. Horoscope, A.D. 284

16 2B.45/F(b), part 10×11 cm After A.D. 284

Broken on the right and bottom edges, although the text is substantially complete. The horoscope is written on the front in a handsome upright hand; on the back are the ends of three lines of indeterminate text. The planetary positions are in order of increasing longitude starting from the ascendant.

	Text	284 December 9, 6th hour of day
☉	♐	♐ 19°
☾	♊	♊ 8°
♄	♑	♑ 29°
♃	♍	♍ 23°
♂	♈	♈ 3°
♀	♐	♐ 13°
☿	♐	♐ 25°
H	♒	♓ !

4260. Horoscope, A.D. 296 (?)

40 5B.116/H(1–5)c 6×11 cm After A.D. 296?

Written in fading ink across the fibres on the back of a document (correspondence addressed to some bankers, other way up) dated A.D. 302. The top, right edge, and bottom of the horoscope are intact, the left edge broken off.

The date of the horoscope is given as Diocletian 13, Phaophi 6 = A.D. 296 October 3, probably hour 12 of day (since there does not seem to be a double date). Only one pairing of a heavenly body with a zodiacal sign survives, Mercury in Sagittarius, and unfortunately this is impossible for Phaophi 6 in any year. Perhaps the double stroke between lines 5 and 6 separates two horoscopes, although one would then have to suppose that a great part of the text is lost on the right.

Notes:

2–3: The full formula for the year would have been Diocletian year 13, Maximian 12, Constantius and Galerius 5. Cf. Bagnall & Worp [1978] 69, [1979] 11.

4261. Horoscope, A.D. 289

28 4 B.60/B(4–5)a | 7×15 cm | After A.D. 289
Plate XII

Written along the fibres in the vacant space at the bottom of a rectangular ticket cut from a receipt dated Diocletian 7 = Maximian 6, Pachon 6 (A.D. 291 May 1), only a year and a half after the nativity date of the horoscope. The back is blank. At the foot of the document, and immediately above and to the right of the horoscope, a hand different from either has written πανεχωτου ϕ[. This might conceivably name the native of the horoscope, but the fact that the text is broken (whereas the horoscope was presumably written after the ticket was cut) suggests that it has some connection with the receipt.

The horoscope bears no date, but the positions of the heavenly bodies, with an uncomfortably large error for Venus, adequately fit A.D. 289 November 18:

	Text	289 November 18, 5th hour of night
☉	♐	♏ 28° !
☾	♋	♋ 17°
♄	♓	♓ 28°
♃	♉	♉ 29°
♂	♐	♐ 5°
♀	♏	♐ 6° !
☿	♏	♏ 7°
H	♌	♌

Notes:

6: An ascendant has already been specified in line 1. It is conceivable that the object in line 6 was intended to be the midheaven, which would be about three signs lower than the ascendant.

4262. Horoscope, A.D. 292/293

85/82(b) | 4×5.5 cm | After A.D. 293

The top left corner of a horoscope, in an upright hand across the fibres although the other side has no text. The month and day, and most positions, are lost.

4263. Horoscope, A.D. 299

102/67(b) 7.5×14 cm After A.D. 299

Written in an upright serifed hand along the fibres on a rectangular ticket; the back is blank. The text is complete. The positions, which do not follow either standard order, fit A.D. 299 June 10 at noon, or one day earlier or later:

	Text	299 June 10, 6th hour of day
☉	♊	♊ 19°
☾	♈	♈ 14°
♄	♌	♌ 8°
♃	♏	♏ 22°
♂	♒	♒ 25°
♀	♊	♊ 23°
☿	♊	♊ 0°
H	♍	♍

4264. Horoscope, A.D. 300

36 4B.110/H(4–6)b 10.5×12 cm After A.D. 300

Written in a practised sloping hand along the fibres on a square ticket, intact except for perforations towards the bottom; the back is blank. The heavenly bodies are presented in the order of the signs that they occupy.

	Text	300 April 18, 4th hour of day
☉	♈	♈ 29°
☾	♍	♎ 2° !
♄	♌	♌ 17°
♃	♉	♉ 3°
♂	♊	♊ 11°
♀	♈	♓ 20° !
☿	♈	♈ 23°
H	♊	♋ !

4265. Horoscope, A.D. 313 and 348

104/69(a) aggregate 15×13.5 cm, debris After A.D. 348

Darkened and badly shattered fragments from a series of horoscopes, written in a practised upright hand along the fibres (the back is blank). Parts of both left and right margins survive, and it would appear also that the top and bottom lines are in part preserved. A horizontal stroke separates the first horoscope (lines 1–5) from the second, which pertained to an unnamed woman of high social status (lines 6–11). It is not clear whether line 12 also belonged to this horoscope, or was a supplementary note.

The heavenly bodies are not listed in either standard order. The month of the first horoscope is restored on the basis of the recorded positions of the sun and Venus. Mercury's and the ascendant's positions are broken off, and the sign for the moon and Saturn cannot be restored since one of them must have been erroneous. Only the moon's position is preserved in the second horoscope.

First horoscope:

	Text	313 July 21, 1st hour of day
☉	♋	♋ 28°
☾	lost	♐ 18°
♄	lost, same as ☾	♑ 22° !
♃	♒	♒ 9°
♂	♋	♋ 24°
♀	♋	♋ 7°
☿	lost	♌ 25°
H	lost	♌

Second horoscope:

	Text	348 October 26 (calculated for 6th hour of day)
☉	lost	♏ 4°
☾	♊	♊ 0°
♄	lost	♈ 0°
♃	lost	♑ 16°

♂	lost	♈ 18°
♀	lost	♎ 22°
☿	lost	♏ 18°
H	lost	?

Notes:

6: There is no room at the end of the line for a name; the last two letters of λαμπροτάτης are raised above the line.

4266. Horoscopes, A.D. 312 and 314

69/24(c) 17×26 cm After A.D. 314

The horoscopes are written in a sloping hand, side by side across the fibres on the back of part of a register with amounts in *litrai* and dates A.D. 310/311 and 311/312, very near the nativity dates of the horoscopes. The first horoscope is intact except for part of the top lines; the second is broken off on the right. An exceptional feature is the 'signature' of the writer, Diapalos; it is not clear whether this was the astrologer who cast the horoscope, or a scribe who copied it out.

In the first horoscope it is not clear whether the heavenly bodies are supposed to be listed in their standard order, since by coincidence this is also the order of the signs they occupy (except for the moon, which is in last place instead of between Mars and Venus, cf. **4269**). The second horoscope apparently follows the order of the signs, so that we can tentatively restore the ascendant as well as Jupiter in line 4. The day number is broken off, but the planetary positions fit A.D. 312 October 16 (= Phaophi 19) or one day earlier or later.

First horoscope:

	Text	314 November 19, 10th hour of day
☉	♏	♏ 28°
☾	♍	♍ 27°
♄	♒	♒ 3°
♃	♓	♓ 10°
♂	entering ♓	♓ 0°
♀	♎	♎ 15°
☿	entering ♏	♏ 30° (moving retrograde)
H	♒	♈ !!

Second horoscope:

	Text	312 October 16, 6th hour of day
☉	♎	♎ 24°
☾	♎	♎ 16°
♄	♑	♑ 8°
♃	♑ ?	♑ 2°
♂	♑	♑ 12°
♀	♐	♐ 11°
☿	♎	♎ 18°
H	♑ ?	♑

Notes:

i 4: The symbol for the ascendant is lined up with the preceding lines. Κρόνοc seems to have been added afterwards, and lines 5–10 aligned with the new margin.

ii 4: Ἄρηc was evidently added afterwards.

4266a. Horoscope, a.d. 313

40 5B.108/K(5–10)c, part 10×13 cm After a.d. 313

Written in a practised upright hand on the front of a rectangular ticket (slight traces of ink on the back). The bottom edge and the top left corner are broken off, the remaining edges intact.

The heavenly bodies are in the order of the signs they occupy. Venus' position is broken off, but from its place in the list, it must have been Gemini or Cancer.

	Text	313 August 1, 2nd hour of day
☉	♌	♌ 9°
☾	♉	♉ 11°
♄	♑	♑ 21°
♃	♒	♒ 8°
♂	♌	♌ 1°
♀	lost	♋ 21°
☿	♍	♍ 5°
H	♌	♍ !

Notes:

1: The regnal year formula for the Alexandrian year A.D. 312/313 would be Maximinus year 9, Constantine 7, Licinius 5 (Bagnall & Worp [1978] 71, [1979] 36–37). The second preserved numeral in this line is written as a single looping stroke joined to the stroke indicating a year number. This does not look right for epsilon, and might perhaps be meant for a six (which would however be an error).

4267. Horoscope, A.D. 318

50 4B.33/J(1–5)b	10×10 cm	After A.D. 318

Written in a wiry, slightly sloping hand across the fibres (!) on a square ticket, unused on the other side. The text is complete except for some breakage on the right.

The heavenly bodies are not listed in either standard order; Saturn's position in Pisces is squeezed in to the right of the other data.

	Text	318 November 8, 9th hour of day
☉	♏	♏ 17°
☾	♏	♏ 4°
♄	♓	♓ 22°
♃	♌	♌ 6°
♂	♊	♊ 3°
♀	♐	♐ 13°
☿	♏	♏ 17°
H	♈	♓ !

4268. Horoscope, A.D. 318

A141/6A	13×7.5 cm	After A.D. 318

Written across the fibres on a rectangular ticket, with its edges intact but abraded and badly stripped in the middle; the other side is blank.

The heavenly bodies are listed in the order of their signs. The day number is broken off, but the recorded positions (except Mercury) fit A.D. 318 September 23/24, or one day earlier.

Text	318 September 23/24, 9th hour of night	
☉	♎	♎ 1°
☾	♒	♒ 25°
♄	♓	♓ 24°
♃	♌	♌ 1°
♂	♊	♊ 6°
♀	♎	♎ 16°
☿	♎	♍ 18° !
H	♌	♌

Notes:

1: The Alexandrian year A.D. 318/319 was the regnal year Constantine 13, Licinius 11, Three Caesars 3. Cf. Bagnall & Worp [1978] 72, [1979] 39.

4269. Horoscope, A.D. 320

37 4B.104/A(1–3)a 7×12 cm After A.D. 320

Written in a slightly sloping hand with a tendency to ligatures, along the fibres on a rectangular ticket; the back is blank.

The heavenly bodies are listed in the order of their signs, except for the moon, which is in the last place.

Text	320 February 17, 2nd hour of day	
☉	♒	♒ 29°
☾	♏	♏ 24°
♄	♈	♈ 10°
♃	♌ entering ♍	♍ 2°
♂	♒	♒ 2°
♀	♒	♒ 23°
☿	♒	♒ 24°
H	♓	♈ !

4269a. Horoscope, a.d. 325

A604/5A 8×11 cm After a.d. 325

Written in a practised upright hand on the front of a rectangular ticket, broken at the bottom but with parts of the remaining edges. The back is blank.

The heavenly bodies are listed in the order of their signs. Hence the broken-off sign occupied by the sun and Mercury must be either Scorpio or Sagittarius, and that of the moon must be either Leo or Virgo.

	Text	325 November 16/17, 10th hour of night
☉	lost (♏?)	♏ 26°
☾	lost (♍?)	♍ 26°
♄	♉	♉ 6°
♃	♒	♒ 11°
♂	♑	♑ 9°
♀	♎	♎ 15°
☿	lost, same as ☉	♐ 6° !
H	♎	♎

4270. Horoscope, a.d. 350

42 5B.72/B(9–11)a 9×17 cm After a.d. 350

Written in a practised upright hand along the fibres on a rectangular ticket, unused on the back, and complete.

The heavenly bodies are listed in the order of their signs, with the moon at the end. Inclusion (by a second hand) of a 'Lot of Marriage' is unusual (cf. **4286** 13), and perhaps indicates that this horoscope was cast for catarchic purposes. Dorotheus II 2 (ed. Pingree, p. 199) gives instructions for finding the Lot of Marriage: the elongation of Venus from Saturn is to be added to the ascendant (for diurnal nativities of men and, apparently, nocturnal nativities of women) or subtracted from the ascendant (for diurnal nativities of women and nocturnal nativities of men). In the present horoscope, Venus is about 308° ahead of Saturn, and by *adding* this to the ascendant, one obtains a position in Scorpio. Since

Ammonas was presumably male, and his nativity nocturnal, we are dealing with a variant of Dorotheus' rule, or a different rule entirely.

	Text	350 March 19/20, 7th hour of night
☉	♓	♓ 29°
☾	♑	♑ 29°
♄	♉	♈ 20°
♃	♓	♓ 15°
♂	♒	♒ 13°
♀	♓	♒ 28° !
☿	♓	♓ 9°
H	♑	♐ !

4271. Horoscope, a.d. 354

A46/2F 9.5×12 cm After a.d. 354

Written in a practised upright hand along the fibres on what was probably a rectangular ticket, now stripped in the middle and broken at the bottom. The back is blank.

The positions of the moon, Mercury, and the ascendant have all been lost either in the stripped lines 4–6 or at the end. The other heavenly bodies have been restored according to the correspondence of the recorded zodiacal signs with the positions recomputed for the given date. The principle of ordering of the heavenly bodies is not clear.

	Text	354 December 31, 12th hour of night
☉	♑	♑ 11°
☾	lost	♑ 15°
♄	♊	♊ 27°
♃	♌	♌ 17°
♂	♌	♌ 27°
♀	♐	♐ 8°
☿	lost	♑ 0°
H	lost	♋

4272. Horoscopes, A.D. 370 and 375

58/A(6), part 6.5×13.5 cm After A.D. 375

A series of horoscopes, written in faded brown ink along the fibres on the back of a ticket cut from part of a document; there are also traces of other writing in a different ink among the lower lines of the horoscope. This horoscope is of special interest for its use of abbreviations and symbols, including the earliest known occurrences of forms of the symbols for Mercury, Venus, Jupiter, and Saturn found in Byzantine manuscripts (cf. pp. 62–63). Months are expressed by their ordinal numbers, although in line 1 the name has been added.

The Lot of Fortune in the first horoscope is computed in the standard way for a diurnal nativity, by adding to the ascendant's longitude the elongation of the moon from the sun (Neugebauer & van Hoesen [1959] 8–9). In the first horoscope the heavenly bodies appear in the order of their signs, but the order in the second horoscope is not clear.

First horoscope:

	Text	370 June 16, 10th hour of day
☉	♊	♊ 24°
☾	♍	♍ 12°
♄	♐	♐ 30°
♃	♏	♏ 15°
♂	♍	♍ 3°
♀	♌	♌ 28°
☿	♊	♊ 7°
H	♏	♏

Second horoscope:

	Text	375 October 26, 1st hour of night
☉	lost	♏ 3°
☾	lost	♉ 10°
♄	♓	♒ 27° !
♃	♉	♉ 16°

♂	lost	♍ 18°
♀	♎	♎ 5°
☿	lost (not ♎?)	♎ 25° !
H	♉	♉

4273. Horoscopes, undated and A.D. 399

104/171(a) 5.5×9 cm After A.D. 399

From a series of horoscopes in a handsome upright hand, broken on the top, bottom, and right; the back is blank. Only the positions of the sun and moon survive from the first horoscope (lines 1–2), so that it cannot be dated. The day number is lost from the second horoscope, but the moon's position in Cancer fits A.D. 399 February 4 (or one day later). A serious error was made in the computation of either Mars' or Venus' longitude; probably both were supposed to be in Capricorn, since Venus' correct sign, Aries, has already been mentioned in line 6.. The heavenly bodies are listed in their standard order.

Text	399 February 4, 12th hour of night	
☉	lost, same as ☿	♒ 16°
☾	♋	♋ 18°
♄	lost, same as H	♐ 15°
♃	♈	♈ 19°
♂	lost, same as ♀	♑ 23°
♀	lost	♈ 2° !
☿	lost	♒ 3°
H	lost	♐

4274. Horoscopes, A.D. 480 and 503

93/Dec. 26/C.1, part 14×9 cm After A.D. 503

Plate XI

The horoscopes were found together with numerous scraps of an astrological text or texts, in part at least in the same sloping hand. A peculiarity that the horoscopes share with those fragments that have text on both sides is that the writing on each side is upside-down with respect to the other side, that is, one must flip along a horizontal axis. On both sides of the horoscopic fragment the text is broken at top and bottom, but right and left margins survive. In the right margin of the side with vertical fibres, written sideways in brown ink, is a word or name]ι̣ωμου; on the other side, similarly situated, there are traces of brown ink that I cannot make sense of as letters. In the horoscope the sun, moon, and planets are represented by abbreviations or symbols (cf. pp. 62–63).

Both horoscopes seem to follow the standard order of the planets, except that Jupiter and Saturn have exchanged places in the first horoscope. The month and day in the first horoscope are broken off; recomputation for A.D. 480 October 24 (or one day later) matches the recorded positions of the outer planets, the sun, and the moon. In the second horoscope, the lost year can be restored from the moon's longitude (confirmed by the given number of days since conjunction in line 3), together with the information from line 4 that Payni 21 was a Sunday.

First horoscope:

	Text	480 October 24, computed for 6th hour of day
☉	♏	♏ 1°
☾	♑	♑ 6°
♄	♎	♎ 10°
♃	♓	♓ 6°
♂	♌	♌ 6°
♀	♐	♏ 19° !
☿	♎	♏ 5° !
H	♑ 6°	♑ 12°

Second horoscope:

	Text	503 June 14/15, computed for 6th hour of night
☉	♊ 21°	♊ 21°
☾	♌ 26°	♌ 23°
♄	♋	♋ 11°
♃	♒	♒ 20°
♂	lost	♋ 3°
♀	lost	♉ 6°
☿	lost	♊ 8°
H	lost	?

Notes:

→ 8: I am unsure of the resolution of the abbreviation, and do not know why Libra is mentioned here.

↓ 1: After]νου are two clear strokes that I cannot identify as a letter or symbol.

4: This seems to be the only known instance of a planetary weekday in a papyrus horoscope. Because of the nocturnal geniture, the text indicated the planets governing both Payni 20 (Saturn, broken off) and Payni 21 (the sun). It is not clear whether there was a word or abbreviation for 'day.'

4275. Horoscope, A.D. 508

68 6B.20/F(1–2)a — 10×19 cm — After A.D. 508

Plate XII

Written in a cursive hand along the fibres on a rectangular ticket, missing a large chunk in the middle (i.e. the last lines of the horoscope). The back is blank. The sun, moon, and planets are represented by symbols or abbreviations (cf. pp. 62–63), and prefaced by an abbreviation to indicate whether each is moving direct or retrograde at the date of the nativity. The positions have been recomputed both by modern theory and Ptolemy's *Handy Tables* (without converting from tropical to sidereal longitudes). The precise agreement with the *Handy Tables* for Jupiter and Saturn, and approximate agreement for the faster planets, shows that Ptolemy's tables were the ultimate basis for the horoscope, though probably through the intermediary of an almanac. The longitude of the ascendant is not what one would find from the *Handy Tables* for the latitude of Lower Egypt, although one finds close to the longitude in the papyrus by using the ascension table for the latitude of Meroe (!). I cannot explain the larger discrepancies for the sun and moon.

	Text	508 February 1/2, 7th hour of night
☉	♒ 13° 11'	♒ 12°
☾	♋ 26° 4'	♋ 30°
♄	♍ 14° 19' retrograde	♍ 15° retrograde
♃	♋ 3° 2' retrograde	♋ 1° retrograde
♂	♑ 7° 9' direct	♑ 7° direct
♀	♑ 21° 36' retrograde	♑ 14° retrograde
☿	♒ 15° 14' retrograde	♒ 6° retrograde
H	♏ 27° 30'	♏ 21°

Notes:

1: The Christian staurogram also appears at the top of the horoscope P. Oxy. XVI 2060 (= No. 478 in Neugebauer & van Hoesen [1959]).

3: The indiction number is correct. Indiction numbers also appear in some of the horoscopes *PSI* I 22–24 (Nos. 366, 370, 376a–b in Neugebauer & van Hoesen [1959]).

4276. Deluxe Horoscope

15 2B.40/C(b) 12.5×20 cm Late second or early third century

Plate IX

The fragment comes from a large-scale prose horoscope in roll format (there is a sheet join 1.5 cm from the left edge). Parts of two columns of texts are extant: the ends of the top lines of column i, and, after a 1 cm intercolumniar gap, the full breadth (9 cm) of the top lines of column ii. The lower part of column ii is represented by a dwindling strip near the left edge of the column. 3 cm top margin and, apparently, 2 cm bottom margin are present, so that the fragment represents the approximate original height of the roll. The writing is a handsome sloping 'severe' hand.

The complete horoscope would have contained a preface and paragraphs concerning the sun, the moon, the five planets, the ascendant, and a selection of other astrologically significant points. Only the paragraph for the ascendant (column ii, lines 1–10) is fairly well preserved. Since it is likely that the ascendant came after the seven heavenly bodies but before the other astrological points, we can estimate that at least three columns of text have been lost before column i. Column i line 1 was the penultimate line of one of the paragraphs (the last line was too short to reach the present left edge), and lines 2–6 belong to the next paragraph. Depending on how prolix the paragraphs for the planets were, there could have been a third paragraph, now entirely lost, below this. We might guess that the subjects were, respectively, Mars, Venus, and Mercury, or just Venus and Mercury. The paragraph following that for the ascendant in column ii most likely pertained to the midheaven, but this is not certain since the choice and order of astrological points was not fixed in the deluxe horoscopes.

The pattern of data given for the ascendant was probably followed more or less for the other objects in the horoscope. The longitude is given by zodiacal sign, degrees, and minutes, and the zodiacal sign itself is described according to its conventional attributes (cf. Appendix L, 3). After some indeterminate text, the planet governing the terms is named (cf. Appendix L, 1), and then a line or so of further information of uncertain nature. A feature of especial interest is the description (column ii, lines 2–6) of the ascendant's derivation from an *anaphorikos* ascribed to Hipparchus, for which see the note *ad. loc.*

Notes:

Column i

1: Probably a description of the planet's location with respect to a 'northern' constellation, e.g. the 'northern fish' of the constellation Pisces. Cf. *P. Lond.* I 130 (= No. 81 in Neugebauer & van Hoesen [1959]) lines 138–139, Venus ἐπὶ τοῦ νοτείου Ἰχθύοc.

2: 'Computed' refers to the longitude of the planet (Venus or Mercury, see introduction above).

3–4: The only sign that is a house of Mars and 'tropical' is Aries. If the planet in question was Venus, the sun would have had to be between Aquarius and Gemini; if Mercury, the sun would have had to be between Pisces and Taurus. In fact we know that the sun was in Aries, since the ascendant at the first hour of day was close to the end of that sign (column ii line 8).

Column ii

1–7: Cf. *P. Lond.* I 98 (= No. 95 in Neugebauer & van Hoesen [1959]) lines 59–61: τὸν δὲ ὡρονόμον ὡροcκόπον προcδεῖ δι' ὅλην τὴν [ἀν]αλογίαν ἐκ τοῦ ἀναφορικοῦ πρὸc κλεψύδραν· ἐκπίπτει ἐν Καρκίνῳ... The term ἀναφορικὸc designates a table of oblique ascensions, correlating the simultaneously rising degrees of the ecliptic and equator. The ascendant would have been computed from such a table by finding the equatorial degree corresponding to the sun's longitude, adding the number of equatorial degrees risen since sunrise, and converting back to the corresponding ecliptic longitude.

The description here is incoherent. It seems certain that the words Ἱππαρχικοῦ cυντάγματοc should be taken with ἀναφορικόc, meaning presumably a Hipparchian collection of tables of ascensions applicable to the several principal latitudinal belts or *klimata*. This was the provenance of the tables used to compute the ascendant. The phrases πρὸc πρώτην ὥραν ἡμέραc and πρὸc τὸ κλίματοc Αἰγύπτου, on the other hand, give two of the variables involved in using the tables: the *klima*, which determines which table must be used, and the number of seasonal hours since sunrise (the third variable, the sun's longitude, is not mentioned). I suppose that κατὰ cύμφωνον τοῦ γνωρίμου τόπου merely means that the *klima* of Egypt is appropriate for the birthplace of the native.

The information that ascension tables were circulating in Roman times under Hipparchus' name is an interesting novelty. It accords with a remark of Pappus of Alexandria (*Collection* VI 109, ed. Hultsch v. 2, 600)) to the effect that Hipparchus in a work 'on the ascension of the twelve signs' demonstrated certain properties of oblique ascensions by numerical means (as opposed to the strictly geometrical reasoning of the 'spherics' tradition). While one may be hesitant to conclude that Hipparchus was the actual author of such tables on the basis of such late witnesses, the hypothesis is not historically implausible *per se*. We actually possess a short work entitled Ἀναφορικὸc by an approximate contemporary of Hipparchus, Hypsicles (middle of second century B.C.), which, although not strictly a table of ascensions, presents a mathematical derivation of such a table appropriate for the latitude of Alexandria based on an arithmetical progression of Babylonian origin. Hipparchus, in whose geographical work the topic of the principal geographical circles of latitude featured prominently, could conceivably have extended the scheme of the Babylonian and Hypsicles' ascension tables to a range of *klimata*.

4277. DELUXE HOROSCOPE

21 3B.27/H(8)a | 30×20.5 cm (fr. 1) | Late second/early third century
2.5×2.5 cm (fr. 2)

The horoscope is written in a good round documentary hand on the front of a roll, the back of which was later used for a register (parts of four columns, in a third-century hand). The top margin (1.5 cm including heading) is preserved for the full width of fr. 1, and the bottom margin (1 cm) for most of its left half. The full width of two columns and the left half of a third survive, with intercolumnia of 2 cm. The text is laid out in paragraphs, each with an indented heading, an enlarged initial capital jutting into the left margin, and *diple obelismene* to separate one paragraph from the next in a column. Preceding the left edge of fr. 1 there would have been an introduction and paragraphs for the sun, moon, and the four planets Saturn through Venus, i.e. enough text for three columns (to one of which the small fr. 2 must have belonged). The right half of fr. 1 is darkened to the extent that col. iii is barely visible, and it is not clear how far down the papyrus the text extends.

The surviving paragraphs pertain to Mercury, the four cardines, and (at least) three lots. Each paragraph has some or all of the following:

1) Identification of the object in question. For Mercury (and presumably the other planets) and the ascendant, this is supplemented with further characterization of the object's cosmic place and function.

2) Identification of the zodiacal sign occupied by the object, with its conventional attributes (cf. Appendix L, 3).

3) Degrees and minutes of the position within the sign.

4) Lord of the sign, term, and *monomoiria* (cf. Appendix L, 1).

5) Lords of the triplicity (cf. Appendix L, 2).

6) Planets with exaltation or depression in the sign (cf. Appendix L, 2).

7) *Dodekatemoria* (sign with attributes, degrees and minutes).

8) Step and wind.

9) Zodiacal place.

10) Sex of the degree.

11) Number of the decan.

There is insufficient astronomical information to make it possible to date the horoscope.

Notes:

Fr. 1, column i

2: Cτίλβων, the 'glittering one,' is one of a set of quasi-rationalistic names for the planets, of which the earliest datable attestations are in two anonymous observation reports cited in Ptolemy's *Almagest* (IX 7, ed. Heiberg v. 2, 264, and IX 10, ed. Heiberg v. 2, 288), pertaining to 265 and 262 B.C. (both, as it happens, for Stilbon = Mercury). The names occur in several deluxe horoscopes: cf. Nos. 46, 81, 95, 137c, 176, 250.2 in Neugebauer & van Hoesen [1959].

2–3: The only other papyrus horoscopes mentioning a 'zone' in connection with a planet are P. Par. 19bis (= No. 137c in Neugebauer & van Hoesen [1959]) col. ii line 2 and **4283**. The reference may be to a scheme associating the elements in their cosmic situations with the heavenly bodies.

5: χερcαῖον characterizes the sign as a 'land animal.' The term is applied variously to certain zodiacal signs (Hübner [1982] 174–175), but the attribution of this property to Capricorn here seems to be unique, and is indeed odd, since the 'goat-fish' is typically thought of as either aquatic or amphibious.

7–8: I do not know of a scheme for *monomoiriai* that assigns Capricorn 6° to Venus, or (i 40–ii 1) Taurus 27° to Mercury. In ii 9 Jupiter for Scorpio 27° fits the rule of Vettius Valens IV 26 (ed. Kroll, 202; ed. Pingree, 193, cf. Neugebauer & van Hoesen [1959] 10), but this may be a coincidence.

8–9: The triplicity Taurus–Virgo–Capricorn is ruled by the moon for nocturnal genitures (cf. Appendix L, 2). Likewise the rulers of the triplicities named in col. i lines 23, 32, and 39, and ii line 8 are consistent with the conventional scheme for night. I know of no parallel for the 'sharers' of the triplicities, which appear to be Saturn for both the triplicity of Taurus and that of Gemini; Mars for the triplicity of Aries; and Jupiter for the triplicity of Cancer.

10–12: The *dodekatemorion* is obtained by projecting the zodiacal circle, starting with the sign in question, on the sign, assigning 2 ¹/₂ ° to each sign. Hence one takes the degrees and minutes, in this instance 6° 22', and multiplies them by 12 to obtain a number of degrees to be counted off from the beginning of the sign (here, Capricorn). The result should have been Pisces 16° 24', but the astrologer seems to have multiplied the minutes of the longitude by 6 instead of 12, thereby obtaining Pisces 14° 12'.

13: The 'step' and 'wind' (i.e. direction) probably go together, although the wind is omitted in the analogous passages col. i lines 24 and 33. The nature of the scheme being used here is not clear. One would expect that the ecliptic is being divided into two equal halves of 12 steps (each step equal to 15°), one half identified as 'northerly' and the other as 'southerly,' with the boundaries at, say, the equinoctial points or the moon's nodes. Four positions ranging from Virgo 7° to Aquarius 22° (i 13, ii 10, ii 22, iii 6) are in fact southerly, and one at Taurus 27° (ii 2) is northerly, and this is consistent with a bisection of the zodiac at the equinoctial points (or points near them). I am unable, however, to find a pattern in the numbers assigned to the steps.

13–14: A sign is given as the 'zodiacal place' here and in i 25. I know of no parallel for this, and do not know what it means or how it was determined.

14–15: The 'second change of direction in anomaly' must refer to one of Mercury's stations, perhaps the one that falls close to first morning visibility (which is only a few days after the other station).

17–18: The conception of the ascendant as a rudder is conventional: see P. Lond. 130 (= No. 81 in Neugebauer & van Hoesen [1959]) line 165; Paul. Alex. ch. 24 (ed. Boer p. 54 line 2); epitome of

Thrasyllus in *Catalogus Codicum Astrologorum Graecorum* 8.3, 101 line 18; epitome of Antiochus, *ibid.* 117 line 1. 'Steersman' develops the same idea.

19–20: It is odd to speak of the motion in longitude of the cardines, as if they were planets rather than the intersections of the stationary horizon and meridian with the revolving zodiac.

21: 'Royal' is a frequent characterization of Leo (and also Aries and Sagittarius); cf. Hübner [1982] 208–209.

24: Numbered 'segments' are mentioned also in i 33, ii 1, 9–10, 21. I cannot identify them, or the principle behind the numbers.

25–26: The Leo 22° and Aquarius 22° (i 33–34) are described as 'feminine.' This does not fit the irregular scheme set out by Firmicus Maternus IV 23, but may simply mean that the number is even; cf. Bouché-Leclercq [1899] 154.

30: For 'damp' and similar terms applied to Aquarius and other signs, see Hübner [1982] 171–173.

Column ii

12–22: The Lot of Fortune for a nocturnal nativity should be as many degrees behind the ascendant as the moon is ahead of the sun (Neugebauer & van Hoesen [1959] 8–9). This cannot, of course, be confirmed in the present instance since the longitudes of the sun and moon are lost.

16: To restore the degrees, we have the information that the Lot of Fortune is in the terms of Venus (Virgo 7°–17°, cf. Appendix L, 1), and in the first decan, i.e. 0–10°. If the *monomoiria* in line 20 was determined by Vettius Valens' scheme (cf. note to i 7–8), the degree must be 7°; otherwise any number from 7° to 9° is possible.

23–32: The Lot of Daimon should be as many degrees ahead of the ascendant as the Lot of Fortune is behind it (Neugebauer & van Hoesen [1959] 9), or in the present instance somewhere in the range Leo 4°–7°. This is consistent with the sign and decan given.

33–iii 9: Two further lots are presented. The name of the first cannot be read, but the second seems to be identified in iii 9 as the Lot of Necessity. Lots of Eros and Necessity are found in two papyrus horoscopes (Nos. 138/161 and 338 in Neugebauer & van Hoesen [1959]). In those horoscopes, the Lots of Eros and Necessity are respectively twice as far from the ascendant as the Lots of Fortune and Daimon. This is compatible with a position in Virgo for the Lot of Eros (iii 2).

4278. Deluxe Horoscope

28 4B.62/E(6–8)b 6.5×10.5 cm Late fourth/early fifth century

The horoscope is on a piece of discoloured parchment, broken on all sides. There are slight traces of writing on the other side. Despite the poor state of preservation, we can see that this was an unusual horoscope. The top six lines give longitudes, in degrees and minutes; unfortunately the names of the heavenly bodies and the zodiacal signs are lost (hence there is also no basis for an exact dating). The remainder of the text seems to contain an interpretation of the horoscope, including remarks about the life-span and manner of death of the native. Such a section would be unparalleled among the documentary horoscopes. Perhaps this is not a genuine personal horoscope at all but rather a fragment of an instructional text.

4279. DELUXE HOROSCOPE

29 4B.42/J(4)a 6×9 cm Late first/early second century

The horoscope is written across the fibres on the back of a piece cut from a land register (early first or late second century); only the bottom portion survives. The structure of the earlier part of the horoscope is not clear; the reference to a 'day' in line 1 is unexpected at what one presumes is part of the section for Venus. Mercury's longitude is given to the degree and minute, with lords of the sign and terms and a specification of the stage of the synodic cycle (during the interval of invisibility between last morning visibility and the following conjunction). The ascendant's position, however, is given only to the zodiacal sign.

Below the horoscope, another hand has written some obscure lines addressing the recipient of the 'ticket' (meaning this very piece of papyrus?). What, if anything, this has to do with the horoscope is not clear to me. Underneath this note, and indented from the left margin, what seems to be a third hand has written down the 'lot of daimon' for the horoscope (cf. note to **4277** ii 23–32).

4280. DELUXE HOROSCOPE

30 4B.37/F(1–3)b 9×5.5 cm Late first/early second century

The bottom left corner of a horoscope (or of a column of text if the horoscope took up several columns), with 0.5 cm margin to the left and 1.5 cm at the bottom, broken on the other sides. The back is blank. Positions were given to the degree (and minute?), with lords of the sign and terms, stage of the synodic cycle, and the number of the decan.

4281. Deluxe Horoscope

46 5B.53/G(1–5)d 12×5.5 cm Second century

The bottom of a column of text, with 2 cm lower margin and 1 cm margin on the left except in line 2, where the first line of a new paragraph jutted about 1.5 cm into the margin. The top is broken; line-ends are barely reached on the right. The horoscope was written across the fibres on the back of a used papyrus, but the extant piece is blank on the front except for a trace of ink.

The text was laid out in paragraphs, with *diple* to separate consecutive sections and the first lines drawn out into the left margin. The paragraph for Saturn must have been fairly near the beginning of the horoscope if it followed the standard sequence. The longitude is given to the degree and minute, with the lords of the sign and terms and an indication of the stage in the synodic cycle. The eclipse mentioned in line 1 may have been an event that took place within a month of the date of nativity; cf. **4282** for the inclusion of nearby syzygies in a horoscope.

4282. Deluxe Horoscope

17 2B.63/G(c) 9×16 cm Late third/early fourth century

Written across the fibres on the back of a petition (third century). What survives is the top of a column of text, with 3 cm upper margin, 1.5 cm intercolumniar space on the left, no vacant space on the right. The horoscope was laid out in paragraphs separated by *diple obelismene*. The first, which is preserved complete, sets out the date of the full moon that preceded the nativity date, the moon's longitude to the degree and minute, and the lords of the sign and terms. Syzygies close to the date of birth are taken into account in several ancient 'literary' horoscopes (cf. Neugebauer & van Hoesen [1959] 174); this is the first example in a papyrus horoscope. The first line of the subsequent paragraph (line 11) is blurred, so that it is not clear whether the conjunction described in the lost sequel was the one preceding or following the nativity.

Opposition is said to have occurred about two hours past sunset on Payni 25/26 (= June 19/20 in any year). During the interval A.D. 200–350, oppositions took place between noon and midnight (Alexandria time) of June 19 only in A.D. 243 (about 4 P.M.), 262 (about 5 P.M.), and 281 (about 2 P.M.). The sidereal longitude of the moon in each instance was approximately Sagittarius 28°.

4283. Deluxe Horoscope

105/117(d) 6×9 cm (fr. 1) Late third/early fourth century
13.5×7 cm (fr. 2) debris

Fragments, broken on all sides, of an elaborate horoscope, written on the front of the papyrus only. Fr. 1 has part of the last lines of Jupiter's paragraph and the first lines of Mars', separated by a *paragraphos*. Fr. 2 preserves most of the section for Mercury. The introductory phrases of Mercury's paragraph resemble **4277** fr. 1 i 1–2. The occupied zodiacal sign is qualified by several adjectives; longitude is stated to the degree and minute, followed by the lords of the sign and terms, exaltations and depressions, triplicity, *monomoiria*, and the number of the decan.

Notes:

Fr. 1

1–3: These lines should belong to Jupiter. Jupiter rules the triplicity Aries–Leo–Sagittarius for nocturnal nativities.

4–5: The phrasing is evidently the same as for Mercury (fr. 2, 1–2), and parallels **4277** fr. 1 i 1–2 (which however has 'the air' in stead of 'the heavens'). The word at the end of line 5 may be ψηφιcθείc (cf. **4277** fr. 1 i 19). For the alternative names of the planets, see note to **4277** fr. 1 i 2.

Fr. 2

4: ἀνθρωποειδὲc is a frequent attribute of several signs including Gemini (Hübner [1982] 133–134). For μουcικὸν and φωναcκὸν I know no exact parallels, but cf. Vettius Valens I 2 (ed. Kroll 7, ed. Pingree 7), where those born under the influence of Gemini are said to be φιλόμουcοι, φωναcκοί.

4284. Deluxe Horoscope

119/71(b) 9×8 cm Late third century

Written in a formal documentary hand on the front of the papyrus only, broken on the top, bottom, and left sides, but with a little margin on the right. This piece is probably from towards the end of an elaborate horoscope; indeed the horizontal stroke in line 14 may mark the very end. The identifiable sections pertain to the Lot of Fortune (9–11) and the Lot of Daimon (12–15), which one would expect to follow the sun, moon, planets, and cardines. (On these lots, see the note to **4277** ii 23–32.) The longitude is specified by sign (with conventional adjectives), degree, and minute, with lords of the sign and terms and the number of the decan.

4285. Deluxe Horoscope

46 5B.51/F(5–7)c | aggregate 15×11 cm | Third century

A horoscope written along the fibres, inside a tabular framework ruled in red; the back of the papyrus is blank. Positions (of unidentifiable heavenly bodies) are given to the degree and minute, with lords of the sign and terms, number of the decan, and other information that I am unable to identify.

4286. Tabular Horoscope

106/84(a) | 9×28 cm | Fifth century

A fragment in wretchedly abraded condition from the top of a horoscope, written in a cursive hand along the fibres in a tabular format, with 3 cm upper margin, broken on the other sides; the back is blank. Triple ruled lines (the middle one in each set in red ink) separate lines that seem to have given the longitudes of the sun, moon, planets, and ascendant. All that survives is a few of the degrees and minutes and part of the identification of the signs in question as exaltations of heavenly bodies. I can make nothing of the lines above the first rulings. Lines 11–13, below the last rulings, give locations for three 'lots', including the Lot of Fortune (cf. note to **4277** ii 12–22) and the Lot of Marriage (cf. **4270** 14). A better preserved example of this variety of tabular horoscope is *P. Oxy.* LXV 4477, for A.D. 430.

4287. Horoscope, A.D. 221–234

20 3B.31/E(6–7)b | 9×6.5 cm | After A.D. 221

A tattered scrap, missing both edges, although the top and bottom lines are both partially extant; the horoscope is written in a practised, slightly sloping hand along the fibres. From line 2 we know that the nativity fell within the reign of Severus Alexander, but the planetary positions are too poorly preserved to allow us to narrow the range to a specific date.

4288. Horoscope

67 6B.16/M(4–5)b 3.5×2.5 cm Second century

Beginnings of four lines from a horoscope written along the fibres in a practised upright hand, with 1 cm left margin, broken on the other sides.

4289. Horoscope

68 6B.19/G(3)d 1.5×4 cm Second or third century

From the middle of a horoscope, written along the fibres in a practised, slightly sloping hand. The numeral in line 5 may be a longitude in degrees.

4290. Horoscope, A.D. 285

102/49(c) 4.5×3.5 cm After A.D. 285?

A fragment of a horoscope written along the fibres in a rough, sloping hand, broken on all sides. Line 1 (merely a speck on the upper edge) was probably the first. After Diocletian's name, the scribe has written τό, as if to introduce a dating by consulship (which would be unusual in a horoscope); but no numeral follows, the text instead continuing as a regnal year formula of Diocletian and Maximian, without further colleagues. This implies a date between A.D. 285/286 and 292/293, and the positions of Jupiter, Mars, and the moon narrow the range to within a day of A.D. 285 September 16 (= Diocletian 2 Thoth 19).

	Text	285 September 16, 1st hour of day
☉	lost	♍ 24°
☾	♍	♍ 23°
♄	lost	♒ 7°
♃	♎	♎ 2°
♂	♍	♍ 22°
♀	lost	♏ 3°
☿	♎	♎ 0°
H	♎	♎

4291. Horoscope, A.D. 285–293

102/58(b) 5×6 cm After A.D. 285

The top lines of a horoscope written along the fibres in a faded, slightly sloping hand, broken on left, right, and bottom. The dating by Diocletian and Maximian requires a year in the range 285/286 through 292/293.

Notes:

3: The traces at the beginning of the line appear to be two diagonal strokes, such as might form part of the sign for ἔτουϲ following the lost numeral.

4292. Horoscope

102/150(a) 12.5×15.5 cm Fifth century

Written across the fibres in a large, practised hand on the back of a business letter (also fifth century). The right half of the horoscope is broken off. I was unable to find a date compatible with the hand on the hypothesis that the signs named in lines 2 and 4 pertain to the planets named in the preceding lines.

4293. Horoscope

102/167(b) 5×8 cm Fourth century

The text on the front of this fragment is the upper left hand corner of a document (official correspondence?) dated by the consulships of Septimius Acyndinus and Populonius, i.e. A.D. 340. The horoscope on the back, in a large practised hand, is oriented perpendicularly to the document, hence also along the fibres, and again it is the top left corner that survives.

4294. Horoscopes, later than A.D. 284

18 2B.69/E(1)b 4.5×4 cm Fourth century

A fragment in a neat upright hand, from the middle of a collection of horoscopes, broken on all sides; the back is blank. Lines 1–2 belong to the end of one horoscope, lines 3–7 to a second one dated by the Era Diocletian (or a true regnal year of Diocletian).

4295. Horoscope

18 2B.69/F(2–4)c 6.5×5 cm Late third or early fourth century

The last lines of a horoscope written in a handsome informal hand on a strip of papyrus, broken on the left and bottom (with no loss of text) as well as the top. The back is blank. The inclusion of a Lots of Fortune and Daimon (for which cf. notes to **4277** fr. 1 ii 12–32) is unusual in a horoscope that exhibits no other 'deluxe' features.

4296. Horoscope

125/25(j) 11.5×6 cm Fourth century

Written in a semicursive hand on the back of a ticket cut from some accounts, and oriented at right angles, i.e. also along the fibres. The top and left sides (relative to the horoscope) are broken off. The ends of the lines survive, with about 2 cm vacant space below line 3; since this is clearly not the end of the horoscope, we must hypothesize a fourth line that ended to the left of the extant fragment. Line 1 is part of the dating formula, and may have been the top line.

4297. Horoscope(?)

6153/C10 15×2.5 cm Third century

Written along the fibres in an upright informal hand on the back of a document (register?), and oriented at right angles to it. There is a vertical line about 2.5 cm to the left of the name Θοώνιος. If this was a horoscope, the format was unusual, since the time seems to have followed at least one of the planetary positions. I can make no sense of line 3.

4298. Horoscope

6153/C11 2.5×7 cm Third century

A small scrap from the middle of a horoscope written across the fibres (though no writing is visible on the back) in a coarse hand. The sign ȹ in line 4 probably stands for the ascendant, since it comes too late for the normal position of the dating formula.

4299. Horoscope

A 188/3F 17×5 cm Second century

On the front, a fragment of accounts of money; the horoscope on the back, in an informal upright hand, is oriented at right angles to this, hence also along the fibres. The top and left are broken off; the surviving portion was the bottom right hand corner, with 3 cm vacant at the bottom and about 4 cm on the right.

4300. Horoscope

A 31/10 | 3×4.5 cm | Third century

A fragment written along the fibres in a fairly large sloping hand, from the middle of the last lines of the horoscope. The top, left, and right sides are broken off; at the bottom, about 1.5 cm margin.

4300a. Horoscope

A 608/32a | 11×16 cm | Third century

A tattered fragment written along the fibres in a large practised hand, broken on all sides. The dating formula and most of the planets' names are lost.

Appendices

Bibliographical Abbreviations

Subject Index

A. Notations and Symbols.

1. Zodiacal signs.

	Symbol	*Longitude of 0°*	*Number from Virgo*
Aries	♈	0°	8
Taurus	♉	30°	9
Gemini	♊	60°	10
Cancer	♋	90°	11
Leo	♌	120°	12
Virgo	♈	150°	1
Libra	♎	180°	2
Scorpio	♏	210°	3
Sagittarius	♐	240°	4
Capricorn	♑	270°	5
Aquarius	♒	300°	6
Pisces	♓	330°	7

2. Planets.

	Symbol
Saturn	♄
Jupiter	♃
Mars	♂
Venus	♀
Mercury	☿
Sun	☉
Moon	☾
Ascending node	☊

3. *Phases.*

	Symbol
Saturn, Jupiter, Mars	
First morning visibility	Γ
First station (morning)	Φ
Acronychal (sunset) rising	Θ
Second station (evening)	Ψ
Last evening visibiity	Ω
Venus, Mercury	
First morning visibility	Γ
Morning station	Φ
Last morning visibility	Σ
First evening visibility	Ξ
Evening station	Ψ
Last evening visibility	Ω

B. Notes on Published Astronomical Papyri.

The purpose of this appendix is to provide an up-to-date checklist of Greek and Demotic astronomical papyri, ostraca, and tablets published elsewhere, incorporating papyri published since Neugebauer [1962a] as well as references to subsequent work on papyri listed there. Numbers **N1**, etc. refer to Neugebauer's checklist, which should be consulted for details of older literature. Only the most recent editions and such discussions as make a significant contribution to the interpretation of the texts are cited here. Palaeographical datings should be regarded with caution, especially for numerical tables.

P. Aberd. 12. **(N1)** Text concerning northern constellations. 2nd/3rd century (by hand), from the Fayum? Discussion (with textual remarks): Neugebauer [1962a] 384.

P. Aberd. 127. Text concerning constellations? 2nd/3rd century (by hand), from the Fayum? Discussion: Neugebauer & van Hoesen [1964] 58.

P. Aberd. 128. **(N2)** Fragment of numerical table with sexagesimal fractions, of undetermed nature. 1st century (by hand), from the Fayum?

Tab. Amst. inv. 1. Sign-entry almanac, 26–24 B.C. Edition: Neugebauer, Sijpesteijn, & Worp [1977]. Discussion: Jones [1993a].

P. Ant. III 141. Procedure text concerning computations of lunar longitudes. 3rd/4th century (by hand), from Antinoopolis. Misidentified by Neugebauer (in the commentary to the edition) as astrological. Discussion: Jones [1998b].

P. Berol. inv. 16511. See **4160** in this volume.

P. Berol. inv. 21226. Monthly almanac for Jupiter, A.D. 52–56, from the Fayum. Edition: Brashear & Neugebauer [1973]. Photograph: Ioannidou [1996] plate 58.

P. Berol. inv. 21236. Template for Mars. Late 1st century B.C. (by hand), from Abusir el Melek. Edition: Neugebauer & Brashear [1976]. Photograph: Ioannidou [1996] plate 61.

P. Berol. inv. 21240. Ephemeris. 3rd century (by hand), from Hermoupolis. Provisional edition with photograph: Ioannidou [1996] 144 and plate 62. Edition: Jones [1999c].

P. Berol. inv. 21359. Sign-entry almanac, 3rd century (by hand), from Oxyrhynchus. Edition: Jones [1999c].

P. dem. Berol. 8279. **(N3)** Sign-entry almanac, 17 B.C.–A.D. 12, from the Fayum. (On back of Greek accounts, A.D. 42). Edition: Neugebauer & Parker [1960–1969] v. 3, 228–232 and plates 66–73. Discussion: van der Waerden [1947]; van der Waerden [1960]; van der Waerden [1972]. A further unpublished fragment, which has not been successfully placed in relation to the main body of the text, has been found by Prof. K.–T. Zauzich (personal communication).

P. dem. Berol. 13146 + 13147. From Abusir el Melek. On front, canon of eclipse predictions (85–74 B.C.), apparently incorporating some observational data. On back, text concerning dates of solstices and equinoxes in Egyptian calendar, 73–70 B.C. Edition: (back) Parker & Zauzich [1981]; (front) Neugebauer, Parker, & Zauzich [1981].

[P. dem. Berol. 23692.] Inventoried inZauzich [1971] 176–177 as an early Ptolemaic astronomical text with columns containing zodiacal signs and numerals. Prof. Zauzich has kindly informed me by letter that the reading of zodiacal signs is incorrect, and the identification of the fragment as astronomical is doubtful. Unpublished.

Bodl. MS Gr. Class. F 7 (P). **(N4)** Sign entry almanac, A.D. 99–102. Edition: Neugebauer [1957]; Neugebauer [1972].

O. Bodl. II 2176. **(N5)** Text on 40-year cycle (?) of Alexandrian calendar dates of Sirius's rising in relation to new moons, with example dated A.D. 257. Discussion: Neugebauer [1975] 946, and commentary to *P. Oxy.* LXV 4473.

O. Bodl. II 2177. **(N6)** Part of a table or list of numbers (degrees and minutes?) followed by a date and longitude. 3rd century (by hand). Edition: Neugebauer [1974] (suggesting that the fragment lists longitudes of Jupiter on Choeac 15 of consecutive years).

P. Brook. **(N7)** A set of small fragments of a table with planetary names and red rulings. Apparently lost; a suggested identification with the grammatical (?) fragment *P. Brook.* 2 (see commentary in that edition) seems to be contradicted by details of Neugebauer's description, Neugebauer [1962a] 385.

P. dem. Carlsberg 1 (partially hieratic). Commentary on the astronomical picture and texts known from the cenotaph of Seti I (1303–1290 B.C.) and Ramses IV (1158–1152 B.C.). 2nd century (by hand and association with P. Carlsberg 9), from Tebtunis?. Edition: Lange & Neugebauer [1940], superseded in part by Neugebauer & Parker [1960–1969] v. 1, 37–80 and plates 36–42 and 44–54.

P. dem. Carlsberg 1a (partially hieratic). Fragments from a text similar to P.

Carlsberg 1, probably written by the same scribe. From Tebtunis? Edition: Neugebauer & Parker [1960–1969] v. 1, 37 and 88–94, and plates 43–54.

P. dem. Carlsberg 9. **(N8)** Schematic new moon dates. Later than A.D. 144 (by contents), from Tebtunis? Edition: Neugebauer & Parker [1960–1969] v. 3, 220–225 and plate 65. Three new fragments: Zauzich [1974]. Photograph of the augmented whole: *P. Carlsberg* v. 1, plate 10.

P. dem. Carlsberg Committee number 10850. Fragment of a text similar to P. Carlsberg 1 and 1a, from Tebtunis? Unpublished. Notice: *P. Carlsberg* v. 1, 8.

P. dem. Carlsberg 31. **(N9)** Schematic eclipse possibilities. 2nd century (by hand), from Tebtunis? Edition: Neugebauer & Parker [1960–1969] v. 3, 241–243 and plate 79A (not B). Discussion: Aaboe [1972].

P. dem. Carlsberg 32. **(N10)** Template for Mercury, Φ to Σ. 2nd century (by hand), from Tebtunis? Edition: Neugebauer & Parker [1960–1969] v. 3, 240–241 and plate 79B (not A).

P. Colker. Computed data, including length of synodic month (Babylonian System B column G), for consecutive syzygies. First half of 1st century (by hand). Photograph and transliteration: Neugebauer [1988]; edition: Jones [1997d].

P. Dub. T.C.D. Pap. inv. F. 7. Ephemeris, probably for April, A.D. 100. Edition: Hoogendijk [1982]. Discussion: Jones [1991] (incorrect interpretation); Jones [1995].

E.E.S. 79/1 (1) b. Numerical table in the format of a template. 2nd century (by hand), from Tebtunis. Edition: Jones [1998c].

E.E.S. 79/82 (1). See below, PSI inv. 75 D.

E.E.S. 79/82 (2). Almanac, longitudes of Mercury. 2nd century (by hand), from Tebtunis. Edition: Jones [1998c].

P. Fay. ined. s.n. (1) (Tait box). Sign-entry almanac, 2nd century, from Tebtunis. Unpublished.

P. Fay. ined. s.n. (2) (Tait box). Sign-entry almanac, A.D. 229–233, from Tebtunis. Unpublished.

P. Fay. ined. G[C]36 (1) (Tait box). Monthly almanac, 2nd century (by hand), from Tebtunis. Unpublished.

P. Fay. ined. G[C]36 (2) (Tait box). Template for moon, from Tebtunis. Unpublished.

P. Flor. 8 (demotic). Concordance between decans and zodiacal signs. Roman period (by hand). Edition: Neugebauer & Parker [1960–1969] v. 3, 252–254 and plate 80C.

P. Flor. 44 (demotic). Text concerning periods (?) involving Jupiter and moon. Perhaps astrological. Roman period (by hand). Edition: Neugebauer & Parker [1960–1969] v. 3, 250–252 and plate 80B.

P. Hamb. inv. 717. Sign-entry almanac, 2nd century. Unpublished.

P. Hamb. inv. 307. Table of unknown purpose, apparently astronomical. Unpublished.

P. Harr. I 60. **(N11)** Almanac-ephemeris for August, A.D. 140. New edition and discussion: Jones [1994a]. *P. Rylands* III 526 does not belong to the same document, as suggested in the original edition.

P. Heid. inv. 34. **(N12)** Five-day almanac for A.D. 348/349 and syzygies and daily lunar longitudes for A.D. 345/346, from a codex. Edition: Neugebauer [1956]. Discussion: Burckhardt [1958], Neugebauer [1975] 1056–1057.

P. Heid. inv. 4144 + *P. Mich.* III 151. **(N13)** Numerical table, interpreted by Jones [1990] as an auxiliary table for use with a modified version of the Babylonian System A scheme for Mars. 3rd century. Edition: Neugebauer [1960]. Discussion: Neugebauer [1975] 946–948; Jones [1990].

P. Hib. I 27. **(N14)** Calendar and parapegma, ca. 300 B.C, from el-Hibeh.

P. Iand. V 84. **(N15)** Text on spherical astronomy. 2nd century (by hand).

[P. Iand. inv. 654.] **(N16)** Concordance of Julian and Alexandrian month names. 6th/7th century (by hand). Not explicitly astronomical. Edition: Gundel [1956].

O. dem. Karnak ODK-NMB no. 2. Fragmentary text of astronomical or astrological contents, referring to zodiacal signs. Ptolemaic? (by hand), from Karnak. Edition: Devauchelle [1987].

P. Laur. IV 144. Sign-entry almanac, A.D. 68–71 (front) and 92–95 (back). Prior edition with photograph: Pintaudi & Neugebauer [1978]. Corrected edition of back, with dating: Jones [1998b].

P. Laur. IV 150 Small fragment of a numerical table ('front', along fibres) and text (both sides), identified in the edition as 'arithmetic-scholastic', but probably astronomical; cf. cτηριγ[on back, line 5. 5th or 6th century (by hand).

P. Lond. III 1278. **(N17)** Fragments from two (?) codices with tables related to the *Handy Tables* and *Almagest*. 2nd/3rd century (by hand). Edition: Neugebauer & Skeat [1958].

P. Lund V 35a. **(N18)** Epoch table for moon, A.D. 59–108, compatible with rules in *P. Rylands* I 27. Edition: Knudtzon & Neugebauer [1947]. Discussion: Neugebauer [1949].

P. Lund V 35b. **(N19)** Monthly almanac, A.D. 119/120. Edition: Knudtzon & Neugebauer [1947]. Possibly from the same manuscript as *P. Tebt.* II 274, which seems to be in the same hand.

P. Mich. III 149. **(N20)** An extensive astrological text, incorporating some astronomical material, e.g. dimensions of planetary epicycles and maximal latitudes, and oblique ascensions. 2nd century (by hand). Discussion: Aaboe [1963]; Neugebauer [1972]; Neugebauer [1975] 805–808.

P. Mich. III 150. **(N21)** Table of syzygies, similar in arrangement to the comparable table in *P. Heid.* inv. 34. 3rd/4th century (by hand). Discussion: Neugebauer [1942b] 251–252; Delporte [1947].

P. Mich. III 151. **(N22)** See above, P. Heid. inv. 4144 + *P. Mich.* III 151.

P. Mich. inv. 29. On the back of an astrological text, parts of two columns of text in a different hand (2nd/3rd century). The contents of col. i appear to be astrological (at any rate not astronomical). Col. ii, however, might belong to a summary of the synodic cycle of an inner planet, broken up into specified numbers of days of visibility and invisibility. Edition and commentary: Ambühl, Markovska, & Milnor [1995].

P. Mich. inv. 1454. **(N23)** Ephemeris for September and October, A.D. 467. Edition: Curtis & Robbins [1935]. Discussion: Fotheringham [1935]; Sloley [1936]; Burckhardt [1958].

[*O. Narm.* I 63–65.] Ostraca from Medinet Madi (Narmuthis) consisting entirely, or almost entirely, of numerals. The editors suggest that these are notes for astronomical calculations; however, the numbers have no obvious astronomical meaning. Edition: Pintaudi & Sijpesteijn [1993] 81–83.

P. Nelson. Monthly almanac, A.D. 101–103. Edition: Nelson [1970]. Discussion: Neugebauer [1971], Jones [1998b].

P. Oslo III 73. **(N24)** Text concerning instrumental measurement of the sun's diameter. 1st/2nd century (by hand).

P. Oslo inv. 1336 (demotic). Numerical table of unknown significance. Roman period (by hand). Edition: Neugebauer & Parker [1960–1969] v. 3, 254–255 and plate 79C.

P. Oxy. I 35. **(N25)** Regnal canon, with schematic lengths of reigns but no totals, Augustus to Decius (A.D. 250). From Oxyrhynchus.

P. Oxy. II 303. **(N26)** Text on lunar anomalistic motion. First century (by hand), from Oxyrhynchus. Edition: Neugebauer [1962a] 386–387.

P. Oxy. III 470. **(N27)** Text on an astronomical or calendrical board (not a game), and on the construction of a water clock. First century (by hand), from Oxyrhynchus. Discussion: West [1989].

[*P. Oxy.* III 539.] **(N28)** Unpublished, now at Columbia University (where it was for a while presumed missing). Neugebauer suspected from the brief published description that this small fragment belonged to *P. Rylands* I 27. It does not, and the contents are not, so far as one can tell, astronomical.

P. Oxy. XXXI 2551 (formerly P. Oxy. inv. 7 B/1959). **(N29)** Regnal canon, with schematic lengths of reigns but no totals, Cyrus to Philip (A.D. 249), and negligible remains of an astronomical (procedure?) text in a different hand, from a codex. From Oxyrhynchus. Edition: Sattler [1962] 39–50. Discussion: Parsons [1963].

P. Oxy. XLVI 3299. Sign-entry almanac, A.D. 217–225, from Oxyrhynchus.

P. Paris 1. **(N30)** Treatise (the so-called 'Ars Eudoxi', although certainly not by Eudoxus) on calendrical and elementary astronomical topics. Early 2nd century B.C. (by astronomical contents and dated documents written around the astronomical text). Discussion and bibliography: Neugebauer [1975] 686–689.

P. Paris inv. 7733. Text on optical phenomena, in part concerning the appearance of the celestial bodies near the horizon. 3rd/2nd century B.C., from Memphis. Edition: Wessely [1891].

P. Petr. III 134. **(N31)** Text mentioning decans and feasts in the Egyptian calendar. 3rd century B.C. (by hand), from Gurob.

P. Princ. inv. GD 7661. Astronomical or astrological treatise. 1st century (by hand). Unpublished.

P. Rylands I 27. **(N32)** Procedure text concerning the construction of epoch tables for the moon. Also rules for extrapolating solstices and equinoxes from Ptolemy's observations in A.D. 139/140 from *Almagest* III 4, and a regnal canon (schematic lengths of reigns and totals from Augustus's reign) from Commodus to Gallus (A.D. 250). Discussion: Neugebauer [1949]; van der Waerden [1958]; Jones [1983] 14–16; Jones [h] (with corrections to edition).

P. Ryl. III 464. **(N33)** Small fragment of a procedure text (originally identified as 'apocryphal gospel'; cf. *P. Rylands* v. 3, xvii). 3rd century (by hand).

P. Ryl. III 522/523. **(N34)** *Handy Tables*, oblique ascensions and table of noteworthy cities (codex). 3rd century (by hand), from the Fayum?.

P. Ryl. III 526. **(N35)** Fragments from a numerical table of unknown significance. 3rd century (by hand), from the Fayum? Cf. *P. Harr.* I 60.

P. Ryl. IV 589 (formerly P. Rylands inv. 666). **(N36)** Text concerning dates of new moons. Early 2nd century B.C., from Philadelphia? Prior edition and discussion: Turner & Neugebauer [1949].

PSI XIII 1296 (Parchment codex) Arithmetical scheme for length of daylight according to Alexandrian calendar months, written (7th/8th century) over erased Coptic text.

PSI XV 1490 (formerly PSI inv. 515). Procedure text concerning tables for the sun. 1st/2nd century (by hand). Awaiting formal publication in *PSI* vol. XV. Preliminary edition: Manfredi [1966].

PSI XV 1491. **(N37)** Text describing the laying out of a table in at least seven columns. 2nd century (by hand). Awaiting publication. Discussion: Neugebauer [1962a] 388; Neugebauer [1975] 946; Jones [1984] 315–316.

PSI XV 1492. **(N38)** Template for Saturn. 2nd century (by hand), from Oxyrhynchus. Awaiting publication. Partial transliteration: Neugebauer [1975] 790–791. Discussion: Jones [1984].

PSI XV 1493. **(N39)** Template for the moon. 2nd century (by hand). Awaiting publication. Partial transliteration: Neugebauer [1975] 822–823. Discussion: Jones [1983] 17–23.

PSI inv. 75 D + EES 79/82 (1). Monthly almanac, A.D. 48–56, from Tebtunis. Edition: Manfredi & Neugebauer [1973], Jones [1998c].

Stobart Tablets, Free Public Museum of Liverpool (demotic). **(N40)** Sign-entry almanac, A.D. 70–133, from Thebes. Edition: Neugebauer & Parker [1960–1969] v. 3, 225–228, 232–240, and plates 74–78. Discussion: van der Waerden [1947]; van der Waerden [1960]; van der Waerden [1972].

P. Stras. inv. gr. 1097. **(N41)** Sign-entry almanac. 2nd century (by hand). Unpublished. Discussion: Neugebauer [1962a] 388.

P. Tebt. II 274. **(N42)** Monthly almanac, A.D. 107–115, from Tebtunis. Edition: Neugebauer [1942a] 241–242. Discussion: Neugebauer & Parker [1960–1969] v. 3, 233–234. Possibly from the same manuscript as *P. Lund* V 35b.

P. Tebt. II 449. **(N43)** Daily longitudes of the moon during Mecheir of an unknown year. 2nd century (by hand), from Tebtunis. Edition: Carp [1975].

P. Vind. D 4876 (demotic). Dates and times of syzygies. Roman period (by hand). Edition: Neugebauer & Parker [1960–1969] v. 3, 243–250 and plate 80A.

P. Vind., Institut für Österr. Geschichtsforschung. **(N44)** Text on weather prognostications, with some astronomical material including a shadow table. 3rd century B.C. (by hand), from the Fayum. Edition: Neugebauer [1962b].

P. Vind. inv. gr. 3231. Column headings from an ephemeris (not a horoscope). 3rd century (by hand). Edition: Neugebauer & Sijpesteijn [1980b].

P. Vind. inv. gr. 36041+26011. Perpetual sign-entry almanac. Late antiquity (by hand). Edition: Neugebauer & Sijpesteijn [1980a]. Discussion: Jones [1994c].

P. Vind. inv. gr. 29370. **(N45a)** Ephemeris, A.D. 489. Edition: Gerstinger & Neugebauer [1962a]. Discussion: Neugebauer [1975] 1057–1058; Jones [1994c].

P. Vind. inv. gr. 29370b. **(N45b)** Ephemeris, A.D. 471. Edition: Gerstinger & Neugebauer [1962a]. Discussion: Jones [1994c].

C. Published Horoscopes from Oxyrhynchus.

Horoscopes are listed in chronological order of nativity date; for multiple horoscopes, the latest date is followed. **GH–3**, etc. refer to the text number in Neugebauer & van Hoesen [1959]; **B1**, etc. refer to the number in Baccani [1992]. Except for *P. Oxy.* XII 1564 and XVI 2060, these editions supersede the publications in *Oxyrhynchus Papyri.*

P. Oxy. IV 804. **(GH–3)** Caesar (Augustus) 27 Phaophi 5 (Alexandrian) = 4 B.C. October 2. A brief note of advice to the native follows the horoscope.

P. Oxy. II 235. **(GH15/22)** Elaborate horoscope of Tryphon, Tiberius [year lost] Phaophi 1 (Alexandrian) = September 28 or 29. The calendrical data imply a year between 15 and 22; unfortunately the astronomical data are contradictory.

P. Oxy. II 307. **(GH46)** Elaborate horoscope. No date given; approximate date 46 January 3.

P. Oxy. XXXI 2555. **(B1)** Elaborate horoscope, Claudius 6 Pachon 18 (Alexandrian) = 46 May 13.

P. Oxy. XLIV 3196. **(B2)** Nero 4 Tybi 16/17 (Alexandrian) = 58 January 11/12.

P. Oxy. III 596. **(GH139 + 149)** (a) Antoninus 13 Mecheir 8/9. (b) Antoninus 2 Phamenoth 30/1 (*sic*). The positions of the heavenly bodies are hard to reconcile with any interpretation of the dates. Marginally acceptable agreement is obtained for Alexandrian calendar dates if the 'Antoninus' of (a) is Marcus Aurelius (hence 173 February 2) whereas that of (b) is Antoninus Pius (hence 139 March 25). But obviously such a spread of dates on one document is dubious.

P. Oxy. XXXI 2556. **(B9)** Antoninus & Verus 1 Mesore 22/23 (Alexandrian) = 161 August 15/16.

P. Oxy. XLVII 3353. **(B10)** Commodus 19 Mecheir 18 (Alexandrian) = 179 February 12.

P. Oxy. III 585. **(GH220)** [Year lost] Mecheir [day lost] (Alexandrian); approximate date 220 February 18.

P. Oxy. XLVI 3298. **(B14)** (a) Horoscope of Apolinarios (*sic*), Antoninus 'the catamite' (i.e. Elagabalus) 2 Pachon 22/23 (Alexandrian) = 219 May 17/18. (b) Horoscope of his son Severos, Gordian 6 Phaophi 7/8 (Alexandrian) = 242 October 4/5. (c)

Horoscope of De[...], Philips 3 [month lost] 20/21 = 245/246, birthdate only. (d) Horoscope of Amm[...], Decius 1 [month lost] 1 = 249/250. (e) Fragmentary horoscope. (f) Horoscope of Sabeinos, son of Iseis, Gordian 6 Tybi 16, birthdate only. (f) Text concerning dream interpretation. I would guess that this document is an astrologer's register rather than a family record.

P. Oxy. XXXVI 2790. **(B15)** (a) No date given; approximate date 257 September 5. (b) No date given; approximate date 255 July 22.

P. Oxy. XII 1563. **(GH258)** Valerian and Gallienus and Cornelius Saloninus 6 Thoth 29 (Alexandrian) = 258 September 26.

P. Oxy. XII 1476. **(GH260)** Horoscope of Sarapammon, Macrinianus and Quietus 1 Phaophi 2 (Alexandrian) = 260 September 29.

P. Oxy. XXXI 2557. **(B16)** Name lost, followed by a symbol that may signify ἑκατοντάρχης, 'centurion'. Probus 4 Thoth 24 (Alexandrian) = 278 September 21.

P. Oxy. XII 1564. **(GH283)** Horoscope of Pichime, Carinus 1 Phamenoth 27 (Alexandrian) = 283 March 23. The papyrus, reported missing by Neugebauer and van Hoesen, p. 62, is at the St. Louis (Missouri) Art Museum; I have been able to confirm the accuracy of the original edition against a photograph at the Ashmolean Museum.

P. Oxy. XII 1565. **(GH293,VIII)** (Diocletian) 9 = (Maximianus) 8 = (Constantius and Galerius) 1 Epagomenae 5 (Alexandrian) = 293 August 28.

P. Oxy. LXV 4477. Name lost, Era Diocletian 146 Epeiph 13 14 (Alexandrian) = 430 July 7. The horoscope is in tabular format.

P. Oxy. XVI 2060. **(GH478)** Horoscope of Anup (not 'Anubis', as Neugebauer and van Hoesen), Era Diocletian 194 Epeiph 5 (Alexandrian) = 478 June 29. The papyrus, reported lost by Neugebauer and van Hoesen, is at the Ashmolean Museum.

D. Regnal Canon.

The schematic totals of regnal years in this table are taken from the regnal canons incorporated in the medieval copies of the *Handy Tables* (e.g *Vat. gr.* 1594 ff. 16^{v}–17^{r}). The numbers under the headings, 'Era Augustus' and 'Era Philip' are for the year *preceding* the first regnal year, so that one simply adds a given regnal year number to the tabulated number corresponding to the emperor in question to obtain the year according to the desired era.

In principle an emperor's first regnal year is the year during which his accession took place, and emperors whose reigns fell entirely within a year are omitted. Almost all the lengths of reigns in the *Handy Tables* canon can be confirmed from other regnal canons, astronomical tables, and horoscopes as the numbers conventionally used for chronological purposes (it does not seem to have made any difference whether one was using the Alexandrian calendar, as in most horoscopes, or the Egyptian calendar, as in most epoch tables). Among the more extended parallel documents are Clemens Alexandrinus, *Strom.* 1.21.144 (Augustus to Commodus), Vettius Valens I 19 (Augustus to Philip, Egyptian calendar), *P. Oxy.* XXXI 2551 (Augustus to Philip), *P. Oxy.* I 35 (Augustus to Decius), and *P. Rylands* I 27 (Antoninus Pius to Gallus). The *Handy Tables* manuscripts give only one year to Decius, while assigning an extra year to either Philip or Gallus; I have substituted the historically correct reigns here, although there is some uncertain evidence that the error was current in antiquity (Sattler [1962] 44–45 note 3).

	Era Augustus	*Era Philip*	*Years of Reign*
Augustus	0	294	43
Tiberius	43	337	22
Gaius	65	359	4
Claudius	69	363	14
Nero	83	377	14
Vespasian	97	391	10
Titus	107	401	3
Domitian	110	404	15
Nerva	125	419	1
Trajan	126	420	19
Hadrian	145	439	21
Antoninus Pius	166	460	23
M. Aurelius, L. Verus, Commodus	189	483	32
Septimius Severus, Caracalla	221	515	25
Elagabalus	246	540	4
Severus Alexander	250	544	13
Maximinus	263	557	3
Gordian	266	560	6
Philip	272	566	6
Decius	278	572	2
Trebonianus Gallus	279	573	2
Gallienus	282	576	15
Claudius Gothicus	297	591	1
Aurelian	298	592	6
Probus	304	598	7
Carus, Carinus, Numerianus	311	605	2
Diocletian	313	607	

E. Tables for Calendrical Conversions.

The following tables are intended to allow easy conversions of dates from the Egyptian to the Alexandrian calendar and *vice versa*, and from the Alexandrian to the Julian calendar and *vice versa*. They assume perfect regularity in the intercalations of the Alexandrian and Julian calendars. For dates from A.D. 8 (era Augustus 38) these assumptions are certainly valid for the Roman and Alexandrian civil calendars actually in use in Egypt. Between 45 B.C. and A.D. 8 the civil Roman calendar was subject to irregular intercalations, leading to discrepancies of as much as three days between civil dates and the modern historian's Julian calendar. Whether the civil Alexandrian calendar was inaugurated in 26/25 B.C. (era Augustus 5) and always intercalated at four-year intervals, or inaugurated in 30/29 B.C. (era Augustus 1) and intercalated in coordination with the Roman civil calendar (as proposed by Snyder [1943]) is a controverted question. Tables for converting between civil Roman, Alexandrian, and Egyptian dates during the reign of Augustus are available in Skeat [1993] (following Snyder's hypothesis) and Hagedorn [1994] (assuming regular four-year intercalations in the Alexandrian calendar).

Table 1: Numerical representation of Egyptian and Alexandrian calendar dates.
For conversion between the Egyptian and Alexandrian calendars it is convenient to count days serially from the beginning of the year instead of by month and day. This number is obtained by adding the number beside the relevant month in the following table to the day within the month; e.g. Pachon 21 becomes 240 + 21 or 261. To convert back, find the largest possible number in the table that can be subtracted from the serial day number; the difference is the day number within the month corresponding to the number that was subtracted.

Thoth	0	Tybi	120	Pachon	240
Phaophi	30	Mecheir	150	Payni	270
Hathyr	60	Phamenoth	180	Epeiph	300
Choeac	90	Pharmuthi	210	Mesore	330
				Epagomenae	360

Table 2: Conversion between Egyptian and Alexandrian calendar dates.
To convert from a given Alexandrian date to the Egyptian calendar: (1) Determine the year according to the era Augustus (Appendix D). (2) Determine the serial day number within the year (Table 1 above). (3) Find the correction in days corresponding to the year in the following table, and *add* it to the serial day number. (4) If the sum is greater than 365, subtract 365 from it, and add 1 to the year number. (5) Convert the sum to month and days using Table 1.

Example: Hadrian 3 ≡ Augustus 148, Mesore 25 (Alexandrian). From Table 1, Mesore 25 is day 355. From Table 2, the correction for Augustus 148 is 35 days. 355 + 35 = 390. Since this is greater than 365, the year becomes Augustus 149 (Hadrian 4), and 390 – 365 = 25, corresponding to Thoth 25 in Table 1. Hence the date in the Egyptian calendar is Hadrian 4, Thoth 25.

To convert from a given Egyptian date to the Alexandrian calendar: (1) Determine the year according to the era Augustus (Appendix D). (2) Determine the serial day number within the year (Table 1 above). (3) Find the correction in days corresponding to the year in the following table. If it is less than the serial day number, *subtract* it from the serial day number. (4) But if the correction is greater than or equal to the serial day number, add 365 to the serial day number before subtracting the correction, and subtract 1 from the year number. If the new year number has a smaller correction in Table 2 below, add 1 to the corrected serial day number. (5) Convert the sum to month and days using Table 1.

Example: Hadrian 4 ≡ Augustus 149, Thoth 25 (Egyptian). From Table 1, Thoth 25 is day 25. From Table 2, the correction for Augustus 149 is 36 days. Since this is greater than 25, we add 365 to 25 and then subtract 36, obtaining 354, and subtract 1 from the year, obtaining Augustus 148 (Hadrian 3). In this instance, Table 2 gives a smaller correction, 35 days, for Augustus 148 than for Augustus 149, so we add 1 day to the 354, obtaining 355. Finally, using Table 1 we find that the equivalent date in the Alexandrian calendar is Hadrian 3, Mesore 25.

		101–104	24	201–204	49	301–304	74	401–404	99
5–8	0	105–108	25	205–208	50	305–308	75	405–408	100
9–12	1	109–112	26	209–212	51	309–312	76	409–412	101
13–16	2	113–116	27	213–216	52	313–316	77	413–416	102
17–20	3	117–120	28	217–220	53	317–320	78	417–420	103
21–24	4	121–124	29	221–224	54	321–324	79	421–424	104
25–28	5	125–128	30	225–228	55	325–328	80	425–428	105
29–32	6	129–132	31	229–232	56	329–332	81	429–432	106
33–36	7	133–136	32	233–236	57	333–336	82	433–436	107
37–40	8	137–140	33	237–240	58	337–340	83	437–440	108
41–44	9	141–144	34	241–244	59	341–344	84	441–444	109
45–48	10	145–148	35	245–248	60	345–348	85	445–448	110
49–52	11	149–152	36	249–252	61	349–352	86	449–452	111
53–56	12	153–156	37	253–256	62	353–356	87	453–456	112
57–60	13	157–160	38	257–260	63	357–360	88	457–460	113
61–64	14	161–164	39	261–264	64	361–364	89	461–464	114
65–68	15	165–168	40	265–268	65	365–368	90	465–468	115
69–72	16	169–172	41	269–272	66	369–372	91	469–472	116
73–76	17	173–176	42	273–276	67	373–376	92	473–476	117
77–80	18	177–180	43	277–280	68	377–380	93	477–480	118
81–84	19	181–184	44	281–284	69	381–384	94	481–484	119
85–88	20	185–188	45	285–288	70	385–388	95	485–488	120
89–92	21	189–192	46	289–292	71	389–392	96	489–492	121
93–96	22	193–196	47	293–296	72	393–396	97	493–496	122
97–100	23	197–200	48	297–300	73	397–400	98	497–500	123

Table 3: Conversion between Alexandrian and Julian calendar dates.
The equivalences in the following table are valid for all dates except those that fall during the interval between an intercalation in the Alexandrian calendar and the subsequent intercalation in the Julian calendar. Hence correction is never needed for dates between Phamenoth 5 and the end of the year in the Alexandrian calendar, and between February 29 and August 28 in the Julian.

We will refer to a Julian year *preceding* a Julian leap-year as a 'pre-leap-year'. For a date between August 30 and December 31 of a pre-leap-year or between January 1 and February 28 of a leap-year, the Alexandrian calendar equivalent is one day *earlier* than the date in the table. (Julian leap-years include 29, 25, 21, 17, 13, 9, and 5 B.C. as well as all years of our era divisible by 4 with no remainder.) August 29 is equivalent to Epagomenae 6 of the old Alexandrian year only in pre-leap-years, otherwise to Thoth 1 of the new Alexandrian year.

For a date between Thoth 1 and Phamenoth 3 of a year *following* an Alexandrian intercalary year, the Julian calendar equivalent is one day *later* than the date in the table. (An Alexandrian year is intercalary if its number according to the era Augustus is divisible by 4 with no remainder.) Phamenoth 4 is equivalent to February 29 only in years in which this correction is made.

A year x in the era Augustus is the year A.D. $(x\text{-}30)/(x\text{-}29)$ or $(31\text{-}x)/(30\text{-}x)$ B.C.; for example Augustus 5 ≡ 26/25 B.C., and Augustus 35 ≡ A.D. 5/6.

Examples: Hadrian 3 ≡ Augustus 148, Mesore 25 (Alexandrian). The Julian equivalent will fall in the years A.D. 118/119. From Table 3 we find August 18, which requires no correction because the given Alexandrian date is later than Phamenoth 5. Since it is also in the second of the two Julian years straddled by the Alexandrian year, the equivalent Julian calendar date is A.D. 119 August 18.

Hadrian 4 ≡ Augustus 149, Phaophi 10 (Alexandrian). The Julian equivalent is in A.D. 119/120. The preceding year, Augustus 148, is an intercalary year and Phaophi 10 falls before Phamenoth 5, so we take the date one day later than that given in the table, i.e. October 8. Since this is within the first of the pair of Julian years, the equivalent Julian calendar date is A.D. 119 October 8.

A.D. 119 October 8. The equivalent Alexandrian year will be either Augustus 148 (A.D. 118/119) or 149 (A.D. 119/120). The following Julian year, A.D. 120, is a leap-year and October 8 is later than September 30, so we take the Alexandrian date one day earlier than the tabulated one, i.e. Phaophi 10. Since October 8 falls within the Julian year that *ends* in the course of the Alexandrian year, the year in question must be A.D. 119/120 rather than A.D. 118/119, and hence the equivalent date in the Alexandrian calendar is Augustus 149 (Hadrian 4), Phaophi 10.

Thoth 1	Aug 29	Phaophi 19	Oct 16	Choeac 7	Dec 3	Tybi 25	Jan 20
Thoth 2	Aug 30	Phaophi 20	Oct 17	Choeac 8	Dec 4	Tybi 26	Jan 21
Thoth 3	Aug 31	Phaophi 21	Oct 18	Choeac 9	Dec 5	Tybi 27	Jan 22
Thoth 4	Sept 1	Phaophi 22	Oct 19	Choeac 10	Dec 6	Tybi 28	Jan 23
Thoth 5	Sept 2	Phaophi 23	Oct 20	Choeac 11	Dec 7	Tybi 29	Jan 24
Thoth 6	Sept 3	Phaophi 24	Oct 21	Choeac 12	Dec 8	Tybi 30	Jan 25
Thoth 7	Sept 4	Phaophi 25	Oct 22	Choeac 13	Dec 9	Mecheir 1	Jan 26
Thoth 8	Sept 5	Phaophi 26	Oct 23	Choeac 14	Dec 10	Mecheir 2	Jan 27
Thoth 9	Sept 6	Phaophi 27	Oct 24	Choeac 15	Dec 11	Mecheir 3	Jan 28
Thoth 10	Sept 7	Phaophi 28	Oct 25	Choeac 16	Dec 12	Mecheir 4	Jan 29
Thoth 11	Sept 8	Phaophi 29	Oct 26	Choeac 17	Dec 13	Mecheir 5	Jan 30
Thoth 12	Sept 9	Phaophi 30	Oct 27	Choeac 18	Dec 14	Mecheir 6	Jan 31
Thoth 13	Sept 10	Hathyr 1	Oct 28	Choeac 19	Dec 15	Mecheir 7	Feb 1
Thoth 14	Sept 11	Hathyr 2	Oct 29	Choeac 20	Dec 16	Mecheir 8	Feb 2
Thoth 15	Sept 12	Hathyr 3	Oct 30	Choeac 21	Dec 17	Mecheir 9	Feb 3
Thoth 16	Sept 13	Hathyr 4	Oct 31	Choeac 22	Dec 18	Mecheir 10	Feb 4
Thoth 17	Sept 14	Hathyr 5	Nov 1	Choeac 23	Dec 19	Mecheir 11	Feb 5
Thoth 18	Sept 15	Hathyr 6	Nov 2	Choeac 24	Dec 20	Mecheir 12	Feb 6
Thoth 19	Sept 16	Hathyr 7	Nov 3	Choeac 25	Dec 21	Mecheir 13	Feb 7
Thoth 20	Sept 17	Hathyr 8	Nov 4	Choeac 26	Dec 22	Mecheir 14	Feb 8
Thoth 21	Sept 18	Hathyr 9	Nov 5	Choeac 27	Dec 23	Mecheir 15	Feb 9
Thoth 22	Sept 19	Hathyr 10	Nov 6	Choeac 28	Dec 24	Mecheir 16	Feb 10
Thoth 23	Sept 20	Hathyr 11	Nov 7	Choeac 29	Dec 25	Mecheir 17	Feb 11
Thoth 24	Sept 21	Hathyr 12	Nov 8	Choeac 30	Dec 26	Mecheir 18	Feb 12
Thoth 25	Sept 22	Hathyr 13	Nov 9	Tybi 1	Dec 27	Mecheir 19	Feb 13
Thoth 26	Sept 23	Hathyr 14	Nov 10	Tybi 2	Dec 28	Mecheir 20	Feb 14
Thoth 27	Sept 24	Hathyr 15	Nov 11	Tybi 3	Dec 29	Mecheir 21	Feb 15
Thoth 28	Sept 25	Hathyr 16	Nov 12	Tybi 4	Dec 30	Mecheir 22	Feb 16
Thoth 29	Sept 26	Hathyr 17	Nov 13	Tybi 5	Dec 31	Mecheir 23	Feb 17
Thoth 30	Sept 27	Hathyr 18	Nov 14	Tybi 6	Jan 1	Mecheir 24	Feb 18
Phaophi 1	Sept 28	Hathyr 19	Nov 15	Tybi 7	Jan 2	Mecheir 25	Feb 19
Phaophi 2	Sept 29	Hathyr 20	Nov 16	Tybi 8	Jan 3	Mecheir 26	Feb 20
Phaophi 3	Sept 30	Hathyr 21	Nov 17	Tybi 9	Jan 4	Mecheir 27	Feb 21
Phaophi 4	Oct 1	Hathyr 22	Nov 18	Tybi 10	Jan 5	Mecheir 28	Feb 22
Phaophi 5	Oct 2	Hathyr 23	Nov 19	Tybi 11	Jan 6	Mecheir 29	Feb 23
Phaophi 6	Oct 3	Hathyr 24	Nov 20	Tybi 12	Jan 7	Mecheir 30	Feb 24
Phaophi 7	Oct 4	Hathyr 25	Nov 21	Tybi 13	Jan 8	Pham 1	Feb 25
Phaophi 8	Oct 5	Hathyr 26	Nov 22	Tybi 14	Jan 9	Pham 2	Feb 26
Phaophi 9	Oct 6	Hathyr 27	Nov 23	Tybi 15	Jan 10	Pham 3	Feb 27
Phaophi 10	Oct 7	Hathyr 28	Nov 24	Tybi 16	Jan 11	Pham 4	Feb 28
Phaophi 11	Oct 8	Hathyr 29	Nov 25	Tybi 17	Jan 12	Pham 4	Feb 29
Phaophi 12	Oct 9	Hathyr 30	Nov 26	Tybi 18	Jan 13	Pham 5	Mar 1
Phaophi 13	Oct 10	Choeac 1	Nov 27	Tybi 19	Jan 14	Pham 6	Mar 2
Phaophi 14	Oct 11	Choeac 2	Nov 28	Tybi 20	Jan 15	Pham 7	Mar 3
Phaophi 15	Oct 12	Choeac 3	Nov 29	Tybi 21	Jan 16	Pham 8	Mar 4
Phaophi 16	Oct 13	Choeac 4	Nov 30	Tybi 22	Jan 17	Pham 9	Mar 5
Phaophi 17	Oct 14	Choeac 5	Dec 1	Tybi 23	Jan 18	Pham 10	Mar 6
Phaophi 18	Oct 15	Choeac 6	Dec 2	Tybi 24	Jan 19	Pham 11	Mar 7

Pham 12	Mar 8	Pharm 30	April 25	Payni 18	June 12	Mesore 6	July 30
Pham 13	Mar 9	Pachon 1	April 26	Payni 19	June 13	Mesore 7	July 31
Pham 14	Mar 10	Pachon 2	April 27	Payni 20	June 14	Mesore 8	Aug 1
Pham 15	Mar 11	Pachon 3	April 28	Payni 21	June 15	Mesore 9	Aug 2
Pham 16	Mar 12	Pachon 4	April 29	Payni 22	June 16	Mesore 10	Aug 3
Pham 17	Mar 13	Pachon 5	April 30	Payni 23	June 17	Mesore 11	Aug 4
Pham 18	Mar 14	Pachon 6	May 1	Payni 24	June 18	Mesore 12	Aug 5
Pham 19	Mar 15	Pachon 7	May 2	Payni 25	June 19	Mesore 13	Aug 6
Pham 20	Mar 16	Pachon 8	May 3	Payni 26	June 20	Mesore 14	Aug 7
Pham 21	Mar 17	Pachon 9	May 4	Payni 27	June 21	Mesore 15	Aug 8
Pham 22	Mar 18	Pachon 10	May 5	Payni 28	June 22	Mesore 16	Aug 9
Pham 23	Mar 19	Pachon 11	May 6	Payni 29	June 23	Mesore 17	Aug 10
Pham 24	Mar 20	Pachon 12	May 7	Payni 30	June 24	Mesore 18	Aug 11
Pham 25	Mar 21	Pachon 13	May 8	Epeiph 1	June 25	Mesore 19	Aug 12
Pham 26	Mar 22	Pachon 14	May 9	Epeiph 2	June 26	Mesore 20	Aug 13
Pham 27	Mar 23	Pachon 15	May 10	Epeiph 3	June 27	Mesore 21	Aug 14
Pham 28	Mar 24	Pachon 16	May 11	Epeiph 4	June 28	Mesore 22	Aug 15
Pham 29	Mar 25	Pachon 17	May 12	Epeiph 5	June 29	Mesore 23	Aug 16
Pham 30	Mar 26	Pachon 18	May 13	Epeiph 6	June 30	Mesore 24	Aug 17
Pharm 1	Mar 27	Pachon 19	May 14	Epeiph 7	July 1	Mesore 25	Aug 18
Pharm 2	Mar 28	Pachon 20	May 15	Epeiph 8	July 2	Mesore 26	Aug 19
Pharm 3	Mar 29	Pachon 21	May 16	Epeiph 9	July 3	Mesore 27	Aug 20
Pharm 4	Mar 30	Pachon 22	May 17	Epeiph 10	July 4	Mesore 28	Aug 21
Pharm 5	Mar 31	Pachon 23	May 18	Epeiph 11	July 5	Mesore 29	Aug 22
Pharm 6	April 1	Pachon 24	May 19	Epeiph 12	July 6	Mesore 30	Aug 23
Pharm 7	April 2	Pachon 25	May 20	Epeiph 13	July 7	Epag 1	Aug 24
Pharm 8	April 3	Pachon 26	May 21	Epeiph 14	July 8	Epag 2	Aug 25
Pharm 9	April 4	Pachon 27	May 22	Epeiph 15	July 9	Epag 3	Aug 26
Pharm 10	April 5	Pachon 28	May 23	Epeiph 16	July 10	Epag 4	Aug 27
Pharm 11	April 6	Pachon 29	May 24	Epeiph 17	July 11	Epag 5	Aug 28
Pharm 12	April 7	Pachon 30	May 25	Epeiph 18	July 12	Epag 6	Aug 29
Pharm 13	April 8	Payni 1	May 26	Epeiph 19	July 13		
Pharm 14	April 9	Payni 2	May 27	Epeiph 20	July 14		
Pharm 15	April 10	Payni 3	May 28	Epeiph 21	July 15		
Pharm 16	April 11	Payni 4	May 29	Epeiph 22	July 16		
Pharm 17	April 12	Payni 5	May 30	Epeiph 23	July 17		
Pharm 18	April 13	Payni 6	May 31	Epeiph 24	July 18		
Pharm 19	April 14	Payni 7	June 1	Epeiph 25	July 19		
Pharm 20	April 15	Payni 8	June 2	Epeiph 26	July 20		
Pharm 21	April 16	Payni 9	June 3	Epeiph 27	July 21		
Pharm 22	April 17	Payni 10	June 4	Epeiph 28	July 22		
Pharm 23	April 18	Payni 11	June 5	Epeiph 29	July 23		
Pharm 24	April 19	Payni 12	June 6	Epeiph 30	July 24		
Pharm 25	April 20	Payni 13	June 7	Mesore 1	July 25		
Pharm 26	April 21	Payni 14	June 8	Mesore 2	July 26		
Pharm 27	April 22	Payni 15	June 9	Mesore 3	July 27		
Pharm 28	April 23	Payni 16	June 10	Mesore 4	July 28		
Pharm 29	April 24	Payni 17	June 11	Mesore 5	July 29		

F. Conversion of Times.

The figure below displays the approximate relationship between seasonal and equinoctial times as a function of solar longitude, according to the table of oblique ascensions for the 3rd Clima through Lower Egypt in the *Handy Tables* (and neglecting the equation of time). Given a date and time in seasonal hours, one can convert to equinoctial hours as follows: (1) Setting a rule vertically aligned with the solar longitude or the Alexandrian date in the figure, locate the given seasonal hour with respect to the curved lines and the scale on the right. (2) The location of this point with respect to the straight horizontal lines and the scale on the left indicates the number of equinoctial hours before or after noon. The reverse procedure leads from equinoctial to seasonal times. (Note that the hour curves represent the *beginnings* of the respective seasonal hours, and that times from the end of the sixth hour of night to the end of the sixth hour of day are before noon.)

The correlation between Alexandrian dates and solar longitudes was derived from the *Handy Tables*, using Theon's formula to convert to sidereal longitudes. It is most accurate for the early third century, but satisfactory for time conversions throughout the Roman period.

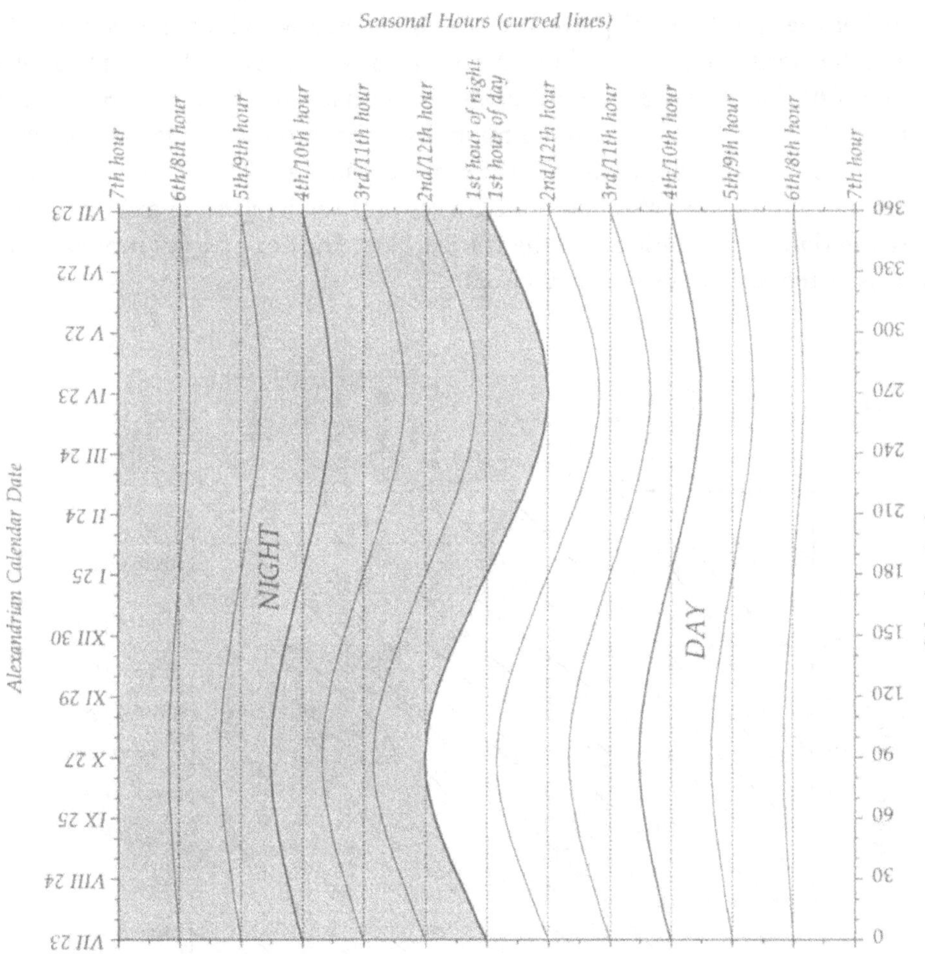

Seasonal Hours (curved lines)
7th hour
6th/8th hour
5th/9th hour
4th/10th hour
3rd/11th hour
2nd/12th hour
1st hour of night
1st hour of day
2nd/12th hour
3rd/11th hour
4th/10th hour
5th/9th hour
6th/8th hour
7th hour
Alexandrian Calendar Date
VII 23
VI 22
V 22
IV 23
III 24
II 24
I 25
XII 30
XI 29
X 27
IX 25
VIII 24
VII 23
NIGHT
DAY
360
330
300
270
240
210
180
150
120
90
60
30
0
Solar Longitude
12
11
10
9
8
7
6
5
4
3
2
1
0
Equinoctial Hours from Noon (straight lines)

G. Determination of Ascendant.

The relationship between solar longitude, time in seasonal hours, and the ascendant point of the ecliptic is displayed in the figure below, which is based on the table of ascensions for the 3rd Clima through Lower Egypt in the *Handy Tables*. Given the sun's longitude (horizontal axis) and the ascendant point (vertical axis), one can read off the time in seasonal hours by reference to the diagonal curves. Conversely, if the solar longitude and the seasonal time are known, one can locate the moment in question using the hour curves and the horizontal axis, and read off the ascendant's longitude along the vertical axis. The hour curves represent the *beginnings* of the respective seasonal hours.

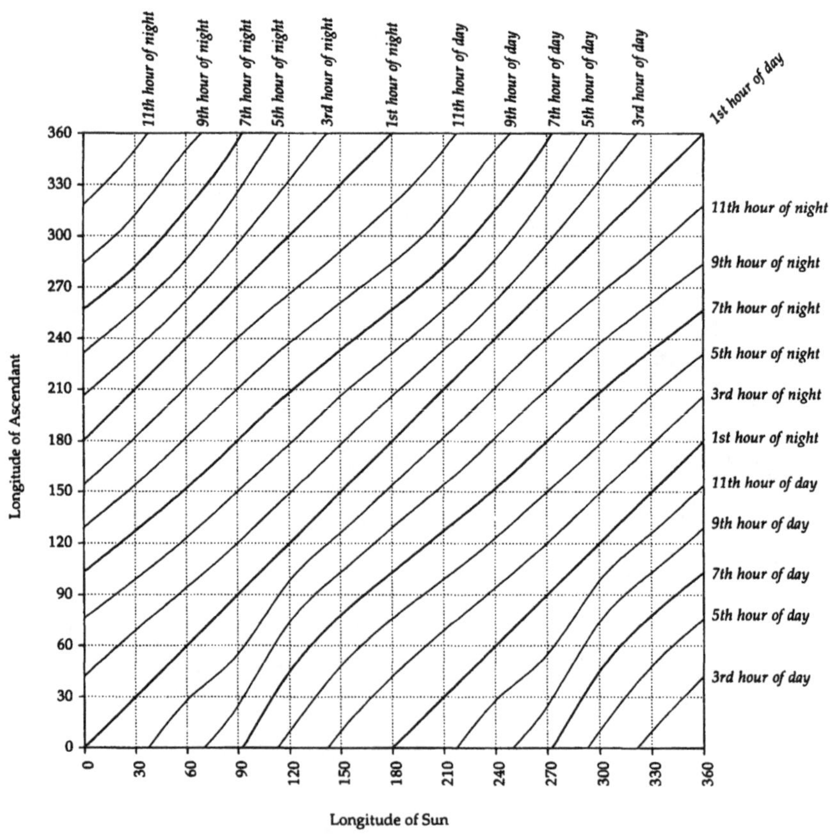

H. The Standard Lunar Scheme.

The Standard Lunar Scheme was a method of computing the moon's position in longitude and argument of latitude (i.e. its longitudinal elongation from the northern limit of its orbit) for any given date. Unlike most of the arithmetical predictive methods of Roman-period astronomy, the operation of the Standard Lunar Scheme is almost entirely understood and reconstructible on the basis of several papyrus texts and tables (Jones [1997a]). In return, it provides a useful tool in dating and analysing other ancient astronomical documents, such as ephemerides.

Given a date (Egyptian or Alexandrian calendar, year according to era Augustus), the corresponding lunar position is determined as follows. One first finds the immediately preceding epoch date from the Table of Epochs (Table 1), noting the associated longitude (in degrees from Aries 0°) and argument of latitude (in 'steps' of 15°). One then counts the number of days elapsed between the epoch date and the given date, and, looking up this number in the first column of the Template (Table 2), one adds the corresponding progress in longitude (column 3) and argument of latitude (column 5) to the epoch values. If the total longitude exceeds 360°, then 360 should be subtracted; and likewise if the total argument of latitude exceeds 24 steps, then 24 should be subtracted.

Positions computed in this way apply to the 'sunset' (effectively 6 P.M.) of the day in question. If the time in question is before sunset, one should calculate the fraction of 24 equinoctial hours between the time and sunset, multiply this by the 'daily progress' tabulated on the *same line* of the Template that was used to find the progress since epoch, and subtract the product from the position at sunset. For times after sunset, calculate the fraction of 24 hours since sunset, and add the proportionate fraction of the 'daily progress' tabulated in the *following line* of the Template to the position at sunset.

It is not known whether standard tables existed for converting the argument of latitude to an actual latitude in degrees from the ecliptic.

Example: Compute the moon's longitude and argument of latitude for Trajan 15, Choeac 1/2 (Alexandrian), end of the 6th hour of night, i.e. 12 P.M.

The year is equivalent to era Augustus 141 (cf. Appendix D). In order not to have to deal with intercalary days, we may prefer to work in the Egyptian calendar. From Appendix E, Tables 1 and 2, we find that the Egyptian calendar equivalent of Choeac 1 is Tybi 5. In the table of Epochs, the immediately preceding epoch date is era Augustus 140, Pachon 9, which is 241 days before the given date. We record and add the numbers in the columns headed 'longitude' and 'argument of latitude' for the epoch date in the Epoch Table and for day 241 in the Template to obtain the positions for the sunset preceding the given time. Since the given time is one quarter of a day after sunset, we further add this fraction of the daily progresses in longitude and argument of latitude tabulated for day 242.

	longitude	*arg. of latitude*
140 Pachon 9	75;38, 6, 5°	3; 7,13,55, 0 steps
day 241	299;49,39,37°	20;50,24,29,45 steps
6 hours	3;17,58,24,15°	0;13,15, 4,26,15 steps
141 Tybi 5, 6 P.M.	18;45,44, 6,15°	0;10,53,29,11,15 steps

The moon's longitude for the given date is thus approximately Aries 18;46°, and its argument of latitude 0;11 steps, or equivalently 2;43°, from the northern limit.

Derivation of Tables 1 and 2:
For a detailed exposition and analysis of the parameters of the Standard Scheme, see Jones [1997a], which incorporates corrections to earlier discussions. The epoch table (Table 1) gives dates and positions in conformity with 'Rule B' in *P. Ryl.* 27. All the dates in the four papyrus epoch tables so far known are in exact agreement with Rule B. For small systematic discrepancies (on the order of a few minutes of arc) in the epoch longitudes and larger deviations in the arguments of latitude, see the commentaries to **4150–4151** and Jones [1997a].

The longitudinal displacements and daily progress in longitude in the Template (Table 2) follow the parameters extracted from **4164** and **4164a**; they are also in agreement with the truncated values in *PSI* 1493 and further confirmed by the procedure text **4136**. The parameters of the latitudinal motion are derived from those of the longitudinal motion by applying a constant nodal motion of –0;3,10,49,15$^{\circ/d}$ and converting the units from degrees to 'steps'. The resulting displacements are once again in agreement with the truncated values in *P.S.I.* 1493. A solid horizontal line in the table marks passage of a minimum in the zigzag functions; a double line marks passage of a maximum. The broken line below day 248 indicates the discontinuity in the computation of the zigzag functions, discussed in the commentary to **4164a**.

	Egyptian			*Alexandrian*		*Longitude*	*Arg. of Latitude*
–2	Phaophi	26				42;18,51,14	9; 3,52, 6,45
–2	Epeiph	4				70; 2,16,10	11;47,20,40,45
–1	Phamenoth	7				97;45,41, 6	14;30,49,14,45
0	Hathyr	10				125;29, 6, 2	17;14,17,48,45
0	Epeiph	18				153;12,30,58	19;57,46,22,45
1	Phamenoth	21				180;55,55,54	22;41,14,56,45
2	Hathyr	24				208;39,20,50	1;24,43,30,45
2	Mesore	2				236;22,45,46	4; 8,12, 4,45
3	Pharmuthi	5				264; 6,10,42	6;51,40,38,45
4	Choeac	8				291;49,35,38	9;35, 9,12,45
4	Mesore	16				319;33, 0,34	12;18,37,46,45
5	Pharmuthi	19	5	Pharmuthi	19	347;16,25,30	15; 2, 6,20,45
6	Mecheir	17	6	Mecheir	17	19;50,10,21	18;16,35,55, 0
7	Phaophi	20	7	Phaophi	20	47;33,35,17	21; 0, 4,29, 0
7	Payni	28	7	Payni	28	75;17, 0,13	23;43,33, 3, 0
8	Phamenoth	1	8	Phamenoth	1	103; 0,25, 9	2;27, 1,37, 0
9	Hathyr	4	9	Hathyr	3	130;43,50, 5	5;10,30,11, 0
9	Epeiph	12	9	Epeiph	11	158;27,15, 1	7;53,58,45, 0
10	Phamenoth	15	10	Phamenoth	14	186;10,39,57	10;37,27,19, 0
11	Hathyr	18	11	Hathyr	17	213;54, 4,53	13;20,55,53, 0
11	Epeiph	26	11	Epeiph	25	241;37,29,49	16; 4,24,27, 0
12	Phamenoth	29	12	Phamenoth	28	269;20,54,45	18;47,53, 1, 0
13	Choeac	2	13	Hathyr	30	297; 4,19,41	21;31,21,35, 0
13	Mesore	10	13	Mesore	8	324;47,44,37	0;14,50, 9, 0
14	Payni	8	14	Payni	6	357;21,29,28	3;29,19,43,15
15	Mecheir	11	15	Mecheir	9	25; 4,54,24	6;12,48,17,15
16	Phaophi	14	16	Phaophi	12	52;48,19,20	8;56,16,51,15
16	Payni	22	16	Payni	20	80;31,44,16	11;39,45,25,15
17	Mecheir	25	17	Mecheir	22	108;15, 9,12	14;23,13,59,15
18	Phaophi	28	18	Phaophi	25	135;58,34, 8	17; 6,42,33,15
18	Epeiph	6	18	Epeiph	3	163;41,59, 4	19;50,11, 7,15
19	Phamenoth	9	19	Phamenoth	6	191;25,24, 0	22;33,39,41,15
20	Hathyr	12	20	Hathyr	9	219; 8,48,56	1;17, 8,15,15
20	Epeiph	20	20	Epeiph	17	246;52,13,52	4; 0,36,49,15
21	Phamenoth	23	21	Phamenoth	19	274;35,38,48	6;44, 5,23,15
22	Hathyr	26	22	Hathyr	22	302;19, 3,44	9;27,33,57,15
23	Thoth	24	23	Thoth	20	334;52,48,35	12;42, 3,31,30
23	Payni	2	23	Pachon	28	2;36,13,31	15;25,32, 5,30
24	Mecheir	5	24	Mecheir	1	30;19,38,27	18; 9, 0,39,30
25	Phaophi	8	25	Phaophi	3	58; 3, 3,23	20;52,29,13,30
25	Payni	16	25	Payni	11	85;46,28,19	23;35,57,47,30
26	Mecheir	19	26	Mecheir	14	113;29,53,15	2;19,26,21,30
27	Phaophi	22	27	Phaophi	17	141;13,18,11	5; 2,54,55,30
27	Payni	30	27	Payni	25	168;56,43, 7	7;46,23,29,30
28	Phamenoth	3	28	Mecheir	28	196;40, 8, 3	10;29,52, 3,30
29	Hathyr	6	29	Phaophi	30	224;23,32,59	13;13,20,37,30
29	Epeiph	14	29	Epeiph	8	252; 6,57,55	15;56,49,11,30
30	Phamenoth	17	30	Phamenoth	11	279;50,22,51	18;40,17,45,30
31	Tybi	15	31	Tybi	9	312;24, 7,42	21;54,47,19,45
32	Thoth	18	32	Thoth	12	340; 7,32,38	0;38,15,53,45
32	Pachon	26	32	Pachon	20	7;50,57,34	3;21,44,27,45
33	Tybi	29	33	Tybi	22	35;34,22,30	6; 5,13, 1,45

	Egyptian			*Alexandrian*		*Longitude*	*Arg. of Latitude*
34	Phaophi	2	34	Thoth	25	63;17,47,26	8;48,41,35,45
34	Payni	10	34	Payni	3	91; 1,12,22	11;32,10, 9,45
35	Mecheir	13	35	Mecheir	6	118;44,37,18	14;15,38,43,45
36	Phaophi	16	36	Phaophi	9	146;28, 2,14	16;59, 7,17,45
36	Payni	24	36	Payni	17	174;11,27,10	19;42,35,51,45
37	Mecheir	27	37	Mecheir	19	201;54,52, 6	22;26, 4,25,45
38	Phaophi	30	38	Phaophi	22	229;38,17, 2	1; 9,32,59,45
38	Epeiph	8	38	Payni	30	257;21,41,58	3;53, 1,33,45
39	Pachon	6	39	Pharmuthi	28	289;55,26,49	7; 7,31, 8, 0
40	Tybi	9	40	Tybi	1	317;38,51,45	9;50,59,42, 0
41	Thoth	12	41	Thoth	3	345;22,16,41	12;34,28,16, 0
41	Pachon	20	41	Pachon	11	13; 5,41,37	15;17,56,50, 0
42	Tybi	23	42	Tybi	14	40;49, 6,33	18; 1,25,24, 0
43	Thoth	26	43	Thoth	17	68;32,31,29	20;44,53,58, 0
43	Payni	4	43	Pachon	25	96;15,56,25	23;28,22,32, 0
44	Mecheir	7	44	Tybi	28	123;59,21,21	2;11,51, 6, 0
45	Phaophi	10	45	Thoth	30	151;42,46,17	4;55,19,40, 0
45	Payni	18	45	Payni	8	179;26,11,13	7;38,48,14, 0
46	Mecheir	21	46	Mecheir	11	207; 9,36, 9	10;22,16,48, 0
47	Phaophi	24	47	Phaophi	14	234;53, 1, 5	13; 5,45,22, 0
47	Mesore	27	47	Mesore	17	267;26,45,56	16;20,14,56,15
48	Pharmuthi	30	48	Pharmuthi	20	295;10,10,52	19; 3,43,30,15
49	Tybi	3	49	Choeac	22	322;53,35,48	21;47,12, 4,15
50	Thoth	6	49	Mesore	30	350;37, 0,44	0;30,40,38,15
50	Pachon	14	50	Pachon	3	18;20,25,40	3;14, 9,12,15
51	Tybi	17	51	Tybi	6	46; 3,50,36	5;57,37,46,15
52	Thoth	20	52	Thoth	9	73;47,15,32	8;41, 6,20,15
52	Pachon	28	52	Pachon	17	101;30,40,28	11;24,34,54,15
53	Mecheir	1	53	Tybi	19	129;14, 5,24	14; 8, 3,28,15
54	Phaophi	4	54	Thoth	22	156;57,30,20	16;51,32, 2,15
54	Payni	12	54	Pachon	30	184;40,55,16	19;35, 0,36,15
55	Mecheir	15	55	Mecheir	3	212;24,20,12	22;18,29,10,15
56	Choeac	13	56	Choeac	1	244;58, 5, 3	1;32,58,44,30
56	Mesore	21	56	Mesore	9	272;41,29,59	4;16,27,18,30
57	Pharmuthi	24	57	Pharmuthi	11	300;24,54,55	6;59,55,52,30
58	Choeac	27	58	Choeac	14	328; 8,19,51	9;43,24,26,30
58	Epagomenae	5	58	Mesore	22	355;51,44,47	12;26,53, 0,30
59	Pachon	8	59	Pharmuthi	25	23;35, 9,43	15;10,21,34,30
60	Tybi	11	60	Choeac	28	51;18,34,39	17;53,50, 8,30
61	Thoth	14	60	Epagomenae	6	79; 1,59,35	20;37,18,42,30
61	Pachon	22	61	Pachon	8	106;45,24,31	23;20,47,16,30
62	Tybi	25	62	Tybi	11	134;28,49,27	2; 4,15,50,30
63	Thoth	28	63	Thoth	14	162;12,14,23	4;47,44,24,30
63	Payni	6	63	Pachon	22	189;55,39,19	7;31,12,58,30
64	Pharmuthi	4	64	Phamenoth	20	222;29,24,10	10;45,42,32,45
65	Choeac	7	65	Hathyr	22	250;12,49, 6	13;29,11, 6,45
65	Mesore	15	65	Epeiph	30	277;56,14, 2	16;12,39,40,45
66	Pharmuthi	18	66	Pharmuthi	3	305;39,38,58	18;56, 8,14,45
67	Choeac	21	67	Choeac	6	333;23, 3,54	21;39,36,48,45
67	Mesore	29	67	Mesore	14	1; 6,28,50	0;23, 5,22,45
68	Pachon	2	68	Pharmuthi	17	28;49,53,46	3; 6,33,56,45
69	Tybi	5	69	Choeac	19	56;33,18,42	5;50, 2,30,45

Egyptian			*Alexandrian*			*Longitude*	*Arg. of Latitude*
70	Thoth	8	69	Mesore	27	84;16,43,38	8;33,31, 4,45
70	Pachon	16	70	Pharmuthi	30	112; 0, 8,34	11;16,59,38,45
71	Tybi	19	71	Tybi	3	139;43,33,30	14; 0,28,12,45
72	Thoth	22	72	Thoth	6	167;26,58,26	16;43,56,46,45
72	Epeiph	25	72	Epeiph	9	200; 0,43,17	19;58,26,21, 0
73	Phamenoth	28	73	Phamenoth	11	227;44, 8,13	22;41,54,55, 0
74	Choeac	1	74	Hathyr	14	255;27,33, 9	1;25,23,29, 0
74	Mesore	9	74	Epeiph	22	283;10,58, 5	4; 8,52, 3, 0
75	Pharmuthi	12	75	Phamenoth	25	310;54,23, 1	6;52,20,37, 0
76	Choeac	15	76	Hathyr	28	338;37,47,57	9;35,49,11, 0
76	Mesore	23	76	Mesore	6	6;21,12,53	12;19,17,45, 0
77	Pharmuthi	26	77	Pharmuthi	8	34; 4,37,49	15; 2,46,19, 0
78	Choeac	29	78	Choeac	11	61;48, 2,45	17;46,14,53, 0
79	Thoth	2	78	Mesore	19	89;31,27,41	20;29,43,27, 0
79	Pachon	10	79	Pharmuthi	22	117;14,52,37	23;13,12, 1, 0
80	Tybi	13	80	Choeac	25	144;58,17,33	1;56,40,35, 0
81	Hathyr	11	81	Phaophi	22	177;32, 2,24	5;11,10, 9,15
81	Epeiph	19	81	Payni	30	205;15,27,20	7;54,38,43,15
82	Phamenoth	22	82	Phamenoth	3	232;58,52,16	10;38, 7,17,15
83	Hathyr	25	83	Hathyr	6	260;42,17,12	13;21,35,51,15
83	Mesore	3	83	Epeiph	14	288;25,42, 8	16; 5, 4,25,15
84	Pharmuthi	6	84	Phamenoth	17	316; 9, 7, 4	18;48,32,59,15
85	Choeac	9	85	Hathyr	19	343;52,32, 0	21;32, 1,33,15
85	Mesore	17	85	Epeiph	27	11;35,56,56	0;15,30, 7,15
86	Pharmuthi	20	86	Phamenoth	30	39;19,21,52	2;58,58,41,15
87	Choeac	23	87	Choeac	3	67; 2,46,48	5;42,27,15,15
87	Epagomenae	1	87	Mesore	11	94;46,11,44	8;25,55,49,15
88	Pachon	4	88	Pharmuthi	14	122;29,36,40	11; 9,24,23,15
89	Phamenoth	2	89	Mecheir	11	155; 3,21,31	14;23,53,57,30
90	Hathyr	5	90	Phaophi	14	182;46,46,27	17; 7,22,31,30
90	Epeiph	13	90	Payni	22	210;30,11,23	19;50,51, 5,30
91	Phamenoth	16	91	Mecheir	25	238;13,36,19	22;34,19,39,30
92	Hathyr	19	92	Phaophi	28	265;57, 1,15	1;17,48,13,30
92	Epeiph	27	92	Epeiph	6	293;40,26,11	4; 1,16,47,30
93	Phamenoth	30	93	Phamenoth	8	321;23,51, 7	6;44,45,21,30
94	Choeac	3	94	Hathyr	11	349; 7,16, 3	9;28,13,55,30
94	Mesore	11	94	Epeiph	19	16;50,40,59	12;11,42,29,30
95	Pharmuthi	14	95	Phamenoth	22	44;34, 5,55	14;55,11, 3,30
96	Choeac	17	96	Hathyr	25	72;17,30,51	17;38,39,37,30
96	Mesore	25	96	Mesore	3	100; 0,55,47	20;22, 8,11,30
97	Payni	23	97	Pachon	30	132;34,40,38	23;36,37,45,45
98	Mecheir	26	98	Mecheir	3	160;18, 5,34	2;20, 6,19,45
99	Phaophi	29	99	Phaophi	6	188; 1,30,30	5; 3,34,53,45
99	Epeiph	7	99	Payni	14	215;44,55,26	7;47, 3,27,45
100	Phamenoth	10	100	Mecheir	17	243;28,20,22	10;30,32, 1,45
101	Hathyr	13	101	Phaophi	19	271;11,45,18	13;14, 0,35,45
101	Epeiph	21	101	Payni	27	298;55,10,14	15;57,29, 9,45
102	Phamenoth	24	102	Mecheir	30	326;38,35,10	18;40,57,43,45
103	Hathyr	27	103	Hathyr	3	354;22, 0, 6	21;24,26,17,45
103	Mesore	5	103	Epeiph	11	22; 5,25, 2	0; 7,54,51,45
104	Pharmuthi	8	104	Phamenoth	14	49;48,49,58	2;51,23,25,45
105	Choeac	11	105	Hathyr	16	77;32,14,54	5;34,51,59,45

Egyptian			*Alexandrian*			*Longitude*	*Arg. of Latitude*
106	Phaophi	9	106	Thoth	14	110; 5,59,45	8;49,21,34, 0
106	Payni	17	106	Pachon	22	137;49,24,41	11;32,50, 8, 0
107	Mecheir	20	107	Tybi	25	165;32,49,37	14;16,18,42, 0
108	Phaophi	23	108	Thoth	28	193;16,14,33	16;59,47,16, 0
108	Epeiph	1	108	Payni	6	220;59,39,29	19;43,15,50, 0
109	Phamenoth	4	109	Mecheir	8	248;43, 4,25	22;26,44,24, 0
110	Hathyr	7	110	Phaophi	11	276;26,29,21	1;10,12,58, 0
110	Epeiph	15	110	Payni	19	304; 9,54,17	3;53,41,32, 0
111	Phamenoth	18	111	Mecheir	22	331;53,19,13	6;37,10, 6, 0
112	Hathyr	21	112	Phaophi	25	359;36,44, 9	9;20,38,40, 0
112	Epeiph	29	112	Epeiph	3	27;20, 9, 5	12; 4, 7,14, 0
113	Pharmuthi	2	113	Phamenoth	5	55; 3,34, 1	14;47,35,48, 0
114	Tybi	30	114	Tybi	3	87;37,18,52	18; 2, 5,22,15
115	Phaophi	3	115	Thoth	6	115;20,43,48	20;45,33,56,15
115	Payni	11	115	Pachon	14	143; 4, 8,44	23;29, 2,30,15
116	Mecheir	14	116	Tybi	17	170;47,33,40	2;12,31, 4,15
117	Phaophi	17	117	Thoth	19	198;30,58,36	4;55,59,38,15
117	Payni	25	117	Pachon	27	226;14,23,32	7;39,28,12,15
118	Mecheir	28	118	Tybi	30	253;57,48,28	10;22,56,46,15
119	Hathyr	1	119	Phaophi	3	281;41,13,24	13; 6,25,20,15
119	Epeiph	9	119	Payni	11	309;24,38,20	15;49,53,54,15
120	Phamenoth	12	120	Mecheir	14	337; 8, 3,16	18;33,22,28,15
121	Hathyr	15	121	Phaophi	16	4;51,28,12	21;16,51, 2,15
121	Epeiph	23	121	Payni	24	32;34,53, 8	0; 0,19,36,15
122	Pachon	21	122	Pharmuthi	22	65; 8,37,59	3;14,49,10,30
123	Tybi	24	123	Choeac	25	92;52, 2,55	5;58,17,44,30
124	Thoth	27	123	Epagomenae	3	120;35,27,51	8;41,46,18,30
124	Payni	5	124	Pachon	6	148;18,52,47	11;25,14,52,30
125	Mecheir	8	125	Tybi	8	176; 2,17,43	14; 8,43,26,30
126	Phaophi	11	126	Thoth	11	203;45,42,39	16;52,12, 0,30
126	Payni	19	126	Pachon	19	231;29, 7,35	19;35,40,34,30
127	Mecheir	22	127	Tybi	22	259;12,32,31	22;19, 9, 8,30
128	Phaophi	25	128	Thoth	25	286;55,57,27	1; 2,37,42,30
128	Epeiph	3	128	Payni	3	314;39,22,23	3;46, 6,16,30
129	Phamenoth	6	129	Mecheir	5	342;22,47,19	6;29,34,50,30
130	Hathyr	9	130	Phaophi	8	10; 6,12,15	9;13, 3,24,30
131	Thoth	7	130	Mesore	11	42;39,57, 6	12;27,32,58,45
131	Pachon	15	131	Pharmuthi	14	70;23,22, 2	15;11, 1,32,45
132	Tybi	18	132	Choeac	17	98; 6,46,58	17;54,30, 6,45
133	Thoth	21	132	Mesore	25	125;50,11,54	20;37,58,40,45
133	Pachon	29	133	Pharmuthi	27	153;33,36,50	23;21,27,14,45
134	Mecheir	2	134	Choeac	30	181;17, 1,46	2; 4,55,48,45
135	Phaophi	5	135	Thoth	3	209; 0,26,42	4;48,24,22,45
135	Payni	13	135	Pachon	11	236;43,51,38	7;31,52,56,45
136	Mecheir	16	136	Tybi	14	264;27,16,34	10;15,21,30,45
137	Phaophi	19	137	Thoth	16	292;10,41,30	12;58,50, 4,45
137	Payni	27	137	Pachon	24	319;54, 6,26	15;42,18,38,45
138	Mecheir	30	138	Tybi	27	347;37,31,22	18;25,47,12,45
139	Choeac	28	139	Hathyr	25	20;11,16,13	21;40,16,47, 0
140	Thoth	1	139	Mesore	3	47;54,41, 9	0;23,45,21, 0
140	Pachon	9	140	Pharmuthi	6	75;38, 6, 5	3; 7,13,55, 0
141	Tybi	12	141	Choeac	8	103;21,31, 1	5;50,42,29, 0

Egyptian			*Alexandrian*			*Longitude*	*Arg. of Latitude*
142	Thoth	15	141	Mesore	16	131; 4,55,57	8;34,11, 3, 0
142	Pachon	23	142	Pharmuthi	19	158;48,20,53	11;17,39,37, 0
143	Tybi	26	143	Choeac	22	186;31,45,49	14; 1, 8,11, 0
144	Thoth	29	143	Mesore	30	214;15,10,45	16;44,36,45, 0
144	Payni	7	144	Pachon	3	241;58,35,41	19;28, 5,19, 0
145	Mecheir	10	145	Tybi	5	269;42, 0,37	22;11,33,53, 0
146	Phaophi	13	146	Thoth	8	297;25,25,33	0;55, 2,27, 0
146	Payni	21	146	Pachon	16	325; 8,50,29	3;38,31, 1, 0
147	Pharmuthi	19	147	Phamenoth	14	357;42,35,20	6;53, 0,35,15
148	Choeac	22	148	Hathyr	17	25;26, 0,16	9;36,29, 9,15
148	Mesore	30	148	Epeiph	25	53; 9,25,12	12;19,57,43,15
149	Pachon	3	149	Phamenoth	27	80;52,50, 8	15; 3,26,17,15
150	Tybi	6	150	Hathyr	30	108;36,15, 4	17;46,54,51,15
151	Thoth	9	150	Mesore	8	136;19,40, 0	20;30,23,25,15
151	Pachon	17	151	Pharmuthi	11	164; 3, 4,56	23;13,51,59,15
152	Tybi	20	152	Choeac	14	191;46,29,52	1;57,20,33,15
153	Thoth	23	152	Mesore	22	219;29,54,48	4;40,49, 7,15
153	Payni	1	153	Pharmuthi	24	247;13,19,44	7;24,17,41,15
154	Mecheir	4	154	Choeac	27	274;56,44,40	10; 7,46,15,15
155	Phaophi	7	154	Epagomenae	5	302;40, 9,36	12;51,14,49,15
155	Mesore	10	155	Epeiph	3	335;13,54,27	16; 5,44,23,30
156	Pharmuthi	13	156	Phamenoth	6	2;57,19,23	18;49,12,57,30
157	Choeac	16	157	Hathyr	8	30;40,44,19	21;32,41,31,30
157	Mesore	24	157	Epeiph	16	58;24, 9,15	0;16,10, 5,30
158	Pharmuthi	27	158	Phamenoth	19	86; 7,34,11	2;59,38,39,30
159	Choeac	30	159	Hathyr	22	113;50,59, 7	5;43, 7,13,30
160	Thoth	3	159	Epeiph	30	141;34,24, 3	8;26,35,47,30
160	Pachon	11	160	Pharmuthi	3	169;17,48,59	11;10, 4,21,30
161	Tybi	14	161	Choeac	5	197; 1,13,55	13;53,32,55,30
162	Thoth	17	161	Mesore	13	224;44,38,51	16;37, 1,29,30
162	Pachon	25	162	Pharmuthi	16	252;28, 3,47	19;20,30, 3,30
163	Tybi	28	163	Choeac	19	280;11,28,43	22; 3,58,37,30
164	Hathyr	26	164	Phaophi	17	312;45,13,34	1;18,28,11,45
164	Mesore	4	164	Payni	25	340;28,38,30	4; 1,56,45,45
165	Pharmuthi	7	165	Mecheir	27	8;12, 3,26	6;45,25,19,45
166	Choeac	10	166	Phaophi	30	35;55,28,22	9;28,53,53,45
166	Mesore	18	166	Epeiph	8	63;38,53,18	12;12,22,27,45
167	Pharmuthi	21	167	Phamenoth	11	91;22,18,14	14;55,51, 1,45
168	Choeac	24	168	Hathyr	14	119; 5,43,10	17;39,19,35,45
168	Epagomenae	2	168	Epeiph	22	146;49, 8, 6	20;22,48, 9,45
169	Pachon	5	169	Phamenoth	24	174;32,33, 2	23; 6,16,43,45
170	Tybi	8	170	Hathyr	27	202;15,57,58	1;49,45,17,45
171	Thoth	11	170	Mesore	5	229;59,22,54	4;33,13,51,45
171	Pachon	19	171	Pharmuthi	8	257;42,47,50	7;16,42,25,45
172	Phamenoth	17	172	Mecheir	6	290;16,32,41	10;31,12, 0, 0
173	Hathyr	20	173	Phaophi	8	317;59,57,37	13;14,40,34, 0
173	Epeiph	28	173	Payni	16	345;43,22,33	15;58, 9, 8, 0
174	Pharmuthi	1	174	Mecheir	19	13;26,47,29	18;41,37,42, 0
175	Choeac	4	175	Phaophi	22	41;10,12,25	21;25, 6,16, 0
175	Mesore	12	175	Payni	30	68;53,37,21	0; 8,34,50, 0
176	Pharmuthi	15	176	Phamenoth	3	96;37, 2,17	2;52, 3,24, 0
177	Choeac	18	177	Hathyr	5	124;20,27,13	5;35,31,58, 0

Egyptian			Alexandrian			Longitude	Arg. of Latitude
177	Mesore	26	177	Epeiph	13	152; 3,52, 9	8;19, 0,32, 0
178	Pharmuthi	29	178	Phamenoth	16	179;47,17, 5	11; 2,29, 6, 0
179	Tybi	2	179	Hathyr	19	207;30,42, 1	13;45,57,40, 0
180	Thoth	5	179	Epeiph	27	235;14, 6,57	16;29,26,14, 0
180	Epeiph	8	180	Pachon	25	267;47,51,48	19;43,55,48,15
181	Phamenoth	11	181	Tybi	27	295;31,16,44	22;27,24,22,15
182	Hathyr	14	182	Thoth	30	323;14,41,40	1;10,52,56,15
182	Epeiph	22	182	Payni	8	350;58, 6,36	3;54,21,30,15
183	Phamenoth	25	183	Mecheir	11	18;41,31,32	6;37,50, 4,15
184	Hathyr	28	184	Phaophi	14	46;24,56,28	9;21,18,38,15
184	Mesore	6	184	Payni	22	74; 8,21,24	12; 4,47,12,15
185	Pharmuthi	9	185	Mecheir	24	101;51,46,20	14;48,15,46,15
186	Choeac	12	186	Phaophi	27	129;35,11,16	17;31,44,20,15
186	Mesore	20	186	Epeiph	5	157;18,36,12	20;15,12,54,15
187	Pharmuthi	23	187	Phamenoth	8	185; 2, 1, 8	22;58,41,28,15
188	Choeac	26	188	Hathyr	11	212;45,26, 4	1;42,10, 2,15
189	Phaophi	24	189	Thoth	8	245;19,10,55	4;56,39,36,30
189	Epeiph	2	189	Pachon	16	273; 2,35,51	7;40, 8,10,30
190	Phamenoth	5	190	Tybi	19	300;46, 0,47	10;23,36,44,30
191	Hathyr	8	191	Thoth	22	328;29,25,43	13; 7, 5,18,30
191	Epeiph	16	191	Pachon	30	356;12,50,39	15;50,33,52,30
192	Phamenoth	19	192	Mecheir	3	23;56,15,35	18;34, 2,26,30
193	Hathyr	22	193	Phaophi	5	51;39,40,31	21;17,31, 0,30
193	Epeiph	30	193	Payni	13	79;23, 5,27	0; 0,59,34,30
194	Pharmuthi	3	194	Mecheir	16	107; 6,30,23	2;44,28, 8,30
195	Choeac	6	195	Phaophi	19	134;49,55,19	5;27,56,42,30
195	Mesore	14	195	Payni	27	162;33,20,15	8;11,25,16,30
196	Pharmuthi	17	196	Mecheir	30	190;16,45,11	10;54,53,50,30
197	Mecheir	15	197	Choeac	27	222;50,30, 2	14; 9,23,24,45
198	Phaophi	18	197	Epagomenae	5	250;33,54,58	16;52,51,58,45
198	Payni	26	198	Pachon	8	278;17,19,54	19;36,20,32,45
199	Mecheir	29	199	Tybi	11	306; 0,44,50	22;19,49, 6,45
200	Hathyr	2	200	Thoth	14	333;44, 9,46	1; 3,17,40,45
200	Epeiph	10	200	Pachon	22	1;27,34,42	3;46,46,14,45
201	Phamenoth	13	201	Tybi	24	29;10,59,38	6;30,14,48,45
202	Hathyr	16	202	Thoth	27	56;54,24,34	9;13,43,22,45
202	Epeiph	24	202	Payni	5	84;37,49,30	11;57,11,56,45
203	Phamenoth	27	203	Mecheir	8	112;21,14,26	14;40,40,30,45
204	Hathyr	30	204	Phaophi	11	140; 4,39,22	17;24, 9, 4,45
204	Mesore	8	204	Payni	19	167;48, 4,18	20; 7,37,38,45
205	Payni	6	205	Pharmuthi	16	200;21,49, 9	23;22, 7,13, 0
206	Mecheir	9	206	Choeac	19	228; 5,14, 5	2; 5,35,47, 0
207	Phaophi	12	206	Mesore	27	255;48,39, 1	4;49, 4,21, 0
207	Payni	20	207	Pharmuthi	30	283;32, 3,57	7;32,32,55, 0
208	Mecheir	23	208	Tybi	3	311;15,28,53	10;16, 1,29, 0
209	Phaophi	26	209	Thoth	5	338;58,53,49	12;59,30, 3, 0
209	Epeiph	4	209	Pachon	13	6;42,18,45	15;42,58,37, 0
210	Phamenoth	7	210	Tybi	16	34;25,43,41	18;26,27,11, 0
211	Hathyr	10	211	Thoth	19	62; 9, 8,37	21; 9,55,45, 0
211	Epeiph	18	211	Pachon	27	89;52,33,33	23;53,24,19, 0
212	Phamenoth	21	212	Tybi	30	117;35,58,29	2;36,52,53, 0
213	Hathyr	24	213	Phaophi	2	145;19,23,25	5;20,21,27, 0

Egyptian			*Alexandrian*			*Longitude*	*Arg. of Latitude*
214	Thoth	22	213	Mesore	5	177;53, 8,16	8;34,51, 1,15
214	Pachon	30	214	Pharmuthi	8	205;36,33,12	11;18,19,35,15
215	Mecheir	3	215	Choeac	11	233;19,58, 8	14; 1,48, 9,15
216	Phaophi	6	215	Mesore	19	261; 3,23, 4	16;45,16,43,15
216	Payni	14	216	Pharmuthi	22	288;46,48, 0	19;28,45,17,15
217	Mecheir	17	217	Choeac	24	316;30,12,56	22;12,13,51,15
218	Phaophi	20	217	Epagomenae	2	344;13,37,52	0;55,42,25,15
218	Payni	28	218	Pachon	5	11;57, 2,48	3;39,10,59,15
219	Phamenoth	1	219	Tybi	8	39;40,27,44	6;22,39,33,15
220	Hathyr	4	220	Thoth	11	67;23,52,40	9; 6, 8, 7,15
220	Epeiph	12	220	Pachon	19	95; 7,17,36	11;49,36,41,15
221	Phamenoth	15	221	Tybi	21	122;50,42,32	14;33, 5,15,15
222	Tybi	13	222	Hathyr	19	155;24,27,23	17;47,34,49,30
223	Thoth	16	222	Epeiph	27	183; 7,52,19	20;31, 3,23,30
223	Pachon	24	223	Phamenoth	30	210;51,17,15	23;14,31,57,30
224	Tybi	27	224	Choeac	3	238;34,42,11	1;58, 0,31,30
225	Thoth	30	224	Mesore	11	266;18, 7, 7	4;41,29, 5,30
225	Payni	8	225	Pharmuthi	13	294; 1,32, 3	7;24,57,39,30
226	Mecheir	11	226	Choeac	16	321;44,56,59	10; 8,26,13,30
227	Phaophi	14	226	Mesore	24	349;28,21,55	12;51,54,47,30
227	Payni	22	227	Pharmuthi	27	17;11,46,51	15;35,23,21,30
228	Mecheir	25	228	Choeac	30	44;55,11,47	18;18,51,55,30
229	Phaophi	28	229	Thoth	2	72;38,36,43	21; 2,20,29,30
229	Epeiph	6	229	Pachon	10	100;22, 1,39	23;45,49, 3,30
230	Pachon	4	230	Phamenoth	8	132;55,46,30	3; 0,18,37,45
231	Tybi	7	231	Hathyr	11	160;39,11,26	5;43,47,11,45
232	Thoth	10	231	Epeiph	19	188;22,36,22	8;27,15,45,45
232	Pachon	18	232	Phamenoth	22	216; 6, 1,18	11;10,44,19,45
233	Tybi	21	233	Hathyr	24	243;49,26,14	13;54,12,53,45
234	Thoth	24	233	Mesore	2	271;32,51,10	16;37,41,27,45
234	Payni	2	234	Pharmuthi	5	299;16,16, 6	19;21,10, 1,45
235	Mecheir	5	235	Choeac	8	326;59,41, 2	22; 4,38,35,45
236	Phaophi	8	235	Mesore	16	354;43, 5,58	0;48, 7, 9,45
236	Payni	16	236	Pharmuthi	19	22;26,30,54	3;31,35,43,45
237	Mecheir	19	237	Choeac	21	50; 9,55,50	6;15, 4,17,45
238	Phaophi	22	237	Mesore	29	77;53,20,46	8;58,32,51,45
238	Mesore	25	238	Payni	27	110;27, 5,37	12;13, 2,26, 0
239	Pharmuthi	28	239	Mecheir	30	138;10,30,33	14;56,31, 0, 0
240	Tybi	1	240	Hathyr	3	165;53,55,29	17;39,59,34, 0
241	Thoth	4	240	Epeiph	11	193;37,20,25	20;23,28, 8, 0
241	Pachon	12	241	Phamenoth	13	221;20,45,21	23; 6,56,42, 0
242	Tybi	15	242	Hathyr	16	249; 4,10,17	1;50,25,16, 0
243	Thoth	18	242	Epeiph	24	276;47,35,13	4;33,53,50, 0
243	Pachon	26	243	Phamenoth	27	304;31, 0, 9	7;17,22,24, 0
244	Tybi	29	244	Hathyr	30	332;14,25, 5	10; 0,50,58, 0
245	Phaophi	2	244	Mesore	8	359;57,50, 1	12;44,19,32, 0
245	Payni	10	245	Pharmuthi	10	27;41,14,57	15;27,48, 6, 0
246	Mecheir	13	246	Choeac	13	55;24,39,53	18;11,16,40, 0
247	Choeac	11	247	Phaophi	11	87;58,24,44	21;25,46,14,15
247	Mesore	19	247	Payni	19	115;41,49,40	0; 9,14,48,15
248	Pharmuthi	22	248	Mecheir	22	143;25,14,36	2;52,43,22,15
249	Choeac	25	249	Phaophi	24	171; 8,39,32	5;36,11,56,15

	Egyptian			Alexandrian		Longitude	Arg. of Latitude
249	Epagomenae	3	249	Epeiph	2	198;52, 4,28	8;19,40,30,15
250	Pachon	6	250	Phamenoth	5	226;35,29,24	11; 3, 9, 4,15
251	Tybi	9	251	Hathyr	8	254;18,54,20	13;46,37,38,15
252	Thoth	12	251	Epeiph	16	282; 2,19,16	16;30, 6,12,15
252	Pachon	20	252	Phamenoth	19	309;45,44,12	19;13,34,46,15
253	Tybi	23	253	Hathyr	21	337;29, 9, 8	21;57, 3,20,15
254	Thoth	26	253	Epeiph	29	5;12,34, 4	0;40,31,54,15
254	Payni	4	254	Pharmuthi	2	32;55,59, 0	3;24, 0,28,15
255	Pharmuthi	2	255	Tybi	30	65;29,43,51	6;38,30, 2,30
256	Choeac	5	256	Phaophi	3	93;13, 8,47	9;21,58,36,30
256	Mesore	13	256	Payni	11	120;56,33,43	12; 5,27,10,30
257	Pharmuthi	16	257	Mecheir	13	148;39,58,39	14;48,55,44,30
258	Choeac	19	258	Phaophi	16	176;23,23,35	17;32,24,18,30
258	Mesore	27	258	Payni	24	204; 6,48,31	20;15,52,52,30
259	Pharmuthi	30	259	Mecheir	27	231;50,13,27	22;59,21,26,30
260	Tybi	3	260	Phaophi	30	259;33,38,23	1;42,50, 0,30
261	Thoth	6	260	Epeiph	8	287;17, 3,19	4;26,18,34,30
261	Pachon	14	261	Phamenoth	10	315; 0,28,15	7; 9,47, 8,30
262	Tybi	17	262	Hathyr	13	342;43,53,11	9;53,15,42,30
263	Thoth	20	262	Epeiph	21	10;27,18, 7	12;36,44,16,30
263	Epeiph	23	263	Pachon	19	43; 1, 2,58	15;51,13,50,45
264	Phamenoth	26	264	Tybi	22	70;44,27,54	18;34,42,24,45
265	Hathyr	29	265	Thoth	24	98;27,52,50	21;18,10,58,45
265	Mesore	7	265	Payni	2	126;11,17,46	0; 1,39,32,45
266	Pharmuthi	10	266	Mecheir	5	153;54,42,42	2;45, 8, 6,45
267	Choeac	13	267	Phaophi	8	181;38, 7,38	5;28,36,40,45
267	Mesore	21	267	Payni	16	209;21,32,34	8;12, 5,14,45
268	Pharmuthi	24	268	Mecheir	19	237; 4,57,30	10;55,33,48,45
269	Choeac	27	269	Phaophi	21	264;48,22,26	13;39, 2,22,45
269	Epagomenae	5	269	Payni	29	292;31,47,22	16;22,30,56,45
270	Pachon	8	270	Phamenoth	2	320;15,12,18	19; 5,59,30,45
271	Tybi	11	271	Hathyr	5	347;58,37,14	21;49,28, 4,45
272	Hathyr	9	272	Thoth	3	20;32,22, 5	1; 3,57,39, 0
272	Epeiph	17	272	Pachon	11	48;15,47, 1	3;47,26,13, 0
273	Phamenoth	20	273	Tybi	13	75;59,11,57	6;30,54,47, 0
274	Hathyr	23	274	Thoth	16	103;42,36,53	9;14,23,21, 0
274	Mesore	1	274	Pachon	24	131;26, 1,49	11;57,51,55, 0
275	Pharmuthi	4	275	Tybi	27	159; 9,26,45	14;41,20,29, 0
276	Choeac	7	276	Thoth	30	186;52,51,41	17;24,49, 3, 0
276	Mesore	15	276	Payni	8	214;36,16,37	20; 8,17,37, 0
277	Pharmuthi	18	277	Mecheir	10	242;19,41,33	22;51,46,11, 0
278	Choeac	21	278	Phaophi	13	270; 3, 6,29	1;35,14,45, 0
278	Mesore	29	278	Payni	21	297;46,31,25	4;18,43,19, 0
279	Pachon	2	279	Mecheir	24	325;29,56,21	7; 2,11,53, 0
280	Mecheir	30	280	Choeac	22	358; 3,41,12	10;16,41,27,15
281	Hathyr	3	280	Mesore	30	25;47, 6, 8	13; 0,10, 1,15
281	Epeiph	11	281	Pachon	2	53;30,31, 4	15;43,38,35,15
282	Phamenoth	14	282	Tybi	5	81;13,56, 0	18;27, 7, 9,15
283	Hathyr	17	283	Thoth	8	108;57,20,56	21;10,35,43,15
283	Epeiph	25	283	Pachon	16	136;40,45,52	23;54, 4,17,15
284	Phamenoth	28	284	Tybi	19	164;24,10,48	2;37,32,51,15
285	Choeac	1	285	Thoth	21	192; 7,35,44	5;21, 1,25,15

Egyptian			Alexandrian			Longitude	Arg. of Latitude
285	Mesore	9	285	Pachon	29	219;51, 0,40	8; 4,29,59,15
286	Pharmuthi	12	286	Mecheir	2	247;34,25,36	10;47,58,33,15
287	Choeac	15	287	Phaophi	5	275;17,50,32	13;31,27, 7,15
287	Mesore	23	287	Payni	13	303; 1,15,28	16;14,55,41,15
288	Payni	21	288	Pharmuthi	11	335;35, 0,19	19;29,25,15,30
289	Mecheir	24	289	Choeac	13	3;18,25,15	22;12,53,49,30
290	Phaophi	27	289	Mesore	21	31; 1,50,11	0;56,22,23,30
290	Epeiph	5	290	Pharmuthi	24	58;45,15, 7	3;39,50,57,30
291	Phamenoth	8	291	Choeac	27	86;28,40, 3	6;23,19,31,30
292	Hathyr	11	291	Epagomenae	5	114;12, 4,59	9; 6,48, 5,30
292	Epeiph	19	292	Pachon	8	141;55,29,55	11;50,16,39,30
293	Phamenoth	22	293	Tybi	10	169;38,54,51	14;33,45,13,30
294	Hathyr	25	294	Thoth	13	197;22,19,47	17;17,13,47,30
294	Mesore	3	294	Pachon	21	225; 5,44,43	20; 0,42,21,30
295	Pharmuthi	6	295	Tybi	24	252;49, 9,39	22;44,10,55,30
296	Choeac	9	296	Thoth	27	280;32,34,35	1;27,39,29,30
297	Phaophi	7	296	Epeiph	30	313; 6,19,26	4;42, 9, 3,45
297	Payni	15	297	Pharmuthi	2	340;49,44,22	7;25,37,37,45
298	Mecheir	18	298	Choeac	5	8;33, 9,18	10; 9, 6,11,45
299	Phaophi	21	298	Mesore	13	36;16,34,14	12;52,34,45,45
299	Payni	29	299	Pharmuthi	16	63;59,59,10	15;36, 3,19,45
300	Phamenoth	2	300	Choeac	19	91;43,24, 6	18;19,31,53,45
301	Hathyr	5	300	Mesore	27	119;26,49, 2	21; 3, 0,27,45
301	Epeiph	13	301	Pharmuthi	29	147;10,13,58	23;46,29, 1,45
302	Phamenoth	16	302	Tybi	2	174;53,38,54	2;29,57,35,45
303	Hathyr	19	303	Thoth	5	202;37, 3,50	5;13,26, 9,45
303	Epeiph	27	303	Pachon	13	230;20,28,46	7;56,54,43,45
304	Phamenoth	30	304	Tybi	16	258; 3,53,42	10;40,23,17,45
305	Tybi	28	305	Hathyr	13	290;37,38,33	13;54,52,52, 0
306	Phaophi	1	305	Epeiph	21	318;21, 3,29	16;38,21,26, 0
306	Payni	9	306	Phamenoth	24	346; 4,28,25	19;21,50, 0, 0
307	Mecheir	12	307	Hathyr	27	13;47,53,21	22; 5,18,34, 0
308	Phaophi	15	307	Mesore	5	41;31,18,17	0;48,47, 8, 0
308	Payni	23	308	Pharmuthi	8	69;14,43,13	3;32,15,42, 0
309	Mecheir	26	309	Choeac	10	96;58, 8, 9	6;15,44,16, 0
310	Phaophi	29	309	Mesore	18	124;41,33, 5	8;59,12,50, 0
310	Epeiph	7	310	Pharmuthi	21	152;24,58, 1	11;42,41,24, 0
311	Phamenoth	10	311	Choeac	24	180; 8,22,57	14;26, 9,58, 0
312	Hathyr	13	311	Epagomenae	2	207;51,47,53	17; 9,38,32, 0
312	Epeiph	21	312	Pachon	5	235;35,12,49	19;53, 7, 6, 0
313	Pachon	19	313	Phamenoth	2	268; 8,57,40	23; 7,36,40,15
314	Tybi	22	314	Hathyr	5	295;52,22,36	1;51, 5,14,15
315	Thoth	25	314	Epeiph	13	323;35,47,32	4;34,33,48,15
315	Payni	3	315	Phamenoth	16	351;19,12,28	7;18, 2,22,15
316	Mecheir	6	316	Hathyr	19	19; 2,37,24	10; 1,30,56,15
317	Phaophi	9	316	Epeiph	27	46;46, 2,20	12;44,59,30,15
317	Payni	17	317	Phamenoth	29	74;29,27,16	15;28,28, 4,15
318	Mecheir	20	318	Choeac	2	102;12,52,12	18;11,56,38,15
319	Phaophi	23	318	Mesore	10	129;56,17, 8	20;55,25,12,15
319	Epeiph	1	319	Pharmuthi	13	157;39,42, 4	23;38,53,46,15
320	Phamenoth	4	320	Choeac	16	185;23, 7, 0	2;22,22,20,15
321	Hathyr	7	320	Mesore	24	213; 6,31,56	5; 5,50,54,15

Egyptian			Alexandrian			Longitude	Arg. of Latitude
322	Thoth	5	321	Payni	21	245;40,16,47	8;20,20,28,30
322	Pachon	13	322	Mecheir	24	273;23,41,43	11; 3,49, 2,30
323	Tybi	16	323	Phaophi	27	301; 7, 6,39	13;47,17,36,30
324	Thoth	19	323	Epeiph	5	328;50,31,35	16;30,46,10,30
324	Pachon	27	324	Phamenoth	8	356;33,56,31	19;14,14,44,30
325	Tybi	30	325	Hathyr	10	24;17,21,27	21;57,43,18,30
326	Phaophi	3	325	Epeiph	18	52; 0,46,23	0;41,11,52,30
326	Payni	11	326	Phamenoth	21	79;44,11,19	3;24,40,26,30
327	Mecheir	14	327	Hathyr	24	107;27,36,15	6; 8, 9, 0,30
328	Phaophi	17	327	Mesore	2	135;11, 1,11	8;51,37,34,30
328	Payni	25	328	Pharmuthi	5	162;54,26, 7	11;35, 6, 8,30
329	Mecheir	28	329	Choeac	7	190;37,51, 3	14;18,34,42,30
330	Choeac	26	330	Phaophi	5	223;11,35,54	17;33, 4,16,45
330	Epagomenae	4	330	Payni	13	250;55, 0,50	20;16,32,50,45
331	Pachon	7	331	Mecheir	16	278;38,25,46	23; 0, 1,24,45
332	Tybi	10	332	Phaophi	19	306;21,50,42	1;43,29,58,45
333	Thoth	13	332	Payni	27	334; 5,15,38	4;26,58,32,45
333	Pachon	21	333	Mecheir	29	1;48,40,34	7;10,27, 6,45
334	Tybi	24	334	Hathyr	2	29;32, 5,30	9;53,55,40,45
335	Thoth	27	334	Epeiph	10	57;15,30,26	12;37,24,14,45
335	Payni	5	335	Phamenoth	13	84;58,55,22	15;20,52,48,45
336	Mecheir	8	336	Hathyr	16	112;42,20,18	18; 4,21,22,45
337	Phaophi	11	336	Epeiph	24	140;25,45,14	20;47,49,56,45
337	Payni	19	337	Phamenoth	26	168; 9,10,10	23;31,18,30,45
338	Pharmuthi	17	338	Tybi	24	200;42,55, 1	2;45,48, 5, 0
339	Choeac	20	339	Thoth	27	228;26,19,57	5;29,16,39, 0
339	Mesore	28	339	Payni	5	256; 9,44,53	8;12,45,13, 0
340	Pachon	1	340	Mecheir	8	283;53, 9,49	10;56,13,47, 0
341	Tybi	4	341	Phaophi	10	311;36,34,45	13;39,42,21, 0
342	Thoth	7	341	Payni	18	339;19,59,41	16;23,10,55, 0
342	Pachon	15	342	Mecheir	21	7; 3,24,37	19; 6,39,29, 0
343	Tybi	18	343	Phaophi	24	34;46,49,33	21;50, 8, 3, 0
344	Thoth	21	343	Epeiph	2	62;30,14,29	0;33,36,37, 0
344	Pachon	29	344	Phamenoth	5	90;13,39,25	3;17, 5,11, 0
345	Mecheir	2	345	Hathyr	7	117;57, 4,21	6; 0,33,45, 0
346	Phaophi	5	345	Epeiph	15	145;40,29,17	8;44, 2,19, 0
346	Mesore	8	346	Pachon	13	178;14,14, 8	11;58,31,53,15
347	Pharmuthi	11	347	Tybi	16	205;57,39, 4	14;42, 0,27,15
348	Choeac	14	348	Thoth	19	233;41, 4, 0	17;25,29, 1,15
348	Mesore	22	348	Pachon	27	261;24,28,56	20; 8,57,35,15
349	Pharmuthi	25	349	Tybi	29	289; 7,53,52	22;52,26, 9,15
350	Choeac	28	350	Phaophi	2	316;51,18,48	1;35,54,43,15
351	Thoth	1	350	Payni	10	344;34,43,44	4;19,23,17,15
351	Pachon	9	351	Mecheir	13	12;18, 8,40	7; 2,51,51,15
352	Tybi	12	352	Phaophi	16	40; 1,33,36	9;46,20,25,15
353	Thoth	15	352	Payni	24	67;44,58,32	12;29,48,59,15
353	Pachon	23	353	Mecheir	26	95;28,23,28	15;13,17,33,15
354	Tybi	26	354	Phaophi	29	123;11,48,24	17;56,46, 7,15
355	Hathyr	24	354	Epagomenae	2	155;45,33,15	21;11,15,41,30
355	Mesore	2	355	Pachon	5	183;28,58,11	23;54,44,15,30
356	Pharmuthi	5	356	Tybi	8	211;12,23, 7	2;38,12,49,30
357	Choeac	8	357	Thoth	10	238;55,48, 3	5;21,41,23,30

Egyptian			*Alexandrian*			*Longitude*	*Arg. of Latitude*
357	Mesore	16	357	Pachon	18	266;39,12,59	8; 5, 9,57,30
358	Pharmuthi	19	358	Tybi	21	294;22,37,55	10;48,38,31,30
359	Choeac	22	359	Thoth	24	322; 6, 2,51	13;32, 7, 5,30
359	Mesore	30	359	Payni	2	349;49,27,47	16;15,35,39,30
360	Pachon	3	360	Mecheir	5	17;32,52,43	18;59, 4,13,30
361	Tybi	6	361	Phaophi	7	45;16,17,39	21;42,32,47,30
362	Thoth	9	361	Payni	15	72;59,42,35	0;26, 1,21,30
362	Pachon	17	362	Mecheir	18	100;43, 7,31	3; 9,29,55,30
363	Phamenoth	15	363	Choeac	16	133;16,52,22	6;23,59,29,45
364	Hathyr	18	363	Mesore	24	161; 0,17,18	9; 7,28, 3,45
364	Epeiph	26	364	Pharmuthi	27	188;43,42,14	11;50,56,37,45
365	Phamenoth	29	365	Choeac	29	216;27, 7,10	14;34,25,11,45
366	Choeac	2	366	Thoth	2	244;10,32, 6	17;17,53,45,45
366	Mesore	10	366	Pachon	10	271;53,57, 2	20; 1,22,19,45
367	Pharmuthi	13	367	Tybi	13	299;37,21,58	22;44,50,53,45
368	Choeac	16	368	Thoth	16	327;20,46,54	1;28,19,27,45
368	Mesore	24	368	Pachon	24	355; 4,11,50	4;11,48, 1,45
369	Pharmuthi	27	369	Tybi	26	22;47,36,46	6;55,16,35,45
370	Choeac	30	370	Thoth	29	50;31, 1,42	9;38,45, 9,45
371	Thoth	3	370	Payni	7	78;14,26,38	12;22,13,43,45
371	Epeiph	6	371	Pharmuthi	5	110;48,11,29	15;36,43,18, 0
372	Phamenoth	9	372	Choeac	8	138;31,36,25	18;20,11,52, 0
373	Hathyr	12	372	Mesore	16	166;15, 1,21	21; 3,40,26, 0
373	Epeiph	20	373	Pharmuthi	18	193;58,26,17	23;47, 9, 0, 0
374	Phamenoth	23	374	Choeac	21	221;41,51,13	2;30,37,34, 0
375	Hathyr	26	374	Mesore	29	249;25,16, 9	5;14, 6, 8, 0
375	Mesore	4	375	Pachon	2	277; 8,41, 5	7;57,34,42, 0
376	Pharmuthi	7	376	Tybi	5	304;52, 6, 1	10;41, 3,16, 0
377	Choeac	10	377	Thoth	7	332;35,30,57	13;24,31,50, 0
377	Mesore	18	377	Pachon	15	0;18,55,53	16; 8, 0,24, 0
378	Pharmuthi	21	378	Tybi	18	28; 2,20,49	18;51,28,58, 0
379	Choeac	24	379	Thoth	21	55;45,45,45	21;34,57,32, 0
380	Phaophi	22	379	Epeiph	24	88;19,30,36	0;49,27, 6,15
380	Payni	30	380	Phamenoth	27	116; 2,55,32	3;32,55,40,15
381	Phamenoth	3	381	Hathyr	29	143;46,20,28	6;16,24,14,15
382	Hathyr	6	381	Mesore	7	171;29,45,24	8;59,52,48,15
382	Epeiph	14	382	Pharmuthi	10	199;13,10,20	11;43,21,22,15
383	Phamenoth	17	383	Choeac	13	226;56,35,16	14;26,49,56,15
384	Hathyr	20	383	Mesore	21	254;40, 0,12	17;10,18,30,15
384	Epeiph	28	384	Pharmuthi	24	282;23,25, 8	19;53,47, 4,15
385	Pharmuthi	1	385	Choeac	26	310; 6,50, 4	22;37,15,38,15
386	Choeac	4	385	Epagomenae	4	337;50,15, 0	1;20,44,12,15
386	Mesore	12	386	Pachon	7	5;33,39,56	4; 4,12,46,15
387	Pharmuthi	15	387	Tybi	10	33;17, 4,52	6;47,41,20,15
388	Mecheir	13	388	Hathyr	8	65;50,49,43	10; 2,10,54,30
389	Phaophi	16	388	Epeiph	16	93;34,14,39	12;45,39,28,30
389	Payni	24	389	Phamenoth	18	121;17,39,35	15;29, 8, 2,30
390	Mecheir	27	390	Hathyr	21	149; 1, 4,31	18;12,36,36,30
391	Phaophi	30	390	Epeiph	29	176;44,29,27	20;56, 5,10,30
391	Epeiph	8	391	Pharmuthi	2	204;27,54,23	23;39,33,44,30
392	Phamenoth	11	392	Choeac	5	232;11,19,19	2;23, 2,18,30
393	Hathyr	14	392	Mesore	13	259;54,44,15	5; 6,30,52,30

Egyptian			Alexandrian			Longitude	Arg. of Latitude
393	Epeiph	22	393	Pharmuthi	15	287;38, 9,11	7;49,59,26,30
394	Phamenoth	25	394	Choeac	18	315;21,34, 7	10;33,28, 0,30
395	Hathyr	28	394	Mesore	26	343; 4,59, 3	13;16,56,34,30
395	Mesore	6	395	Pharmuthi	29	10;48,23,59	16; 0,25, 8,30
396	Payni	4	396	Mecheir	27	43;22, 8,50	19;14,54,42,45
397	Mecheir	7	397	Phaophi	29	71; 5,33,46	21;58,23,16,45
398	Phaophi	10	397	Epeiph	7	98;48,58,42	0;41,51,50,45
398	Payni	18	398	Phamenoth	10	126;32,23,38	3;25,20,24,45
399	Mecheir	21	399	Hathyr	13	154;15,48,34	6; 8,48,58,45
400	Phaophi	24	399	Epeiph	21	181;59,13,30	8;52,17,32,45
400	Epeiph	2	400	Phamenoth	24	209;42,38,26	11;35,46, 6,45
401	Phamenoth	5	401	Hathyr	26	237;26, 3,22	14;19,14,40,45
402	Hathyr	8	401	Mesore	4	265; 9,28,18	17; 2,43,14,45
402	Epeiph	16	402	Pharmuthi	7	292;52,53,14	19;46,11,48,45
403	Phamenoth	19	403	Choeac	10	320;36,18,10	22;29,40,22,45
404	Hathyr	22	403	Mesore	18	348;19,43, 6	1;13, 8,56,45
405	Thoth	20	404	Payni	16	20;53,27,57	4;27,38,31, 0
405	Pachon	28	405	Mecheir	18	48;36,52,53	7;11, 7, 5, 0
406	Mecheir	1	406	Phaophi	21	76;20,17,49	9;54,35,39, 0
407	Phaophi	4	406	Payni	29	104; 3,42,45	12;38, 4,13, 0
407	Payni	12	407	Phamenoth	2	131;47, 7,41	15;21,32,47, 0
408	Mecheir	15	408	Hathyr	5	159;30,32,37	18; 5, 1,21, 0
409	Phaophi	18	408	Epeiph	13	187;13,57,33	20;48,29,55, 0
409	Payni	26	409	Phamenoth	15	214;57,22,29	23;31,58,29, 0
410	Mecheir	29	410	Hathyr	18	242;40,47,25	2;15,27, 3, 0
411	Hathyr	2	410	Epeiph	26	270;24,12,21	4;58,55,37, 0
411	Epeiph	10	411	Phamenoth	29	298; 7,37,17	7;42,24,11, 0
412	Phamenoth	13	412	Choeac	2	325;51, 2,13	10;25,52,45, 0
413	Tybi	11	413	Thoth	29	358;24,47, 4	13;40,22,19,15
414	Thoth	14	413	Payni	7	26; 8,12, 0	16;23,50,53,15
414	Pachon	22	414	Mecheir	10	53;51,36,56	19; 7,19,27,15
415	Tybi	25	415	Phaophi	13	81;35, 1,52	21;50,48, 1,15
416	Thoth	28	415	Payni	21	109;18,26,48	0;34,16,35,15
416	Payni	6	416	Mecheir	24	137; 1,51,44	3;17,45, 9,15
417	Mecheir	9	417	Phaophi	26	164;45,16,40	6; 1,13,43,15
418	Phaophi	12	417	Epeiph	4	192;28,41,36	8;44,42,17,15
418	Payni	20	418	Phamenoth	7	220;12, 6,32	11;28,10,51,15
419	Mecheir	23	419	Hathyr	10	247;55,31,28	14;11,39,25,15
420	Phaophi	26	419	Epeiph	18	275;38,56,24	16;55, 7,59,15
420	Epeiph	4	420	Phamenoth	21	303;22,21,20	19;38,36,33,15
421	Pachon	2	421	Tybi	18	335;56, 6,11	22;53, 6, 7,30
422	Tybi	5	422	Thoth	21	3;39,31, 7	1;36,34,41,30
423	Thoth	8	422	Pachon	29	31;22,56, 3	4;20, 3,15,30
423	Pachon	16	423	Mecheir	2	59; 6,20,59	7; 3,31,49,30
424	Tybi	19	424	Phaophi	5	86;49,45,55	9;47, 0,23,30
425	Thoth	22	424	Payni	13	114;33,10,51	12;30,28,57,30
425	Pachon	30	425	Mecheir	15	142;16,35,47	15;13,57,31,30
426	Mecheir	3	426	Phaophi	18	170; 0, 0,43	17;57,26, 5,30
427	Phaophi	6	426	Payni	26	197;43,25,39	20;40,54,39,30
427	Payni	14	427	Mecheir	29	225;26,50,35	23;24,23,13,30
428	Mecheir	17	428	Hathyr	2	253;10,15,31	2; 7,51,47,30
429	Phaophi	20	428	Epeiph	10	280;53,40,27	4;51,20,21,30

Egyptian			Alexandrian			Longitude	Arg. of Latitude
429	Mesore	23	429	Pachon	7	313;27,25,18	8; 5,49,55,45
430	Pharmuthi	26	430	Tybi	10	341;10,50,14	10;49,18,29,45
431	Choeac	29	431	Thoth	13	8;54,15,10	13;32,47, 3,45
432	Thoth	2	431	Pachon	21	36;37,40, 6	16;16,15,37,45
432	Pachon	10	432	Tybi	24	64;21, 5, 2	18;59,44,11,45
433	Tybi	13	433	Thoth	26	92; 4,29,58	21;43,12,45,45
434	Thoth	16	433	Payni	4	119;47,54,54	0;26,41,19,45
434	Pachon	24	434	Mecheir	7	147;31,19,50	3;10, 9,53,45
435	Tybi	27	435	Phaophi	10	175;14,44,46	5;53,38,27,45
436	Thoth	30	435	Payni	18	202;58, 9,42	8;37, 7, 1,45
436	Payni	8	436	Mecheir	21	230;41,34,38	11;20,35,35,45
437	Mecheir	11	437	Phaophi	23	258;24,59,34	14; 4, 4, 9,45
438	Choeac	9	437	Mesore	26	290;58,44,25	17;18,33,44, 0
438	Mesore	17	438	Pharmuthi	29	318;42, 9,21	20; 2, 2,18, 0
439	Pharmuthi	20	439	Tybi	2	346;25,34,17	22;45,30,52, 0
440	Choeac	23	440	Thoth	5	14; 8,59,13	1;28,59,26, 0
440	Epagomenae	1	440	Pachon	13	41;52,24, 9	4;12,28, 0, 0
441	Pachon	4	441	Tybi	15	69;35,49, 5	6;55,56,34, 0
442	Tybi	7	442	Thoth	18	97;19,14, 1	9;39,25, 8, 0
443	Thoth	10	442	Pachon	26	125; 2,38,57	12;22,53,42, 0
443	Pachon	18	443	Tybi	29	152;46, 3,53	15; 6,22,16, 0
444	Tybi	21	444	Phaophi	2	180;29,28,49	17;49,50,50, 0
445	Thoth	24	444	Payni	10	208;12,53,45	20;33,19,24, 0
445	Payni	2	445	Mecheir	12	235;56,18,41	23;16,47,58, 0
446	Phamenoth	30	446	Choeac	10	268;30, 3,32	2;31,17,32,15
447	Choeac	3	446	Mesore	18	296;13,28,28	5;14,46, 6,15
447	Mesore	11	447	Pharmuthi	21	323;56,53,24	7;58,14,40,15
448	Pharmuthi	14	448	Choeac	24	351;40,18,20	10;41,43,14,15
449	Choeac	17	448	Epagomenae	2	19;23,43,16	13;25,11,48,15
449	Mesore	25	449	Pachon	4	47; 7, 8,12	16; 8,40,22,15
450	Pharmuthi	28	450	Tybi	7	74;50,33, 8	18;52, 8,56,15
451	Tybi	1	451	Thoth	10	102;33,58, 4	21;35,37,30,15
452	Thoth	4	451	Pachon	18	130;17,23, 0	0;19, 6, 4,15
452	Pachon	12	452	Tybi	21	158; 0,47,56	3; 2,34,38,15
453	Tybi	15	453	Thoth	23	185;44,12,52	5;46, 3,12,15
454	Thoth	18	453	Payni	1	213;27,37,48	8;29,31,46,15
454	Epeiph	21	454	Phamenoth	29	246; 1,22,39	11;44, 1,20,30
455	Phamenoth	24	455	Choeac	2	273;44,47,35	14;27,29,54,30
456	Hathyr	27	455	Mesore	10	301;28,12,31	17;10,58,28,30
456	Mesore	5	456	Pharmuthi	13	329;11,37,27	19;54,27, 2,30
457	Pharmuthi	8	457	Choeac	15	356;55, 2,23	22;37,55,36,30
458	Choeac	11	457	Mesore	23	24;38,27,19	1;21,24,10,30
458	Mesore	19	458	Pharmuthi	26	52;21,52,15	4; 4,52,44,30
459	Pharmuthi	22	459	Choeac	29	80; 5,17,11	6;48,21,18,30
460	Choeac	25	460	Thoth	2	107;48,42, 7	9;31,49,52,30
460	Epagomenae	3	460	Pachon	10	135;32, 7, 3	12;15,18,26,30
461	Pachon	6	461	Tybi	12	163;15,31,59	14;58,47, 0,30
462	Tybi	9	462	Thoth	15	190;58,56,55	17;42,15,34,30
463	Hathyr	7	462	Epeiph	18	223;32,41,46	20;56,45, 8,45
463	Epeiph	15	463	Phamenoth	21	251;16, 6,42	23;40,13,42,45
464	Phamenoth	18	464	Hathyr	24	278;59,31,38	2;23,42,16,45
465	Hathyr	21	464	Mesore	2	306;42,56,34	5; 7,10,50,45

	Egyptian			Alexandrian		Longitude	Arg. of Latitude
465	Epeiph	29	465	Pharmuthi	4	334;26,21,30	7;50,39,24,45
466	Pharmuthi	2	466	Choeac	7	2; 9,46,26	10;34, 7,58,45
467	Choeac	5	466	Mesore	15	29;53,11,22	13;17,36,32,45
467	Mesore	13	467	Pharmuthi	18	57;36,36,18	16; 1, 5, 6,45
468	Pharmuthi	16	468	Choeac	21	85;20, 1,14	18;44,33,40,45
469	Choeac	19	468	Mesore	29	113; 3,26,10	21;28, 2,14,45
469	Mesore	27	469	Pachon	1	140;46,51, 6	0;11,30,48,45
470	Pharmuthi	30	470	Tybi	4	168;30,16, 2	2;54,59,22,45
471	Mecheir	28	471	Hathyr	2	201; 4, 0,53	6; 9,28,57, 0
472	Hathyr	1	471	Epeiph	10	228;47,25,49	8;52,57,31, 0
472	Epeiph	9	472	Phamenoth	13	256;30,50,45	11;36,26, 5, 0
473	Phamenoth	12	473	Hathyr	15	284;14,15,41	14;19,54,39, 0
474	Hathyr	15	473	Epeiph	23	311;57,40,37	17; 3,23,13, 0
474	Epeiph	23	474	Phamenoth	26	339;41, 5,33	19;46,51,47, 0
475	Phamenoth	26	475	Hathyr	29	7;24,30,29	22;30,20,21, 0
476	Hathyr	29	475	Mesore	7	35; 7,55,25	1;13,48,55, 0
476	Mesore	7	476	Pharmuthi	10	62;51,20,21	3;57,17,29, 0
477	Pharmuthi	10	477	Choeac	12	90;34,45,17	6;40,46, 3, 0
478	Choeac	13	477	Mesore	20	118;18,10,13	9;24,14,37, 0
478	Mesore	21	478	Pharmuthi	23	146; 1,35, 9	12; 7,43,11, 0
479	Payni	19	479	Mecheir	21	178;35,20, 0	15;22,12,45,15
480	Mecheir	22	480	Phaophi	24	206;18,44,56	18; 5,41,19,15
481	Phaophi	25	480	Epeiph	2	234; 2, 9,52	20;49, 9,53,15
481	Epeiph	3	481	Phamenoth	4	261;45,34,48	23;32,38,27,15
482	Phamenoth	6	482	Hathyr	7	289;28,59,44	2;16, 7, 1,15
483	Hathyr	9	482	Epeiph	15	317;12,24,40	4;59,35,35,15
483	Epeiph	17	483	Phamenoth	18	344;55,49,36	7;43, 4, 9,15
484	Phamenoth	20	484	Hathyr	21	12;39,14,32	10;26,32,43,15
485	Hathyr	23	484	Epeiph	29	40;22,39,28	13;10, 1,17,15
485	Mesore	1	485	Pharmuthi	1	68; 6, 4,24	15;53,29,51,15
486	Pharmuthi	4	486	Choeac	4	95;49,29,20	18;36,58,25,15
487	Choeac	7	486	Mesore	12	123;32,54,16	21;20,26,59,15
488	Phaophi	5	487	Payni	10	156; 6,39, 7	0;34,56,33,30
488	Payni	13	488	Mecheir	13	183;50, 4, 3	3;18,25, 7,30
489	Mecheir	16	489	Phaophi	15	211;33,28,59	6; 1,53,41,30
490	Phaophi	19	489	Payni	23	239;16,53,55	8;45,22,15,30
490	Payni	27	490	Mecheir	26	267; 0,18,51	11;28,50,49,30
491	Mecheir	30	491	Phaophi	29	294;43,43,47	14;12,19,23,30
492	Hathyr	3	491	Epeiph	7	322;27, 8,43	16;55,47,57,30
492	Epeiph	11	492	Phamenoth	10	350;10,33,39	19;39,16,31,30
493	Phamenoth	14	493	Hathyr	12	17;53,58,35	22;22,45, 5,30
494	Hathyr	17	493	Epeiph	20	45;37,23,31	1; 6,13,39,30
494	Epeiph	25	494	Phamenoth	23	73;20,48,27	3;49,42,13,30
495	Phamenoth	28	495	Hathyr	26	101; 4,13,23	6;33,10,47,30
496	Tybi	26	496	Thoth	24	133;37,58,14	9;47,40,21,45
497	Thoth	29	496	Payni	2	161;21,23,10	12;31, 8,55,45
497	Payni	7	497	Mecheir	4	189; 4,48, 6	15;14,37,29,45
498	Mecheir	10	498	Phaophi	7	216;48,13, 2	17;58, 6, 3,45
499	Phaophi	13	498	Payni	15	244;31,37,58	20;41,34,37,45
499	Payni	21	499	Mecheir	18	272;15, 2,54	23;25, 3,11,45
500	Mecheir	24	500	Phaophi	21	299;58,27,50	2; 8,31,45,45

day	*longitude*	*daily progress*	*arg. of latitude*	*daily progress*
1	11;42,10,37	11;42,10,37	0;47, 1,25,45	0;47, 1,25,45
2	23;37,11,14	11;55, 0,37	1;34,54,11,30	0;47,52,45,45
3	35;45, 1,51	12; 7,50,37	2;23,38,17,15	0;48,44, 5,45
4	48; 5,42,28	12;20,40,37	3;13,13,43, 0	0;49,35,25,45
5	60;39,13, 5	12;33,30,37	4; 3,40,28,45	0;50,26,45,45
6	73;25,33,42	12;46,20,37	4;54,58,34,30	0;51,18, 5,45
7	86;24,44,19	12;59,10,37	5;47, 8, 0,15	0;52, 9,25,45
8	99;36,44,56	13;12, 0,37	6;40, 8,46, 0	0;53, 0,45,45
9	113; 1,35,33	13;24,50,37	7;34, 0,51,45	0;53,52, 5,45
10	126;39,16,10	13;37,40,37	8;28,44,17,30	0;54,43,25,45
11	140;29,46,47	13;50,30,37	9;24,19, 3,15	0;55,34,45,45
12	154;33, 7,24	14; 3,20,37	10;20,45, 9, 0	0;56,26, 5,45
13	168;49,18, 1	14;16,10,37	11;18, 2,34,45	0;57,17,25,45
14	183;18,18,38	14;29, 0,37	12;16,11,20,30	0;58, 8,45,45
15	197;54,26,15	14;36, 7,37	13;14,48,34,15	0;58,37,13,45
16	212;17,43,52	14;23,17,37	14;12,34,28, 0	0;57,45,53,45
17	226;28,11,29	14;10,27,37	15; 9,29, 1,45	0;56,54,33,45
18	240;25,49, 6	13;57,37,37	16; 5,32,15,30	0;56, 3,13,45
19	254;10,36,43	13;44,47,37	17; 0,44, 9,15	0;55,11,53,45
20	267;42,34,20	13;31,57,37	17;55, 4,43, 0	0;54,20,33,45
21	281; 1,41,57	13;19, 7,37	18;48,33,56,45	0;53,29,13,45
22	294; 7,59,34	13; 6,17,37	19;41,11,50,30	0;52,37,53,45
23	307; 1,27,11	12;53,27,37	20;32,58,24,15	0;51,46,33,45
24	319;42, 4,48	12;40,37,37	21;23,53,38, 0	0;50,55,13,45
25	332; 9,52,25	12;27,47,37	22;13,57,31,45	0;50, 3,53,45
26	344;24,50, 2	12;14,57,37	23; 3,10, 5,30	0;49,12,33,45
27	356;26,57,39	12; 2, 7,37	23;51,31,19,15	0;48,21,13,45
28	8;16,15,16	11;49,17,37	0;39, 1,13, 0	0;47,29,53,45
29	20; 4, 8,53	11;47,53,37	1;26,25,30,45	0;47,24,17,45
30	32; 4,52,30	12; 0,43,37	2;14,41, 8,30	0;48,15,37,45
31	44;18,26, 7	12;13,33,37	3; 3,48, 6,15	0;49, 6,57,45
32	56;44,49,44	12;26,23,37	3;53,46,24, 0	0;49,58,17,45
33	69;24, 3,21	12;39,13,37	4;44,36, 1,45	0;50,49,37,45
34	82;16, 6,58	12;52, 3,37	5;36,16,59,30	0;51,40,57,45
35	95;21, 0,35	13; 4,53,37	6;28,49,17,15	0;52,32,17,45
36	108;38,44,12	13;17,43,37	7;22,12,55, 0	0;53,23,37,45
37	122; 9,17,49	13;30,33,37	8;16,27,52,45	0;54,14,57,45
38	135;52,41,26	13;43,23,37	9;11,34,10,30	0;55, 6,17,45
39	149;48,55, 3	13;56,13,37	10; 7,31,48,15	0;55,57,37,45
40	163;57,58,40	14; 9, 3,37	11; 4,20,46, 0	0;56,48,57,45
41	178;19,52,17	14;21,53,37	12; 2, 1, 3,45	0;57,40,17,45
42	192;54,35,54	14;34,43,37	13; 0,32,41,30	0;58,31,37,45
43	207;25, 0,31	14;30,24,37	13;58,47, 3,15	0;58,14,21,45
44	221;42,35, 8	14;17,34,37	14;56,10, 5, 0	0;57,23, 1,45
45	235;47,19,45	14; 4,44,37	15;52,41,46,45	0;56,31,41,45
46	249;39,14,22	13;51,54,37	16;48,22, 8,30	0;55,40,21,45
47	263;18,18,59	13;39, 4,37	17;43,11,10,15	0;54,49, 1,45
48	276;44,33,36	13;26,14,37	18;37, 8,52, 0	0;53,57,41,45
49	289;57,58,13	13;13,24,37	19;30,15,13,45	0;53, 6,21,45
50	302;58,32,50	13; 0,34,37	20;22,30,15,30	0;52,15, 1,45
51	315;46,17,27	12;47,44,37	21;13,53,57,15	0;51,23,41,45
52	328;21,12, 4	12;34,54,37	22; 4,26,19, 0	0;50,32,21,45
53	340;43,16,41	12;22, 4,37	22;54, 7,20,45	0;49,41, 1,45
54	352;52,31,18	12; 9,14,37	23;42,57, 2,30	0;48,49,41,45
55	4;48,55,55	11;56,24,37	0;30,55,24,15	0;47,58,21,45
56	16;32,30,32	11;43,34,37	1;18, 2,26, 0	0;47, 7, 1,45

day	*longitude*	*daily progress*	*arg. of latitude*	*daily progress*
57	28;26, 7, 9	11;53,36,37	2; 5,49,35,45	0;47,47, 9,45
58	40;32,33,46	12; 6,26,37	2;54,28, 5,30	0;48,38,29,45
59	52;51,50,23	12;19,16,37	3;43,57,55,15	0;49,29,49,45
60	65;23,57, 0	12;32, 6,37	4;34,19, 5, 0	0;50,21, 9,45
61	78; 8,53,37	12;44,56,37	5;25,31,34,45	0;51,12,29,45
62	91; 6,40,14	12;57,46,37	6;17,35,24,30	0;52, 3,49,45
63	104;17,16,51	13;10,36,37	7;10,30,34,15	0;52,55, 9,45
64	117;40,43,28	13;23,26,37	8; 4,17, 4, 0	0;53,46,29,45
65	131;17, 0, 5	13;36,16,37	8;58,54,53,45	0;54,37,49,45
66	145; 6, 6,42	13;49, 6,37	9;54,24, 3,30	0;55,29, 9,45
67	159; 8, 3,19	14; 1,56,37	10;50,44,33,15	0;56,20,29,45
68	173;22,49,56	14;14,46,37	11;47,56,23, 0	0;57,11,49,45
69	187;50,26,33	14;27,36,37	12;45,59,32,45	0;58, 3, 9,45
70	202;27,58,10	14;37,31,37	13;44,42,22,30	0;58,42,49,45
71	216;52,39,47	14;24,41,37	14;42,33,52,15	0;57,51,29,45
72	231; 4,31,24	14;11,51,37	15;39,34, 2, 0	0;57, 0, 9,45
73	245; 3,33, 1	13;59, 1,37	16;35,42,51,45	0;56, 8,49,45
74	258;49,44,38	13;46,11,37	17;31, 0,21,30	0;55,17,29,45
75	272;23, 6,15	13;33,21,37	18;25,26,31,15	0;54,26, 9,45
76	285;43,37,52	13;20,31,37	19;19, 1,21, 0	0;53,34,49,45
77	298;51,19,29	13; 7,41,37	20;11,44,50,45	0;52,43,29,45
78	311;46,11, 6	12;54,51,37	21; 3,37, 0,30	0;51,52, 9,45
79	324;28,12,43	12;42, 1,37	21;54,37,50,15	0;51, 0,49,45
80	336;57,24,20	12;29,11,37	22;44,47,20, 0	0;50, 9,29,45
81	349;13,45,57	12;16,21,37	23;34, 5,29,45	0;49,18, 9,45
82	1;17,17,34	12; 3,31,37	0;22,32,19,30	0;48,26,49,45
83	13; 7,59,11	11;50,41,37	1;10, 7,49,15	0;47,35,29,45
84	24;54,28,48	11;46,29,37	1;57,26,31, 0	0;47,18,41,45
85	36;53,48,25	11;59,19,37	2;45,36,32,45	0;48,10, 1,45
86	49; 5,58, 2	12;12, 9,37	3;34,37,54,30	0;49, 1,21,45
87	61;30,57,39	12;24,59,37	4;24,30,36,15	0;49,52,41,45
88	74; 8,47,16	12;37,49,37	5;15,14,38, 0	0;50,44, 1,45
89	86;59,26,53	12;50,39,37	6; 6,49,59,45	0;51,35,21,45
90	100; 2,56,30	13; 3,29,37	6;59,16,41,30	0;52,26,41,45
91	113;19,16, 7	13;16,19,37	7;52,34,43,15	0;53,18, 1,45
92	126;48,25,44	13;29, 9,37	8;46,44, 5, 0	0;54, 9,21,45
93	140;30,25,21	13;41,59,37	9;41,44,46,45	0;55, 0,41,45
94	154;25,14,58	13;54,49,37	10;37,36,48,30	0;55,52, 1,45
95	168;32,54,35	14; 7,39,37	11;34,20,10,15	0;56,43,21,45
96	182;53,24,12	14;20,29,37	12;31,54,52, 0	0;57,34,41,45
97	197;26,43,49	14;33,19,37	13;30,20,53,45	0;58,26, 1,45
98	211;58,32,26	14;31,48,37	14;28,40,51,30	0;58,19,57,45
99	226;17,31, 3	14;18,58,37	15;26, 9,29,15	0;57,28,37,45
100	240;23,39,40	14; 6, 8,37	16;22,46,47, 0	0;56,37,17,45
101	254;16,58,17	13;53,18,37	17;18,32,44,45	0;55,45,57,45
102	267;57,26,54	13;40,28,37	18;13,27,22,30	0;54,54,37,45
103	281;25, 5,31	13;27,38,37	19; 7,30,40,15	0;54, 3,17,45
104	294;39,54, 8	13;14,48,37	20; 0,42,38, 0	0;53,11,57,45
105	307;41,52,45	13; 1,58,37	20;53, 3,15,45	0;52,20,37,45
106	320;31, 1,22	12;49, 8,37	21;44,32,33,30	0;51,29,17,45
107	333; 7,19,59	12;36,18,37	22;35,10,31,15	0;50,37,57,45
108	345;30,48,36	12;23,28,37	23;24,57, 9, 0	0;49,46,37,45
109	357;41,27,13	12;10,38,37	0;13,52,26,45	0;48,55,17,45
110	9;39,15,50	11;57,48,37	1; 1,56,24,30	0;48, 3,57,45
111	21;24,14,27	11;44,58,37	1;49, 9, 2,15	0;47,12,37,45

day	*longitude*	*daily progress*	*arg. of latitude*	*daily progress*
112	33;16,27, 4	11;52,12,37	2;36,50,36, 0	0;47,41,33,45
113	45;21,29,41	12; 5, 2,37	3;25,23,29,45	0;48,32,53,45
114	57;39,22,18	12;17,52,37	4;14,47,43,30	0;49,24,13,45
115	70;10, 4,55	12;30,42,37	5; 5, 3,17,15	0;50,15,33,45
116	82;53,37,32	12;43,32,37	5;56,10,11, 0	0;51, 6,53,45
117	95;50, 0, 9	12;56,22,37	6;48, 8,24,45	0;51,58,13,45
118	108;59,12,46	13; 9,12,37	7;40,57,58,30	0;52,49,33,45
119	122;21,15,23	13;22, 2,37	8;34,38,52,15	0;53,40,53,45
120	135;56, 8, 0	13;34,52,37	9;29,11, 6, 0	0;54,32,13,45
121	149;43,50,37	13;47,42,37	10;24,34,39,45	0;55,23,33,45
122	163;44,23,14	14; 0,32,37	11;20,49,33,30	0;56,14,53,45
123	177;57,45,51	14;13,22,37	12;17,55,47,15	0;57, 6,13,45
124	192;23,58,28	14;26,12,37	13;15,53,21, 0	0;57,57,33,45
125	207; 2,54, 5	14;38,55,37	14;14,41,46,45	0;58,48,25,45
126	221;28,59,42	14;26, 5,37	15;12,38,52,30	0;57,57, 5,45
127	235;42,15,19	14;13,15,37	16; 9,44,38,15	0;57, 5,45,45
128	249;42,40,56	14; 0,25,37	17; 5,59, 4, 0	0;56,14,25,45
129	263;30,16,33	13;47,35,37	18; 1,22, 9,45	0;55,23, 5,45
130	277; 5, 2,10	13;34,45,37	18;55,53,55,30	0;54,31,45,45
131	290;26,57,47	13;21,55,37	19;49,34,21,15	0;53,40,25,45
132	303;36, 3,24	13; 9, 5,37	20;42,23,27, 0	0;52,49, 5,45
133	316;32,19, 1	12;56,15,37	21;34,21,12,45	0;51,57,45,45
134	329;15,44,38	12;43,25,37	22;25,27,38,30	0;51, 6,25,45
135	341;46,20,15	12;30,35,37	23;15,42,44,15	0;50,15, 5,45
136	354; 4, 5,52	12;17,45,37	0; 5, 6,30, 0	0;49,23,45,45
137	6; 9, 1,29	12; 4,55,37	0;53,38,55,45	0;48,32,25,45
138	18; 1, 7, 6	11;52, 5,37	1;41,20, 1,30	0;47,41, 5,45
139	29;46,12,43	11;45, 5,37	2;28,33, 7,15	0;47,13, 5,45
140	41;44, 8,20	11;57,55,37	3;16,37,33, 0	0;48, 4,25,45
141	53;54,53,57	12;10,45,37	4; 5,33,18,45	0;48,55,45,45
142	66;18,29,34	12;23,35,37	4;55,20,24,30	0;49,47, 5,45
143	78;54,55,11	12;36,25,37	5;45,58,50,15	0;50,38,25,45
144	91;44,10,48	12;49,15,37	6;37,28,36, 0	0;51,29,45,45
145	104;46,16,25	13; 2, 5,37	7;29,49,41,45	0;52,21, 5,45
146	118; 1,12, 2	13;14,55,37	8;23, 2, 7,30	0;53,12,25,45
147	131;28,57,39	13;27,45,37	9;17, 5,53,15	0;54, 3,45,45
148	145; 9,33,16	13;40,35,37	10;12, 0,59, 0	0;54,55, 5,45
149	159; 2,58,53	13;53,25,37	11; 7,47,24,45	0;55,46,25,45
150	173; 9,14,30	14; 6,15,37	12; 4,25,10,30	0;56,37,45,45
151	187;28,20, 7	14;19, 5,37	13; 1,54,16,15	0;57,29, 5,45
152	202; 0,15,44	14;31,55,37	14; 0,14,42, 0	0;58,20,25,45
153	216;33,28,21	14;33,12,37	14;58,40,15,45	0;58,25,33,45
154	230;53,50,58	14;20,22,37	15;56,14,29,30	0;57,34,13,45
155	245; 1,23,35	14; 7,32,37	16;52,57,23,15	0;56,42,53,45
156	258;56, 6,12	13;54,42,37	17;48,48,57, 0	0;55,51,33,45
157	272;37,58,49	13;41,52,37	18;43,49,10,45	0;55, 0,13,45
158	286; 7, 1,26	13;29, 2,37	19;37,58, 4,30	0;54, 8,53,45
159	299;23,14, 3	13;16,12,37	20;31,15,38,15	0;53,17,33,45
160	312;26,36,40	13; 3,22,37	21;23,41,52, 0	0;52,26,13,45
161	325;17, 9,17	12;50,32,37	22;15,16,45,45	0;51,34,53,45
162	337;54,51,54	12;37,42,37	23; 6, 0,19,30	0;50,43,33,45
163	350;19,44,31	12;24,52,37	23;55,52,33,15	0;49,52,13,45
164	2;31,47, 8	12;12, 2,37	0;44,53,27, 0	0;49, 0,53,45
165	14;30,59,45	11;59,12,37	1;33, 3, 0,45	0;48, 9,33,45
166	26;17,22,22	11;46,22,37	2;20,21,14,30	0;47,18,13,45

day	longitude	daily progress	arg. of latitude	daily progress
167	38; 8,10,59	11;50,48,37	3; 7,57,12,15	0;47,35,57,45
168	50;11,49,36	12; 3,38,37	3;56,24,30, 0	0;48,27,17,45
169	62;28,18,13	12;16,28,37	4;45,43, 7,45	0;49,18,37,45
170	74;57,36,50	12;29,18,37	5;35,53, 5,30	0;50, 9,57,45
171	87;39,45,27	12;42, 8,37	6;26,54,23,15	0;51, 1,17,45
172	100;34,44, 4	12;54,58,37	7;18,47, 1, 0	0;51,52,37,45
173	113;42,32,41	13; 7,48,37	8;11,30,58,45	0;52,43,57,45
174	127; 3,11,18	13;20,38,37	9; 5, 6,16,30	0;53,35,17,45
175	140;36,39,55	13;33,28,37	9;59,32,54,15	0;54,26,37,45
176	154;22,58,32	13;46,18,37	10;54,50,52, 0	0;55,17,57,45
177	168;22, 7, 9	13;59, 8,37	11;51, 0, 9,45	0;56, 9,17,45
178	182;34, 5,46	14;11,58,37	12;48, 0,47,30	0;57, 0,37,45
179	196;58,54,23	14;24,48,37	13;45,52,45,15	0;57,51,57,45
180	211;36,33, 0	14;37,38,37	14;44,36, 3, 0	0;58,43,17,45
181	226; 4, 2,37	14;27,29,37	15;42,38,44,45	0;58, 2,41,45
182	240;18,42,14	14;14,39,37	16;39,50, 6,30	0;57,11,21,45
183	254;20,31,51	14; 1,49,37	17;36,10, 8,15	0;56,20, 1,45
184	268; 9,31,28	13;48,59,37	18;31,38,50, 0	0;55,28,41,45
185	281;45,41, 5	13;36, 9,37	19;26,16,11,45	0;54,37,21,45
186	295; 9, 0,42	13;23,19,37	20;20, 2,13,30	0;53,46, 1,45
187	308;19,30,19	13;10,29,37	21;12,56,55,15	0;52,54,41,45
188	321;17, 9,56	12;57,39,37	22; 5, 0,17, 0	0;52, 3,21,45
189	334; 1,59,33	12;44,49,37	22;56,12,18,45	0;51,12, 1,45
190	346;33,59,10	12;31,59,37	23;46,33, 0,30	0;50,20,41,45
191	358;53, 8,47	12;19, 9,37	0;36, 2,22,15	0;49,29,21,45
192	10;59,28,24	12; 6,19,37	1;24,40,24, 0	0;48,38, 1,45
193	22;52,58, 1	11;53,29,37	2;12,27, 5,45	0;47,46,41,45
194	34;36,39,38	11;43,41,37	2;59,34,35,30	0;47, 7,29,45
195	46;33,11,15	11;56,31,37	3;47,33,25,15	0;47,58,49,45
196	58;42,32,52	12; 9,21,37	4;36,23,35, 0	0;48,50, 9,45
197	71; 4,44,29	12;22,11,37	5;26, 5, 4,45	0;49,41,29,45
198	83;39,46, 6	12;35, 1,37	6;16,37,54,30	0;50,32,49,45
199	96;27,37,43	12;47,51,37	7; 8, 2, 4,15	0;51,24, 9,45
200	109;28,19,20	13; 0,41,37	8; 0,17,34, 0	0;52,15,29,45
201	122;41,50,57	13;13,31,37	8;53,24,23,45	0;53, 6,49,45
202	136; 8,12,34	13;26,21,37	9;47,22,33,30	0;53,58, 9,45
203	149;47,24,11	13;39,11,37	10;42,12, 3,15	0;54,49,29,45
204	163;39,25,48	13;52, 1,37	11;37,52,53, 0	0;55,40,49,45
205	177;44,17,25	14; 4,51,37	12;34,25, 2,45	0;56,32, 9,45
206	192; 1,59, 2	14;17,41,37	13;31,48,32,30	0;57,23,29,45
207	206;32,30,39	14;30,31,37	14;30, 3,22,15	0;58,14,49,45
208	221; 7, 7,16	14;34,36,37	15;28,34,32, 0	0;58,31, 9,45
209	235;28,53,53	14;21,46,37	16;26,14,21,45	0;57,39,49,45
210	249;37,50,30	14; 8,56,37	17;23, 2,51,30	0;56,48,29,45
211	263;33,57, 7	13;56, 6,37	18;19, 0, 1,15	0;55,57, 9,45
212	277;17,13,44	13;43,16,37	19;14, 5,51, 0	0;55, 5,49,45
213	290;47,40,21	13;30,26,37	20; 8,20,20,45	0;54,14,29,45
214	304; 5,16,58	13;17,36,37	21; 1,43,30,30	0;53,23, 9,45
215	317;10, 3,35	13; 4,46,37	21;54,15,20,15	0;52,31,49,45
216	330; 2, 0,12	12;51,56,37	22;45,55,50, 0	0;51,40,29,45
217	342;41, 6,49	12;39, 6,37	23;36,44,59,45	0;50,49, 9,45
218	355; 7,23,26	12;26,16,37	0;26,42,49,30	0;49,57,49,45
219	7;20,50, 3	12;13,26,37	1;15,49,19,15	0;49, 6,29,45
220	19;21,26,40	12; 0,36,37	2; 4, 4,29, 0	0;48,15, 9,45
221	31; 9,13,17	11;47,46,37	2;51,28,18,45	0;47,23,49,45

day	*longitude*	*daily progress*	*arg. of latitude*	*daily progress*
222	42;58,37,54	11;49,24,37	3;38,58,40,30	0;47,30,21,45
223	55; 0,52,31	12; 2,14,37	4;27,20,22,15	0;48,21,41,45
224	67;15,57, 8	12;15, 4,37	5;16,33,24, 0	0;49,13, 1,45
225	79;43,51,45	12;27,54,37	6; 6,37,45,45	0;50, 4,21,45
226	92;24,36,22	12;40,44,37	6;57,33,27,30	0;50,55,41,45
227	105;18,10,59	12;53,34,37	7;49,20,29,15	0;51,47, 1,45
228	118;24,35,36	13; 6,24,37	8;41,58,51, 0	0;52,38,21,45
229	131;43,50,13	13;19,14,37	9;35,28,32,45	0;53,29,41,45
230	145;15,54,50	13;32, 4,37	10;29,49,34,30	0;54,21, 1,45
231	159; 0,49,27	13;44,54,37	11;25, 1,56,15	0;55,12,21,45
232	172;58,34, 4	13;57,44,37	12;21, 5,38, 0	0;56, 3,41,45
233	187; 9, 8,41	14;10,34,37	13;18, 0,39,45	0;56,55, 1,45
234	201;32,33,18	14;23,24,37	14;15,47, 1,30	0;57,46,21,45
235	216; 8,47,55	14;36,14,37	15;14,24,43,15	0;58,37,41,45
236	230;37,41,32	14;28,53,37	16;12,33, 1, 0	0;58, 8,17,45
237	244;53,45, 9	14;16, 3,37	17; 9,49,58,45	0;57,16,57,45
238	258;56,58,46	14; 3,13,37	18; 6,15,36,30	0;56,25,37,45
239	272;47,22,23	13;50,23,37	19; 1,49,54,15	0;55,34,17,45
240	286;24,56, 0	13;37,33,37	19;56,32,52, 0	0;54,42,57,45
241	299;49,39,37	13;24,43,37	20;50,24,29,45	0;53,51,37,45
242	313; 1,33,14	13;11,53,37	21;43,24,47,30	0;53, 0,17,45
243	326; 0,36,51	12;59, 3,37	22;35,33,45,15	0;52, 8,57,45
244	338;46,50,28	12;46,13,37	23;26,51,23, 0	0;51,17,37,45
245	351;20,14, 5	12;33,23,37	0;17,17,40,45	0;50,26,17,45
246	3;40,47,42	12;20,33,37	1; 6,52,38,30	0;49,34,57,45
247	15;48,31,19	12; 7,43,37	1;55,36,16,15	0;48,43,37,45
248	27;43,24,56	11;54,53,37	2;43,28,34, 0	0;47,52,17,45
249	39;26,59,33	11;43,34,37	3;30,35,35,45	0;47, 7, 1,45
250	51;23,24,10	11;56,24,37	4;18,33,57,30	0;47,58,21,45
251	63;32,38,47	12; 9,14,37	5; 7,23,39,15	0;48,49,41,45
252	75;54,43,24	12;22, 4,37	5;57, 4,41, 0	0;49,41, 1,45
253	88;29,38, 1	12;34,54,37	6;47,37, 2,45	0;50,32,21,45
254	101;17,22,38	12;47,44,37	7;39, 0,44,30	0;51,23,41,45
255	114;17,57,15	13; 0,34,37	8;31,15,46,15	0;52,15, 1,45
256	127;31,21,52	13;13,24,37	9;24,22, 8, 0	0;53, 6,21,45
257	140;57,36,29	13;26,14,37	10;18,19,49,45	0;53,57,41,45
258	154;36,41, 6	13;39, 4,37	11;13, 8,51,30	0;54,49, 1,45
259	168;28,35,43	13;51,54,37	12; 8,49,13,15	0;55,40,21,45
260	182;33,20,20	14; 4,44,37	13; 5,20,55, 0	0;56,31,41,45
261	196;50,54,57	14;17,34,37	14; 2,43,56,45	0;57,23, 1,45
262	211;21,19,34	14;30,24,37	15; 0,58,18,30	0;58,14,21,45
263	225;56, 3,11	14;34,43,37	15;59,29,56,15	0;58,31,37,45
264	240;17,56,48	14;21,53,37	16;57,10,14, 0	0;57,40,17,45
265	254;27, 0,25	14; 9, 3,37	17;53,59,11,45	0;56,48,57,45
266	268;23,14, 2	13;56,13,37	18;49,56,49,30	0;55,57,37,45
267	282; 6,37,39	13;43,23,37	19;45, 3, 7,15	0;55, 6,17,45
268	295;37,11,16	13;30,33,37	20;39,18, 5, 0	0;54,14,57,45
269	308;54,54,53	13;17,43,37	21;32,41,42,45	0;53,23,37,45
270	321;59,48,30	13; 4,53,37	22;25,14, 0,30	0;52,32,17,45
271	334;51,52, 7	12;52, 3,37	23;16,54,58,15	0;51,40,57,45
272	347;31, 5,44	12;39,13,37	0; 7,44,36, 0	0;50,49,37,45
273	359;57,29,21	12;26,23,37	0;57,42,53,45	0;49,58,17,45
274	12;11, 2,58	12;13,33,37	1;46,49,51,30	0;49, 6,57,45
275	24;11,46,35	12; 0,43,37	2;35, 5,29,15	0;48,15,37,45
276	35;59,40,12	11;47,53,37	3;22,29,47, 0	0;47,24,17,45

day	longitude	daily progress	arg. of latitude	daily progress
277	47;48,57,49	11;49,17,37	4; 9,59,40,45	0;47,29,53,45
278	59;51, 5,26	12; 2, 7,37	4;58,20,54,30	0;48,21,13,45
279	72; 6, 3, 3	12;14,57,37	5;47,33,28,15	0;49,12,33,45
280	84;33,50,40	12;27,47,37	6;37,37,22, 0	0;50, 3,53,45
281	97;14,28,17	12;40,37,37	7;28,32,35,45	0;50,55,13,45
282	110; 7,55,54	12;53,27,37	8;20,19, 9,30	0;51,46,33,45
283	123;14,13,31	13; 6,17,37	9;12,57, 3,15	0;52,37,53,45
284	136;33,21, 8	13;19, 7,37	10; 6,26,17, 0	0;53,29,13,45
285	150; 5,18,45	13;31,57,37	11; 0,46,50,45	0;54,20,33,45
286	163;50, 6,22	13;44,47,37	11;55,58,44,30	0;55,11,53,45
287	177;47,43,59	13;57,37,37	12;52, 1,58,15	0;56, 3,13,45
288	191;58,11,36	14;10,27,37	13;48,56,32, 0	0;56,54,33,45
289	206;21,29,13	14;23,17,37	14;46,42,25,45	0;57,45,53,45
290	220;57,36,50	14;36, 7,37	15;45,19,39,30	0;58,37,13,45
291	235;26,37,27	14;29, 0,37	16;43,28,25,15	0;58, 8,45,45
292	249;42,48, 4	14;16,10,37	17;40,45,51, 0	0;57,17,25,45
293	263;46, 8,41	14; 3,20,37	18;37,11,56,45	0;56,26, 5,45
294	277;36,39,18	13;50,30,37	19;32,46,42,30	0;55,34,45,45
295	291;14,19,55	13;37,40,37	20;27,30, 8,15	0;54,43,25,45
296	304;39,10,32	13;24,50,37	21;21,22,14, 0	0;53,52, 5,45
297	317;51,11, 9	13;12, 0,37	22;14,22,59,45	0;53, 0,45,45
298	330;50,21,46	12;59,10,37	23; 6,32,25,30	0;52, 9,25,45
299	343;36,42,23	12;46,20,37	23;57,50,31,15	0;51,18, 5,45
300	356;10,13, 0	12;33,30,37	0;48,17,17, 0	0;50,26,45,45
301	8;30,53,37	12;20,40,37	1;37,52,42,45	0;49,35,25,45
302	20;38,44,14	12; 7,50,37	2;26,36,48,30	0;48,44, 5,45
303	32;33,44,51	11;55, 0,37	3;14,29,34,15	0;47,52,45,45

I. Conversion of Longitudes.

Computed longitudes in modern astronomical tables such as Tuckerman's are tropical, implying a tropical year of approximately 365;14,32 days. Ptolemy's tables were also intended to generate tropical longitudes, but were based on a longer estimate of the tropical year, 365;14,48 days. However, most longitudes in the papyri are sidereal, and the schemes they are based on assume a sidereal year length somewhere between 365;15,30 and 365;15,40 days (the modern value is approximately 365;15,23 days, which is incidentally very close to Ptolemy's sidereal year).

In his Shorter Commentary on the *Handy Tables* (chapter 12, ed. Tihon pp. 236–237), Theon of Alexandria sets out a formula used by 'the ancient astrologers' (οἱ παλαιοὶ τῶν ἀποτελεσματικῶν) to convert Ptolemy's tropical longitudes to sidereal. Given a Ptolemaic tropical longitude λ_t computed for year x in the era Augustus, the sidereal longitude λ_s is determined thus:

$$\lambda_s = \lambda_t + 8° - \left(\frac{x+128}{80}\right)° \tag{1}$$

Theon's formula, which implies a sidereal year close to the value of 365;15,33,46 days attested in the solar epoch table **4148**, was actually used to obtain the longitudes in many almanacs of the third and fourth century.

Similar approximative formulas can be used for conversion of tropical longitudes computed by modern theory to the ancient frames of reference. Given a modern tropical longitude λ_m, these corrections produce rough Ptolemaic and sidereal equivalents:

$$\lambda_t = \lambda_m - 0;10° - \frac{x}{230}° \tag{2}$$

$$\lambda_s = \lambda_m + 6;15° - \frac{x}{60}° \tag{3}$$

J. Zodiacal Signs of Saturn and Jupiter.

The table below lists years, within the range from –49 (i.e. 50 B.C.) to A.D. 550, during which Saturn and Jupiter occupied each possible pair of zodiacal signs. It provides an index to Tuckerman's tables for the sake of dating horoscopes and other sets of simultaneous planetary positions.

The longitudes, which are sidereal, were computed using the *Handy Tables* and Theon's formula (cf. Appendix I). Because of variations in the ancient methods of computation, the longitude of one or both planets for a given date according to the *Handy Tables* can sometimes fall one zodiacal sign before or after the sign occupied according to another ancient method of computation. If the years listed under the attested signs do not yield an acceptable date for a text, one should check for years in which either Jupiter or Saturn was close to the end of the preceding sign, or close to the beginning of the following one.

Saturn in Aries:

Jupiter	
Aries	-5/-4, 54/55, 113/114, 172/174, 232/233, 291/292, 350/351
Taurus	-5/-4, 55/56, 114/115, 173/174, 233
Gemini	-4/-3, 55/56, 436, 495/496
Cancer	319, 378, 436/437, 496/497
Leo	142, 201, 260, 319/320, 378/379, 437/438, 496/498
Virgo	-35, 24, 83, 142/143, 201/202, 260/261, 319/321, 379/380, 438/439, 497/498
Libra	-35/-34, 24/25, 83/84, 142/144, 202/203, 261/262, 320/321
Scorpio	-34/-33, 25/26, 84/85, 143/145, 203
Sagittarius	-33/-32, 26/27, 85/86, 145, 466, 525
Capricorn	-32, 289, 348, 407/408, 466, 467, 525/526
Aquarius	171, 230/231, 289/290, 349/350, 408/409, 467/468, 526/527
Pisces	53/54, 112/113, 172/173, 231/232, 290/291, 350/351, 409/410, 468/469, 528

Saturn in Taurus:

Aries	232, 291/292, 350/351, 410/411, 469/470, 528/529
Taurus	115, 174, 233/234, 292/293, 351/352, 411/412, 470/471, 529/530
Gemini	-3, 56/57, 115/116, 174/175, 233/235, 293/294, 352/353, 411/412
Cancer	-3/-2, 56/58, 116/117, 175/176, 234/235
Leo	-2/-1, 57/58, 498
Virgo	321, 380, 439, 498/499
Libra	144, 203, 262, 321/322, 380/381, 439/440, 498/499
Scorpio	-33, 26, 85, 144/145, 203/204, 262/263, 321/322, 381/382, 440/441, 499/500
Sagittarius	-33/-32, 27, 86, 145/146, 204/205, 263/264, 323
Capricorn	-32/-31, 27/28, 86/87, 146/147, 205/206, 265
Aquarius	-31/-30, 28/29, 88, 147, 527
Pisces	-30, 350, 409, 468/469, 527/528

Saturn in Gemini:

Aries	-29/-28, 30/31, 89/90, 149, 208
Taurus	-28, 31, 470/471, 529/530
Gemini	293/294, 352/353, 411/413, 471/472, 530/531
Cancer	117, 176, 235/236, 294/295, 353/354, 412/414, 472/473, 531/532
Leo	-1, 58/59, 117/118, 176/177, 235/237, 295/296, 354/355, 413/414
Virgo	-1/0, 58/60, 118/119, 177/178, 236/237
Libra	0/1, 59/60, 500
Scorpio	2, 323, 382, 441, 500
Sagittarius	205, 264, 323/324, 382/383, 441/442, 501/502
Capricorn	28, 87, 146, 205/206, 264/265, 324/325, 383/384, 442/443, 502
Aquarius	-31, 29, 88/89, 147/148, 206/207, 266, 325/326, 384/385
Pisces	-30/-29, 29/30, 89/90, 148/149, 207/208, 267, 326

Saturn in Cancer:

Jupiter	
Aries	30, 89/90, 149/150, 208/209, 267/268, 327/328, 386/387, 445/446, 505
Taurus	-28/-27, 31/32, 90/91, 150/151, 209/210, 268/269, 327/328, 387
Gemini	-28/-26, 32/33, 91/92, 150/151, 210
Cancer	-27/-26, 33, 472/473, 531/532
Leo	296, 355, 413/414, 473/474, 532/533, 591/592
Virgo	60, 119, 178, 237/238, 296/297, 355/356, 414/415, 473/475, 533/534
Libra	1, 60/61, 119/120, 178/179, 237/239, 297/298, 356/357, 415/416, 475
Scorpio	1/2, 60/62, 120/121, 179/180, 238/239
Sagittarius	2/3, 62, 122
Capricorn	4, 63, 122, 384, 443, 502/503
Aquarius	266/267, 325/326, 384/385, 443/444, 503/504
Pisces	89, 148, 207, 267/268, 326/327, 385/386, 444/445, 504

Saturn in Leo:

Aries	6, 327/328, 386/387, 445/446, 505/506
Taurus	150, 209, 268/269, 327/329, 387/388, 446/447, 505/507
Gemini	150/152, 210/211, 269/270, 328/330, 388/389, 447/448, 506/507
Cancer	-26/-25, 33/34, 92/93, 151/153, 211/212, 270/271, 329/330, 389, 448
Leo	-26/-25, -24, 34/35, 93/94, 152/153, 212, 271
Virgo	-25/-23, 35, 475, 533/534
Libra	298, 357, 416, 475/476, 534/535
Scorpio	121, 180, 239, 298/299, 357/358, 416/417, 476/477, 535/536
Sagittarius	3, 62/63, 121/122, 180/181, 240/241, 299/300, 358/359, 417/418, 477
Capricorn	3/4, 63/64, 122/123, 181/182, 241, 300, 360
Aquarius	5/6, 64/65, 123/124, 183, 242, 301
Pisces	6, 65, 124, 183, 445, 504/505

Saturn in Virgo:

Aries	6/7, 66/67, 125/126, 184/185, 244, 303, 362
Taurus	7/8, 66/67, 126, 185, 244, 507
Gemini	8, 67, 388/389, 447/448, 506/507
Cancer	330, 389/390, 448/449, 507/508
Leo	153/154, 212/213, 271/272, 330/331, 389/391, 449/450, 508/509
Virgo	-24/-23, 35/36, 94/95, 153/155, 213/214, 272/273, 331/332, 390/391, 450
Libra	-24/-22, 36/37, 95/96, 154/156, 214/215, 273/274, 333
Scorpio	-23/-21, 37/38, 96/97, 156, 477, 536
Sagittarius	-21, 300, 359, 418/419, 477/478, 536/537
Capricorn	123, 182, 241/242, 300/301, 359/360, 419/420, 478/479, 537/538
Aquarius	5, 64/65, 123/124, 182/183, 242/243, 301/302, 360/361, 420, 479, 539
Pisces	6/7, 65/66, 124/125, 183/184, 243/244, 302/303, 361/362, 421, 480, 539

Saturn in Libra:

Jupiter	
Aries	244/245, 303/304, 362/363, 422/423, 481/482, 540/541
Taurus	67/68, 126/127, 185/186, 244/246, 304/305, 363/364, 422/423, 482, 541
Gemini	8/9, 67/69, 127/128, 186/187, 245/246, 305, 364/365, 423/424
Cancer	-49, 9/10, 68/70, 128/129, 187/188, 246/247, 305/306, 365
Leo	-49/-48, 10/11, 69/70, 128/129, 188, 247
Virgo	-48, 11, 332, 391/392, 450/451, 509/510
Libra	155, 214/215, 273/274, 332/333, 391/393, 451/452, 510/511
Scorpio	-22, 37/38, 96/97, 155/156, 215/216, 274/275, 333/334, 393/394, 452/453, 511/512
Sagittarius	-21/-20, 38/39, 97/98, 157/158, 216/217, 275/276, 334/335, 394
Capricorn	-20/-19, 39/40, 98/99, 158, 217, 538
Aquarius	-19, 40, 361, 420/421, 479/480, 538/539
Pisces	-18, 41, 303, 361/362, 421/422, 480/481, 539/540

Saturn in Scorpio:

Aries	-17/-16, 42/43, 101/102, 161, 220, 279
Taurus	-17/-16, 43, 102, 161, 424, 482/483, 541/542
Gemini	-16, 305/306, 364/365, 423/424, 483/484, 542/543
Cancer	70, 128/129, 187/188, 246/247, 306/307, 365/366, 424/425, 483/485, 543-544
Leo	11, 70, 129/130, 188-189, 247-248, 306/308, 366/367, 425/426, 484/485, 544
Virgo	-48/-47, 11/12, 70/71, 129/131, 189/190, 248/249, 307/308, 367, 426, 485
Libra	-47/-46, 12/13, 71/72, 130/132, 190, 249
Scorpio	-46/-45, 13/14, 72/73, 394, 453, 511/512
Sagittarius	276, 335, 394/395, 453/454, 512/513
Capricorn	99, 158/159, 217/218, 276/277, 336/337, 395/396, 454/455, 514
Aquarius	-19/-18, 40/41, 99/100, 159/160, 218/219, 277/278, 337, 396, 455, 515
Pisces	-18/-17, 41/42, 100/101, 160/161, 219/220, 278/279, 338, 397, 456

Saturn in Sagittarius:

Aries	102, 161/162, 220/221, 279/280, 339/340, 398/399, 457/458, 516/517
Taurus	-16/-15, 43/44, 102/103, 162/163, 221/222, 280/281, 339/340, 399, 458, 517
Gemini	-16/-14, 4445, 103/104, 162/163, 222/223, 281, 340, 399
Cancer	-15/-14, 45/46, 104/105, 163/164, 222/223, 282, 485, 544
Leo	-14/-13, 45/46, 105, 164, 223, 486, 544/545
Virgo	-13, 308/309, 367/368, 426/427, 485/487, 545/546
Libra	131/132, 190/191, 249/250, 308/310, 368/369, 427/428, 486/488, 546/547
Scorpio	-45, 13/14, 72/73, 132/133, 191/192, 250/251, 310/311, 369/370, 428/429, 487/488, 547
Sagittarius	-45/-44, 14/15, 73/75, 133/134, 192/193, 251/252, 311, 370
Capricorn	-44/-43, 15/16, 75, 134, 193, 514/515
Aquarius	-43, 16, 338, 397, 455/456, 515/516
Pisces	161, 220, 279, 338/339, 397/398, 456/457, 516/517

Saturn in Capricorn:

Jupiter	
Aries	-41/-40, 18/19, 78, 137, 196, 517/518
Taurus	-40, 19, 78, 400, 458/459, 518
Gemini	223, 282, 341, 400/401, 459/460, 518/519
Cancer	46, 105, 164, 223/224, 282/283, 341/342, 400/402, 460/461, 519/520
Leo	-13, 46/47, 105/106, 164/165, 223/225, 283/284, 342/343, 401/402, 461, 520
Virgo	-13/-12, 46/48, 106/107, 165/166, 224/225, 284, 343, 402
Libra	-12/-11, 47/48, 107/108, 166/167, 225/226, 488, 547
Scorpio	-11/-10, 49, 311, 370, 429, 488/489, 547/548
Sagittarius	252, 311/312, 370/371, 429/430, 489/490, 548/549
Capricorn	75/76, 134/135, 193/194, 253/254, 312/313, 371/372, 431/432, 490/491, 549/550
Aquarius	-42, 16/17, 76/77, 135/136, 194/195, 254/255, 313/314, 372/373, 432
Pisces	-42/-41, 17/18, 77/78, 136/137, 195/196, 255, 314

Saturn in Aquarius:

Aries	-40, 19, 78/79, 137/138, 196/197, 255/257, 315/316, 374/375, 433/435, 493
Taurus	-40/-39, 19/20, 78/80, 138/139, 197/198, 256/257, 316/317, 375/376, 434
Gemini	-39/-38, 20/21, 79/80, 139/140, 198/199, 257/258, 316
Cancer	-39/-37, 21/22, 80/81, 139, 402, 461, 520
Leo	-38, 284, 343, 402/403, 461/462, 520/521
Virgo	226, 285, 343/344, 402/404, 462/463, 521/522
Libra	49, 108, 167, 226/227, 285/286, 344/345, 403/404, 463/464, 522/523
Scorpio	-10, 49/50, 108/109, 167/168, 227/228, 286/287, 345/346, 404/405
Sagittarius	-10/-9, 50/51, 109/110, 168/169, 228
Capricorn	-8/-7, 51, 110, 491, 550
Aquarius	314, 373, 432/433, 491/492, 550
Pisces	137, 196, 255/256, 314/315, 373/374, 433/434, 492/493

Saturn in Pisces:

Aries	-5, 375, 434, 493/494
Taurus	257, 316/317, 375/376, 434/435, 494/495
Gemini	140, 199, 257/258, 317/318, 376/377, 435/436, 494/496
Cancer	-37, 22, 81, 140/141, 199/200, 258/259, 317/319, 377/378, 436/437, 495/496
Leo	-37/-36, 22/23, 81/82, 140/142, 200/201, 259/260, 318/319
Virgo	-37/-35, 23/24, 82/83, 141/142
Libra	-36/-35, 24, 405, 464, 523
Scorpio	-35/-34, 228, 287, 346, 405/406, 464/465, 523/524
Sagittarius	51, 110, 169, 228/229, 287/288, 346/347, 406/407, 465/466, 524/525
Capricorn	-8/-7, 51/52, 110/111, 169/171, 229/230, 288/289, 347/349, 407, 466
Aquarius	-7/-6, 52/53, 111/112, 171/172, 230/231, 289/290, 349
Pisces	-6/-5, 53/54, 112/113, 172

K. Solar Longitudes and Calendar Dates.

In the Julian and Alexandrian calendars, the sun's longitude remains within about a degree of a constant value for the same date from year to year, with only a very slow shift. The Tables 1 and 2 give approximate sidereal longitudes of the sun computed according to the *Handy Tables* for the first day of each calendar month about the beginning of the reign of Augustus (30 B.C.). The sidereal longitudes tend to decrease by about 1° in a century.

Since four Egyptian calendar years fall short of four Alexandrian years by one day, a given month and day in the Egyptian calendar will have its Alexandrian counterpart 25 days earlier after a century. The solar longitude therefore also decreases at nearly the same rate. Table 3 gives approximate sidereal longitudes computed by the *Handy Tables* for the first day of each Egyptian month, at intervals of 100 years from the beginning of the reign of Augustus.

1. Julian calendar

January 1	286
February 1	317
March 1	345
April 1	15
May 1	44
June 1	73
July 1	102
August 1	131
September 1	161
October 1	191
November 1	223
December 1	254

2. Alexandrian calendar

Thoth 1	159
Phaophi 1	189
Hathyr 1	220
Choeac 1	250
Tybi 1	281
Mecheir 1	312
Phamenoth 1	342
Pharmuthi 1	11
Pachon 1	40
Payni 1	68
Epeiph 1	96
Mesore 1	125

3. Egyptian calendar

	Aug. 0	*100*	*200*	*300*	*400*	*500*
Thoth 1	160	135	110	85	61	36
Phaophi 1	190	164	139	114	89	65
Hathyr 1	221	194	168	143	118	93
Choeac 1	251	225	199	173	147	122
Tybi 1	282	256	229	203	177	151
Mecheir 1	313	287	260	233	207	181
Phamenoth 1	343	317	291	264	238	211
Pharmuthi 1	12	347	321	295	268	242
Pachon 1	41	16	351	325	299	273
Payni 1	69	44	20	355	329	303
Epeiph 1	97	73	48	23	358	333
Mesore 1	126	101	77	52	27	2

L. Astrological Houses, Terms, and Attributes.

1. Rulers of the signs and terms (Egyptian System).

Sign	*House Ruler*	*Terms*				
Aries (♈)	♂	0°–6°: ♃	6°–12°: ♀	12°–20°: ☿	20°–25°: ♂	25°–30°: ♄
Taurus (♉)	♀	0°–8°: ♀	8°–14°: ☿	14°–22°: ♃	22°–27°: ♄	27°–30°: ♂
Gemini (♊)	☿	0°–6°: ☿	6°–12°: ♃	12°–17°: ♀	17°–24°: ♂	24°–30°: ♄
Cancer (♋)	☾	0°–7°: ♂	7°–13°: ♀	13°–19°: ☿	19°–26°: ♃	26°–30°: ♄
Leo (♌)	☉	0°–6°: ♃	6°–11°: ♀	11°–18°: ♄	18°–24°: ☿	24°–30°: ♂
Virgo (♍)	☿	0°–7°: ☿	7°–17°: ♀	17°–21°: ♃	21°–28°: ♂	28°–30°: ♄
Libra (♎)	♀	0°–6°: ♄	6°–14°: ☿	14°–21°: ♃	21°–28°: ♀	28°–30°: ♂
Scorpio (♏)	♂	0°–7°: ♂	7°–11°: ♀	11°–19°: ☿	19°–24°: ♃	24°–30°: ♄
Sagittarius (♐)	♃	0°–12°: ♃	12°–17°: ♀	17°–21°: ☿	21°–26°: ♄	26°–30°: ♂
Capricorn (♑)	♄	0°–7°: ☿	7°–14°: ♃	14°–22°: ♀	22°–26°: ♄	26°–30°: ♂
Aquarius (♒)	♄	0°–7°: ☿	7°–13°: ♀	13°–20°: ♃	20°–25°: ♂	25°–30°: ♄
Pisces (♓)	♃	0°–12°: ♀	12°–16°: ♃	16°–19°: ☿	19°–28°: ♂	28°–30°: ♄

2. Exaltations and depressions, and rulers of triplicities and decans.

Sign	*Exaltation*	*Depression*	*Triplicity*		*Decan*		
			day	*night*	1	2	3
Aries (♈)	☉	♄	☉	♃	♂	☉	♀
Taurus (♉)	☾		♀	☾	☿	☾	♄
Gemini (♊)			♄	☿	♃	♂	☉
Cancer (♋)	♃	♂	♀	♂	♀	☿	☾
Leo (♌)			☉	♃	♄	♃	♂
Virgo (♍)	☿	♀	♀	☾	☉	♀	☿
Libra (♎)	♄	☉	♄	☿	☾	♄	♃
Scorpio (♏)		☾	♀	♂	♂	☉	♀
Sagittarius (♐)			☉	♃	☿	☾	♄
Capricorn (♑)	♂	♃	♀	☾	♃	♂	☉
Aquarius (♒)			♄	☿	♀	☿	☾
Pisces (♓)	♀	☿	♀	♂	♄	♃	♂

3. *Common attributes of the signs.*

Masculine signs:	Aries, Gemini, Gemini, Leo, Libra, Sagittarius, Aquarius
Feminine signs:	Taurus, Cancer, Virgo, Scorpio, Capricorn, Pisces
Tropical signs:	Aries, Cancer, Libra, Capricorn
Solid signs:	Taurus, Leo, Scorpio, Aquarius
Bicorporal signs:	Gemini, Virgo, Sagittarius, Pisces

Bibliographical Abbreviations

Aaboe [1955] A. Aaboe, 'On the Babylonian Origin of Some Hipparchian Parameters.' *Centaurus* 4 (1955) 122–125.

Aaboe [1958] A. Aaboe, 'On Babylonian Planetary Theories.' *Centaurus* 5 (1958) 209–277.

Aaboe [1963] A. Aaboe, 'On a Greek Qualitative Planetary Model of the Epicyclic Variety.' *Centaurus* 9 (1963) 1–10.

Aaboe [1972] A. Aaboe, 'Remarks on the Theoretical Treatment of Eclipses in Antiquity.' *Journal for History of Astronomy* 3 (1972) 105–118.

Aaboe & Sachs [1966] A. Aaboe and A. Sachs, 'Some Dateless Computed Lists of Longitudes of Characteristic Planetary Phenomena from the Late-Babylonian Period.' *Journal of Cuneiform Studies* 20 (1966) 1–33.

Ambühl, Markovska, & Milnor [1995] A. Ambühl, D. Markovska, and K. Milnor, 'P. Mich. Inv. 29: Two Astrological Treatises.' *Zeitschrift für Papyrologie und Epigraphik* 105 (1995) 229–236 and plate VIII.

Baccani [1989] D. Baccani, 'Appunti per oroscopi negli ostraca di Medinet Madi.' *Analecta Papyrologica* 1 (1989) 67-77.

Baccani [1992] D. Baccani, *Oroscopi Greci: Documentazione papirologica*. Ricerca Papirologica 1. Messina, Sicania, 1992.

Baccani [1995] D. Baccani, 'Appunti per oroscopi negli ostraca di Medinet Madi (II).' *Analecta Papyrologica* 7 (1995) 63–72.

Bagnall [1993] R. S. Bagnall, *Egypt in Late Antiquity*. Princeton, Princeton University Press, 1993.

Bagnall & Worp [1978] R. S. Bagnall and K. A. Worp, *The Chronological Systems of Byzantine Egypt*. Studia Amstelodamensia ad Epigraphicam, Ius Antiquum et Papyrologicam Pertinentia 8. Zutphen, 1978.

Bagnall & Worp [1979] R. S. Bagnall and K. A. Worp, *Regnal Formulas in Byzantine Egypt*. Bulletin of the American Society of Papyrologists, Supplement 2. Missoula, 1979.

Bickerman [1980] E. J. Bickerman, *Chronology of the Ancient World*. London, 1968.

Biscottini [1966] M. V. Biscottini, 'L'archivio di Tryphon, tessitore di Oxyrhynchos.' *Aegyptus* 46 (1966) 60-90 and 186-292.

Blanchard [1974] A. Blanchard, *Sigles et abréviations dans les papyrus documentaires grecs: recherches de paléographie*. University of London, Institute of Classical Studies Bulletin Supplement 30. London, 1974.

Boeckh [1863] A. Boeckh, *Ueber die vierjährigen Sonnenkreise der Alten, vorzüglich den Eudoxischen. Ein Beitrag zur Geschichte der Zeitrechnung und des Kalenderwesens der Aegypter, Griechen und Römer*. Berlin, 1863.

Bouché-Leclercq [1899] A. Bouché-Leclercq, *L'astrologie grecque*. Paris, 1899.

Bowen & Goldstein [1988] A. C. Bowen and B. R. Goldstein, 'Meton of Athens and Astronomy in the Late Fifth Century B.C.' *A Scientific Humanist: Studies in Memory of Abraham Sachs*, ed. E. Leichty, M. deJ. Ellis, and P. Gerardi. Occasional Publications of the Samuel Noah Kramer Fund 9. Philadelphia, 1988. 39–81.

Brashear & Neugebauer [1973] W. Brashear and O. Neugebauer, 'Zwei Berliner Papyri: Ein Horoskop und eine Jupiter-Tafel.' Österreichische Akademie der Wissenschaften, phil.-hist. Klasse, *Anzeiger* 110 (1973) 306–312.

Brashear & Jones [1999] W. Brashear and A. Jones, 'An Astronomical Table Containing Jupiter's Synodic Phenomena.' *Zeitschrift für Papyrologie und Epigraphik* 125 (1999) 206–210.

Britton & Jones [2000] J. P. Britton and A. Jones, 'A New Babylonian Planetary Model in a Greek Source.' *Archive for History of Exact Sciences* (forthcoming).

Britton & Walker [1991] J. P. Britton and C. B. F. Walker, 'A 4th century Babylonian model for Venus: B.M. 33552.' *Centaurus* 34 (1991) 97-118.

Bulmer-Thomas [1974] I. Bulmer-Thomas, 'Menelaus of Alexandria.' *Dictionary of Scientific Biography* 9 (1974) 296–302.

Bulmer-Thomas [1978] I. Bulmer-Thomas, 'Menelaus of Alexandria.' *Dictionary of Scientific Biography* 15 (1978) 420–421. (*addenda to* Bulmer-Thomas [1974])

Burckhardt [1958] J. J. Burckhardt, 'Zwei griechische Ephemeriden.' *Osiris* 13 (1958) 73–76.

Bureth [1964] P. Bureth, *Les Titulatures impériales dans les papyrus, les ostraca et les inscriptions d'Égypte (30 a.C. – 284 p.C.)*. Bruxelles, 1964.

Carp [1975] T. C. Carp, 'Five Tebtunis Papyri.' *Zeitschrift für Papyrologie und Epigraphik* 16 (1975) 47–62.

Clarysse [1997] W. Clarysse, 'Greek Accents on Egyptian Names.' *Zeitschrift für Papyrologie und Epigraphik* 119 (1997) 177–184.

Cumont [1935] F. Cumont, 'Les noms des planètes et l'astrolatrie chez les Grecs.' *L'Antiquité Classique* 4 (1935) 5–43.

Curtis & Robbins [1935] H. D. Curtis and F. E. Robbins, 'An Ephemeris of 467 A.D.' *Publications of the Observatory of the University of Michigan* 6.9 (1935) 77-100.

Delporte [1947] E. Delporte, 'Papyrus Michigan III 150.' *Chronique d'Égypte* 22 (1947) 73–83.

Devauchelle [1987] D. Devauchelle, 'Cinq ostraca démotiques de Karnak.' *Cahiers de Karnak* 8 (1987)137-142.

Fotheringham [1924] J. K. Fotheringham, 'The Metonic and Callippic Cycles.' *Monthly Notices of the Royal Astronomical Society* 84 (1924) 383–392.

Fotheringham [1935] J. K. Fotheringham, review of Curtis & Robbins [1935], *Classical Review* 49 (1935) 242.

Gerstinger & Neugebauer [1962] H. Gerstinger and O. Neugebauer, 'Eine Ephemeride für das Jahr 348 oder 424 n. Chr. in den PER, Pap. Graec. Vindob. 29370.' Österreichische Akademie der Wissenschaften, Phil.-hist. Klasse, *Sitzungsberichte* 240.2 (1962) 3–25.

Gignac [1976–1981] F. T. Gignac, *A Grammar of the Greek Papyri of the Roman and Byzantine Periods*. Testi e documenti per lo studio dell' antichità 55. 2 vols. Milano, 1976–1981.

Goldstein & Bowen [1989] B. R. Goldstein and A. C. Bowen, 'On Early Hellenistic Astronomy: Timocharis and the First Callippic Calendar.' *Centaurus* 32 (1989) 272–293.

Goldstine [1973] H. H. Goldstine, *New and Full Moons 1001 B.C. to A.D. 1651*. Memoirs of the American Philosophical Society 94. Philadelphia, 1973.

Gundel [1956] H. Gundel, 'Monatsnamen. Bruchstück einer vergleichenden Liste.' *Archiv für Papyrusforschung* 16 (1956) 13–19.

Halma [1822-1825] N. Halma, *Commentaire de Théon d'Alexandrie sur le livre III de l'Almageste de Ptolemée* [sic]; *Tables manuelles des mouvemens des astres*. Paris, 1822. *Tables manuelles astronomiques de Ptolemée et de Théon*. II, III. Paris, 1823 and 1825.

Hoogendijk [1982] F. A. J. Hoogendijk, 'Fragment of a Greek Planetary Table.' *Zeitschrift für Papyrologie und Epigraphik* 48 (1982) 135-141 and plate vi.

Houlden & Stephenson [1986] M. A. Houlden and F. R. Stephenson, *A Supplement to the Tuckerman Tables*. Memoirs of the American Philosophical Society 170, Philadelphia, 1986.

Huber [1957] P. Huber, 'Zur täglichen Bewegung des Jupiter nach babylonischen Texten.' *Zeitschrift für Assyrologie* 52 = N.F. 18 (1957) 265–303.

Hübner [1982] W. Hübner, *Die Eigenschaften der Tierkreiszeichen in der Antike*. Sudhoffs Archiv Beihefte 22, Wiesbaden, 1982.

Ioannidou [1996] G. Ioannidou, *Catalogue of Greek and Latin Literary Papyri in Berlin (P. Berol. inv. 21101–21229, 21911)*. Berliner Klassikertexte 9, Mainz am Rhein, 1996.

Irani [1955] R. A. K. Irani, 'Arabic Numeral Forms.' *Centaurus* 4 (1955) 1–12.

James [1982] T. G. H. James, *Excavating in Egypt: The Egypt Exploration Society 1882–1982*. Chicago, University of Chicago Press, 1982.

Jones [1983] A. Jones, 'The Development and Transmission of 248-Day Schemes for Lunar Motion in Ancient Astronomy.' *Archive for History of Exact Sciences* 29 (1983) 1-36.

Jones [1984] A. Jones, 'A Greek Saturn Table.' *Centaurus* 27 (1984) 311-317.

Jones [1990a] A. Jones, 'Babylonian and Greek Astronomy in a Papyrus Concerning Mars.' *Centaurus* 33 (1990) 97–114.

Jones [1990b] A. Jones, *Ptolemy's First Commentator*. Transactions of the American Philosophical Society 80.7, Philadelphia, 1990.

Jones [1991a] A. Jones, 'A Second-Century Greek Ephemeris for Venus.' *Archives Internationales d'Histoire des Sciences* 41 (1991) 3–12.

Jones [1991b] A. Jones, 'Hipparchus's Computations of Solar Longitudes.' *Journal for the History of Astronomy* 22 (1991) 101–125.

Jones [1991c] A. Jones, 'The adaptation of Babylonian Methods in Greek Numerical Astronomy.' *Isis* 82 (1991) 441–453.

Jones [1993a] A. Jones, 'The Date of the Astronomical Almanac Tab. Amst. inv. no. 1.' *Chronique d'Égypte* 68 (1993) 178–185.

Jones [1993b] A. Jones, 'Evidence for Babylonian Arithmetical Schemes in Greek Astronomy.' *Die Rolle der Astronomie in den Kulturen Mesopotamiens. Beiträge zum 3. Grazer Morgenländischen Symposion (23.–27. September 1991)*, ed. H. D. Galter. Grazer Morgenländische Studien 3. Graz, 1993. 77–94.

Jones [1994a] A. Jones, 'An Ephemeris for A.D. 140: P. Harris I.60.' *Zeitschrift für Papyrologie und Epigraphik* 100 (1994) 59–63.

Jones [1994b] A. Jones, 'The Place of Astronomy in Roman Egypt.' *The Sciences in Greco Roman Society*, ed. T. D. Barnes. Edmonton, 1994. (*Apeiron* 27.4). 25–51.

Jones [1994c] A. Jones, 'Two Astronomical Papyri Revisited.' *Analecta Papyrologica* 6 (1994) 111–126.

Jones [1995] A. Jones, 'On the Planetary Table, Dublin TCD Pap. F. 7.' *Zeitschrift für Papyrologie und Epigraphik* 107 (1995) 255–258.

Jones [1997a] A. Jones, 'Studies in the Astronomy of the Roman Period. I. The Standard Lunar Scheme.' *Centaurus* 39 (1997) 1–36.

Jones [1997b] A. Jones, 'Studies in the Astronomy of the Roman Period. II. Tables for Solar Longitude.' *Centaurus* 39 (1997) 211–229.

Jones [1997c] A. Jones, 'On the Reconstructed Macedonian and Egyptian Lunar Calendars.' *Zeitschrift für Papyrologie und Epigraphik* 119 (1997) 157–166.

Jones [1997d] A. Jones, 'A Greek Papyrus Containing Babylonian Lunar Theory.' *Zeitschrift für Papyrologie und Epigraphik* 119 (1997) 167–172 and plate II.

Jones [1998a] A. Jones, 'Studies in the Astronomy of the Roman Period. III. Planetary Epoch Tables.' *Centaurus* 40 (1998) 1–41.

Jones [1998b] A. Jones, 'Notes on astronomical papyri.' *Zeitschrift für Papyrologie und Epigraphik* 121 (1998) 203–210.

Jones [1998c] A. Jones, 'Three astronomical tables from Tebtunis.' *Zeitschrift für Papyrologie und Epigraphik* 121 (1998) 211–218.

Jones [1999a] A. Jones, 'A Classification of Astronomical Tables on Papyrus.' *Ancient Astronomy and Celestial Divination*, ed. N. M. Swerdlow. Cambridge (U.S.A.), MIT Press (forthcoming).

Jones [1999b] A. Jones, 'Astrologers of Oxyrhynchus.' *Oxyrhynchus: A City and its Texts* (forthcoming).

Jones [1999c] A. Jones, 'Two Astronomical Tables: P. Berol. 21240 and 21359.' *Zeitschrift für Papyrologie und Epigraphik* 125 (1999) 201-205.

Knudtzon & Neugebauer [1947] E. J. Knudtzon and O. Neugebauer, 'Zwei astronomische Texte.' *K. Humanistiska Vetenskapssamfundet i Lund, Årsberättelse (Bulletin de la Société Royale des Lettres de Lund)* 1946-1947, II, 77-88.

Kubitschek [1915] W. Kubitschek, *Die Kalenderbücher von Florenz, Rom und Leyden*. Denkschriften d. kaiserl. Akad. d. Wissenschaften in Wien, Philos.-histor. Kl. 57.3 (1915).

Kugler [1907-1935] F. X. Kugler, *Sternkunde und Sterndienst in Babel*. 2 vols. and 3 Ergänzungshefte (the last by J. Schaumberger). Münster, Aschendorff, 1907-1935.

Lange & Neugebauer [1940] H. O. Lange and O. Neugebauer, *Papyrus Carlsberg No. I. Ein hieratisch-demotischer kosmologischer Text*. Kgl. Danske Vidensk. Selskab, Histor.-filol. Skrifter 1.2 (1940).

Manfredi [1966] M. Manfredi, 'Presentazione di un testo astronomico e discussione di un documento di Antinoe.' *Atti dell' XI Congresso Internazionale di Papirologia, Milano 2–8 Settembre 1965*. Milano, 1966. 237–243.

Manfredi & Neugebauer [1973] M. Manfredi and O. Neugebauer, 'Greek Planetary Tables from the Time of Claudius.' *Zeitschrift für Papyrologie und Epigraphik* 11 (1973) 101–114.

Nelson [1970] C. A. Nelson, 'Astronomical Table.' *Bulletin of the American Society of Papyrologists* 7 (1970) 35–37.

Neugebauer [1941] O. Neugebauer, 'On a Special Use of the Sign "Zero" in Cuneiform Astronomical Texts.' *Journal of the American Oriental Society* 61 (1941) 213–215.

Neugebauer [1942a] O. Neugebauer, 'Egyptian Planetary Texts.' *Transactions of the American Philosophical Society* NS 32.2 (1942) 209-250.

Neugebauer [1942b] O. Neugebauer, 'On Some Astronomical Papyri and Related Problems of Ancient Geography.' *Transactions of the American Philosophical Society* NS 32.2 (1942) 251-263.

Neugebauer [1943] O. Neugebauer, 'Demotic Horoscopes.' *Journal of the American Oriental Society* 63 (1943) 115-127.

Neugebauer [1949] O. Neugebauer, 'The Astronomical Treatise P. Ryl. 27.' Kgl. Danske Vidensk. Selskab, *Hist.-fil. Medd.* 32.2 (1949).

Neugebauer [1955] O. Neugebauer, *Astronomical Cuneiform Texts.* 3 vols. London, Lund Humphries, [1955].

Neugebauer [1956] O. Neugebauer, 'An Astronomical Almanac for the Year 348/9 (P. Heid. Inv. No. 34).' Kgl. Danske Vidensk. Selskab, *Hist.-fil. Medd.* 36.4 (1956).

Neugebauer [1957a] O. Neugebauer, 'A Greek Planetary Table.' *Chronique d'Égypte* 32 (1957) 269–272.

Neugebauer [1957b] O. Neugebauer, *The Exact Sciences in Antiquity.* 2nd ed. Providence, Brown University Press, 1957.

Neugebauer [1960] O. Neugebauer, 'A New Greek Astronomical Table (P. Heid. Inv. 4144 +, P. Mich. 151).' Kgl. Danske Vidensk. Selskab., *Hist.-filos. Medd.* 39.1 (1960).

Neugebauer [1962a] O. Neugebauer, 'Astronomical Papyri and Ostraca: Bibliographical Notes.' *Proceedings of the American Philosophical Society* 106 (1962) 383–391.

Neugebauer [1962b] O. Neugebauer, 'Über griechische Wetterzeichen und Schattentafeln. Institut für Österr. Geschichtsforschung Pap. Graec. Nr. 1.' Österreichische Akademie der Wissenschaften, Phil.-hist. Klasse, *Sitzungsberichte* 240.2 (1962) 27–44.

Neugebauer [1971] O. Neugebauer, 'On a Fragment from a Planetary Table.' *Zeitschrift für Papyrologie und Epigraphik* 7 (1971) 267–274.

Neugebauer [1972a] O. Neugebauer, 'Planetary Motion in P. Mich. 149.' *Bulletin of the American Society of Papyrologists* 9 (1972) 19–22.

Neugebauer [1972b] O. Neugebauer, 'The Date of the Planetary Tables Bodleian MS. Gr. Class. F 7 (P).' *Chronique d'Égypte* 93 (1972) 224–226.

Neugebauer [1974] O. Neugebauer, 'A Table for Positions of Jupiter (Ostr. Tait II 2177).' *Chronique d'Égypte* 97 (1974) 128–129.

Neugebauer [1975] O. Neugebauer, *A History of Ancient Mathematical Astronomy.* 3 vols. Berlin/Heidelberg/New York, Springer, 1975.

Neugebauer [1988] — O. Neugebauer, 'A Babylonian Lunar Ephemeris from Roman Egypt.' *A Scientific Humanist: Studies in Memory of Abraham Sachs*, ed. E. Leichty, M. deJ. Ellis, and P. Gerardi. Occasional Publications of the Samuel Noah Kramer Fund 9. Philadelphia, 1988. 301–304.

Neugebauer & Brashear [1976] — O. Neugebauer and W. Brashear, 'An Ephemeris for Mars.' *Zeitschrift für Papyrologie und Epigraphik* 20 (1976) 117–118.

Neugebauer & Parker [1960–1969] — O. Neugebauer and R. A. Parker, *Egyptian Astronomical Texts*. 3 vols. Providence, Brown University Press, 1960–1969.

Neugebauer & Parker [1968] — O. Neugebauer and R. A. Parker, 'Two Demotic Horoscopes.' *Journal of Egyptian Archaeology* 54 (1968) 231-235.

Neugebauer, Parker, & Zauzich [1981] — O. Neugebauer, R. A. Parker, and K.-T. Zauzich, 'A Demotic Lunar Eclipse Text of the First Century B.C.' *Proceedings of the American Philosophical Society* 125 (1981) 312–327.

Neugebauer & Pingree [1970–1971] — O. Neugebauer and D. Pingree, *The Pañcasiddhāntikā of Varāhamihira*. Danske Vidensk. Selskab, Histor.-Filos. Skrifter 6.1 and 6.2 (1970–1971).

Neugebauer & Sijpesteijn [1980a] — O. Neugebauer and P. J. Sijpesteijn, 'A New Version of Greek Planetary Tables.' *Zeitschrift für Papyrologie und Epigraphik* 37 (1980) 280–293.

Neugebauer & Sijpesteijn [1980b] — P. J. Sijpesteijn & O. Neugebauer, 'Fragment of an Astronomical Text.' *Zeitschrift für Papyrologie und Epigraphik* 39 (1980) 163–164.

Neugebauer, Sijpesteijn, & Worp [1977] — O. Neugebauer, P. J. Sijpesteijn, and K. A. Worp, 'A Greek Planetary Table.' *Chronique d'Égypte* 52 (1977) 301–310.

Neugebauer & Skeat [1958] — O. Neugebauer and T. C. Skeat, 'The Astronomical Tables, P. Lond. 1278.' *Osiris* 13 (1958) 93–113.

Neugebauer & van Hoesen [1959] — O. Neugebauer and H. B. van Hoesen, *Greek Horoscopes*. Memoirs of the American Philosophical Society 48. Philadelphia, 1959.

Neugebauer & van Hoesen [1964] — O. Neugebauer and H. B. van Hoesen, 'Astrological Papyri and Ostraca: Bibliographical Notes.' *Proceedings of the American Philosophical Society* 108 (1964) 57–72.

Newton [1977] — R. Newton, *The Crime of Claudius Ptolemy*. Baltimore, Johns Hopkins University Press, 1977.

O. Bodl. — *Greek Ostraca in the Bodleian Library at Oxford and Various Other Collections*. 3 vols. Ed. J. G. Tait and C. Préaux. Egypt Exploration Society, Graeco-Roman Memoirs 21, 33, 43. London, 1930–1964.

O. Narm. — *Ostraka greci da Narmuthis (OgN I)*. Ed. R. Pintaudi and P. J. Sijpesteijn. Quaderni di Medinet Madi 2. Pisa, 1993.

Oates et al. [1992] J. F. Oates, R. S. Bagnall, W. H. Willis, and K. A. Worp, *Checklist of Editions of Greek and Latin Papyri, Ostraca and Tablets*. 4th ed. Bulletin of the American Society of Papyrologists, Supplement 7. Atlanta, Scholars Press, 1992.

P. Aberd. *Catalogue of Greek and Latin Papyri and Ostraca in the Possession of the University of Aberdeen*. Ed. E. G. Turner. Aberdeen Univ. Studies 116. Aberdeen, 1939.

P. Ant. *The Antinoopolis Papyri*. 3 vols. Ed. C. H. Roberts, J. W. B. Barns, and H. Zilliacus. Egypt Exploration Society, Graeco-Roman Memoirs 28, 51, 110. London, 1950–1967.

P. Brook. *Greek and Latin Papyri, Ostraca, and Wooden Tablets in the Collection of the Brooklyn Museum*. Ed. J. C. Shelton. Papyrologica Florentina 22. Firenze, Gonnelli, 1992.

P. Carlsberg *The Carlsberg Papyri* 1. *Demotic Texts from the Collection*. Ed. P. J. Frandsen, with contributions by K.-Th. Zauzich, W. J. Tait and M. Chauveau. Copenhagen, Museum Tusculanum Press, 1991.

P. Harr. I *The Rendel Harris Papyri of Woodbrooke College, Birmingham*. Ed. J. E. Powell. Cambridge, 1936.

P. Hib. *The Hibeh Papyri*. 2 vols. Ed. B. P. Grenfell, A. S. Hunt, E. G. Turner, and M.-Th. Lenger. Egypt Exploration Society, Graeco-Roman Memoirs 7, 32. London, 1906–1955.

P. Iand. *Papyri Iandanae*. 8 vols. Ed. C. Kalbfleisch *et al.* Leipzig, 1912–1938.

P. Laur. *Dai Papiri della Biblioteca Medicea Laurenziana*. 5 vols. Ed. R. Pintaudi and G. M. Browne. Papyrologica Florentina 1, 2, 5, 12, 13. Firenze, Gonnelli, 1976–1984.

P. Lond. *Greek Papyri in the British Museum*. 7 vols. Ed. F. G. Kenyon *et. al.* London, 1893–1974.

P. Lund *Aus der Papyrussammlung der Universitätsbibliothek in Lund*. 5 installments. Ed. A. Wifstrand, K. Hanell, E. J. Knudtzon, and O. Neugebauer. Published in *K. Humanistiska Vetenskapssamfundet i Lund, Årsberättelse (Bulletin de la Société Royale des Lettres de Lund)* 1934/1935 – 1951/1952. (See Knudtzon & Neugebauer [1947]).

P. Oslo *Papyri Osloenses*. 3 vols. Ed. S. Eitrem and L. Amundsen. Oslo, 1925–1936.

P. Oxy. *The Oxyrhynchus Papyri*. 65 vols. Ed. B. P. Grenfell, A. S. Hunt, *et al.* Egypt Exploration Society, Graeco-Roman Memoirs. London, 1898–1998.

P. Ryl. *Catalogue of the Greek and Latin Papyri in the John Rylands Library, Manchester*. 4 vols. Ed. A. S. Hunt *et al.* Manchester, 1911–1952.

PSI — *Papiri greci e latini*. 14 vols. Various editors. Pubblicazioni della Società Italiana per la ricerca dei papiri greci e latini in Egitto. Firenze, 1912–1957.

P. Tebt. — *The Tebtunis Papyri*. 4 vols. Various editors. Egypt Exploration Society, Graeco-Roman Memoirs 4, 52, 23, 25, 64. London, 1902–1976.

Pack [1965] — R. A. Pack, *The Greek and Latin Literary Texts from Greco-Roman Egypt*. 2nd ed. Ann Arbor, University of Michigan Press, 1965.

Parker & Zauzich [1981] — R. A. Parker and K.-T. Zauzich, 'The Seasons in the First Century B.C.' *Studies Presented to Hans Jakob Polotsky*. East Gloucester, Pirtle & Polsen, 1981. 472–479.

Parsons [1963] — Review of Sattler [1962]. *Classical Review* N.S. 13 (1963) 364–365.

Pintaudi & Neugebauer [1978] — R. Pintaudi & O. Neugebauer, 'Pap, Laurentiana III/423: A Planetary Table.' *Zeitschrift für Papyrologie und Epigraphik* 30 (1978) 211–218.

Pintaudi & Sijpesteijn [1993] — R. Pintaudi & P. J. Sijpesteijn, *Ostraka Greci da Narmuthis (OGN I)*. Quaderni di Medinet Madi 2, Pisa, Giardini, 1993.

Rawlins [1987] — D. Rawlins, 'Ancient Heliocentrists, Ptolemy, and the Equant.' *American Journal of Physics* 55 (1987) 235–239.

Rochberg [1988] — F. Rochberg-Halton, *Aspects of Babylonian Celestial Divination: The Lunar Eclipse Tablets of Enūma Anu Enlil*. Archiv für Orientforschung, Beiheft 22, Horn, Verlag Ferdinand Berger & Söhne, 1988.

Sachs & Hunger [1988–] — A. J. Sachs and H. Hunger, *Astronomical Diaries and Related Texts from Babylonia*. Denkschriften d. Österr. Akad. d. Wissenschaften, Phil.-hist. Kl. 195, 210, … Wien, 1988–.

Samuel [1972] — A. E. Samuel, *Greek and Roman Chronology: Calendars and Years in Classical Antiquity*. Handbuch der Altertumswissenschaft I.7, München, C. H. Beck'sche Verlagsbuchhandlung, 1972.

Sattler [1962] — P. Sattler, *Studien aus dem Gebiet der Alten Geschichte*. Wiesbaden, Steiner, 1962.

Sezgin [1974] — F. Sezgin, *Geschichte des Arabischen Schrifttums. 5. Mathematik bis ca. 430 H.* Leiden, Brill, 1974.

Skeat [1982] — T. C. Skeat, 'The Length of the Standard Papyrus Roll and the Cost-Advantage of the Codex.' *Zeitschrift für Papyrologie und Epigraphik* 45 (1982) 169–175.

Sloley [1936] — R. W. Sloley, review of Curtis & Robbins [1935], *Journal of Egyptian Archaeology* 22 (1936) 218–219.

Stahlman [1960] — W. D. Stahlman, *The Astronomical Tables of Codex Vaticanus Graecus 1291*. Thesis, Brown University, 1960.

Tihon [1968] A. Tihon, 'Le calcul de la longitude de Vénus d'après un texte anonyme du *Vat. gr.* 184.' *Bulletin de l'Institut historique belge de Rome* 52 (1968) 51–82.

Tihon [1978] A. Tihon, *Le "Petit Commentaire" de Théon d'Alexandrie aux Tables Faciles de Ptolémée (Histoire du texte, édition critique, traduction)*. Studi e testi 282. Città del Vaticano, Biblioteca Apostolica Vaticana, 1978.

Tihon [1992] A. Tihon, 'Les *Tables Faciles* de Ptolémée dans les manuscrits en onciale (ixe–xe siècles).' *Revue d'histoire des textes* 22 (1992) 47–87 and plates I–V.

Toomer [1980] G. J. Toomer, 'Hipparchus's Empirical Basis for his Lunar Mean Motions.' *Centaurus* 24 (1980) 97-109.

Toomer [1984] G. J. Toomer, *Ptolemy's Almagest*. London, Duckworth, 1984.

Tuckerman [1962] B. Tuckerman, *Planetary, Lunar, and Solar Positions 601 B.C. to A.D. 1 at Five-day and Ten-day Intervals*. Memoirs of the American Philosophical Society 56. Philadelphia, 1962.

Tuckerman [1964] B. Tuckerman, *Planetary, Lunar, and Solar Positions A.D. 2 to A.D. 1649 at Five-day and Ten-day Intervals*. Memoirs of the American Philosophical Society 59. Philadelphia, 1964.

Turner [1952] E. G. Turner, 'Roman Oxyrhynchos.' *Journal of Egyptian Archaeology* 38 (1952) 78–93.

Turner [1956] E. G. Turner, 'Scribes and Scholars of Oxyrhynchos.' *Akten des VIII. Internationalen Kongresses für Papyrologie, Wien 1955*. Mitteilungen aus der Papyrussammlung der Oesterreichischen Nationalbibliothek, N.S. 5. Wien, 1956. 141-146.

Turner & Neugebauer E. G. Turner and O. Neugebauer, 'Gymnasium Debts and New Moons.' *Bulletin of the John Rylands Library* 32 (1949) 80–96.

van der Waerden [1947] B. L. van der Waerden, 'Egyptian "Eternal Tables."' Koninkl. Nederl. Akad. van Wetenschappen, *Proceedings* 50 (1947) 536–547 and 782–788.

van der Waerden [1958] B. L. van der Waerden, 'The astronomical papyrus Ryland 27.' *Centaurus* 5 (1958) 177–191.

van der Waerden [1960] B. L. van der Waerden, 'Babylonische Methoden in ägyptischen Planetentafeln.' *Vierteljahrsschrift d. Naturforsch. Ges. in Zürich* 105 (1960) 97–144.

van der Waerden [1972] B. L. van der Waerden, 'Aegyptische Planetenrechnung.' *Centaurus* 16 (1972) 65–91.

van der Waerden [1984a] B. L. van der Waerden, 'Greek Astronomical Calendars II: Callippos and his Calendar.' *Archive for History of Exact Sciences* 29 (1984) 115–124.

van der Waerden [1984b] — B. L. van der Waerden, 'Greek Astronomical Calendars III: The Calendar of Dionysios.' *Archive for History of Exact Sciences* 29 (1984) 125–130.

van der Waerden [1988] — B. L. van der Waerden, 'The Motion of Venus in Greek, Egyptian and Indian Texts.' *Centaurus* 31 (1988) 105–113.

Wessely [1891] — C. Wessely, 'Bruchstücke einer optischen Schrift aus dem Alterthum.' *Wiener Studien* 13 (1891) 312–323.

West [1989] — S. West, 'Astrologoumena.' *Zeitschrift für Papyrologie und Epigraphik* 77 (1989) 30–32.

Wirth [1974] — P. B. Wirth, 'Die Venustheorie von Vasiṣtha in der Pañcasiddha3ntika3 des Varāha Mihira.' *Centaurus* 18 (1974) 29–43.

Zauzich [1971] — *Ägyptische Handschriften*, ed.. E. Luddeckens. Verzeichnis der orientalischen Handschriften in Deutschland 19. Teil. 2. Wiesbaden, Franz Steiner Verlag, 1971.

Zauzich [1974] — K.-Th. Zauzich, 'Drei neue Fragmente zu Pap. Carlsberg 9.' *Enchoria* 4 (1974) 157–158 and plate 12.

Subject Index

Aaboe 98
ACT 17 n. 1, 17–18
Agrippa 73–75
Alexandria 15, 69, 73, 84, 89–90, 102
almanac-ephemerides *see* ephemerides
almanacs
  monthly 40, 44–45, 176–177
  sign-entry 42–44, 176
    four-column format 176, 204–205
  five-day 45–46, 177
  perpetual 44, 176, 214
  other formats 46
  dating 50–53, 57–58
  computation 55
anomaly (*see also*: sun, moon, planets) 15–17, 38, 119
Antiochus 286
Apollonius 15
archives 58–60
  'Group A' 59–60, 160
arithmetical astronomy 17–33, 60, 67–68, 84–86, 113–118, 175–176, 187–188, 192–193
ascendant 10, 33, 285–286, 320
ascensions 33, 103–105, 283
Ashmolean 1
astrologers 5–8, 57–60, 249, 272
astrology 5–8, 10–11
  astrological literature 1
  astrological papyri 5, 279–280
  catarchic astrology 5, 42, 58, 176
  general astrology 58, 68, 91
astronomical papyri, classification of 4–5
astronomy, theoretical 5, 35, 67–80, 95–99
Augustus 12–13, 120, 216

Babylon 8
Babylonian astronomy (*see also:* System A, System B) 1, 8, 44, 53, 68–69, 92, 99
  planetary theory and tables 17–33, 37, 55, 78, 82, 114–118, 123–150, 157–159
  lunar theory and tables 46, 90, 97–99, 178
  ascensions 283
  Diaries 76
  mathematics 9
Baccani 8
Bahnasa *see* Oxyrhynchus

Blest, Islands of the 104
boards as medium of writing 1 n. 2, 43

calendars 11–14
  Alexandrian 11–13, 37, 40–41, 45, 176, 178, 250, 312–317
  Athenian (Callippic) 13, 68, 87, 89–90, 94
  Babylonian 22, 114
  Bithynian 75
  Dionysian 75–77
  Egyptian ('Old' Egyptian) 11–12, 312–314
    survival in astronomy 12, 16, 22, 36–37, 94–95, 114, 120–122, 125, 133, 137, 139–140, 142, 146, 148–149, 178, 250
  Julian 12–13, 315–317
  Roman 12–13, 40–41, 75, 176, 178
Callippic dates 13–14, 68, 87–90
catarchic astrology *see* astrology
Christians 8, 282
Clemens Alexandrinus 310
codices 38, 40–41, 43, 45–46, 60–61

dates 11–15
  double dates for nights 14, 44
decans 11, 351
demotic astronomy and astrology 1 n. 3, 5–6, 8, 61, 68, 87–88
depressions 10–11, 351
Diapalos 272
Diocletian 250
Dionysian calendar, observers using 75–77, 80
dodekatemorion 285
Domitian 75
Dorotheus 276–277

eclipses 68–69, 75, 87–99, 101
  canons 35, 58, 68, 87–94
  cycles 90–91
  inclinations (προcνεύcειc) 68, 87, 91–92
ecliptic 9
  astronomical divisions 9–10
  astrological divisions 10–11
Elegabalus 257
emperors *see* years
ephemerides 13, 40–42, 47–48, 175–176

almanac-ephemerides 40–41, 175
dating 50–53, 181–182
epoch tables 17, 20, 35–37, 54–55, 113–118
equation of time 104–105
Eternal Tables 16
exaltations 10–11, 351

Firmicus Maternus 286
five-day almanacs *see* almanacs
fractions 9

Galba 14
general astrology *see* astrology
geographical tables (*see also:* Ptolemy, *Handy Tables*) 163
goal-year periods 44, 78, 176
Greek letter phenomena *see* planets
Grenfell and Hunt 1, 4, 7, 56–57, 59

Handy Tables (*see also:* Ptolemy, *Handy Tables*) 38–40, 61, 119, 165–168, 236–238
Hipparchus 15, 68, 71, 79, 90, 93–94, 97–99, 151–152, 283
horoscopes 1, 5–8, 10, 46–47
dating 50–53, 57, 344–348
deluxe 8, 10–11, 47, 249, 258–259, 282–290
grammar and orthography 250
multiple 249
tabular 290
without astronomical data 267
hours, equinoctial and seasonal 14–15, 102–105, 205, 318–319
houses, astrological 11
Hypsicles 283

Indian astronomy 1, 15–16, 22 n. 1, 25, 35, 68, 82, 114
indiction 282

Julius Caesar 13
Jupiter
anomalistic period 72, 77–78
observations 69–80
'Phaethon' 70, 77
latitude 79
tables 118, 145–148, 157–158

kinematic astronomy 15–17, 38–40, 68–69, 82–83, 108–109, 117–119, 151
klima 103–104, 161–162, 165, 283

latitudes, astronomical 9–10, 68, 100–101
longitude, astronomical 9
sidereal and tropical 49, 117, 175–176, 343
Theon's formula 45, 49, 55, 176, 343
equinoctial and solstitial points 117, 151–152
lots
lot of fortune 278, 286, 289–290, 293
lot of daimon 286, 287, 289, 293
lots of eros and necessity 286
lot of marriage 276–277, 290

Macrobius 13
Mars
latitude 101
retrogradation 25, 141–145
tables 118, 137–145, 155–157
mean motion 15–17, 38, 119, 169–171, 228, 237, 244
Medinet Madi *see* Narmuthis
Menelaus 35, 67, 69–80
Mercury
pushes 25, 115, 129–130
'Stilbon' 77, 285
System A in Babylonian astronomy and papyri 114–115
tables 118, 123–134, 232–235
midheaven 10
modern theory 49, 249–250
monomoiria 11, 285–286
monthly almanacs *see* almanacs
months (*see also:* moon) 12–13
Athenian 13
Egyptian 12
represented by numerals 176
lunar 13
moon
anomalistic month 18
anomaly 18
kinematic models 119, 237
lunar diameter as unit 75, 80
Standard Scheme 18–19, 33, 35, 48, 54, 68, 84–86, 95, 113, 116, 120–123, 136, 153–155, 175, 178–187, 192, 195
tables for 321–342
symbol 62–63, 280
tables 18–19, 119, 169, 228

Narmuthis 5, 8
Nerva 75 n. 2
Neugebauer 1, 4, 61, 114, 119, 171
numbers, notation 9

observations 35, 67, 70–80, 101
omens *see* astrology, general
ostraca (*see also:* papyri) 1 n. 2, 5
Oxyrhynchus 4–8, 249
excavations 1, 4–5, 56, 59
provenance of papyri 4, 56–57
Oxyrhynchus papyri
published series 1, 7
unpublished papyri and inventory numbers 4, 7, 56–57, 59

Pappus 105, 283
papyri, ostraca, tablets
Bodl. MS Gr. Class. F 7 (P) 14
O. Bodl. Tait 2176 110
P. Aberd. 12 35
P. Aberd. 127 35
P. Ant. 141 35
P. Berol. 8279 55, 58, 114
P. Berol. 13146+13147 14 , 35, 87–91
P. Berol. 16511 145
P. Berol. 21226 14, 216
P. Carlsberg 31 96
P. Colker 46
P. Dubl. TCD F.7 175
P. Harris 60 175
P. Heid. inv. 34 46
P. Heid. inv. 4144 + P. Mich. 151 232
P. Iand. 84 35
P. Laur. 144 60
P. Lond. 98 35, 283
P. Lond. 110 250
P. Lond. 130 8, 47, 283, 285
P. Lond. 1278 119
P. Lund 35a 17, 58, 113
P. Mich. 150 46
P. Mich. inv. 1454 40 n. 2, 175, 190
P. Nelson 14
P. Oslo 73 5, 35
P. Oxy. 35 263, 310
P. Oxy. 303 35
P. Oxy. 988 263
P. Oxy. 2060 8
P. Oxy. 2551 263, 310
P. Oxy. 3298 257
P. Oxy. 3299 58
P. Oxy. 3353 263
P. Oxy. 4133 5, 35, 80
P. Oxy. 4135 35, 182
P. Oxy. 4136 35
P. Oxy. 4137 13–14, 35, 58
P. Oxy. 4138a 35
P. Oxy. 4139 35
P. Oxy. 4140 35
P. Oxy. 4142 35
P. Oxy. 4143 35
P. Oxy. 4144 35
P. Oxy. 4148 15, 18, 58, 152, 187
P. Oxy. 4149 58
P. Oxy. 4150 61
P. Oxy. 4151 61
P. Oxy. 4152 54
P. Oxy. 4153 54
P. Oxy. 4154 23
P. Oxy. 4155 233
P. Oxy. 4158 54, 58, 135, 140
P. Oxy. 4160 14
P. Oxy. 4160a 30
P. Oxy. 4161 30, 54
P. Oxy. 4162 18, 33, 36, 61, 152
P. Oxy. 4163 18, 33, 61, 109, 120
P. Oxy. 4164 61
P. Oxy. 4164a 185
P. Oxy. 4167 164
P. Oxy. 4175 13, 62, 100, 179
P. Oxy. 4177 63, 141
P. Oxy. 4177a 155
P. Oxy. 4184 62
P. Oxy. 4198
P. Oxy. 4199–4203a 62
P. Oxy. 4199 58, 60
P. Oxy. 4203a 183
P. Oxy. 4204 46
P. Oxy. 4205 45
P. Oxy. 4212 45
P. Oxy. 4216 46
P. Oxy. 4218 46
P. Oxy. 4219 46
P. Oxy. 4225 46
P. Oxy. 4230–4235 46
P. Oxy. 4241 84
P. Oxy. 4263 51
P. Oxy. 4265 8
P. Oxy. 4266 8
P. Oxy. 4272 62
P. Oxy. 4274 62
P. Oxy. 4275 5, 8, 62
P. Oxy. 4276 8
P. Oxy. 4277 8, 11, 47, 289
P. Oxy. 4282 288
P. Oxy. 4283 285
P. Oxy. 4286 276
P. Oxy. 4473 110
P. Oxy. 4477 290
P. Par. 19 250
P. Par. 19bis 285
P. Ryl. 27 19, 35, 180, 186, 263, 310
PSI 1490 35
PSI 1491 35
PSI 1492 117
PSI 1493 116, 153
P. Tebt. 274 14
P. Vind. 3231 175
P. Vind. 29370 175, 190
P. Vind. 29370b 175, 190
P. Vind. 36041+26011 fr. n. 176, 214
P. Warren 21 257
Stobart Tablets 55, 58, 114, 176
Tab. Amst. inv. 1 14, 89
Paulus of Alexandria 285
perpetual almanacs *see* almanacs
planets

anomaly 20, 117
astrological attributes 10–11
Greek letter phenomena 19–21
symbols 62–63, 182, 217–218, 300
kinematic models 119, 167
latitude 68, 101, 105–108, 238–239
order 42, 47, 250
symbols 62–63, 278, 280–281, 299
Pliny 99 n. 1
Plutarch 75
pontifices 13
precession (*see also:* longitude) 113
prediction 5, 5 n. 1
procedure texts 35, 67–69, 80–86, 109–110, 152–153
Ptolemy 1, 48, 79, 95, 113, 116, 119, 151–152
Almagest 14–16, 39, 53, 67–69, 79–80, 90–94, 97–98, 166–167, 237
observation reports 73–75, 84, 89–90, 94, 118–119
star catalogue 71, 80, 92–93
Geography 163
Handy Tables 102–108, 118–119, 160–168, 170
in papyri 38–40, 44–45, 49, 53, 55, 59–61, 160–165, 175–177, 190, 228, 281
used to date papyri 49–50
anonymous third-century commentator 97
calendrical tables 13
geographical tables 102, 104
instruction manuals 35, 68–69, 102–108
mean motion tables 228, 237
planetary latitude tables 105–108
regnal canon 14–17, 102, 310–311
star catalogue 93–94
time conversions 102–105
Phaseis 91
Planetary Hypotheses 15
Tetrabiblos 10–11, 91–92
Ptolemy Philadelphus 76

regnal canons 14
rolls 40, 43, 45, 61
Rome 75, 102

Saturn
tables 117–118, 148–150, 158–160
sexagesimal numerals 9, 15
Sextilis 178
sidereal longitude *see* longitudes
sign-entry almanacs *see* almanacs
Sirius 110
Solinus 13
solstice 113, 117, 120
Standard Scheme *see* moon
steps 10–11, 19, 37–38, 101, 122, 285
Strabo 99

sun
anomaly 18, 119, 151
latitude 119, 170–171
longitude, table for 349–350
symbol 62–63, 280
tables 18, 113, 117, 119–121, 150–153, 170–171
symbols in papyri 61–63
synodic arc 21, 54, 82
System A
planets 19–29, 54, 114, 117–118, 123–134, 137–148, 157
moon 69, 97–99, 227
System B
planets 25–30, 54, 114, 148–150
moon 46, 97–99
syzygies 46
in horoscope 288

Tebtynis 8
templates 33, 37–38, 54–55, 81, 115–118, 150–160, 244
temples 8
terms, astrological 11
Theon, associate of Ptolemy 73, 75
Theon of Alexandria 119
shorter commentary on Handy Tables 40, 102, 105, 343
Theon's formula *see* longitude
Theon of Smyrna 73, 109, 119, 170–171
Thrasyllus 286
Timocharis 68, 84, 89
trigonometry 16–17, 117, 151
triplicities 11, 285, 351
tropical longitude *see* longitude

Uruk 69, 99

Varāhamihira 25, 82, 114
Vasiṣṭha 82, 182
velocity schemes 31–33, 81–84
Venus
calculation of latitude 105–108
tables 135–137, 237
velocity scheme 81–84, 182
Vettius Valens 8, 16, 42, 101, 109, 135–136, 285–286, 289, 310

Waerden, van der 55, 82–83, 114
week, planetary 187,281
winds 11, 42, 101, 107–108, 285
Wirth 83
women 8, 271

years, regnal (*see also:* calendars) 11, 13–14, 199, 250

zero 9, 61–62, 86, 170
zigzag function 25–30, 68, 84–86, 115–116, 148–150, 153
zodiacal signs 9
  astrological attributes 9, 351–352
  represented by numerals 9, 42, 45, 176
  symbols and abbreviations 62, 178, 299

Volume II
Texts, Translations, and Indexes

Part I

Theoretical and Instructional Texts

4133. Menelaus of Alexandria (?), Treatise on Planetary Theory

Col. i

λήνηc. ἐπ' αὐτοῦ ἄρα ἦν τοῦ δι̣ὰ μέcω⟨ν⟩ ἢ βραχεῖ βορειότεροc· ἦν̣ δὲ καὶ ἐξ ἴcου τῷ ἀcτέρι. μετὰ δὲ ἔτη $\overline{\text{τμα}}$ Αἰγυπτιακὰ καὶ ἡμέραc $\overline{\text{πζ}}$ ἡμεῖc ἐτηρήcαμεν τι $\overline{\text{η}}$ τῆc Τραιανοῦ καίcαροc ἡγεμονίαc πρὸ μιᾶc καλ̣ανδῶν̣ Ἰ̣ανουαρίων Μεχεὶρ ζ εἰc τὴν $\overline{\text{η}}$. τῆc̣ διὰ τοῦ νοτίου ὄνου καὶ τοῦ νοτίου τῶν̣ περὶ̣ τὸ̣ νεφ̣έ̣λιον [Φάτν]η̣ν τετραπ̣λ̣ε̣ύρου̣ πρ̣[ο]ηγουμέ[νω]ν̣ εὐθείαc πρ̣ὸ̣[c] ἄ̣ρκτ̣[ου]c̣ ἅμα κα[ὶ π]ρ̣ὸc δυcμὰc εἶχεν c[ε]ληνιακῆ[c] διαμέτρου τῷ ἡ̣μίcει, ἀπὸ δὲ τοῦ νοτείου ὄνου̣ πρὸc ἄ̣ρ̣κτουc ἅμ̣α̣ καὶ̣ πρὸc ἀν̣[α]τολὰc δ̣ι̣ειcτήκε̣[ι] μ̣ίαν cεληνι̣α̣κὴν δ̣ι̣ά̣μετρ[ον] ἢ καὶ π̣α̣λ̣ι̣ν̣ πέμπτο̣ν μέ[ροc] μιᾶc διαμέτρου. ὑπε̣[λ]είπε[το] ἄ̣ρ̣α τοῦ̣ νο̣τίου ὄνου τ̣[. .]τ̣ον [. . . . .] . [. . .] . cεληνιακῆc̣ διαμέ[τρου τοῦ δι]ὰ μέcων βρα̣χὺ βο[ρειότεροc . . . . . .]μενου[.] πάλιν

— — — — — — — — — — — — — —

1 *sc.* (cε)λήνηc   μεcω̅   3 *l.* $\overline{\text{τμδ}}$ 5 τι: *l.* ἔτει?   14 *l.* νοτίου

4133. Menelaus of Alexandria (?), Treatise on Planetary Theory

Col. i

moon. Hence it was right on the ecliptic or a little to the north. It was also in line with the star. After 341 (*sic*, for 344) Egyptian years and 87 days, we made an observation in the 8th (year) of the emperor Trajan's rule, the day before the Calends of January, Mecheir 7/8. (Jupiter) was, from the straight line through the southern Asellus and the southern of the leading (stars) of the quadrilateral around the nebula Praesepe, to the north and also to the west half a lunar diameter; and, from the southern Asellus, it stood away to the north and also to the east one lunar diameter; or in addition a fifth of one diameter. Hence (Jupiter) was to the east of the southern Asellus by [...] lunar diameter, [...] a little to the north of [the] ecliptic.

...

— — — — — — — — — — — — — — — —

Col. ii

(traces of initial letters)

τ[ ]
μ̣[ ]
ω̣[ ]
ε̣ντ ̣ [ ]
νω ̣ [ ]
αcο ̣ [ὑ-]
πελ[είπετο cεληνιακῆc διαμέ-]
τρου [ ]
τῷ τε[τά]ρ̣[τῳ]
τοῦ νοτί[ου]
ρα cεληνν̣[ ]
δαψιλῶc [ ]
cιοc προc̣[τη-]
ρήcαc ἐφα̣[ ]
αc Φαεθο̣[ ]
δὲ κατὰ̣ ̣ [ ]
ετ̣αι δε ̣α̣[ἐ-]
πόμεν̣α̣ [ ]
cεληνια̣[κῆc διαμέτρου]
ωc ἔγγιc[τα]
— — — — — — — — — — — — — —

Col. ii

1–8 …
9 …
10 …
…
…
…
…
15 was to the east [by … of a lunar]
diameter […]
a quarter […]
the southern […]
a lunar [diameter? …]
20 abundantly […]
…
having observed […]
Phaethon (*scil.* Jupiter) […]
tenth? […]
25 …
trailing? […]
of a lunar [diameter …]
approximately […]

4134. Procedure Text for a Planet?

Fr. 1

— — — — — — — — — — — — — — — — — — — —

[*c. 7 letters*]ο̣ν̣[ ] ̣[
[*c. 7 letters*]α̣τα γίνοντα̣[ι] μοῖρ̣[αι
[̣ ̣ ̣ ̣ ̣ ἀ]φαιρῶ ἀπὸ τῶν ι̅α̣̅ [
[*c. 7 letters*]λείπεται α̣̅ [̣ ̣ ̣ ̣] ̣[
[με]ρ̣ίζω παρὰ [τῶν] ρ̣̅α̣̅ [ ] ̣[
ται 𐆊 𐆊 λ̅ζ̣̅ κ̅ε̅. τα̣ῦ̣τα εἰς[ ]ρα ̣[
ἀπὸ τῶν 𐆊 𐆊 𐆊 ε̅ ἐπὶ τὸν α ̣[
τ̣α̣[̣ ̣]νόμε[νο]ν. γίνε̣[ται] ρ̣̅ζ̅ 𐆊 [
[ἀ]πὸ τῶν 𐆊 𐆊 α̅ λ̅ τῆς πρ̣[οσ]θ̣έ̣[σεως
[̣ ̣]πης ἐπὶ τὸν ρ̅ γενόμ̣ε̣ν̣ο̣ν [
𐆊 β ιζ, ὁμο̣ῦ προ⟨σ⟩θέσε̣ω̣ς ι ̣[
ἐπὶ δὲ τῆς ἀ[ν]τ̣ι̣[στρ]οφῆς ἐ̣φίσ̣[ταμαι
[ἐ]ξ̣ῆς αὔξην[*c. 7 letters*] ταύτας ε[
να ἐπὶ τὸν ι̣[ ] ̣η̅ μέσος ̣[
[̣ ̣]α̣τα̣ α̣̅ ν̣̅ ς̅ μ̣̅. ταῦτα ἐκτί̣θ[εται
[*c. 10 letters*]σ̣τ̣η̣ π̣ρ̣ό̣σθεσι[ς
[̣ ̣ ̣ ̣] τ̣ὰ̣ γενόμενα ἐπὶ τὸν α[
[*stripped*] ̣ ̣[
[̣ ̣] ̣ ̣[̣ ̣] ταῦτα ἀπὸ τ̣ῶν ι̅α̅ [
[μοῖρ]αι α̅ ν̣̅ζ̅ ν̣̅ κ̅. ταῦτα μερ̣[ίζω
[*c. 8 letters* γ]ί̣ν̣εται 𐆊 α̅ κ̣̅ [
[*c. 9 letters*] καὶ ἀπὸ τῶν 𐆊 [
[*c. 10 letters*] ̣ ἀπαρτ̣ιζο̣[
[*c. 10 letters*] ̣ ̣ καὶ ἀ̣πὸ [
[*c. 9 letters*]ς 𐆊 𐆊 η̅ κ̅. γίνε[ται
[̣ ̣ ̣] 𐆊̣ 𐆊̣ ̣ν̅, ὁμοῦ προ⟨σ⟩θέ̣[σεως
[̣ ̣ ̣] ι̅ς̅ λ̅γ̅ κ̅α̅ ἐ̣λ̣αχίστης α[

⇓

5 ρ̣̅α̣̅: the stroke above the numeral is not visible

4134. Procedure Text for a Planet?

Fr. 1

— — — — — — — — — — — — — — — — — — — —

...
[...]... there result [...] degrees [
[...] I subtract from the 11 [
[...] there remains 1 [
I divide by 101 [
result? 0 0 37 25. This into [
from 0 0 0 5 to the ...[
... there results 107? 0 [
from the 0 0 1 30 of the increment [
... to 100 resulting? [
0 2 17, total increment ...[
at the turning back I stop [
next increase [...] these ...[
... by the ...[...] x8 mean ...[
... 1 5x 6 40. These are recorded [
[...]... increment [
[...] the result by the ...[
[...]...[
[...]... these from 11 [
[...] 1 57 5x 20 degrees. These I divide [
[...] there results 0 1 2x [
[...] and from 0 [
[...]... completed [
[...]... and from [
[...]... 0 0 8 20. There results [
[...] 0 0 x 50, total increment [
[...] 16 33 21 of the minimum ...[

⇓

⇑

[. . .] μ̣οῖραι ι̅α̅. ἐφίσταμαι α[
[. . . ō] ō ō ε̣̅. ταῦτα παρὰ τ[
[. . .]αcτερην γίνεται ζ̅ μ̅[̣
[*c. 9 letters*] ἐ̣φίσταμαί τε [
[̣] ̣ [̣]c̣ι̣ν[̣] ̣ ̣ α̅ ι̅ε̅. ταῦτα ε̣[

Fr. 2

— — — — — — — — — —
] ̣ [
]ν ι̅ζ̅ κ[
] κ̣̅γ̅ β̣[ ]ντρο̣[
] ̣ c̣η̣[
ō] ọ̄ δ̅ ι̅ ι̣[
] π̣ρ[οcθε]cεωc̣ [
] ̣ ζ̣ καὶ γίνεται ō [
] ἐκ τοῦ ὁμαλοῦ μ̣[
ὁ]μοῦ γίνεται̣ [
] ̣ ̣ι̣ τ̣αc λ̣[οι]π̣ὰ[c
]α̣c⟦α⟧\`ο´μεθα τη[
] ἐλαχίcτηc κει̣[νήcεωc?
— — — — — — — — — —

11 α crossed out, ο written above 12 *l.* κινήcεωc?

⇑

[...] 11 degrees. I stop ...[
[... 0] 0 0 5. These by the [
[...]... there results 7 4[x
[...] and I stop [
[...]... 1 15. These by [

Fr. 2

...
...]... 17 ...[
...] 23 ...[
...
... 0] 0 4 10 ...[
...] increment [
...]x7 and there results 0 [
...] from the uniform ...[
...] together there results [
...]... the remainder [
...] we ... [
...] the minimum motion? [

4135. Procedure Text for Venus

\- -

[ ] ̣[ ]
[ἐν] ἡμέραις [ξ κιν–]
[ήcεται μοί(ρας) οδ, ἑκάcτη]ϲ̣ ἡμέρας α [ιδ· καὶ]
[ἐν ἡμέραις ξ ογ] μοί(ρας), ἑκάcτης ἡ̣[μέρας]
[α ιγ· ἔπειτα ἐν ἡ]μέραις ξ μοί(ρας) οβ, [ἑκάc–]
[της ἡμέρας α ιβ· ἔπ]ε̣ιτα ἐν ἄλλαι[c ἡμέ–]
[ραις μ μοί(ρας) λ, ἑκάcτ]ης ἡμέρας ∻ μ[ε· καὶ]
[ἔπειτα ἐν ἡμέραι]c κ μοί(ρας) ι, ἑκάcτ[ης ἡμέρ–]
[ας ∻ λ· ἔπειτα ἐ]ν ἡμέραις γ μοί(ρας) [∻ ιη,]
[ἑκάcτης ἡμέ(ρας) ∻] ϛ̣ καὶ cτηρίcι ἑcπέ̣[ριος.]
[αἱ πᾶcαι ἡμ]έ̣(ραι) c̣μγ, μοῖ(ραι) cνθ ιη. [καὶ]
[cτρέψεται τ]ὴ̣ν̣ κείν̣η̣c̣ι̣ν̣ εἰ̣c̣ [τὰ]
[προηγούμεν]α̣ ζωιδίων· κ[ά]πε̣[ιτα]
[ἐν ἡμέραις ̣ ̣ ⟦ογ⟧ μοί(ρας) κ̣, ἑ̣κάcτη[c]
[ἡμέ(ρας) ̣ ̣ ̣ · καὶ ἐν] ἡμέραις ξ μοί(ρας) δ̣, ἑ̣[κάc–]
[της ἡμέ(ρας) ̣ ̣ ̣ ·] ἔ̣πε̣ιτα ἐν ἡμέρ[αις]
[̣ ̣ μοί(ρας) ̣, ἑκάcτη]ϲ̣ ἡμέρας μοί(ρας) ∻ λ̣[̣]
[*c. 8 letters* ἑκάc]τ̣[ης] ἡ̣μέρ(ας) ∻ ̣ ̣.
[*c. 6 letters* αἱ πᾶcαι] ἡ̣μέραι λη καὶ̣
[μοῖ(ραι) ιη ϛ. καὶ cτρέψε]ται τὴν κείν̣η(cιν)
[εἰc τὰ ἑπόμενα ζ]ω̣ιδίων· καὶ ἐ̣̣ν̣
[ἄλλαις ἡμέραις γ] μ̣ο̣ί̣(ρας) ∻ ιη, ἑκάc̣τ̣η̣ς̣
[ἡμέρας ∻ ϛ· ἐν ἡμέ]ραις κ μοί(ρας) ι̣, ἑ̣κ̣ά̣–
[cτης ἡμέ(ρας) ∻ λ· ἐν ἡ]μ̣έραις μ μοί(ρας) λ̣,
[ἑκάcτης ἡμέ(ρας) ∻ με·] ἔπειτα ἐν ἡμέ(ραις)
[ξ οβ, ἑκάcτης ἡμέ(ρας)] α ιβ· ξ μοί(ρας) ογ,
[ἑκάcτης ἡμέ(ρας) α ιγ·] ἔπειτ' ἐν ἡμέ(ραις)

⇓

4 μο : so in subsequent lines   10 *l.* cτηρίcει   11 ημε : so in subsequent lines except 18
12 *l.* κίνηcιν   14 ογ is crossed out, and an M-shaped symbol is written above the horizontal stroke
18 ημερ   20 κειν^{η} : *l.* κίνηcιν

4135. Procedure Text for Venus

[...] [... in 60] days [it will move 74°], each day 1;[14°; and] [in 60 days, 73]°, each day [1;13°; then in] 60 days, 72°, [each day 1;12°]; then in another [40 days, 30°], each day 0;4[5°; and]; [then in] 20 days, 10°, each [day 0;30°; then] in 3 days, [0;18]°, [each day 0;]6$^{(\circ)}$; and it will be at evening station. [Total] 243 days, 259;18°. [And it will turn] its motion to [the leading] signs; and then [in xx days], 20$^{?\circ}$, each [day 0;xx$^{\circ ?}$; and in] 60$^{!}$ days, 4°, each [day 0;xx°]; then in [xx] days, [x°], each day 0;3x°; [...] [...] each day 0;x$^{(\circ)}$ [... Total] 38 days and [18;6°. And it will turn back] its motion [to the trailing] signs; and in [another 3 days], 0;18°, each [day 0;6°; in] 20 days, 10°, each [day 0;30°; in] 40 days, 30°, [each day 0;45°]; then in [60] days, [72°, each day] 1;12$^{(\circ)}$; (in) 60 (days) 73°, [each day 1;13°]; then in [60] days,

⇓

[ξ μοί(ραϲ) οδ, ἑκάϲτηϲ] ἡμέ(ραϲ) α ιδ· καὶ
[δύϲ]εται ἑῷιοϲ. αἱ π[ᾶϲ]αι ἡμέραι
[ϲμγ,] μοῖ(ραι) ϲνθ ιη. ἔπειτα κι[ν]ή[ϲ–]
[ε]ται [ἐ]ν ἡμέραιϲ ξ μοί(ραϲ) [οε, ἑκάϲτηϲ]
[ἡμέρ]αϲ α ιε· καὶ ἀνατελεῖ ἑϲπέ[ριοϲ.]
[ταῖϲ] τῆϲ διεξαγωγῆϲ ἡμέραι[ϲ φπδ]
[μοί(ραϲ) ϲι]ε λ εἰϲ τὰ ἑπόμε[να ζωιδ–]
[ίων] πορεύεται. ἀρχὴ δὲ [*c. 6 letters*]
[*c. 5 letters*] τῶν Τειμοχάριδο[ϲ *c. 4 letters*]
[*c. 7 letters*]εωϲ. ἔϲτι δ' ἀν[*c. 5 letters*]
[*c. 7 letters*]⌐ Ταύρου δ ζ [*c. 5 letters*]
[*c. 8 letters*]ϲ (*vac.*)
[*c. 10 letters* ψ]ηφου (*vac.*)
[*c. 12 letters*]εια κατακι[
[*c. 12 letters*] . . . . [
[*c. 12 letters*]ϲιαϲ ἀποκ[
[*c. 12 letters*]εται ιδ ἀπο[
[*c. 12 letters*]α γ(ίνεται) ϛ . . ‾ ι[
— — — — — — — — — — — — — — — — — — —

36 *l.* Τιμοχάριδοϲ   41–45 abraded   45 ⌐┼

[74°, each] day $1;14^{(\circ)}$; and it will set in the morning. Total [243] days, 259;18°. Then it will move in 60 days, [75]°, [each] day $1;15^{(\circ)}$; and it will rise in the evening. In the [584] days of the whole cycle, it travels [21]5;30° towards the trailing [signs]. The beginning [...] [...] of Timocharis [...] [...] It is [...] [...] Taurus $4;7^{?(\circ)}$ [[...] ... [...] computation$^{?}$ [...] ... [[...] ... [[...] ... [[...] $14^{?}$ [[...] there results $6;0,x^{?}$ [

4136. Procedure Text for the Moon

ὡc δεῖ εἰδέναι τὴν κατὰ κζ–[cτιχ(α) καὶ κη–cτιχ(α) ἐπιcυν–] αγομέ(νην). πολυπλαcί(αcον) τὴν μέcη[ν κίν(ηcιν) ἐ]πὶ τὰ κ΄η΄–[cτι]χ(α)· [τ]αῦ[τ]α γ(ίνεται) τξη νς ις [ις . .] . . καὶ πο[λυπλ(αcίαcον) τὴ]ν κη–cτιχ(ων) πρ(όcθεcιν) ∻ ε μ[γ ἐπὶ] ζ ὅ ἐcτιν d΄ τῶν κη–[cτί]χ(ων)· γ(ίνεται) ∻ μ ∻ α. ταῦ[τα ἐκ τῶ]ν ἐκ–κειμένων κ(αταλείπεται) τξη ις ιε ις, [ὅ ἐc]τιν ἡ ἐπὶ παcῶν κη–cτίχ(ων) ἐπιcυναγομ(ένη).

ἐπὶ δὲ τῶν κζ–cτ(ιχων) ἀφέλε ἀπὸ τῶ[ν τξη] ις ιε ις τὸν ἐλάχιcτον, μοί(ραc) ιλ μβ ι λζ, [καὶ γ(ίνονται)] μοῖ(ραι) τνς λδ δ λθ. καὶ πρόcθε[c τούτ]ων τὴν ὑπεροχὴν τῆc ἀφ(αιρέcεωc) τῶ[ν κζ–cτιχ(ων) πα]ρὰ τὴν πρ(όcθεcιν) τῶν κη–cτιχ(ων), ∻ α κδ· γ(ίνεται) τν[ς λε κη] λθ. τηλικαύτη ἐcτὶν ἡ ἐπιcυναγομ(ένη) τοῦ γ̄. ἐπεὶ δὲ αὕτη λείπει πρὸc τὰc τξ μοίραc γ κδ λα κα, ἀπὸ τούτων ἀφ(έλε) τὴν α κδ ∻· κατα–λείπεται τοῦ ε̄ cελιδ(ίου) γ κγ ζ κα. πάλιν ἀπὸ τῆc ∻ α κδ ἀφέλε τῆc αὐξήc(εωc) τὸ ρ΄ι΄, ∻ ∻ ζ· κ(αταλείπεται) ∻ α ιζ. ταῦτα πρ(οcτίθει) τῆι ἐπιλείψει τοῦ ε̄ cελιδ(ίου) πρὸc τὸν κύκλον, μοί(ραιc) γ κδ ζ κα, καὶ ἔcται τοῦ ζ̄ cελ(ιδίου) γ κδ κδ κδ. τούτοιc προcήχθω ὅλη ἡ ὑπεροχή, ∻ λ κδ· γ(ίνονται) τοῦ ἐνάτου γ κε μη κδ. ὁμοίωc προcκείcθω . . [*c. 8 letters* ι]΄[α]΄ [cελ(ιδίου) *c. 8 letters*]

— —

1 κζ–cτιχ(α) and κη–cτιχ(α) written throughout as, e.g., (line 8) or (line 7): here the superscript is lost 2 αγομ^{ε}; πολυπλαcι 3 : so lines 5, 12, 21 4 : so lines 12, 18 5 *l.* ∻ μ α 6 7 επιcυναγο: so line 13 9 (*bis*): so line 19; *l.* ια 11 αφ : so line 15 16 cελ : so 19 17 αυξη^{c}? 20 cε^{λ}; *l.* κδ κδ κα 22 *l.* κα

4136. Procedure Text for the Moon

How one should determine the total for 27-[line and 28-line (columns).] Multiply the mean [movement] by the 28 lines; there results 368;56,16,[16$^{\circ?}$.] And multiply the amount added for 28-line (columns), 0;5,4[3, by] 7, which is $^1/_4$ of the 28 lines; there results 0;40,1. (Subtracting) this from the foregoing, there remains 368;16,15,16, [which] is the total for all 28-line (columns).

For the 27-line (columns), subtract from the [368];16,15,16 the minimum (course), 11;42,10,37°, [and there results] 356;34,4,39°. And add to these the excess of the amount subtracted for [27-line (columns)] over the amount added for 28-line (columns), 0;1,24; there results 35[6;35,28],39. This is the size of the total for the third (column). And since this falls short of 360° by 3;24,31,21, subtract from this the 1,24,0; there remains for the fifth column 3;23,7,21. Again, from the 0;1,24 subtract $^1/_{110}$ of the increment, (namely) 0;0,7; there remains 0;1,17. Add this to the amount by which the fifth column falls short of a circle, (namely) 3;23,7,21°, and there will be for the seventh column 3;24,24,21. Add to this the entire excess, 0;1,24; there results for the ninth (column) 3;25,48,21. In the same way let there be added ...[... the eleventh column ...]

4137. Predictions of Lunar Eclipses, A.D. 56 and 57

– –

. . ἐ̣γ̣[λ(είψει)] δ̣ὲ̣ ἀπὸ νότο̣[υ *c. 11 letters*] [ἔ]λαϲϲον ἢ τὰ δύο μέ̣[ρη τῆϲ διαμέ(τρου), καὶ] τ̣ῆϲ ἐγλείψεωϲ ἔϲτ̣[αι ὁ πᾶϲ χρόνοϲ] π̣ερὶ ὥραϲ γ̄. ἡ μὲν ο[ὖν πρόϲν(ευϲιϲ) πρώτου] ἐγλ(ελοιπότοϲ) ἔϲται μεταξὺ μ̣[εϲημβρίαϲ καὶ] ἀνατολ(ῆϲ)· περιϲτήϲε̣[ι δὲ πλεῖϲτον] ἐγλελοιπὸϲ πρόϲν̣(ευϲιν) [ὡϲ πρὸϲ μεϲημβ(ρίαν)·] ἔϲχατον δ' ἀναπ[ληρούμενον ὡϲ] μεταξὺ μεϲημ(βρίαϲ) καὶ δ̣[ύϲεωϲ. κατὰ] δ̣ὲ τὸ πλεῖϲτον ἐγλ(ελοιπὸϲ) [προϲνεύϲει] ἡ̣ διχοτομοῦϲα τ̣[ὰ τῆϲ ϲελήνηϲ] τ̣μήματα διάμ[ετροϲ ὡϲ πρὸϲ] μεϲημ(βρίαν). ἔϲται δὲ ἡ [ϲελ(ήνη) ἐν τοῖϲ Διδύ–] μοι̣ϲ̣ μεταξὺ τοῦ δ[εξιοῦ *c. 5–7 letters*] [τ]ο̣ῦ̣ ἡγουμένου δ[ιδύμου καὶ] [το]ῦ̣ ἐν τῷ ὀμφαλ[ῷ τοῦ ἑπομέ–] [νου] διδύμου ἀϲτέ[ροϲ.] δ]ι̣[ὰ δ]ὲ̣ μ̣ηνῶν ϛ̄, (ἔτει) γ´ κ̣[*c. 5 letters* κατὰ] δ̣ὲ Κάλλιπ(πον) τῆϲ ϛ̄ πε̣[ριό(δου) (ἔτει) ϛ̄, διχό–] μηνοϲ μηνὸϲ Θαργ[ηλιῶνοϲ, μη–] ν̣ὸϲ δ' Αἰγυπτίο(υ) τοῦ Ἐπ[εὶφ ᾱ εἰϲ β̄,] [ἔ]ϲται ἡ παϲϲέλ(ηνοϲ) τοῦ με̣[*c. 8 letters*]

– –

1 εγ^{λ}? (only traces of the bottoms of the letters visible): *l.* ἐκλείψει : sim. lines 3, 5, 7, 10   5 εγ^{λ} 6 ανατο^{λ} 7 προϲ^{ν} 9 μεϲημ^{β} 10 εγ^{λ} 13 μεϲημ^{β} 14 μεταξ^{υ} 17 διδυμο^{υ} 18 L ; γ´ κ̣ : ΓΙΚ   19 καλλι^{π} 21 αιγυπτι^{ο} 22 παϲϲε^{λ}: *l.* πανϲέληνοϲ

4137. Predictions of Lunar Eclipses, A.D. 56 and 57

... [The moon's disk?] will be eclipsed from the south by less than two thirds [of its diameter]. [The total duration] of the eclipse will be about three hours. The [inclination at first] obscuration will be between south [and] east; the [maximum] obscuration will turn the inclination [to the south;] and the final emersion between south and west. [At] the maximum obscuration the diameter bisecting the [moon's] segments [will incline towards] the south. The [moon] will be [in] Gemini between the right? [...] [...] of the leading twin [and] the star in the navel of the [trailing] twin.

At an interval of 6 months, in the year 3? [..., according] to Callippus [the 6th year] of period 6, mid-month of the month Thargelion, and in the Egyptian month Epeiph [(night of days) 1/2,] the full moon will be [...]

4138. On Eclipse Prediction

— — — — — — — — — — — — — —

[ ]β̣αί̣νει̣[ ]
[ ] Δ̣ιδύμω[ν]
[ ]βιβ̣άζων α̣[ ]
[ ] ̣ ̣ τμ̣ήμ̣[α-]
τα δε̣ ἰ̣c̣η̣μ̣ε̣ρ̣ι̣ν̣ὰ ⸓ ν̅ κ̅ε̣̅ τ̣[οῦ]
μέcου τῆc ἐκλεί̣ψεώc ἐ[cτι]
χρόνου. ἐὰν γὰρ τὸ νυχθήμ[ε-]
ρον διέλω εἰc ξ λεπτὰ, ἀπ[ο-]
λ̣αβὼν δὲ ἀπὸ τοῦ μεc̣ο̣ν̣υ̣[κτί-]
ο̣υ τὰ παρακείμενα ⸓ ν̅ κ̅ε̣̅,
λοιπὰ καταλείπεταί μοι π̣[ρὸ]
με̣c̣ο̣ν̣υκτ̣ί̣ο̣υ̣ τ̣ῆ̣c μελλ[ού-]
c̣ηc̣ [νυκτὸc ̣ ̣ ̣] Μεχε̣[ὶρ]

— — — — — — — — — — — — — —

5 ⸓: written ⁊ here and in 10

4138. On Eclipse Prediction

[...]scends [...] [...] Gemini [...] node[...] [...] xx. The equinoctial divisions, 0;50,25, are of the middle time of the eclipse. For if I divide the day-and-night into 60 sixtieths, and counting off from midnight the 0;50,25 under discussion, there is left to me as remainder before midnight of the following [night ...] Mecheir

4138a. TREATISE ON ECLIPSE PREDICTION

Col. i

] ̣ω τοῦ πλα–
]ασεων πα–
] ἐν τῇ πε–
]εν γὰρ εν
ὅ]λοις κύκλοις
]ν ἐπελαμ–
] ̣τοιουδο–
]γ̣ε̣ι̣ν̣[̣ ̣ ̣ ̣]
— — — — —

Col. ii

λ̣[ης] τ̣ῆς ἐγλε̣[ιπτικῆς] π̣[ε–]
ριφερείας, ὅθεν̣ [ἤρξα]τ̣ὸ π̣[α–]
ραγεί̣ν̣εται ἡ̣ c[ύνεγ]γ̣υς, ἐ[μ–]
φανὲς ὅτι ἐπεὶ [̣ ̣ ̣ ̣]κ̣ον ἡ̣[μι–]
κυκλίου ἐπιλ̣α̣μ̣β̣[άν]ει τ[̣ ̣ ̣]
ταῖς πενταμήνοις ὅλο[υ, ̣ ̣ ̣ ἀ–]
φαιρεῖται ἐν ὅλῃ τῇ τ̣ῶ̣[ν ἐ–]
γλείψεων περιόδῳ τ̣[οcαῦ–]
[τα ἡμ]ι̣κύκλια ἡ cελήν[η ̣ ̣]
[*c. 8 letters*] ὅcα ἐ̣cτὶν ἐν τῇ
[περιόδῳ ἐξ]αμη̣νιαῖα καὶ πε[ν–]
[ταμηνιαῖ]α δ̣ι̣αcτήματ̣α̣.
[*c. 10 letters*]υ̣ ἀ̣ναcτρ̣[̣ ̣ ̣]
[*c. 11 letters*] περιοδ ̣[̣]
[*c. 9 letters* ἡ]μ̣ικυκλι̣[̣ ̣]
[*c. 11 letters*] ̣ ̣ ̣[̣ ̣ ̣] ̣[̣]
— — — — — — — — — —

1 *scil.* ὅλης; *l.* ἐκλειπτικῆς : sim. lines 7–8 2–3 *l.* παραγίνεται

4138a. Treatise on Eclipse Prediction

Col. i

(Text too broken for translation)

Col. ii

entire circumference of the ecliptic, it comes back to where it began or very close, it is manifest that since … takes up … a whole[?] semicircle in[?] the (groups of) five months, the moon deducts[?] in the entire period of eclipses as many semicircles […] as there are six-month and five-month intervals in the period. … revolving backward … … period … … semicircle(s[?]) … …

4139. Treatise on Lunar Periods

] ε̣ὑρη̣[ ]λ̣α̣ ̣ ̣
] ̣ κοc ,ϛ̅χ̅ϙ̅ε̅ ἢ
,δ̅υ̅]ο̣̅ζ̅ καὶ ὡρῶν Ƹ
]ι̣ ἐφύλαξαν αἱ
χρ]ό̣νου ἐκλείψειc
χ]ρόνου αἰεὶ γει̣– Ƹ
] ὕ̣cτερο̣ν δὲ του Ƹ
]α̣ντο[ ]ορχη– Ƹ̣
] ἀποκαταcτα– Ƹ
] ἔφαcαν ἐμπε–
] ̣ αρτων ἐγ̣ ̣ ̣ [

– – – – – – – – – – – –

2 ,ϛ : ϛ 6 *l.* γί[νονται] ?

4139. Treatise on Lunar Periods

[…] discovered …
[…] longitude? 6695 or
[… 44]77 [days] and [x] hours
[…] … preserved …
[…] time, eclipses
[…] time, always …
[…] later than this?
[…] … the Orchenoi?
[…] periodic restitutions
[…] they said …
[…] … […]

4140. Tabulated Lunar Phenomena

i	ii
[. . .]	[καταβαί]ν̣ο̣[υcα
	μφ̣ τ̣[ῆ]c̣ γενέc̣ε̣ω[c
	cελήνη οὖcα ἐν Ἰχ̣[θύcι
φοα	τὰ βόρια ἣ καὶ cυν̣[
	νοc ἀντὶ τοῦ ἀν̣α̣τ̣[
	cυνῆπ̣τ̣εν ἡ cελή̣[νη
φοβ	ἀναβαίνο̣υcα τὰ βόρι[α
	λοιπῶν ἀcτέρων [
	cελήνη οὖc̣α̣ [

4 l. βόρεια : so line 7

4141. On Latitudes of a Planet

]. .[
]α̅ μο(ίραc) τ̣[οῦ Ταύρου
ἀναβ]αίνει τ̣[ὰ βόρεια
]α̣̅ μο(ίραc) τοῦ Λ̣[έοντοc
κατα]βαίνει τ[ὰ βόρεια
]α̣̅ μο(ίραc) τοῦ Σκ̣[ορπίου
καταβ]α̣ί̣[ν]ει τ[ὰ νότια

2 μ̊: so lines 4, 6

4140. Tabulated Lunar Phenomena

i	ii
	descending? [... of the geniture [...
571	The moon being in Pisces? [the north or even at the node? [... against? the eastern? [
572	The moon reached [ascending the north [the rest of the stars [
	The moon being [

4141. On Latitudes of a Planet

]...[
]1° of [Taurus
] ascends? [the north?
]1° of Leo [
] descends? the [north?
]1° of Scorpio [
] descends? the [south?

4142. Instructions for Ptolemy's *Handy Tables*

Fr. 1

— — — — — — — — — — — — — — — —

ι̣[ ]ε̣ ̣[ ]
κε̣ ̣ ̣[ ] ̣ ̣[ ]
̣ ̣ ̣[ ] ̣ ̣ ̣[ ]
̣ ̣[π]ολυπλαcί̣α̣[cον]
[ἐπὶ τὰc ἤτοι] πρὸ μεcημ–
βρ̣ι̣α̣c̣ ἢ μετὰ μεcημβρ̣ί̣[αν]
ὥρ̣(αc) κ(αὶ) τὸν γινόμεν(ον) ἀριθμ̣(ὸν)
[ἐ]κ̣ τοῦ πολυπλαcιαcμοῦ π̣(αρά)–
[β]α̣λ̣ε̣ π(αρὰ) τ(ὸν) $\overline{ιε}$. ὅ̣cον ἐὰν γ̣(ίνεται)
τ̣ὸ̣ τ̣(ῶν) ὡρ̣(ῶν) πλῆ̣θ̣ο̣c, τοcαύτα̣c
ἕξομ(εν) ἰ̣c̣η̣μεριν̣ὰc ἤτοι πρὸ
[μ]εcημβ̣[ρί]α̣c ἢ μετὰ μεcημβρίαν̣.
[̣]προ̣ ̣[ ] ̣[ ] ̣αμεν ̣ ̣ ̣ ̣
ἐκ π(αρα)[δείγματοc. ὑπο]κ̣ε̣ί̣cθω γ̣ὰ̣[ρ]
ὑ̣(πο)δείγματοc χ̣[ά]ρ̣ι̣ν̣ τὰc δο̣[θεί–]
c̣αc κ(αι)ρικὰc ἡ̣μερινὰ[c] ὥρ(αc)
[ἀ]ν̣αλῦcαι ε[ἰc] τ̣ὰc ἰcημ̣[ερινάc,]
τ̣οῦ (ἡλίου) ἐ̣πέχοντοc ὡ̣cα[ύτωc]
[Ζ]υγοῦ μο(ίραc) $\overline{ϛ}$ $\overline{κϛ}$ [*c. 8 letters*]
[το]ῦ οἰκείου κλ[ίματοc ̣ ̣]

Compendia regularly employed: ꝁ = και (also as part of word), ϧ = δια, ꝕ = ὡρα etc., μ̃ = μοῖρα etc., ⳁ = γίνεται etc., δ′ = δὲ, μ′ = μεν (also as part of word), π\ = παρα or παρ–, α′ = ἀπο or ἀπ–, ε′ = ἐπι or ἐπει, κ′ = κατα or κατ–, ′ = –ουc or –ον, \ = –ων

9 l. ἂν 10 ωρ\ 15 υ′ ? 18 ☉

4142. INSTRUCTIONS FOR PTOLEMY'S *HANDY TABLES*

Fr. 1

– – – – – – – – – – – – – – – – –

... ... multiply [by the] hours [whether] before noon or after noon, and the number resulting from the multiplication divide by 15. As great as is the number of hours, we will have this many equinoctial (hours), either before noon or after noon. ... by means of an example[?]. For let it be required for example to convert the given seasonal hours of day into equinoctial (hours), with the sun again occupying Libra 6;26° ... the appropriate *clima* ...

Fr. 2, col. i

(*traces of 6 lines*)
[. . .] κ̣(ατ̣ὰ) δ̣(ιά)μ̣ετ̣ρ̣(ον) χ̣ρ̣(όν̣ο̣υ̣c) . . [*c. 16 letters*] ὡ̣ρ̣(ῶν)
[. . . .]ι . ἐκ τῆc τῶν β̣ [*c. 16 letters*]c̣
[ὥρ(αc)] ἰcημεριν̣ὰ̣c ἀ̣[*c. 16 letters*]τ̣ου
[. . .] . . . π̣α̣ρ̣έ̣χ̣ειν [ὑποκε]ι̣c̣θ̣ω̣ [τὰc δοθεί]c̣αc
[ἰcη]μ̣ε̣ρ̣ιν̣ὰc ὥρ(αc) δ νδ ἀ̣ν̣α̣λ̣ύ̣cα̣[ι εἰc τὰc πρὸ μ]εc̣–
[ημβρ]ίαc ἢ̣ μ̣ετα̣ [μεcημβρίαν] . ουc̣
[. . .] . . . γ(ίνεται) [. . .] . . τ̣ . [*c. 10 letters*] . τ(ῶ̣ν̣?) ἰ̣cη–
[μεριν(ῶν?) . . . . .] . . μ̣ . [. . . .] π̣(αρα)βάλλε̣ π(αρὰ)
[τοὺc αὐτοὺ]c χρόνο̣υc̣ τ̣οὺc π(αρα)κειμ̣έν̣ουc
[τοῖc τοῦ Ζυγ]ο̣ῦ̣ μ̣ο(ίραιc) ϛ κϛ οἳ ἦcαν ιδ μ̣δ. τ̣α̣ῦ̣τα̣ ο̣ῦν
[*c. 7 letters*] . . ογ μ̣ π(αρ)έ̣[χ]ο̣υcι π̣(αρὰ) τ̣ὰc ιδ μ̣δ
[ὥραc κ(αι)ρι]κ̣ὰ̣c̣ ε. καὶ ἀνελύcαμ(εν) τ̣ὰc δο–
[θείcαc ἰcη]μ̣ε̣ρ̣[ι]ν̣ὰ̣c δ̣ ∟′ γ′ ιε′ ὥ̣ρ(αc) εἰc τὰc κ(αι)ρ–
[ικὰc . . . . . .] . . [.] . . . π(αρα)δείγματ̣ι καὶ ε .
] . [ ] . ε̣θ̣α̣[ ] ἕξομ̣[εν]
] μ̣ε̣θ̣ώ̣δευμε̣ν δε[.]
] . α̣ . καθ' (ὃν) δε . .
(*traces of approximately 20 lines*)

7 χ ?

Fr. 2, col. i

(*traces of 6 lines*)

… diametrically opposite time-degrees … … from the … of the 2$^?$ … equinoctial [hours] … … provide an example, to convert the given 4;54 equinoctial hours to [seasonal hours] either before noon or after noon … … there results … of the equinoctial … divide by the same time-degrees corresponding to Libra 6;26°, which were 14;44. Hence these … 73;40 divided by 14;44 yield 5 seasonal [hours]. And we have converted the $4\,{}^1/_2\,{}^1/_3\,{}^1/_{15}$ equinoctial hours to seasonal … for example … … we will obtain … … we proceed … …

(*traces of approximately 20 lines*)

– –

Fr. 2, col. ii

χρόνους ι̅η̅ μ̅. ἐπεὶ οὖν ἡ δοθεῖcα κ(αι)ρικὴ ὥρ(α) β̅ πρὸ δ̅ ὡρ(ῶν) ἐcτὶ κ(αι)ρικ(ῶν) τῆc μεcημ̣βρίαc, πο̣⟨λ⟩̣υπλαcιά[ζ]ω τὰc δ̅ ὥρ(αc) ἐ(πὶ) τ(οὺc) ι̅η̅ μ̅, καὶ γ(ίνεται) ο̅δ̅ μ̅. ταύταc παραβαλ(ὼν) π(αρὰ) τ(οὺc) ι̅ε̅, γ(ίνεται) ἰcημεριναὶ ἔνγιcτα ὥ̣ρ̣(αι) ε̅. ἡ ἄρα δ̣ο̣θ̣[ε]ῖcα ἐν Ῥώμῃ κ(αι)ρικὴ ὥ̣ρ(α) β̅ ἐ[cτὶν] π[ρὸ ε̅] ὡρ(ῶν) ἰcημεριν(ῶν) τῆc ἐν Ῥώ̣[μῃ] μεc̣η̣μ̣[β]ρ̣ίαc, τῆc δ(ὲ) προτέραc με̣[c]η̣μ̣β̣[ρίαc ἔ]c̣ται μετὰ ι̅θ̅ ὥρ(αc) ἰcημερινάc. το̣ύ̣τ(ων) γ̣ε̣γ̣[ε]ν̣ημέν(ων) λαμβάνω τ̣ὸ μῆκοc τοῦ δ(ιὰ) Ῥώμη[c π(αρ)]αλλήλ̣ο̣υ̣ κ(αὶ) τ̣οῦ δ(ι') Ἀλεξανδρείαc· ἔ̣cτιν δ(ὲ) τοῦ μ̣(ὲν) δ(ιὰ) Ῥώμηc μο(ῖραι) λ̅ϛ̅ κ̅, τοῦ δ(ὲ) δ(ι') Ἀλεξανδρείαc μο(ῖραι) ξ̅ λ̅. κ(αὶ) φανερ(ὸν) ἐ̣κ̣ τούτ(ων) ὅ̣τι ὁ δ(ιὰ) Ῥώμηc μεcημβρινὸc δ̣υcμικώτερόc ἐcτι τοῦ δ(ι') Ἀλεξανδ̣ρ̣ε̣ίαc μεcημβρινοῦ· τὸ γὰρ μῆκόc ἐcτι̣ν̣ ὡc εἴρηται ἡ ἀ(πο)χὴ ἣν ἀ(π)έχει ὁ μεcημβρινὸc τοῦ δυcμικωτάτου. ἔcτιν οὖν ἡ τούτ(ων) τ(ῶν) μή[κων] ὑ̣περοχὴ ἐν μο(ίραιc) κ̅δ̅ ι̅, αἵτινεc̣ π(αρα)βληθε[ῖc]α̣ι π(αρὰ) τ(οὺc) ι̅ε̅ π(αρ)έχουcιν ὥρ(αc) ἰ̣c̣ημ̣ε̣ρινὰ̣[c] τ̣ὰc̣ α̅ κ(αὶ) τρία πεμπταμώρια ὥc ἔνγιc[τα. τ]α̣ύταc ἀφαιρῶ ἀ(πὸ) τ(ῶν) ε̅ ὡρ(ῶν) ἰ̣cημεριν(ῶν) τῆc ἐ̣ν Ῥώμῃ μεcημβρίαc, ἐ(πει)δήπερ οἱ ἐν [το]ῖc δυ⟦cμ⟧`τ´ικωτέροιc τόποιc π̣ρ̣ό̣τ̣ερο̣ι̣ εἰ̣cι τ(ῶν) ἐν τοῖc̣ ἀνατο̣λικω̣τ̣έροιc. ε̣ἰ̣ γ̣ὰ̣ρ ἦν ἀ̣ν̣ά̣[το]λικώ̣[τερ]ο̣c ὁ δ(ιὰ) Ῥώ̣μ̣[ηc] μ̣εcη̣[μβ(ρινόc), κ(ατὰ)] τὸ ἀ̣ν̣[άπα]λ̣ιν ἂ̣ν προc̣[ε]τ̣ίθου[ν. ἀφαί]ρεθειc(ῶν) οὖν ὡρ(ῶν) ἰcημεριν[(ῶν) α̅ κ(αὶ)] γ̅ πεμπ̣τ̣ημορί(ων) ἀ(πὸ) τῶν ε̅ ὡ̣ρ(ῶν) ἰcημ(ερινῶν) [τῆc ἐν Ῥώ]μ̣η μ̣ε̣c̣ημβρίαc, κ(ατὰ)λείπεται [ὧρ(αι) τρεῖc] κ̣(αὶ) β̅ πεμπ̣τ̣ημ̣όρια. πρὸ τοcούτ̣[ων ἄρα]

⇓

5 *l.* ἔγγιcτα : so line 21   20–21 *l.* πεμπτημόρια   23 cμ crossed out, and τ written above
27 *l.* προcετίθην

Fr. 2, col. ii

18;40 time-degrees. Hence since the given second seasonal hour is 4 seasonal hours before noon, I multiply the 4 hours by 18;40, and there results 74;40. This I divide by 15, and there result approximately 5 equinoctial hours. Hence the given second seasonal hour at Rome is [5] equinoctial hours [before] noon at Rome, and it will be 19 equinoctial hours after the preceding noon. This being so, I take the longitude of the parallel[!] through Rome and (the longitude) of that through Alexandria: that of the (meridian) through Rome is 36;20°, while that of the (meridian) through Alexandria is 60;30°. And it is obvious from this that the meridian through Rome is west of the meridian through Alexandria. For the longitude, as has been stated, is the distance of the meridian from the most westerly (meridian). Hence the difference between these longitudes in degrees is 24;10, which, divided by 15 yield approximately 1 $^3/_5$ equinoctial hours. I subtract these from the 5 equinoctial hours (before) noon at Rome, since those (people[?]) who are in more westerly localities are earlier than those who are in more easterly (localities). For if the (meridian) through Rome had been more easterly, on the contary I would add. Hence when 1 $^3/_5$ equinoctial hours have been subtracted from the 5 equinoctial hours (before) noon at Rome, there remain [3] $^2/_5$ [hours]. This is the number (of hours), therefore,

⇓

τ̣ῆc ἐν ’Αλεξ(ανδρείαc) μεcημ(βρίαc) ἡ δο̣θε̣ῖ[cα]
κ(αι)ρική ὥρ(α) β̣̅, τουτέcτι μετὰ κ̅ κ(αὶ) γ̣̅ [πεμ–]
πτημο̣ρια τῆc προτέρου τῆc μεcη[μβρίαc.]
κ(αὶ) ἀ̣νελύcαμ(εν) τὰc δοθείcαc ἐν 'Ρώμῃ κ̣(αι)[ρικὰc]
ὥρ(αc) β̅ εἰc τὰc ἰcημερινάc, κ(αὶ) ἐγένετ̣[ο ὥρ(α) ἰcημε–]
ρι̣ν̣ὴ α̅. κ(αὶ) προcεθήκαμ̣(εν) τὴν τ(ῶν) δύο τόπ̣[ων]
δ(ια)φοράν, τουτέcτι τὴν ἰcημερινὴν [ὥρ(αν) α̅]
κ̣(αὶ) τρία πεμπτημόρια, κ(αὶ) ἐγένετο ἐν ['Αλεξαν–]
δ̣ρείᾳ̣ ὧρ(αι) ἰcημεριναὶ β̅ κ(αὶ) τρία πεμ[πτημό–]
[ρ]ια, αἵτινέc ⟨ἐc⟩τι τῆc προτέραc μ̣[εcημβρί–]
αc μετὰ ὥρ(αc) ἰcημ̣[ε]ρ̣ινὰc κ̅ κ(αὶ) γ̅ πεμπτη̣(μόρια).
κ(αὶ) ἐκ τού̣[του φανερ](ὸν) ὅτι καθόλου ταῖc [δοθεί–]
c̣αῖc ὥρ(αιc) κ̣(α̣ι̣)[ρικαῖc ταῖ]c̣ ἐν τοῖc δυτικ[οῖc]
[πρ]ο̣c̣τιθ̣[έαμεν τὴν τ(ῶν)] δύο τόπ(ων) δ(ια)φο̣ρ̣[άν]

42 πεμπτη

Fr. 3, column i

κ(αὶ) [ἐ]ξηκοcτῶν οἷc διοίcει τὰ κ(ατὰ) [. . . . . .]
ἐ̣π̣οχ̣ῆc εἰc ἐκεῖν(ον) τ(ὸν) χρόνο[ν . . .]α̣
⟨ὁ⟩μαλὰ νυχθήμερα τῶν φαιν̣ο̣μ̣έ̣[νων]
κ(αὶ) . . . . . ἐὰν ἀ(πὸ) τ(ῶν) ἀνωμάλ̣(ων) ν̣υ̣–
χ̣θ[ημέρ(ων)] τὰ ὁμαλὰ προαιρ̣ώμ̣[ε]–
θα λ[α]μ[β]άνειν, προcθήcομ̣(εν) ὡ̣c̣ ἔ–
φ̣η̣ν̣ [τ]ὰ̣ δ(ια)φορὰ ἑξηκοcτὰ τ̣οῖc̣
ἀ̣ν̣[ωμ]α̣λοῖc, κ(αὶ) ἕξομ(εν) τὰ ὁμαλά.
κ(αὶ) ἐὰ[ν] κ̣(ατὰ̣) τὸ ἀνάπαλιν ἀ(πὸ) τ(ῶν) ὁμαλ(ῶν)
[τὰ ἀν]ώμαλα λαμ[βά]νωμ(εν), ἀφελ̣(ὼν)
[ἀ(πὸ) τ(ῶν) ὁμ]αλ(ῶν) τὰ διαφορὰ (ἑξηκοcτὰ) τὰ π(αρα)κείμενα
[τῇ ἡλιακῇ] μο(ίρᾳ) ἐν τῷ ἐπ’ ὀρθῆc τῆc cφαί̣–

⇓

11 ξ́

by which the given second seasonal hour precedes noon at Alexandria, i.e. $20\ ^{3}/_{5}$ (hours) after the preceding noon. And we have converted the given 2 seasonal hours at Rome to equinoctial, and there resulted 1 equinoctial [hour]. And we added the difference between the two localities, i.e. [1] $^{3}/_{5}$ equinoctial hours, and there resulted $2\ ^{3}/_{5}$ equinoctial hours at Alexandria, i.e. $20\ ^{3}/_{5}$ equinoctial hours after the preceding noon. And it is obvious from this that in general to the seasonal hours … in the west we add the difference between the two localities

Fr. 3, column i

and sixtieths by which the uniform day-and-nights … [from?] the epoch up to that time … differ from the apparent (day-and-nights) and … If from the non-uniform day-and-nights we choose to obtain the uniform ones, we add as I said? the sixtieths of difference to the non-uniform (day-and-nights), and we will obtain the uniform ones. And if contrariwise from the uniform ones we would obtain the uniform ones, subtracting [from the] non-uniform ones the sixtieths corresponding [to the solar] degree in the table of *sphaera recta*,

⇓

[ρας κα]ν̣ο̣νίῳ, κ(αὶ) ἕξομ(εν) τὰ ἀνώμ̣α̣–
[λα . . . . .]α φαιν̣όμενα τὰ κ̣α̣τ̣ὰ
[*c. 7 letters*]αμεν̣α̣ χάριν τοῦ δύ̣ναc̣–
[θαι τ(οὺc) ἀνωμά]λ(ουc) χρόν(ουc) μετ̣α̣–
[βάλλειν εἰc τὰ ὁμαλ]ὰ̣ νυχθήμερα
[*c. 15 letters* τ]η̣ τ(ῶν) καν̣όν(ων) π̣ρα–
[γματείᾳ̣ *c. 7 letters*] ἐ̣π̣έχ(ον) cυνε̣c̣τα–
. [.] τ̣ὰ̣ ὁ̣μ̣α̣λ̣ὰ̣
21–24 *traces of 4 lines*

— — — — — — — — — — — — — — — —

Fr. 3, column ii

ἐπεὶ οὖν ἡ̣ δ̣ο̣θ̣ε̣ῖ̣c̣α̣ ‾ . ὥ̣ρ(α) π̣ρ̣ὸ̣ [.]
ἐcτι τῆc [μεcημ]β̣ρ̣ί̣α̣[c], π̣ο̣λ̣υ̣π̣λ̣α̣[cι–]
άζω τὰ [*c. 8 letters*] ἐ̣(πὶ) τὰ [. .] . [κ(αὶ)]
γ(ίνονται) $\overline{\kappa\theta}$ μ̣ο̣(ῖραι) . [ ] . [.] . . . . ο̣β̣ . [. .]
π̣(α̣ρ̣ὰ̣) $\overline{\iota\varepsilon}$ γ(ίνονται) ὧρ(αι) [*c. 7 letters*] . . . . λ(ον?) . . . [. .]
[.] ὥ̣ρ(αc) ἰcημερ[ινὰc . .] $\overline{\varepsilon}$ [π]ρ̣ὸ̣ μ̣[. . .]
μ̣ε̣c̣ημ̣[β]ρ̣ί̣[αc . . . . .] . γενήc̣ε[ται αἱ]
δοθεῖ̣cα̣ι̣ [*c. 7 letters* ἰ]c̣η̣μ̣ε̣ρ̣ι̣ν̣[αὶ . .]
9–24 (*traces of c. 16 lines*)

— — — — — — — — — — — — — — — —

1 The bar above the lost numeral seems to have been prolonged over the sign for ὥρα

and we will obtain the non-uniform
... apparent ...
for the sake of being able
to change non-uniform times
into uniform day-and-nights
... the (instructions for) use of the tables
...
... the uniform
21–24 (*traces of 4 lines*)

— — — — — — — — — — — — — — — —

Fr. 3, column ii

Since the given hour x is x (hours)
before noon, I multiply
... by ...
there result 29 degrees ...
[divided] by 15, there result ... hours ...
... equinoctial hours ...
noon ... there will result the
given ... equinoctial ...
9–24 (traces of c. 16 lines)

— — — — — — — — — — — — — — — —

4143. Instructions for Ptolemy's *Handy Tables*

Recto (along the fibres)

̣[ ]̣η̣ν̣ ̣κ̣α̣ ̣ ̣α̣[
ε[ἰcενεχ]θ̣έ̣ντοc ἀριθμ̣[οῦ
ἐ̣γ̣ρά̣[ψ]α̣μεν ἐκ̣ τ̣ο̣ῦ̣ ̄ ̣[
ἐ̣κ̣λ̣ο̣[γ]ι̣ζ̣ομ̣έ̣ν̣ο̣υ̣. ἔπ̣ε̣ι̣τ̣[α
μεν̣ τὸν ἀπ̣ὸ τοῦ̣ μήκ[ουc τοῦ μέc–
cου̣ πλάτουc λ̣η̣φ̣θ̣έ̣ν̣[τα ἀριθμὸν
πρὸc τ̣ὰ δ̣ι̣α̣φ̣ ̣[
τὴν διάcταcιν [
[τ]ῆ β̄. καὶ ἐὰν μ̣ὲ[ν τυγχάνωcι οἱ
ἄ̣νεμ̣ο̣ι̣ [ὁ]μ̣ο̣γ̣ε̣ν̣[εῖc, προcθέν–
τεc ἑκάτ̣ερα π̣ρ̣[ὸc
λέξ̣ομεν τοcοῦ[τον ἀπέχειν τοῦ
διὰ μ̣έ̣cων π̣ρ̣ὸ̣[c
ἐ̣ὰν δὲ ἀνο̣μογ[ενεῖc τυγχάνωcι,
ἀφελόν̣τ̣εc τ̣ὸ̣ν̣ [ἐλάccονα
τῆc διαcτά[cεωc
τ̣ο̣c̣ο̣ῦτον το̣[ῦ διὰ μέcων πρὸc τὸν
ἄ̣ν̣ε̣μο̣ν̣ [
λ̣ε̣ ̣ ̣[
τ̣ω ̣[
̣ ̣ ̣κ̣[
αυτ̣ ̣[
θη̣ ̣[
την̣[
c̣η̣ ̣[
του ̣[
δʹ ᾱ ̣[

— — — — — — — —

4143. Instructions for Ptolemy's *Handy Tables*

Recto (along the fibres)

...
the number that was entered ...
we wrote down from the xth [column ...]
computed. Next ...
the ... from the longitude ...
obtained from [the column?] of mean latitude ...
...
the distance ...
the 2nd (latitude?). And if ...
directions are of the same kind ...
adding both to [each other? ...]
we will say this is the amount [by which the planet is away from the]
ecliptic in the [same?] direction ...
But if [the directions] are of opposite kinds, ...
subtracting [the smaller ...]
of the distance ...
...
direction ...
...
...
...
...
...
...
...
...
4th?, (namely?) 1 ...
– – – – – – – –

Verso (across the fibres)

τῶ]ν̣ (ἐξηκοστῶν). ταῦτα ἐπ̣[ε̣ὶ . . . .]
ἐν τοῖc ἐ]πιπροcθετικ̣[οῖc] . . . c̣
ἐξ]η̣κ̣ο̣cτὰ τῆc ἐλα[χί]c̣τηc
] ἃ̣ ἧ̣ν̣ ∻ α̣̅, ἐφ' ἃ̣c̣ δεῖ τὰ̣
]α ἐ̣ξ̣ηκοcτὰ πολυπλα–
πολυ]πλαcιάζω οὖν τὰ λ̣̅α̣̅
] κ̣α̣ὶ̣ ἔ̣τ̣ι̣ π̣ο̣λ̣υπλαcιαζο–
]α̣νεν κ̣αὶ̣ μ̣ερ̣ι̣–
] ̣ ξ̅ ἅπαξ ξ̅ ἐν τοῖ̣[c
]τ̣ο̣ῖ̣c̣ τοῦ [μ]έcου
] ̣ ἐπ̣ι̣π̣ροcθετι–
]ν̣ο̣ν̣ τ̣ὰ α̣̅ κ̅θ̅. καὶ
] ἡ̣ cχ̣έ̣cιc τοῦ πλα–
π]ρ̣ὸc τὸ πλάτοc
] ἐν τῷ ζ̣ῳδιακῷ
] ̣ ̣ ̣ τ̣οῦ ἐπι–
] ̣ ρ̅κ̣̅β̅ ι̅ καὶ τὸ(ν)
]χ̣ β̅ καὶ τὴν
προcθ]ή̣κην
] ̣ ψ̅μ̅β̅
λοι]π̣αὶ
] δ̣ηλ̣ο̣υ̣–
π]λάτου[c
]α̣ι̣c κ̅α̅
] ̣ ζ̣. τ̣αυ–
] ̣ ̣ ̣ c̣
] μ̣έ̣c̣ο̣υ̣
] ̣ ι̣

– – – – – – – – – – – – – – – –

1 ξ

Verso (across the fibres)

... the sixtieths. Since these [were found?] in the additive ones, ... sixtieths of the minimum [distance], which were 0;1, by which one should multiply the ... sixtieths. Hence I multiply 31? [by xx], and moreover we? multiply ... and we? divide ... 60 once 60 in the ... the [numbers] of the mean [latitude, xx];xx, which are additive ... 1;29. And ... the situation of the latitude [of the planet?] with respect to the latitude [of the epicycle? ...] in the ecliptic. [Moreover? we add] the [number?] of the epicycle, 122;10, and that [of the planet, ...]... 2, and the quantity to be added [that is appropriate for the planet?], [270, and there results?] 742;xx. [After subtracting complete circles?] the remainder [is 22;xx ...] indicate ... [We enter with these in the table?] of latitude ... 21 ... 7th (column?). These ... ... mean [latitude?] ...

4144. Treatise on Kinematic Models

]τ̣ ̣ ̣ ̣υτη ἀνωμ̣[αλ
ἤχθ]ω̣ γ̣ὰρ̣ κά̣θ̣ε̣τ̣[οc
]c̣αυcτον̣ και ̣ ̣[

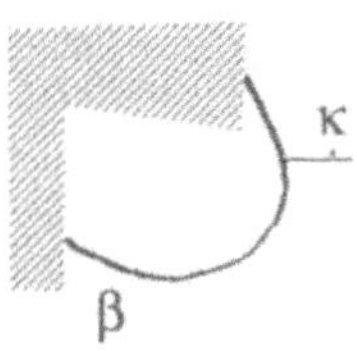

κιν]ήcεωc̣ αὐτῆc ̣ ̣ ̣ερ̣ ̣ ̣μ̣η[
ἀνώ]μ̣αλ⟦η⟧`ο´c̣ ἐφ̣[ε]ξῆc̣ ἀ̣π̣’ α̣[ὐ]τ̣ῶ̣ν̣ [
]ν. λέγω `δὲ´ τοῦτον̣ ἐχ[̣ ̣] π̣ρ̣ωτ̣ο̣[
τ]ῆc πρώτηc φορ[ᾶc ̣ ̣]τ̣ωτοι[
] ̣ ἤδη αἱ φοραὶ cυν̣ελήλ̣[υ]θ̣α[ν
ἀνω]μάλωc κε̣ι̣νήcεται̣ δ[̣] ̣[
] κ̣αὶ ἀκμὴ κα̣[ὶ] ἄνεcιc τῆc φ[ορᾶc
ἀνω]μάλοc φο̣ρὰ μ̣ια[̣]ε ̣[̣] ̣ ̣ ̣ ̣[
] ̣ ̣μ̣ ̣[ ] ̣τι[
]ο̣ν ` ̣και ̣[ ]´ οιc̣[ ]κ̣οιc μ[
]α̣ι̣τ̣ ̣[

5 η crossed out and ο written above 6 δε written above the line 9 *l.* κινήcεται
13 και ̣[ ] written above (indeterminate number of letters lost)

4144. Treatise on Kinematic Models

———————————————

]... anomal[
] For let a perpendicular be drawn [
]...[

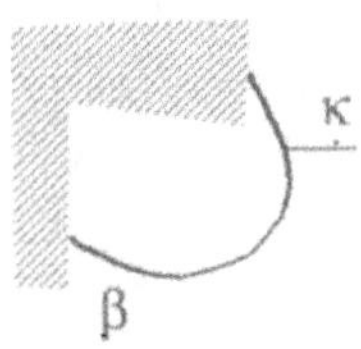

] its motion ...[
] nonuniform. Next, from them [
].... I say that ... ...[
] the first motion ...[
]... by this time the motions have come together [
] will move nonuniformly ...[
] both peak and relaxation of the motion [
] nonuniform motion ...[
]...[
]...[
]...[

———————————————

4145. Procedure Text

Front — — — — — — — — — — — —

]. [ ]. . ιαυ.ρ[
ἐ]πὶ τῷ ι̅γ̣̅ γ(ίνεται) ν̣̅β̅. ὁμοί̣[ως
προς]τίθω ο̅ζ̅ καὶ ν̅β̅. γ(ίνεται) [ρ̅κ̅θ̅
τ]ῶ̣ν ο̅ζ̅ γ(ίνεται) κα´´
]ι̣μετριταιζ̣.[
προς]τίθω ι̅γ̣̅ καὶ ο̅ζ̅ [
]c̣ μ̅ε̅. γ(ίνεται) β̣̅ . [
— — — — — — — — — — — —

Back — — — — — — — — —

.[
[ ]
μ̣[
.[ ]. ν̣τ̣ω̣[
τ̣α̣ δεκατρ[ὶ]c̣ μ[
ἀπαρτίζομεν ε[
ι̅α̅ γ(ίνεται) ο̅ζ̅. καὶ ζ̅ θ̣[
λ̣η εἴπομεν τ̣[
τ̣οῦ μεγάλου̣ [
τ]ο̣υτέστιν τῷ [
]ζ̣ κα[ὶ] ι̅α̅´´ [
]. [
— — — — — — — — —

4145. Procedure Text

Front

— — — — — — — — — — —

]...[
] times 13. There results 52. Similarly [
] I add 77 and 52. There results [129
] from? 77. There results
]...[
] I add 13 and 77 [
] 45. There results 2 ...[

— — — — — — — — — — —

Back

— — — — — — — —

...[
[...]
...[
...[
thirteen ...[
we complete ...[
11. There results 77. And 7 ...[
... we said ...[
of the great [
that is, the [
]7 and 11 [
]...[

— — — — — — — —

4146. Procedure Text

fr. 1

] ε̣ [
τ]ῶ̣ν περιοδικ[ῶν
]λ̣ε̣τ̣α̣ κυνὸϲ̣ [
]ε̣ι̣ζοντα το[
]ε̣̅ δὲ αἱ̣ π̣ρ̣ο̣[
]ε̣ιζονται ε̣ὐ̣να[
] ̣ ωϲ δὲ γί̣νετ[αι
]λ̣οϲ νυν ἀρξο[
τὴ]ν Θωὺθ α̅ νε̣ο[μηνίαν
ἐπ]ιτολὴ καὶ ϲυν[
ϲελ]ήνηϲ κατα[
α]υτη ἄρξει τω̣[
]ου πρωτω ̣ [
]η̣λοϲ ἔϲτα[ι
]ω̣ν ̄ ̣ θεῶ̣ν̣ [
— — — — — — — —

fr. 2

— — — —
] ̣ ̣ ̣ [
]νηϲ̣[
]α̣[ ]ϲ ̣ [
]ο̣λε[
]α̣να ̣ [
]ν̣τω ̣ [
]νε ̣ ̣ [
]ϲων δε[
]ταγμα ε ̣ [
]κ̣ατεχ̣ει̣[
]τινο̣θε[
]α̣ϲεδ̣η ̣ [
— — — —

8 *l.* Θὼθ

(Text too broken for translation)

4147. Procedure Text

fr. 1, col. i

] ̣ ̣
]ι̣ ̣
]το
] ̣ου
] καὶ λοι-
]φέρηται
]αϲαμ-
]δ̣ευτε-
]ι̣
]ϲελ-

fr. 1, col. ii

]ν̣[
]τητων̣[
]ϲ̣τ̣ι̣ν̣ δε ̣[
 ̣[ ]ι̣ ̣[ ] ̣ ̣ Λέο̣ντο[ϲ
ο̣ ̣μ[ ] ̣υ̣ ̣ Αἰγο̣[
κριο̣ ̣ ̣ (traces)
ἀριθμὸϲ ἑκάϲτου ε[
[μ]ερ̣ιζόμε̣νοϲ εἰϲ [
ἀ̣πὸ̣ ἀνα[τ]ο̣λῆ̣ϲ μ̣έχρι δ̣[ύϲ]ε̣[ωϲ
οὕτωϲ ϲελήνηϲ ἀπόδε̣[ι]ξιϲ [
ξιϲων ἔχει ἐν ἑκάϲτῳ οι ̣[
τοῦ ζῳδίου ἐϲτίν ⁄.

fr. 2

] ̣ ̣ ̣[
] ̣ ̣τ̣αμ[
] ̣ν̣[ ] ̣ω[
τὴ]ν ϲελή̣ν̣[ην
] ̣ι̣δ̣εδε̣ ̣[
]α̣ν̣ δ̣ὲ̣ ἐ̣π̣ὶ τῆϲ ̣[
] (traces) ἥμιϲυ [
] (traces) ρ̣εων ̣[

fr. 3

] ̣ ̣[
]νοϲαγο̣[
]ϲ/ τα[
] ̣ου ῑ ο̣δ̣ ̣[
Ϲ]κορπ̣[

(Text too broken for translation)

Part II

Primary Tables

4148. Epochs of the Sun, a.d. 161–237

i	ii	iii	iv	v	⇒
		['Αντωνῖνοc καὶ Λούκι]οc			
		α̣	Μ̣ε̣ϲ̣ορή	κ̣ ιϛ [ν]ε μ	
		β	Μεϲορή	κ λβ κ̣θ̣ κϛ	
		γ	Μεϲ[ο]ρή	κ μη γ ιβ	
	[μ]ζ	δ	Μεϲορή	κα γ λϛ νη	
		ε	Μεϲορή	κα ιθ ι̣ μδ	
		ϛ	Μεϲορή	κ̣α λδ μδ λ	
		ζ	Μεϲορή	κα̣ ν ι̣η ιϛ	
	[μη]	η	Μεϲορή	κβ̣ ε νβ β	
		θ	Μεϲορή	κβ κα κ̣ε μη	
		ι	Μεϲορή	κβ λε ν̣θ̣ λδ	
		ι̣α	Μεϲορή	κβ νβ [λ]γ κ	
	[μθ]	ι̣β	Μεϲορή	κγ η ζ̣ ϛ	
— — —		[ι]γ	Μ̣ε̣ϲ̣ορή	κ[γ] κγ μ̣ νβ	
]ϛ		ι̣δ	Μεϲορή	κγ λθ [ι]δ̣ λη	
]ϛ		ιε	Μεϲορή	κγ νδ [μ]η̣ [κ]δ	
] ̣ζ	ν	ιϛ	Μ̣ε̣ϲ̣ορή	κ̣[δ] ι̣ [κβ] ι̣	
]η		ιζ	Μεϲορή	κ[δ κε νε ν]ϛ̣	
]θ		ιη	Μεϲορή	κ[δ μα κ]θ̣ μ̣β̣	
— — —		ι̣θ	Μεϲορή	κ̣[δ νζ] γ̣ κη	
	[να]	κ	Μεϲορή	κ̣[ε ιβ λ]ζ̣ ιδ	
		κα	Μεϲορή	[κε κη ι]α ✻	
		κ̣β	Μεϲορή	[κε μγ] μ̣δ μϛ	
		[κγ]	Μ̣εϲορή	[κε νθ ι]η λβ	
	[νβ]	[κδ]	Μεϲορή̣	[κϛ ιδ] ν̣β ιη	
		[κε]	Μεϲορ[ή]	[κϛ λ κ]ϛ δ	
		[κϛ]	Μ̣ε̣ϲ̣[ορή]	[κϛ με] ν̣θ ν	
		[κζ]	[Μεϲορή]	[κζ α] λ̣γ λϛ	
	[νγ]	[κη]	[Μεϲορή]	[κζ ιζ] ζ̣ κβ	
		[κθ]	[Μεϲορή]	[κζ λβ] μ̣α η	
		[λ]	[Μεϲορή]	[κζ μη] ιδ νδ	
		[λα]	[Μεϲορή]	[κη γ] μ̣η̣ μ̣	
	[νδ]	[λβ]	[Μεϲορή]	[κη ιθ] κ̣β κϛ	

11 *l.* κβ λϛ νθ λδ

4148. Epochs of the Sun, A.D. 161–237

i	ii	iii	iv	v	⇒
		[Antoninus and] Lucius			
		1	Mesore	20 16 [5]5 40	
		2	Mesore	20 32 29 26	
		3	Mesore	20 48 3 12	
	[4]7	4	Mesore	21 3 36 58	
		5	Mesore	21 19 10 44	
		6	Mesore	21 34 44 30	
		7	Mesore	21 50 18 16	
	[48]	8	Mesore	22 5 52 2	
		9	Mesore	22 21 25 48	
		10	Mesore	22 35! 59 34	
		11	Mesore	22 52 [3]3 20	
	[49]	12	Mesore	23 8 7 6	
— — —		[1]3	Mesore	2[3] 23 40 52	
]6		14	Mesore	23 39 [1]4 38	
]6		15	Mesore	23 54 [4]8 [2]4	
]x7	50	16	Mesore	2[4] 10 [22] 10	
]8		17	Mesore	2[4 25 55 5]6	
]9		18	Mesore	2[4 41 2]9 42	
— — —		19	Mesore	2[4 57] 3 28	
	[51]	20	Mesore	2[5 12 3]7 14	
		21	Mesore	[25 28 1]1 0	
		22	Mesore	[25 43] 44 46	
		[23]	Mesore	[25 59 1]8 32	
	[52]	[24]	Mesore	[26 14] 52 18	
		[25]	Mesore	[26 30 2]6 4	
		[26]	Mesore	[26 45] 59 50	
		[27]	[Mesore]	[27 1] 33 36	
	[53]	[28]	[Mesore]	[27 17] 7 22	
		[29]	[Mesore]	[27 32] 41 8	
		[30]	[Mesore]	[27 48] 14 54	
		[31]	[Mesore]	[28 3] 48 40	
	[54]	[32]	[Mesore]	[28 19] 22 26	

⇐ vi	vii	viii	ix ⇒
	Cεουῆρ(οc) κ(αὶ) 'Αν[τωνῖνοc]		
	α	Μεcορή	[κη λδ νϛ ιβ]
	β	Μεcορή	[κη ν κθ νη]
	γ	Μεcορή	[κθ ϛ γ μδ]
νε	δ	Μεcορή	[κθ κα λζ λ]
	ε	Μεcορή	[κθ λζ ια ιϛ]
	ϛ	Μεcο[ρή]	[κθ νβ με β]
	ζ	Μεcο[ρή]	[λ η ιη μη]
νϛ	η	Μεcο[ρή]	[λ κγ νβ λδ]
	θ	Μ[εcορή]	[λ λθ κϛ κ]
	ι	Μ[εcορή]	[λ νε ∘ ϛ]
	ια	'Επ̣[αγόμεναι]	[α ι λγ νβ]
νζ	ιβ	'Ε̣[παγόμεναι]	[α κϛ ζ λη]
	ιγ	'Ε̣[παγόμεναι]	[α μα μα κδ]
	ιδ	['Επαγόμεναι]	[α νζ ιε ι]
	ιε	'Ε̣[παγόμεναι]	[β ιβ μη νϛ]
νη	ιϛ	'Ε̣[παγόμεναι]	[β κη κβ μβ]
	ιζ	['Επαγόμεναι]	[β μγ νϛ κη]
	ιη	['Επαγόμεναι]	[β νθ λ ιδ]
	ιθ	['Επαγόμεναι]	[γ ιε δ ∘]
νθ	κ´	['Επαγόμεναι]	[γ λ λζ μϛ]
	κα	['Επαγόμεναι]	[γ μϛ ια λβ]
	κβ	'Ε̣[παγόμεναι]	[δ α με ιη]
	κγ	'Ε̣[παγόμεναι]	[δ ιζ ιθ δ]
ξ´	κδ	'Ε̣[παγόμεναι]	[δ λβ νβ ν]
	κε	'Ε̣[παγόμεναι]	[δ μη κϛ λϛ]
'Αντω[νῖνοc]			
	α	['Επαγόμεναι]	[ε δ ∘ κβ]
	β	['Επαγόμεναι]	[ε ιθ λδ η]
ξα	γ	['Επαγόμεναι]	[ε λε ζ νδ]
	δ	['Επαγόμεναι]	[ε ν μα μ]
'Αλέ[ξανδροc]			
	α	[Θώθ]	[α ϛ ιε κϛ]
	β̣	[Θώθ]	[α κα μθ ιβ]

1 cεουηρό

⇐ ⇒

vi	vii	viii	ix
	Severus and Antoninus		
	1	Mesore	[28 34 56 12]
	2	Mesore	[28 50 29 58]
	3	Mesore	[29 6 3 44]
55	4	Mesore	[29 21 37 30]
	5	Mesore	[29 37 11 16]
	6	Mesore	[29 52 45 2]
	7	Mesore	[30 8 18 48]
56	8	Mesore	[30 23 52 34]
	9	Mesore	[30 39 26 20]
	10	Mesore	[30 55 0 6]
	11	Epagomenae	[1 10 33 52]
57	12	Epagomenae	[1 26 7 38]
	13	Epagomenae	[1 41 41 24]
	14	[Epagomenae]	[1 57 15 10]
	15	Epagomenae	[2 12 48 56]
58	16	Epagomenae	[2 28 22 42]
	17	[Epagomenae]	[2 43 56 28]
	18	[Epagomenae]	[2 59 30 14]
	19	[Epagomenae]	[3 15 4 0]
59	20	[Epagomenae]	[3 30 37 46]
	21	[Epagomenae]	[3 46 11 32]
	22	Epagomenae	[4 1 45 18]
	23	Epagomenae	[4 17 19 4]
60	24	Epagomenae	[4 32 52 50]
	25	Epagomenae	[4 48 26 36]
	Antoninus		
	1	[Epagomenae]	[5 4 0 22]
	2	[Epagomenae]	[5 19 34 8]
61	3	[Epagomenae]	[5 35 7 54]
	4	[Epagomenae]	[5 50 41 40]
	Alexander		
	1	[Thoth]	[1 6 15 26]
	2	[Thoth]	[1 21 49 12]

⇐	x	xi	xii	xiii
	[ξβ]	[γ]	[Θώθ]	[α λζ κβ νη]
		[δ]	[Θώθ]	[α νβ νϛ μδ]
		[ε]	[Θώθ]	[β η λ λ]
		[ϛ]	[Θώθ]	[β κδ δ ιϛ]
	[ξγ]	[ζ]	[Θώθ]	[β λθ λη β]
		[η]	[Θώθ]	[β νε ια μη]
		[θ]	[Θώθ]	[γ ι με λδ]
		[ι]	[Θώθ]	[γ κϛ ιθ κ]
	[ξδ]	[ια]	[Θώθ]	[γ μα νγ ϛ]
		[ιβ]	[Θώθ]	[γ νζ κϛ νβ]
		[ιγ]	[Θώθ]	[δ ιγ ▾ λη]
	Μαξιμῖ[voc]			
		α̣	Θώ[θ]	[δ κη λδ κδ]
	[ξε]	[β]	Θώ̣[θ]	[δ μδ η ι]
		γ	Θ[ώθ]	[δ νθ μα νϛ]
	Γ̣ορ[δι]α̣[voc]			
		α̣	[Θώθ]	[ε ιε ιε μβ]

— — — — — — — — — — — — — — —

⇐

x	xi	xii	xiii
[62]	[3]	[Thoth]	[1 37 22 58]
	[4]	[Thoth]	[1 52 56 44]
	[5]	[Thoth]	[2 8 30 30]
	[6]	[Thoth]	[2 24 4 16]
[63]	[7]	[Thoth]	[2 39 38 2]
	[8]	[Thoth]	[2 55 11 48]
	[9]	[Thoth]	[3 10 45 34]
	[10]	[Thoth]	[3 26 19 20]
[64]	[11]	[Thoth]	[3 41 53 6]
	[12]	[Thoth]	[3 57 26 52]
	[13]	[Thoth]	[4 13 0 38]
Maximinus			
	1	Thoth	[4 28 34 24]
[65]	[2]	Thoth	[4 44 8 10]
	3	Thoth	[4 59 41 56]
Gordianus			
	1	[Thoth]	[5 15 15 42]

– – – – – – – – – – – – – – –

4149. EPOCHS OF THE MOON, A.D. 96–166

Front

i	ii	iii	iv
[Θώ]θ̣	ι̣α̣	Ζ̣υ̣γ̣[ο]ῦ̣	κ̣γ̣ [
[Πα]χ̣ῶ̣[ν]	ιθ̣	Ϲκο̣ρ̣(πίου)	κα [
[Τῦβι]	κ̣β	Το̣ξ[ό(του)]	ιθ [
[Θώθ]	κ̣ε̣	Αἰγο̣κ̣(έρῳ)	ιϛ [
[Παῦν]ι	γ̣	Ὑδρο̣(χόου)	ι̣δ̣ [
[Μεχεί]ρ̣	ε	Ἰχθύ̣ϲ̣ι̣	ι̣β̣ κ̣[
[Φαῶφ]ι̣	η	Κρει(οῦ)	ι ϛ [
[Μεϲορή]	ια	Ταύρ(ο̣υ̣)	ιβ λθ̣ [
[Φαρμοῦθι]	ι̣δ̣	Δ̣ι̣δ[ύ(μων)]	κ̣ κ̣[
[Χοίακ]	ι̣ζ̣	Κ̣α̣ρκ(ίνου)	η ϛ [
[Μεϲορή]	κ̣ε̣	[	
[	*8 lines lost*		]
	]	Κ̣ρ̣ε̣[ιῷ]	[
	]	Ταύρ(ῳ)	[
	]	[Διδύ(μοιϲ)]	[
	]	Καρκί(νῳ)	ιγ [
	]	Λέον(τι)	ια [
	]	Παρθ(ένῳ)	η̣ [
	]	Ζυγῷ	ϛ̣ [
	]	Ϲκ[ορ(πίῳ)]	[
	]	Τ̣[οξό(τῃ)]	[

Names of zodiacal signs abbreviated by truncation 7 *l.* Κριοῦ (so line 20)

4149. Epochs of the Moon, A.D. 96–166

Front

i	ii	iii	iv
Thoth	11	Libra	23 [
Pachon	19	Scorpio	21 [
[Tybi]	22	Sagittarius	19 [
[Thoth]	25	Capricorn	16 [
Payni	3	Aquarius	14 [
Mecheir	5	Pisces	12 2[x
Phaophi	8	Aries	10 6 [
[Mesore]	11	Taurus	12 39 [
[Pharmuthi]	14	Gemini	20 2[x
[Choeac]	17	Cancer	8 6 [
Mesore]	25	[	
[	*8 lines lost*		]
	]	Aries	[
	]	Taurus	[
	]	[Gemini]	[
	]	Cancer	13 [
	]	Leo	11 [
	]	Virgo	8 [
	]	Libra	6 [
	]	Scorpio	[
	]	Sagittarius	[

Back (across the fibres)

i	ii	iii
'Αντωνίνου κα]ὶ Οὐή̣ρ̣ου		
(ἔτος) α̣	Τῦβι̣	[
β̣	Θ̣ώθ	κ̣β̣
	Π̣αχ̣ών̣	[
γ̣	Μ̣εχεί̣ρ	[
[δ]	Φαῶφι	[
	Π̣α̣οῖ̣νι	κ̣[
[ε]	[Με]χ̣ε̣ί̣ρ	[
ϛ̣	Φ̣α̣ῶ̣[φ]ι̣	[

2 ∟ 7 *l.* Παῦνι

Back

	i	ii	iii
Antoninus] and Verus			
	Year 1	Tybi	[
	2	Thoth	22
		Pachon	[
	3	Mecheir	[
	[4]	Phaophi	[
		Payni	2[x
	[5]	Mecheir	[
	6	Phaophi	[

4150. Epochs of the Moon, a.d. 187–198, and Template

Front

i	ii	iii	iv	v
]	Ὑδρο(χόῳ)	ις	κ̣ ̣	[
]	Ἰχθύ(ϲι)	ιδ	ι	[
]	Κριῷ	ια̣	ν ̣	[
]	Ταύρ(ῳ)	θ	λϛ̣	̣[
δ̣	Δ̣ιδύ(μοιϲ)	ζ	κ	ϛ̣
ι]β̣	Κ̣αρ̣κ(ίνῳ)	ε̣	γ	θ̣
]	Λ̣έον̣(τι)	β	[	
]	Παρ̣θ̣(ένῳ)	ε̣	[	
]	Ζυγ̣ῷ	γ̣	[	
]	Cκορ̣(πίῳ)	ō	[	
]	Cκο̣ρ̣(πίῳ)	κ̣[η		
]	Τ[ο]ξό(τῃ)	κϛ̣	[	
]	Ἐγόκ(ερῳ)	κ[δ		
]	Ὑδρο(χόῳ)	κ̣[α		
]	Ἰ̣χ̣θύ(ϲι)	[		
]	Κ̣ρ̣ι̣ῷ	[		

Names of zodiacal signs abbreviated by truncation   13 *l.* Αἰγόκερῳ

4150. Epochs of the Moon, a.d. 187–198, and Template

Front

i	ii	iii	iv	v
]	Aquarius	16	2x	[
]	Pisces	14	10	[
]	Aries	11	5x	[
]	Taurus	9	36	x[
4	Gemini	7	20	6?
1]2	Cancer	5	3	9?
]	Leo	2	[	
]	Virgo	5	[	
]	Libra	3	[	
]	Scorpio	0	[	
]	Scorpio	2[8		
]	Sagittarius	26	[	
]	Capricorn	2[4		
]	Aquarius	2[1		
]	Pisces	[		
]	Aries	[		

Back

i	ii	iii
] κ̣ζ̣	[	
] δ̣	[	
μ]α̣	[	
] ̣ ̣	[	
] νε̣	[cλε	
λ]β̣	[	
μ]ε̣ θ	[	
] μ̣ϛ̣	[	
] κ̣γ̣	[	
] ̣ ̣	c̣[μ	
[		]
[		]
]		κ̣β̣ [
]		κγ [
]	cμε	ō ι̣[
]		α [
]		α [
]		β̣ [

Back

i	ii	iii
] 27	[	
] 4	[	
4]1	[	
]xx	[	
] 55	[235	
3]2	[	
4]5 9	[	
] 46	[	
] 23	[	
]xx	2[40	
[		]
[		]
]		22 [
]		23 [
]	245	0 1[x
]		1 [
]		1 [
]		2 [

4151. Epochs of the Moon, a.d. 217–254

[-III]	[-II]	[-I]	[0]	i	ii	iii	iv	v ⇒
	['Αντωνίνου]							
	[α]	[Χοί(ακ)]	[ια]	[Διδύμ(ων)]	[κη]	[ō]	[κ]	[νη]
	[ ]	[Μεϲ(ωρή)]	[ιθ]	[Καρκί(νου)]	[κε]	[μγ]	[κγ]	[μα]
	[β]	[Φαρ(μοῦθι)]	[κβ]	[Λέοντ(οϲ)]	[κγ	[κϛ]	[β]	[κδ]
[ξα]	[γ]	[Χοί(ακ)]	[κε]	[Παρθέ(νου)]	[κα]	[ι]	[ε[	ζ]
	[ ]	['Επα(γόμεναι)]	[γ]	[Ζυγοῦ]	[ιη]	[νγ]	ζ̣	ν̣
	[δ]	[Παχ(ών)]	[ϛ]	[Cκορπ(ίου)]	[ιϛ]	[λ]ζ̣	ι̣	[λγ]
	[α]	[Τῦβ(ι)]	[θ]	[Τοξότ(ου)]	[ιδ]	κ ̣	ιγ	ιϛ
	[β]	[Θώθ]	[ιβ]	[Αἰγόκ(ερω)]	[ιβ]	γ	ιε	νθ
	[ ]	[Παχ(ών)]	[κ]	['Υδροχ(όου)]	[θ]	μζ	ιη	μβ
[ξβ]	[γ]	[Τῦβ(ι)]	[κγ]	['Ιχθύω(ν)]	[ζ]	λ	κα	κε
	[δ]	[Θώθ]	[κϛ]	[Κριοῦ]	[ε]	ιδ	ọ̄	η̣
[					] ̣ 'Α̣λ̣ε̣ξ̣[άνδρου]			
	[ ]	[Παῦ(νι)]	[δ]	[Ταύρο(υ)]	β	ν̣ζ̣	β̣	[νβ]
	[ε]	[Φαρ(μοῦθι)]	[β]	[Διδύμ(ων)]	ε	λ̣ ̣ ̣	μ̣η̣	[ ]
	[ϛ]	[Χοί(ακ)]	[ε]	[Καρκ]ί̣(νου)	γ	ιδ	ιγ	[ ]
	[ ]	[Μεϲ(ωρή)]	[ιγ]	[Λέο]ν̣τ̣(οϲ)	ō	νζ̣	λη	[ ]
[ξγ]	[ζ]	[Φαρ(μοῦθι)]	[ιϛ]	[ ]	κη	μα̣	γ	[ ]
	[η]	[Χοί(ακ)]	[ιθ]	[Παρθέ(νου)]	κ̣ϛ̣	κδ	κ̣η	[ ]

Names of zodiacal signs and months apparently abbreviated by truncation

4151. Epochs of the Moon, a.d. 217–254

[-III]	[-II]	[-I]	[0]	i	ii	iii	iv	v ⇒
	[Antoninus]							
	[1]	[Choeac]	[11]	[Gemini]	[28]	[0]	[20]	[58]
	[ ]	[Mesore]	[19]	[Cancer]	[25]	[43]	[23]	[41]
	[2]	[Pharmuthi]	[22]	[Leo]	[23]	[26]	[2]	[24]
[61]	[3]	[Choeac]	[25]	[Virgo]	[21]	[10]	[5]	[7]
	[ ]	[Epagomenae]	[3]	[Libra]	[18]	[53]	7	50
	[4]	[Pachon]	[6]	[Scorpio]	[16]	[3]7	10	[33]
	[1]	[Tybi]	[9]	[Sagittarius]	[14]	20	13	16
	[2]	[Thoth]	[12]	[Capricorn]	[12]	3	15	59
	[ ]	[Pachon]	[20]	[Aquarius]	[9]	47	18	42
[62]	[3]	[Tybi]	[23]	[Pisces]	[7]	30	21	25
	[4]	[Thoth]	[26]	[Aries]	[5]	14	0	8
[					]… Alexander			
	[ ]	[Payni]	[4]	[Taurus]	2	57	2	[52]
	[5]	[Pharmuthi]	[2]	[Gemini]	5	30	48	[ ]
	[6]	[Choeac]	[5]	Cancer	3	14	13	[ ]
	[ ]	[Mesore]	[13]	Leo	0	57	38	[ ]
[63]	[7]	[Pharmuthi]	[16]	[ ]	28	41	3	[ ]
	[8]	[Choeac]	[19]	[Virgo]	26	24	28	[ ]

⇐ vi vii viii ix x

] ᾱc̣μ̣[

] ̣ ̣[

Γ̣ọρδ[ιανοῦ

vi	vii	viii	ix	x
	[ ]	ς	Ἁ̣[θύ(ρ)]	[θ]
		ς̣	Ἐ̣[πε(ίφ)]	[ιζ]
	ξ̣ζ̣	α̣	[Φαμ(ενώθ)]	[κ]
		β̣	[Ἁθύ(ρ)]	[κγ]
			[Μεc(ωρή)]	[α]
		[γ]	[Φαρ(μοῦθι)]	[δ]
		δ	Χ̣[οί(ακ)]	[ζ]
			Μ̣ε̣c̣(ωρή)	[ιε]
	ξη	ε	Φ̣α̣[ρ̣(μοῦθι)]	[ιη]
	⟦Δ̣ε̣κ̣´⟧	⟦α⟧	⟦Χ̣⟧[ ]	[ ]
		ς	Χ̣[οί(ακ)]	[κα]
		ς	Με̣[c(ωρή)]	[κθ]
	Δεκ(ίου)	α	Παχ̣(ών)	[β]
	Γ̣[άλλου]	α̣	Μεχ(είρ)	λ̣
	[ξθ]	[β]	Ἁθύρ	γ̣
		[ ]	Ἐπε(ίφ)	ια̣
		[γ]	Φ̣αμ(ενώθ)	ι̣δ̣
	[Γαλλιηνοῦ]	[α]	Ἁ̣θ̣ύ̣(ρ)	ι̣[ζ]

⇐ vi	vii	viii	ix	x
		]...[		
		]...[		
		Gordian		
	[ ]	6	Hathyr	[9]
		6	Epeiph	[17]
	67	1	[Phamenoth]	[20]
		2	[Hathyr]	[23]
			[Mesore]	[1]
		[3]	[Pharmuthi]	[4]
		4	Choeac	[7]
			Mesore	[15]
	68	5	Pharmuthi	[18]
	⟦Decius⟧	⟦1⟧	⟦Choeac⟧	[ ]
	...	6	Choeac	[21]
		6	Mesore	[29]
	Decius	1	Pachon	[2]
	G[allus]	1	Mecheir	30
	[69]	[2]	Hathyr	3
		[ ]	Epeiph	11
		[3]	Phamenoth	14
	[Gallienus]	[1]	Hathyr	1[7]

– – – – – – – – – – – – – –

13 crossed out

4152. Epochs of Mercury

Front

i	ii
[Ἰ]χ̣θύ(ων)	ιβ λϛ
[Δ]ι̣δύ(μων)	ιη λϛ
[Cκ]ορ(πίου)	ι 𐆊
[Ὑδ]ρο(χόου)	κϛ μ
[Δι]δ̣ύ(μων)	δ 𐆊
[Ζυ]γο(ῦ)	κβ μ
[Ὑδ]ρο(χόου)	ι κϛ [μ
[Τα]ύ̣ρ(ου)	ιθ κδ
[Παρ]θ(ένου)	κη κ̣
[Αἰ]γό(κερω)	κδ [ιγ κ
[Τα]ύ̣ρ(ου)	δ μ̣[η

Names of zodiacal signs abbreviated by truncation

Back (offsets)

i	ii
Δι̣δύ̣(μων)	κ̣[
Ζ̣υ̣γ̣ο̣(ῦ)	ιδ̣ [
Ὑ̣δ̣ρο̣(χόου)	ι̣β̣ [
Δ̣ι̣δ̣ύ̣(μων)	γ [
Π̣αρθ̣(ένου)	κε̣ [
Αἰγό̣(κερω)	κ ̣ [
Ταύρ̣(ου)	ιδ [
Παρθ̣(ένου)	θ [
Αἰγό(κερω)	[
Κρ̣ι̣ο̣(ῦ)	[
Λ̣έον̣(τοc)	[

4152. Epochs of Mercury

Front

i	ii
Pisces	12 36
Gemini	18 36
Scorpio	10 0
Aquarius	26 40
Gemini	4 0
Libra	22 40
Aquarius	10 26 [40
Taurus	19 24
Virgo	28 20
Capricorn	24 [13 20
Taurus	4 4[8

Back (offsets)

i	ii
Gemini	2[x
Libra	14 [
Aquarius	12 [
Gemini	3 [
Virgo	25 [
Capricorn	2x [
Taurus	14 [
Virgo	9 [
Capricorn	[
Aries	[
Leo	[

4153. Epochs of Mercury, A.D. 206–215

i	ii	iii	iv
	[		
	[		
νη	[		
	[		
	[		
	[		
		Μ̣[εχείρ]	[
		Π[αῦνι]	[
	ιη	Φ[αῶφι]	[
		Τ[ῦβι]	[
		Π[αχών]	[
	ιθ	Θώ̣[θ]	[
		Τῦ̣[βι]	[
		Παχώ̣ν	ϛ̣ λ̣ ̣ [
ν̣θ	κ	Θώθ	η μβ̣ [
		Χοίακ	κβ ̣[
		Φαρμο(ῦθι)	ιε [
		Μεσορ(ή)	κθ ̣[
	κα	Χοίακ	ϛ λ[̣
		Φαμεν(ώθ)	κ̣δ̣ [
		Μ[εσορ(ή)]	[
	κβ	Ἁθ̣[ύρ]	[
		Φα[μεν(ώθ)]	[
		Ἐπ[είφ]	[
	κγ	Ἁθύρ̣	[
		Μεχ[είρ]	[

Names of zodiacal signs abbreviated by truncation

4153. Epochs of Mercury, a.d. 206–215

i	ii	iii	iv
	[		
	[		
58	[		
	[		
	[		
	[		
		Mecheir	[
		Payni	[
	18	Phaophi	[
		Tybi	[
		Pachon	[
	19	Thoth	[
		Tybi	[
		Pachon	6 3x [
59	20	Thoth	8 42 [
		Choeac	22 x[
		Pharmouthi	15 [
		Mesore	29 x[
	21	Choeac	6 3[x
		Phamenoth	24 [
		Mesore	[
	22	Hathyr	[
		Phamenoth	[
		Epeiph	[
	23	Hathyr	[
		Mecheir	[

4154. Epochs of Mercury

i	ii
Μ̣εχε[ίρ]	[
Π̣αῦνι	[
Θώθ	κη̣ μβ [
Τῦβι	ιη νβ ̣[
Παχών	ιε β ̣[
Θώθ	ιβ ιβ [
Τῦβι	β κβ [
[Φ]αρμοῦ(θι)	κη λβ [
Μεσορ(ή)	ō μβ [
Χοίακ	ιε νβ ̣[
Φαρμοῦ(θι)	ιβ ̣ ̣[
Μεσορ(ή)	ια ιβ ̣[
Ἀθύρ	κθ κβ ̣[
Φ̣αμεν̣ώ(θ)	κ̣ε̣ λ̣β [

Names of zodiacal signs abbreviated by truncation (in line 12, the ρ is embellished with a curved tail above). 9 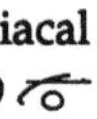 11 ιβ  12 *l.* ιδ ιβ

4154. Epochs of Mercury

i	ii
Mecheir	[
Pauni	[
Thoth	28 42 [
Tybi	18 52 ...[
Pachon	15 2 ...[
Thoth	12 12 [
Tybi	2 22 [
Pharmuthi	28 32 [
Mesore	0 42 [
Choeac	15 52 ...[
Pharmuthi	12 2? ...[
Mesore	11! 12 ...[
Hathyr	29 22 ...[
Phamenoth	25 32 [

4155. Epochs of Mercury

Fr. 1

i	ii	iii	
]	[ ]	Δ̣[ιδύμ(οιϲ)]	[
]	ι̣ζ	Ζ[υγῷ]	[
]	ιγ	Αἰγό̣[κε(ρῳ)]	[
]	λε	Ταύ̣[ρῳ]	[
]	[ι]η̣	Παρθ[έ(νῳ)]	[
]	ι̣δ	Αἰ̣γ[όκε(ρῳ)]	[
]	λδ	Τα̣[ύρῳ]	[
]	ιη	Π̣α̣ρ̣θ̣[έ(νῳ)]	[
]	ιε	Τοξό[τ(ῃ)]	[
]	λ̣	Κρει̣ῷ	[
]	[ ]	Λέον[τ(ι)]	[
]	[ ]	Τοξό̣[τ(ῃ)]	[
]	[ ]	[Ἰχθύ(ϲι)]	[
]	[ ]	[Καρ]κ̣[ίν(ῳ)]	[
]	[ ]	[Ϲκο]ρ̣π(ίῳ)	ι̣[
]	[ ]	Ἰ]χ̣θύϲ(ι)	ιβ̣ [
]	κ̣β	Κ̣α̣ρ̣κ̣ί(νῳ)	ι[
]	ι̣ζ	Ϲ̣κ̣ο̣[ρ(πίῳ)]	α̣ [
]	ιε	Ὑ̣δρ̣οχ(όῳ)	κ[
]	κϛ	Διδύμ(οιϲ)	κ̣[
]	ιζ	Ζυ̣γῷ	ι̣[
]	ιβ	Ὑδ[ρ]οχ(όῳ)	ζ̣ [
]	λβ	Δ̣ι̣[δ]ύ̣μ̣(οιϲ)	[
]̣ νβ λ	ιη	Π̣α̣[ρθέν(ῳ)]	[

Names of zodiacal signs abbreviated by truncation 10 *l.* Κριῷ

4155. Epochs of Mercury

Fr. 1

i	ii	iii	
]	[ ]	Gemini	[
]	17	Libra	[
]	13	Capricorn	[
]	35	Taurus	[
]	[1]8	Virgo	[
]	14	Capricorn	[
]	34	Taurus	[
]	18	Virgo	[
]	15	Sagittarius	[
]	30	Aries	[
]	[ ]	Leo	[
]	[ ]	Sagittarius	[
]	[ ]	[Pisces]	[
]	[ ]	Cancer	[
]	[ ]	Scorpio	1[x
]	[ ]	Pisces	12? [
]	22	Cancer	1[x
]	17	Scorpio	1 [
]	15	Aquarius	2[x
]	26	Gemini	2[x
]	17	Libra	1[x
] 0	12	Aquarius	7 [
]	32	Gemini	[
]x 52 30	18	Virgo	[

Fr. 2

i	ii	iii	iv
]	[ ]	Ἰ̣χθύϲ(ι)	[
]	[ ]	Κ̣α̣ρ̣κί(νῳ)	[
]	[μ]β̣	Ζυγῷ	[
]̣	λζ̣	Ὑδρ̣ο̣χ̣ό̣(ῳ)	θ̣ [
]̣ λ̣	κε	Διδύμο(ιϲ)	κ [
]	[λ]θ̣	Ζυγῷ	α̣ [
]	μδ̣	Αἰγόκε(ρῳ)	[
]	[ ]	[Διδύμο(ιϲ)]	[
]	[ ]	[Παρθέν(ῳ)]	[
]̣ λ	με	Αἰγόκ(ερῳ)	α κβ λ̣
]̣ ō	κβ	Ταύρῳ̣	ιζ ν ọ̄
]	λβ	Λέοντ(ι)	κθ ιε ọ̄
]	μζ	Τοξότ(ῃ)	ιε ιε ọ̄
]	κ̣[ ]	[Κρι]ῷ̣	κ̣ϛ̣ κ ọ̄

Fr. 2

i	ii	iii	iv
]	[ ]	Pisces	[
]	[ ]	Cancer	[
]	[4]2	Libra	[
]x	37	Aquarius	9 [
]x 30	25	Gemini	20 [
]	[3]9	Libra	1 [
]	44	Capricorn	[
]	[ ]	[Gemini]	[
]	[ ]	[Virgo]	[
]x 30	45	Capricorn	1 22 30
]x 0	22	Taurus	17 50 0
]	32	Leo	29 15 0
]	47	Sagittarius	15 15 0
]	2[x]	Aries	26 20 0

4156a. Epochs of Mercury

i	ii	iii	iv	v	vi
Μεσορή]	[ιε]	[θ]	[Κ]α̣[ρ]κ̣(ίνου)	ō	λ̣
Χ]ο̣ί̣(ακ)	ō	ιθ	Ζυγο(ῦ)	ιθ	ō
Φαμενώθ]	κδ	λθ	Ὑδροχ(όου)	ια	μ
Ἐπείφ]	κδ	λθ	Διδύμ(ων)	ι	ọ̄
Ἀθύρ]	[ιγ]	[μθ]	[Ζυγοῦ]	β̣	λ̣
Φαμενώθ]	[ϛ]	[ιθ]	[Αἰγόκερω]	κγ	κ̣
Ἐπείφ]	[δ]	[α λ]	[Ταύρου]	[ιθ]	[κβ λ
Φαῶφι]	[κζ]	[ιθ]	Π̣[αρ]θ̣έ̣(νου)	ι̣ϛ̣	[ō
Μεχείρ]	[ιζ]	[νθ]	[Αἰγόκερω]	[ε]	[ō
Παῦνι]	[ιγ]	[λθ]	[Κριοῦ]	κ̣θ̣	[ō
Φαῶφι]	[ι]	[μθ]	[Λέοντος]	[κ]θ̣	λ
Μεχείρ]	[ō]	[νθ]	[Τοξότου]	[ιη]	[ō
Παχών]	[κζ]	θ̣	Κ̣ρ̣[ιοῦ]	ι̣β̣	λ̣

Names of zodiacal signs and months abbreviated by truncation. 4 διδυ^{μ}

4156a. Epochs of Mercury

i	ii	iii	iv	v	vi
Mesore]	[15]	[9]	Cancer	0	30
Choeac	0	19	Libra	19	0
Phamenoth]	24	39	Aquarius	11	40
Epeiph]	24	39	Gemini	10	0
Hathyr]	[13]	[49]	[Libra]	2	30
Phamenoth]	[6]	[19]	[Capricorn]	23	20
Epeiph]	[4]	[1 30]	[Taurus]	[19]	[22 30
Phaophi]	[27]	[19]	Virgo	16	[0
Mecheir]	[17]	[59]	[Capricorn]	[5]	[0
Payni]	[13]	[39]	[Aries]	29	[0
Phaophi]	[10]	[49]	[Leo]	[2]9	30
Mecheir]	[0]	[59]	[Sagittarius]	[18]	[0
Pachon]	[27]	9	Aries	12	30

4156b. Epochs of Mercury

	i	ii	iii	iv	v
	]	̣	̣	[	
	]ε̣	ιβ̣	θ̣	Π̣[αρθ(ένου)]	[
3	]̣	κθ	λϛ	Αἰ̣[γόκ(ερω)]	[
	]	κ̣	ν̣	Διδύμ(ων)	[
	]	κϛ̣	μζ	Παρθ(ένου)	θ̣ [

Names of zodiacal signs abbreviated by truncation.

4156c. Epochs of Mercury

	i	ii	
	]̣	Ζυγο(ῦ)	[
	]̣[̣]	Ὑδρο(χόου)	[
3	] ι̣δ̣	Διδύ(μων)	[
	]	Ζυγο(ῦ)	[
	]	Αἰγ̣ό(κερω)	[
	]	Ταύρ(ου)	[

Names of zodiacal signs abbreviated by truncation.

4156b. Epochs of Mercury

	i	ii	iii	iv	v
	]	x	x	[	
	]...	12	9	Virgo	[
3	]...	29	36	Capricorn	[
	]	20	50	Gemini	[
	]	26	47	Virgo	9? [

4156c. Epochs of Mercury

	i	ii	
	]x	Libra	[
	]x[x]	Aquarius	[
3	] 14?	Gemini	[
	]	Libra	[
	]	Capricorn	[
	]	Taurus	[

Names of zodiacal signs abbreviated by truncation.

4157. Epochs of Venus

i	ii	iii	iv
].		φ̣[ ].	[
]α̣		χι ζ	Παῦν(ι)

2 παυν

4157a. Epochs of Venus, a.d. 106–121

i	ii	iii	iv
		c̣ύνοδ̣[ο]c ἑῴ̣α [	
] λ̣δ	ι	Ἁθύρ	ϛ̣ [
]ζ̣	ια	Παῦνι	ι̣ε̣ . .
]ϛ	ιγ	Τῦβι	κγ̣ κβ
] ν̣η	ιδ	Μεσ̣ο̣(ρή)	κδ ν̣η̣
]	ιϛ	Φαρ̣μ̣(οῦθι)	[.]. μ
].	ιη	Ἁ[θύρ]	[.] μ̣
]	ι̣θ̣	[Π]α̣[ῦνι]	[. .] .
]	β̣	[Τῦ]βι	[. .] .
]	γ̣	[Μ]εσο̣(ρή)	[ ].
]	ε̣	[Φα]ρ̣μ̣(οῦθι)	. [ ]α̣
]	[ζ]	[Ἁθ]ῦ̣ρ̣	ϛ̣ κ̣.

Names of months abbreviated by truncation.

4157. Epochs of Venus

i	ii	iii	iv
]...		5[xx]x	[
]...		610 7	Payni

4157a. Epochs of Venus, a.d. 106–121

i	ii	iii	iv
		Morning conjunction [	
] 34	10	Hathyr	6 [
]7?	11	Payni	15 xx [
]6?	13	Tybi	23 22
] 58	14	Mesore	24 58
]	16	Pharmuthi	[x]x 40
]x	18	Hathyr	[x] 40
]	19	Payni	[xx] x
]	2	Tybi	[xx] x
]	3	Mesore	[ ]x
]	5	Pharmuthi	x [ ]1
]	[7]	Hathyr	6 2x

4158. Epochs of Mars, A.D. 271–325

Fr. 1

i	ii	iii	iv
[μϛ ιδ δ]	Τ̣ῦβ̣[ι]	[ι] ν̣ε κ	
[λδ κδ δ]	Μεχ[είρ]	ιε ιθ κδ	ξε
[λα ιδ δ]	Φαμ[ε(νώθ)]	[ι]ϛ λγ κη	
[λθ ζ] κδ	Φαρμ(οῦθι)	κε μ νβ	[ ]
[μζ ▾] μδ	Παῦν(ι)	[ι]β̣ μα̣ λ̣ϛ	[ ]
[ξθ νδ] δ	Μεσο(ρή)	[κβ λ]ε̣ μ	[ ]
[οζ ιδ] δ̣	Ἁθύρ	δ [μθ μ]δ	[ ]
[νβ ιδ] δ	Χ̣οίακ	κ[ζ γ μη]	[ ]
[μ ιδ] δ	[Μ]εχί(ρ)	ζ [ιζ νβ]	[ ]
[λα ιδ] δ	[Φαμε(νώθ)]	[η λα νϛ]	[ ]
[λε ιδ] δ	[Φαρμ(οῦθι)]	[ιγ μϛ ▾]	[ ]
[μα ιδ] δ̣	Π̣[αχώ(ν)]	[κε ▾ δ]	[ ]
[νθ ιδ] δ	Ἐ[πείφ]	[κδ ιδ η]	[ ]
[πβ νθ] δ	Φα[ῶφ(ι)]	[ιβ ιγ ιβ]	[ ]
[ξ νθ] δ	Χοίακ̣	[ιγ ιβ ιϛ]	[ ]
[μϛ δ] δ	Τ̣[ῦ]βι	[κθ ιϛ κ]	[ ]
[λα ιδ] δ̣	[Με]χ̣(είρ)	[λ λ κδ]	[ ]
[λα κ μδ]	[Φαρ]μ̣(οῦθι)	[α να η]	[ ]
[μα ιδ δ]	[Παχ]ώ̣(ν)	[ιγ ε ιβ]	[ ]
[να] κζ [κδ]	[Ἐπεί]φ̣	δ̣ [λβ λϛ]	[ ]
ο̣ϛ̣ λδ δ̣	[Θώ]θ̣	[ιϛ ϛ μ]	[ ]
ο̣β̣ μ̣δ δ	[Ἁθύ]ρ̣	[κη κ μδ]	[ ]
μ̣ζ̣ ι̣δ̣ δ̣	[Τ]ῦ̣βι	ι̣ε̣ λ̣[δ μη]	[ ]
[λ]ϛ ν̣δ δ̣	Μεχ(είρ)	κβ κ[η] ν̣β	
[λα ι]δ δ̣	Φαμε(νώθ)	κγ μβ νϛ	
[λζ] κζ κδ	Παχώ(ν)	α ι κ	

Names of months abbreviated by truncation. 9 *l.* Μεχείρ 22 *l.* οβ λδ δ

4158. EPOCHS OF MARS, A.D. 271–325

Fr. 1

i	ii	iii	iv
[46 14 4]	Tybi	[10] 55 20	
[34 24 4]	Mecheir	15 19 24	65
[31 14 4]	Phamenoth	[1]6 33 28	
[39 7] 24	Pharmuthi	25 40 52	[ ]
[47 0] 44	Payni	[1]2 41 36	[ ]
[69 54] 4	Mesore	[22 3]5 40	[ ]
[77 14] 4	Hathyr	4 [49 4]4	[ ]
[52 14] 4	Choeac	2[7 3 48]	[ ]
[40 14] 4	Mecheir	7 [17 52]	[ ]
[31 14] 4	[Phamenoth]	[8 31 56]	[ ]
[35 14] 4	[Pharmuthi]	[13 46 0]	[ ]
[41 14] 4	Pachon	[25 0 4]	[ ]
[59 14] 4	Epeiph	[24 14 8]	[ ]
[82 59] 4	Phaophi	[12 13 12]	[ ]
[60 59] 4	Choeac	[13 12 16]	[ ]
[46 4] 4	Tybi	[29 16 20]	[ ]
[31 14] 4	Mecheir	[30 30 24]	[ ]
[31 20 44]	Pharmuthi	[1 51 8]	[ ]
[41 14 4]	Pachon	[13 5 12]	[ ]
[51] 27 [24]	Epeiph	4 [32 36]	[ ]
76 34 4	Thoth	[16 6 40]	[ ]
72 44' 4	Hathyr	[28 20 44]	[ ]
47 14 4	Tybi	15 3[4 48]	[ ]
[3]6 54 4	Mecheir	22 2[8] 52	
[31 1]4 4	Phamenoth	23 42 56	
[37] 27 24	Pachon	1 10 20	

Fr. 2

i	ii
Πα[χώ(ν)]	[
Π̣αῦν̣(ι)	[
Μεσο(ρή)	θ̣ [
Φ̣α̣ῶ̣φ(ι)	κα̣ [
Τ̣ῦ̣βι	δ [
Μεχί(ρ)	λ̣ [
[Φ]α̣με(νώθ)	κ̣η̣ [
[Φαρ]μ̣(οῦθι)	κ̣[
[Π]α̣χ̣[ώ(ν)]	ϛ̣ [

6 *l.* Μεχείρ

Fr. 3

]γ̣ ιδ η̣ [
]α̣ νη ι̣[β
]γ μβ [
]δ̣ ις [
] λ κ[δ
] να [

4159. Epochs of Mars, A.D. 225–235

[0]	i	ii	iii	iv
[ε]	[Φ]α̣ρ̣(μοῦθι)	κ̣[		
[ζ]	[Ἐ]πείφ	[		
[θ]	Μ̣εσο(ρή)	κη	[	
[ιβ]	Φαῶ(φι)	α	α	ι̣[
[Μαξ]ιμείνο[υ]				
[α]	[Ἁ]θύρ	β	ιε̣	[
			] ̣[	

1 φ]α̣ρ̣ 3 μ̣εσο 4 φα^{ω} 5 *l.* Μαξιμίνου

Fr. 2

i	ii
Pachon	[
Payni	[
Mesore	9$^{?}$ [
Phaophi	21$^{?}$ [
Tybi	4 [
Mecheir	30$^{?}$ [
Phamenoth	28$^{?}$ [
Pharmuthi	2$^{?}$[x
Pachon	6$^{?}$ [

Fr. 3

]3 14 8 [
]1 58 1[2
]3 42 [
]4 16 [
] 30 2[4
] 51 [

4159. Epochs of Mars, a.d. 225–235

[0]	i	ii	iii	iv
5]	Pharmuthi	2[x		
7]	Epeiph	[		
9]	Mesore	28	[	
12]	Phaophi	1	1	1[x
Maximinus				
1]	Hathyr	2	15	[
			]x[	

4159a. Epochs of Mars, a.d. 101–133

i	ii	iii	iv	v
[ ]	[ ]	Κ̣[α]ρ̣[κ(ίνου)]	[ ]	[ ]
[κ]θ̣	μ̣[̣]	Λ̣έο[ν(τος)]	[ ]	[ ]
η	θ	Παρ̣θ̣(ένου)	κ̣β̣	[ ]
κζ	κγ	Ϲκο̣ρ̣(πίου)	ιγ	[ ]
θ	λζ	Αἰγ[ό(κερω)]	λ	ō
κβ	ϛ	Κρ[ι]ο̣(ῦ)	ζ̣	λ
ιγ	λε	Τα[ύ]ρ(ου)	κε	ō
κ	λθ	Διδ̣ύ̣(μων)	λ	ō
κα	ν̣γ	Κα[ρκ(ίνου)]	λ	ō
κϛ	ι̣ζ̣	Π̣α̣[ρθ(ένου)]	ε̣	μ̣[
η	μα	Ζ[υγο(ῦ)]	ι̣η	νγ
η	ιε	Τ[οξό(του)]	κγ̣	κ
κθ	μθ	Ἰ[χθύ(ων)]	ια	ι̣ε̣
κη	μγ̣	Τ̣[αύρ(ου)]	ζ	λ̣
ιβ	μ̣ζ̣	[Διδύ(μων)]	[ ]	[ ]
[ ]	[ ]	[Κα]ρκ(ίνου)	[ ]	[ ]

Names of zodiacal signs abbreviated by truncation.

4159a. Epochs of Mars, a.d. 101–133

i	ii	iii	iv	v
[ ]	[ ]	Cancer	[ ]	[ ]
[2]9	4[x]	Leo	[ ]	[ ]
8	9	Virgo	22	[ ]
27	23	Scorpio	13	[ ]
9	37	Capricorn	30	0
22	6	Aries	7	30
13	35	Taurus	25	0
20	39	Gemini	30	0
21	53	Cancer	30	0
26	17	Virgo	5?	4[0?
8	41	Libra	18	53
8	15	Sagittarius	23?	20
29	49	Pisces	11	15
28	43	Taurus	7	30
12	47	[Gemini]	[ ]	[ ]
[ ]	[ ]	Cancer	[ ]	[ ]

4160. Epochs of Jupiter, a.d. 30–79

i	ii	iii	iv	v
]	α			̣[
] κδ κα ιε		(ἔτοc) ιϛ	Μ[εc]ο(ρή)	[ι]θ νε ιβ ̣[
]ϛ̣ ιε		(ἔτοc) ιη	Θωύθ	κ λβ νδ κ[
] λ̣		(ἔτοc) ιθ	Φαῶφ(ι)	κε β μγ [
]̣ δ ι̣ε		(ἔτοc) κ	Ἁθύρ	κη η γ [
] κ ō		(ἔτοc) κα	Τῦβι	ō ō κζ λ̣[
]̣ ọ̄		(ἔτοc) κβ	Μεχ[(είρ)]	ō ̣ ̣ νζ [
]γ με̣	(ἔ̣τ̣ο̣c̣) κγ τὸ \κ(αὶ)/	Γαίου	Φαμ[ε(νώθ)]	α νη [
]η με̣		(ἔτοc) β	Φαρμο(ῦθι)	δ ι μ̣[
]̣ λ̣		(ἔτοc) γ	Παχώ(ν)	η μ[
]	[ ]	(ἔτοc) δ	Π̣α̣ῦ̣ν̣ι̣	ι̣δ κ[
]	[ ]	Κλαυδ(ίου)	Ἐπείφ	κα̣ [
]	[ ]	(ἔτοc) β	Μ̣[εcο(ρή)]	κ̣ ̣ [

14–26 (*lost*)

i	ii	iii	iv	v
]	[ ]	(ἔτοc) δ̣	[Χοίακ]	[
]	[ ]	(ἔτοc) ε	[Τῦβι]	ιδ κ ̣ [
]	[ ]	(ἔτοc) ϛ	Μ̣ε̣χ[εί(ρ)]	ιε η [
]	[ ]	(ἔτοc) ζ	Φ̣α̣με̣(νώθ)	ιϛ ν̣[
]	[ ]	(ἔτοc) η	Φ̣α̣ρ̣μ̣(οῦθι)	ιθ λ̣[
]	[ ]	(ἔτοc) θ	Παχώ(ν)	κ̣δ [
]	[ ]	(ἔτοc) ι	Ἐπείφ	ọ̄ λ̣[
]	[ ]	(ἔτοc) ια	Μεcο(ρή)	ζ κ̣[
]	[ ]	(ἔτοc) ιγ	Θωύθ	θ ζ [
]	[ ]	[Γ]ά̣λ̣βα	Φαῶφ(ι)	ιγ μ[
]	[ ]	[Ο]ὐεc()	Ἁθύρ	ιη ι[
]	[ ]	(ἔτοc) β	Χοίακ	κ κ̣[
]	[ ]	(ἔτοc) γ	Τῦβ̣ι	κα [

⇓

Month names abbreviated by truncation. ἔτοc = L

2 *l.* Θώθ (also 35, 46)   8 τὸ^κ   12 κλαυ^δ   37 *scil.* Οὐεcπαcιανοῦ

4160. Epochs of Jupiter, a.d. 30–79

i	ii	iii	iv	v
]	1			...[
] 24 21 15		year 16	Mesore	[1]9 55 12 x[
]6 15		year 18	Thoth	20 32 54 2[x
] 30		year 19	Phaophi	25 2 43 [
] x4 15		year 20	Hathyr	28 8 3 [
] 20 0		year 21	Tybi	0 0 27 3[x
]x 0		year 22	Mecheir	0 xx 57 [
]3 45	year 23 =	Gaius	Phamenoth	1 58 [
]8 45		year 2	Pharmuthi	4 10 4[x
]x 30		year 3	Pachon	8 4[x
]	[ ]	year 4	Payni	14 2[x
]	[ ]	Claudius	Epeiph	21 [
]	[ ]	year 2	Mesore	2x [

14–26 (*lost*)

]	[ ]	year 4	[Choeac]	[
]	[ ]	year 5	[Tybi]	14 2x [
]	[ ]	year 6	Mecheir	15 8 [
]	[ ]	year 7	Phamenoth	16 5[x
]	[ ]	year 8	Pharmuthi	19 3[x
]	[ ]	year 9	Pachon	24 [
]	[ ]	year 10	Epeiph	0 3[x
]	[ ]	year 11	Mesore	7 2[x
]	[ ]	year 13	Thoth	9 7 [
]	[ ]	Galba	Phaophi	13 4[x
]	[ ]	Vespasian	Hathyr	18 1[x
]	[ ]	year 2	Choeac	20 2[x
]	[ ]	year 3	Tybi	21 [

⇓

⇑

i	ii	iii	iv	v
]	[ ]	[(ἔτοc) δ]	Μεχεί(ρ)	[
]	[ ]	[(ἔτοc) ε]	Φ̣αμε̣(νώθ)	[
]	[ ]	[(ἔτοc) ϛ]	Φ̣αρμ(οῦθι)	κ̣ζ̣ [
]	[ ]	[(ἔτοc) ζ]	Π̣α̣ῦ̣ν(ι)	α ν[
]	[ ]	[(ἔτοc) η]	[Ἐ]π̣ε̣ίφ	η μ̣[
]	[ ]	[(ἔτοc) θ]	[Με]c̣ο̣(ρή)	ιε̣ . . [
]	[ ]	[Τίτου]	[Θω]ύθ	ιζ̣ [
]	[ ]	[(ἔτοc) β]	[Φαῶ]φ̣(ι)	κ̣[

4160a. EPOCHS OF JUPITER, A.D. 6–13

i	ii	iii
λϛ μθ λζ]	(ἔτουc) [λϛ]	[Παχών
λ]ε̣ α λζ	(ἔτουc) λ[ζ]	[Παῦνι
λ]γ ι̣γ̣ λζ	(ἔτουc) λη	[Ἐπείφ
λ]α κ̣ε̣ λζ	(ἔτουc) λθ	Μ̣[εcορή
κ]θ̣ λζ̣ [λζ]	(ἔτουc) μα	[Θ]ωύ[θ
λ ι λζ]	[(ἔτουc) μ]β	Φ̣[αῶφι

1 ʟ (so subsequent lines)   5 *l.* Θώθ

⥣

i	ii	iii	iv	v
]	[ ]	[year 4]	Mecheir	[
]	[ ]	[year 5]	Phamenoth	[
]	[ ]	[year 6]	Pharmuthi	27 [
]	[ ]	[year 7]	Payni	1 5[x
]	[ ]	[year 8]	Epeiph	8 4[x
]	[ ]	[year 9]	Mesore	15 xx [
]	[ ]	[Titus]	Thoth	17 [
]	[ ]	[year 2]	Phaophi	2[x

4160a. Epochs of Jupiter, A.D. 6–13

i	ii	iii
36 49 37]	Year [36]	[Pachon
3]5 1 37	Year 3[7]	[Payni
3]3 13 37	Year 38	[Epeiph
3]1 25 37	Year 39	Mesore
2]9 37 [37]	Year 41	Thoth
30 10 37]	[Year 4]2	Phaophi

4161. Epochs of Saturn, A.D. 23–92

i	ii	iii	iv	v	vi	vii	viii
[θ]	[η]	[ιβ]	[ζ]	[ια]	[ ]	β̣	[θ
[ι]	[η]	[κϛ]	[ζ]	[κδ]	[ ]	γ̣	[θ
[ια]	[θ]	[ι]	[η]	[η]	ζ̣	δ	[ι
[ιβ]	[θ]	[κδ]	[η]	κα		ε	ι
[ιγ]	ι	η	[θ]	ε̣		ϛ	ι[α
[ιδ]	ι	κγ̣	θ	ι̣θ̣		ζ	ια̣
[ιε]	ια	ζ	ι	γ̣		η	ι[β
[ιϛ]	ια	κα	ι	ιζ		θ	ιβ̣
[ιζ]	ιβ	ε̣	ια	α		ια̣	α̣
[ιη]	ιβ	ιθ	ια	ιδ		[ιβ]	α̣
[ιθ]	ιγ	β	ια	κζ̣	[		
[κα]	α	ι̣α̣	ι̣[β]	[ι]	[		
[κβ]	α̣	κ̣[δ]	[ιβ]	[κγ]	[		
Γ̣α̣[ί	ου		]				
α	β	δ	[α]	[ε]	[		
β	β	κ̣	α̣	[ιζ]	[		
γ̣	γ̣	β̣	[				

4161. Epochs of Saturn, a.d. 23–92

i	ii	iii	iv	v	vi	vii	viii
[9]	[8]	[12]	[7]	[11]	[ ]	2	[9]
[10]	[8]	[26]	[7]	[24]	[ ]	3	[9]
[11]	[9]	[10]	[8]	[8]	7?	4	[10]
[12]	[9]	[24]	[8]	21		5	10
[13]	10	8	[9]	5		6	1[1]
[14]	10	23	9	19		7	11
[15]	11	7	10	3		8	1[2]
[16]	11	21	10	17		9	12
[17]	12	5	11	1		11	1
[18]	12	19	11	14		[12]	1
[19]	13	2	11	27	[		
[21]	1	11?	12	[10]	[		
[22]	1	2[4]	[12]	[23]	[		
Gaius							
1	2	4	[1]	[5]	[		
2	2	20?	1	[17]	[		
3	3	2	[				

4162. Template for the Sun

Recto

i	ii	iii		iv	v	vi ⇒
ρ̣ν̣	ι̣β̣	[λ]γ̣			ι̣	[̣]
	[ι]γ	λ̣		[ρπ]	ι̣[α]	[̣]
	ι̣δ̣	κζ			[ια]	ν̣η̣
	ιε	κδ			ι̣β̣	ν̣ ̣
	[ι]ς	κ̣α̣			ιγ̣	νβ̣
ρ̣ν̣ε̣	ι̣ζ	ιη			ιδ̣	μ̣θ̣
	[ι]η̣	[ι]ε̣		[ρπε]	ιε̣	μ̣ ̣
	ι̣θ̣	ιβ̣			ι̣ς	μ̣β̣
	κ̣	θ			ι̣ζ̣	λ̣θ̣
	κ̣α	ς			ι̣η	λς
ρξ	κβ	γ			ι̣θ̣	λγ
	κγ	ō		[ρ]ϙ̣	κ̣	λ̣
	κγ	νζ			κα	κ̣ζ̣
	κδ	νδ̣			κβ̣	κ̣δ̣
	κε̣	να̣			κ̣[γ]	κα̣
ρξε	κς	[μ ̣]			κ[δ]	ιζ
	κζ	μδ̣		ρ̣ϙ̣[ε]	[κ]ε	[ι]δ
	κη	μα			[κ]ς̣	ια
	κθ	λη			[κ]ζ̣	η̣
Διδύ̣μ̣[οις]					κη̣	ε
	[ō]	[λ ̣]			κθ̣	β̣
ρο	[α]	[λ ̣]		[c]	κθ̣	[ν]θ
	β̣	[κ ̣]		[Καρ]κί(νῳ)		
	γ̣	[κ ̣]			ō	νς
	δ̣	[κ ̣]			α	νγ
	ε	ιθ̣			β	ν
ροε	ς	[ι ̣]			γ	μς̣
	ζ	ιγ		c̣ε̣	δ	[μ]γ̣
	η	ι			ε̣	[̣ ̣]
	θ̣	ζ			ς̣	[λ]ζ̣

23 καρ]κι

4162. Template for the Sun

Recto

i	ii	iii	iv	v	vi ⇒
150	12	[3]3		10	[x]
	[1]3	30	**[180]**	1[1]	[x]
	14	27		[11]	58
	15	24		12	5x
	[1]6	21		13	52
155	17	18		14	49
	[1]8	[1]5	**[185]**	15	4x
	19	12		16	42
	20	9		17	39
	21	6		18	36
160	22	3		19	33
	23	0	**[1]90**	20	30
	23	57		21	27
	24	54		22	24
	25	51		2[3]	21
165	26	[4x]		2[4]	17
	27	44	**19[5]**	[2]5	[1]4
	28	41		[2]6	11
	29	38		[2]7	8
Gemini				28	5
	[0]	[3x]		29	2
170	[1]	[3x]	**[200]**	29	[5]9
	2	[2x]	**Cancer**		
	3	[2x]		0	56
	4	[2x]		1	53
	5	19		2	50
175	6	[1x]		3	46
	7	13	**205**	4	[4]3
	8	10		5	[xx]
	9	7		6	[3]7

Recto, continued

vii	viii	ix	x	xi	xii
	ζ	λδ		[ϛ]	[ι ̣]
	η	λ̣[̣]		[ζ]	[ι ̣]
ϲι	θ	κ̣[̣]	[ϲμ]	[η]	[̣]
	ι	[κ ̣]		[θ]	[̣]
	ια̣	[κ ̣]		[ι]	[̣]
	ι̣β̣	ι̣ ̣		[ια]	[̣]
	ι̣γ̣	ι̣ϛ		[ιβ]	[̣]
[ϲιε]	[ιδ]	ι̣ ̣	[ϲμε]	[ιβ]	[ν ̣]
	ιε	ι		ι̣γ̣	[ν ̣]
	ιϛ	ζ		ι̣δ̣	[ν ̣]
	ιζ	ε		ι̣ε̣	[ν ̣]
	ιη	β		ι̣ϛ̣	[μ]θ̣
ϲ̣κ̣	ι̣η̣	νθ	ϲν	ιζ̣	μζ
	ι̣θ̣	νϛ		ι̣[η]	μ̣ε̣
	κ	ν̣δ		ι[θ]	μδ
	κα	να		[κ]	μ̣β
	κβ̣	μη		κ̣α̣	μ̣
ϲκε	κγ̣	με̣	ϲν[ε]	κ̣β̣	λη
	κ̣δ̣	μ̣γ̣		κγ̣	λ̣ϛ
	κε	μ		κδ̣	λε
	κϛ	λη		[κε]	[λ]γ̣
	κζ̣	λ̣ε		[κϛ]	[λ ̣]
ϲ̣λ̣	κη̣	λβ	[ϲξ]	[κζ]	[̣ ̣]
	κθ̣	λ		[κη]	[κ ̣]
Λέ̣οντ(ι)	ō	κ[̣]		[κθ]	[κ ̣]
	α	κε	[Παρθένῳ]		
	β	κβ		[ō]	[κ ̣]
ϲλε	γ̣	ιθ		[α]	[κ ̣]
	δ	ιζ̣	[ϲξε]	[β]	[κ ̣]
	ε	ιε		[γ]	[̣ ̣]

23 λ̣εοντ

Recto, continued

vii	viii	ix	x	xi	xii
	7	34		[6]	[1x]
	8	3[x]		[7]	[1x]
210	9	2[x]	**[240]**	[8]	[x]
	10	[2x]		[9]	[x]
	11	[2x]		[10]	[x]
	12	1x		[11]	[x]
	13	16		[12]	[x]
[215]	[14]	1x	**[245]**	[12]	[5x]
	15	10		13	[5x]
	16	7		14	[5x]
	17	5		15	[5x]
	18	2		16	[4]9
220	18	59	**250**	17	47
	19	56		1[8]	45
	20	54		1[9]	44
	21	51		[20]	42
	22	48		21	40
225	23	45	**25[5]**	22	38
	24	43		23	36
	25	40		24	35
	26	38		[25]	[3]3
	27	35		[26]	[3x]
230	28	32	**[260]**	[27]	[xx]
	29	30		[28]	[2x]
Leo	0	2[x]		[29]	[2x]
	1	25	**[Virgo]**		
	2	22		[0]	[2x]
235	3	19		[1]	[2x]
	4	17	**[265]**	[2]	[2x]
	5	15		[3]	[xx]

Verso

i	ii	iii	iv	v	vi
	[δ]	ι̣ ̣		[δ]	[̣]
	[ε]	κ̣α̣		ε̣	[̣]
	[ϛ]	[κ ̣]		ϛ̣	[̣]
[τλ]	[ζ]	[κ]γ̣		ζ	ϛ̣
	[η]	κ̣δ̣	[τξ]	η	η
	[θ]	κ̣ε̣		θ̣	[̣]
	[ι]	κζ		[ι]	[ι ̣]
	[ια]	[κ]η		[ια]	[ι ̣]
τλε	ι̣β̣	κθ		ι̣β̣	ιδ
	ι̣γ	λα	τξε	ιγ	ιϛ
	ιδ	λβ̣		ι̣γ̣	λ
	ιε̣	λ̣δ̣			
	ιϛ	λε			
τμ̣	ιζ	λ̣[̣]			
	ι̣η̣	λη̣			
	ι̣θ̣	λθ̣			
	κ̣	μ̣α̣			
	[κ]α̣	μβ̣			
τμ̣ε̣	[κ]β̣	μδ			
	[κ]γ̣	μϛ̣			
	[κδ]	μζ			
	[κε]	μ̣η			
	[κϛ]	[̣ ̣]			
[τν]	[κζ]	[ν ̣]			
	[κη]	[ν ̣]			
	[κθ]	[ν ̣]			
[Τοξότη]					
	[ō]	[ν] ̣			
	[α]	[ν] ̣			
[τνε]	[β]	[ν ̣]			

Verso

i	ii	iii		iv	v	vi
	[4]	1x			[4]	[x]
	[5]	21			5	[x]
	[6]	[2x]			6	[x]
[330]	[7]	[2]3			7	6
	[8]	24		**[360]**	8	8
	[9]	25			9	[x]
	[10]	27			[10]	[1x]
	[11]	[2]8			[11]	[1x]
335	12	29			12	14
	13	31		**365**	13	16
	14	32			13	30
	15	34				
	16	35				
340	17	3[x]				
	18	38				
	19	39				
	20	41				
	[2]1	42				
345	[2]2	44				
	[2]3	46				
	[24]	47				
	[25]	48				
	[26]	[xx]				
[350]	[27]	[5x]				
	[28]	[5x]				
	[29]	[5x]				
[Sagittarius]						
	[0]	[5x]				
	[1]	[5]x				
[355]	[2]	[5x]				

4163. Template for the Sun, and Procedure Text

Recto

[0]	i	ii	iii
	[Ταύρ]ῳ		
	[ō ν]ζ		κδ̣ [μβ]
	[α ν]δ̣		κε̣ [λθ]
[τε]	[β ν]α̣	τλ	κ[ϛ λϛ]
	[γ μ]η̣		κ[ζ λγ]
	[δ μ]ε̣		κη λ
	[ε μ]β̣		κθ κζ
	[ϛ λ]θ̣		Διδύμω⟨ν⟩
[τι]	[ζ λ]ϛ̣	τλ̣[ε]	α̣ κ
	[η λ]γ		[β ι]ζ
	[θ λ]		γ ιδ̣
	[ι κ]ζ̣		δ ια̣
	[ια κ]δ̣		ε η̣
[τιε]	[ιβ κ]α̣	τμ	ϛ δ
	[ιγ ι]η̣		ζ α
	[ιδ ι]ε̣		ζ ⟨ν⟩η
	[ιε ι]β̣		η νε
	[ιϛ] θ̣		θ νβ
[τκ]	[ι]ζ ϛ	τμ̣ε	ι μθ
	[ιη] γ		ια με
	[ιθ] ọ̄		[ι]β μβ
	[ιθ ν]ζ		ι̣γ λθ
	[κ ν]δ̣		ι̣[δ λ̣]
[τκε]	[κα ν]α̣	τν	ι̣[ε λ̣]
	[κβ μ]η		ι̣[ϛ ̣ ̣]
	[κγ] μ̣ε		ι̣[ζ κ̣]

8 διδυμω̅

Verso, table

i	ii
	[ιη] κγ
	[ιθ] κ
[τνε]	[κ ι]ζ̣
	κα κδ
	κβ κα
	κγ η̣
	κδ [̣]
τξ	κε̣ α̣
κε νη	κϛ̣ νε
	κζ νβ
	κη μη
τξε	κθ με
̣ ̣ ̣ ̣ια λ ō	
	λ ō

4 *l.* κα ιδ 5 *l.* κβ ια

9 The entry κε νη was omitted in error before this line, then added to the left of the subsequent entry.

4163. Template for the Sun, and Procedure Text

Recto

[0]	i	ii	iii
	Taurus		
	[0 5]7		24 [42]
	[1 5]4		25 [39]
[305]	[2 5]1	330	2[6 36]
	[3 4]8		2[7 33]
	[4 4]5		28 30
	[5 4]2		29 27
	[6 3]9		Gemini
[310]	[7 3]6	33[5]	1 20
	[8 3]3		[2 1]7
	[9 30]		3 14
	[10 2]7		4 11
	[11 2]4		5 8
[315]	[12 2]1	340	6 4
	[13 1]8		7 1
	[14 1]5		7 ⟨5⟩8
	[15 1]2		8 55
	[16] 9		9 52
[320]	[1]7 6	345	10 49
	[18] 3		11 45
	[19] 0		[1]2 42
	[19 5]7		13 39
	[20 5]4		1[4 3x]
[325]	[21 5]1	350	1[5 3x]
	[22 4]8		1[6 xx]
	[23] 45		1[7 2x]

Verso, table

i	ii
	[18] 23
	[19] 20
[355]	[20 1]7
	21 24
	22 21
	23 8
	24 [x]
360	25 1
25 58	26 55
	27 52
	28 48
365	29 45
... 30 0	
	30 0

Verso, text surrounding table

ἐὰν ὀ̣ [
τῆc το[
δ̣έξαι[
ε̣κ̣και[
κατ̣η̣[
 ̣μ̣η̣α̣[
cιc ἐπα[προ–
c̣θέτω ι[
 ̣εναι[
 ̣ζ̣ετ(αι) βα[
 ̣ ̣ καὶ ̣[
ο̣π̣ ̣[ποίη–
cον ἐπ[ὶ διέκ–
βαλε τ̣ὴ̣[ν λοιπὴν
καὶ οὗπ[ερ ἂν καταλήξῃ, ἐκεῖ
ἔcται ἡ [οἷ–
ον ἐπὶ τ[οῦ
ζ (ἔτουc) Ο̣ὐαλ̣ε̣[ριανοῦ τὰ
π̣λήρη ἔτη [
δ̣ὲ̣ δέκα γ(ίνεται) $\overline{\text{cϙη}}$. π̣α̣ρ̣ὰ̣ τὸ ν̣[λοι–
π̣αὶ $\overline{\text{ιγ}}$. τὰ $\overline{\text{ιγ}}$ ἐπὶ $\overline{\text{ιθ}}$ γ(ίνεται) $\overline{\text{cμ[ζ}}$. παρὰ τὰ $\overline{\lambda}$
γ(ίνεται)] $\overline{\text{η}}$ λοιπαὶ ζ α̣ ̣[̣] ἡ [c]ύνο̣[δοc
$\overline{\text{η̣}}$ ἐcτιν. ἐὰν δ̣[ὲ] ̣[πανcέλη–
νοc ζητήcα[ ] ̣ ̣[
ὤ]φελε ὥρ(αc) $\overline{\text{ια}}$ d′· λοιπαὶ ἡμέρα[ι
 ̣] ̣ ̣ ε ̣β̣[̣] ̣ ̣ ̄ ̣ ̣ γ(ίνεται) $\overline{\text{ϛ̣}}$. εἰc ζ̣ [
ὡcα]ύτωc κ(αὶ) οἱ λ̣ο[ιποὶ
πα]νcεληνο̣ ̣ ̣ [
] τ̣ῆc cυνόδ[ου
]α̣ ̣ ̣ κατὰ μ[
]ο̣c οἱ λ̣[

10 ζετ′ 15 οϋπ[ερ 18 ϛ 20 ϯ (so in subsequent lines) 25 ῳ 27 ϗ

Verso, text surrounding table

If …[… …[… …[… …[… …[… …[… …[… let there be added …[… …[… …[… xx and …[… …[… multiply by [… count off the [remainder… and wherever [it falls, there will be the [… For example, for the [… year 7 of Valerianus [the completed years [… ten, there results 298. Divided by the …[… remainder 13. The 13 times 19, there results 24[7. Divided by 30, there results] 8, remainder 7. … the conjunction … is $8^{?}$. If […]…[… full moon we$^{?}$ seek […]…[… subtract $11\,^{1}/_{4}$ hours. There remain […] days [… …]…[…]… x … there results $6^{?}$. into $7^{?}$ [… …] similarly the rest$^{?}$ [… …] full moon [… …] of the conjunction [… …]…[… …]…[…

4164. Template for the Moon

[-IV]	[-III]	[-II]	[-I]	[0]	i	ii	iii	iv	v	⇒
						]ν̣ο̣ι	̣[			
					]	ϲ̣υ̣ναγό(μεναι)			̣[	
[ια]	[μβ]	[ι]	[λζ]		[ια]	[μβ]	ι	λ̣ζ		
[ια]	[νε]	[ō]	[λζ]		[κγ]	[λζ]	ι̣α̣	ι̣δ		
[ιβ]	[ζ]	[ν]	[λζ]		[λε]	[με]	α̣	να		
[ιβ]	[κ]	[μ]	[λζ]		[μη]	[ε]	[μβ]	κ̣η̣		
[ιβ]	[λγ]	[λ]	[λζ]	[ε]	[ξ]	[λθ]	[ιγ]	ε̣		
[ιβ]	[μϛ]	[κ]	[λζ]		[ογ]	[κε]	[λγ]	μ̣β̣		
[ιβ]	[νθ]	[ι]	[λζ]		[πϛ]	[κδ]	[μ]δ̣	ιθ		
[ιγ]	[ιβ]	[ō]	[λζ]		[ϙθ]	[λϛ]	μ̣δ̣	νϛ		
[ιγ]	[κδ]	[ν]	[λζ]		[ριγ]	α̣	λε	λγ		
[ιγ]	[λζ]	[μ]	[λζ]	[ι]	[ρκϛ]	λ̣θ	ιϛ	ι		
[ιγ]	[ν]	[λ]	[λζ]		[ρμ]	κ̣θ	μϛ	μζ		
[ιδ]	[γ]	[κ]	[λζ]		[ρνδ]	[λ]γ̣	ζ	κδ		
[ιδ]	[ιϛ]	[ι]	[λζ]		[ρξη]	[μ]θ	ιη	α		
[ιδ]	[κθ]	[ō]	[λζ]		[ρπγ]	[ι]η̣	ιη	λη̣		
[				]						
[ιδ]	[λϛ]	[ζ]	[λζ]	[ιε]	[ρ]ϙ̣ζ	νδ	κϛ̣	ι[ε]		
[ιδ]	[κγ]	[ιζ]	[λζ]		ϲ̣ιβ	ιζ	μγ	ν̣[β]		
[ιδ]	[ι]	[κζ]	[λζ]		[ϲ]κ̣ϛ̣	κ̣η	ι̣α̣	[κθ]		
[ιγ]	[νζ]	[λζ]	[λζ]		ϲ̣μ	κ̣ε̣	μ̣[θ]	[ϛ]		
[ιγ]	[μδ]	[μζ]	[λζ]		[ϲνδ]	[ι]	[λϛ]	[μγ]		

– –

2 ϲ̣υ̣ναγ°

4164. Template for the Moon

[-IV]	[-III]	[-II]	[-I]	[0]	i	ii	iii	iv	v	⇒
						]...[				
					]	running total ...[				
[11]	[42]	[10]	[37]		[11]	[42]	10	37		
[11]	[55]	[0]	[37]		[23]	[37]	11	14		
[12]	[7]	[50]	[37]		[35]	[45]	1	51		
[12]	[20]	[40]	[37]		[48]	[5]	[42]	28		
[12]	[33]	[30]	[37]	[5]	[60]	[39]	[13]	5		
[12]	[46]	[20]	[37]		[73]	[25]	[33]	42		
[12]	[59]	[10]	[37]		[86]	[24]	[4]4	19		
[13]	[12]	[0]	[37]		[99]	[36]	44	56		
[13]	[24]	[50]	[37]		[113]	1	35	33		
[13]	[37]	[40]	[37]	[10]	[126]	39	16	10		
[13]	[50]	[30]	[37]		[140]	29	46	47		
[14]	[3]	[20]	[37]		[154]	[3]3	7	24		
[14]	[16]	[10]	[37]		[168]	[4]9	18	1		
[14]	[29]	[0]	[37]		[183]	[1]8	18	38		
[				]						
[14]	[36]	[7]	[37]	[15]	[1]97	54	26	1[5]		
[14]	[23]	[17]	[37]		212	17	43	5[2]		
[14]	[10]	[27]	[37]		[2]26	28	11	[29]		
[13]	[57]	[37]	[37]		240	25	4[9]	[6]		
[13]	[44]	[47]	[37]		[254]	[10]	[36]	[43]		

- -

⇐ vi	vii	viii	ix	[X]
[ιδ]	[λ]	[κδ]	[λζ]	
[ιδ]	[ιζ]	[λδ]	[λζ]	
[ιδ]	[δ]	[μδ]	[λζ]	[με]
[ιγ]	[να]	[νδ]	[λζ]	
[ιγ]	[λθ]	[δ]	[λζ]	
[ιγ]	κ̣ς̣	[ιδ]	[λζ]	
[ι]γ̣	ιγ	[κδ]	[λζ]	
ι̣γ̣	ō	λ̣[δ]	[λζ]	[ν]
ιβ	μζ	μ̣[δ]	[λζ]	
ιβ	λδ	νδ̣	[λζ]	
ιβ	κβ	δ	[λζ]	
ιβ	θ	ιδ	[λζ]	
[ια]	νς	κ̣δ̣	[λζ]	[νε]
[ια]	μγ	λδ	λζ̣	
[πλ]ήρης [ ]				
[ι]α̣	νγ	λς	λζ	
[ι]β̣	ς	κς	λζ	
ι̣β	ιθ	ι̣ς	λζ	
ι̣β	λ̣β̣	ς̣	λζ	[ξ]
ι̣β̣	μ̣δ̣	ν̣ς	λζ	

— — — — — — — — — —

vi	vii	viii	ix	[X]
[14]	[30]	[24]	[37]	
[14]	[17]	[34]	[37]	
[14]	[4]	[44]	[37]	[45]
[13]	[51]	[54]	[37]	
[13]	[39]	[4]	[37]	
[13]	26	[14]	[37]	
[1]3	13	[24]	[37]	
13	0	3[4]	[37]	[50]
12	47	4[4]	[37]	
12	34	54	[37]	
12	22	4	[37]	
12	9	14	[37]	
[11]	56	24	[37]	[55]
[11]	43	34	37	
[ ] complete? [				]
[1]1	53	36	37	
[1]2	6	26	37	
12	19	16	37	
12	32	6	37	[60]
12	44	56	37	

⇐

4164a. TEMPLATE FOR THE MOON

i	ii	iii	
ι̣β̣ β ζ λζ		ν̣[θ να ε κς]	
ι̣β̣ ιδ νζ λ̣[ζ]		[ο]β̣ ς γ [γ]	
ι̣β̣ κζ μζ λζ	[cπ]	πδ [λγ ν μ]	
ι̣β̣ μ λζ λζ		ϙζ ιδ κ[η ιζ]	
ι̣β̣ νγ κζ λζ		ρ̣ι ζ νε [νδ]	
ιγ ς ιζ λζ		ρκγ ιδ ι[γ λα]	
ιγ ιθ ζ λζ		ρλς λγ κ[α η]	
ιγ λα νζ λζ	c̣[πε]	ρν ⟨ε⟩ ιη με̣	
ιγ μδ μζ λζ		ρξγ ν ς̣ [κβ]	
ιγ νζ λζ λζ		ροζ̣ μζ [μγ νθ]	
ιδ ι κζ λζ		ρϙα νη [ια λς]	
ιδ κγ κζ λζ		cς κα κ̣[θ κγ]	
ιδ λς ζ λζ	c̣ϙ̣	cκ νζ λς ν	
[ιδ κ]θ̣ ọ̄ λ̣ζ		c̣[λε κ̣ς̣ λ̣ζ̣ [κζ]	
[ιδ ι]ς̣ ι̣ λ̣ζ		c̣μ̣θ̣ μβ μ̣η̣ [δ]	
ι̣δ̣ γ̣ κ λ̣ζ		c̣ξ̣γ̣ μς [η μα]	
[ιγ] ν̣ λ̣ λ̣ζ		cο̣ζ̣ λ̣ς̣ λ̣[θ ιη]	
[ιγ] λ̣ζ μ̣ λζ	cϙε	c̣ϙ̣α̣ ι̣δ̣ ι̣θ̣ ν̣ε̣	
[ιγ κδ] ν̣ λ̣ζ		τ̣δ λθ̣ ι λβ	
[ιγ ι]β̣ ō λ̣ζ		τ̣ιζ να ια θ̣	
[ιβ] ν̣θ̣ ι̣ λ̣ζ̣		τλ ν κ̣α μς	
[ι]β̣ μ̣ς̣ κ̣ λζ		τ̣μ̣γ̣ [λς μβ κγ]	
[ιβ] λ̣γ λ̣ λ̣ζ	τ	τνς ι ι̣γ̣ ō	
[ιβ κ μ] λ̣ζ̣		η λ νγ λζ	
[ιβ] ζ̣ ν̣ λ̣ζ̣		κ λ̣η μδ ⟨ι⟩δ	
ι̣α̣ ν̣ε̣ ọ̄ λζ̣		λ̣β λ̣γ̣ μδ ν̣α̣	

3 Contents of col. iii are shifted approx. 2 letters to the right, probably because the index number in col. ii was too wide for the available space. 10 The entire line, omitted in error, was squeezed in the space above line 11, with no ruling between. 15ff Contents of cols. i–iii and rulings shifted to right of upper part of table as shown.

4164a. TEMPLATE FOR THE MOON

i	ii	iii
12 2 7 37		5[9 51 5 26]
12 14 57 3[7]		[7]2 6 3 [3]
12 27 47 37	[280]	84 [33 50 40]
12 40 37 37		97 14 2[8 17]
12 53 27 37		110 7 55 [54]
13 6 17 37		123 14 1[3 31]
13 19 7 37		136 33 2[1 8]
13 31 57 37	2[85]	150 <5> 18 45
13 44 47 37		163 50 6 [22]
13 57 37 37		177 47 [43 59]
14 10 27 37		191 58 [11 36]
14 23 27 37		206 21 2[9 23]
14 36 7 37	290	220 57 36 50

i	ii	iii	
[14 2]9 0 37		2[35 26 37 [27]	
[14 1]6 10 37		249 42 48 [4]	
14 3 20 37		263 46 [8 41]	
[13] 50 3 037		277 36 3[9 18]	
[13] 37 40 37	295	291 14 19 55	
[13 24] 50 37		304 39 10 32	
[13 1]2 0 37		317 51 11 9	
[12] 59 10 37		330 50 21 46	
[1]2 46 20 37		343 [36 42 23]	
[12] 33 30 37	300	356 10 13 0	
[12 20 40] 37		8 30 53 37	
[12] 7 50 37		20 38 44 <1>4	
11 55 0 37		32 33 44 51	

4165. Template for Mars

fr. 1

i	ii	iii	iv	v
	[Καρκίνου καὶ] Λ̣[έο]ντος			
	[ō ō λβ] λ̣	ō ō λβ λ	ν̣ε	ō [κ]ε μδ [λ]
	[ō α ō] λ̣	ō α λγ ō		ō κ̣ϛ ιβ [λ]
	[ō α κη] λ̣	ō γ α λ		ō κϛ μ [λ]
	[ō α νϛ] λ̣	ō δ νη ō		ō κζ η [λ]
[ε]	[ō β κδ λ]	ō ζ κβ λ		ō κζ λϛ [λ]
	[ō β νβ λ]	ō ι ιε ō	ξ	ō κη δ [λ]
	[ō γ κ λ]	ō ιγ λε λ		ō κη λβ̣ [λ]
	[ō γ μη λ]	ō ιζ κδ̣ ọ̄		[ō] κ̣θ̣ ọ̄ [λ]
	ọ̄ [δ ιϛ λ]	ọ̄ κ̣α̣ [μ λ]		
ι	ọ̄ [δ μδ λ]	[ō κϛ κε ō]		
	ō ε̣ [ιβ λ]	[ō λα λζ λ]		
	ō ε̣ [μ λ]	[ō λζ ιη ō]		
	ọ̄ ϛ [η λ]	[ō μγ κϛ λ]		
	(lost)			
	ō ιζ̣ [κ λ]	[ε λ ν λ]		
	ō ιζ̣ [μη λ]	[ε μη λθ ō]		
	ō ιη [ιϛ λ]	[ϛ ϛ νε λ]		
μ̣	ō ιη̣ [μδ λ]	[ϛ κε μ ō]		
	[ō] ι̣[θ ιβ λ]	[ϛ μδ νβ λ]		

fr. 2

Καρκ[ίνου	
α	ō κ[
	ō κη̣ [
	ō κζ̣ [
	ō [
ε̣	ọ̄ [

4165. Template for Mars

fr. 1

i	ii	iii	iv	v
	[Cancer and] Leo			
	[0 0 32] 30	0 0 32 30	55	0 [2]5 44 [30]
	[0 1 0] 30	0 1 33 0		0 26 12 [30]
	[0 1 28] 30	0 3 1 30		0 26 40 [30]
	[0 1 56] 30	0 4 58 0		0 27 8 [30]
[5]	[0 2 24 30]	0 7 22 30		0 27 36 [30]
	[0 2 52 30]	0 10 15 0	60	0 28 4 [30]
	[0 3 20 30]	0 13 35 30		0 28 32 [30]
	[0 3 48 30]	0 17 24 0		[0] 29 0 [30]
	0 [4 16 30]	0 21 [40 30]		
10	0 [4 44 30]	[0 26 25 0]		
	0 5 [12 30]	[0 31 37 30]		
	0 5 [40 30]	[0 37 18 0]		
	0 6 [8 30]	[0 43 26 30]		
	(lost)			
	0 17 [20 30]	[5 30 50 30]		
	0 17 [48 30]	[5 48 39 0]		
	0 18 [16 30]	[0 0 55 30]		
40	0 18 [44 30]	[0 25 40 0]		
	[0] 1[9 12 30]	[0 44 52 30]		

fr. 2

Cancer [	
1	0 2[x
	0 28 [
	0 27 [
	0 [
5	0 [

4165a. Template for Jupiter

fr. 1

i	ii	iii
	[ια λα λ]β̣ [ō]	[ō ς ιη κδ]
	[ια μγ] ζ̣ [ιβ]	[ō ς θ λς]
	[ια νδ] λ̣γ̣ λ̣[ς]	[ō ς ō μη]
	[ιβ] ε̣ να ιβ	[ō ε νβ ō]
[ν̄]	[ιβ] ι̣ζ̣ ō ō	ō ε̣ [μγ ιβ]
	[ιβ] κ̣η̣ ō ō	ō ε̣ [λδ κδ]
	[ιβ λη να̣ ι̣β̣	ō ε̣ κ̣[ε λς]
	ιβ με λγ λς	ō ε ι̣ς̣ [μη]
	ιγ ō ζ ιβ	ō ε̣ η ō
ν̄ε̄	ιγ ι λβ ō	ō δ ν̣θ̣ [ι]β̣
	ιγ κ μη ō	ọ̄ δ̣ ν̣ κ̣δ̣
	[ι]γ λ νε ιβ	ọ̄ δ̣ μ̣[α λς]
	[ιγ] μ̣ νγ λ̣[ς]	ō δ λβ μ̣[η]
	[ιγ ν] μγ [ι]β̣	ō δ κδ̣ [ō]
[ξ]	[ιδ ō κδ ō]	ō δ ιε [ιβ]
	[ιδ θ νς ō]	ō δ ς κ̣[β]
	[ιδ ιθ ιθ ιβ]	ō γ νζ [λς]
	[ιδ κη λγ λς]	ọ̄ γ μη [μη]
	[ιδ λζ λθ ιβ]	ọ̄ γ μ ọ̄
[ξε]	[ιδ μς λς ō]	[ō] γ̣ λ̣α̣ ι̣β̣

fr. 2

ō [
ō ι[
ō ι[
ō [
ō ι[
ō [ι]ε [
ō ιε [
ō ιε ̣[
ō ι ̣ [
ō ι̣ ̣ [

4165a. Template for Jupiter

fr. 1

i	ii	iii
	[11 31 3]2 [0]	[0 6 18 24]
	[11 43] 7 [12]	[0 6 9 36]
	[11 54] 33 3[6]	[0 6 0 48]
	[12] 5 51 12	[0 5 52 0]
[50]	[12] 17 0 0	0 5 [43 12]
	[12] 28 0 0	0 5 [34 24]
	12 38 51 12	0 5 2[5 36]
	12 45 33 36	0 5 16 [48]
	13 0 7 12	0 5 8 0
55	13 10 32 0	0 4 59 [1]2
	13 20 48 0	0 4 50 24
	[1]3 30 55 12	0 4 4[1 36]
	[13] 40 53 3[6]	0 4 32 4[8]
	[13 50] 43 [1]2	0 4 24 [0]
[60]	[14 0 24 0]	0 4 15 [12]
	[14 9 56 0]	0 4 6 2[2]
	[14 19 19 12]	0 3 57 [36]
	[14 28 33 36]	0 3 48 [48]
	[14 37 39 12]	0 3 40 0
[65]	[14 46 36 0]	[0] 3 31 12

fr. 2

0 [
0 1[x
0 1[x
0 [
0 1[x
0 [1]5 [
0 15 [
0 15 x[
0 1x [
0 1x [

4166. Template for Saturn with Jottings

Front

— — — — — — — — —

. ō .[
. . . ō δ
(λοιπαὶ) ō ō ν νδ λβ
. ō ō δ
γ̣ί̣(νεται) ō ō νδ νδ λβ . .
τ̣ο̣ῦ̣ ō ō κζ κζ . .
(λοιπαὶ) ō ζ μγ κζ ι̣ς̣

ζ ι̣ς

2 ›γ‹ ō δ 3 (λοιπαὶ): ɔ (so line 7) 4 ꝭ ō ō δ 5 ⌐

4166. Template for Saturn with Jottings

Front

— — — — — — — — —

… 0 …

… 0,4

remainder 0;0,50,54,32

0;0,4

there results 0;0,54,54,32 …

the[?] 0;0,27,27,xx

remainder 0;7,43,27,16

7,16[?]

Back

i	ii	iii	iv
[ō ς κγ κζ ις]	[ε η ι] νδ [ō]		[ō μς μ]
[ō ς κζ κζ ις]	[ε] ι̣[δ λη] κα ι̣[ς]		[ō νγ κ]
[ō ς λα κζ ις]	[ε] κ̣[α θ] μ̣η λ̣[β]		[α ō ō]
[ō ς λε κζ ις]	[ε] κ̣[ζ με] ιε [μη]	[ι]	[α ς μ]
[ō ς λθ κζ ι]ς	[ε] λ̣δ̣ [κδ μγ δ]		[α ιγ κ]
[ō ς μγ] κζ ις	ε̣ μ̣α̣ [η ι κ]		[α κ ō]
[ō ς μ]ζ κζ ις	ε μζ ν̣ε̣ λ̣ζ̣ [λς]		[α κς μ]
[ō ς ν]α̣ κζ ις	ε νδ μ̣ζ̣ δ ν̣β̣		α̣ [λγ κ]
[ō] ς ν̣ε̣ κζ ις	ς α μβ̣ λβ η	ι̅ε̅	α [μ ō]
[ō] ς̣ ν̣θ κζ ις	ς η μα νθ κδ		α [μς μ]
[ō] ζ γ κζ ις	ς ιε με κς μ		α ν̣[γ κ]
[ō] ζ ζ κζ ις	ς κβ νβ νγ νς		β ō [ō]
[ō] ζ ια κζ ις	ς λ δ κα ιβ		β ς [μ]
[ō] ζ ιε κζ ις	ς λζ ιθ μη κη	κ̅	β ι̣γ [κ]
[ō] ζ ιθ κζ ις	ς μδ λθ ιε μδ		β κ [ō]
[ō] ζ κγ κζ ις	ς νβ β μγ ō		β κ̣[ς μ]
[ō] ζ κ̣ζ κζ ις	ς νθ λ ι ις		β [λγ κ]
[ō ζ λ]α κζ ις	ζ ζ α̣ λζ λβ		β [μ ō]
[ō ζ λ]ε κζ ις	ζ ιδ λ̣ζ̣ δ μη	κ̅ε̣̅	β [μς μ]
[ō ζ λ]θ κζ ις	ζ κβ ι̣ς̣ λ̣β δ		β [νγ κ]
[ō ζ] μ̣γ κζ ις	ζ κθ ν̣θ̣ νθ κ		γ [ō ō]
			γ [ς μ]
			γ [ιγ κ]
		λ̅	γ [κ ō]
		Διδ[ύμοις	
		Ὑδρο̣[χόῳ	
		τ̣ [	
		Αἰγ̣[όκερῳ	
		μ̣ ̣[	
		̣[	

Back

i	ii	iii	iv
[0 6 23 27 16]	[5 8 10] 54 [0]		[0 46 40]
[0 6 27 27 16]	[5] 1[4 38] 21 1[6]		[0 53 20]
[0 6 31 27 16]	[5] 2[1 9] 48 3[2]		[1 0 0]
[0 6 35 27 16]	[5] 2[7 45] 15 [48]	[10]	[1 6 40]
[0 6 39 27 1]6	[5] 34 [24 43 4]		[1 13 20]
[0 6 43] 27 16	5 41 [8 10 20]		[1 20 0]
[0 6 4]7 27 16	5 47 55 37 [36]		[1 26 40]
[0 6 5]1 27 16	5 54 47 4 52		1 [33 20]
[0] 6 55 27 16	6 1 42 32 8	15	1 [40 0]
[0] 6 59 27 16	6 8 41 59 24		1 [46 40]
[0] 7 3 27 16	6 15 45 26 40		1 5[3 20]
[0] 7 7 27 16	6 22 52 53 56		2 0 [0]
[0] 7 11 27 16	6 30 4 21 12		2 6 [40]
[0] 7 15 27 16	6 37 19 48 28	20	2 13 [20]
[0] 7 19 27 16	6 44 39 15 44		2 20 [0]
[0] 7 23 27 16	6 52 2 43 0		2 2[6 40]
[0] 7 27 27 16	6 59 30 10 16		2 [33 20]
[0 7 3]1 27 16	7 7 1 37 32		2 [40 0]
[0 7 3]5 27 16	7 14 37 4 48	25	2 [46 40]
[0 7 3]9 27 16	7 22 16 32 4		2 [53 20]
[0 7] 43 27 16	7 29 59 59 20		3 [0 0]
			3 [6 40]
			3 [13 20]
		30	3 [20 0]
		Gemini? [	
		Aquarius? [	
		…[	
		Capricorn? [	
		…[	
		…[	

— — — — — — —

4167. Ptolemy, *Handy Tables*

fr. 1, left

i	ii	iii	iv	v	vi	vii	viii	ix	x
				Ταύρου			[Διδύμων]		
[ὡριαῖοι] [χρ(όνου)]		[ἀνα]φ̣ο̣ρ̣α̣ί̣		ὡριαῖο[ι] [χρ(όνου)]		[ἀναφοραί]		[ὡριαῖ]οι χρ(όνου)	
		]	λ	ιζ	γ	λ̣ϛ	λ̣ϛ	ιη	μβ
		]	[λ]ε	ιζ	ζ̣	[λζ]	κε̣	ιη	μδ
	]	[ι]ζ	μ̣α	ιζ	[ι]	[λη]	ι̣δ	ιη̣	μϛ
	]	ι̣η̣	ιζ	ιζ	ιδ	λ̣θ̣	γ	ιη̣	μθ
	]	ιθ̣	νγ	ιζ	ιη̣	[λθ]	ν̣β	ι̣η	να
	]	ιθ̣	κθ	ιζ	κα	μ̣	μ̣β̣	ι̣η	νγ
]	κ]η̣	κ̣	ε̣	ιζ	κε̣	μ̣α̣	λ̣β̣	ιη	νϛ
]	λ̣β	κ	μα	ιζ	κθ	μ̣β	κ̣β̣	ιη̣	νθ̣
]	λϛ	κα	ιζ	ιζ	λβ	μ̣γ	ιβ	ιθ	ō
]	μ	κα	νγ	ιζ	λϛ̣	μδ̣	β	ιθ	β
]	μδ	κβ	λβ	ιζ	λθ	με	νζ	ι̣θ	δ̣
]	μη	κγ	ια	ιζ	μγ	με	νγ̣	ιθ	[
]	νβ	κγ	ν	ιζ	μϛ̣	μϛ	μ̣θ̣	ιθ	[
]	νϛ	κδ	κθ	ιζ	ν	μζ	με̣	ιθ̣	[
]	ō	κε	η	ιζ	νγ	μη	μα	[	
ϛ̣	δ	κε	νζ	ιζ	νζ	μθ̣	λζ	[	
ϛ	η	κϛ	λζ	ιη	ō	ν	λ[γ]	[	
ϛ	ιβ	κζ	ζ	ιη	γ	να	κ[θ]	[	
ϛ	ιϛ	κζ	μζ	ιη	ζ	νβ	κε	[	
]	κ̣	κη	κ̣ζ	ιη̣	ι	νγ	[κ]α̣	[	
				]	[ι]γ̣	νδ	[		
				[ι]η̣	[ϛ]	νε̣	[		
			]	ιη	[				
		]	[κ]γ̣	ι̣η̣	[				
			].	ι[η]	[				
		]	[ν]α	ιη	[				
	]	[λ]γ	λε	ιη	[				
	]	λδ	ιθ	ιη	[				
	]	λε	γ	ιη	[				
	]	λε	μζ	ιη	[λθ]	ξγ	μ[ε]	[	

⇒

2 ꝉ (*tris*)

4167. Ptolemy, *Handy Tables*

fr. 1, left

	i	ii	iii	iv	v	vi	vii	viii	ix	x ⇒
			Taurus				Gemini			
2 (cont.)	[Hourly] [degrees]		Ascensions		Hourly [degrees]		[Ascensions]		Hourly degrees	
			]	30	17	3	36	36	18	42
			]	[3]5	17	7	[37]	25	18	44
		]	[1]7	41	17	[10]	[38]	14	18	46
		]	18	17	17	14	39	3	18	49
		]	19	53	17	18	[39]	52	18	51
		]	19	29	17	21	40	42	18	53
	]	2]8	20	5	17	25	41	32	18	56
	]	32	20	41	17	29	42	22	18	59
	]	36	21	17	17	32	43	12	19	0
	]	40	21	53	17	36	44	2	19	2
	]	44	22	32	17	39	45	57	19	4
	]	48	23	11	17	43	45	53	19	[
	]	52	23	50	17	46	46	49	19	[
	]	56	24	29	17	50	47	45	19	[
	]	0	25	8	17	53	48	41	[	
	6	4	25	57	17	57	49	37	[	
	6	8	26	37	18	0	50	3[3]	[	
	6	12	27	7	18	3	51	2[9]	[	
	6	16	27	47	18	7	52	25	[	
	]	20	28	27	18	10	53	[2]1	[	
					]	[1]3	54	[		
				]	[1]8	[6]	55	[		
				]	18	[				
			]	[2]3	18	[				
				]	1[8]	[				
			]x	[5]1	18	[				
		]	[3]3	35	18	[				
		]	34	19	18	[				
		]	35	3	18	[				
		]	35	47	18	[39]	63	4[5]	[	

fr. 1, right

⇐ i	ii	iii	iv	v	vi	vii	viii	ix	x	xi
	[Καρκίνου]				[Λέοντος]					
ἀναφο̣[ραί]		[ὡριαῖοι] [χρόνοι]		[ἀναφοραί]		[ὡριαῖοι] [χρόνοι]		[ἀναφοραί]		
ξ̣δ	ν[γ]	[ιθ]	[κ]β	ρα	[λη]	[ιη]	λ̣ζ̣	ρ̣μ̣[α]	[	
ξ̣ς	α̣	[ιθ]	κα	ρβ	[ν]ζ̣	[ιη]	λδ	ρμβ	[	
ξζ	θ̣	[ιθ]	κα	ρδ	ι̣ς	ιη	λα	ρμδ	ι̣[̣]	[
ξη	[ι]ζ̣	[ιθ]	κ	ρ̣ε	λε	ιη	κη	ρμε	λ̣[̣]	[
ξθ	κς	[ιθ]	κ	ρς	νδ	ιη	κε	ρμς	[	
ο	λε	[ιθ]	ιθ	ρη	ιγ	ιη	κβ	ρμη	[	
οα	μδ	[ιθ]	ιθ	ρθ	λβ	ιη	ιθ	ρμθ	λ̣[̣]	[
οβ	νγ	ιθ	ιη	ρι	να̣	ιη	ις	ρν	ν̣[̣]	[
οδ	β	[ι]θ̣	ιη	ριβ	ια	ιη	ιγ	ρνβ	[	
οε	ια	ι[θ]	ιζ	ριγ	λα	ιη	ι	ρνγ	[	
ος	κδ	ι̣[θ]	ις	ριδ	να	ιη	ζ	ρνδ̣	[	
]	[λ]ζ	[ιθ]	ιδ	ριε	ια	ιη	γ	ρν[ς]	ι̣β	[
		]	ιγ	ριζ	λα	ιη	[illegible]	ρνζ	λβ	ις
		ι]θ̣	ια	ριη	να	ιζ	νζ	ρνη	νβ	ις
		]	ι	ρκ	ια	ι̣ζ	νγ	ρξ	ιβ	ις
			]	ρκα	λα	ιζ	ν	ρξα	λβ̣	ι[ε
			]	ρκβ̣	[να]	ι̣ζ̣	μς	ρξβ	να	ι[ε
						]	μ̣γ	ρ̣ξ̣δ	ι	ι̣[ε
							]	[ρξ]ε	κ̣[̣	

fr. 1, right

⇐

i	ii	iii	iv	v	vi	vii	viii	ix	x	xi
[Cancer]				[Leo]						
Ascensions		[Hourly] [degrees]		[Ascensions]		[Hourly] [degrees]		[Ascensions]		
64	5[3]	[19]	[2]2	101	[38]	[18]	37	14[1]	[	
66	1	[19]	21	102	[5]7	[18]	34	142	[	
67	9	[19]	21	104	16	18	31	144	1[x]	[
68	[1]7	[19]	20	105	35	18	28	145	3[x]	[
69	26	[19]	20	106	54	18	25	146	[	
70	35	[19]	19	108	13	18	22	148	[	
71	44	[19]	19	109	32	18	19	149	3[x]	[
72	53	19	18	110	51	18	16	150	5[x]	[
74	2	[1]9	18	112	11	18	13	152	[	
75	11	1[9]	17	113	31	18	10	153	[	
76	24	1[9]	16	114	51	18	7	154	[	
]	[3]7	[19]	14	115	11	18	3	15[6]	12	[
		]	13	117	31	18	0	157	32	16
		1]9	11	118	51	17	57	158	52	16
		]	10	120	11	17	53	160	12	16
			]	121	31	17	50	161	32	1[5
			]	122	[51]	17	46	162	51	1[5
						]	43	164	10	1[5
							]	[16]5	2[x	

fr. 2a

i	ii	iii	iv	v
]	τ̣λ̣α̣	λ̣β̣	[	
]	τλβ	ιζ	ι̣γ	[
]	τλγ	β	ιγ	ε
]	τλγ	μζ	ιγ	ζ
].	τλδ	λ	ιγ	ι
]	τλε	ιβ	ιγ	ιβ
].	τ̣λ̣ε	ν̣δ	ιγ	ιε
]	τλς	λ̣ς̣	ιγ	ιη
]	τ̣λ̣ζ̣	ι̣η	ιγ	κ

fr. 2b

i	ii	iii
ι]η̣	ιη	[
μα	ιη	κη
δ	ιη	κε
κ̣ζ̣	ιη̣	κ̣[

fr. 2a

i	ii	iii	iv	v
]	331	32	[	
]	332	17	13	[
]	333	2	13	5
]	333	47	13	7
]x	334	30	13	10
]	335	12	13	12
]x	335	54	13	15
]	336	36	13	18
]	337	18	13	2

fr. 2b

i	ii	iii
1]8	18	[
41	18	28
4	18	25
27	18	2[x

fr. 3, left

i	ii
]	ε̣̄
]	κοινά
]	ξ´
]	γ
]	ϛ̣
]	θ̣
]	[ι]β̣
]	[ιε]
]	[ιη]
]	[κα]
]	[κδ]
]	κζ
]	λ
]	λγ
]	λε
]	λη
]	μ
]	μβ
]	μ̣δ
ι̣[ε]	μϛ
ιϛ	μη
ιζ	ν̣
]	νβ
]	νδ
]	νε
]	[ν]ϛ
]	[ν]ζ

fr. 3, right

iii	iv	v	vi	vii	viii	ix	x	xi
	]	ō	ν[δ]	[				
[κ]α̣	[τλ]θ̣	ō	νβ	ō	[			
[κ]δ	τλϛ	ō	ν	ō	[			
[κ]ζ	τλγ	ō	μθ	ō	[			
λ̣	τλ	ō	μδ	ọ̄	[			
λγ	τκζ	ō	μα	ọ̄	[			
λϛ	τκδ	ō	λη	ọ̄	[			
λθ	τκα	ō	λδ	[				
μβ	τιη	ō	λα	[				
με	τιε	ō	κζ	ō	[			
μη	τιβ	ō	κγ	ō	ι̣[δ]	[		
να	τθ	ō	ιθ	ō	ιδ	α	[	
νδ	τϛ	ō	ιε	ō	ιδ	α	[	
νζ	τγ	ō	ια	ō	ιδ	α	[	
ξ	τ	ō	ζ	ō	ιδ	α	[	
ξγ	cϥζ	ō	β	ō	ιε	α	[	
		προcτίθει						
ξϛ	cϥδ	ō	β	ō	ιε̣	β	[	
ξθ	cϥα	ō	η	ọ̄	ιε	β	γ	ō
οβ	cπη	ō	ιδ	ō	ιϛ	β	ε	[
οε̣	c̣π̣ε̣	ọ̄	κ	ō	ιϛ	β	ζ	ọ̄
ο̣η̣	c̣πβ	ō	κ̣ε	ō	ιζ	β	θ	ō
πα	coθ	ō	λ	ō	ιζ	β	ια	ō
πδ	c̣ọϛ	ō	λε	ō	ιη	β	ιγ	ō
πζ	coγ	ō	μ	ō	ιη	β	ιε	ō
ϥ	co	ō	μδ	ō	ιθ	β	ιη	ō

left ii 3 *scil.* ἑξηκοcτά

fr. 3, left

i	ii
]	5
]	Common
]	60ths
]	3
]	6
]	9
]	[1]2
]	[15]
]	[18]
]	[21]
]	[24]
]	27
]	30
]	33
]	35
]	38
]	40
]	42
]	44
1[5]	46
16	48
17	50
]	52
]	54
]	55
]	[5]6
]	[5]7

fr. 3, right

iii	iv	v	vi	vii	viii	ix	x	xi
	]	0	5[4]	[				
[2]1	[33]9	0	52	0	[			
[2]4	336	0	50	0	[			
[2]7	333	0	49	0	[			
30	330	0	44	0	[			
33	327	0	41	0	[			
36	324	0	38	0	[			
39	321	0	34	[				
42	318	0	31	[				
45	315	0	27	0	[			
48	312	0	23	0	1[4]	[		
51	309	0	19	0	14	1	[	
54	306	0	15	0	14	1	[	
57	303	0	11	0	14	1	[	
60	300	0	7	0	14	1	[	
63	297	0	2	0	15	1	[	
		additive						
66	294	0	2	0	15	2	[	
69	291	0	8	0	15	2	3	0
72	288	0	14	0	16	2	5	[
75	285	0	20	0	16	2	7	0
78	282	0	25	0	17	2	9	0
81	279	0	30	0	17	2	11	0
84	276	0	35	0	18	2	13	0
87	273	0	40	0	18	2	15	0
90	270	0	44	0	19	2	18	0

fr. 4

i	ii	iii	iv	v	vi	vii	viii
					ὡρῶν ιε̅		ὡ̣[ρῶν ιε ∟
					κλίμα ε̅		[κλίμα ς̅
	]	ι̣β̣	[ζ]	κε	ζ	λη	[
]	ζ	ιβ	ζ	κε	ζ	μ	ζ̣
ζ̣	ζ	ιθ̣	ζ	λη	η	β	η̣
[ι]η̣	ζ	μ̣ε̣	η	ιη	θ	ς	θ̣
μ̣α	η	κβ	θ	ιθ	ι	μ̣[γ]	[
κ̣ε	θ	κη	ι	[			
[ι]β	ι	κγ	ιβ	[			
[κ]ζ	ι	μη	ι̣	[			
	]	ιβ	ι[̣]	[. .]	[. .]	κε	[
	]	θ	θ̣	[. .]	[. .]	η	ι[β
	]	κς	η	[. .]	[. .]	ζ	θ̣
]	ζ̣	κγ	ζ̣	[. .]	[. .]	δ	η

fr. 5

	ι
	θ
	ζ
4	ς

fr. 6a

	i	ii
	ιδ	θ
	ιβ	λγ
	ιβ	ν
4	ιβ	ιη

fr. 6b

	i	ii
	ι̣[ ]	κ̣α̣
	ζ	ις
3	κα	ιγ

fr. 4

i	ii	iii	iv	v	vi	vii	viii
					5 Hours Clima 5		5 1/2 Hours [Clima 6
	]	12	[7]	25	7	38	[
]	7	12	7	25	7	40	7
7	7	19	7	38	8	2	8
[1]8	7	45	8	18	9	6	9
41	8	22	9	19	10	4[3]	[
25	9	28	10	[			
[1]2	10	23	12	[			
[2]7	10	48	1x	[			
	]	12	1[x]	[xx]	[xx]	25	[
	]	9	9	[xx]	[xx]	8	1[2
	]	26	8	[xx]	[xx]	7	9
]	7	23	7	[xx]	[xx]	4	8

fr. 5

10
9
7
6

fr. 6a

ii

i	
14	9
12	33
12	50
12	18

fr. 6b

i	ii
1[x]	21
7	16
21	13

fr. 7

	i	ii	iii
	. . .	ἀ̣[ν]α̣[τολῆc	
2	Κ̣ρ̣ιοῦ	ιθ	λϛ

fr. 8

	i		ii	
	διαφο]ρ̣(ὰ)		ἀριθμοὶ̣	
	προcθέc]εωc		κοινοί	
3			ρνα	

fr. 9

i	ii
[ι]θ	cλγ
κδ	cλγ
κθ	cλ[γ]
λδ	cλγ
[λ]η̣	cλγ

fr. 7

	i	ii	iii
	. . .	rising	
2	Aries	19	36

fr. 8

	i		ii	
	Difference in increment		Common numbers	
3			151	

fr. 9

	i	ii
	[1]9	233
	24	233
	29	23[3]
	34	233
5	[3]8	233

4168. Ptolemy, *Handy Tables*

fr. 1, front

i	ii	iii	iv	v	vi
]	☾ πλα–		☉ καὶ ☾	ἀριθ(μοί)	☉
]	τοϲ		ἀριθ(μοί)		λ[όξωϲιϲ
]	δ	νθ	ϙγ	ϲξζ	[
μ]γ	δ	νη	ϙϛ	ϲξδ	[
λ]ϛ	δ	νϛ	ϙθ	ϲξα	γ
]	δ	νδ	ρβ	ϲνη	δ̣
]	δ	ν	ρε	ϲνε	ϛ̣
]	δ	με	ρη	ϲνβ	ζ̣
ι]α̣	δ	μ	ρια	ϲμθ	[
]	δ	λδ	ριδ	ϲμϛ	[
]	δ	κζ	ριζ	ϲμγ	[
]	δ	κ	ρκ	ϲμ	[
]	δ	ια	ρκγ	ϲλζ	[
ϛ	δ	γ	ρκϛ	ϲλδ	[
ι]θ̣	γ	νγ	ρκθ	ϲλα	[
κ]θ̣	γ	μγ	ρλβ	ϲκη	[
λ]ζ̣	γ	λβ	ρλε	ϲκε	[
]	γ	κ	ρλη	ϲ[κβ	
μ]ϛ	γ	η̣	ρμα	ϲ[ιθ	
μ]ε̣	β	νε	ρμδ	ϲ̣[ιϛ	
μ]γ	β	μγ	ρμζ	ϲ[ιγ	
μ̣	β	λ	ρν	ϲ[ι	
λ]ε̣	β	ιϛ	ρνγ	ϲ[ζ	
.	β	β	ρνϛ	ϲ[δ	
]	α	μη	ρνθ	ϲ[α	
]	α	λγ	ρξβ	ρ[ϙη	
α̣	α	ιζ	ρξε	ρ̣[ϙε	
]	α	γ	ρξη̣	[ρϙβ	
λ]η	ō	μη	ροα	ρπ[θ	
].	ō	λβ	ροδ	ρπ[ϛ	
]	ō	ιϛ	ροζ	[ρπγ	
]	ō	ō	ρπ	[ρπ	
			κα.[		
			νο[		

1 ☉ = ἡλίου ; ☾ = ϲελήνηϲ ; αριθ′ (so line 2)

4168. Ptolemy, *Handy Tables*

fr. 1, front

i	ii	iii	iv	v	vi
]	Moon's latitude		Sun, Moon numbers	Numbers	Sun's declination
]					
]	4	59	93	267	[
4]3	4	58	96	264	[
3]6	4	56	99	261	3
]	4	54	102	258	4
]	4	50	105	255	6
]	4	45	108	252	7
1]1	4	40	111	249	[
]	4	34	114	246	[
]	4	27	117	243	[
]	4	20	120	240	[
]	4	11	123	237	[
6	4	3	126	234	[
1]9	3	53	129	231	[
2]9	3	43	132	228	[
3]7	3	32	135	225	[
]	3	20	138	2[22	
4]6	3	8	141	2[19	
4]5	2	55	144	2[16	
4]3	2	43	147	2[13	
40	2	30	150	2[10	
3]5	2	16	153	20[7	
x	2	2	156	20[4	
]	1	48	159	20[1	
]	1	33	162	1[98	
1	1	17	165	1[95	
]	1	3	168	[192	
3]8	0	48	171	18[9	
]x	0	32	174	18[6	
]	0	16	177	[183	
]	0	0	180	[180	
			...[		
			...[		

fr. 1, back

] ὧρ(αι) [

] ̣[παραλ]λ̣ά̣[cc]ει κα̣[τά

	i	ii	iii	iv
Καρκίν]ου		Λέοντο̣c		
	α	β	γ	δ
	ι̣c̣η̣μερ̣(ιναὶ) ̣ ὧρ(αι)	μῆκ(οc)	πλάτ(οc)	ἀποc(τήματοc) π(ερι)φέ(ρεια)
	ϭ ̣	νδ	ε νό(τιον)	ō
	ς	να	γ	ρκ
	ε	μη	β βό(ρειον)	ϙε
	δ	μγ	ε	ϙ
	γ	λδ	ς	πε
	β	κδ	ζ	πε
	α	ιβ	ς	πς
		α	δ νό(τιον)	πζ
	α	ιγ	ō	πη
	β	κε	δ	πε
	γ	λδ	θ	π
	δ	μα	ιδ	ο
	ε	μα	ιη	ν
	ς	με	κα	ε
	ϭ	μδ	κε	ō
]		Ὑδρηχό[ου		
]	α̣	β	γ	δ
	]	μῆκ(οc)	πλάτ(οc)	ἀπο[c(τήματοc)
	]			π(ερι)φέ(ρεια)
	]	μα	κε νό(τιον)	ō
	]	μβ	κη	ρξβ
	]	λζ	λα	ρμδ
	]	λη	λγ	ρκγ
	]	ιη	λδ	ρθ
	]	ις	λβ	ϙς
	]	ζ	λ	πε
	]	ιθ	κζ	οζ
	]	λα	κγ	οβ
	]	λθ	ιη	ξα
	]	μς	ιβ̣	ν̣β̣
	]	να̣	ζ	λη
	]	ν̣α	ε	ō

1 ϕ (so line 6) 5 ι̣c̣η̣μερ̣ (the first two letters written to the left of the ruling on the left side of column i); μηκʹ (so line 24); πλατʹ (so line 24); αποc (so line 24?) 6 πʹφεʹ (so line 25)

fr. 1, back

	i	ii	iii	iv
		]...[		
	]...[...]...[			
Cancer		Leo		
	1	2	3	4
	Equinoctial hours	Longitude	Latitude	Arc of distance away
	☌	54	5 south	0
	6	51	3	120
	5	48	2 north	95
	4	43	5	90
	3	34	6	85
	2	24	7	85
	1	12	6	86
		1	4 south	87
	1	13	0	88
	2	25	4	85
	3	34	9	80
	4	41	14	70
	5	41	18	50
	6	45	21	5
	☌	44	25	0
]		Aquarius		
]	1	2	3	4
	]	Longitude	Latitude	Arc of
	]			distance away
	]	41	25 south	0
	]	42	28	162
	]	37	31	144
	]	38	33	123
	]	18	34	109
	]	16	32	96
	]	7	30	85
	]	19	27	77
	]	31	23	72
	]	39	18	61
	]	46	12	52
	]	51	7	38
	]	51	5	0

fr. 2, front

i	ii	iii	iv	v	vi
			Κρόνου	[	
α	β	γ	δ	[	
ἀριθμοὶ κοινοί		ἐκκ(έντρου)	πρ(οcθ)αφαί(ρεcιc)	ἐξηκ(οcτὰ) ἀφαιρ(έcεωc)	δι̣[αφόρα ἀφαιρ(έcεωc)
ξα	cϥθ	ε	λγ	κθ	ō [
ξβ	cϥη	ε	λζ	κη	ō [
ξγ	cϥζ	ε	μ	κζ	[
ξδ	cϥϛ	ε̣	μδ	κϛ	[
ξε	cϥε	ε	μζ	κε	ō [
ξϛ	cϥ[δ]	ε̣	ν	κδ	ō [
ξζ	cϥγ̣	ε	νγ	κγ	ō [
ξη	cϥ̣[β]	ε	νϛ	κβ	ō [
ξθ	cϥ[α]	ε	νθ	κα	ọ̄ [
ο	c̣ϥ̣	ϛ	β	κ	ō [
οα	cπ̣θ̣	ϛ	ε	ιθ	ọ̄ [
οβ	cπ̣η	ϛ	ζ	ιη	[
ογ	cπ̣[ζ]	ϛ	ι	ιϛ	[
οδ	cπϛ̣	ϛ	ιβ	ιε	[
οε	cπε	ϛ	ιδ	ιδ	[
οϛ	cπδ̣	ϛ	ιϛ	ιγ	[
οζ	cπγ	ϛ	ιη	ιβ	[
οη	cπβ	ϛ	ιθ	ια	[
ο̣θ	cπα	ϛ	κα	θ	[
π	cπ	ϛ	κβ	η	[
πα	cοθ	ϛ	κγ	ζ	[
πβ̣	cοη	ϛ	κε	ϛ	[
πγ	cοζ	ϛ	κϛ	ε	[
πδ	cοϛ	ϛ	κζ	δ	[
πε	cοε	ϛ	κη	γ	[
πϛ	cοδ	ϛ	κη	β	[
πζ	cο[γ]	ϛ	κθ	α	[
				προc(τίθει)	[
	]	ϛ	λ	α	[
	]	ϛ	λ	β	[
	]	ϛ	λα	γ	[

3 εκκ′ ; εξηκ′ 4 /ρ̸αφαι′ ; αφαιρ′ (*bis*) 31a προc

fr. 2, front

i	ii	iii	iv	v	vi	
			Saturn	[		
1	2	3	4	[		
Common numbers		Correction of eccentre	subtraction	Sixtieths of subtraction	Difference in	
61	299	5	33	29	0	[
62	298	5	37	28	0	[
63	297	5	40	27	[	
64	296	5	44	26	[	
65	295	5	47	25	0	[
66	29[4]	5	50	24	0	[
67	293	5	53	23	0	[
68	29[2]	5	56	22	0	[
69	29[1]	5	59	21	0	[
70	290	6	2	20	0	[
71	289	6	5	19	0	[
72	288	6	7	18	[	
73	28[7]	6	10	16	[	
74	286	6	12	15	[	
75	285	6	14	14	[	
76	284	6	16	13	[	
77	283	6	18	12	[	
78	282	6	19	11	[	
79	281	6	21	9	[	
80	280	6	22	8	[	
81	279	6	23	7	[	
82	278	6	25	6	[	
83	277	6	26	5	[	
84	276	6	27	4	[	
85	275	6	28	3	[	
86	274	6	28	2	[	
87	27[3]	6	29	1	[	
				Additive	[	
	]	6	30	1	[	
	]	6	30	2	[	
	]	6	31	3	[	

fr. 2, back

i	ii	iii	iv	v
ε	ς		ζ	
διαφ]όρ(α)	ἐπικύκ(λου)		διαφόρ(α)	
ἀφα]ιρ(έσεως)	πρ(οσθ)αφαίρ(εσις)		πρ(οσθ)αφαιρ(έσεως)	
]	ς	ιβ	ō	κγ
]	ς	ιβ	ō	κγ
]	ς	ιβ	ō	κγ
]	ς	ιγ	ō	κγ
]	ς	ιγ	ō	κγ
]	ς	ιγ	ō	κγ
]	ς	ιγ	ō	κγ
]	ς	ιγ	ō	κδ
]	ς	ιβ	ō	κδ
]α	ς	ιβ	ō	κδ
]α	ς	ιβ	ō	κδ
]	ς	ια	ō	κδ
]	ς	ια	ō	κδ
]	ς	ι	ō	κδ
]	ς	θ	ō	κδ
]	ς	η	ō	κε
]	ς	ζ	ō	κε
]	ς	ε	ō	κε
]	ς	δ	ō	κε
]	ς	β	ō	κε
]̣	ς	α	ō	κδ
]̣	ε	νθ	ō	κδ
]̣	ε	νζ	ō	κδ
]	ε	νε	ō	κδ
]	ε	νγ	ō	κδ
]̣	ε	να	ō	κδ
]	ε	μη	ō	κ̣δ
]	ε	μς	ō	[ ]
]̣	ε	μγ	ō	[ ]
]̣	ε	μ	ō	[ ]

2 διαφ]ορ′ ; επικυκ′ ; διαφορ′   3 αφα]ιρ′ ; /πραφαιρ′ (*bis*)

fr. 2, back

i	ii	iii	iv	v
5	6		7	
Difference in subtraction	Correction of epicycle		Difference in correction	
]	6	12	0	23
]	6	12	0	23
]	6	12	0	23
]	6	13	0	23
]	6	13	0	23
]	6	13	0	23
]	6	13	0	23
]	6	13	0	24
]	6	12	0	24
]1	6	12	0	24
]1	6	12	0	24
]	6	11	0	24
]	6	11	0	24
]	6	10	0	24
]	6	9	0	24
]	6	8	0	25
]	6	7	0	25
]	6	5	0	25
]	6	4	0	25
]	6	2	0	25
]x	6	1	0	24
]x	5	59	0	24
]x	5	57	0	24
]	5	55	0	24
]	5	53	0	24
]x	5	51	0	24
]	5	48	0	24
]	5	46	0	[ ]
]x	5	43	0	[ ]
]x	5	40	0	[ ]

fr. 3, front

i	ii	iii
νζ̣	[	
νς	ō	κβ
νς	ō	κγ
νε	ō	κδ
ν̣[ε]	ō	κς

fr. 3, back

i	ii	iii
]	ι̣	να
η	ι̣α̣	ς
ι	[ι]α̣	κα̣
ιβ	[ι]α̣	λ[.

fr. 4, front

]	κδ
].	κε
]	κς
]	κζ
]	κη

fr. 4, back

i	ii
β	ιβ
β	ιδ
β	ις̣
β	ι̣[.
β	[
β	[
β	[

fr. 5

]α̣
]α̣
κα
μ̣α
α

fr. 3, front

i	ii	iii
57	[	
56	0	22
56	0	23
55	0	24
5[5]	0	26

fr. 3, back

i	ii	iii
]	10	51
8	11	6
10	[1]1	21
12	[1]1	3[x

fr. 4, front

]	24
]x	25
]	26
]	27
]	28

fr. 4, back

i	ii
2	12
2	14
2	16
2	1[x
2	[
2	[
2	[

fr. 5

]1
]1
21
41
1

fr. 6, front

i	ii	iii	iv
]	θ	ι̣	[
]	ι	ια	[
ιβ	ια	ιβ	ια
ιγ	ιβ	ιγ	ιβ
]	ιγ	ιδ	ιγ̣

fr. 6, back

i	ii	iii	iv
	[		
	Φρυ̣γ[ίαϲ		
	Ἀπάμ̣[εια		
	ϲοζ	ρ̣κ̣[	
]ν	Γαλλίαϲ	Ϲυρίαϲ	Δ̣[
] ̣ν			
] ̣ν	Κιλικία	Δα̣μ̣α̣[ϲ]κ̣ό̣ϲ	

fr. 7

] ̣ειτα ̣ ̣[
] ̣τωνεν̣η̣μ̣[

fr. 6, front

i	ii	iii	iv
]	9	10	[
]	10	11	[
12	11	12	11
13	12	13	12
]	13	14	13

fr. 6, back

i	ii	iii	iv
...	[		
	in Phrygia		
	Apamea		
	277$^{?}$	12[x$^{?}$	
...	in Gallia	in Syria	D[
...			
...	Cilicia	Damascus	

fr. 7 (too fragmentary for translation)

4169. PTOLEMY, *HANDY TABLES*

Front

i	ii	iii	iv	v	
ν]α̣	τθ	β	[̣]	ιθ	α
νβ]	τη	β	[̣]	ιζ	α
ν]γ̣	τζ	β	ιβ	ις	α
ν]δ̣	τς	β	[ι]δ	ιδ̣	α
ν]ε	τε	β	ις	ιγ	α
νς]	τδ	β	ιθ̣	ιβ	α
ν]ζ̣	τγ	β	κ[̣	ια	α
ν]η̣	τβ	β	κγ̣	θ	α
νθ]	τα	β	κε̣	η	α
ξ]	τ	β	κς̣	ζ	α

Back

i	ii	iii	iv	v
ι̣δ	ιθ	ις̣	α	κβ̣
ις	ιθ	κε	α	κγ
ιη	ι̣θ̣	λε	α	κγ̣
ιθ̣	ιθ̣	μ̣δ̣	α	κε
κα	ιθ	νγ	α	κς
κγ	κ	β	α	κζ
κδ	κ	ι	α	κη
κ̣ς	κ	ιη̣	α	κθ
κ̣η̣	κ	κς	α	λ
κ̣θ	κ	λγ	α	λα̣

4169. Ptolemy, *Handy Tables*

Front

i	ii	iii	iv	v	
5]1	309	2	[x]	19	1
52]	308	2	[x]	17	1
5]3	307	2	12	16	1
5]4	306	2	[1]4	14	1
5]5	305	2	16	13	1
56]	304	2	19	12	1
5]7	303	2	2[x	11	1
5]8	302	2	23	9	1
59]	301	2	25	8	1
60]	300	2	26	7	1

Back

i	ii	iii	iv	v
1]4	19	19	1	22
16	19	25	1	23
18	19	35	1	23
19	19	44]	1	25
21	19	53	1	26
23	20	2	1	27
24	20	10	1	28
26	20	18	1	29
28	20	26	1	30
29	20	33	1	31

4170. PTOLEMY, *HANDY TABLES*

Front

i	ii	iii	iv	v	vi
			]	αὐτοῦ (ἀςτέρος)	
			]	cιδ	μ
			]	ρ̣κ̣γ	ιη
			]	λα	ν̣ε
			]	τ̣	λ̣γ
				]	ια̣
		]	λ̣η̣	ρι̣ζ̣	μ̣η
		]	ιη	ρ̣κς	κ̣ς
		]	νη	cϙε	δ
		]	λθ	cγ	μβ
		]	[ι]θ̣	ριβ	κ
		]	ν̣θ̣	κ̣	νη
		]	μ̣	c̣π̣θ̣	λς
[c]ξ̣[θ]	[̣ ̣]	[̣ ̣]	κ̣	ρ̣ϙη	ι̣γ
ρ̣λε	[κ]β̣	[̣ ̣]	τ̣	[ρς]	[ν]α̣
α	α	τ̣ιγ	[̣ ̣]	[̣ ̣]	κ̣θ
cκς	λθ	τ̣[ζ]	[̣ ̣]	[̣ ̣]	ς̣
ϙβ	ιη̣	τ̣[α]	[		
τιζ	νς̣	c̣[ϙδ]	[		
ρ]π̣γ	λ̣ε̣	cπ̣η̣	[		
μ]θ̣	ιγ	cπβ	[		
c]ο̣δ	νβ	coε	[		
]	λ	cξθ	κ̣[		
]	θ	cξ̣γ	γ	[	
]	μ̣[η]	cν̣[ς]	[		
]	κς̣	c̣ν̣	[		
]	δ	c̣μ̣δ	[		
]	μ̣γ	cλζ	μ[δ]	[	
	]	[c]λ̣α	κ[ε]	[	
	]	[c]κ̣ε̣	ε	[	
	]	c̣ιη̣	[		
	]	ι̣β	[		

1 *

4170. Ptolemy, *Handy Tables*

Front

i	ii	iii	iv	v	vi
			]	The planet itself	
			]	214	40
			]	123	18
			]	31	55
			]	300	33
				]	11
		]	38	117	48
		]	18	126	26
		]	58	295	4
		]	39	203	42
		]	[1]9	112	20
		]	59	20	58
		]	40	289	36
[2]6[9]	[xx]	[xx]	20	198	13
135	[2]2	[xx]	0	[106]	[5]1
1	1	313	[xx]	[xx]	29
226	39	30[7]	[xx]	[xx]	6
92	18	30[1]	[		
317	56	2[94]	[		
1]83	35	288	[		
4]9	13	282	[		
2]74	52	275	[		
]	30	269	20[		
]	9	263	3	[	
]	4[8]	25[6]	[		
]	26	250	[		
]	4	244	[		
]	43	237	4[4]	[	
	]	[2]31	2[5]	[	
	]	[2]25	5	[	
	]	218	[		
	]	12	[		

Back

i	ii	iii
ρ̣[να]	[	
ρ[νβ]	[	
ρ[νγ]	[	
[ρνδ]	[	
ρ[νε]	[	
ρ̣[νϛ]	[	
ρ̣[ν]ζ̣	cγ̣	β̣
ρ]ν̣η	c̣β	β̣
ρν]θ̣	cα̣	α̣
ρξ]	c	α̣
ρξ]α̣	ρ̣ϙ̣θ	α̣
ρ]ξ̣β	ρϙη	.
ρ]ξγ	ρϙζ̣	.
ρ̣ξδ	ρϙϛ̣	[
ρξε	ρϙε	[
ρ̣ξϛ	ρϙ̣δ	[
ρ]ξ̣ζ	ρϙ̣γ	[
ρ]ξη	ρϙβ̣	[
ρ̣ξθ	ρϙα	[
ρο	ρϙ̣	[
ρ̣ο̣α	[	
ρ̣ο̣β	[	
ρ̣ο̣γ̣	[ρ]π̣[ζ]	[
[ροδ]	ρ̣[πϛ]	[

— — — — — — — — — —

Back

i	ii	iii
1[51]	[	
1[52]	[	
1[53]	[	
[154]	[	
1[55]	[	
1[56]	[	
1[5]7	203	2
1]58	202	2
15]9	201	1
160]	200	1
16]1	199	1
1]62	198	x
1]63	197	x
164	196	[
165	195	[
166	194	[
1]67	193	[
1]68	192	[
169	191	[
170	190	[
171	[	
172	[	
173	[1]8[7]	[
[174]	1[86]	[

4171. PTOLEMY, *HANDY TABLES*

Front

	i	ii
		Κ[ριοῦ
	μοῖ–	ἀνα–
	ραι	φορα̣[ί
4	(ἡλίου)	

— — — — — — — — — — —

4 ☉

Back

	Κα]ρ̣κίνου
	ὡρῶν
	χρόνος
4	ι̣ζ̣ [

— — — — — —

4171. PTOLEMY, *HANDY TABLES*

Front

	i	ii
		Aries?
	deg- rees of	Asc- ensions
4	the Sun	

— — — — — — — — — — —

Back

	Cancer	
	Time of hours	
4	17	[

— — — — — —

4172. Handy Tables

Fr. 1

i	ii	iii
]	η̣	[
]	ι	ε
]	ιβ	ζ
]	ιδ	θ
]	ιε	νδ
]	ις	νβ
]δ	ις	λα
]	ιε	δ
]	ιγ	β̣
]	ι	μ̣[̣]
] ̣α̣	η	ν[̣]

Fr. 2

i	ii	iii	iv
		ὥρ(αι) κ̅[	
̣	κ̣	Καρκί(νου)	ι̣[
ε	[̣]η̣	Λέον(τος)	ι̣[
ζ	ι̣η	Παρθέ(νου)	ι̣[
θ̣	ι̣ε̣	Ζυγοῦ	[
ια	ιδ	Cκορπί(ου)	[
ιβ	νε	Τοξότ(ου)	[
ιγ	μζ	Αἰγόκε(ρῳ)	[
ιγ	κ̣ε	Ὑδρο̣χ̣(όου)	[
ια	ν̣ζ	Ἰχθύω̣(ν)	[
θ	ν̣ζ	Κρειοῦ	[
ζ	μζ	Ταύρ[ου	
ς	̣ ̣	Διδύ̣μ(ων)	[

Names of zodiacal signs abbreviated by truncation. 11 *l.* Κριοῦ

4172. Handy Tables

Fr. 1

i	ii	iii
]	8	[
]	10	5
]	12	7
]	14	9
]	15	54
]	16	52
]4	16	31
]	15	4
]	13	2
]	10	4[x]
]x1	8	5[x]

Fr. 2

i	ii	iii	iv
		2[x] hours [	
x	2x	Cancer	1[x?
5	[x]8	Leo	1[x?
7	18	Virgo	1[x?
9	15	Libra	[
11	14	Scorpio	[
12	55	Sagittarius	[
13	47	Capricorn	[
13	25	Aquarius	[
11	57	Pisces	[
9	57	Aries	[
7	47	Taurus	
6	xx	Gemini	[

4173. Handy Tables

Front, left

i	ii	iii	iv	v	vi	vii ⇒
]	ọ̄	[				
]	α	[				
ϛ]	β	ō	λ̣γ̣	[		
θ̣	γ	ō	ν	ζ	κ̣[.]	[
ι]β̣	δ	α	ζ	θ	νη̣	[
ι]ε	ε	α	κγ	ιβ	κη	ϛ
ι]η	ϛ	α	μ	ιδ	νζ	μγ̣
κ]α̣	ζ	α	νζ	ιζ	κζ	κ
κδ	η	β	ιδ	ιθ	νϛ	νζ
κ̣ζ	θ	β	λ	κβ	κϛ	λε
λ	ι	β	μζ	κδ	νϛ	ιβ
λγ	ια	γ	δ	κζ	κε	μθ
λϛ	ιβ	γ	κα	κθ	νε	κϛ

Front, right

⇐ viii	ix	x	xi	xii	xiii	xiv	xv
[							]
	]	ι̣η̣	ι̣[η]	κ̣θ̣	μ̣β̣	ϙ̣[γ]	[
	]	λϛ	λϛ	νθ	κ̣ε̣	ρπ̣[ϛ]	[
	]	ν̣ε	νε	κθ	η̣	c[οθ]	[
	]	[ι]γ̣	ο̣γ	νη	[να]	ι̣[β]	[
	]	λ̣β	ϙβ	κη	λδ	ρϛ	ọ̄
ϙ̣δ̣	[ιθ]	ν̣	ρι	νη	ιζ	ρϙθ	[
ρι	γ	θ̣	ρκθ	κη	ō	cϙβ	κ̣[δ
ρκε	μϛ	κζ	ρμζ	νζ	μγ	κε	λ̣[ϛ
ρμα	κθ	μϛ	ρξϛ	κζ	κϛ	ριη	μ̣[η
ρνζ	ιγ	δ	ρπδ	νζ	θ	cιβ	ọ̄
ροβ	νϛ	κβ	cγ	κϛ	νβ	τε	ιβ
ρπη	λθ	μα	cκα	νϛ	λε	λη	κδ

4173. Handy Tables

Front, left ⇒

i	ii	iii	iv	v	vi	vii
]	0	[				
]	1	[				
6]	2	0	33	[		
9	3	0	50	7	2[8]	[
1]2	4	1	7	9	58	[
1]5	5	1	23	12	28	6
1]8	6	1	40	14	57	43
2]1	7	1	57	17	27	20
24	8	2	14	19	56	57
27	9	2	30	22	26	35
30	10	2	47	24	56	12
33	11	3	4	27	25	49
36	12	3	21	29	55	26

Front, right ⇐

viii	ix	x	xi	xii	xiii	xiv	xv
[							]
	]	18	1[8]	29	42	9[3]	[
	]	36	36	59	25	18[6]	[
	]	55	55	29	8	2[79]	[
	]	[1]3	73	58	[51]	1[2]	[
	]	32	92	28	34	16	0
94	[19]	50	110	58	17	199	[
110	3	9	129	28	0	292	2[4
125	46	27	147	57	43	25	3[6
141	29	46	166	27	26	118	4[8
157	13	4	184	57	9	212	0
172	56	22	203	26	52	305	12
188	39	41	221	56	35	38	24

Back, left ⇒

i	ii	iii	iv	v	vi	vii
[						]
]	ō	ō	β	ō	μη̣	[
]	ō	ō	β	ō	ν	[
]	ọ̄	ō	β	ō	ν[β]	[
κ]η̣	ō	ō	β	ō	ν[δ]	[
κ]θ	ō	ō	β	ō	ν̣ς̣	ι̣ε
λ̣	ō	ō	β	ō	ν̣η	ις

Back, right ⇐

viii	ix	x	xi	xii	xiii	xiv	xv
					]	ιδ	[
				]	λ̣η	ιδ	[
	]	[μ]α̣	ιγ	ς	ε	ιε	κ̣[δ
	]	μ̣	ιγ	λζ	λα	ις	α̣
]	δ	λ̣θ	ιδ	η	νη̣	ις	λ̣[η
β	ιθ	λη	ιδ	μ	κε	ιζ	ι̣[ε
β	κδ	λ̣η̣	ιε	ια	να	ιζ	ν̣[β

Back, left

i	ii	iii	iv	v	vi	vii	⇒
[						]	
]	0	0	2	0	48	[	
]	0	0	2	0	50	[	
]	0	0	2	0	5[2]	[	
2]8	0	0	2	0	5[4]	[	
2]9	0	0	2	0	56	15	
30	0	0	2	0	58	16	

Back, right

⇐ viii	ix	x	xi	xii	xiii	xiv	xv
					]	14	[
				]	38	14	[
	]	[4]1	13	6	5	15	2[4
	]	40	13	37	31	16	1
]	4	39	14	8	58	16	3[8
2	19	38	14	40	25	17	1[5
2	24	38	15	11	51	17	5[2

4173a. Handy Tables

Fr. 1, front

i	ii
[λη]	[τ]κ̣β̣
[λθ]	τκα
[μ]	τκ̣
[μ]α̣	τιθ
[μβ]	τ̣ι̣η̣
[μγ]	[τιζ]
[μ]δ̣	τ̣[ις]
[μ]ε	τ[ιε]
[μ]ς	τ[ιδ]
[μ]ζ	τ[ιγ]
[μη]	τ[ιβ]
[μ]θ̣	τ[ια]
[ν]	τ̣[ι]

Fr. 1, back

i	ii	iii
] ̣	α̣	[
]	α	νβ
]	α	ν
]η	α	μθ̣
]ζ̣	α	μζ̣
]	[α]	[
]	α̣	[
]	α̣	μβ̣
]	α̣	μ
]	α̣	λθ̣
]	α̣	λζ̣
]	[ ]	λ ̣

4173a. Handy Tables

Fr. 1, front

i	ii
[38]	[3]22
[39]	321
[40]	320
[4]1	319
[42]	318
[43]	[317]
[4]4	3[16]
[4]5	3[15]
[4]6	3[14]
[4]7	3[13]
[48]	3[12]
[4]9	3[11]
[50]	3[10]

Fr. 1, back

i	ii	iii
]x	1	[
]	1	52
]	1	50
]8	1	49
]7	1	47
]	[1]	[
]	1	[
]	1	42
]	1	40
]	1	39
]	1	37
]	[ ]	1x

Fr. 2, front

i	ii	iii
ρνε̣	[cε]	[
ρνς	[cδ]	[
ρ̣[νζ]	[cγ]	[
[ρνη]	[cβ]	[
[ρν]θ̣	cα̣	.
[ρξ]	c	[
[ρξ]α̣	ρϙθ	.
[ρξ]β̣	ρϙη	ọ̄
[ρξγ]	ρϙ̣ζ̣	[
[ρξδ]	[ρϙς]	[
[ρξε]	[ρϙε]	[
[ρξς]	[ρϙδ]	[
[ρξζ]	[ρϙγ]	[
[ρξη]	[ρϙβ]	[
[ρξθ]	[ρϙα]	[
[ρο]	[ρϙ]	[
[ροα]	ρπθ	ọ̄
[ροβ]	ρπη	[
[ρογ]	ρπζ	ọ̄
[ροδ]	ρπς	ō
[ροε]	ρπε	ō
[ρος]	ρπδ	ō
[ροζ]	[ρπ]γ	ō

Fr. 2, back

i	ii	iii	iv
[γ]	[ ]	[γ]	κ̣δ̣
[δ]	[ ]	[δ]	ε
[ε]	[ ]	[δ]	μς
[ε]	[ ]	[ε]	[
[ς]	[ ]	[ς]	[
[ζ]	[ ]	ς̣	[
[η]	η	ζ	[
[η]	νγ	η	[
[θ]	λη	η	[
[ι]	κγ	θ	[
[ια]	η̣	ι	[
[ια]	[ ]	[ ]	[
[ιβ]	[ ]	[ ]	[
[ιγ]	[ ]	[ ]	[
[ιδ]	[ ]	[ ]	[
[ιδ]	[ ]	[ ]	[
[ιε]	[ ]	[ ]	[
[ις]	[ ]	[ ]	[
[ιζ]	[ι]α̣	ι̣[ ]	[
[ιζ]	[ν]η̣	ι̣[ ]	[
ι̣η	μδ	ι[ ]	[
[ι]θ	λα	ιζ̣	[
κ	ιζ	ιη̣	[
κα	δ	ι .	[
κ̣α̣	ν̣	[ ]	[

Fr. 2, front

i	ii	iii
155	[205]	[
156	[204]	[
1[57]	[203]	[
[158]	[202]	[
[15]9	201	x
[160]	200	[
[16]1	199	x
[16]2	198	0
[163]	197	[
[164]	[196]	[
[165]	[195]	[
[166]	[194]	[
[167]	[193]	[
[168]	[192]	[
[169]	[191]	[
[170]	[190]	[
[171]	189	0
[172]	188	[
[173]	187	0
[174]	186	0
[175]	185	0
[176]	184	0
[177]	[18]3	0

Fr. 2, back

i	ii		iii	iv
[3]	[ ]		[3]	24
[4]	[ ]		[4]	5
[5]	[ ]		[4]	46
[5]	[ ]		[5]	[
[6]	[ ]		[6]	[
[7]	[ ]		6	[
[8]	8		7	[
[8]	53		8	[
[9]	38		8	[
[10]	23		9	[
[11]	8		10	[
[11]	[ ]		[ ]	[
[12]	[ ]		[ ]	[
[13]	[ ]		[ ]	[
[14]	[ ]		[ ]	[
[14]	[ ]		[ ]	[
[15]	[ ]		[ ]	[
[16]	[ ]		[ ]	[
[17]	[1]1		1[x]	[
[17]	[5]8		1[x]	[
18	44		1[x]	[
[1]9	31		17	[
20	17		18	[
21	4		1x	[
21	50		[ ]	[

Fr. 3, front

	i	ii	iii
	] κοιν[οί]		
	]α̣	ϲ̣ ̣[θ]	
	]β̣	ϲ[̣η]	
	]γ	ϲ[̣ζ]	
5	δ]	ϲ̣[̣ϛ]	
	ε]	ϲ̣[̣ε]	

Fr. 3, back

	i	ii	iii	iv
	] ̣	̣[		
	]β̣	[		
	]β	[		
	]β	̣[		
5	]β	[		
	]β	[		

Fr. 4, front

		i	ii
		ιε	[ ]
		ιϛ	ιθ
3		ιζ	ιη

Fr. 4, back

	i	ii	iii	iv
	ν̣[ζ]	[ ]		
	ν̣[η]	[ ]		
	ν[θ]	[ ]		
4	ξ	ọ̄		

Fr. 5

	κ
2	ō

Fr. 3, front

	i	ii	iii
	Common [numbers]		
	]1	2x[9]	
	]2	2[x8]	
	]3	2[x7]	
5	4]	2[x6]	
	5]	2[x5]	

Fr. 3, back

	i	ii	iii	iv
	]...	...[		
	]2	[		
	]2	[		
	]2	x[		
5	]2	[		
	]2	[		

Fr. 4, front

	i	ii
	15	[ ]
	16	19
3	17	18

Fr. 4, back

	i	ii	iii	iv
	5[7]	[ ]		
	5[8]	[ ]		
	5[9]	[ ]		
4	60	0		

Fr. 5

	20
2	0

4174. Table of Lunar Mean Motions

Recto?

i	ii	iii
	[Αἰγό(κερω)]	ι̣
	[Τ]α̣ύ̣[ρ(ου)]	ι[θ
	Ζυγο(ῦ)	ια
	Ὑ[δ]ρο(χόου)	κ
	`] ̣ι̣α̣´	
	Διδύ(μων)	κθ̣
	Σκορ(πίου)	η̣
	Κριο(ῦ)	ō
	Λέο̣ν̣(τος)	ι̣β̣
	Τοξό(του)	κ̣α̣
	Κρι̣ο̣(ῦ)	κζ̣
	[Παρ]θ(ένου)	ιθ
	[Αἰγό(κερω)]	κη

Names of zodiacal signs abbreviated by truncation. 4a inserted in space above line 5.

Verso?

	i	ii
]	κ̣β̣	β̣
]	κγ	ιε̣
]	κδ	κη
]	κε	μ̣β
]	κϛ	νε
]	κ̣[ζ]	[ο]η
]	κη	π̣α̣
]	κθ	ϙδ
]	λ	ρ̣η̣
Χ]ο̣ί̣α̣κ̣		
]	α̣	ρκα
]	β	ρ̣λδ
]	γ̣	ρμζ̣
]	δ̣	ρ̣ξ̣
]	ε̣	ρ̣ο̣[̣]

4174. Table of Lunar Mean Motions

Recto?

i	ii	iii
	[Capricorn]	10
	Taurus	1[9
	Libra	11
	Aquarius	20
	. . .	
	Gemini	29
	Scorpio	8
	Aries	0
	Leo	12
	Sagittarius	21
	Aries	27
	Virgo	19
	[Capricorn]	28

Verso?

	i	ii
]	22	2
]	23	15
]	24	28
]	25	42
]	26	55
]	2[7]	[7]8
]	28	81
]	29	94
]	30	108
Choeac		
]	1	121
]	2	134
]	3	147
]	4	160
]	5	17[x]

4174a. Table of Solar Mean Motions

i				ii				iii				⇒
ō	ō	ō]	ō	ō	ō	ọ̄	ọ̄	ọ̄	ō	ō	ō	
ō	νθ	η]	ι̣ς	ō	νθ	ε̣	μ̣[η]	ọ̄	ν̣θ	θ	λ	
α	νη	ις]	λ̣β	α	νη	ια	λ̣ς̣	α̣	ν̣η̣	ι̣θ	ō	
β	νζ	κδ]	μη	β	νζ	ιζ	κ̣δ̣	β	ν̣ζ̣	κη	λ	
γ	νς	λγ]	δ	γ	νς	κγ	ιβ	γ	[νς]	λ̣η	ō	
δ	νε	μα]	κ̣	δ	νε	κη̣	ō	δ	[νε]	μ̣ζ	λ	
			]	ε	νδ	λδ̣	μ̣η̣	ε̣	[νδ]	ν̣ζ	ō	
– – –	– – –	– – –	– – –	– – –	– – –	– – –	]	ς	[νδ]	ζ̣	λ	
							]	ζ̣	ν̣γ̣	ι̣ς̣	ō	
								η	νβ]	κ̣ς	α	
								θ	να	λ]ε̣	ọ̄	
								ι	ν	με]	λ̣	
								ια	μθ	νδ]	ọ̄	
								ιβ	μθ	δ]	λ̣	
								ιγ	μη	ιγ]	ọ̄	
								– – –	– – –	– – –	– – –	

10 *l.* η νβ κς λ

4174a. Table of Solar Mean Motions

i				ii				iii				⇒
0	0	0]	0	0	0	0	0	0	0	0	0	
0	59	8]	16	0	59	5	4[8]	0	59	9	30	
1	58	16]	32	1	58	11	36	1	58	19	0	
2	57	24]	48	2	57	17	24	2	57	28	30	
3	56	33]	4	3	56	23	12	3	[56]	38	0	
4	55	41]	20	4	55	28	0	4	[55]	47	30	
			]	5	54	34	48	5	[54]	57	0	
— — —	— — —	— — —	— — —	— — —	— — —	— — —	]	6	[54]	7	30	
							]	7	53	16	0	
								8	52]	26	1'	
								9	51	3]5	0	
								10	50	45]	30	
								11	49	54]	0	
								12	49	4]	30	
								13	48	13]	0	
								— —	— —	— —	— —	

⇐

iv	v					vi		
α̣	[ō]	β	κζ	ν	μ̣	ō	β	κ[
β	[ō]	{κ}δ	νε	μα	κ	ō	δ	[
γ	[ō]	ζ	κγ	λβ	ō	ọ̄	ζ̣	[
δ	[ō]	θ	να	κα	μ	[ō		
ε	[ō]	ιβ	ιθ	ιγ	κ̣	[ō		
ς	[ō]	ιδ	μζ	δ	[ō	ō		
ζ	[ō]	ιζ	ιδ	νδ	μ̣	[ō		
η	ọ̄	ιθ	μβ	με	κ	[ō		
θ	ō	κβ	ιζ	λς	ọ̄	ọ̄		
ι	ō	κδ	λη	κς	μ̣	ọ̄		
ια	ō	κς	ς	ιζ	κ	ọ̄		
[ιβ]	ọ̄	κθ	λδ	η̣	ō	ọ̄		
[ιγ]	ọ̄	[λβ	α	νη	μ	ō		

4 *l.* ō θ να κβ μ ō 11 *l.* ō κζ ς ιζ κ ō

⇐

iv	v					vi		
1	[0]	2	27	50	40	0	2	2[x
2	[0]	{2}4	55	41	20	0	4	[
3	[0]	7	23	32	0	0	7	[
4	[0]	9	51	21'	40	[0		
5	[0]	12	19	13	20	[0		
6	[0]	14	47	4	[0	0		
7	[0]	17	14	54	40	[0		
8	0	19	42	45	20	[0		
9	0	22	17	36	0	0		
10	0	24	38	26	40	0		
11	0	26'	6	17	20	0		
[12]	0	29	34	8	0	0		
[13]	0	[32	1	58	40	0		

Part III

Ephemerides and Almanacs

4175. Almanac-Ephemeris, 24 b.c.

Left:

i	ii	iii	iv	v	vi	vii	viii	ix	x	xi ⇒
				[Μεϲωρή]						
				[ ] Π̣α̣(ρθένου) [̣ ἑϲπ(έριοϲ)]						
				ιε Ζυ(γοῦ) ϛ						
] ̣				λ Ϲκο(ρπίου) ϲυνάψ(ει)						
ϲυνά]ψει				β Κα(ρκίνου) ϲυνά(ψει) κζ Λέ(οντοϲ) ϲυνά(ψει)						
				ιβ Πα(ρθένου) ϲυνά(ψει) κη Πα(ρθένου) ια θ̣ ε̣ϲ̣π̣(έριοϲ)						
[ ]						α	Κρι(οῦ)	κϛ		
ε						β	Ταύ(ρου)	η		
ι̣η̣						γ		ιθ		
[ ]						δ		λ		
[ ]						ε	Δι(δύμων)	ιβ		
[ ]						ϛ		κγ		
[ ]						ζ	Κα(ρκίνου)	ε		
[ ]				**Ϲεξ(τίλιοϲ)**		η		ιη		
[ ]						θ̣		[ ]		
[ ]						[ι]	[Λέ(οντοϲ)]	[ ]		
[ ]						[ια]		[ ]		
[ ]						[ιβ]	[Πα(ρθένου)]	[ ]		
[ ]						[ιγ]		[ ]		
[ ]				**[νῶν(αι)]**		[ιδ]	[Ζυ(γοῦ)]	[ ]		
[ ]						[ιε]		[ ]		

---

2 Π̣α̣(ρθένου): probably the same compendium as in line 6 3 [illegible] 4 [illegible]; ϲυναψ
5 [illegible]; [illegible]; [illegible]; ϲυνα 6 [illegible] (*bis*); ϲυνα; ε̣ϲ̣π̣? 8 [illegible] 9 [illegible] 12 [illegible] 14 [illegible]
15 ϲεξ (written in pink)

4175. Almanac-Ephemeris, 24 b.c.

Left:

[Mesore]

i	ii	iii	iv	v	vi	vii	viii	ix	x	xi ⇒
				[xx] Virgo [evening setting?]						
				15 Libra 6						
				30 Scorpio boundary						
]...				2 Cancer boundary 27 Leo boundary						
boundary				12 Virgo boundary 28 Virgo 11 evening setting?						
[xx]						1	Aries	26		
5						2	Taurus	8		
18						3		19		
[xx]						4		30		
[xx]						5	Gemini	12		
[xx]						6		23		
[xx]						7	Cancer	5		
[xx]				**Sextilis**		8		18		
[xx]						9		[xx]		
[xx]						[10]	[Leo]	[xx]		
[xx]						[11]		[xx]		
[xx]						[12]	[Virgo]	[xx]		
[xx]						[13]		[xx]		
[xx]				**[nones]**		[14]	[Libra]	[xx]		
[xx]						[15]		[xx]		

Right:

⇐	xii	xiii	xiv	xv	xvi
		ζ Θ[ώθ			
		[			
		κδ Ζυ(γοῦ) ιδ ̣ ἑ̣[ϲπ(έριοϲ)			
		ιζ Ϲ̣κ̣ο̣(ρπίου) [			
		ιζ Πα(ρθένου) ϲ̣υ̣[νά(ψει)			
		ι[̣ Πα(ρθένου) ̣ ἑϲπ(έριοϲ)			
				α	[Κ]α̣(ρκίνου)
				[β]	[Λέ(οντοϲ)]
				[γ]	[ ]
		Ϲεπτ(έμβριοϲ)		δ	[Πα(ρθένου)]
				ε	[ ]
				ϛ	[Ζυ(γοῦ)]
				ζ	
		νῶν(αι)		η	
				θ	Ϲ̣κ̣ο̣(ρπίου)
				ι	
				ια	[Τοξότου]
				ιβ	[ ]
				ιγ	[Αἰγόκερω]
				ιδ	
				ιε	Ὑ̣δ̣(ροχόου)
		[εἰδοί]		ιϛ	
				[ι]ζ̣	Ἰ̣χ̣(θύων)
				ιη̣	[ ]
				ι̣θ	[ ]

1 Θ[ώθ: written in pink 3 ♎︎; ̣ ἑ̣[ϲ(περία): very uncertain; the first letter could be ε or ο with something suspended above. 4 Ϲκο(ρπίου): as in v 4? 5 ♍︎ 8 [K]α̣(ρκίνου): as in v 5? 11 ϲεπτ (in pink) 15 νων (in pink) 16 Ϲ̣κ̣ο̣(ρπίου): as in v 4? 22 compendium uncertain 24 compendium uncertain

Right:

⇐ xii xiii xiv xv xvi

7 Thoth

xii	xiii	xiv	xv	xvi
	[			
	24 Libra 14 evening setting?			
	17 Scorpio [			
	17 Virgo boundary			
			1	Cancer
			[2]	[Leo]
			[3]	[ ]
	September		4	[Virgo]
			5	[ ]
			6	[Libra]
			7	
	nones		8	
			9	Scorpio
			10	
			11	[Sagittarius]
			12	[ ]
			13	[Capricorn]
			14	
			15	Aquarius
	[ides]		16	
			17	Pisces
			18	[ ]
			19	[ ]

4176. Almanac-Ephemeris, a.d. 111

i	ii	iii
α̅ Ἰχθύ]ϲι̣ ιβ κ̣		
α̅ Αἰγόκερῳ] κζ ι̣̅[ ]		Ὑ̣δ̣ρ̣ο̣χ̣(όῳ) ╤
α̅ Κριῷ] ιγ̣		
‾ Cκορπίῳ] ╤	κ̣ε̣̅	Τ̣ο̣ξ̣ό̣τ̣(η) ╤
‾ Τοξότῃ] ╤	κ̣α̣̅ δ . .	Τ̣ο̣ξ̣ό̣τ̣ῃ κ̣ .

Κρ]ι̣ο̣ῦ̣	ιε θ̣	
	κη̣ κ̣α	
Ταύρ]ου	ια̣ κ̣α̣	ν̣υ̣κ̣τ̣(ὸϲ) α̣̅
	κδ ϛ	
Δ]ι̣δ̣ύ̣μω(ν)	ϛ μ	πρὸ ἡλίου
	ιθ ╤	
Κα]ρ̣κ̣ί̣νο(υ)	α θ	ἡμέραϲ ‾ .
	ιγ γ̣	
	κδ με̣	
Λ]έοντι	ϛ μ̣	πρὸ ἡλίου
	ιθ ε	
Π]α̣ρθένου	α κ̣ε	ἡμέρ(αϲ) θ γ
	ιγ νθ	
	κϛ̣ με	
Ζ̣υγοῦ	θ̣ μ̣δ̣	ἡμέρα̣ϲ̣ δ̣ ∟
	κ̣[β] ν̣ϛ	
C]κ̣ορπίῳ	ϛ̣ κα	ἡμέρ(αϲ) δʹ
	[ιθ] ν̣θ	
Τ̣οξότῃ	[γ] μ̣θ	ἡμέ̣ρ(αϲ) ε δ̣ʹ . .
	[ιζ ν]β̣	
Α]ἰγόκερῳ	[β] η̣	ἡμέρ(αϲ) ε̣ [
	[ιϛ λζ]	
Ὑ]δροχό̣ῳ	α [ιγ]	ἡ̣μ̣έ̣ρ̣(αϲ) [
	[ιε λζ]	

2 υ̣δ̣ρ̣ο̣χ̣   4 τ̣ο̣ξ̣ο̣τ̣?   5 δ . . : abbreviation of δύϲιϲ?   9 νυκτ̅   11 δ]ι̣δ̣υ̣μω̅   13 κα]ρ̣κ̣ι̣νο̅   18 ημερ̅ (so lines 23, 25, 27, 29)

4176. Almanac-Ephemeris, a.d. 111

i	ii	iii
1] Pisces 12° 2x'?		
1 Capricorn] 27° 1x Aquarius 0°		
1 Aries] 13°		
x Scorpio] 0° 25 Sagittarius 0°		
x Sagittarius] 0° 21 setting? Sagittarius 2x°		
Aries	15 9	
	28 21	
Taurus	11 21	night 1?
	24 6	
Gemini	6 40	before sun(rise)
	19 0	
Cancer	1 9	day x?
	13 3	
	24 45	
Leo	6 40	before sun(rise)
	19 5	
Virgo	1 25	day 9 ⅓
	13 59	
	26 45	
Libra	9 44	day 4? ½
	2[2] 56	
Scorpio	6 21	day 4
	[13] 59	
Sagittarius	[3] 49	day 5 ¼...?
	[17 5]2	
Capricorn	[2] 8	day 5? [
	[16 37]	
Aquarius	1 [13]	day [
	[15 37]	

4177. Almanac-Ephemeris, A.D. 121–127

Fragment 1, upper table, left

i	ii					⇒
ιε̣	Κ̣ρ̣ι̣(οῦ) ō ι̣η̣					
ι	Ταύ(ρου) ō ιη					
ε	Διδ(ύμων) ō ι̣η̣	λ	Καρ(κίνου)	ō λ̣ς̣		
κε̣	Λέον(τος) α α					

Fragment 1, lower table, left

i	ii	iii	iv	v	vi	⇒
	[Φ]α̣[μ(ενώθ)]			Φ̣α̣[ρ]μ̣(οῦθι)		
[α]	Cκο(ρπίου)	ιβ	α̣	Τοξ(ότου)	ιη	
[γ]	Τοξ(ότου)	θ	β̣	Αἰγ(όκερω)	[β]	
ε̣	Αἰγ(όκερω)	ζ	[δ]	[Ὑδρ(οχόου)]	[α]	
ζ	Ὑδρ(οχόου)	ς	[ζ]	[Ἰχθ(ύων)]	[ιδ]	
θ	Ἰχθ(ύων)	ε	[θ]	[Κρι(οῦ)]	[ιβ]	
ιβ	Κρι(οῦ)	γ	[ια]	[Ταύ(ρου)]	θ̣	
ιδ	Ταύ(ρου)	γ	[ιγ]	[Διδ(ύμων)]	ε	
ις	Διδ(ύμων)	θ	[ις]	[Καρ(κίνου)]	ιγ	
ιη	Καρ(κίνου)	δ	[ιη]	Λ̣έο(ντος)	ζ	
κα	Λέο(ντος)	ι	κα	Π̣αρ(θένου)	ιγ	
κγ	Παρ(θένου)	δ	κγ	Ζυγ(οῦ)	ζ	
κς	Ζυγ(οῦ)	ιβ	κε̣	Cκο(ρπίου)	α	
κη	Cκο(ρπίου)	ζ	κη̣	Τοξ(ότου)	ιδ	
λ	Τοξ(ότου)	ε	λ	Α̣ἰγ(όκερω)	ιβ	

Names of zodiacal signs abbreviated by truncation.

4177. ALMANAC-EPHEMERIS, A.D. 121–127

Fragment 1, upper table, left

i	ii	⇒
15	Aries 0 18?	
10	Taurus 0 18	
5	Gemini 0 18? 30 Cancer 0 36?	
25	Leo 1 1	

Fragment 1, lower table, left

i	ii	iii	iv	v	vi
	Phamenoth			Pharmouthi	
[1]	Scorpio	12	1	Sagittarius	18
[3]	Sagittarius	9	2	Capricorn	[2]
5	Capricorn	7	[4]	[Aquarius]	[1]
7	Aquarius	6	[7]	[Pisces]	[14]
9	Pisces	5	[9]	[Aries]	[12]
12!	Aries	3	[11]	[Taurus]	9
14	Taurus	3	[13]	[Gemini]	5
16	Gemini	9	[16]	[Cancer]	13
18	Cancer	4	[18]	Leo	7
21	Leo	10	21	Virgo	13
23	Virgo	4	23	Libra	7
26	Libra	12	25	Scorpio	1!
28	Scorpio	7	28	Sagittarius	14
30	Sagittarius	5	30	Capricorn	12

⇒

Fragment 1, upper table, right

⇐ iii	iv
α//	Δ̣[ιδ(ύμων)
ι̣ε	Δ̣[ιδ(ύμων)
γ//	Διδ̣(ύμων) [
δ	Παρ(θένου) [

1 // = last evening visibility 3 // = last morning visibility

Fragment 1, lower table, right

⇐ vii	viii	ix	x	xi
	[Π]αχ̣(ών)			Π̣[αῦ(νι)
α	Αἰγ(όκερω)	κϛ	α	Ἰχθ(ύων)
β	Ὑδρ(οχόου)	ια̣	γ	Κρι(οῦ)
δ	Ἰχθ(ύων)	ι̣	ϛ	Ταύ(ρου)
ϛ	Κ̣ρ̣ι̣(οῦ)	η̣	η	Διδ(ύμων)
η	[Ταύ(ρου)]	[ε]	[ι]	Καρ(κίνου)
ι	Δ̣ι̣δ(ύμων)	α̣	ι̣[γ]	Λέο(ντος)
ιγ	Καρ(κίνου)	θ̣	ι̣ε̣	Παρ(θένου)
ιε	Λέο(ντος)	γ̣	ι̣η̣	Ζυγ(οῦ)
ιη	Παρ(θένου)	θ̣	κ̣	Cκο(ρπίου)
κ	Ζ̣υγ(οῦ)	δ̣	κ̣	[Τοξ(ότου)]
κγ	Cκο(ρπίου)	ιβ̣	κδ̣	[Αἰγ(όκερω)]
κε	Τοξ(ότου)	θ	κϛ̣	[Ὑδρ(οχόου)]
κζ	Αἰγ(όκερω)	ζ̣	κθ̣	Ἰ̣χ̣θ̣(ύων)
κθ	Ὑδρ(οχόου)	ϛ		

Fragment 1, upper table, right

⇐ iii	iv
1//	Gemini [
15	Gemini [
3//	Gemini [
4	Virgo [

Fragment 1, lower table, right

⇐ vii	viii	ix	x	xi
	Pachon			Payni
1	Capricorn	26	1	Pisces
2	Aquarius	11	3	Aries
4	Pisces	10	6	Taurus
6	Aries	8	8	Gemini
8	[Taurus]	[5]	[10]	Cancer
10	Gemini	1	1[3]	Leo
13	Cancer	9	15	Virgo
15	Leo	3	18	Libra
18	Virgo	9	20	Scorpio
20	Libra	4	22	[Sagittarius]
23	Scorpio	12	24	[Capricorn]
25	Sagittarius	9	26	[Aquarius]
27	Capricorn	7	29	Pisces
29	Aquarius	6		

Fragment 2, upper table

i
– – – – – – –
Ὑδ]ρ̣(οχόου)] β̣
]
]
]
]
]
]

Fragment 2, lower table

i	ii	iii	iv
		Μ̣ε̣[cωρ]η	
κ]α̣	α	Παρ(θένου)	κδ
γ̣	β	Ζυγ(οῦ)	ε
θ]	ε	Cκο(ρπίου)	ιβ
γ̣	ζ	Τοξ(ότου)	ζ
ι]α̣	θ	Αἰγ(όκερω)	γ̣
ζ]	ιβ	Ὑδρ(οχόου)	ι̣δ̣
ε̣	ιδ	Ἰχθ(ύων)	ιβ
γ̣	ις	Κρι(οῦ)	ια
β]	ιη	Ταύ(ρου)	ι
ιδ]	κ	Διδ(ύμων)	η
ιβ]	κβ	Καρ(κίνου)	δ
θ]	κδ	Λέο(ντος)	α
ε]	κζ	Παρ(θένου)	η
ιβ]	κθ	Ζυγ(οῦ)	β
	β	Cκο(ρπίου)	η
	δ	Τοξ(ότου)	γ

Fragment 2, upper table

i
Aquarius 2?
]
]
]
]
]
]

Fragment 2, lower table

i	ii	iii	iv
		Mesore	
2]1	1	Virgo	24
3	2	Libra	5
9]	5	Scorpio	12
3	7	Sagittarius	7
1]1	9	Capricorn	3
7]	12	Aquarius	14
5	14	Pisces	12
3	16	Aries	11
2?]	18	Taurus	10
14]	20	Gemini	8
12]	22	Cancer	4
9]	24	Leo	1
5]	27	Virgo	8
12]	29	Libra	2
	2	Scorpio	8
	4	Sagittarius	3

Fragment 3, upper table

	i	ii
	]̣	[
	]α	Z[υγ(οῦ)
3	]α	Z[υγ(οῦ)

Fragment 3, lower table

	i	ii	iii	iv
		Θώθ̣		
5	[α]	Cκο(ρπίου)	ιζ	α
	[β]	Τοξ(ότου)	γ̣	β
	[δ]	Αἰγ(όκερω)	α	δ
	[ζ]	[Ὑ]δρ(οχόου)	ια	ϛ̣
	[θ]	[Ἰχθ(ύων)]	ζ̣	[θ

Fragment 3, upper table

	i	ii
	]x	[
	]1	Libra [
3	]1	Libra [

Fragment 3, lower table

	i	ii	iii	iv
		Thoth		
5	[1]	Scorpio	17'	1
	[2]	Sagittarius	3	2
	[4]	Capricorn	1	4
	[7]	Aquarius	11	6
	[9]	[Pisces]	7	[9

4177a. Ephemeris, a.d. 245

i	ii	iii	iv
[Ὑδροχόῳ]	[ε]	ιδ	
	[ιθ]	ν	
[Ἰχθύϲι]	[δ]	ιθ	[
	[ι]η̣	λε	
Ḳρ̣ι̣ῷ̣	β̣	λη	[
	ιϛ	κθ	
Ταύρῳ	ō	ϛ	[
	ιγ	λα	
	κϛ	μγ	[
Δ]ιδύμ(οιϲ)	θ	μβ	
	κβ	κη	
Κ]αρκίν(ῳ)	ε	β	ι̣[
	ιζ	κβ	
	κθ	λ	α̣ [νυκτόϲ
[Λέ]ο̣ν̣[τι]	[ια]	μ̣ε	
	[κ]γ̣	η	
[Παρθένῳ]	[ε]	ε	β̣ [ἡμέραϲ
	[ι]ζ̣	ιδ	
	[κθ]	λδ	α̣ [νυκτόϲ
[Ζυγῷ]	[ιβ]	ια	
	[κ]δ̣	νθ	

7 ō : [illegible]   10 δ]ιδυμ`   12 κ]α̣ρκιν

4177a. Ephemeris, a.d. 245

i	ii	iii	iv
[Aquarius]	[5]	14	
	[19]	50	
[Pisces]	[4]	19	[
	[1]8	35	
Aries	2	38	[
	16	29	
Taurus	0	6	[
	13	31	
	26	43	[
Gemini	9	42	
	22	28	
Cancer	5	2	1[x?
	17	22	
	29	30	1 [of night
Leo	[11]	45!	
	[2]3	8	
[Virgo]	[5]	5	2 [of day
	[1]7	14	
	[29]	34!	1 [of night
[Libra]	[12]	11	
	[2]4	59	

4178. Ephemeris, a.d. 261

i	ii	iii	iv	v	vi	vii
ιε]	κα	[κη]	[Ὑδροχ̣όῳ̣]	[ιβ]	μ̣ς̣	[
ις	κβ	κθ̣		κς̣	ια	[
ιζ	κγ	λ α	Ἰχθύϲι	θ	κα	[
ιη	κδ	β		[κ]β̣	ι̣.̣	[
ιθ	κε	γ	Κρ̣ειῷ̣	ε	γ	ἡ̇μέ(ραϲ) [
κ]	κς	δ		ιζ	λς̣	[
κα]	κζ	ε		κθ	νδ̣	νυ(κτὸϲ) α̣ [
κβ]	κη̣	ς	Τ̣αύρῳ̣	ι̣β̣	α̣	[
κγ]	κθ	ζ		[κ]γ̣	ν̣δ	[
κδ]	λ	η	Διδύμοιϲ	ε	λ̣η̣	ἡ̇μ(έραϲ) β [
	Φαμε̣νώθ					
κε]	α	θ		ιζ	λ̣ε	[
κς]	β	ι		θ̣	[μ]δ̣	νυ(κτόϲ) ε̣ [
κζ]	γ	ια	Καρκ̣[ίν]ῳ̣	ιβ	[	
κη]	δ	ιβ		κδ	[	

5 *l.* Κριῷ ; ἡμ̄ε (so line 10) 7 ν̄υ (so line 12)

4178. Ephemeris, a.d. 261

i	ii	iii	iv	v	vi	vii
15]	21	[28]	Aquarius	[12]	46	[
16	22	29		26	11	[
17	23	30/1	Pisces	9	21	[
18	24	2		[2]2	1x	[
19	25	3	Aries	5	3	day [
20]	26	4		17	36	[
21]	27	5		29	54	night 1? [
22]	28	6	Taurus	12	1	[
23]	29	7		[2]3	54	[
24]	30	8	Gemini	5	38	day 2 [
	Phamenoth					
25]	1	9		17	35	[
26]	2	10		9	[4]4	night 5? [
27]	3	11	Cancer	12	[	
28]	4	12		24	[	

4179. Ephemeris, A.D. 348

	i	ii	iii	iv	v	vi	vii	viii ⇒
	[καλ()]	[α]	[ϛ]	[ιζ]	[Τοξότ(ῃ)]	[ε ιη]	[ἡμ(έραϲ) ̣]	
	[ϛ]	[β]	[ζ]	[ιη]		[ιη νε]		
	[ε]	[γ]	[η]	[ιθ]	[Αἰγόκε(ρῳ)]	[β]κ̣	ἡ̣μ̣(έραϲ) η	
	[δ]	[δ]	[θ]	[κ]		[ιε] λβ		
	[γ]	[ε]	[ι]	[κα]		[κη] λ̣α̣	ν̣υ̣(κτόϲ) γ	
	[α]	[ϛ]	[ια]	[κβ]	[Ὑδροχ(όῳ)]	[ια] ιζ		
C	[νῶναι]	[ζ]	[ιβ]	[κγ]		[κγ] να		
	[η]	[η]	[ιγ]	κ̣δ	Ἰχθύ(ϲι)	[ϛ] ια	ἡμ(έραϲ) α	
	[ζ]	[θ]	ι̣δ	κε		ιη ιθ		
	[ϛ]	ι	ιε	κϛ	Κριῷ	ō ιδ	ἡμ(έραϲ) ια̣ L	
	[ε]	ι̣α	ιϛ	κζ		ια [νϛ]		
	[δ]	ι̣β	ιζ	κη		κγ να		
	[γ]	ιγ	ιη	κθ	Ταύρῳ	ε νθ	ἡμ(έραϲ) α̣	
C	[α]	ιδ	ιθ	λ		ιη ιθ		ϲύνοδ(οϲ)
	[εἰδοί]	ι̣ε	κ	α	Διδ̣ύμ(οιϲ)	ō νγ	ἡμ(έραϲ) ι δʹ	Ταύρῳ
	[ιζ]	ιϛ	κα	β̣		ιγ λθ		κ̣γ̣ λ̣ ̣
								ν̣υ̣(κτόϲ) ι̣
	[ιϛ]	[ι]ζ̣	κ̣β̣	γ		κϛ λη	νυ(κτόϲ) ϛ	
	[ιε]	ιη	κγ	δ	Καρκίν(ῳ)	θ ν		
	[ιδ]	ιθ	κδ	ε		κγ ιε	νυ(κτόϲ) ιβ̣	
	[ιγ]	κ	κε	ϛ	Λέοντι	ϛ νγ̣		
C	[ιβ]	[κ]α̣	κ̣ϛ̣	ζ		κ [μγ]		
	ι̣α̣	[κβ]	[κζ]	η̣	[Π]αρθέ̣ν(ῳ)	δ μ̣ζ	ἡ̣μ(έραϲ) δ̣	
	ι	κγ̣	κ̣η̣	θ		ιθ γ		
	θ̣	κ̣δ̣	[κθ]	[ι]	Ζυγῷ	γ λβ	ἡμ(έραϲ) ε L	
	[η]	κ̣η̣	λ̣	[ι]α̣		ιη η		
	[ζ]	κ̣ϛ̣	α̣	ιβ	Ϲκορπ(ίῳ)	β λα	ἡμ(έραϲ) ζ	
	[ϛ]	κ̣ζ̣	β	ιγ		ιϛ μβ		
C	[ε]	κη	γ	ιδ	Τοξότ(ῃ)	ō λ̣θ̣	ἡμ(έραϲ) ι̣α̣	
	[δ]	κ̣θ	δ	ιε		ιδ κδ		παν(ϲέληνοϲ)
	[γ]	[λ]	ε	ιϛ		κη ν[ϛ]	νυ(κτόϲ) ̣ Ḷ	Τοξότῃ
	[α]	λα	ϛ	ι̣ζ	Αἰγόκερῳ	ια ιε		ϛ ν̣δ
								ἡμ(έραϲ) β̣ δʹ

Names of zodiacal signs abbreviated by truncation. Other regular abbreviations: $\overset{\mu}{\eta}$ = ἡμέραϲ, ν^{υ} = νυκτόϲ 14 ϲυνοδ 29 παν(

4179. Ephemeris, A.D. 348

	i	ii	iii	iv	v	vi	vii	viii ⇒
	[kal.]	[1]	[6]	[17]	[Sagittarius]	[5 18]	[day x]	
	[6]	[2]	[7]	[18]		[18 55]		
	[5]	[3]	[8]	[19]	[Capricorn]	[2] 20	day 8	
	[4]	[4]	[9]	[20]		[15] 32		
	[3]	[5]	[10]	[21]		[28] 31	night 3	
	[1]	[6]	[11]	[22]	[Aquarius]	[11] 17		
☾	[nones]	[7]	[12]	[23]		[23] 51		
	[8]	[8]	[13]	24	Pisces	[6] 11	day 1	
	[7]	[9]	14	25		18 19		
	[6]	10	15	26	Aries	0 14	day 11 $^1/_2$	
	[5]	11	16	27		11 [56]		
	[4]	12	17	28		23 51		
	[3]	13	18	29	Taurus	5 59	day 1	
☾	[1]	14	19	30		18 19		conjunction
	[ides]	15	20	1	Gemini	0 53	day 10 $^1/_4$	Taurus
	[17]	16	21	2		13 39		23 3x
								night 10
	[16]	[1]7	22	3		26 38	night 6	
	[15]	18	23	4	Cancer	9 50		
	[14]	19	24	5		23 15	night 12	
	[13]	20	25	6	Leo	6 53		
☾	[12]	[2]1	26	7		20 [43]		
	11	[22]	[27]	8	Virgo	4 47	day 4	
	10	23	28	9		19 3		
	9	24	[29]	[10]	Libra	3 32	day 5 $^1/_2$	
	[8]	28	30	[1]1		18 8		
	[7]	26	1	12	Scorpio	2 31	day 7	
	[6]	27	2	13		16 42		
☾	[5]	28	3	14	Sagittarius	0 39	day 11	
	[4]	29	4	15		14 24		full moon
	[3]	[30]	5	16		28' 5[6]	night x $^1/_2$	Sagittarius
	[1]	31	6	17	Capricorn	11 15		6 54
								day 2 $^1/_4$

⇐	ix	x	xi	xii	xiii	xiv		xv	xvi
	[ ]	[ō] μ	κ̣β̣ λ̣	κς [. .]	[ ]	[ ]		[καλ()]	[α]
	[ ]	[ō μ]ς	κ̣β̣ λ̣	κς μη	κζ [. .]	[ ]		[δ]	[β]
	[ ]	[ō ν]β̣	κβ̣ λ̣	κζ λβ	κη [. .]	[ ]		[γ]	[γ]
	[ ]	[ō ν]η	κβ λα	κη ις	κθ̣ δ	[ ]	☾	[α]	[δ]
5	[ ]	[ ]	κβ λα̣	κθ ọ̄	κ̣θ νς	[ ]		[νῶναι]	[ε]
	[ ]	[ ]	κ̣β λα̣	κ̣θ̣ μγ	ō μ̣[.]	[θ λ]α̣		[η]	[ς]
6a				Ἴ̣χ̣θ̣υ̣c̣ι̣					
	[ ]	[ ]	κβ λα̣	ō κζ̣	α [μ]ε̣	ι̣α̣ [κ]ε		[ζ]	[ζ]
	[ ]	[ ]	[κβ] λ̣α̣	[α . .]	β μα	ιγ̣ [ι]θ̣		[ς]	[η]
	[ ]	[ ]	[κβ λ]α̣	α̣ νδ	γ λζ	ι̣ε ιγ		[ε]	[θ]
10	κ̣[. . .]	[ ]	[ ]	[β] λη	δ λγ	ιζ ζ		[δ]	[ι]
	κ[. . .]	[ ]	[ ]	[γ] κ̣α̣	ε̣ κθ	ιθ̣ ō	☾	γ	[ια]
	κα [. .]	[ ]	[ ]	δ̣ δ	ς̣ κε̣	[κ ν]δ		α̣	[ιβ]
12a						cυνοδικ(όc)			
	κβ [. .]	[ ]	[κβ] λ̣α̣	δ μζ	ζ κ	κβ μη		ε̣ἰ̣[δοί]	[ιγ]
	[ ]	[ ]	κβ λ̣[α]	[ε] λ	η ιζ	κδ μγ		ιη̣	ιδ̣
15	[ ]	[ ]	[κ]β λα̣	[ς] ιγ	θ ιε	κς λς		ιζ	ι̣ε̣
	[ ]	[ ]	κβ λ̣[α]	[ς] ν̣ς	ι ιδ	κη κθ		ις	ις
16a						Διδύμ(οιc)			
	[ ]	[ ]	[κβ λα]	[ζ λ̣]θ	ια ιγ	ọ̄ κγ	☾	ι̣ε̣	ιζ
	[ ]	[ ]	[κβ λα]	[η .].	ιβ̣ ιγ	[β .].		[ιδ]	ι̣η̣
	[κ]ζ̣ νδ	β̣ [. .]	[κ]β̣ λα	[ ]	ιγ ιγ	δ̣ [. .]		[ιγ]	[ιθ]
20	[κ]η̣ να	β [. .]	κ̣β λ	[ ]	[ιδ] ιδ	ς̣ ς		ιβ	κ
	[κ]θ μθ	β [. .]	κβ κη	[ ]	[ιε ι]δ	[η] ō		ια	κα
21a	Δ̣ι̣δύμ(οιc)								
	ō με	[ ]	[ ]	[ ]	[ις] ιδ	θ νγ		[ι]	[κβ]
	α μβ	[ ]	[ ]	[ ]	[ιζ] ιδ	ια μς		[θ]	[κγ]
	β λ̣[θ]	β̣ ν̣η̣	[κβ] κβ	ι̣γ [. .]	[ιη] ιε	ιγ λθ		[η]	[κδ]
25	γ λς	γ δ	κβ ιθ̣	ι̣δ̣ [. .]	[ιθ] ι̣ε	ιε λβ	☾	ζ̣	[κε]
	δ λγ̣	[ ]	κ̣β̣ ις	ι̣ε̣ ς	[κ] ιζ	ιζ κε		ς̣	[κς]
	ε λ	[ ]	[κβ] ι̣γ̣	ι̣ε̣ μ̣.	κα̣ ι̣ς̣	ιθ ις		ε̣	[κζ]
	ς κζ	γ̣ [. .]	κβ η̣	ις κ̣.	κβ ιζ	κα θ		δ̣	[κη]
	ζ κ̣γ	γ . .	κ̣β̣ ε̣	ι̣ζ̣ ς	κγ κ	κγ β		γ	κθ̣
30	η κα	γ̣ λ̣β̣	κ̣β̣ β̣	[ι]ζ̣ μ̣ζ	κδ κγ	κδ ν̣ε		[α]	[λ]
	θ ι̣ζ̣	γ λζ	κα νη	[ι]η κη	κ̣ε̣ κ̣ς	κς μ̣η			

12a cυνοδικ 16a διδυμ 21a διδυμ

⇐	ix	x	xi	xii	xiii	xiv		xv	xvi
	[ ]	[0] 40	22 30	26 [xx]	[ ]	[ ]		[kal.]	[1]
	[ ]	[0 4]6	22 30	26 48	27 [xx]	[ ]		[4]	[2]
	[ ]	[0 5]2	22 30	27 32	28 [xx]	[ ]		[3]	[3]
	[ ]	[0 5]8	22 31	28 16	29 4	[ ]	☾	[1]	[4]
5	[ ]	[ ]	22 31	29 0	29 56	[ ]		[nones]	[5]
	[ ]	[ ]	22 31	29 43	0 4[x]	[9 3]1		[8]	[6]
6a				Pisces					
	[ ]	[ ]	22 31	0 27	1 [4]5	11 [2]5		[7]	[7]
	[ ]	[ ]	[22] 31	[1 xx]	2 41	13 [1]9		[6]	[8]
	[ ]	[ ]	[22 3]1	1 54	3 37	15 13		[5]	[9]
10	2[x xx]	[ ]	[ ]	[2] 38	4 33	17 7		[4]	[10]
	2[x xx]	[ ]	[ ]	[3] 21	5 29	19 0	☾	3	[11]
	21 [xx]	[ ]	[ ]	4 4	6 25	[20 5]4		1	[12]
12a						conjunction			
	22 [xx]	[ ]	[22] 31	4 47	7 20	22 48		ides	[13]
	[ ]	[ ]	22 3[1]	[5] 30	8 17	24 43		18	14
15	[ ]	[ ]	[2]2 31	[6] 13	9 15	26 36		17	15
	[ ]	[ ]	22 3[1]	[6] 56	10 14	28 29		16	16
16a						Gemini			
	[ ]	[ ]	[22 31]	[7 3]9	11 13	0 23		15	17
	[ ]	[ ]	[22 31]	[8 x]x	12 13	[2 x]x	☾	[14]	18
	[2]7 54	2 [xx]	[2]2 31	[ ]	13 13	4 [xx]		[13]	[19]
20	[2]8 51	2 [xx]	22 30	[ ]	[14] 14	6 6		12	20
	[2]9 49	2 [xx]	22 28	[ ]	[15 1]4	[8] 0		11	21
21a	Gemini								
	0 45	[ ]	[ ]	[ ]	[16] 14	9 53		[10]	[22]
	1 42	[ ]	[ ]	[ ]	[17] 14	11 46		[9]	[23]
	2 3[9]	2 58	[22] 22	13 [xx]	[18] 15	13 39		[8]	[24]
25	3 36	3 4	22 19	14 [xx]	[19] 15	15 32	☾	7	[25]
	4 33	[ ]	22 16	15 6	[20] 17	17 25		6	[26]
	5 30	[ ]	[22] 13	15 4x	21 16	19 16		5	[27]
	6 27	3 [xx]	22 8ʾ	16 2x	22 17	21 9		4	[28]
	7 23	3 xx	22 5	17 6	23 20	23 2		3	29
30	8 21	3 32	22 2	[1]7 47	24 23	24 55		[1]	[30]
	9 17	3 37	21 58	[1]8 28	25 26	26 48			

4180. Ephemeris, a.d. 465

Recto [0]	i	ii	iii	iv	v ⇒
[ ]	[ ]	[ ]	[ ]	[ ]. .	[ ]
[ ]	[ ]	[ ]	[ ]	[ ]	[ ]
[ ]	[ ]	[ ]	[ ]	[ ]	[ ]
[ ]	[ ]	[ ]	[ ]	[ ]	[ ]
[α]	[ε]	[ ]	[c]τερ(εά)	[Λέοντ(ι)]	[
[β]	[δ]	[ ]	[ ] cτερ(εά)		[
[γ]	[γ]	[ ]	[ ]ηι δίcωμ(οc)	Πα[ρθ(ένῳ)]	[
[δ]	[α]	[ ]	[κ]ακὴ δίcωμ(οc)		[
[ε]	[καλ(ὸ)]	[ ]	[μ]έcη δίcωμ(οc)		[
[ϛ]	[δ]	[ ]	[μ]έcη τροπικ(ή)	Ζ̣[υγῷ]	[
[ζ]	[γ]	[ ]	[μ]έcη τροπικ(ή)		[
[η]	[α]	[ ]	[cτερ(εά)]	[Cκορπ(ίῳ)]	[
[θ]	[νῶναι]	[α]	[cτερ(εά)]		[
[ι]	[η]	[β]	[δίcωμ(οc)]	[Τοξότ(ῃ)]	[
[ια]	[ζ]	[γ]	[ ]ηι δίcωμ(οc)		[
[ιβ]	[ϛ]	[δ]	[τρο]π̣ι̣κ̣(ὴ) ἀποφρά̣c̣	Αἰγόκ(ερῳ) ὥρ(ᾳ) δ ἡ̣μ̣έ̣ρ(αc)	[
[ιγ]	[ε]	[ε]	[τρ]ο̣π̣ι̣κ̣(ὴ) ἀποφρά̣c̣		[
[ιδ]	[δ]	[ϛ]	[c]τερ(εά)	Ὑ̣δρ̣υ̣χ(όῳ) ὥρ(ᾳ) . . .	[
[ιε]	[γ]	[ζ]	[cτ]ε̣ρ̣(εά)		ι̣η̣ ιη̣
[ιϛ]	[α]	[η]	[δίcωμ(οc)]	[Ἰχ]θ̣(ύcι) ὥ̣ρ̣(ᾳ) . [ ]	β ε
[ιζ]	[εἰδοί]	[θ]	[δίcωμ(οc)]		ιε ν.
[ιη]	[ιη]	[ι]	[δίcωμ(οc)]		κθ μ̣.
[ιθ]	[ιζ]	[ι]α̣	[ ] τ̣ρ̣ο̣π̣ι̣κ̣(ή)	[Κριῷ ὡρ(ᾳ)] α̣ νυκ(τόc)	ιγ ιϛ
[κ]	[ιϛ]	ιβ̣	μέcη̣ τροπικ(ή)		[
[κα]	[ιε]	ιγ	μέc̣η cτερ(εά)	Τ̣[αύρ(ῳ)] ὥρ(ᾳ) ζ νυκ(τόc)	θ ιβ
[κβ]	ιδ	ι̣[δ]	φαύλη [cτερ(εά)]		κα μ̣.
[κγ]	ι̣γ	ι̣ε	μέcη δίcωμ̣(οc)	Δ̣ι̣δ̣[ύμ(οιc) ὡρ(ᾳ) .] ἡ̣μ̣έ̣ρ̣(αc)	δ ι̣[
[κδ]	ιβ̣	ιϛ	κακὴ δ̣ί̣c̣ω[μ(οc)]		ιϛ κ[
[κε]	ι̣α̣	ιζ	ἐκλ̣ε̣ι̣π(τικὴ) ἀποφρ̣ά̣c̣		[
[κϛ]	ι̣	ιη	ἐκλειπ(τικὴ) μ̣έ̣c̣η	Καρ̣[κ(ίνῳ)]	[
[κζ]	[θ]	ι̣θ	μέcη τροπι̣κ(ή)		[
[κη]	[η]	κ̣	μέc̣η c̣τ̣ερ(εά)	[Λ]έο̣ν̣τ̣(ι) ὥρ(ᾳ) . ἡ̣μ̣έ̣ρ(αc)	[
[κθ]	[ζ]	[κ]α̣	μέ̣c̣[η cτε]ρ(εά)		[
[λ]	[ϛ]	[κβ]	[cτερ(εά)]		κ[
[cύνοδοc Ἀθὺρ η εἰ]c θ ὥρ(ᾳ) ζ ḍ νυ̣κ(τὸc) [Cκορπίῳ μοίρ(ᾳ) . .					
[πανcέληνοc] κβ εἰc κγ ὥρ(ᾳ) ε̣ [. νυ]κ̣(τὸc) Τ[αύρῳ μοίρ(ᾳ) . .					

Regular abbreviations: ⳨ = ὥρᾳ, ημερ = ἡμέρας, ν$\overset{κ}{υ}$ = νυκτός, στερ = στερεά, δισωμ = δίσωμος, τροπ$\overset{κ}{ι}$ = τροπική, εκλειπ = ἐκλειπτική 16 αιγ$\overset{κ}{ο}$ (abbreviations of other names in col. iv are unclear)

4180. Ephemeris, A.D. 465

Recto

[0]	i	ii	iii	iv	v ⇒
[ ]	[ ]	[ ]	[ ]	...	[ ]
[ ]	[ ]	[ ]	[ ]	[ ]	[ ]
[ ]	[ ]	[ ]	[ ]	[ ]	[ ]
[ ]	[ ]	[ ]	[ ]	[ ]	[ ]
[1]	[5]	[ ]	[ ] solid	[Leo]	[
[2]	[4]	[ ]	[ ] solid		[
[3]	[3]	[ ]	... bicorporal	Virgo [ ]	[
[4]	[1]	[ ]	bad bicorporal		[
[5]	[kal.]	[ ]	mean bicorporal		[
[6]	[4]	[ ]	mean tropical	Libra [ ]	[
[7]	[3]	[ ]	mean tropical		[
[8]	[a]	[ ]	[solid]	[Scorpio]	[
[9]	[nones]	[1]	[solid]		[
[10]	[8]	[2]	[bicorporal]	[Sagittarius]	[
[11]	[7]	[3]	... bicorporal		[
[12]	[6]	[4]	tropical unlucky	Capricorn, hour 4 of day	[
[13]	[5]	[5]	tropical unlucky		[
[14]	[4]	[6]	[ ] solid	Aquarius, hour x ...	[
[15]	[3]	[7]	[ ] solid		18 18
[16]	[1]	[8]	[bicorporal]	Pisces, hour x [ ]	2 5
[17]	[ides]	[9]	[bicorporal]		15 5x
[18]	[18]	[10]	[bicorporal]		29 4x
[19]	[17]	[1]1	[ ] tropical	[Aries, hour] 1 of night	13 16
[20]	[16]	12	mean tropical		[
[21]	[15]	13	mean solid	Taurus, hour 7 of night	9 12
[22]	14	1[4]	poor [solid]		21 4x
[23]	13	15	mean bicorporal	Gemini, hour x] of day	4 1[x
[24]	12	16	bad bicorporal		16 2[x
[25]	11	17	ecliptic unlucky		[
[26]	10	18	ecliptic mean	Cancer [ ]	[
[27]	[9]	19	mean tropical		[
[28]	[8]	20	mean solid	Leo, hour x of day	[
[29]	[7]	[2]1	mean solid		[
[30]	[6]	[22]	[solid]		2[x
[Conjunction Hathyr 8] to 9, hour 7 $^1/_4$ of night, [Scorpio xx°					
[Full moon] 22 to 23, hour 5 [...] of night, Taurus [xx°					

Recto

⇐	vi	vii	viii
	ἥ̣[λιοc]	Κρόνο̣c̣	Ζεὺc
	[ ]	Ὑδρυχ̣(όῳ)	Παρθέν̣(ῳ)
	[ ]	ἑcπέρι(οc)	ἑῷοc
4	[ ]	φάc̣[ιc]	[

— — — — — — — — — — — — — — — —

Abbreviations in cols. vii and viii are unclear.

Recto

⇐ vi vii viii

	Sun	Saturn	Jupiter
	[ ]	Aquarius	Virgo
	[ ]	evening	morning
4	[ ]	visibility	[

– – – – – – – – – – – – – – – – – –

Verso [–II]	[–I]	[0]	i	ii	iii ⇒
[ ]	[ ]	[ ]	. . [ ]	ζῴ[δι]ον	μοῖρ[αι
[ ]	[ ]	[ ]	. . καταρχή	. . [ ]	[ ]
[ ]	[ ]	[ ]		κ[ ]	[ ]
[ ]	[ ]	[ ]		. [ ]	[ ]
[α]	[ε]	[κγ]	[ ] . . .	[Παρθ(ένῳ)]	[
[β]	[δ]	[κδ]	[ ]		[
[γ]	[γ]	[κε]	[ ]	[Ζυγῷ]	[
[δ]	[α]	[κϛ]	[ ]		[
[ε]	[καλ()]	[κζ]	[ ]	[Cκορπ(ίῳ)]	[
[ϛ]	[δ]	[κη]	[ ]		[
[ζ]	[γ]	[κθ]	[ ]	[Τοξότ(ῃ)]	[
[η]	[α]	[ ]	[ ]		[
[θ]	[νῶναι]	[ ]	[ ]		[
[ι]	[η]	[ ]	[ ]	[Αἰγόκ(ερῳ)]	[
[ια]	[ζ]	[ ]	[ ]		[
[ιβ]	[ϛ]	[ ]	[ ]	[Ὑδρυχ(όῳ)]	ιγ μ .
[ιγ]	[ε]	[ ]	[ ]		κη κε
[ιδ]	[δ]	[ ]	[ ]	[Ἰχθύcι]	ιβ να
[ιε]	[γ]	[ ]	[ ]		κϛ [μ]α
[ιϛ]	[α]	[ ]	[ ]	[Κριῷ] ὥρ(α) . νυκ(τόc)	θ γ
[ιζ]	[εἰδοί]	[ ]	[ ]		κγ .
[ιη]	[ιθ]	[ ]	[ ]	[Ταύρῳ ὥρ(α)] . ἡμέρ(αc)	ϛ ιδ
[ιθ]	[ιη]	[ ]	[ ]		ιη μ[
[κ]	[ιζ]	[ ]	[ ]	[Διδύμ(οιc)] ὥρ(α) ι ἡμέρ(αc)	α .
[κα]	[ιϛ]	[ ]	[ ]		ιγ ι .
[κβ]	[ιε]	[ ]	[ ]		κε ι .
[κγ]	[ιδ]	[ ]	[ ]	[Καρκ(ίνῳ) ὥρ(α)] η . νυκ(τόc)	ζ ι .
[κδ]	[ιγ]	[ ]	[ ]		κ[
[κε]	[ιβ]	[ ]	[ ]	[Λέοντ(ι) ὥρ(α) .] ἡμέρ(αc)	[
[κϛ]	[ια]	[ ]	[ ]		[
[κζ]	[ι]	[ ]	[ ]		[
[κη]	[θ]	[ ]	[ ]	[Παρθ(ένῳ) ὥρ(α) . ἡμ]έρ(αc)	ϛ [
[κθ]	[η]	[ ]	[ ]		ιη λ . .
[λ]	[ζ]	[ ]	[ ]	[Ζυγῷ ὥρ(α) ἡ]μέρ(αc)	α κε
[cύνοδοc Χοιὰκ η ὥρ(α) . ἡμέ]ρ(αc) Τοξότῃ μοίρ(αιc) [					
[πανcέληνοc κβ εἰc κγ ὥρ(α) . νυκ(τὸc)] Διδύμοιc μοίρ(αιc) κζ [					

35 μοιρ (so line 36)

Verso [–II]	[–I]	[0]	i	ii	iii ⇒
[ ]	[ ]	[ ]	...	sign	degrees
[ ]	[ ]	[ ]	appraisal of day	...	[ ]
[ ]	[ ]	[ ]		...	[ ]
[ ]	[ ]	[ ]		...	[ ]
[1]	[5]	[23]	[ ]...	[Virgo]	[
[2]	[4]	[24]	[ ]		[
[3]	[3]	[25]	[ ]	[Libra]	[
[4]	[1]	[26]	[ ]		[
[5]	[kal.]	[27]	[ ]	[Scorpio]	[
[6]	[4]	[28]	[ ]		[
[7]	[3]	[29]	[ ]	[Sagittarius]	[
[8]	[1]	[ ]	[ ]		[
[9]	[nones]	[ ]	[ ]		[
[10]	[8]	[ ]	[ ]	[Capricorn]	[
[11]	[7]	[ ]	[ ]		[
[12]	[6]	[ ]	[ ]	[Aquarius]	13 4x
[13]	[5]	[ ]	[ ]		28 25
[14]	[4]	[ ]	[ ]	[Pisces]	12 51
[15]	[3]	[ ]	[ ]		26 [4]1
[16]	[1]	[ ]	[ ]	Aries, hour x of night	9! 3
[17]	[ides]	[ ]	[ ]		23 x
[18]	[19]	[ ]	[ ]	[Taurus, hour] x of day	6 14
[19]	[18]	[ ]	[ ]		18 4[x
[20]	[17]	[ ]	[ ]	[Gemini,] hour 10 of day	1 x
[21]	[16]	[ ]	[ ]		13 1x
[22]	[15]	[ ]	[ ]		25 1x
[23]	[14]	[ ]	[ ]	[Cancer, hour] 8 ... of night	7 1x
[24]	[13]	[ ]	[ ]		2[x
[25]	[12]	[ ]	[ ]	[Leo, hour x] of day	[
[26]	[11]	[ ]	[ ]		[
[27]	[10]	[ ]	[ ]		[
[28]	[9]	[ ]	[ ]	[Virgo, hour x] of day	6 [
[29]	[8]	[ ]	[ ]		18 3x
[30]	[7]	[ ]	[ ]	[Libra, hour x] of day	1 25
[Conjunction, Choeac 8, hour x] of day, Sagittarius [xx]°					
[Full moon, 22 to 23, hour x of night,] Gemini 27° [					

Verso

iv	v	vi	vii	viii	ix
[ ]	[ ]	η̣ [ ]	[ ]	[ ]	[
[ ]	[ις λ]γ	η κζ̣	[ ]	[ ]	[
[ ]	[ι]ς λδ	η μα	κ̣[ ]	[ ]	[
[ ]	[ι]ς λε	η νε	κ̣[ ]	[ ]	[
[ ]	[ις] λς	θ̣ θ	κζ ν̣[ ]	[ ]	[
[ ]	ι̣ς λζ	θ κγ	κη ις	[ ]	[
[ ]	[ι]ς̣ λη	θ λς	κη λς̣	[ ]	[
[ ]	[ ]	[ ]	[ ]	[ ]	[
[ ]	[ ]	[ ]	[ ]	[ ]	[
[ ]	[ ]	[ ]	[ ]	[ ]	[
[ ]	[ις] μ̣ς	ι̣ λ̣	[ ]	[ ]	[
ι̣ς λε	ι̣ς μη	ι μδ	ọ̄ [ ]	[ ]	[
ιζ λς	κ̣ς ν̣	ι νη	ọ̄ [ ]	[ ]	[
ιη λ̣η̣	ι̣ς ν̣α̣	ια ι̣β̣	ọ̄ [ ]	[ ]	[
ιθ μ̣	ι̣ς νδ	ι̣[ ]	[ ]	[ ]	[
κ μ̣ .	[ ]	[ ]	[ ]	[ ]	[
κα̣ [ ]	[ ]	[ ]	[ ]	[ ]	[
[ ]	[ ]	[ ]	[ ]	[ ]	[
[ ]	[ ]	[ ]	[ ]	[ ]	[
[ ]	[ι]ζ η̣	ι̣β λ̣[ ]	[ ]	[ ]	[
[ ]	ι̣ζ ια	ιβ μ̣[ ]	β ιζ	κ̣β̣ μ̣δ̣	κ̣θ [
					Τοξότ(η)
					ὥρ(α̣) ε̣ [
[ ]	ι̣ζ ιδ	ιγ .	β κε	κ[γ] ν̣η	ọ̄ . [
[ ]	ι̣ζ ι̣ .	ι̣[ ]	β̣ λ̣ζ̣	[κε] ι̣γ	α̣ . [
[ ]	[ ]	[ ]	β̣ μθ	κφ̣ .	β ν̣[
[ ]	[ ]	[ ]	β̣ νθ̣	κζ μβ̣	δ .
[ ]	[ ]	[ ]	γ ς̣	κη ν̣ . .	ε̣ [
				Ὑδρυ̣χ̣(όῳ)	
				ὥ̣ρ̣(α̣) . . [	
[ ]	[ ]	[ ]	γ . .	ọ̄ ι̣β	[
[ ]	[ ]	[ ]	γ̣ .	α̣ κ̣ς	[
δ̣ . [ ]	[ ]	[ ]	γ . β̣	[ ]	[
ε̣ [ ]	[ ]	[ ]	γ̣ . .	[ ]	[

Verso

⇐

iv	v	vi	vii	viii	ix
[ ]	[ ]	8 [ ]	[ ]	[ ]	[
[ ]	[16 3]3	8 27	[ ]	[ ]	[
[ ]	[1]6 34	8 41	2[x]	[ ]	[
[ ]	[1]6 35	8 55	2[x]	[ ]	[
[ ]	[16] 36	9 9	27 5[x]	[ ]	[
[ ]	16 37	9 23	28 16	[ ]	[
[ ]	[1]6 38	9 36	28 36	[ ]	[
[ ]	[ ]	[ ]	[ ]	[ ]	[
[ ]	[ ]	[ ]	[ ]	[ ]	[
[ ]	[ ]	[ ]	[ ]	[ ]	[
[ ]	[16] 46	10 30	[ ]	[ ]	[
16 35	16 48	10 44	0 [ ]	[ ]	[
17 36	16 50	10 58	0 [ ]	[ ]	[
18 38	16 51	11 12	0 [ ]	[ ]	[
19 40	16 54	1[x]	[ ]	[ ]	[
20 4x	[ ]	[ ]	[ ]	[ ]	[
21 [ ]	[ ]	[ ]	[ ]	[ ]	[
[ ]	[ ]	[ ]	[ ]	[ ]	[
[ ]	[ ]	[ ]	[ ]	[ ]	[
[ ]	[1]7 8	12 3[x]	[ ]	[ ]	[
[ ]	17 11	12 4[x]	2 17	22 44	29 [
					Sagittarius
					hour [x
[ ]	17 14	13 x	2 25	2[3] 58	0 x[
[ ]	17 1x	1[x]	2 37	[25] 13	1 x[
[ ]	[ ]	[ ]	2 49	26 x	2 5[x
[ ]	[ ]	[ ]	2 59	27 42	4 x
[ ]	[ ]	[ ]	3 7	28 5x	5 [
				Aquarius	
				hour xx[	
[ ]	[ ]	[ ]	3 xx	0 12	[
[ ]	[ ]	[ ]	3 x	1 26	[
4 x[ ]	[ ]	[ ]	3 x2	[ ]	[
5 [ ]	[ ]	[ ]	3 xx	[ ]	[

4181. Ephemeris, a.d. 161

fr. 1

i	ii	iii	iv
Ζ̣εύc̣	Ἄ̣[ρ]η̣c̣	[Ἀφροδίτη]	[Ἑρμῆc]
Κριο(ῦ)	Λ̣[έ]ον̣(τοc)	[Κριοῦ]	[Ταύρου]
ἀ]νατολ(ικόc)	ἀ̣ν̣ώ̣μ̣(αλοc) β̣	[ ]	[ ]
[ ]	[ ]	[ ]	[ ]
[ ]	[ ]	[ ]	[ ]
[ ]	[ ]	[ ]	[ ]
ζ κδ	[ ]	[ ]	[ ]
ζ λε	β̣ [ ]	[ ]	[ ]
ζ με	γ [ ]	[ ]	[ ]
ζ̣ νη	γ [ ]	[ ]	[ ]
η θ	δ̣ [ ]	[ ]	[ ]
η̣ κ̣	[ ]	[ ]	[ ]
[ ]	[ ]	[ ]	[ ]
[ ]	ζ [ ]	[ ]	α λ̣γ
[ ]	ζ μς	ι̣ ν̣β	β γ
[ ]	η [ι]η̣	ι̣β̣ δ̣	δ μ
[ ]	η̣ ̣ ̣	ιγ ις̣	ς κγ
[ ]	[ ]	ιδ̣ κη	η ι̣γ
[ ]	[ ]		δύcεω[c]
[ ]	[ ]	ιε μ	ι ζ
[ ]	[ ]	ις νβ	ιβ ι̣ ̣
[ ]	[ ]	ιη̣ δ	ιδ λ̣
[ ]	[ ]	[ιθ] ι̣ς̣	ις λε
[ ]	[ ]	[ ]	ιη μ
[ ]	[ ]	[ ]	[κ] με
[ ]	[ ]	[ ]	[κβ] ν
[ ]	[ ]	[ ]	[κδ νε]
[ ]	[ ]	[ ]	[κζ ō]
[ ]	[ ]	[ ]	[κθ] ε̣
			[Καρ]κ̣(ίνου)
[ ]	[ ]	[ ]	[α] ι̣
[ ]	[ ]	[ ]	[γ ι]ε
[ ]	[ ]	[ ]	[ε] κ̣
[ ]	[ ]	[ ]	[ζ κ]ε
[ ]	[ ]	[ ]	[θ] λ̣

1 text in red   2 κριο; λεον?   3 text in red; ανατο (with λ written above); α̣ν̣ω̣μ̣?   9 *l.* ζ μς?   18a in red

4181. EPHEMERIS, A.D. 161

fr. 1

i	ii	iii	iv
Jupiter	**Mars**	**[Venus]**	**[Mercury]**
Aries	Leo	[Aries]	[Taurus]
morning	**non-uniform? 2?**	[ ]	[ ]
[ ]	[ ]	[ ]	[ ]
[ ]	[ ]	[ ]	[ ]
[ ]	[ ]	[ ]	[ ]
7 24	[ ]	[ ]	[ ]
7 35	2 [ ]	[ ]	[ ]
7 45	3 [ ]	[ ]	[ ]
7 58	3 [ ]	[ ]	[ ]
8 9	4 [ ]	[ ]	[ ]
8 20	[ ]	[ ]	[ ]
[ ]	[ ]	[ ]	[ ]
[ ]	7 [ ]	[ ]	1 33
[ ]	7 46	10 52	2! 3
[ ]	8 [1]8	12 4	4 40
[ ]	8 xx	13 16	6 23
[ ]	[ ]	14 28	8 13
			setting
[ ]	[ ]	15 40	10 7
[ ]	[ ]	16 52	12 1x
[ ]	[ ]	18 4	14 30
[ ]	[ ]	[19] 16	16 35
[ ]	[ ]	[ ]	18 40
[ ]	[ ]	[ ]	[20] 45
[ ]	[ ]	[ ]	[22] 50
[ ]	[ ]	[ ]	[24 55]
[ ]	[ ]	[ ]	[27 0]
[ ]	[ ]	[ ]	[29] 5
			Cancer
[ ]	[ ]	[ ]	[1] 10
[ ]	[ ]	[ ]	[3 1]5
[ ]	[ ]	[ ]	[5] 20
[ ]	[ ]	[ ]	[7 2]5
[ ]	[ ]	[ ]	[9] 30

fr. 2

i	ii	iii	iv	v	vi
[νῶναι]	[ιγ]	[κ]ζ̣	Δ̣[ιδ(ύμοις)]	[ ]	[
η	ι̣δ̣	κ̣η̣		[ ]	[
ζ	ι̣ε̣	κθ̣	Κ̣α̣[ρ(κίνῳ)]	[ ]	[
ς	ις	λ		[ ]	[
ε	ιζ	α	Λ̣έ̣[ο(ντι)]	[ ]	[
δ	ιη	β		[ ]	[
γ	ιθ	γ̣	Π̣α̣[ρ(θένῳ)]	[ ]	[
α	κ	δ		[ ]	[
ε̣ἰ̣δ̣ο̣ί̣	κα	ε	Ζ̣υ̣γ̣(ῷ)	[ ]	[
ιζ	κβ	ς̣		[ ]	[
ις	κγ	ζ		[ ]	[
ιε	κδ	η	Cκ̣ο̣(ρπίῳ)	[ ]	[
ιδ	κ̣ε̣	[θ]		[ ]	[
[ιγ]	[κς]	[ι]	Τ̣οξ(ότῃ)	[ ]	[
[ιβ]	[κζ]	[ια]		[ ]	[
[ια]	[κη]	[ιβ]	Α̣[ἰ]γ̣(όκερῳ)	[ ]	[
[ι]	[κθ]	[ιγ]		[ ]	[
[θ]	[λ]	[ιδ]	Ὑ̣δ̣ρ̣(οχόῳ)	[ ]	[
[η]	[α]	[ιε]		[ ]ς	[
[ζ]	[β]	[ις]		κγ̣ ̣γ̣	[
[ς]	[γ]	[ιζ]	[Ἰ]χθ(ύσι)	[ ] ̣β	ὥρ(α) ̣ [
[ε]	[δ]	[ιη]		ιζ μ	[

Names of zodiacal signs abbreviated by truncation. 9 ε̣ι̣δ̣ο̣ι̣ in red 21 ϕ

fr. 2

i	ii	iii	iv	v	vi
[nones]	[13]	[2]7	Gemini	[ ]	[
8	14	28		[ ]	[
7	15	29	Cancer	[ ]	[
6	16	30		[ ]	[
5	17	1	Leo	[ ]	[
4	18	2		[ ]	[
3	19	3	Virgo	[ ]	[
1	20	4		[ ]	[
ides	21	5	Libra	[ ]	[
17	22	6		[ ]	[
16	23	7		[ ]	[
15	24	8	Scorpio	[ ]	[
14	25	[9]		[ ]	[
[13]	[26]	[10]	Sagittarius	[ ]	[
[12]	[27]	[11]		[ ]	[
[11]	[28]	[12]	Capricorn	[ ]	[
[10]	[29]	[13]		[ ]	[
[9]	[30]	[14]	Aquarius	[ ]	[
[8]	[1]	[15]		[ ]6	[
[7]	[2]	[16]		23 x3	[
[6]	[3]	[17]	Pisces	[ ] x2	hour x [
[5]	[4]	[18]		17 40	[

4182. Ephemeris

Recto

]κ̣β̅ λεπ(τοῖc) ν̅ε̅

i	ii	iii	iv
Ἄ]ρ̣ηc	Ἀφροδ(ίτη)	Ἑρμῆ(c)	[ ]α̣
]τ	Αἰγ[ό]κ̣ε̣(ρῳ)	Ἴχθ(υcι)	. .
].	ἑῷ̣α	ἑῷο̣c̣	[
]	[ ] . [ ]	[ ]	[

1 λεπ′ 2–3 abbreviations by truncation?

Verso

i	ii	iii	iv
Μ̣	Φ	κ	cελή̣ν(ῆc)
Α	Α	.	ζῴδι[ον]
Ρ̣	Ρ	.	[ ]

3

1 cεληṿ′

4182. Ephemeris

Recto

]22;55

i	ii	iii	iv
Mars	Venus	Mercury	...
...	Capricorn	Pisces	...
...	morning	morning	[
]	...	[ ]	[

Verso

	i	ii	iii	iv
	M	PH	...	moon's
	A	A	...	sign
3	R	R	...	[ ]

4183. Ephemeris

fr. 1

Recto?

i	ii	iii
Μάρ–]	Φα–	κ̣α̣–
τι–	με–	τὰ
οϲ	νώθ	ϲ[ε–
		λῆ̣–
		[νην

fr. 1

Verso?

i	ii
] ̣ ̣[ ]	[
]μετα̣ ̣ ̣	ρ̣ι̣[
]π̣ερι	ι̣ ̣ ̣[
]ω̣	̣ ̣[

fr. 2

η̣[ ] ̣[
ἐπὶ μήκ̣[ει
ε ̣β̣ ̣[
πιπτ̣ω̣[
ἀρχὴ ἐν [
Καρκίν[
Ζεὺ̣[ϲ

4183. Ephemeris

fr. 1

Recto?

i	ii	iii
March	Phamenoth	Lunar

fr. 1

Verso? (not translatable)

fr. 2

...
in longitude? [
...
falling? [
beginning in [
Cancer [
Jupiter? [

4184. Ephemeris

i	ii	iii	iv
α̣	[ ]	[ ]	[
β	.	θ	λ̣ .
γ		κγ	μ .
δ	Ξ̲	ζ	μθ̣

4184a. Ephemeris, A.D. 293

i	ii	iii
[κα]		κθ̣ [
[κβ]	Αἰγόκ(ερῳ)	ια [
[κγ]		κγ [
κδ	Ὑδροχ(όῳ)	ε [
κε		ιζ̣ [
κϛ	Ἰχθύ(cι)	ō [
κζ		ιβ [
κη		κε [
κθ	Κριῷ	η [
λ		κα̣ [

2 αιγο$^{\kappa}$ 4 ϋδρο$^{\chi}$ 6 ϊχθυ

4184. Ephemeris

i	ii	iii	iv
1	[ ]	[ ]	[
2	...	9	3x
3		23	4x
4	三	7	49

4184a. Ephemeris, A.D. 293

i	ii	iii
[21]		29 [
[22]	Capricorn	11 [
[23]		23 [
24	Aquarius	5 [
25		17 [
26	Pisces	0 [
27		12 [
28		25 [
29	Aries	8 [
30		21 [

4185. SIGN-ENTRY ALMANAC, A.D. 129–133

fr. 1

i			ii	iii		
.	.	[̣]	[ ]	[		
.	.	δ			(ἔ̣τ̣ο̣c̣)	
		]			ιδ	
		]			—	
		]		η	η̣	[
		]			κ̣ ̣	[
		]		ε̣	θ̣	[
		]		ι̣α̣	κ[	

2 L

fr. 2

i			ii			iii			iv	
		]	[ζ	̣]ε̣	ζ	η	κ	η	[	
		]	[η]	ι̣ ̣	[η]	ι̣	β̣	θ	[	
γ	. .	γ]	[θ	. .	θ]	ι̣α̣	κ̣ε̣	ι̣	.	[
	κ̣η̣	δ̣	[ι	.	ι]	——			[	
δ̣	κ̣δ̣	ε	[ια]	α̣	ια	α	α	α	[	
ε̣	κ̣	ς		κ̣ζ̣	ιβ		η̣	β	δ̣	[
[ς	̣]	ζ̣	——			β	β	γ	ε̣	[
ζ̣	.	η̣	α̣	γ	α̣	γ̣	ε̣	δ	ς̣	[
	κ̣θ̣	θ̣		κ̣β̣	β̣		κθ	ε̣	ζ	[
η	κ̣δ̣	ι̣	β̣	ιβ	γ	δ	κ̣[̣]	ς̣	η	[
θ	ιθ̣	ια	δ	ις	δ̣	ε	ι̣ ̣	ζ	[	
[ι	̣ ̣]	ιβ	ε̣	β	ε	ς	ιγ	η	[	
[ια	. .	α]		κα	ς	ζ	ις	θ̣	[	
[ιβ	. .	β]	ς	ι	ζ	ι̣α̣	ι ̣	[ι]	[	
——			η	κβ̣	η	ι̣β̣	[̣ ̣]	ι̣[α]	[	
α̣	α̣	ιβ̣	[θ]	.	[θ]	[		]	[	

4185. Sign-Entry Almanac, a.d. 129–133

fr. 1

i			ii		iii		
x	x	[x]	[	]	[		
x	x	4				(year)	
		]				14	
		]				—	
		]			8	8	[
		]				2x	[
———	———	———			———	———	———
		]			5	9	[
		]			11	2[x	

fr. 2

i			ii			iii			iv	
		]	[7	x]5	7	8	20	8	[	
		]	[8]	1x	[8]	10	2	9	[	
3	xx	3]	[9	xx	9]	11	25	10	x	[
	28	4	[10	x	10]	———			[	
4	24	5	[11]	1	11	1	1	1	[	
5	20	6		27	12		8	2	4	[
[6	x]	7	———			2	2	3	5	[
7	x	8	1	3	1	3	5	4	6	[
	29	9		22	2		29	5	7	[
8	24	10	2	12	3	4	2[x]	6	8	[
9	13	11	4	16	4	5	1x	7	[	
[10	xx]	12	5	2	5	6	13	8	[	
[11	xx	1]		21	6	7	16	9	[	
[12	xx	2]	6	10	7	11	1x	[10]	[	
———			8	22	8	12	[xx]	1[1]	[	
1	1	12	[9]	x	[9]	[		]	[	

4186. Sign-Entry Almanac, a.d. 142–146

i			ii			iii			iv		
ς (ἔτοc)] μ̣α̣			ζ (ἔτοc) μ̣δ̣			η (ἔτοc) μ̣[			[θ (ἔτοc)		
	α]			[α]			[α]			[α	
α	α	η]	[α	α]	η̣	[α	α]	θ̣	[α	α	θ
	β]		[ια	ς]	θ̣		[β]		[ιβ	κ	ι
α	α	α]		[β]		α̣	α	β	[		
	γ]		[α	.	β]	[.	..]	γ̣		[β	
α	α	ιβ]	[ς	ε	γ]	[.	..	δ]	α̣	[α	γ
β	.]	α̣	[ι	ι	β]	[..	..	γ]	γ̣	δ̣	[δ
γ	..	β]		[γ]			γ̣		ζ	ια̣	[ε
ε	.	γ]	[α	..	θ]	[α	.].	[α]		γ	
ς	..]	δ̣	γ̣	κ̣ζ̣	η̣	[γ]	δ̣	β	α	α	[θ
.	.]δ̣	ε	ζ	δ	θ	δ̣	[ι]ς̣	γ	β	β̣	[ι
θ]	ι̣	ς	η	κγ	ι	ε	κη	δ	θ	ιε̣	[ια
ι	.]η̣	ζ	ι	ιε̣	ια	ζ	ι̣	ε	ια	θ	[ιβ
ιβ]	ς̣	η	ιβ	α	ιβ	η	ιη	ς		δ̣	
	δ]			δ		θ	κς	ζ	α	ι[.	ια
α	..	ι]β̣	α	α	ια	ια	ζ̣	[η]	β	β	[ιβ
β]	ι̣γ̣	α̣	β	ις	ιβ	ιβ	ιη	θ		κ[ζ	α
		]	γ	κδ	α		δ		γ	[..	β
		]	δ	ιθ̣	β̣	α	ιγ	α	[δ	..	γ
		]	ε	ιδ	γ	β	ς	β	ε	θ̣	[δ
		]	ς	ς	δ	γ	ε	γ	ς	β̣	[ε
		]	ζ	α	ε		κζ	δ		κ̣[.	ς
		]		κη	ς	δ	κβ	ε	ζ	.[.	ζ
		]	η	κγ	ζ	ε	ιη̣	ς	η	[..	η
		]	θ̣	ι̣ζ̣	η	ς	κ̣η̣	ζ	θ	[.	θ
		]	ι̣	ι̣α̣	θ̣	η	ια̣	ς	ι	δ̣	[ι
		]	ι̣α̣	ε̣	ι	θ	ις	ζ		κ̣[.	ια
		]		κ̣θ̣	ια	ι	κη	η	ι[α	..	ιβ
		]	[ι]β	κγ̣	ιβ	ια	κδ	θ	[		
		]		ε̣		[ι]β̣	κ̣[.	ι]	[		

1 S (*bis*)

4186. Sign-Entry Almanac, a.d. 142–146

i			ii			iii			iv		
6th year] 41$^{?}$			7th year 44$^{?}$			8th year 4[x$^{?}$			[9th year …		
	1]			[1]			[1]			[1	
1	1	8]	[1	1]	8	[1	1]	9	[1	1	9
	2]		[11	6]	9		[2]		[12	20	10
1	1	1]		[2]		1	1	2	[		
	3]		[1	x	2]	[x	xx]	3		[2	
1	1	12]	[6	5	3]	[x	xx	4]	1	[1	3
2	x]	1	[10	10	2]	[xx	xx	3]	3	4	[4
3	xx	2]		[3]			3		7	11	[5
5	x	3]	[1	xx	9]	[1	x]x	[1]		3	
6	xx]	4	3	27	8	[3]	4	2	1	1	[9
x	x]4	5	7	4	9	4	[1]6	3	2	2	[10
9]	10	6	8	23	10	5	28	4	9	15	[11
10	x]h	7	10	15	11	7	10	5	11	9	[12
12]	6	8	12	1	12	8	18	6		4	
	4]			4		9	26	7	1	1[x	11
1	xx	1]2	1	1	11	11	7	[8]	2	2	[12
2]	13	1	2	16	12	12	18	9		2[7	1
		]	3	24	1		4		3	[xx	2
		]	4	19	2	1	13	1	[4	xx	3
		]	5	14	3	2	6	2	5	9	[4
		]	6	6	4	3	5	3	6	2	[5
		]	7	1	5		27	4		2[x	6
		]		28	6	4	22	5	7	x[x	7
		]	8	23	7	5	18	6	8	[xx	8
		]	9	17	8	6	28	7	9	[x	9
		]	10	11	9	8	11	6	10	4	[10
		]	11	5	10	9	16	7		2[x	11
		]		29	11	10	28	8	1[1	xx	12
		]	[1]2	23	12	11	24	9	[		
		]		5		[1]2	2[x	10]	[		

4187. Sign-Entry Almanac, a.d. 160/161

	α (ἔτοc)	
α	α	[γ
ϛ	κη	[η
ι	κε	[θ
α	κδ	[ι
γ̣	ιε	[θ
ϛ	κε	[ι
η̣	ιγ	[ια
ι̣	ε	[ιβ
ι̣α	κζ	[α
α]	ιη	[γ

— — — — —

1 L

4187. SIGN-ENTRY ALMANAC, A.D. 160/161

Year 1		
1	1	[1
6	28	[8
10	25	[9
1	24	[10
3	15	[9
6	25	[10
8	13	[11
10	5	[12
11	27	[1
1]	18	[3

— — — — —

4188. Sign-Entry Almanac, A.D. 161–165

Ο]ὐ̣ή̣ρο[υ τ]ῶ̣ν κ̣υ̣ρ̣ί̣ων cεβαc̣τ̣ῶν

i			ii			iii			iv		⇒
β	ι̣	β	ι̣α	.	α̣	ζ	.	η		κϛ	δ
	κ̣θ̣	γ̣	ι̣β̣	ι .	β	θ	.	θ	γ	κ	ε
γ̣	ιη	δ̣	γ	η̣	β		κ̣β	[ι]	δ	ιε	ϛ
ε	.	ε̣	δ	ιδ	γ̣	ι̣	[ι]β	[ι]α	ε	ι	ζ
ϛ	.	[ϛ]	ε̣	θ̣	[δ]	[ια]	. .	ιβ	ϛ	ζ	η
	[. .]	ζ̣	ϛ̣	[.	ε]	. .	(ἔτοc)	δ	ζ	ζ̣	θ
		]		κ̣[. .	ϛ]	[.	. .]	ε̣	η	ζ	η´
					]	[. .	. .]	δ[ʹ]	ι̣	θ̣	θ
					]	[. .	. .]	ι̣[β]	ι̣α	[. .	ι]
								]	[ιβ	. .	ια]
								]	[α	.	α]
								]	[	. .	β]
								]	[β	.	γ]
								]	[		]
								]	[		]
								]	[δ	. .	δ]
								]	[ε	.]	ε̣
								]	[	. .]	ϛ̣
								]	[ϛ	.]	ζ̣
								]	[η	. .]	η

7 L

4188. Sign-Entry Almanac, a.d. 161–165

… and] Verus, lords Augusti

i			ii			iii			iv			⇒
2	10	2	11	x	1	7	x	8		26	4	
	29	3	12	1x	2	9	x	9	3	20	5	
3	18	4	3	8	2		22	[10]	4	15	6	
5	x	5	4	14	3	10	[1]2	[1]1	5	10	7	
6	x	[6]	5	9	[4]	[11]	xx	12	6	7	8	
	[xx]	7	6	[x	5]	… (Year) 4			7	7	9	
		]		2[x	6]	[x	xx]	5	8	7	8′	
					]	[xx	xx]	4[']	10	9	9	
					]	[xx	xx]	1[2]	11	[xx	10]	
								]	[12	xx	11]	
								]	[1	x	1]	
								]	[	xx	2]	
								]	[2	x	3]	
								]	[		]	
								]	[		]	
								]	[4	xx	4]	
								]	[5	x]	5	
								]	[	xx]	6	
								]	[6	x]	7	
								]	[8	xx]	8	

⇐ v			vi			vii
(ἔτοϲ) ε						
β	ιη	ε̣	α	α	α	[
α	α	ιβ		κβ	β	̣[
β	λ	ιβ	β	ι	γ	[
ϛ	ιζ	ια´	δ	ε	δ	[
η	ιζ	ιβ	ε	ι	ε	[
ι̣	ι[̣]	α̣		κ̣ζ̣	ϛ̣	[
[ιβ	̣ ̣]	β	ϛ	β̣	ζ	̣[
[α]	̣	ι̣β̣	ζ	ϛ̣	[ϛ]´	[
	κ̣η̣	α̣		̣ ̣	ζ̣	[
[β	̣ ̣	β]	[— — — — — —			
[γ	̣ ̣	γ]	[			
[δ	̣	δ]	[			
[ε	̣	ε]	[			
[	̣ ̣	ϛ]	[			
ϛ	κβ	ζ	[			
ζ	ιζ	η	[			
η	ι	θ	[			
θ	ε	ι	[			
	κθ	ια	[			
ι	κα	ιβ	[			
ια	ι̣θ̣	α	[			
ιβ	ιδ	β̣	[			

— — — — — — —

1 ι

⇐

v			vi			vii
Year 5						
2	18	5	1	1	1	[
1	1	12		22	2	x[
2	30	12	2	10	3	
6	17	11′	4	5	4	[
8	17	12	5	10	5	[
10	1[x]	1		27	6	[
[12	xx]	2	6	2	7	x[
[1]	x	12	7	6	[6]′	[
	28	1		xx	7	[
[2	xx	2]	[			
[3	xx	3]	[			
[4	x	4]	[			
[5	x	5]	[			
[	xx	6]	[			
6	22	7	[			
7	17	8	[			
8	10	9	[			
9	5	10	[			
	29	11	[			
10	21	12	[			
11	19	1	[			
12	14	2	[			

4188a. Sign-Entry Almanac, A.D. 201–208

Along the fibres

i	ii	iii
]	ιη ζ	[
]	η ιζ η	ι̣[
]	θ ζ θ	[
]	κ̣ . ι	α̣ [
]	ι̣ . ι̣α̣	β̣ [
]	ιβ ιθ ιβ	γ̣ [
]		δ̣ [
]	ιβ//	ε̣ [
. . ε̣	[α] α θ	[
ϛ̣ ιε ϛ	α α β	[
ζ ζ ζ	β ιβ γ	η̣ [
η δ η	β κ ϛ	θ̣ [
κ̣ϛ̣ θ	γ κε ζ	ι [
θ κ̣ ι	ε ιδ η	ι̣[
] . ι̣α̣	ϛ κδ [θ]	[

Col. ii, 8: ιβ (but not the following diagonal strokes) overwritten in red ink.

4188a. Sign-Entry Almanac, a.d. 201–208

Along the fibres

i	ii	iii
]	18 7	[
]	8 17 8	1[x
]	9 7 9	[
]	2x 10	1 [
]	10 x 11	2 [
]	12 19 12	3 [
]		4 [
]	(year) 12	5 [
x x 5	[1] 1 9	[
6 15 6	1 1 2	[
7 7 7	2 12 3	8 [
8 4 8	2 20 6	9 [
26 9	3 25 7	10 [
9 20 10	5 14 8	1[x
]x 11	6 24 [9]	[

Across the fibres

iv	v
ι]	θ κα ι̣
ι]α̣	ι ιε ι̣α̣
ιβ]	ια .[. ιβ]
] α̣	ιβ .[. α]
	κη β
[ις/]/	β ι β
	κθ [γ]
ι]α̣	γ ιη δ
ι]	δ̣ κα ε
ι]α̣	ς ια ς
]η̣	κη ζ
] ι	ζ ια η
] ια	θ ζ θ
]	[ι] ς ι
]	[..] ι̣α̣

Across the fibres

iv	v
10]	9 21 10
1]1	10 15 11
12]	11 x[x 12]
] 1	12 x[x 1]
	28 2
[(year) 16]	2 10 2
	29 [3]
1]1	3 18 4
10]	4 21 5
1]1	6 11 6
]8	28 7
] 10	7 11 8
] 11	9 7 9
]	[10] 6 10
]	[xx] 11

4189. Sign-Entry Almanac, a.d. 136–144 (?)

Recto:

i			ii		
ε̣	[. .	ζ]	[		]
ϛ̣	κζ	ϛ	[		]
η	ε	ζ̣	[		]
θ	ιϛ	η	ϛ̣	[	]
ι	ιβ	θ	ζ	ε	η
ια	ζ	ι		κθ	θ
ιβ	α	ια	η	κδ	ι
	κϛ	ιβ	θ	ιη	ια
α	γ	ιβ	ι	ιϛ	ιβ
	ϛ	α	ια	η	[α]
	κθ	β	ιβ	ιϛ	β̣
β	ιε	γ	α̣	[ ]	ι̣β̣
δ	κ	δ	.	[	α]
ε̣	θ	ε	.	. .	[β]
	κθ	ϛ	[ ]	ιζ	[γ]
ϛ̣	ιβ	ζ	δ	ιε̣	δ
η	ι̣δ	η		λ	ε
θ	α	θ	ε̣	ιη	ϛ
	[ι]ζ	ι	ϛ	δ	ζ
[ι	.]	ι̣α		κα̣	ϛ
[	]	ιβ	ζ	. .	[ζ]
[		ι]α̣	[		]

4189. Sign-Entry Almanac, a.d. 136–144 (?)

Recto:

i			ii		
5	[xx	7]	[		]
6	27	6	[		]
8	5	7	[		]
9	16	8	6	[	]
10	12	9	7	5	8
11	7	10		29	9
12	1	11	8	24	10
	26	12	9	18	11
1	3	12	10	16	12
	6	1	11	8	[1]
	29	2	12	16	2
2	15	3	1	[ ]	12
4	20	4	x	[	1]
5	9	5	x	xx	[2]
	29	6	[ ]	17	[3]
6	12	7	4	15	4
8	14	8		30	5
9	1	9	5	18	6
	[1]7	10	6	4	7
[10	x]	11		21	6
[	]	12	7	xx	[7]
[		1]1	[		]

fr. 1, verso:

i	ii			iii		
]	[		]	[ζ	.]	ζ
]		κ̣ς̣	[ζ]	[η]	.	η̣
]	ς̣	κ̣β	η		κθ	θ
ε	ζ	ιζ	θ	ι	η	ι
ς̣	η	ιβ	ι		ιζ	ια
]	θ	ις	ια	ια	ιβ	ιβ
]	ι	κε	ιβ	ιβ	ς	α
]	[ιβ]	. .	ια	$\overline{\text{ιγ}}$	α	β
]	[	]α̣	α	α	ζ	β
]	[		]β̣	β	ιγ	γ
]	[		]	δ	ιη	δ̣
]	[		]	ε	ιε	ε
]	[		]		κδ̣	ς
]	[ ]	.	δ	ς̣	ιδ	ζ
]	ε̣	ι̣δ̣	ε	[η	ι]α̣	η̣
]	ς̣	α̣	ς		κη	θ
]		ι̣ζ	ζ	[θ]	ιε	ι
]	[ ]	.	[ ]	[ι]	β̣	ι̣α̣
]	[		]	[. .]	.	[.]

fr. 1, verso:

i	ii			iii		
]	[		]	[7	x]	7
]		26	[7]	[8]	x	8
]	6	22	8		29	9
5	7	17	9	10	8	10
6	8	12	10		17	11
]	9	16	11	11	12	12
]	10	25	12	12	6	1
]	[12]	xx	11	13	1	2
]	[	]1	1	1	7	2
]	[		]2	2	13	3
]	[		]	4	18	4
]	[		]	5	15	5
]	[		]		24	6
]	[ ]	x	4	6	14	7
]	5	14	5	[8	1]1	8
]	6	1	6		28	9
]		17	7	[9]	15	10
]	[ ]	x	[ ]	[10]	2	11
]	[		]	[xx]	x	[x]

4190. Sign-Entry Almanac, A.D. 241–243

Recto:

ε (ἔτος)			
[α]	α	ιβ̣	
[β]	β	α	ε
[ζ]	ζ η	ιβ	ε
[ι]	λ	α	ε
[α]	α	ε	
[ϛ]	ε ϛ	ϛ	α
[α]	α̣	ι	
	κ̣γ̣ κδ̣	ια̣	[ ]
[ε]	ι̣ ια	[ι]	[ ]
[ζ]	ι̣ ια̣	[ια]	[ ]
[θ]	ι̣ζ̣	[ιβ]	[ ]

1 ϛ″

Verso:

ϛ (ἔτος) Γορδιανο[ῦ			
α	α	α	
α	α	ϛ	
ϛ	ιϛ ιζ	ζ	ϛ
α	α	β	
[β]	[β]	γ	γ̣
[γ]	[ιβ ι]γ	δ	ϛ̣
[δ]	[κα κβ]	ε	ια
[ε]	[λ]	ϛ̣	ε
[ζ]	[η]	[ζ]	ια̣

1 ϛ″

4190. Sign-Entry Almanac, a.d. 241–243

Recto:

Year 5			
[1]	1	12	
[2]	2	1	5
[7]	7/8	12	5
[10]	30	1	5
[1]	1	5	
[6]	5/6	6	1
[1]	1	10	
	23/24	11	[ ]
[5]	10/11	[10]	[ ]
[7]	10/11	[11]	[ ]
[9]	17	[12]	[ ]

Verso:

Year 6 of Gordianus			
1	1	1	
1	1	6	
6	16/17	7	6
1	1	2	
[2]	[2]	3	3
[3]	[12/1]3	4	6
[4]	[21/22]	5	11
[5]	[30]	6	5
[7]	[8]	[7]	11

4191. Sign-Entry Almanac, A.D. 236–243

fr. 1, recto

i			ii			iii			iv		
	α′′]			[β′′]			[γ′′]			[δ]′′	
α	α	ια]	[α	α	ια]	[α	α	ι]α̣	[α]	α	ιβ
═══	═══	═══	═══	═══	═══	[.	. .	ι]β̣	═══	═══	═══
α	α	α]	[α	α	β]	═══	═══	═══	α	α	δ
γ	.	β]	δ̣	ι̣γ̣	γ	[α	α]	γ̣	ε	ις	ε
η	. .	α]	θ̣	ι	β	δ	λ	δ	═══	═══	═══
ια	.	β]	ι̣β	ιη	γ	ια	ιγ	γ	α	θ	β
═══	═══	═══	═══	═══	═══	ιβ	κε	δ	β	κγ	γ
α]	α̣	η	α	κ̣ε	β	═══	═══	═══	δ	δ	δ
.	]ζ	ζ	γ	ιβ	γ	α	α	θ	ε	ιγ	ε
.	]θ̣	η	δ	[. .	δ]	ʼ	κδ	ι	ς	κα	ς
ε	.].	θ	ς	[.	ε]	γ	ις	θ	ζ̣	κθ	ζ
ζ	.]ς	ι	ζ	ιδ	ς	ς̣	[.].	[ι]	[θ]	θ̣	η̣
θ	. .	ι]α̣	η	κ. .	[ζ]	η	ιγ	ι̣[α]	[ι	. .	θ]
ια	.]	ιβ	ι	β	η	ι	ε	ι̣[β]	[ιβ	.	ι]
ιβ]	ι̣θ	α	ια	ιε	θ	ια	κζ	α̣	[		]
═══	═══	═══	═══	═══	═══	═══	═══	═══	═══	═══	═══
α	.]	β	α	ς	β	α	α	α	α	[α	α]
β	.]	γ̣	γ	ι̣[.]	γ	β	ιζ	β	β	[.	β]
δ	.]	δ̣	δ	β	δ	γ	ε	γ		κ̣[.	γ]
	. .]	ε		κ. .	ε̣		. .	δ̣	γ̣	[. .	δ]
ε	. .]	ς̣	ς	κα	ς	[δ]	ιβ	ε	ε̣	κ̣β	ε
ζ	. .]	ζ̣	ζ	ια	ζ	ς	η	ς	ς	ια̣	ς
η	.]	η̣		κ̣ζ	η	ζ	δ	ζ		κδ	ζ
	. .]	θ	[η	ι]δ	θ		[. .]	η̣	ζ̣	ιδ̣	η̣
θ	.]	ι̣	θ	[.	ι]	η	[. .]	θ̣	θ̣	[. .	θ]
ια	. .	ι]α	ια	ιβ	ια̣	[		]	[		]
		]		κ̣η̣	ιβ	[		]	[		]
		]	[ιβ]	. .	[α]	[		]	[		]

10 θ corrected from α (?) by overwriting.

4191. Sign-Entry Almanac, a.d. 236–243

fr. 1, recto:

i			ii			iii			iv		
	1]			[2]			[3]			[4]	
1	1	11]	[1	1	11]	[1	1	1]1	[1]	1	12
═══	═══	═══	═══	═══	═══	[x	xx	1]2	═══	═══	═══
1	1	1]	[1	1	2]	═══	═══	═══	1	1	4
3	x	2]	4	13	3	[1	1]	3	5	16	5
8	xx	1]	9	10	2	4	30	4	═══	═══	═══
11	x	2]	12	18	3	11	13	3	1	9	2
═══	═══	═══	═══	═══	═══	12	25	4	2	23	3
1]	1	8	1	25	2	═══	═══	═══	4	4	4
x	]7	7	3	12	3	1	1	9	5	13	5
x	]9	8	4	[xx	4]		24	10	6	21	6
5	x]x	9	6	[x	5]	3	16	9	7	29	7
7	x]6	10	7	14	6	6	[x]x	[10]	[9]	9	8
9	xx	1]1	8	2x	[7]	8	13	1[1]	[10	xx	9]
11	x]	12	10	2	8	10	5	1[2]	[12	x	10]
12]	19	1	11	15	9	11	27	1	[		]
═══	═══	═══	═══	═══	═══	═══	═══	═══	═══	═══	═══
1	x]	2	1	6	2	1	1	1	1	[1	1]
2	x]	3	3	1[x]	3	2	17	2	2	[x	2]
4	x]	4	4	2	4	3	5	3		2[x	3]
	xx]	5		2x	5		xx	4	3	[xx	4]
5	xx]	6	6	21	6	[4]	12	5	5	22	5
7	xx]	7	7	11	7	6	8	6	6	11	6
8	x]	8		27	8	7	4	7		24	7
	xx]	9	[8	1]4	9		[xx]	8	7	14	8
9	x]	10	9	[x	10]	8	[xx]	9	9	[xx	9]
11	xx	1]1	11	12	11	[		]	[		]
		]		28	12	[		]	[		]
		]	[12]	xx	[1]	[		]	[		]

fr. 1, verso:

i			ii			iii		
	ε´´]			[ς´´]			[α´´]	
α	α	ιβ]	[α	α	α]	[α	α	α]
β	ι[.	α]	═══	═══	═══	[.	..	β]
ς	ι[.	ιβ]	[α	α	ς]	═══	═══	═══
ια	α	α̣	[ς	..	ζ]	[α	α	ζ]
═══	═══	═══	═══	═══	═══	[ς	..	η]
α	α	ε	α	ε	γ	[ια	..	θ]
ς	β̣	ς	γ̣	ιδ	δ	═══	═══	═══
═══	═══	═══	δ	κγ̣	ε̣	α	α̣	[ια
α	κα	ια	ε̣	..	ς̣	β	ι̣η	[ιβ
ς̣	ι̣ζ	ι	ζ	ς	ζ̣	ς	ιζ	ι̣[α
ζ̣	ιζ	ια	[η	..]	η̣	η	ιζ	ι[β
[θ	..	ιβ]	[θ	..]	θ̣	ι	ιη	α̣
ια	.]	α̣	[ια	..	ι]	ιβ	δ	[β
ι]β̣	κε	β	[ιβ]	κ̣θ̣	ια	═══	═══	═══
═══	═══	═══	═══	═══	═══	α	ι.	[β
α	ια	α	α	η	α	β	ζ	[γ
β	β	β		κζ	β	γ	[..	β
	κ̣β̣	γ	β	κζ	γ		ι[.	γ
γ̣	ια	δ	δ̣	κα̣	δ	δ̣	.[.	δ
δ	α	γ	ε̣	θ	ε		[..	ε
ʼ	κα	δ		κς	ς	ε̣	[..	ς
ε	ιζ	ε	ς	ιγ̣	ζ̣	[ς	.	ζ
ς	.	ς	η	ις	η		[..	ς
	..]	ζ̣	θ	β	θ̣	ζ̣	[..	ζ
		]		ιη	ι	η̣	[.	η
		]	ι̣	ζ	ια	[		
		]	ιβ	ιη	ιβ̣	[		
		]	ἐ̣π̣	[.	α]	[		

27 ἐπ (*scil.* ἐπαγόμεναι): the traces are uncertain.

fr. 1, verso:

i			ii			iii		
	5]			[6]			[1]	
1	1	12]	[1	1	1]	[1	1	1]
2	1[x	1]	===	===	===	[x	xx	2]
6	1[x	12]	[1	1	6]	===	===	===
11	1	1	[6	xx	7]	[1	1	7]
===	===	===	===	===	===	[6	xx	8]
1	1	5	1	5	3	[11	xx	9]
6	2	6	3	14	4	===	===	===
===	===	===	4	23	5	1	1	[11
1	21	11	5	xx	6	2	18	[12
6	17	10	7	6	7	6	17	1[1
7	17	11	[8	xx]	8	8	17	1[2
[9	xx	12]	[9	xx]	9	10	18	1
11	x]	1	[11	xx	10]	12	4	[2
1]2	25	2	[12]	29	11	===	===	===
===	===	===	===	===	===	1	1x	[2
1	11	1	1	8	1	2	7	[3
2	2	2		27	2	3	[xx	2
	22	3	2	27	3		1[x	3
3	11	4	4	21	4	4	x[x	4
4	1	3	5	9	5		[xx	5
ʼ	21	4		26	6	5	[xx	6
5	17	5	6	13	7	[6	x	7
6	x	6	8	16	8		[xx	6
	xx]	7	9	2	9	7	[xx	7
		]		18	10	8	[x	8
		]	10	7	11	[		
		]	12	18	12	[		
		]	ep	[x	1]	[		

fr. 2, recto

	κζ̣	[θ
θ̣	ιβ	ι̣
ι̣	ϛ	ια
ιβ	ια	ιβ
	κ̣θ̣	α

fr. 2, verso:

	i			ii		
	ια	̣ ̣	ια]	θ̣	[̣	ι
	ιβ	̣ ̣]	ι̣β̣	ι̣α	[̣ ̣	ια
		̣ ̣]	α̣		[̣ ̣	ιβ
4				ιβ̣	[̣ ̣	α

fr. 2, recto:

– –	– –	– –
	27	[9
9	12	10
10	6	11
12	11	12
	29	1

fr. 2, verso:

i			ii		
– –	– –	– –	– –	– –	– –
11	xx	11]	9	[x	10
12	xx]	12	11	[xx	11
	xx]	1		[xx	12
			12	[xx	1

4192. Sign-Entry Almanac, A.D. 276–280

recto:

[0]	i	ii	iii	iv	v
α]	α̣	β		α	[α
β]	[ι]ς	γ̣	ζ̣	α	[α
α]	α̣	δ		δ	[
δ]	[δ]	ε	β̣	θ̣	[
α]	α̣	β		ι̣β̣	[
	ς̣ ζ̣	γ	α̣	α	α̣
β]	[ι]ζ̣	δ	ι̣β	[α]	ι̣ε̣ ι̣ς
γ]	[κς] κ̣ζ̣	ε̣	.	γ̣	θ
ε]	ε̣	[ς]	ι̣α̣	ι̣	[
ς]	[ι]γ̣	ζ̣	.	ι̣[β]	[
]	[ ]	η̣	[ ]	[α]	α̣
]	[ ]	θ̣	[ ]	[ ]	[
]	[ ]	[ι]	[ ]	[ ]	[
]	[ ]	[ι]α̣	[ ]	[ ]	[

verso:

[0]	i	ii	iii	iv	v	vi
[α]	[α]	γ		α	α	[γ
[ε]	[ις]	δ	δ	β	[ι]ε ις	[δ
[θ]	[ιη]	γ	ιβ	α	α	[η
[α]	α̣	ς̣		θ	ιδ	θ̣
[δ]	κα	ζ	.	α	α	ι[β
[θ]	ς	η	β	β	ι̣γ ιδ	[α
[α]	α	γ		δ	ε ς	[β
	κε	δ	ιβ	ς	ε ς	[γ
[γ]	ε̣	ε	ε	θ	κ	[β
[δ]	[ι]γ ιδ	ς	ια̣	ι	κζ κ̣[η]	[γ
[ε]	[κ]β κγ	ζ	.	ιβ	λ̣	[δ
[ζ]	[γ̣]	η	.	α	[α]	[ια

4192. Sign-Entry Almanac, a.d. 276–280

recto:

[0]	i	ii	iii	iv	v
1]	1	2		1	[1
2]	[1]6	3	7	1	[1
1]	1	4		4	[
4]	[4]	5	2	9	[
1]	1	2		12	[
	6/7	3	1	1	1
2]	[1]7	4	12	[1]	15/16
3]	[26]/27	5	x	3	9
5]	5	[6]	11	10	[
6]	[1]3	7	x	1[2]	[
]	[ ]	8	[ ]	[1]	1
]	[ ]	9	[ ]	[ ]	[
]	[ ]	[10]	[ ]	[ ]	[
]	[ ]	[1]1	[ ]	[ ]	[

verso:

[0]	i	ii	iii	iv	v	vi
[1]	[1]	3		1	1	[3
[5]	[16]	4	4	2	[1]5/16	[4
[9]	[18]	3	12	1	1	[8
[1]	1	6		9	14	9
[4]	21	7	x	1	1	1[2
[9]	6	8	[2	2	13/14	[1
[1]	1	3		4	5/6	[2
	25	4	12	6	5/6	[3
[3]	5	5	5	9	20	[2
[4]	[1]3/14	6	11	10	27/2[8]	[3
[5]	[2]2/23	7	x	12	30	[4
[7]	[3]	8	x	1	[1]	[11

4193. Sign-Entry Almanac, a.d. 195–203

Folio A, recto

i			ii			iii			iv		
ι̣[		]									
[		]									
α	η̣	[ϛ]									
δ	ιδ	[ζ]									
θ	[̣	η]									
α	α̣	[β]									
	κ̣ζ̣	[γ]									
γ	ζ̣	[δ]	ι̣β	ιδ̣	[ια]						
δ	[ι]ζ	ε	α	ζ	α				ε̣	[̣	ε]
ε̣	κ̣ϛ	ϛ		κζ	β					κ̣ε̣	[ϛ]
ζ̣	η	[ζ]	β	ιγ	γ				ϛ̣	ιγ̣	ζ
η	θ̣	η	δ	κ	δ				η	ιϛ	η
θ̣	κβ	θ	ε	ι	ε̣				θ	β̣	θ
ια	γ̣	ι		κδ	[ϛ]					ιη	ι
ιβ	̣ ̣	[ι]α	ϛ	ιδ	ζ				ι	ϛ	ια
			η	ιϛ	η	[ε]	κ̣ζ	ϛ	ιβ	ιβ̣	ιβ̣
			θ	γ̣	θ	[ϛ	̣]γ̣	ζ			
				̣ ̣	ι						
			[ι	̣]	ι̣α̣						
			[ιβ	̣ ̣]	ι̣β̣						

4193. Sign-Entry Almanac, a.d. 195–203

Folio A, recto

i			ii			iii			iv		
1[x		]									
[		]									
1	8	[6]									
4	14	[7]									
9	[x	8]									
1	1	[2]									
	27	[3]									
3	7	[4]	12	14	[11]						
4	[1]7	5	1	7	1				5	[x	5]
5	26	6		27	2					25	[6]
7	8	[7]	2	13	3				6	13	7
8	9	8	4	20	4				8	16	8
9	22	9	5	10	5				9	2	9
11	3	10		24	[6]					18	10
12	xx	[1]1	6	14	7				10	6	11
			8	16	8	[5]	27	6	12	12	12
			9	3	9	[6	x]3	7			
				xx	10						
			[10	x]	11						
			[12	xx]	12						

Folio A, verso

i			ii			iii			iv		
									[δ	.]	δ
										[κ]δ̣	ε̣
									ε̣	ιβ	ς
									ζ̣	ε	ζ
									[η	.]	η̣
										[.].	θ
									[θ]	ς̣	ι
						[β	.].	γ	[ια	ι]ς̣	ια
						[γ]	ιδ	δ	ιβ	[.]	ιβ
[ς	..]	ζ̣	[		]	[δ]	η	ε		ι.	α
ζ̣	κ̣ς	η	.	[	]	[ε]	γ̣	ς			
θ	κ̣	θ	[		]		[κ]η̣	ζ			
ι	ε	ι	[		]	ς̣	κ̣β	η			
ια	α	ια	.	[	]	ζ̣	ιζ̣	θ			
[α	.]	γ̣	θ	[..	ι]	[η	..]	ι			
.	..	δ̣	ια	ι̣[.]	ι̣[α]	[θ	.].	ια			
ς	η	ε	ιβ	ις	ιβ	ιβ	.	ι̣			
[ζ]	ι̣δ	ς	ε̣π̣[α̣]ʹ	[.	α]		..	ια̣			

18 *scil.* ἐπαγόμεναι

Folio A, verso

i			ii			iii			iv		
									[4	x]	4
										[2]4	5
									5	12	6
									7	5	7
									[8	x]	8
										[x]x	9
									[9]	6	10
						[2	x]x	3	[11	1]6	11
						[3]	14	4	12	[x]	12
[6	xx]	7	[		]	[4]	8	5		1x	1
7	26	8	x	[	]	[5]	3	6			
9	20	9	[		]		[2]8	7			
10	5	10	[		]	6	22	8			
11	1	11	x	[	]	7	17	9			
[1	x]	3	9	[xx	10]	[8	xx]	10			
x	xx	4	11	1[x]	1[x]	[9	x]x	11			
6	8	5	12	16	12	12	x	10			
[7]	14	6	Epag. [x		1]		xx	11			

Folio B, recto

i		
α	[. .	δ]
γ	θ	[ε]
δ	ι[.	ς]
ε	κ[.	ζ]
ζ	δ	[η]
η	ιζ	θ
θ	κ̣ς̣	ι
ια	θ	[ι]α̣
επα´	β	ιβ
α	α	ιβ
β	ιβ	α
γ	ζ	β
δ̣	β̣	γ
	[κ]ζ̣	δ̣
ε	κ̣β̣	ε
[ς	. .]	ς
ζ̣	[.]	ζ

ii		
α̣	[α	β]
β	ι	α̣
	ιζ̣	[β]
γ̣	ιγ	[γ]
	κζ̣	δ̣
ε	ιζ	ε
ς	κ	ς
ζ	η	ζ
	κς	η
η	ιβ	θ
θ	.	ι̣
ια	ζ	ια̣

iii		
[.	]δ	α
[.	].	β
[.	]η̣	γ̣
[α	. .]	γ
[β	.].	δ
[γ	.]	ε
δ̣	α̣	ς
ε	ζ	ζ
ς̣	κ̣η̣	ς
η	ε	ζ
θ	ις	η
[ι	. .]	θ
[ια	.]	ι̣

iv		
[ε	.]	δ̣
	κβ	ε
ς	ιε	ς̣
ζ	α	ζ
	ιη	η̣
η	η	θ
ι	ιβ	ι
	κς	ια
ια̣	ια	ιβ
ιβ	θ̣	α̣
	κθ	ιβ

11 *scil.* ἐπαγόμεναι

Folio B, recto

	i			ii			iii			iv	
									— —	— —	—
						— —	— —	—	[5	x]	4
— —	—	— —				[x	]4	1		22	e5
1	[xx	4]				[x	]x	2	6	15	6
3	9	[5]				[x	]8	3	7	1	7
4	1[x	6]	— —	— —	— —	[1	xx]	3		18	8
5	2[x	7]	a	[1	2]	[2	x]x	4	8	8	9
7	4	[8]	2	10	1	[3	x]	5	10	12	10
8	17	9		17	[2]	4	1	6		26	11
9	26	10	3	13	[3]	5	7	7	11	11	12
11	9	[1]1		27	4	6	28	6	12	9	1
Epag. 2		12	5	17	5	8	5	7		29	12
1	1	12	6	20	6	9	16	8	— —	— —	— —
2	12	1	7	8	7	[10	xx]	9			
3	7	2		26	8	[11	x]	10			
4	2	3	8	12	9	— —	— —	—			
	[2]7	4	9	x	10						
5	22	5	11	7	11						
[6	xx]	6	— —	— —	— —						
7	[x]	7									
— —	— —	—									

Folio B, verso

	i			ii			iii			iv	
ε̣	.	ζ̣	α̣	[	]				[	]	α̣
ς̣	ι̣α̣	η̣	[		]				[		β]
ζ̣	κ̣.	θ̣	[		]				[	]	ιβ̣
[θ	.]	ι̣	.	[	]				[		α]
[ι]	κ̣α̣	ι̣α̣	ε	.[.	ε]				[	]	β̣
ιβ	.	ι̣β̣	ς	η	[ς]				[	]	γ
α	ι̣ς̣	α		κ̣η̣	ζ				[	]	δ̣
β̣	.	β̣	ζ̣	ι.	η̣	[	].	ε̣	ε̣	ς̣	[ε]
[γ	.]	γ̣	θ	κ̣	[θ]	[α]	α̣	α	ς	α	ς
	[.].	δ̣	ι	α	ι	[γ]	ι̣α	β		ιθ	ζ
δ	κα̣	ε		ι̣ε̣	ια	[δ]	κ̣α̣	γ	η	..	η
ε	ις	ς	ια	α̣	[ιβ]	[ε	]α	δ̣	θ	ζ	θ
						[	].	ε	[	..	ι
						[		ς]	.	.[.	ια]
						[	]ς̣	ζ	ι̣.	.	ι̣β̣

Folio B, verso

i			ii			iii			iv		
5	x	7	1	[	]				[	]	1
6	11	8	[		]				[		2]
7	2x	9	[		]				[	]	12
[9	x]	10	x	[	]				[		1]
[10]	21	11	5	x[x	5]				[	]	2
'12	x	12	6	8	[6]				[	]	3
1	16	1		28	7				[	]	4
2	x	2	7	1x	8	[	]x	5	5	6	[5]
[3	x]	3	9	20	[9]	[1]	1	1	6	1	6
	[x]x	4	10	1	10	[3]	11	2		19	7
4	21	5		15	11	[4]	21	3	8	xx	8
5	16	6	11	1	[12]	[5	]1	4	9	7	9
						[	]x	5	[	xx	10
						[		6]	x	x[x	11]
						[	]6	7	1x	x	12

4194. Sign-Entry Almanac, A.D. 282–286

Recto:

i	ii	iii	iv	v
]	[ ]	[ ]	ιβ	ι[
α̣	α	α		ι[
	ιη	β	α̣	̣[
β̣	ζ	γ	β	α
δ	ι̣δ̣	δ	ιβ	[
ε	α β̣	ε	ιβ	[
	ιη	ϛ	ε	[
ϛ	ϛ	ζ	β̣	[
η	ιγ	[η]	α̣	δ̣
	κθ	θ	ιβ	
θ	ιε	ι	η	ε̣
ι	β γ	ια	ι	ζ̣
ι̣β	ι ια	ιβ	ζ	η̣
	κη	α	β	
				θ̣

4194. Sign-Entry Almanac, a.d. 282–286

Recto:

i	ii	iii	iv	v
]	[ ]	[ ]	12	1[x
1	1	1		1[x
	18	2	1	x[
2	7	3	2	1
4	14	4	12	[
5	1/2	5	12	[
	18	6	5	[
6	6	7	2	[
8	13	[8]	1	4
	29	9	12	
9	15	10	8	5
10	2/3	11	10	7
12	10/11	12	7	8
	28	1	2	
				9

Verso:

i	ii	iii	iv	v
]	[ι]α̣	[ ]	[ ]	[
]	ιβ	[ ]	[ ]	[
]		[ ]	[ ]	[
]	α	α	[α]	[
]	β̣	ιβ ιγ	β̣	ε
]	γ	γ	γ	ζ
]		κ κα	δ	δ
]	δ	η θ	ε	β
]	ς	ις ιζ	ς	γ
]	ζ	γ	ζ	ζ
]		ιθ	η	β
]	η	ς	θ	ς̣
]	ι	ιγ ι̣δ	ι	α
] ̣		λ	ια	ζ̣
]α̣	ια	ις ιζ	ιβ	β
]α̣	ιβ	ε	α	γ̣
]β̣				

Verso:

i	ii	iii	iv	v
]	[1]1	[ ]	[ ]	[
]	12	[ ]	[ ]	[
]		[ ]	[ ]	[
]	1	1	[1]	[
]	2	12/13	2	5
]	3	3	3	7
]		20/21	4	4
]	4	8/9	5	2
]	6	16/17	6	3
]	7	3	7	7
]		19	8	2
]	8	6	9	6
]	10	13/14	10	1
]x		30	11	7
]1	11	16/17	12	2
]1	12	5	1	3
]2				

4195. Sign-Entry Almanac, a.d. 300–304

Recto:

[0]	i	ii	iii	iv	v
α]	[ ]	[ ]	[ ]	α̣	α
ζ]	[ ]	[ι]β̣	ς	α	α̣
ι]	[ ]	α	ι	β	κ̣[
α]	[α]	δ		ζ	κ[
β]	[κζ] κη	ε	ς	α	α̣
ζ]	[ιε]	ς	ι[ ]	β	κ̣[
ιβ]	[κθ] λ	ε	ς	ε	δ̣
α]	[α]	α		ζ	ι̣[
β]	[ ]	β̣	[ ]	[ ]	[

Verso:

[–I]	[0]	i	ii	iii	iv
α]	[α]	[α]		[α]	[α
ε]	[ ]	[β]	α	[α]	[
ε]	[ ]	[α]	ς	α	α̣
α]	[α]	[ζ]		η̣	ιβ ιγ
η]	[ ]	[η]	β	α	α
α]	[α]	α̣		η	ς ζ
α]	[ ]	[β]	ζ	ι	α [β
γ]	[ ]	[γ]	ιβ	[ι]α̣	ι̣θ̣ [κ

4195. Sign-Entry Almanac, a.d. 300–304

Recto:

[0]	i	ii	iii	iv	v
1]	[ ]	[ ]	[ ]	1	1
7]	[ ]	[1]2	6	1	1
10]	[ ]	1	10	2	2[x
1]	[1]	4		7	2[x
2]	[27]/28	5	6	1	1
7]	[15]	6	1[x]	2	2[x
12]	[29]/30	5	6	5	4
1]	[1]	1		7	1[x
2]	[ ]	2	[ ]	[ ]	[

Verso:

[–I]	[0]	i	ii	iii	iv
1]	[1]	[1]		[1]	[1
5]	[ ]	[2]	1	[1]	[
5]	[ ]	[1]	6	1	1
1]	[1]	[7]		8	12/13
8]	[ ]	[8]	2	1	1
1]	[1]	1		8	6/7
1]	[ ]	[2]	7	10	1/[2
3]	[ ]	[3]	12	[1]1	19/[20

4196. Sign-Entry Almanac, A.D. 218–220

Recto

i	ii
ε̣	β̣
ϛ	ε
ζ	η
η	ια
θ̣	[ ]
ι	[ ]
ια	[ι]α̣
ι̣β̣	[ ]

Verso

i	ii
γ	[
δ	ια̣
ε	ζ̣
ϛ	ε [ϛ
ι	ι [ια
ια	ιβ̣
ιβ̣	θ ι̣
ἐ̣[πα]´	ε

8 *scil.* ἐπαγόμεναι

4196. Sign-Entry Almanac, a.d. 218–220

Recto

i	ii
5	2
6	5
7	8
8	11
9	[ ]
10	[ ]
11	[ ]
12	[ ]

Verso

i	ii
3	[
4	11
5	7
6	5/[6
10	10/[11
11	12
12	9/10
Epag.	5

4196a. Sign-Entry Almanac

Along the fibres

1	ιβ

Across the fibres

	i	ii
	ια	ζ
	ι	α
	ια	θ
4	[ι]β	ς̣

4196a. Sign-Entry Almanac

Along the fibres

1	12

Across the fibres

	i	ii
	11	7
	10	1
	11	9
4	[1]2	6

4197. Perpetual Sign-Entry Almanac

Recto:

i				ii			⇒
						ο	
[νγ	α	.	ιβ]	ν]θ	β̣	ε̣	β
	[β	. .	α]		γ	ιθ̣	γ
	[δ	. .	β]		ε̣	.	δ
	[ια	. .	γ]		ϛ	[ι]γ̣	ε
[νδ	α	.	δ]		ζ	κγ	ϛ
	[β	. .	ε]		θ	δ	ζ
	[γ	. .	ϛ]		ι	ιγ	η
	[ε	.	ζ]		[ι]β̣	β	θ
	[ϛ	. .	η]	ξ̣	[ζ]	.	ι
	[ζ	. .	θ]		[η]	[. .]	ια
	[θ	. .	ι]		[ι]	[. .]	ιβ
	[ι	. .]	ι̣α̣		ι̣β̣	α̣	α
	ι̣β̣	θ̣	ιβ	ξ̣α̣	α	ιϛ	β
[νε]	β̣	γ	α		β	λ	γ
	γ	κα	β		δ	ι	[δ]
	ε	η	γ		ε	κ[.	ε]
	ζ̣	γ̣	δ		ζ	α	[ϛ]
[νϛ	α	.]	ε̣		η	η	[ζ]
	[β	. .	ϛ]		θ	ιε̣	η̣
	[δ	.	ζ]		ι̣α	λ	θ
	[ε	. .	η]		ιβ	ιη	ι̣
	ζ̣	β̣	θ	ξβ	α	α	ι
	η	γ̣	ι		η̣	β	ια
	[ι]	ι̣	ια		θ	κ̣γ	ι̣β
	[ια	.]	ιβ		ια	. .	α̣
[νζ	α]	.	α		. .	γ̣	β̣
	[β]	κϛ	β	ξγ	β̣	ι̣ϛ	γ̣
	[δ]	ιϛ	γ		γ	κ̣γ	δ
	[ε]	κ̣ε	δ		ε	α	ε
	[ζ]	.	ε		ϛ	θ̣	ϛ
	[η]	κ̣γ̣	ϛ̣		ζ	ιε	ζ
	[ι]	.	ζ		η	κγ	η
[νη	δ	. .	η]		ι	ιη	θ
	[ϛ	. .	θ]		ια	ιδ	ι
	[η	.	ι]	ξδ	α̣	δ	ια
	[θ	. .	ια]		γ̣	ϛ	ιβ
	[ια]	. .	ι̣β̣		ι̣α̣	.	α̣
	[ιβ]	κε	α		ι̣β̣	γ̣	β̣

4197. Perpetual Sign-Entry Almanac

Recto:

	i				ii		⇒
						o	
[53	1	x	12]	5]9	2	5	2
	[2	xx	1]		3	19	3
	[4	xx	2]		5	x	4
	[11	xx	3]		6	[1]3	5
[54	1	x	4]		7	23	6
	[2	xx	5]		9	4	7
	[3	xx	6]		10	13	8
	[5	x	7]		[1]2	2	9
	[6	xx	8]	60	[7]	x	10
	[7	xx	9]		[8]	[xx]	11
	[9	xx	10]		[10]	[xx]	12
	[10	xx]	11		12	1	1
	12	9	12	61	1	16	2
[55]	2	3	1		2	30	3
	3	21	2		4	10	[4]
	5	8	3		5	2[x	5]
	7	3	4		7	1	[6]
[56	1	x]	5		8	8	[7]
	[2	xx	6]		9	15	8
	[4	x	7]		11	30	9
	[5	xx	8]		12	18	10
	7	2	9	62	1	1	10
	8	3	10		8	2	11
	[10]	10	11		9	23	12
	[11	x]	12		11	xx	1
[57	1]	x	1		xx	3	2
	[2]	26	2	63	2	16	3
	[4]	16	3		3	23	4
	[5]	25	4		5	1	5
	[7]	x	5		6	9	6
	[8]	23	6		7	15	7
	[10]	x	7		8	23	8
[58	4	xx	8]		10	18	9
	[6	xx	9]		11	14	10
	[8	x	10]	64	1	4	11
	[9	xx	11]		3	6	12
	[11]	xx	12		11	x	1
	[12]	25	1		12	3	2

Recto (continued):

⇐ iii iv

ο							
ξε	α	κδ	γ	ο	β	ιβ	α
	γ	α̣	δ		γ	λ	β
	δ	[.]	ε		ε̣	κη	γ̣
	ε	κ̣	ς		ια	κζ	δ
	ς	κζ	ζ̣	οα	α̣	λ̣	ε
	η	δ	η̣		γ	ιβ̣	ς
	θ	ιδ	θ̣		ε̣	γ	ζ
	ι̣	λ	ι		ς̣	.	η
	[ιβ]	ις	ια		ζ̣	ι̣.	θ̣
ξς	[α]	.	ιβ		[θ]	..	ι
	ι̣	α̣	α		[ι]	ια	ια̣
	ια	κγ	β		ιβ̣	α	ιβ̣
ξζ	α	β	γ	οβ	α	ιη̣	α
	β	ιγ	δ̣		γ	ι̣.	β
	γ	κα	ε		δ̣	κ[.]	γ
	δ	λ	ς		ς	γ	δ̣
	ς	ζ̣	ζ		ζ	ι.	ε̣
	ζ	ιε̣	η̣		θ	κθ̣	ς
	[η]	κγ̣	θ	ογ̣	α	λ	ς
	ι	ς	ι		γ	κ̣.	ζ
	ια	κζ	ια		ε	.	η̣
ξη	α̣	ια̣	[ιβ]		ς̣	[..	θ]
	β̣	[ ]ε̣	[α]		[.	..	ι]
	δ̣	[..	β]		[.]	.	[ια]
	η̣	[..	α]		[.	..	ιβ]
	ι	ε̣	[β]	[οδ]	[.	..	β]
	ιβ	ια	[γ]		[.]	.	[γ]
ξθ	α	κα̣	δ̣		ε	[..	δ]
	γ	ι	[ε]		[.	..	ε]
	δ	ιγ	[ς]		[.	..	ς]
	ε	κς	[ζ]		[.	..]	ζ
	ς	α	[η]		ια̣	..	[η]
	η	ια	[θ]				
	θ	ιδ	[ι]				
	ια	α	[ια]				
	ιβ	κδ	[ιβ]				

Recto (continued):

⇐

iii

o			
65	1	24	3
	3	1	4
	4	[x]	5
	5	20	6
	6	27	7
	8	4	8
	9	14	9
	10	30	10
	[12]	16	11
66	[1]	x	12
	10	1	1
	11	23	2
67	1	2	3
	2	13	4
	3	21	5
	4	30	6
	6	7	7
	7	15	8
	[8]	23	9
	10	6	10
	11	27	11
68	1	11	[12]
	2	[ ]5	[1]
	4	[xx	2]
	8	[xx	1]
	10	5	[2]
	12	11	[3]
69	1	21	4
	3	10	[5]
	4	13	[6]
	5	26	[7]
	6	1	[8]
	8	11	[9]
	9	14	[10]
	11	1	[11]
	12	24	[12]

iv

70	2	12	1
	3	30	2
	5	28	3
	11	27	4
71	1	30	5
	3	12	6
	5	3	7
	6	x	8
	7	1x	9
	[9]	xx	10
	[10]	11	11
	12	1	12
72	1	18	1
	3	1x	2
	4	2[x]	3
	6	3	4
	7	1x	5
	9	29	6
73	1	30	6
	3	2x	7
	5	x	8
	6	[xx	9]
	[x	xx	10]
	[x]	x	[11]
	[x	xx	12]
[74]	[x	xx	2]
	[x]	x	[3]
	5	[xx	4]
	[x	xx	5]
	[x	xx	6]
	[x	xx]	7
	11	xx	[8]

Verso:

i ii ⇒

ο̣	Ἄ̣ρης				. Ἀφροδεί[τη]		
οε	ε̣	κα̣	θ̣	α̣	α	γ	ιβ
	ζ̣	ι̣ζ̣	ι̣			κη	α
	η	κζ	ι̣α		β̣	κγ	β
	ι	κ	ιβ		γ	ιζ	γ
	ιβ	ιθ	α		δ	ιβ	δ
ος	α	κγ	β		ε	γ	ε
	γ	α	γ			λ̣	ς
	δ	.	δ		ς̣	.	ζ
	ς	[̣]	ε		ζ̣	[. .]	η
	ζ	[̣]	ς		η̣	.	θ̣
	η̣	η̣	ζ		θ	.	ι
	θ	κβ̣	η			λ	ια
	ια	ιβ	θ		ι	κγ	ιβ
[οζ]	α	ια	ι		ια	κ	α̣
	δ	ιδ	θ		ι̣β̣	ιε	[β]
	ς	ιγ	ι	β̣	α	ι̣	[γ]
	η	ι	ια		β	ιβ	[δ]
	ι̣	.	ιβ		[ς]	.	[ε]
	ι̣α̣	ιθ	α̣		[ζ]	. .	[ς]
[οη	α	.	β]		[η]	η	ζ
	[β	. .	γ]		[θ]	.	η
	[γ	. .	δ]			κ̣ζ	θ̣
	[ε	.	ε]		[ι	. .]	ι
	[ς	. .	ς]		[ι]α̣	ιζ	[ι]α̣
	[ζ	. .	ζ]		ιβ	ια	ι̣β
	θ̣	[δ	η]	[γ]	α	α	[α]
	ι̣	[ι]ε	[θ]			κ. .	β̣
	ι̣α̣	[. .	ι]		β̣	.	γ̣
ο̣θ̣	α	ιε	[ια]		γ	ι̣δ̣	δ̣
	β	ιβ̣	[ιβ]		δ	ζ̣	ε̣
	ι	κγ	α̣		ε	β̣	ς̣
	ιβ	ιθ	[β]			κς	ζ
					ς	κ̣α	η
					ζ	ις	θ
					η	ια	ι
					θ	θ	[ι]α̣

1 *l.* Ἀφροδίτη

Verso:

i ii ⇒

o	Mars				Venus		
75	5	21	9	1	1	3	12
	7	17	10			28	1
	8	27	11		2	23	2
	10	20	12		3	17	3
	12	19	1		4	12	4
76	1	23	2		5	3	5
	3	1	3			30	6
	4	x	4		6	x	7
	6	[x]	5		7	[xx]	8
	7	[x]	6		8	x	9
	8	8	7		9	x	10
	9	22	8			30	11
	11	12	9		10	23	12
[77]	1	11	10		11	20	1
	4	14	9		12	15	[2]
	6	13	10	2	1	1x	[3]
	8	10	11		2	12	[4]
	10	x	12		[6]	x	[5]
	11	19	1		[7]	xx	[6]
[78	1	x	2]		[8]	8	7
	[2	xx	3]		[9]	x	8
	[3	xx	4]			27	9
	[5	x	5]		[10	xx]	10
	[6	xx	6]		[1]1	17	[1]1
	[7	xx	7]		12	11	12
	9	[4	8]	[3]	1	1	[1]
	10	[1]5	[9]			2x	2
	11	[xx	10]		2	x	3
79	1	15	[11]		3	14	4
	2	12	[12]		4	7	5
	10	23	1		5	2	6
	12	19	[2]			26	7
					6	21	8
					7	16	9
					8	11	10
					9	9	[1]1

Verso (continued):

⇐ iii iv

δ	α	[. .	ιβ]	[ς	
	β	.	[α]	[	
	γ	.	[β]	[	
	δ̣	.	[γ]	[	
		.	[δ]	[	
	ε	κα	[ε]	[	
	ς	ι.	[ς]	[	
	ζ	.	[ζ]	[	
	η̣	.	[η]	[	
		κ̣η	θ	[	
	θ̣	κα	ι		[
	ι	κγ	ια		[
	ια	ι	ιβ		[̣
	ιβ	δ	α	ζ	δ̣
		κθ	β		[ε
[ε]	α	ιη	γ		[ς
	β	ις	δ		[
	γ	η	ε̣		[ζ
	δ	ς	ς		η
	ε	κα	ε		θ
	ς	ι	ς		[ι
	η	ζ	ζ		[ια
	θ	ιη̣	η		[
	ι	ιβ	θ		[ιβ
	ια	ε	ι	η	[
		λ	ι̣α̣		[
	ιβ	κη	ι̣β̣		[

Verso (continued):

⇐ iii iv

4	1	[xx	12]	[6	
	2	x	[1]	[	
	3	x	[2]	[	
	4	x	[3]	[	
		x	[4]	[	
	5	21	[5]	[	
	6	1x	[6]	[	
	7	x	[7]	[	
	8	x	[8]	[	
		28	9	[	
	9	21	10		[
	10	23	11		[
	11	10	12		x[
	12	4	1	7	4
		29	2		[5
[5]	1	18	3		[6
	2	16	4		[
	3	8	5		[7
	4	6	6		8
	5	21	5		9
	6	10	6		[10
	8	7	7		[11
	9	18	8		[
	10	12	9		[12
	11	5	10	8	[
		30	11		[
	12	28	12		[

4198. PERPETUAL SIGN-ENTRY ALMANAC

Fr. 1, along the fibres

i	ii	iii	iv		v
	[ ]	ιζ̣	[ς]		
	ζ	θ	[ζ]		
	η	ζ	η		κη̣
		[κ]ε̣	θ		
	θ	[ ]η	ι		
	ι̣[α]	[ ]η	ια		
	ιβ	[ ]	ιβ		
		κη	α		
[ ]	α	[ ]	β̣		
	[γ]	[ ]	γ̣		
	[δ]	[ ]	δ		
		[ ]	ε		
	[ε]	ιζ	ς		
	[ζ]	δ	ζ		
		λ	η		
	[η]	ιζ	θ		
	[θ]	ς	ι		

Fr. 2, along the fibres

i	ii
]	ε̣
ι̣ζ̣	ς
ζ	ζ
κδ	η̣
.	θ
]	ι
]	ι̣α

4198. Perpetual Sign-Entry Almanac

Fr. 1, along the fibres

i	ii	iii	iv		v
	[ ]	17	[6]		
	7	9	[7]		
	8	7	8		28
		[2]5	9		
	9	[ ]8	10		
	1[1]	[ ]8	11		
	12	[ ]	12		
		28	1		
[ ]	1	[ ]	2		
	[3]	[ ]	3		
	[4]	[ ]	4		
		[ ]	5		
	[5]	17	6		
	[7]	4	7		
		30	8		
	[8]	17	9		
	[9]	6	10		

Fr. 2, along the fibres

i	ii
]	5
17	6
7	7
24	8
x	9
]	10
]	11

Fr. 1, across the fibres

i	ii	iii	iv
	ϛ	[ ]	[ ]
	ζ	[ ]	[ ]
	η	[ ]	θ̣
	[ι]	[ ]	ι
	[ι]α̣	ι̣β̣	ια
		κζ	ι̣[β]
	ι̣β	ι	[α]
[ ]	β	ι̣ζ	[β]
	[γ]	ε	γ̣
		κδ	δ̣
	δ̣	ιβ̣	ε̣
	ϛ̣	η	ϛ̣
	ζ̣	.	ζ
		κα	η̣
	η̣	θ	[θ]
	ι̣	ιε̣	[ι]
	ια	α	ι̣α̣
		κδ	ι̣β̣
	ιβ	ϛ	α
[ ]	β	.	β

Fr. 2, across the fibres

i	ii	iii
]	ζ̣	β̣
	κ̣θ	γ
γ]	ι	δ
ε]	ι̣δ̣	ε
	λ	[
ϛ]	ι̣ .	[
]	.	[

Fr. 1, across the fibres

i	ii	iii	iv
	6	[ ]	[ ]
	7	[ ]	[ ]
	8	[ ]	9
	[10]	[ ]	10
	[1]1	12	11
		27	1[2]
	12	10	[1]
[ ]	2	17	[2]
	[3]	5	3
		24	4
	4	12	5
	6	8	6
	7	x	7
		21	8
	8	9	[9]
	10	15	[10]
	11	1	11
		24	12
	12	6	1
[ ]	2	x	2

Fr. 2, across the fibres

i	ii	iii
]	7	2
	29	3
3]	10	4
5]	14	5
	30	[
6]	1x	[
]	x	[

4199. MONTHLY ALMANAC, JUPITER, 14 B.C. – 6 B.C.

Recto

i	ii	iii
	[ι]ζ	
[α]	[ ]	Λέο(ντος) [
[β]	[ ]	Παρ(θένου) [
[γ]	[ ]	Παρ(θένου) [
[δ]	[ ]	Παρ(θένου) [
[ε]	[ ]	Παρ(θένου) [
[s]	[ ]	[Πα]ρ̣(θένου) [
[ζ]	[ ]	Λέο(ντος) [
[η]	[ ]	Λέο(ντος) κ̣ ̣[
[θ]	[ ]	Λέο(ντος) [
[ι]	[ ]	Παρ(θένου) ⸓̣ [
[ια]	[ ]	Παρ(θένου) [
[ιβ]	[ ]	Παρ(θένου) [
	ιη	
[α]	[ ]	Πα̣ρ̣(θένου) [
[β]	[ ]	[Παρ(θένου)
[γ]	[ ]	[Πα]ρ̣(θένου) [
[δ]	.	Ζ̣υ̣γ̣(οῦ) [⸓
[ε]	.	[Ζ]υ̣γ̣(οῦ) α̣ κ̣[
[s]	ιβ	Ζ̣υ̣γ(οῦ) ε λβ̣ ς̣[τ(ηριγμῷ α´)
[ζ]	[ ]	[Ζ]υ̣γ(οῦ) λ ἀκ[ρ(όνυχος)
[η]	[ ]	Παρ(θένου) κθ λγ
[θ]	[ ]	Παρ̣(θένου) κε κε ς[τ(ηριγμῷ β´)
ι	ιβ	Παρ(θένου) κε λ
ια	α	Παρ(θένου) κs μ[
ιβ	α	Ζυγ(οῦ) ⸓ s

⇓

Names of zodiacal signs are abbreviated by truncation to the first three letters. Other regular abbreviations: ςτ′ = ςτηριγμῷ α´, ςτ′′ = ςτηριγμῷ β´, ακρ′ = ἀκρόνυχος, ⸓ = first or last visibility.

4199. Monthly Almanac, Jupiter, 14 B.C. – 6 B.C.

Recto

i	ii	iii
	[1]7	
[1]	[ ]	Leo [
[2]	[ ]	Virgo [
[3]	[ ]	Virgo [
[4]	[ ]	Virgo [
[5]	[ ]	Virgo [
[6]	[ ]	Virgo [
[7]	[ ]	Leo [
[8]	[ ]	Leo 2x [
[9]	[ ]	Leo [
[10]	[ ]	Virgo 0 [
[11]	[ ]	Virgo [
[12]	[ ]	Virgo [
	18	
[1]	[ ]	Virgo [
[2]	[ ]	[Virgo
[3]	[ ]	Virgo [
[4]	x	Libra [0
[5]	x	Libra 1 2[x
[6]	12	Libra 5 32 1st station
[7]	[ ]	Libra[1] 30 acronychal
[8]	[ ]	Virgo 29 33
[9]	[ ]	Virgo 25 25 2nd station
10	12	Virgo 25 30
11	1	Virgo 26 4x
12	1	Libra 0 6

⇓

Recto (continued)

i	ii	iii
	ιθ	
α	ιη	Ζυγ(οῦ) ια α̣ [
β	ια	Ζυγ(οῦ) ιs ι ↗
γ̣	α	Ζυγ(οῦ) κα̣ ᵕ [
δ	ια	Cκο̣(ρπίου) ᵕ .
ε	ιδ	[C]κ̣[ο(ρπίου)] γ α̣
s	ι̣[ ]	[C]κ[ο(ρπίου)] s .
ζ	α	Cκ[ο(ρπίου)] ζ ṣ cτ(ηριγμῷ α´)
η	λ	Cκο̣(ρπίου) δ κṣ ἀκρ(όνυχος)
θ	θ	Ζυ̣γ(οῦ) λ
ι	λ	Ζυγ(οῦ) κs κγ cτ(ηριγμῷ β´)
ια	α̣	[Ζυγ(οῦ)] κθ κζ
ιβ	ιε	Cκο(ρπίου) ᵕ .
	κ	
α	κ	C̣κ̣ο(ρπίου) ε̣ κ̣ .
β	.	Cκο(ρπίου) ι̣γ̣ κs
γ	[ ]	Cκο(ρπίου) [
δ	[ ]	[

Verso

[0]	i	ii
[α]	[ ]	[Τοξ(ότου)] μ̣θ
[β]	[ ]	[Τοξ(ότου)]
[γ]	[ ]	[Τοξ(ότου)] ↗
[δ]	[ ]	[Τοξ(ότου)].
[ε]	[ ]	[Τοξ(ότου)] Αἰγό(κερω) ᵕ̣ [ ]
[s]	[ ]	[Αἰγό(κερω)].
[ζ]	[ ]	[Αἰγό(κερω)].
[η]	[ ]	[Αἰγό(κερω)] λ̣β
[θ]	[ ]	[Αἰγό(κερω)]ζ cτ(ηριγμῷ α´)
[ι]	[ ]	[Αἰγό(κερω)] κ
[ια]	[ ]	[Αἰγό(κερω)]. ἀκρ(όνυχος)
[ιβ]	[ ]	[Αἰγό(κερω)]

⇓

Recto (continued)

i	ii	iii
	19	
1	18	Libra 11 1 [
2	11	Libra 16 10
3	1	Libra 21 0 [
4	11	Scorpio 0 x
5	14	Scorpio 3 1
6	1[x]	Scorpio 6 x
7	1	Scorpio 7 6 1st station
8	30	Scorpio 4 26 acronychal
9	9	Libra 30
10	30	Libra 26 23 2nd station
11	1	[Libra] 29 27
12	15	Scorpio 0 x
	20	
1	20	Scorpio 5 2x
2	x	Scorpio 13 26
3	[ ]	Scorpio [
4	[ ]	[

Verso

[0]	i	ii
[1]	[ ]	[Sagittarius] 49
[2]	[ ]	[Sagittarius]
[3]	[ ]	[Sagittarius]
[4]	[ ]	[Sagittarius]x
[5]	[ ]	[Sagittarius] Capricorn 0 [ ]
[6]	[ ]	[Capricorn]x
[7]	[ ]	[Capricorn]x
[8]	[ ]	[Capricorn] 32
[9]	[ ]	[Capricorn]7 1st station
[10]	[ ]	[Capricorn] 20
[11]	[ ]	[Capricorn]x acronychal
[12]	[ ]	[Capricorn]

⇓

Verso (continued)

⥣

[0]	i	ii
	[κβ]	
[α]	[ ]	[ ] ϲ̣τ̣(ηριγμῷ β´)
[β]	[ ]	[ ]
[γ]	[ ]	[ ]
[δ]	[ ]	[ ]. . [ ]
[ε]	[ ]	[Αἰγ(όκερω)] ⟋ . [ ] κ̣[ ]
[ϛ]	[ ]	[Ὑδρ(οχόου) .] ι̣α̣ ⟋
[ζ]	[ ]	[Ὑδρ(οχόου)] η κϛ
[η]	[ ]	[Ὑδρ(οχόου)] ιε ιε
[θ]	[ ]	[Ὑδρ(οχόου)] ιθ ν
[ι]	[ ]	[Ὑ]δ̣ρ̣(οχόου) κβ ϲτ(ηριγμῷ α´)
[ια]	[ ]	[Ὑ]δρ(οχόου) κα με
[ιβ]	[ ]	[Ὑδ]ρ̣(οχόου) ι . [ ] ἀκρ(όνυχος)
	[κγ]	
[α]	[ ]	[Ὑδρ(οχόου)] . . κϛ
[β]	[ ]	[Ὑδρ(οχόου)] . . ϲτ(ηριγμῷ β´)
[γ]	[ ]	[Ὑδρ(οχόου)] ιδ
[δ]	[ ]	[Ἰχθ(ύων)] νδ
[ε]	[ ]	[Ἰχ]θ̣(ύων) ι̣ ε
[ϛ]	[ ]	[Ἰ]χ̣θ̣(ύων) [ ] κγ . ⟋ λ
[ζ]	[ ]	[Ἰ]χθ(ύων) . . ⟋ λ
[η]	[ ]	[Ἰ]χθ(ύων) κ̣ κϛ
[θ]	[ ]	Ἰχθ(ύων) κ̣α̣ θ
[ι]	[ ]	[Ἰ]χ̣[θ(ύων)] κ̣ζ κϛ
[ια]	[ ]	[Ἰχθ(ύων)] ϲ̣τ̣(ηριγμῷ α´)
[ιβ]	[ ]	Ἰχθ(ύων) . . ε
	κ̣δ	
[α]	[ ].	Ἰχθ(ύων) κγ λ ἀκρ(όνυχος)
[β]	[ ]	Ἰχθ(ύων) κα η
[γ]	[ ]	Ἰχθ(ύων) ιζ ν ϲτ(ηριγμῷ β´)
[δ]	[ ]	Ἰχθ(ύων) ι̣ . . β̣ ζ
[ε]	[ ]	Ἰχθ(ύων) κ ζ [ ]
[ϛ]	[ ]	Κρι(οῦ) [ ]

Verso (continued)

⇑

[0]	i	ii
	[22]	
[1]	[ ]	[ ] 2nd station
[2]	[ ]	[ ]
[3]	[ ]	[ ]
[4]	[ ]	[ ]...[ ]
[5]	[ ]	[Capricorn] ⌒ 6 [ ] 2[x]
[6]	[ ]	[Aquarius x] 11 ⌒
[7]	[ ]	[Aquarius] 8 26
[8]	[ ]	[Aquarius] 15 15
[9]	[ ]	[Aquarius] 19 50
[10]	[ ]	Aquarius 22 1st station
[11]	[ ]	Aquarius 21 45
[12]	[ ]	Aquarius 1x [ ] acronychal
	[23]	
[1]	[ ]	[Aquarius] xx 26
[2]	[ ]	[Aquarius] xx 2nd station
[3]	[ ]	[Aquarius] 14
[4]	[ ]	[Pisces] 54
[5]	[ ]	Pisces 10 5
[6]	[ ]	Pisces [ ] 23 x ⌒ λ
[7]	[ ]	Pisces xx ⌒ λ
[8]	[ ]	Pisces 20 26
[9]	[ ]	Pisces 21 9
[10]	[ ]	Pisces 27 26
[11]	[ ]	[Pisces] 1st station
[12]	[ ]	Pisces x x5
	24	
[1]	[ ]x	Pisces 23 30 acronychal
[2]	[ ]	Pisces 21 8
[3]	[ ]	Pisces 17 50 2nd stations
[4]	[ ]	Pisces 1x x 2 7
[5]	[ ]	Pisces 20 7 [ ]
[6]	[ ]	Aries [ ]

– – – – – – – – – – – – – – – – – –

4200. Monthly Almanac, Jupiter

Recto

i		ii
]	[δ]	[ ]. . [
]	[ε]	[ ] ε̣ μο(ίραιϲ) ζ̣ [λ(επτοῖϲ) . .
]	ϛ̣	α̣ [ε μο(ίραιϲ)] γ λ̣(επτοῖϲ) νγ
]	ζ	α ε μο(ίραιϲ) θ λ(επτοῖϲ) νβ
]	η	α ε μο(ίραιϲ) ιγ λ(επτοῖϲ) λβ
]//	θ	κ ϲ(τηριγμῷ) α´ ε μο(ίραιϲ) κε λ(επτοῖϲ) . .
]	ι	α ε [. . μο(ίραιϲ) . . λ(επτοῖϲ) .]γ̣
]	ια	ζ ἀκρῴ(νυχοϲ) ε̣ μ̣ο̣(ίραιϲ) θ λ(επτοῖϲ) λ̣. .
]	ιβ	α ε μο(ίραιϲ) ϛ λ(επτοῖϲ) κ

Regular abbreviations: μ̊ = μοίραιϲ , λ̊ = λεπτοῖϲ , c = ϲτηριγμῷ , ακρω = ἀκρώνυχοϲ (*l.* ἀκρόνυχοϲ)

Verso

i		ii	
[θ	. . ζ] μ̣ο̣(ίραιϲ) ιϛ̣ [λ(επτοῖϲ) . .]	[	
[ι	. .] ζ̣ μο(ίραιϲ) κα λ(επτοῖϲ) ε̣		[
[ια	. . ϲ(τηριγμῷ)] α̣´ ζ μο(ίραιϲ) κ̣γ̣ λ̣(επτοῖϲ) ν̣η̣	α	α [
[ιβ	. .] ζ μο(ίραιϲ) κγ λ(επτοῖϲ) κ	β	κ̣ [
	(ἔτουϲ) ξ	γ	α [
α	ιγ ἀκ̣ρ̣ῴ̣(νυχοϲ) ζ μ̣ο̣(ίραιϲ) [. . λ(επτοῖϲ) .]η̣	δ	κδ [
β	α ζ μο(ίραιϲ) ιδ λ(επτοῖϲ) νζ	ε	θ [
γ	η ϲ(τηριγμῷ) β´ ζ μο(ίραιϲ) ιβ λ(επτοῖϲ) κ̣ζ	ϛ	ιϛ̣ [

4200. Monthly Almanac, Jupiter

Recto

i	ii	
]	[4]	[ ]...[
]	[5]	[ ] (sign) 5, 7 degrees [xx minutes
]	6	1 [(sign) 5], 3 [degrees] 53 minutes
]	7	1 (sign) 5, 9 degrees 52 minutes
]	8	1 (sign) 5, 13 degrees 32 minutes
]...	9	20 1st station (sign) 5, 25[!] degrees x minutes
]	10	1 (sign) 5, [xx degrees x]3 [minutes
]	11	7 acronychal (sign) 5, 9 degrees 3x minutes
]	12	1 (sign) 5, 6 degrees 20 minutes

Verso

i		ii	
[9	xx (sign) 7], 16 degrees [xx] minutes	[	
[10	xx] (sign) 7, 21 degrees 5 minutes	[	
[11	xx] 1st [station] (sign) 7, 23 degrees 58 minutes	1	1 [
[12	xx] (sign) 7, 23 degrees 20 minutes	2	2x [
	Year 60	3	1 [
1	13 acronychal (sign) 7, [xx] degrees [x]8 [minutes]	4	24 [
2	1 (sign) 7, 14 degrees 57 minutes	5	9 [
3	8 2nd station (sign) 7, 12 degrees 27 minutes	6	16 [

4201. MONTHLY ALMANAC, A.D. 126–128

i	ii
ε] α β̣ ζ̣ .	[ε
ϛ] ϛ̣ β̣ η̣ λ̣ c̣(τηριγμῷ) $\overline{\alpha}$	[ϛ] α̣ θ . [
ζ] κβ β̣ δ ἀκρώ(νυχος)	ζ α̣ θ . [
η] α β̣ γ	η α θ . [
θ] κϛ β̣ . . . κ	θ ιβ ι ▾̣ [
ι] α β̣ ▾ ϛ	ι α̣ ᵒ⁄ ι ϛ̣ [
ι̣]α α β̣ ▾ ν̣η	ια α ι ι[
ι]β̣ α̣ β̣ β̣ μ̣α	ιβ α ι [
α] (*illegible traces*)	α α . [
β] (*illegible traces*)	β η [

2 c̣(τηριγμῷ): apparently c 3 ακρω (*l.* ἀκρόνυχος) 6 ᵒ⁄ = last (?) visibility

(Cols. iii–iv too damaged for transcription.)

4201. MONTHLY ALMANAC, A.D. 126–128

i	ii
5] 1 (sign) 2, 7 x	[5
6] 6 (sign) 2, 8 30 1st station	[6] 1 (sign) 9, x [
7] 22 (sign) 2, 4 acronychal	7 1 (sign) 9, x [
8] 1 (sign) 2, 3	8 1 (sign) 9, x [
9] 26 (sign) 2, ...	9 12 (sign) 10, 0 [
10] 1 (sign) 2, 0 6	10 1 ♂ (sign) 10, 6 [
1]1 1 (sign) 2, 0 58	11 1 (sign) 10, 1[x
1]2 1 (sign) 2, 2 41	12 1 (sign) 10 [
1] (*illegible traces*)	1 1 ...[
2] (*illegible traces*)	2 8 [

4202. Monthly Almanac, a.d. 152–155

⇒

i	ii	iii
[Φαμε(νὼθ)]	[ ]	[ ]
[Φαρμο(ῦθι)]	[ ]	[ ] .
[Παχὼν]	[ ]	[Λέο(ντοc)] η .
[Παῦνι]	[ ]	[Λέ]ο(ντοc) θ . .
[Ἐπεὶφ]	[ ]	Λέ[ο(ντοc)] κα . δ
[Μεcορὴ]	[ ]	Λ[έ]ο(ντοc) κ . α
[Θὼθ]	[ ]	[Λ]έο(ντοc) [ ] .
[Φαῶφι]	[ ]	[ ] .
[Ἁθὺρ]	.	[ ] . .
Χοίακ	[ ]	[ ] . . .
Τῦβι	[ ]	[ ] . . λ c(τηριγμῷ) [α΄]
Μεχεὶρ	[ ]	[ ] . .
[Φ]αμε(νὼθ)	α	[Π]αρ(θένου) α . . .
[Φ]αρμο(ῦθι)	κδ	Παρ(θένου) ∗ λ c(τηριγμῷ) β΄
[Παχὼν]	.	Παρ(θένου) ∗ λη
[Παῦνι]	.	Παρ(θένου) α ιδ
[Ἐπεὶφ]	.	Παρ(θένου) γ ϛ
[Μεcορὴ]	.	Παρ(θένου) η .
[Θὼθ]	[ ]	[Π]αρ(θένου) [ ]
[Φαῶφι]	[ ]	[Παρ(θένου)] . .
[Ἁθὺρ]	[ ]	[Π]αρ(θένου) [ ] λ .
[Χοίακ]	[ ]	Πα[ρ(θένου)] μα
[Τῦβι]	[ ]	[Παρ(θένου)] κα λε
[Μεχεὶρ]	[ ]	Παρ(θένου) κ κθ
[Φαμε(νὼθ)]	[ ]	[Π]αρ(θένου) ιζ ν[ ]
[Φαρμο(ῦθι)]	[ ]	[Π]αρ(θένου) ιε ιδ
[Παχὼν]	[ ]	[Παρ(θένου)] ιβ . c(τηριγμῷ) β΄
[Παῦνι]	[ ]	[Π]αρ(θένου) ιγ ε
[Ἐπεὶ]φ	α	Παρ(θένου) ιδ β
[Μεcορὴ]	.	Παρ(θένου) κ νζ

Names of zodiacal signs regularly truncated to first three letters; long month names in col. i apparently also truncated. ∗ = cτηριγμῷ ; ℓ = first or last visibility

4202. Monthly Almanac, a.d. 152–155

⇒

i	ii	iii
[Phamenoth]	[ ]	[ ]
[Pharmuthi]	[ ]	[ ] x
[Pachon]	[ ]	[Leo] 8 x
[Payni]	[ ]	Leo 9 xx
[Epeiph]	[ ]	Leo 21 x4
[Mesore]	[ ]	Leo 2x x1
[Thoth]	[ ]	Leo [ ] x
[Phaophi]	[ ]	[ ] x
[Hathyr]	x	[ ] xx
Choeac	[ ]	[ ] ...
Tybi	[ ]	[ ] xx 30 1st station
Mecheir	[ ]	[ ] x x
Phamenoth	1	Virgo 1 x ...
Pharmuthi	24	Virgo 0 30 1st station
[Pachon]	x	Virgo 9 38
[Payni]	x	Virgo 1 14
[Epeiph]	x	Virgo 3 6
[Mesore]	x	Virgo 8 x
[Thoth]	[ ]	Virgo [ ]
[Phaophi]	[ ]	[Virgo] x x
[Hathyr]	[ ]	Virgo [ ] 3x
[Choeac]	[ ]	Virgo [ ] 41
[Tybi]	[ ]	[Virgo] 21 35
[Mecheir]	[ ]	Virgo 20 29
[Phamenoth]	[ ]	Virgo 17 5[x]
[Pharmuthi]	[ ]	Virgo 15 14
[Pachon]	[ ]	[Virgo] 12 x 2nd station
[Payni]	[ ]	Virgo 13 5
Epeiph	1	Virgo 14 2
[Mesore]	x	Virgo 20 57

⇐ ⇒

iv	v
[ ]	[Καρ(κίνου)] . . . [ ]
[ ]	[Καρ(κίνου)] . . [ ]
[ ]	[Λ]έο(ντος) ⸋ β
α	Λέο(ντος) α . .
.	Λέο(ντος) ιε ν ∠
α	Λέο(ντος) ι . . . .
[ ]	Λ[έο(ντος)] . ζ κ .
[ ]	Λ[έο(ντος)] . ϛ ζ
[ ]	[Λέο(ντος)] . . . ϛ
[ ]	[Λέο(ντος)] . .
[ ]	[Λέο(ντος)] . .
[ ]	[Λέο(ντος)] . . . . .
.	Λέο(ντος) [ ] .
.	Λέο(ντος) . . . c(τηριγμῷ) β´
α	Λέο(ντος) . . . .
α	Παρ(θένου) ⸋ ϛ
.	Παρ(θένου) γ νθ
[ ]	Παρ(θένου) η . [ ]
[ ]	Παρ(θένου) ϛ $\overline{κγ}$ ∠/ ιβ
.	Παρ(θένου) [ ] ζ
.	Ζυγ(οῦ) ⸋ .
α	Ζυγ(οῦ) α δ
. .	Ζυγ(οῦ) . . . δ
[ ]	Ζυ[γ(οῦ)] . . β
[ ]	Ζυγ(οῦ) . . . .
α	Παρ(θένου) κϛ
κϛ	Παρ(θένου) κγ κα c(τηριγμῷ) β´
ϛ	Παρ(θένου) κδ κα
α	Παρ(θένου) κϛ μζ
[ ]	Ζυγ(οῦ) . .

⇐ ⇒

iv	v
[ ]	[Cancer] … [ ]
[ ]	[Cancer xx [ ]
[ ]	Leo 0 2
1	Leo 1 xx
x	Leo 15 50 /
1	Leo 1x …
[ ]	Leo x7 20 …
[ ]	Le] x6 7
[ ]	[Leo] xx x6
[ ]	[Leo] x x
[ ]	[Leo] x x
[ ]	[Leo] …
x	Leo [ ] x
x	Leo … 2nd station
1	Leo …
1	Virgo 0 6
x	Virgo 3 59
[ ]	Virgo 8 x[ ]
[ ]	Virgo 6 23 // 12
x	Virgo [ ] 7
x	Libra 0 x
1	Libra 1 x4
xx	Libra …
[ ]	Libra … 2
[ ]	Libra …
1	Virgo 26
26	Virgo 23 21 2nd station
6	Virgo 24 21
1	Virgo 26 47
[ ]	Libra xx

⇐

vi	vii
[ ]	[ ]
[ ]	[ ]
[ ]	[ ]
.	[Δ]ιδ(ύμων) [ ]
[ ]	Καρ(κίνου) . [ ]
α	Καρ(κίνου) [ ]
α	Λ[έο(ντοϲ)] ϛ [ ]
ιδ	Παρ(θένου) ▾ . [ ]
α	Π[α]ρ(θένου) [ι]α κ. .
α	Ζυγ(οῦ) ▾ λ[.
α	Ζυγ(οῦ) ιδ [ ]
δ	Ϲκορ(πίου) ▾ [ ].
. .	Ϲκ[ορ(πίου)] κε [ ] ϲ(τηριγμῷ) α´ .
α	Ϲκορ(πίου) ιθ νε
. .	Ϲκορ(πίου) ις λς
ιδ	Ϲκορ(πίου) δ ϲ(τηριγμῷ) β´
ια	Ϲκ[ο]ρ(πίου) ζ .
κα	Ϲκορ(πίου) κ . . γ̄ Τοξ(ότου)
α	Τοξ(ότου) β με
α	Αἰγ(όκερω) ▾ μα
ιγ	Ὑδρ(οχόου) ▾ λζ
κϛ	Ἰχθ(ύων) ▾ γ
α	Ἰχθ(ύων) δ ι. .
ζ	Κρι(οῦ) ▾ λη
ις	Τα[ύ(ρου)] ▾ ζ
α	[Τ]αύ(ρου) θ ιδ
γ	Διδ(ύμων) ▾ δ κ̄s̄ ℓ ιζ
ιδ	[Κα]ρ(κίνου) ▾ ιδ
κγ	[Λέο(ντοϲ) ▾ .]´´
ε	[Λέο(ντοϲ) .]s . . β̄ [Παρ(θένου) ▾]

⇐

vi	vii
[ ]	[ ]
[ ]	[ ]
[ ]	[ ]
x	Gemini [ ]
[ ]	Cancer x [ ]
1	Cancer [ ]
1	Leo 6 [ ]
14	Virgo 0 x[ ]
1	Virgo [1]1 2x
1	Libra 0 3[x
1	Libra 14 [ ]
4	Scorpio 0 [ ]x
xx	Scorpio 25 [ ] 1st station
1	Scorpio 19 55
xx	Scorpio 16 36
14	Scorpio 4 2nd station
11	Scorpio 7 x
21	Scorpio 2x xx **3** Sagittarius
1	Sagittarius 2 45
1	Capricorn 0 41
13	Aquarius 0 37
26	Pisces 0 3
1	Pisces 4 1x
7	Aries 0 38
16	Taurus 0 7
1	Taurus 9 14
3	Gemini 0 4 **26** ℓ 17
14	Cancer 0 14
23	[Leo 0 x]...
5	[Leo x]6 ... **2** [Virgo 0]

4203. Monthly Almanac, Saturn, a.d. 215/216

Recto

] ἔτοϲ α[
] ἔτοϲ [
Παρθ]έ̣νῳ α ις̣ [
Παρθ]έ̣νῳ ιγ κ̣ [
Παρθ]έ̣νῳ ις ζ [
Παρθ]έ̣νῳ θ κ[
Παρθ]έ̣νῳ θ κζ [
Παρθ]ένῳ κ κθ̣ [
Παρθέ]ν̣ῳ̣ ι̣ε̣ ̣[
— — — — — — —

Verso

Ϲεουη]ρ̣οῦ κ̅δ̅ α ̣[
] ἔτοϲ [
Ζυγ]ῷ̣ α ιγ [
Ζυγ]ῷ̣ ε ι̣θ ⟋ [
Ζυγῷ̣] η ν̇ζ [
Ζυγ]ῷ̣ ια ις [
Ζυγ]ῷ ιγ ιζ [
Ζυγ]ῷ̣ ιδ νθ̣ [
— — — — — — — —

4 ⟋ = first visibility

4203. Monthly Almanac, Saturn, a.d. 215/216

Recto

] year ...[	
] year	[
] Virgo	1 16 [
] Virgo	13 20 [
] Virgo	16 7 [
] Virgo	9 2[x
] Virgo	9 27 [
] Virgo	20 29 [
] Virgo	15 x[

— — — — — — —

Verso

] Severus 24 ...[	
] year	[
] Libra	1 13 [
] Libra	5 19 / [
Libra]	8 57 [
] Libra	11 16 [
] Libra	13 17 [
] Libra	14 59 [

— — — — — — — —

4203a. Monthly Almanac, Mercury

] κ̣α̣̅ ιγ ⊽̣ [
] ̣ ̣̅ ̣ Cκο(ρπίου) [
] ̣ ι̣α̅ Τοξ(ότου) [
] ○○/ [
] κγ̅ ι̣ϛ̣ κ [
]̣̅ Ἰχθ(ύων) ⊽̣ ζ̣ [
] κθ̣̅ ζ λ [
] ○○/ [
] ○○/ ια̅ Τ̣[αύ(ρου)
] ●●/ ια̅ Κα̣[ρ(κίνου)
] ι̣α̣ ●●/ [
]γ̣ ○○/ ι̣[

] κ̣ϛ κ̣γ̣̅ Ζυγ(οῦ) [
] ⊽ θ κ̅ ̣ ι̣θ̣ [
] ̣ϛ λ̣ ●●/ κδ̅ ι̣[
] λϛ [
] ̣ κ ○○̣/ κε̅ Ὑδ[ρ(οχόου)
]ζ̣ ̣ ι̣̅ ̣ Ἰχθ(ύων) [
] ̣θ̣ ●●/ [
]ε ̣̅ ια νβ [
] ⊽̣ λβ ιζ̅ Δι̣δ(ύμων) ⊽ ν̣[
] κη [
] ●●/ κζ̅ ιε [
] ̣ κγ̅ ια ϛ [

Names of zodiacal signs regularly truncated to their first three letters. ○○/ = first or last morning visibility; ●●/ = first or last evening visibility.

4203a. MONTHLY ALMANAC, MERCURY

] **21** 13 0 [
]... **xx** Scorpio [

] ... **11** Sagittarius [
] 〃 [

] **23** 16 20 [
]x Pisces 0 7 [

] **29** 7 30 [
] 〃 [

] 〃 **11** Taurus [
] 〃 **11** Cancer [

] 11 〃 [
]2 〃 **1**[x

] 26 **23** Libra [
] 0 9 **2x** 19 [

]x6 30 〃 **24** 1[x
] 36 [

]x 20 〃 **25** Aquarius [
]7 ...**1x** Pisces [

] x9 〃 [
]5 x 11 52 [

] 0 32 **17** Gemini 0 5[x
] 28 [

] 〃 **27** 15 [
]x **23** 11 6 [

4204. ALMANAC, A.D. 54

i ⇒

]φ̣ω̣ ̣η ̣αια̣
]ν̣ ζῴδιον ἀπὸ
] ̣ ̣ ̣ Τιβε̣ρίου
] ̣ατο ̣
] ̣ος
Κρόνο]ς̣
] ̣ ̄ δύ̣νει Κ̣[ρ]ι̣ῶι
] ̣ ̄ ἀνατέλλει ἐν Κ̣ρ̣ι̣ῶ̣ι̣
] ̣ ̣ ̣ ̣ ̣ Κριῶι
Ζεύς]
δύ]νει Κριῶι̣
ἀν]α̣τέλλει Κριῶι δ̣̄
στηρί]ζει ἐν Κριῶι
]
]ε̣ταιτη

4204. ALMANAC, A.D. 54

i ⇒

]...
]... sign from
]... Tiberius
]...
]...
] Saturn
] x sets in Aries
]x rises in Aries
]... Aries
Jupiter]
] sets in Aries
] rises in Aries 4?
] is stationary in Aries
]
]...

⇐ ii ⇒

Ἄρη̣[c]
Φα̣μ̣ε̣ν̣[ὼθ] ̣εταιτη
Πα[χὼν]
Πα[ῦνι]
Ἐπ[εὶφ [ ] Πα]ρ̣θ̣έ̣[νῳ]
Μ̣ε[c]ω̣ρ̣[ὴ] ⟨ι⟩ Ζυγῷ
Ἀ[φ]ροδί̣τ̣η
Φαμ[εν]ὼθ ⟨ϛ⟩ Ὑδρηχό̣[ῳ]
⟨κα⟩ Ἰχθύcι
Φα̣[ρμο]ῦθι ⟨κη⟩ Κριῷ
Παχ̣[ὼν [ ]] Ταύρῳ
Πα[ῦνι [ ]] Διδύμ(οιc)
Ἐπ[εὶφ [ ]] Καρκίνωι
Μ[εcωρὴ [ ]] Λέοντοc
Ἐπα̣[γομ]ένων Πάρθενοc

15 *l.* Παρθένῳ

⇐ ii ⇒

Mars
Phamenoth [ ]...
Pachon [ ]
Payni [ ]
Epeiph xx in Virgo
Mesore 1x in Libra
Venus
Phamenoth 6 in Aquarius
21 in Pisces
Pharmuthi 28 in Aries
Pachon [xx] in Taurus
Payni [xx] in Gemini
Epeiph [xx] in Cancer
Mesore [x]x in Leo
Epagomenae in Virgo

⇐ iii

Ἑρμῆς
Φαμενὼθ ι̣γ̣ Ἰχθύων
κη Κριῶι
Φαρμ[ο]ῦθι ιγ Ταύρωι
κη̣ Διδύμων
Π̣α̣χὼν κβ Καρκίνωι ς
[Π]α̣ῦ̣ν̣ι̣ ζ̣ δ̣ύ̣ν̣ε̣ι̣ ἑσπέρᾳ
[ ]
κα στηριγμὸς Καρκίνο(υ) α
Ἐπεὶφ θ ἀνατ̣έλλει ἑῴ̣[ος . . . .]α
κδ Καρκί̣ν̣ωι ιζ
Μεσωρ̣ὴ̣ ̣ [Λ]έοντος
.[. Πάρ]θενος
.[ἀνατέλλ]ει ἑσπέρᾳ Παρ–
[θένου] ια ∓ μ̣

9 καρκιν^{o} 13 *l.* Παρθένῳ

⇐ iii

Mercury
Phamenoth 13 in Pisces
28 in Aries
Pharmuthi 13 in Taurus
28 in Gemini
Pachon in Cancer 6
Payni 7 evening setting
[ ]
21 station Cancer 1
Epeiph 9 morning rising ...
24 Cancer 17
Mesore x Leo
x[x] Virgo
...[ ] evening rising in Virgo
[ ] 11 0 $40^{?}$

4205. Five-Day Almanac, a.d. 257/258

Recto

[ε	Ϲκ(ορπίου)	ι]α̣	ιε
ι	Ϲ̣κ̣(ορπίου)	ιζ	μ
ιε	Ϲκ(ορπίου)	κδ	ō
κ	Τοξ(ότου)	ō	ιθ
κε̣	Τ̣[ο]ξ̣(ότου)	ϛ̣	λ̣ζ
λ̣	Τοξ(ότου)	ιβ	νη
Χοίακ			
ε	Τοξ̣(ότου)	[ι]θ	κ

Abbreviations: ϲκ = Ϲκορπίου, τοξ' = Τοξότου.

Verso

i				ii			
Ἐπείφ				[			
ε	Λέ(οντοϲ)	ιθ	λε	[			
ι	Λέ(οντοϲ)	κγ	δ	Φαρμο̣[ῦθι			
[ι]ε	Λέ(οντοϲ)	κε	νε	α	Ἰχ(θύων)	κζ̣	[
[κ]	Λέ(οντοϲ)	κη	κγ	ϛ	Κρ(ιοῦ)	ζ	[
κε	Παρ(θένου)	ō	ια	ια	Κρ(ιοῦ)	[ιϛ	
λ	Παρ̣(θένου)	α	ε	ι̣ϛ̣	Κ̣ρ̣(ιοῦ)	κ̣[ϛ	
Μεϲορή				κα	Ταύ(ρου)	ε̣	[
ε	Παρ(θένου)	α	ι	κϛ	Ταύ(ρου)	ιδ	[
ι	Πα]ρ̣(θένου)	ō	ιε	Π̣αχών			

Abbreviations: λε = Λέοντοϲ, παρ' = Παρθένου, ιχ' = Ἰχθύων, κρ' = Κριοῦ, ταυ' = Ταύρου.

4205. Five-Day Almanac, a.d. 257/258

Recto

5	Scorpio	1]1	15
10	Scorpio	17	40
15	Scorpio	24	0
20	Sagittarius	0	19
25	Sagittarius	6	37
30	Sagittarius	12	58
Choeac			
5	Sagittarius	[1]9	20

Verso

i				ii			
Epeiph				[			
5	Leo	19	35	[			
10	Leo	23	4	Pharmuthi			
[1]5	Leo	25	55	1	Pisces	27	[
[20]	Leo	28	23	6	Aries	7	[
25	Virgo	0	11	11	Aries	[16	
30	Virgo	1	5	16	Aries	2[6	
Mesore				21	Taurus	5	[
5	Virgo	1	10	26	Taurus	14	[
10]	Virgo	0	15	Pachon			

4205a. Five-Day Almanac

Along the fibres

]. .	.	
] κα	/α[	
] κ̣δ	Θώ[θ	
Μεc]ορ(ή)	ι	Λ̣[έοντι
] κ̣α α	κ	Λ̣[έοντι
] κγ	λ	[
]	Φα̣[ῶφι	

4]ορ΄

Across the fibres

i	ii	iii
[Τοξ](ότη)	κ. ι̣θ̣	[
[Τοξ](ότη)	ιζ ι̣β	[
[Τ]ο̣ξ(ότη)	ιδ α	.[
[Τ]οξ(ότη)	. νθ	[
[.].	[ ]	[

1–4 τοξ΄

4205a. Five-Day Almanac

Along the fibres

]xx	...	
] 21	/1? [	
] 24	Thoth	
Mesore	10	Leo?
] 21 1	20	Leo?
] 23	30	[
]	Phaophi	

Across the fibres

i	ii	iii
Sagittarius	2x 19	[
Sagittarius	17 12	[
Sagittarius	14 1	...[
Sagittarius	x 59	[
...	[ ]	[

4206. Five-Day Almanac, a.d. 272–274

Fr. 1, recto

i				ii			
Φα]μ̣ενώ̣θ				Θ̣ώ̣[θ			
α	ϛ]	ι̣α	μθ̣	α	β	ιε̣	[
ϛ	ϛ]	ιε	μ̣ϛ	ϛ	β	κ̣α	κ̣
ια	ϛ]	ι̣θ	μα	ια	β	κ̣ζ	ι̣[
ι]ϛ	[ϛ]	κ̣γ	λε̣	ιϛ	γ	γ	ι̣[
κ̣α	ϛ	κζ	κζ	κα̣	γ	θ̣	[
κ̣ϛ	[ζ]	α̣	κ:	κ̣ϛ	γ̣	[ιδ	
Φαρμ̣[οῦθι]				Φ̣α̣ῶ̣φ̣[ι			
α	ζ̣	ε̣	[ι]δ̣	α	[γ]	κ̣	[
ϛ	ζ	θ̣	ϛ̣	ϛ	γ̣	κ̣[ϛ	
ια	ζ	ι̣β̣	νζ̣	ια	δ	β̣	[
ιϛ	ζ	ιϛ	μζ̣	ι̣ϛ	δ̣	ζ̣	[
κα	ζ	κ:	λϛ̣	κ̣α̣	δ̣	[ιγ	
κϛ	ζ	κδ	ιθ	κ̣ϛ	δ̣	[ιη	
Παχών							
α	ζ	κ[η					
α]	η̣	α̣	[				

Fr. 2, recto

i				ii			
				κ]ϛ	θ	κϛ̣	[
[Τῦβι]				Ἐπεί̣φ			
α	ϛ	α]	μθ	α̣	ι	β	λ̣[
ϛ	ϛ	ō]	λ̣	ϛ̣	ι	η	λ̣[
ια	ε	κ]η	[ι]γ	ια	ι	ιδ	[
ιϛ	ε	κ]ε	λ̣ζ	ι̣ϛ	ι	κ	κ̣[
κα	ε	κ]β	μδ	κα̣	ι	[κϛ	
κϛ	ε]	κ̣	ι̣θ	κϛ	ια̣	[β	
Μεχε]ίρ				Μ̣ε[σορή			
α	ε	ιη]	λβ	α	[ια	η	

4206. Five-Day Almanac, A.D. 272–274

Fr. 1, recto

i				ii			
Phamenoth				Thoth			
1	6]	11	49	1	2	15	[
6	6]	15	46	6	2	21	20
11	6]	19	41	11	2	27	1[x
1]6	[6]	23	35	16	3	3	1[x
21	6	27	27	21	3	9	[
26	[7]	1	20	26	3	[14	
Pharmuthi				Phaophi			
1	7	5	[1]4	1	[3]	20	[
6	7	9	6	6	3	2[6	
11	7	12	57	11	4	2	[
16	7	16	47	16	4	7	[
21	7	20	36	21	4	[13	
26	7	24	19	26	4	[18	
Pachon							
1	7	2[8					
1]	8	1	[				

Fr. 2, recto

i				ii			
				2]6	9	26	[
[Tybi]				Epeiph			
1	6	1]	49	1	10	2	3[x
6	6	0]	30	6	10	8	3[x
11	5	2]8	[1]3	11	10	14	[
16	5	2]5	37	16	10	20	2[x
21	5	2]2	44	21	10	[26	
26	5]	20	19	26	11	[2	
Mecheir				Mesore			
1	5	18]	32	1	[11	8	

Fr. 1, verso

i				ii			
				Ἄρεω̣[ϲ			
Φαμ]ε̣ν̣ώθ				Θώθ			
α	β]	ιθ	μγ	α	ι	κβ	κ[δ
ϛ	β]	ιθ	κβ	ϛ	ι	κε	ζ̣
ια	β]	ιη	νζ	ια	ι	κζ	μδ
ιϛ	β]	ιη	κθ	ιϛ	ια	ō	ιζ
κα	β]	ιζ	νθ	κα	ια	β	μ̣η̣
κϛ	β]	ι̣ζ̣	κγ	κ̣ϛ	ια	ε	ι̣γ
Φαρ]μοῦθι				Φαῶφι̣			
α	β	ιϛ]	μ̣ϛ̣	α̣	ια	ζ	λ̣α̣
ϛ	β	ιϛ]	η	ϛ	ια	θ̣	[
ια	β	ιε]	λα	ια	ια	ι̣α̣	[μ]α̣
ιϛ	β	ιδ]	ν̣β	ι̣ϛ	ια	ι̣γ̣	μ̣α̣
[κα	β	ιδ	. .]	κα̣	ια	ι̣ε̣	λ̣
κϛ	β	ιγ]	λϛ̣	κ̣ϛ̣	ι̣α	ι̣ζ̣	β̣
				Ἁ]θ̣ύρ̣			
				α	ια]	ι̣η̣	κ̣ϛ

F. 2, verso

i				ii			
κϛ]	α̣	κ̣[η					
Ἐπεί]φ̣				Τ̣[ῦβι			
α	α	κη	ν]α	[α]	β	[ιη	
ϛ	α	κθ]	ι̣β̣	[ϛ]	β	[ιη	
ια	α	κθ]	κ̣ϛ	[ια]	β	[ιθ	
ιϛ	α	κθ	μ]γ	[ιϛ	β	ιθ	
κα	β]	ọ̄	γ̣	κ̣α̣	β	[ιθ	
κϛ	β	ō	κ]ϛ	κϛ	β	[κ	
Μεϲορ]ή̣				Μεχ̣[είρ			
α	β	ō]	μ̣ζ	α	β	κ̣	[
ϛ	β	α	ι]δ̣	ϛ̣	β	[κ	
ια]	β̣	[κ		]	.	[	

Fr. 1, verso

i				ii			
				Mars			
Phamenoth				Thoth			
1	2]	19	43	1	10	22	2[4
6	2]	19	22	6	10	25	7
11	2]	18	57	11	10	27	44
16	2]	18	29	16	11	0	17
21	2]	17	59	21	11	2	48
26	2]	17	23	26	11	5	13
Pharmuthi				Phaophi			
1	2	16]	46	1	11	7	31
6	2	16]	8	6	11	9	[
11	2	15]	31	11	11	11	[4]1
16	2	14]	52	16	11	13	41
[21	2	14	xx]	21	11	15	30
26	2	13]	36	26	11	17	2
				Hath]yr			
				1	11]	18	26

Fr. 2, verso

i				ii			
26]	1	2[8					
Epeiph				Tybi			
1	1	28	5]1	[1]	2	[18	
6	1	29]	12	[6]	2	[18	
11	1	29]	26	[11]	2	[19	
16	1	29	4]3	[16	2	19	
21	2]	0	3	21	2	[19	
26	2	0	2]6	26	2	[20	
Mesore				Mecheir			
1	2	0]	47	1	2	20	[
6	2	1	1]4	6	2	[20	
11]	2	[20		]	x	[	

4207. Five-Day Almanac, a.d. 289–291

Recto

].	ς̣	[
]γ̣	ιạ	β
].	ις	β
]δ	κ̣ạ	β
].	κς	.
	Ἀ̣θ̣ύ̣[ρ	
]	α	[
]	ς̣	β̣

Verso

[ιθ]	μ̣θ̣
κ	κ̣θ̣
κα	θ
κα	με
κβ	κς
[Πα]χών	
κγ	ζ
κγ	μ̣[
κδ	κ̣[
κ̣ε	[
κ̣[	

4207. Five-Day Almanac, a.d. 289–291

Recto

]x	6	[
]3	11	2
]x	16	2
]4	21	2
]x	26	x
	Hathyr	
]	1	[
]	6	2

Verso

[19]	49
20	29
21	9
21	45
22	26
Pachon	
23	7
23	4[x
24	2[x
25	[
2[x	

4208. Five-Day Almanac, a.d. 295/296

i	ii			
	ιϛ	[κη		
	κα	κθ̣	ε̣	
	κϛ	α΄	⊤̣	ε
	Μεσορή			
]γ	α	α	ϛ	
]β̣	ϛ	β̣	ζ̣	
]δ̣	ια	γ̣	θ̣	
]δ̣	ιϛ	δ	ι̣γ̣	
	κα	ε	ιζ	
	κϛ	ϛ	κβ	
] .	Ἐπαγό(μεναι)			
] . ϛ	α	α΄	ζ	κϛ
	ε]	α̣΄	η̣	ι̣η̣

11 επαr

4208. Five-Day Almanac, a.d. 295/296

i	ii			
	16	[28		
	21	29	5	
	26	1	0	5
	Mesore			
]3	1	1	6	
]2	6	2	7	
]4	11	3	9	
]4	16	4	13	
	21	5	17	
	26	6	22	
]x	Epagomenae			
] x6	1	1	7	26
	5]	1	8	18

4209. Five-Day Almanac, a.d. 295/296

i	ii	iii
ἐφή]μ̣εροι τ[ῶν] πέν̣[τε πλανωμένων		
Κρόν̣[ου]	Κ̣[ρόνου]	[Διόc]
Μ̣εχείρ̣	Ἐ̣[πείφ]	[Χοίακ]
α ι´ κα νγ	α̣ [ια β . .]	[α ιβ κη .]
ϛ κα [. .]	[ϛ γ] .	ϛ κη ιβ̣
ια κα [. .]	ια γ̣ μ̣ζ	ια κη κδ
ιϛ κα̣ [. .]	ιϛ δ̣ ιζ	ιϛ κη κζ
κα κα̣ [. .]	κα δ [. .]	ι̣ζ̣ κη κζ c´α´
κδ̣ κα [. .] c´]β̣´	κϛ ε̣ [. .]	[κα] κη κε
κϛ̣ κ̣α̣ ι̣ϛ̣		[κϛ κη κ]
Φαμενώθ [		
α κα ιη [		
ϛ κα κ [		
[ια κα κ]ζ̣ [		

4209. Five-Day Almanac, a.d. 295/296

Positions? [of the] five? [planets

i	ii	iii
Saturn	Saturn	[Jupiter]
Mecheir	Epeiph	[Choeac]
1 10 21 53	1 [11 2 xx]	[1 12 28 x]
6 21 [xx]	[6 3 x]x	6 28 1x
11 21 [xx]	11 3 47	11 28 24
16 21 [xx]	16 4 17	16 28 27
21 21 [xx]	21 4 [xx]	17 28 27 Φ
24 21 [xx] Ψ	26 5 [xx]	[21] 28 25
26 21 16		[26 28 20]
Phamenoth [		
1 21 18 [		
6 21 20 [		
[11 21 2]7 [		

4210. Five-Day Almanac, A.D. 300–302

Fr. 1, recto

Τῦβι		
α	ε	κα
ϛ	ε	ιγ
ια	ε	γ̣
ιϛ	δ	ν̣ ̣
κα	δ	λ[̣
κϛ	δ̣	ιζ̣
[Μεχείρ]		
α̣	δ̣	̣
ϛ	γ	λζ
ια	γ	ιγ
ιϛ	β	ν

Fr. 1, verso

i	ii		
	ι]ϛ̣	δ	νη
]	κ̣α	ε	νζ
]	κ̣ϛ	ϛ	νϛ
]	Π̣αχών		
]	α̣	ζ	ν
]	ϛ	η	μδ
	ι]α̣	θ	λγ
]	ιϛ	ι]	κ̣
] ̣ ̣ ̣	κ̣[α	ια	ε
] ̣ ̣ιμη	κϛ	[ια	μϛ
]	Παῦνι		
]	α	ιβ	κα
]	ϛ	ιβ	νγ
]	ι̣α	ιγ	κβ
	ιϛ	ιδ]	α̣

4210. Five-Day Almanac, a.d. 300–302

Fr. 1, recto

Tybi		
1	5	21
6	5	13
11	5	3
16	4	5x
21	4	3[x
26	4	17
[Mecheir]		
1	4	0
6	3	37
11	3	13
16	2	50

Fr. 1, verso

i	ii		
	1]6	4	58
]	21	5	57
]	26	6	56
]	Pachon		
]	1	7	50
]	6	8	44
	1]1	9	33
]	16	10]	20
]...	2[1	11	5
]...	26	[11	46
]	Payni		
]	1	12	21
]	6	12	53
]	11	13	22
	16	14]	1

Fr. 2

] ιη [
Φαμεν]ώ̣θ
α̣ α ι̣[ε] η
ϛ ιδ μϛ
ι̣α ιδ ιθ
ι̣ϛ ιγ νγ̣
κα ι̣γ̣ λ̣
κϛ ιγ ζ
Φ̣αρμοῦθι
α] ιβ μγ
ϛ] ιβ κδ
ια] ιβ ϛ
ιϛ] ια μ̣ .
κα ια λ]γ̣
— — — — —

Fr. 2

] 18 [

Phamen]oth

1	1 1[5]	8
6	14	46
11	14	19
16	13	53
21	13	30
26	13	7

Pharmuthi

1]	12	43
6]	12	24
11]	12	6
16]	11	4x
21	11	3]3

— — — — —

4211. Five-Day Almanac, A.D. 306/307

Recto			Verso		
[Θώθ]			Φαμεν̣[ώθ		
α	γ΄] ☿	λ̣ζ̣	α	γ΄ ις	κ
ς]	α̣	γ̣	ς	ις	ιη
ια]	α	λβ	ια	ις	ιδ
ις]	β	β	ις	ις	ς
κα]	β	λγ	κα	ιε	νζ̣
κς]	γ	ε	κς	ιε	μ[ς
[Φα]ῶ̣φι			Φαρμ(οῦθι)		
α]	γ̣	λς	α	ιε	λα
ς]	δ̣	ι	ς	ιε	ι
ια]	δ	μδ	ια	ιδ	νς
ις]	ε	κ	ις	ιδ	λζ
κα]	ε	νδ	κα	ιδ	ιε
κς]	ς	λγ	κς	ιγ	νγ
Ἀθύ]ρ̣					

4211. Five-Day Almanac, a.d. 306/307

Recto

[Thoth]		
1	3] 0	37
6]	1	3
11]	1	32
16]	2	2
21]	2	33
26]	3	5
[Pha]ophi		
1]	3	36
6]	4	10
11]	4	44
16]	5	20
21]	5	54
26]	6	33
Hathy]r		

Verso

Phamen[oth		
1	3 16	20
6	16	18
11	16	14
16	16	6
21	15	57
26	15	4[6
Pharm(uthi)		
1	15	31
6	15	10
11	14	56
16	14	37
21	14	15
26	13	53

4212. Five-Day Almanac, a.d. 217/218

i			ii			iii		
[		]	[		]	[	]η̣[	]
[Φαμενώθ]			['Επείφ]			Φαρμ(οῦθι)		
[ι		]	[ι		]	ε̣	ιδ	μ[ ]
[κ]	.[ ]	.[ ]	[κ		].	ι	κ	ν
[λ]	γ	κζ	λ	ς	μς	ιε	κς	λε
[Φα]ρμ(οῦθι)			Μεσορ(ή)			κ	ι β	ιζ
[ι]	ε	θ	ι	θ	ζ	κ̣ε̣	ζ	νθ
[κ]	ς	μ	κ	[ι]γ̣	κ̣ζ̣	λ̣	ι[γ]	νζ
[λ]	ζ	μθ	λ	ι̣η̣	λ̣δ̣	Παχώ(ν)		
Π̣αχ(ών)			. . . .	.	̣	ε	[ι]θ	νδ
[ι]	η	μ̣γ̣	η. . ( )			ι	κ̣ε	η
[κ		]	. Θ̣ώθ			ιε	ια̣	α̣λθ
[λ		]	ε	.[	]	κ̣	[ς]	κδ̣
[		]	[		]	κ̣[ε]	[ι]β̣	[.]

2 φαρ^{μ} (so line 6) 6 μεσορʹ 9 παχω̅ 10 πα^{χ}

4213. Five-Day Almanac, a.d. 301/302

	ις	ιθ	[
	κα	κ	[
	κς	κβ	[
Χοίακ			[
	α	κε	κ[
	ς	κη	[
		].[	

4212. Five-Day Almanac, a.d. 217/218

i			ii			iii		
[		]	[		]	[	]...[	]
[Phamenoth]			[Epeiph]			Pharmuthi		
[10		]	[10		]	5	14	4[x]
[20]	x[x]	x[x]	[20		]x	10	20	50
[30]	3	27	30	6	46	15	26	35
Pharmuthi			Mesore			20	10 2	17
[10]	5	9	10	9	7	25	7	59
[20]	6	40	20	[1]3	27	30	1[3]	57
[30]	7	49	30	18	34	Pachon		
Pachon			...	x	0	5	[1]9	54
[10]	8	43	...			10	25	8
[20		]	... Thoth			15	11 1	39
[30		]	5	x[x	]	20	[6]	24
[		]	[		]	2[5]	[1]2	[ ]

4213. Five-Day Almanac, a.d. 301/302

	16	19	[
	21	20	[
	26	22	[
Choeac			[
	1	25	2[x
	6	28	[
		]x[	

4214. Syzygies

	Λ̣[έον(τος)]	[
	Π̣α̣ρ̣θ̣(ένου)	[
	Ζυγο(ῦ)	ι̣ . [
	Cκορ(πίου)	ια μ̣[
	Τοξό(του)	θ ν̣[
	Αἰγό(κερω)	η [
	Ὑδρο(χόου)	ς̣ ζ̣ [
	Ἰχθύ(ων)	δ̣ [

Names of zodiacal signs abbreviated by truncation.

4215. Mean Motions of the Moon, A.D. 287/288

] .	. [
cπ̣ζ μθ	λς λβ
οβ κζ	ρξζ ια
με μ̣ζ	ρκζ μη
ο̣η̣ ν̣β	cλε ι

4214. Syzygies

Leo	[
Virgo	[
Libra	1x [
Scorpio	11 4[x
Sagittarius	9 5[x
Capricorn	8 [
Aquarius	6 7? [
Pisces	4 [

4215. Mean Motions of the Moon, a.d. 287/288

]...		...[	
287	49	36	32
72	27	167	11
45	47	127	48
78	52	235	10

Part IV

Miscellaneous Tables

4216. Arithmetical Table: Thirtieths of 40;33

ii	iii
β̣ μ̣β̣	[β]
δ̣ γ	[γ]
ε̣ κδ	[δ]
ς μ̣ε	ε̣
η̣ ς	ς
θ̣ κζ	ζ
[ι] μ̣η̣	η
ι̣β̣ [θ]	θ
ι̣γ λ̣α	[ι]
ιδ νβ	ια
ις ις	ιβ
ιζ λδ	ιγ
ιη̣ νε	ιδ
κ ις	ιε
κα λζ	ις
κβ νθ	ιζ
κδ κ	ιη
κ̣ε μα	ιθ
κζ β	κ
κη κς	κα
κθ μδ	κβ
λα ε	κγ
λβ κς	κδ
λγ μζ	κ̣[ε]
λε η	κ[ς]
λς λ	κζ̣
λζ να	κη
λθ ιβ	κθ
μ λγ̣	λ

4216. Arithmetical Table: Thirtieths of 40;33

ii	iii
2 42	[2]
4 3	[3]
5 24	[4]
6 45	5
8 6	6
9 27	7
[10] 48	8
12 [9]	9
13 31	[10]
14 52	11
16 16	12
17 34	13
18 55	14
20 16	15
21 37	16
22 59	17
24 20	18
25 41	19
27 2	20
28 26	21
29 44	22
31 5	23
32 26	24
33 47	2[5]
35 8	2[6]
36 30	27
37 51	28
39 12	29
40 33	30

4217. Arithmetical Table: Templates for Mercury?

[0]	i	ii	iii	iv	v	vi	vii	viii	ix
[λα]		λ		κθ		κη		κζ	
[○]	[α]	○̣	β̣	○	β	○	β	○	[β]
[○]	[ζ]	○̣	η	○	η	○	η	○	[η]
[○]	[ιζ]	[○]	[ι]η̣	○	ιη	○	ιθ̣	○	[κ]
[○]	λ̣	○	λβ	○	λγ	○	λδ	○	[λε]
[○]	μ̣η̣	○̣	ν	○	να	○	νγ̣	○	ν̣[ε]
[α]	[θ]	α̣	ιβ	α	ιδ	α	ιζ	α̣	κ̣
[α]	[λδ]	α	λη	α	μα	α	με	α	[μη]
[β]	[γ]	[β]	η	β	ιβ	β	ιζ	β̣	κ[β]
[β]	[λϛ]	[β]	μβ	β	μζ	β	νγ	γ̣	○̣
[γ]	[ιγ]	[γ]	κ	γ	κϛ	γ	[λδ]	[γ]	[μβ]
[γ]	[νδ]	[δ]	β	δ	ι	δ	ιθ	δ	κ̣[η]
[δ]	[λη]	[δ]	μη	δ	νζ	ε	ν	ε	κ̣
[ε]	[κζ]	[ε]	λη	ε	μθ	ϛ	β	ϛ	[ιε]
[ϛ]	[ιθ]	[ϛ]	λβ	ϛ	με	ζ	○	ζ	[ιε]
[ζ]	[ιε]	[ζ]	λ	ζ	με	η	[β]	η	[κ]
[η]	[ιε]	[η]	λβ	η	μθ	θ	η̣	θ̣	[κη]
[θ]	[ιθ]	[θ]	λη	θ	νζ	ι	ιθ̣	ι̣	[μβ]
[ι]	[κζ]	[ι]	μ̣η̣	ια	ι̣	ια	λδ	ιβ̣	[○]
[ια]	[λη]	[ιβ]	[β]	[ιβ]	[κϛ]	ιβ	νγ	[ιγ]	[κβ]
[ιβ]	[νδ]	[ι]γ̣	κ	ιγ	μζ	ιδ	ιζ	[ιδ]	[μη]
[ιδ]	[ιγ]	[ι]δ̣	μ̣β	ιε	ιβ	ιε	με	[ιϛ]	[κ]
[ιε]	[λϛ]	[ι]ϛ̣	η	ιϛ	μα	ιζ	ι̣ζ	[ιζ]	[νε]
[ιζ]	[γ]	[ι]ζ̣	λη	ιη	ιβ	ιη	νγ	[ιθ]	[λε]
[ιη]	[λδ]	[ι]θ̣	ιβ	ιθ	να	κ	λδ̣	[κ]	[κ]
[κ]	[θ]	[κ]	ν	κα	λγ	κβ	ιθ̣	[κγ]	[η]
[κα]	[μη]	[κβ]	λβ̣	κγ	ιη	κδ	η	[κε]	[β]
[κγ]	[λ]	[κ]δ̣	ιη	κε	η	κϛ	β	[κζ]	[○]
[κε]	[ιζ]	[κ]ϛ̣	η	κζ	β	κη	○		
[κζ]	[ζ]	[κ]η̣	β	κθ	○				
[κθ]	[α]	[λ]	○						
[λα]	[○]								

1 numerals in red ink   v 12 ϊ (so vi 18, presumably also v 19, viii 18)

4217. Arithmetical Table: Templates for Mercury?

[0]	i	ii	iii	iv	v	vi	vii	viii	ix
[31]		30		29		28		27	
[0]	[1]	0	2	0	2	0	2	0	[2]
[0]	[7]	0	8	0	8	0	8	0	[8]
[0]	[17]	[0]	[1]8	0	18	0	19	0	[20]
[0]	30	0	32	0	33	0	34	0	[35]
[0]	48	0	50	0	51	0	53	0	5[5]
[1]	[9]	1	12	1	14	1	17	1	20
[1]	[34]	1	38	1	41	1	45	1	[48]
[2]	[3]	[2]	8	2	12	2	17	2	2[2]
[2]	[36]	[2]	42	2	47	2	53	3	0
[3]	[13]	[3]	20	3	26	3	[34]	[3]	[42]
[3]	[54]	[4]	2	4	10	4	19	4	2[8]
[4]	[38]	[4]	48	4	57	5	50	5	20
[5]	[27]	[5]	38	5	49	6	2	6	[15]
[6]	[19]	[6]	32	6	45	7	0	7	[15]
[7]	[15]	[7]	30	7	45	8	[2]	8	[20]
[8]	[15]	[8]	32	8	49	9	8	9	[28]
[9]	[19]	[9]	38	9	57	10	19	10	[42]
[10]	[27]	[10]	48	11	10	11	34	12	[0]
[11]	[38]	[12]	[2]	[12]	[26]	12	53	[13]	[22]
[12]	[54]	[1]3	20	13	47	14	17	[14]	[48]
[14]	[13]	[1]4	42	15	12	15	45	[16]	[20]
[15]	[36]	[1]6	8	16	41	17	17	[17]	[55]
[17]	[3]	[1]7	38	18	12	18	53	[19]	[35]
[18]	[34]	[1]9	12	19	51	20	34	[20]	[20]
[20]	[9]	[20]	50	21	33	22	19	[23]	[8]
[21]	[48]	[22]	32	23	18	24	8	[25]	[2]
[23]	[30]	[2]4	18	25	8	26	2	[27]	[0]
[25]	[17]	[2]6	8	27	2	28	0		
[27]	[7]	[2]8	2	29	0				
[29]	[1]	[30]	0						
[31]	[0]								

4217a–c. Arithmetical Tables: Templates for Mercury?

(a)

i	ii
[λ η]	κθ̣ [λζ]
[λ μα]	λ ζ̣
λα ι	λ [λγ]
λα λϛ	λ ν̣[ϛ]
[λ]α̣ νη	λα [ιε]
[λβ] ιζ	λα λ̣[α]
[λβ] λβ	λα̣ μ̣[δ]
[λβ] μδ	λ[α νβ]
[λβ] νγ	λ̣[α νη]
[λβ] νη̣	[λβ ō]
[λγ] ọ̄	

(c)

i	ii
]θ	
] κε	
]̣ νε	ọ̄ [

(b)

i	ii	iii	iv	v
]ε̣				
]β	α [νη ιδ]			
]α̣	[γ να ϛ]	[α νη ιη]		
]̣	ε λ̣η̣ [λϛ]	[γ ν νη]	α νη̣ [ζ]	
] ε	ζ κα μ̣[γ]	[ε λη ō]	γ ν [κα]	[
] λ	η νζ κ̣[η]	[ζ ιθ κδ]	[ε] λ̣[ϛ] μ̣	[
]	[ι κη ν]	η̣ [νε ι]	ζ̣ ιζ δ	[
]	[ια νδ ν]	ι κ[ε ι]η̣	η να λε	ζ̣ [
]	[ιγ ι]ε κη	ια μθ μη	ι κ ι̣α̣	[
]	[ιδ λ] μ̣β̣	ιγ η μ	ια μβ [νβ]	[
		(7 lines lost)		
		κ̣ ιϛ̣ [μη]		
		κ μδ νη̣		
		κα ζ λ		
		κ̣α̣ κ̣δ̣ [κδ]		

4217a–c. Arithmetical Tables: Templates for Mercury?

(a)

	i	ii
	[30 8]	29 [37]
	[30 41]	30 7
	31 10	30 [33]
	31 36	30 5[6]
5	[3]1 58	31 [15]
	[32] 17	31 3[1]
	[32] 32	31 4[4]
	[32] 44	3[1 52]
	[32] 53	3[1 58]
10	[32] 58	[32 0]
	[33] 0	

(c)

	i	ii
	]9	
	] 25	
3	]x 55	0 [

(b)

	i	ii	iii	iv	v
	]5				
	]2	1 [58 14]			
	]1	[3 51 6]	[1 58 18]		
	]x	5 38 [36]	[3 50 58]	1 58 [7]	
5	] 5	7 21 4[3]	[5 38 0]	3 50 [21]	[
	] 30	8 57 2[8]	[7 19 24]	[5] 3[6] 40	[
	]	[10 28 50]	8 [55 10]	7 17 4	[
	]	[11 54 50]	10 2[5 1]8	8 51 35	7 [
	]	[13 1]5 28	11 49 48	10 20 11	[
10	]	[14 30] 42	13 8 40	11 42 [52]	[
11–17			*(7 lines lost)*		
18			20 16 [48]		
			20 44 58		
20			21 7 30		
			21 24 [24]		

4218. Arithmetical Table: Fractions of 30

] ̣ραι

i	ii
ε̣	ō
[ι]	ō
[ι]ε	ō
κ	ō
κε	ō
λ	ō
̣]	
γ	με
ζ	λ
ια	ιε
ιε	ō
ι̣η	με
κα	λ
κϛ	ι̣ε
λ	ō
] λ	
β	λ
ε	ō
ζ	λ
ι̣	ọ̄

17 λ in red ink

4218. Arithmetical Table: Fractions of 30

i	ii
]...	
5	0
[10]	0
[1]5	0
20	0
25	0
30	0
...]	
3	34
7	30
11	15
15	0
18	45
21	30
26	15
30	0
] **30**	
2	30
5	0
7	30
10	0

– – – – – – – – –

4219. Numerical Table

i		ii	
δ	ε	ε	ϛ̣
η	θ	θ	ι
ιβ	ιγ	ιγ	ι[δ]
ιϛ	ιζ	ιζ	ιη̣
κ̣	κα	κα	κ̣[β]
κδ	κε	κε	α̣
κ̣η̣	δ	κθ	ε
γ̣	η	δ	θ̣
ζ̣	ιβ	η	[̣ ̣]

Second numeral in each cell in red ink.

4219. Numerical Table

i		ii	
4	**5**	5	**6**
8	**9**	9	**10**
12	**13**	13	**1[4]**
16	**17**	17	**18**
20	**21**	21	**2[2]**
24	**25**	25	**1**
28	**4**	29	**5**
3	**8**	4	**9**
7	**12**	8	**[xx]**

4220. HANDY TABLES

Fr. 1

i	ii	iii
[ρξβ]	λ̣δ̣	κ̣β
[c]γ̣	ιβ	νζ
[cμ]γ̣	να	λ̣γ̣
c̣πδ	λ	η̣
τ̣[κ]ε	η	μ[δ]
ε̣	μζ	ι[θ]
μ̣ς̣	κ̣ς̣	[νε]
π̣ζ	δ	[λ]
ρ̣κζ	μ⟨γ⟩	[ς]
[ρ]ξη	κα	μα
c̣θ	ō	ιζ
[c]μ̣θ	λη	νβ
c̣ϙ̣	ιζ	κη
[τλ]	ν̣ς̣	γ
[ι]α̣	λ̣δ̣	λθ
[ν]β	ιγ̣	ιδ
[ϙ]ζ	να̣	ν
[ρ]λ̣γ	λ̣	κε̣
[ρο]δ̣	θ	α
[cι]δ̣	μζ	[λς]
[cν]ε̣	κς	[ιβ]
[cϙ]ς̣	δ̣	[μζ]

Fr. 2

i	ii
θ̣	[ϙζ]
ι̣	ρ̣[θ]
ι̣α̣	ρκ[α]
ι̣β̣	ρλδ̣
ι̣γ̣	ρμς
ι̣δ̣	ρνη
ι̣ε̣	ρο
[ι]ς̣	ρπβ
ι̣ζ̣	ρϙε
[ι]η̣	cζ̣
[ι]θ	cιθ
κ̣	cλα

4220. Handy Tables

Fr. 1

i	ii	iii
[162]	34	22
[20]3	12	57
[24]3	51	33
284	30	8
3[2]5	8	4[4]
5	47	1[9]
46	26	[55]
87	4	[30]
127	4<3>	[6]
[1]68	21	41
209	0	17
[2]49	38	52
290	17	28
[330]	56	3
[1]1	34	39
[5]2	13	14
[9]7	51	50
[1]33	30	25
[17]4	9	1
[21]4	47	[36]
[25]5	26	[12]
[29]6	4	[47]

Fr. 2

i	ii
9	[97]
10	1[09]
11	12[1]
12	134
13	146
14	158
15	170
[1]6	182
17	195
[1]8	207
[1]9	219
20	231

Fr. 3, along the fibres

μς νθ̣
μζ ̣
μς νς
μς νβ
μς μη
μς μβ
μς λδ
μς̣ [
μς̣ [
μ[
μ[
μ ̣ [
ε̣ ̣ [

Fr. 3, across the fibres

i	ii
α] ζ̣	β̣ [
]α̣ η	β̣ [
]α̣ θ	β [
]α̣ ι	β [
]α [ι]α	β̣ [
]α ιβ	[
]α̣ ιγ	[
]α̣ ιδ	[
α ι]ε̣	[
α ι]ς̣	β [
α ι]ζ̣	β [
α ι]η	β̣ [
]	β [
]	β̣ [

Fr. 3, along the fibres

— — —

46 59
47 0
46 56
46 52
46 48
46 42
46 34
46 [
46 [
4[x
4[x
4x [
...[

— — —

Fr. 3, across the fibres

— — — — —

i	ii
1] 7	2 [
]1 8	2 [
]1 9	2 [
]1 10	2 [
]1 [1]1	2 [
]1 12	[
]1 13	[
]1 14	[
1 1]5	[
1 1]6	2 [
1 1]7	2 [
1 1]8	2 [
]	2 [
]	2 [

— — — — —

4221. Longitudes

Fr. 1

i	ii	iii
]		κ̣[γ ○ ○
]		κγ̣ ν̣ϛ̣ [μ
]		κδ νγ̣ [κ
]		κε ν [○
]		κϛ μ[ϛ μ
]		κ[ζ μγ κ
]		κη μ ○
] μ		κθ λϛ μ
] κ̣	Cκορ(πίου)	○ [λγ κ
]		α λ̣ [○
]		β κ̣[ϛ μ

9 cκορ᾽

Fr. 2

i	ii
	κ[
	κ̣[
Τ̣αύρο(υ)	○ [

3 ταυρο

Fr. 3

μ] ○̣
λ]ϛ μ
λ]γ κ
] λ̣ ○
κ]ϛ μ
κ]γ κ
] ι̣ ○

4221. Longitudes

Fr. 1

i	ii	iii
]		2[3 0 0
]		23 56 [40
]		24 53 [20
]		25 50 [0
]		26 4[6 40
]		2[7 43 20
]		28 40 0
] 40		29 36 40
] 20	Scorpio	0 [33 20
]		1 30 [0
]		2 2[6 40

Fr. 2

	i	ii
		2[x
		2[x
3	Taurus	0 [

Fr. 3

40] 0
3]6 40
3]3 20
] 30 0
2]6 40
2]3 20
] 10 0

4222. Table with Zodiacal Signs and Latitudes (?)

Fr. 1

i	ii	iii
[Τ]α̣[ύροc]	[βο]	[
Δίδυμο̣ι̣	β̣[ο]	[
Κάρκιν(οc)	β[ο]	[
Λέων	βο̣	[
Πάρθεν(οc)	βο	γ̣ [
Ζυγόc	νο	𐆊̣ [
Cκόρπιοc	νο	γ̣ [
Τοξό̣τ̣ηc	νο	ϛ̣ [
Αἰγό̣κ̣ε(ρωc)	νο	ζ̣ [
Ὑδρηχ(όοc)	νο	ϛ̣ [
Ἰχθύεc	νο	γ̣ [
ταβ[		

3 καρκιν 5 παρθεν 9 αιγοκε 10 υδρηχ, *l.* Ὑδροχόοc

Fr. 2

i	ii
]	ζῴδια
].	Τοξότη[c
]	Cκόρπι̣[οc

4222. Table with Zodiacal Signs and Latitudes (?)

Fr. 1

i	ii	iii
Taurus	N	[
Gemini	N	[
Cancer	N	[
Leo	N	[
Virgo	N	3 [
Libra	S	0 [
Scorpio	S	3 [
Sagittarius	S	6 [
Capricorn	S	7 [
Aquarius	S	6 [
Pisces	S	3 [
...[		

Fr. 2

i	ii
]	signs
]...	Sagittarius
]	Scorpio

4223. Sinusoidal Function

[0]	i	ii	iii	iv	v	vi
	[ ]ρ̣ε̣	ια			[ ]	[ ]
	ζ̣	κη			[ ]	[ ]
	ζ	κθ			[ ]	[ ]
	ζ̣	κθ			[ ]	[ ]
	ζ̣	κθ			[ ]	[ ]
[ϙε]	ζ̣	κθ		[ρμ]	[ ]	[ ]
	[ζ]	λ			[ ]	[ ]
	ζ̣	κθ			[ ]	[ ]
	ζ̣	κθ			[ ]	[ ]
	ζ̣	κ̣θ̣			[ ]	[ ]
[ρ]	ζ̣	κθ		ρμε	[ ]	[ ]
	ζ̣	κζ			[ ]	[ ]
	ζ̣	κϛ			[ ]	[ ]
	[ζ]	κε			[ ]	[ ]
	[ζ]	κδ			[ ]	[ ]
[ρε]	[ζ]	κγ		ρν	[ ]	[ ]
	[ζ]	κβ			[ ]	[ ]
	[ζ]	κ̣			[ ]	[ ]
	[ζ]	ι̣η̣			[ ]	[ ]
	[ζ]	ι̣[̣]			[ ]	[ ]
[ρι]	[ζ]	ι̣δ̣		ρνε̣	[ ]	[ ]
	[ζ]	[ ]			[ ]	[ ]
	[ζ]	[ ]			[ ]	[ ]
	[ζ]	[ ]			[ ]	[ ]
	[ζ]	[ ]			[ ]	[ ]
[ριε]	[ ]	[ ]		ρξ	[ ]	[ ]
	[ ]	[ ]			[ ]	[ ]
	[ ]	[ ]			[β]	μ[̣]
	[ ]	[ ]			β̣	λ̣
	[ ]	[ ]			β	κζ

⇓

4223. Sinusoidal Function

[0]	i	ii	iii	iv	v	vi
	[ ].	.			[ ]	[ ]
	7	28			[ ]	[ ]
	7	29			[ ]	[ ]
	7	29			[ ]	[ ]
	7	29			[ ]	[ ]
[95]	7	29		[140]	[ ]	[ ]
	[7]	30			[ ]	[ ]
	7	29			[ ]	[ ]
	7	29			[ ]	[ ]
	7	29			[ ]	[ ]
[100]	7	29		145	[ ]	[ ]
	7	27			[ ]	[ ]
	7	26			[ ]	[ ]
	[7]	25			[ ]	[ ]
	[7]	24			[ ]	[ ]
[105]	[7]	23		150	[ ]	[ ]
	[7]	22			[ ]	[ ]
	[7]	20			[ ]	[ ]
	[7]	18			[ ]	[ ]
	[7]	1[x]			[ ]	[ ]
[110]	[7]	14		155	[ ]	[ ]
	[7]	[ ]			[ ]	[ ]
	[7]	[ ]			[ ]	[ ]
	[7]	[ ]			[ ]	[ ]
	[7]	[ ]			[ ]	[ ]
[115]	[ ]	[ ]		160	[ ]	[ ]
	[ ]	[ ]			[ ]	[ ]
	[ ]	[ ]			[2]	4[x]
	[ ]	[ ]			2	3x
	[ ]	[ ]			2	27

⇓

[0]	i	ii	iii	iv	v	vi
[ρκ]	[ ]	[ ]		ρ̣ξ̣ε̣	β	ιθ̣
	[ ]	[ ]			β̣	ια̣
	[ ]	[ ]			β̣	α̣
	[ ]	[ ]			α	νδ
	[ ]	[ ]			α	μδ
[ρκε]	[ ]	[ ]		ρο	α	λ̣δ̣
	[ ]	[ ]			α	[ ]
	[ ]	[ ]			α	[ ]
	[ ]	[ ]			α	[ ]
	[ ]	[ ]			ō	[ ]
[ρλ]	[ ]	[ ]		ροε	ō	[ ]
	[ ]	[ ]			ō	[ ]
	[ ]	[ ]			ō	[ ]
	[ ]	[ ]			ō	[ ]
	[ ]	[ ]			ō	[ ]
[ρλε]	[ ]	[ ]		ρπ	ō	[ ]

[0]	i	ii	iii	iv	v	vi
[120]	[ ]	[ ]		165	2	19
	[ ]	[ ]			2	11
	[ ]	[ ]			2	1
	[ ]	[ ]			1	54
	[ ]	[ ]			1	44
[125]	[ ]	[ ]		170	1	34
	[ ]	[ ]			1	[ ]
	[ ]	[ ]			1	[ ]
	[ ]	[ ]			1	[ ]
	[ ]	[ ]			0	[ ]
[130]	[ ]	[ ]		175	0	[ ]
	[ ]	[ ]			0	[ ]
	[ ]	[ ]			0	[ ]
	[ ]	[ ]			0	[ ]
	[ ]	[ ]			0	[ ]
[135]	[ ]	[ ]		180	0	[ ]

4224. Sinusoidal Function

Fr. 1

i	ii	iii	iv	v
		[ιθ]	[κ]δ	ι̣[θ]
		[ιθ]	[κ]γ̣	νη
	[ρ]	[ι]θ	κγ	ιζ
		[ι]θ	κβ	ιε
		[ι]θ	κ	νγ
		ιθ	ιθ	ι
		ιθ	ιζ̣	ϛ
	ρε	ιθ	ιε̣	μ[α]
]κ̣		ιθ	ια	νϛ
]ζ̣				

Fr. 2

i	ii
[.]ε̣	β̣
μ̣ζ	λθ
ϛ	γ
κδ	κζ
μ̣β	[..]

Fr. 3

i	ii
γ̣	[..]
δ	ι
ε	ι[.]
ϛ	ιη̣
ζ	κ[.]
η	κζ̣
θ̣	κ[.]

4224. Sinusoidal Function

Fr. 1

i	ii	iii	iv	v
		[19]	[2]4	1[9]
		[19]	[2]3	58
	[100]	[1]9	23	17
		[1]9	22	15
		[1]9	20	53
		19	19	10
		19	17	6
	105	19	15'	4[1]
]20		19	11	56
]7				

Fr. 2

i	ii
[x]5	2
47	39
6	3
24	27
42	[xx]

Fr. 3

i	ii
3	[xx]
4	10
5	1[x]
6	18
7	2[x]
8	27
9	2[x]

4225. Numerical Table

i	ii	iii	
[ε]	[ ]	[ ]	η̣
[ϛ]	[ ]	[̣]δ	ε
[ζ]	[ ]	ζ	ϛ
[η]	[ ]	δ	β
[θ]	[ ]	ι̣δ	θ
[ια]	.	ζ	ι .
[ιβ]	ϛ	ιβ	ι
ιγ	ιγ	ιβ	ι
ιδ	ιζ	θ̣	β
ιε	ϛ	[ ]	γ
ιϛ	δ	[ ]	ϛ
ιζ	ζ	[ ]	ιϛ
ιη	ιϛ	[ ]	θ
ιθ	ε	[ ]	[
κ	δ̣	ι̣ζ̣	[
κα	α	ε̣	[
κβ	ιθ	ϛ	[
κγ	θ	θ	[
κδ	η	ιϛ	[
κε	ε	ζ	[
κϛ	κδ̣	ε	[
κζ	ϛ	ιδ̣	[
κη	ζ	.	[
κθ	ε	[ ]	[
λ	ιδ	[ ]	[

4225. Numerical Table

i	ii	iii	
[5]	[ ]	[ ]	8
[6]	[ ]	[x]4	5
[7]	[ ]	7	6
[8]	[ ]	4	2
[9]	[ ]	14	9
[11]	x	7	1x
[12]	6	12	10
13	13	12	10
14	17	9	2
15	6	[ ]	3
16	4	[ ]	6
17	7	[ ]	16
18	16	[ ]	9
19	5	[ ]	[
20	4	17	[
21	1	5	[
22	19	6	[
23	9	9	[
24	8	16	[
25	5	7	[
26	24	5	[
27	6	14	[
28	7	x	[
29	5	[ ]	[
30	14	[ ]	[

4226. Table with Planets, Zodiacal Signs, and Symbols

Fr. 1

i	ii	iii
[ ]	[ ]	Π[άρθ(ενοϲ)]
[ ]	[ ]	Ζ[υγό(ϲ)]
[ ]	[ ] .	Ϲ[κόρ(πιοϲ)]
[ ]	[ ] ○	Το̣[ξότ(ηϲ)]
[ ]	[ ] ○	Αἰγ̣[όκ(ερωϲ)]

i	ii	iii
[Κρό]ν̣οϲ		Ζεύ[ϲ]
[Πάρθ]ε̣(νοϲ)	○ ○	Ἰχθύ̣(εϲ)
[Ζυγό]ϲ̣	○ ○	Κ̣ρ̣ι[όϲ]
[Ϲκό]ρ̣(πιοϲ)	○ ○	Ταύ[ρο(ϲ)]
[Τοξότ(ηϲ)]	• •	Δίδυ̣(μοι)
[Αἰγόκ(ερωϲ)]	○ ○	Κ̣ά̣ρ̣κ̣(ινοϲ)
[Ὑδροχ(όοϲ)]	○ •	Λ̣έ̣ω[ν]
[Ἰχθύ(εϲ)]	○ ○	Π̣ά̣ρ̣[θε(νοϲ)]
[Κριόϲ]	• ○	Ζυγ̣[όϲ]
[Ταύρο(ϲ)]	[ ]	Ϲκό[ρ(πιοϲ)]
[Δίδυ(μοι)]	[ ]	Τοξ[ότ(ηϲ)]
[Κάρκ(ινοϲ)]	[ ]	Α[ἰγόκ(ερωϲ)]
[Λέων]	[ ]	Ὑδρ̣[οχ(όοϲ)]

4226. Table with Planets, Zodiacal Signs, and Symbols

Fr. 1

i	ii		iii
[ ]	[ ]		Virgo
[ ]	[ ]		Libra
[ ]	[ ] ...		Scorpio
[ ]	[ ] ○		Sagittarius
[ ]	[ ] ○		Capricorn
Saturn			Jupiter
Virgo	○ ○		Pisces
Libra	○ ○		Aries
Scorpio	○ ○		Taurus
[Sagittarius]	• •		Gemini
[Capricorn]	○ ○		Cancer
[Aquarius]	○ •		Leo
[Pisces]	○ ○		Virgo
[Aries]	• ○		Libra
[Taurus)]	[ ]		Scorpio
[Gemini]	[ ]		Sagittarius
[Cancer]	[ ]		Capricorn
[Leo]	[ ]		Aquarius

Fr. 2

i		ii	iii		iv	v
		[῎Α]ρη[ϲ]			[᾽Αφ]ρ[οδίτη]	
[ ]		[Ζ]υγόϲ	* *		Ζ[υ]γόϲ	[
[ ]		[Ϲκό]ρ(πιοϲ)	○ •		[Ϲκόρ(πιοϲ)]	○ [
[ ]		[Τοξό]τ(ηϲ)	○		[Το]ξότ(ηϲ)	[
[ ]		[Αἰγό]κ(ερωϲ)	* •		Α[ἰ]γόκ(ερωϲ)	[
[ ]		[῾Υδρ]οχ(όοϲ)	* •		῾Υδρο[χ(όοϲ)]	[
[ ]		[Ι]χθύ(εϲ)	○ ○		᾽Ιχθύ(εϲ)	[
*		Κριόϲ	* •		[Κριόϲ]	[
[ ] ○		Ταύρο(ϲ)	*		[Ταύρο(ϲ)]	[
[ ]		Δίδυ(μοι)	○ ○		[Δίδυ(μοι)]	[

Fr. 3

i	ii
Ϲκόρ(πιοϲ)	[
[Τ]οξότ(ηϲ)	○ ○
[Α]ἰγόκ(ερωϲ)	○
[῾Υ]δροχ(όοϲ)	○
᾽Ιχθύ(εϲ)	*
[Κ]ριόϲ	○

Fr. 4

i	ii
[Λέω]ν	• •
[Π]άρθ(ενοϲ)	• [
Ζυγό(ϲ)	*
Ϲκόρ(πιοϲ)	• •
[Τοξότ(ηϲ)]	• •

Fr. 5

Κρόν[οϲ]	
Κριόϲ	
[Τ]αύρ(οϲ)	
[Δί]δυ(μοι)	

Fr. 6

*
2 • [

Fr. 2

i		ii	iii		iv	v
		Mars			Venus	
[ ]		Libra	* *?		Libra	[
[ ]		Scorpio	○ ●		[Scorpio]	○ [
[ ]		Sagittarius	○		Sagittarius	[
[ ]		Capricorn	* ●		Capricorn	[
[ ]		Aquarius	* ●		Aquarius	[
[ ]		Pisces	○ ○		Pisces	[
*		Aries	* ●		[Aries]	[
[ ] ○		Taurus	*		[Taurus]	[
[ ]		Gemini	○? ○		[Gemini]	[

Fr. 3

i	ii
Scorpio	[
Sagittarius	○ ○
Capricorn	○
Aquarius	○
Pisces	*
Aries	○

Fr. 4

i	ii
Leo	● ●?
Virgo	● [
Libra	*
Scorpio	● ●
[Sagittarius]	● ●?

Fr. 5

Saturn	
Aries	
Taurus	
Gemini	

Fr. 6

*
● [

4227. Sexagesimals

λγ λε [
λα μα [
κθ μζ κθ ιε [
κζ να λε λδ̣ [
κε νδ κθ μγ̣ [
κ̣γ̣ κ̣ε̣ ọ̄ ε̣ λ[

4228. Sexagesimals

i	ii	iii	
[ρκζ β μβ]	[ϲλβ νζ ιη]	ϲπς̣ νζ ι̣[η]	[γ κς β]
[ρκδ κη κη]	ϲλε [λα λβ]	ϲπγ λα λβ̣	[γ κε μς]
[ρκα νγ] ν̣η	ϲλη ς β	ϲπ° ς β	[γ κε λ]
[ριθ ιθ ι]β	ϲμ μ μη̣	ϲος μ μη	γ̣ [κε ιδ]
[ριζ μ]δ̣ ις	ϲμγ ιε ν	ϲογ ιε ν	γ̣ [κδ νη]
[ριδ η] νβ	ϲμε να η	ϲξθ να η	γ κ[δ μβ]
[ρια λ]γ̣ ιη	ϲμη κς μβ̣	ϲξς κς μβ	γ κ[δ κς]
[ρη ν]ζ̣ κη	ϲνα β λβ̣	ϲξγ β λβ	γ κ̣[δ ι]
[ρς κα κ]β	ϲνγ [λη] λ̣η̣	ϲ̣νθ λη λ̣η̣	γ [κγ νδ]
[ργ με] ọ̄	ϲνς ι̣ε̣ ọ̄	ϲ̣νς ιε ọ̄	γ̣ κ̣[γ λη]

4227. Sexagesimals

33 35 [
31 41 [
29 47 29 15 [
27 51 35 34 [
25 54 29 43 [
23 55 0 5 3[x

4228. Sexagesimals

i	ii	iii	
[127 2 42]	[232 57 18]	286 57 1[8]	[3 26 2]
[124 28 28]	235 [31 32]	283 31 32	[3 25 46]
[121 53] 58	238 6 2	280 6 2	[3 25 30]
[119 19 1]2	240 40 48	276 40 48	3 [25 14]
[116 4]4 16	243 15 50	273 15 50	3 [24 58]
[114 8] 52	245 51 8	269 51 8	3 2[4 42]
[111 3]3 18	248 26 42	266 26 42	3 2[4 26]
[108 5]7 28	251 2 32	263 2 32	3 2[4 10]
[106 21 2]2	253 [38] 38	259 38 38	3 [23 54]
[103 45] 0	256 15 0	256 15 0	3 2[3 38]

4229. Sexagesimals

Before corrections			Corrected
i	ii	iii	ii
] μθ̣	δ̣ [ ]	[	[ ]
]ζ λ	δ μη νη	α .[	δ μθ ιδ
]θ̣ δ	δ μη μβ	α [	δ μη κϛ
] μϛ	δ μη κϛ	α̣ [	δ μζ λη̣
] .ϛ	δ μη ι̣	α [	δ μϛ ν
]	δ μζ ν̣ .	α [	δ μϛ .
] .	δ μ̣[ζ . .]	α [	[ ]
] .	δ μ[ζ . .]	[	[ ]
] .	δ μζ̣ .[.]	[	[ ]
]	δ̣ μ̣ϛ̣ .[.]	[	[ ]
]	δ μϛ λ̣ .	[	[ ]
]	δ μϛ ιθ̣	α [	δ μα ιδ
]	δ με̣ ν̣γ̣	α̣ [	δ μ κϛ
]	δ̣ μ̣ε̣ λ̣ .	[	δ [λ]θ . .

4229. SEXAGESIMALS

Before corrections i	ii	iii	Corrected ii
] 49	4 [ ]	[	[ ]
]7 30	4 48 58	1 x[	4 49 14
]9 4	4 48 42	1 [	4 48 26
] 46	4 48 26	1 [	4 47 38
] x6	4 48 10	1 [	4 46 50
]	4 47 5x	1 [	4 46 x
]x	4 4[7 xx]	1 [	[ ]
]x	4 4[7 xx]	[	[ ]
]x	4 47 x[x]	[	[ ]
]	4 46 x[x]	[	[ ]
]	4 46 3x	[	[ ]
]	4 46 19	1 [	4 41 14
]	4 45 53	1 [	4 40 26
]	4 45 3x	[	4 [3]9 xx

4230. Tabular Scraps

(1), along the fibres

	i	ii
	α̣	γ̣
	α	γ
3	α	γ

across the fibres

	i	ii	iii
	]		[
	]	β	ᴗ λ ̣ [
	]̣	β	δ ι κ̣̄[
4	]	β	ζ [

(2)

	[β]	ζ̣
	[β]	ιε
	β	κγ
	β	λα
5	β̣	λθ
	[β]	μϛ
	[β]	νδ

(3)

	i	ii
	λθ	λ̣[
	μζ	α̣ [
3	νε	λ [

4230. TABULAR SCRAPS

(1), along the fibres

	i	ii
	1	3
	1	3
3	1	3

across the fibres

	i	ii	iii
	]		[
	]	2	0 3x [
	]x	2	4 10 ...[
4	]	2	7 [

(2)

	[2]	7
	[2]	15
	2	23
	2	31
5	2	39
	[2]	46
	[2]	54

(3)

	i	ii
	39	3[x
	47	1 [
3	55	30 [

4231. Tabular Scraps

(1) Template

i	ii
κ̣ε	ο̄ [
	ο̄ [
	ο̣̄ [
	ο̣̄ [
	ο̣̄ [
λ	[
	[
	[
	[

(2) Mean motions?

i	ii	iii	
[κ]ϛ̣	[ ]	[ ]	[
κζ	[ ]	[ ]	[
κη	ρλζ	νζ	λ̣[
κθ	ρπγ	δ	ι̣[
λ	cκη	ι	μ̣[

4231. Tabular Scraps

(1) Template

i	ii
25	0 [
	0 [
	0 [
	0 [
	0 [
30	[
	[
	[
	[

(2) Mean motions?

i	ii	iii	
[2]6	[ ]	[ ]	[
27	[ ]	[ ]	[
28	137	57	3[x
29	183	4	1[x
30	228	10	4[x

4231a. Tabular Scraps

(1)

i	ii
ε	μα̣
ε	νζ
ϛ	ιβ
ϛ	κ̣ .
ϛ	[
ϛ̣	[

(2)

Along the fibres

i	ii
] .	κ̣ .
]ε	ϛ
]γ̣	α̣
]α	ιβ
]μ̣θ̣	[ ]

Across the fibres

	i	ii
	η	λ̣[
	η	.[
3	η	κ̣[

4231a. Tabular Scraps

(1)

i	ii
5	41
5	57
6	12
6	2x
6	[
6	[

(2)

Along the fibres

i	ii
]x	2x
]5	6
]3	1
]1	12
]49	[ ]

Across the fibres

i	ii
8	3[x
8	x[
8	2[x

4232. Tabular Scraps

(1)

Fr. 1

i	ii	iii	iv
['Ιχ]θ̣(ύεϲ)	γ	'Ιχθ(ύεϲ)	δ̣
['Ιχ]θ(ύεϲ)	δ	'Ιχθ(ύεϲ)	ε
['Ιχ]θ(ύεϲ)	ε	[Κρ]ιό(ϲ)	α
[Κρι]ό(ϲ)	α	[Κρι]ό̣(ϲ)	β
[Κρ]ιό(ϲ)	β	Κ̣[ριό̣(ϲ)]	γ
[Κριό(ϲ)]	γ	.[ ]	[α?]
[ ]	α	.[ ]	[ ]

Fr. 2

α
β
α
α

Names of zodiacal signs abbreviated by truncation.

(2)

Fr. 1

] λ̣ λζ
] κ̣ζ λδ
] κ̣δ λα
] κ̣α κη
ι]η̣ κε
ι]ε κβ
ι]β̣ ι̣ . .

Fr. 2

] ν̣ε ν̣ . [
] . μζ κθ λ
]θ δ ιε
] . λθ ō

Fr. 3

[ ]
[ι]ε ι . [
ιϛ κ[
ιζ .
ιη [

4232. Tabular Scraps

(1)

Fr. 1

i	ii	iii	iv
Pisces	3	Pisces	4
Pisces	4	Pisces	5
Pisces	5	Aries	1
Aries	1	Aries	2
Aries	2	Aries	3
[Aries]	3	...	[1?]
[ ]	1	...	[ ]

Fr. 2

1
2
1
1

(2)

Fr. 1

] 30 37
] 27 34
] 24 31
] 21 28
1]8 25
1]5 22
1]2 1x

Fr. 2

] 55 5x [
]x 47 29 30
]9 4 15
]x 39 0

Fr. 3

[ ]
[1]5 1x [
16 2[x
17 x
18 [

(3)

i	ii
[Κριοῦ]	ιη
Ταύρου	ις
Διδύμων	ιε
[Καρκί]ν̣ου	ιδ
[Λέοντο]c	ιβ

– – – – – – – – – – –

(4)

– – –

κ
κε
λ
λε
μ
με

– – –

(3)

i	ii
[Aries]	18
Taurus	16
Gemini	15
Cancer	14
Leo	12

(4)

20
25
30
35
40
45

4233. Tabular Scraps

(1)

i	ii	iii
]	κγ	λ̣[
]ς̣	λε	κ[
] ̣	μζ	ι̣[
]	νθ	[
] ̣	ι̣	[

(2)

[ι]γ λ̣[
ιγ λ̣[
ιγ λα
ιγ λ
ιγ κ[
[ι]γ̣ [

(3)

ο̄
λ
ο̄
λ
ο̄
λ

4233. Tabular Scraps

(1)

i	ii	iii
]	23	3[x
]6	35	2[x
]x	47	1[x
]	59	[
]x	10	[

(2)

[1]3 3[x
13 3[x
13 31
13 30
13 2[x
[1]3 [

(3)

0
30
0
30
0
30

4234. Tabular Scraps

(1)

i	ii	iii
]	[ ]	β̣
]	.	γ
]	ϛ̣	δ
]	γ̣	ε̣
]	ε	ō
]	ϛ	α
]	ō	β
]	α	γ
.	γ	ō
ε̣	δ	α
.	ε	β
]	ϛ̣	γ
β̣	.	ō
]	γ	α
]	δ	β̣
]	ε	γ̣
]	ϛ	ō
]	. .	
]	. . .	δ̣ .[
]	[ ]	.[

4234. TABULAR SCRAPS

(1)

i	ii	iii
]	[ ]	2
]	x	3
]	6	4
]	3	5
]	5	0
]	6	1
]	0	2
]	1	3
x	3	0
5	4	1
x	5	2
]	6	3
2	x	0
]	3	1
]	4	2
]	5	3
]	6	0
]	...	
]	...	...[
]	[ ]	...[

(2)

Fr. 1

– – – –

] ọ̄
]γ̣ κ
]ς̣ μ
] ō
]γ̣ κ
]ς μ
] ō

– – – –

Fr. 2

– – – –

κς μ̣
ō ō
]γ̣ κ

– – – –

4235. Tabular Scraps

(1)

– – – – –

i	ii
]β̣	υν̣[
]ζ̣	φι̣[

– – – – –

(2)

– – – – –

i	ii
]	ιγ
]̣	ιδ
]	ις
]	[̣]̣

– – – – –

(2)

Fr. 1

	]0
	]3 20
	]6 40
	]0
5	]3 20
	]6 40
	]0

Fr. 2

	26 40
	0 0
3	]3 20

4235. Tabular Scraps

(1)

	i	ii
	]2	45[x
2	]7	51[x

(2)

	i	ii
	]	13
	]x	14
	]	16
4	]	[x]x

Part V

Horoscopes

4236. Horoscope, A.D. 63

ἀγαθ]ῆ τύχῃ. γένεcιc Ἀπ[
(ἔτουc) ι] Νέρωνοc, Ἁθὺρ κ̄η̄, ὥραc [‾ ̣
τῆc ἡ]μέραc.
ἥλιοc] ἐν Τοξότηι μοί(ρᾳ) [‾ ̣, ζῳδίῳ καὶ
ὁρίοιc Δι]ό̣c.
cελήνη] ἐν Δ̣ιδύμ(οιc) μοί(ρᾳ) [‾ ̣ ̣, ζῳδίῳ Ἑρμοῦ,
ὁρίοι]c̣ Ἀφροδε(ίτηc).
Κρόνοc] ἀνατολικ̣(ὸc) ἑῷοc [ἐν Λέοντι μοί(ρᾳ)
‾ ̣ ̣, ζῳδίῳ] ἡ̣λίου, ὁρίοιc ἰδ[ίοιc.
Ζεὺc ἐ]ν̣ Α̣ἰ̣[γ]ό̣κ̣[ε]ρ̣ῳ μοί(ρᾳ) [‾ ̣ ̣ ζῳδίῳ Κρόνου
— — — — — — — — — — — — — — —

2 ∟ 4 μ^{οι} (so 6, 10) 6 διδυ̣^{μ} 7 αφροδε (a superscript may be lost), *l.* Ἀφροδίτηc
8 ανατολικ 10 α̣ι̣[γ]ο̣κ̣[ε]ρ̣^{ω}

4236a. Horoscope, A.D. 80

Ζεὺc ἐν Διδ(ύμοιc)
Κρόνο(c) Ὑδρηχ(όῳ)
Ἑρμῆc Παρθέν(ῳ)
οἱ λοιποὶ ἐν Λέοντι

1 δι^{δ} 2 κρο^{ν}; υδρη^{χ}, *l.* Ὑδροχόῳ 3 παρθε^{ν} 4 λεοντι̅

4236. Horoscope, a.d. 63

With good fortune. Nativity of Ap[
Year 10 of] Nero, Hathyr 28, hour [x
of the] day.
Sun] in Sagittarius [x°, house and
terms] of Jupiter.
Moon] in Gemini [xx]°, [house of Mercury,
Terms of Venus.
Saturn] rising in the east [in Leo
xx°, house] of the sun, his own terms.
Jupiter in] Capricorn [xx]°, [house of Saturn...

4236a. Horoscope, a.d. 80

Jupiter in Gemini.
Saturn in Aquarius.
Mercury in Virgo.
The rest in Leo.

4237. Horoscope, A.D. 84

(ἔτους) δ Δομιτ̣ιανοῦ, μηνὸς Θὼθ
$\overline{θ}$, ὥρ̣ᾳ $\overline{ϛ}$ ἡ̣μέρας.
ἥλ̣ις ἐ̣ν̣ Παρθ(ένῳ) μ̣ο̣ί̣(ρ̣ᾳ) $\overline{ι̣α̣}$, οἴ̣κ̣ῳ Ἑρμ̣ο̣ῦς,
ὁρίο̣ις Ἀφ̣ροδεί̣τ̣ης.
σελήνη ἐν Κριῷ̣ μ̣ο̣ί̣(ρ̣ᾳ) $\overline{κγ}$ L, οἴκῳ
Ἄρεως, ὁρίοις Κρόνου.
Κρόνος ἐν Κ̣[ρι]ῷ μοί(ρᾳ) $\overline{κδ}$, οἴκῳ
[Ἄρεως, ὁρίοις Ἄρεως.]
— — — — — — — — — — — — — — — —

1 L   3 *l.* ἥλιος; $\overset{οι}{μ}$ (so 5, 7); *l.* Ἑρμοῦ   4 *l.* Ἀφροδίτης

4238. Horoscope, A.D. 117

— — — — — — —
σ̣ε̣λ̣[ήν]η̣ [
Κρόνος Διδύμ[οις
Ζε̣ὺς Καρκίν̣[ῳ
Ἄρης Τα̣[ύ]ρ̣ῳ
Ἀ̣φροδ̣είτη̣ Λ̣έον[τι
Ἑρ]μῆς Λέον[τι
ὡροσκόπο̣ς Σ̣κ̣[ορπίῳ
μεσουράνημα ̣[

5 *l.* Ἀφροδίτη

4237. Horoscope, a.d. 84

Year 4 of Domitian, month Thoth 9, hour 6 of day. Sun in Virgo 11°, house of Mercury, terms of Venus. Moon in Aries 23½°, house of Mars, terms of Saturn. Saturn in Aries 24°, house [of Mars, terms of Mars.]

— — — — — — — — — — — — — — — — —

4238. Horoscope, a.d. 117

— — — — — — —

Moon [Saturn in Gemini. Jupiter in Cancer. Mars in Taurus. Venus in Leo. Mercury in Leo. Ascendant in Scorpio. Midheaven in …

4239. Horoscope, A.D. 130

ἀγαθῇ τύχῃ. γ]έ̣νεϲ[ιϲ ̣ ̣] ̣ ̣ ̣ ̣πο[̣ ̣] ̣α. ιδ (ἔτουϲ) Ἁδραιανοῦ
Αἰλίου τοῦ κυρ]ίου, Τῦβι̣ ιθ, ὥρα̣ϲ ε̣ ἡμέραϲ καθ' Ἑλήνου
ἥλιοϲ ἐν Αἰγό]κειρῳ [μ]οιριῶν κ̅. Ἄρ̣ηϲ `Ἄρ̣ηϲ´ ἐν Αἰγόκε̣ρῳ μοιριῶν ι̅
λεπτῶν ̣ ̣ . Ἑ]ρ̣μῆϲ ἐν Αἰγόκερῳ γ̅ λεπτῶν γ̅. ϲελήνη ἐν
Λέοντι μοιρι]ῶ̣ν θ̅ λε̣π̣τ̣ῶν κε̣̅. Ζε̣ὺϲ ἐν Λέοντι μοιρ̣ιῶν θ̅.
Κρόνοϲ ἐν Ϲκορπ]ί̣ῳ μοιρ̣ιῶ[ν ̣ ̣ λ]ε̣π̣τ̣ῶ̣ν β̅. Ἀφροδείτη ἐν Ὑδρο̣χ̣[ό]ῳ̣ μ̣ο̣ι̣ρ̣ιῶ̣ν̣

– –

1 ʟ ; *l.* Ἁδριανοῦ 2 *l.* Ἕλληναϲ 3 αρηϲ, though seemingly written correctly in the text, is written again in the space above 3 *l.* Αἰγόκερῳ ; *l.* μοιρῶν (*bis*, so 5, 6 *bis*) 6 *l.* Ἀφροδίτη

4240. Horoscopes, A.D. 150, 183, and undated

– – – – – – – – – – – – –

γ̣έ̣ν̣[εϲ]ι̣[ϲ
(ἔτουϲ) ι̅δ̅ Αἰλίου Ἀ̣[ντωνίνου,
Φαῶφι κ̅δ̅, ὥρ(α) ι̣[̣] ν̣[υκτόϲ,
ἧλιϲ Ἑρμῆϲ Ἀφροδείτ[η Ϲκορπίῳ
Κρόνοϲ Καρκίνῳ
Ζε̣ὺϲ Ταύρῳ
Ἄρηϲ Αἰγόκερῳ
Ϲελήνη Διδύμοι[ϲ
ὡροϲκοπεῖ Ζυγ[όϲ

γ̣[ένεϲ]ι̣[ϲ] Ϲ̣αραπο̣ῦτο̣[ϲ
(ἔτουϲ) κγ Αὐρηλί[ου
Ἀντωνίνου Καίϲαρ[οϲ
κυρί[ο]υ, Π̣[α]ῦ̣ν̣ι̣ ̣[̣] ὥρ̣(α) [
ἡμέραϲ

γένεϲιϲ Α ̣[
ἔ̣(τουϲ) [

– – – – – – – – – – – – –

2 ʟ (so 11, 15) 3 ϕ (so 13) 4 *l.* ἥλιοϲ ; *l.* Ἀφροδίτη 13 π̣[α]ῦ̣ν̣ι̣

4239. Horoscope, a.d. 130

Good Fortune.] Nativity of ..., year 14 of Hadrian Aelius the] lord, Tybi 19, hour 5? of day according to the Greeks. Sun in] Capricorn 20°. Mars in Capricorn 10° xx'.] Mercury in Capricorn 3?° 3'. Moon in Leo] 9° 25'. Jupiter in Leo 9°. Saturn in] Scorpio [xx]° 2'. Venus in Aquarius [x]°...

4240. Horoscopes, a.d. 150, 183, and undated

Nativity [of ... Year 14 of Aelius Antoninus, Phaophi 24, hour 1[x] of night. Sun, Mercury, Venus [in Scorpio.]

Saturn	in Cancer.
Jupiter	in Taurus.
Mars	in Capricorn.
Moon	in Gemini.
Ascendant	Libra.

Nativity of Sarapous. Year 23 of Aurelius Antoninus Caesar the lord, Payni xx, hour [x] of day.

Nativity of A[...] Year [...]

4241. Horoscopes, a.d. 162, 188, 191, and undated

] (ἔτους) κη Κọ[μμόδου
] .κ̣α̣. ̣ ̣ ς̄ .[
] Ἑρμ(ῆς) Λέο(ντι) Ἀφ[ροδίτη
]
] . . . .ρ. .() (ἔτους) γ
] Θ̣ω̣ὺ̣θ ε̄ ὥρ(ᾳ) ς̄
] (Ἄρης?) Ζε(ὺς) Διδύ(μοις)
Ἑρ]μ̣(ῆς) Λέ(οντι) ὡροσκ(όπος) Σκο(ρπίῳ)
]. . . . . .ε()
] (σελήνη) Σκ[ο](ρπίῳ) Ἑρμ̣(ῆς) Ζε(ὺς) Ζυγ(ῷ)
] Διδύ(μοις) Κρ(όνος) Τ̣ο̣ξ(ότῃ)
]
] Ἑ̣ρμ̣(ῆς) Ζυγ(ῷ) Ζ̣ε̣ὺ̣[ς] Σκο(ρπίῳ)

1 L (so line 5)   3 $\varepsilon\rho^{\mu}$ (so lines 10, 12, and, apparently, 8); $\lambda\varepsilon^{o}$ (so line 8)   5 uncertain traces of bottoms of letters, followed by ρ, then a smudged letter with a superscript   6 [symbol]   7 [symbol]; ζ^{ε} (so line 10); δ̣ι̣δ̄ (so line 11)   8 $\omega\rho o\sigma^{\kappa}$; $\sigma\kappa^{o}$ (so line 12 and, apparently, 10)   9 [symbol] (the first symbol perhaps is ⲫ)   10 ☾; $\zeta\upsilon^{\gamma}$ (so line 12)   11 [symbol]; [symbol]

4242. Horoscope, a.d. 212

γένεσις Πτολεμαίου. (ἔτους) κʹ Σεουήρου καὶ Ἀντωνίνου,
Παχὼν κ̄β̄ (ὥρᾳ) γ̄ ἡμέραις.
(ὡροσκόπος) ὡροσκόπος, Ζεὺς Καρκίνῳ.
ἥλις, Ἑρμῆς Ταύρῳ.
Ἀφροδίτη Κρειῷ.
Ἄρης, Κρόνος Λέοντι.
σελήνη Τοξότῃ.
)———

1 L   2 ⲫ; *l.* ἡμέρας   3 ⲫ   4 *l.* ἥλιος   5 *l.* Κριῷ

4241. Horoscopes, a.d. 162, 188, 191, and undated

] year 28 of Commodus,
] ... 6 ...
] Mercury in Leo; Venus
]
] ... year 3
] Thoth 5, hour 6.
] Mars?, Jupiter in Gemini.
] Mercury in Leo. Ascendant in Scorpio.
]...
] moon in Scorpio. Mercury, Jupiter in Libra.
] in Gemini. Saturn in Sagittarius.
]
] Mercury in Libra. Jupiter in Scorpio.

4242. Horoscope, a.d. 212

Nativity of Ptolemaios. Year 20 of Severus and Antoninus,
Pachon 22, hour 3 of day.
Ascendant, Jupiter in Cancer.
Sun, Mercury in Taurus.
Venus in Aries.
Mars, Saturn in Leo.
Moon in Sagittarius.

4243. Horoscope, a.d. 217

— — — — — — — —

Ἄρηϲ Ἑρμῆϲ̣
ἥ̣λ̣ι̣ο̣ϲ Ταύρῳ
Ἀφροδείτ(η) Ἰχθύϲι
ὡροϲκόπον Τ̣[ο]ξ̣ότη
Ζ̣ε̣ὺ̣ϲ̣ Αἰγόκαιρῳ
Κρόνοϲ Ζυγῷ̣
ϲελήνη⟨ν⟩ Λέοντι
8–11 (*illegible traces*)

2 αφροδει^{τ}, *l.* Ἀφροδίτη   4 *l.* ὡροϲκόποϲ   5 *l.* Αἰγόκερῳ   7 ϲεληνη̅, *l.* ϲελήνη

4244. Horoscope, a.d. 217

— — — — — — — —

̣ ̣[

γένεϲιϲ Ἀλι ̣ ̣[
τ̣ο̣ῦ̣ κορύφου Θὼθ̣ [
δ̅ ἡμέραϲ ̣ ̣ ̣[
Ζεὺϲ Τοξότη̣ [
Ἄρηϲ Ἀφροδε[ίτη

— — — — — — — —

6 *l.* Ἀφροδίτη

4243. Horoscope, a.d. 217

– – – – – – – –
Mars, Mercury,
sun in Taurus.
Venus in Pisces.
Ascendant in Sagittarius.
Jupiter in Capricorn.
Saturn in Libra.
Moon in Leo.
8–11 (*illegible traces*)

4244. Horoscope, a.d. 217

– – – – – – – –
. . .

Nativity of Ali… [
the catamite, Thoth [
4 of day …[
Jupiter in Sagittarius. [
Mars, Venus [
– – – – – – – –

4245. Deluxe horoscope, A.D. 218

Χοίακ] α̣ ε̅ἰ̅ς̣̅ [β
] .ς̣κ̣[
]α̣ρ̣α̣τ̅[
] ἡ̣λ̣ί̣ο̣υ̣ ἦν [
ἐν ζ]ῳδίῳ ἀρρεν̣[ι]κ̣ῷ̣ δισώμῳ Τ̣[ο]ξό̣[τῃ
μοιρ]ῶ̣ν ζ̅ λεπ̣τ̣ῶ̣ν ν̣̅ε̅ οἴκ̣ῳ Δι[ὸ]ς ὁρ̣[ίοις] Δ̣ι̣ὸ̣ς̣
ἐπὶ τοῦ] πρώτου δεκα̣νοῦ̣. ἡ δὲ θεὰ κ̣αὶ σελας̣φ(όρος)
. . ἦν] ἐν ζῳδίῳ θηλυκῷ δισώμῷ Παρθέν̣ῳ
μοιρῶ]ν ϛ̅ λεπτῶν ν̅α̣̅ [οἴκ]ῳ Ἑρμοῦ ὁρ̣ίοις [Ἑρ-
μ]ο̣υ ἀφαιροῦσα τῷ φ[ ]αθουσα τ̣ο̣ῖς ἀριθ[μοῖς
π̣λάτος ἔχουσα ἐν [τῷ ‾ . βαθμ]ῷ ἐπὶ το̣ῦ̣ α̅ δεκ̣[ανοῦ.
ὁ δὲ] τ[ο]ῦ Κρόνου ἦν ἐν [ζῳδίῳ στερε]ῷ̣ Σκορπίῳ
μ̣ο̣ιρῶν ι̅α̅ λεπτ(ῶν) λ̣̅[‾ οἴκ̣ῳ Ἄρεως ὁρίο]ις Ἑρμοῦ
ἀν]ατολ[ι]κὸς ἑῷος ἐ̣[πὶ τοῦ β̅ δ]εκα̣ν̣[ο]ῦ̣.
ὁ δὲ] τ̣ο̣ῦ Διὸς ἦν ἐν ζ[ῳδίῳ ἀρρενι]κ̣ῷ̣ ς̣[τερεῷ
Ὑδ]ροχ[ό]ῳ μοιρῶν δ̅ [λεπτῶν ‾ . . οἴ]κ[ῳ Κρόνου
ὁρ]ίοις Ἑρμοῦ ἀνατ[ολικὸς ἑῷο]ς̣ ἐ̣[πὶ τοῦ α̅ δεκανοῦ.
ὁ δὲ τ]οῦ Ἄ̣ρεως ἦν ο .[ἐν ζῳδίῳ] ἀ̣[ρρενικῷ
στ]ε[ρ]εῷ Ὑδροχῷ [μοιρῶν ‾ . . λε]π̣τ̣(ῶν) [‾ . .
ο]ἴ̣[κ]ῳ̣ Κρόνου ὁρίο[ις ἐπὶ το]ῦ̣ τρ[ίτου δεκανοῦ.
ὁ δ]ὲ̣ τῆς Ἀφροδείτη̣[ς ἀστὴρ ἦν ἐ]ν ζ[ῳδί]ῳ ἀρ-
ρ]ενικῷ τροπικῷ [Ζυγῷ μοιρ]ῶ̣ν [. .] λ̣επτ(ῶν) μ̅ε̅
ο]ἴκῳ ἰδίῳ ὁρίοι̣[ς ἐπὶ τοῦ] α̣̅ [δε]κανοῦ.
ὁ] δ̣ὲ̣ τοῦ Ἑρμοῦ̣ ἀ̣στ[ὴρ ἦν ἐν] ζ̣[ῳδ]ίῳ θηλυ-
κ̣ῷ̣ ς̣τερεῷ̣ Σ̣κ̣ορπ̣ί̣[ῳ μοι]ρ̣ῶ̣ν κ̣̅ε̅ λεπτ(ῶν) ν̅
οἴκῳ Ἄρεως ὁρίοις Κ[ρόνο]υ ἐπὶ τ̣οῦ τρίτου
δ̣ε̣κ̣ανοῦ προπορε[υό]μ̣ε̣ν̣ο̣ς̣ τ̣[οῦ Κ]ρ̣ονου
ἐν τῇ ἑῴα̣ φάσει ἐπ̣ὶ [δ]ύ̣σιν ἐρχ̣[όμενος.
ὁ] δ̣ὲ̣ τῆς γενέσεως ὡρο[ς]κόπος [

7 σελας̣ with φ written above   13 λεπ' (so lines 19, 22, 25)   19 *l.* Ὑδροχόῳ   21 *l.* Ἀφροδίτης

4245. Deluxe horoscope, a.d. 218

... Choeac] 1/[2 ...

...

...

... of the sun was ... [in] a masculine bicorporal sign, Sagittarius, at 7 degrees 55 minutes, house of Jupiter, terms of Jupiter, [at] the first decan. The goddess and light-bearing (moon?) [was] in a feminine bicorporal sign, Virgo, at 6 degrees 51 minutes, house of Mercury, terms of Mercury, diminishing the ... and ...ing the numbers, having its latitude in step [x], in the 1st decan. [The (star)] of Saturn was in a solid? [sign], Scorpio, at 11 degrees 3[x] minutes, [house of Mars], terms of Mercury, rising in the east, in the [2nd] decan. [The (star)] of Jupiter was in a masculine solid sign, Aquarius, at 4 degrees [xx minutes], house of Saturn, terms of Mercury, rising in the east, in [the 1st decan.] [The (star)] of Mars was in a masculine solid [sign], Aquarius, [at xx degrees xx] minutes, house of Saturn, terms of [...], in the third decan. [The star] of Venus [was] in a masculine tropical sign, [Libra], at [xx] degrees 45 minutes, (Venus') own house, terms of [...], in the 1st decan. [The] star of Mercury was in a feminine, solid sign, Scorpio, at 25 degrees 50 minutes, house of Mars, terms of Saturn, in the third decan, advancing ahead of Saturn, approaching setting in its morning visibility. [The] ascendant of the nativity...

4246. Horoscope, a.d. 223

(ἔτουϲ) β Ἀλεξ̣άνδρου, Τῦβι ε̄ εἰϲ ϛ̄, ὥραϲ
ζ ν̣υ̣κ̣τὸϲ. ὡ̣ροϲκόπον̣ Ϲκορπίῳ. Κρόνοϲ̣,
Ἑρμῆϲ Τοξό[τ]ῃ. ἥλιον Αἰγόκερῳ.
Ἀφρ[ο]δείτῃ Ὑδρηχῷ. Ἄρην ⟦Ἰχθύϲι⟧
Κρ[ι]ῷ. ϲε̣λ̣ήνη Ταύρῳ. Ζεὺϲ
] Διδύμοιϲ.

1 L   2 *l.* ὡρόϲκοποϲ   3 *l.* ἥλιοϲ   4 *l.* Ἀφροδίτη ; *l.* Ὑδροχόῳ ; *l.* Ἄρηϲ ; ιχθυϲι crossed out

4247. Horoscopes, a.d. 223

— — — — — — — — — — — — — — — —
] ζ̣ ὥρ(ᾳ) ε [νυκτόϲ
ἥλι]ϲ̣ [Ἑρμῆϲ] Ἀφρο[δίτῃ Τοξότῃ
Κρόνοϲ Ἐγόκ̣[ερῳ
ϲ]ελήνῃ̣ Ἰχθύ[ϲι
Ἄραιϲ Ϲκορπίῳ
Ζεὺϲ ὡρ(οϲκόποϲ) ⟦ζ̣⟧ `Λ´έοντι
⟨ἀγ⟩άθῃ τύχῃ

(m. 2) ̣ ἔτουϲ γ Ἀλεξά̣[νδ]ρ̣ο̣υ̣ Χοίακ ιϛ̄ εἰϲ ιζ̄
ὥρ(ᾳ) ε νυκτόϲ.
ἥλιοϲ (m. 3) ˊ (m. 2) Ἑρμῆϲ (m. 3) ˊ (m. 2) Ἀφρ̣[ο]δ̣ι̣τ̣ῃ Τοξότῃ (m. 3) ˊ
Κρόνοϲ Αἰγόκερῳ (m. 3) ˊ
(m. 2) Ϲ̣ελήνῃ Ἰχθύϲι (m. 3) ↝ Ὑδρηχόῳ ˊ
(m. 2) Ἄρηϲ Ϲκορπίῳ (m. 3) ↝ Τοξότῃ ˊ
(m. 2) Ζεὺϲ ὡ̣ροϲκό(ποϲ) Λέοντι (m. 3) ˊ
(m. 3) Ἀ̣φ̣ρ̣ο̣[δε]ί̣τ̣ῃ Ϲκορπίῳ ˊ

1 φ (so line 9)   2 *l.* ἥλιοϲ   3 *l.* Αἰγόκερῳ   5 *l.* Ἄρηϲ   6 φ ; ζ (?) crossed out, and λ written below   8 at the beginning of the line, a symbol with a long stroke descending below and hooked to the right, the top part broken off   12 *l.* Ὑδροχόῳ   15 *l.* Ἀφροδίτῃ

4246. Horoscope, a.d. 223

Year 2 of Alexander, Tybi 5/6, hour 7 of night. Ascendant in Scorpio. Saturn, Mercury in Sagittarius. Sun in Capricorn. Venus in Aquarius. Mars ⟦in Pisces⟧ in Aries. Moon in Taurus. Jupiter [...] in Gemini.

4247. Horoscopes, a.d. 223

– – – – – – – – – – – – – – – –

... hour 5 [of night.] Sun, [Mercury], Venus in Sagittarius. Saturn in Capricorn. Moon in Pisces. Mars in Scorpio. Jupiter, ascendant in Leo. Good fortune.

(m. 2) Year 3 of Alexander, Choeac 16/17, hour 5 of night. Sun (m. 3) ˊ (m. 2) Mercury (m. 3) ˊ (m. 2) Venus in Sagittarius (m. 3) ˊ Saturn in Capricorn. (m. 3) ˊ (m. 2) Moon in Pisces. (m. 3) ⸓ In Aquarius ˊ (m. 2) Mars in Scorpio. (m. 3) ⸓ In Sagittarius ˊ (m. 2) Jupiter, ascendant in Leo. (m. 3) ˊ (m. 3) Venus in Scorpio. ˊ

4248. Horoscope, A.D. 229

(ἔ̣τ̣ο̣υ̣ϲ̣) η, Παχὼν κ̣[β,
ὥ̣ρ(α) δ̅ περὶ ἄκρο̣[ν.
ἥ̣λιϲ, Ἑρμῆϲ Τα[ύρῳ.
Ἀ̣φροδίτη Διδ[ύμοιϲ.
Κ̣ρόνοϲ Εἰκθύ[ϲι.
Ζ]εὺϲ Αἰγόκερ[ῳ.
supr. Παχὼ]ν̣ κ̅α̅ Παχ̣ὼ̣ν̣ κ̅β̅
Ἄρ]ηϲ, ϲελήνη [Λέοντι.
ὡ]ρ̣οϲκόποϲ ⟦δ̣⟧[
Κ̣αρκίνῳ ι̣̅δ̣̅ [

1 (ἔτουϲ): a raised diagonal stroke at the edge of the papyrus; the symbol itself is broken off
2 ῳ̸ 3 *l.* ἥλιοϲ 5 *l.* Ἰχθύϲι 9 δ apparently crossed out

4249. Horoscope, A.D. 172 (?)

(ἔτουϲ) ι̅β̣̅ Κο̣[μμόδου
Παῦνι η̣̅ [
κατὰ̣ τοὺϲ Αἰ̣[γυπτίουϲ
ὥρα̣ γ̅ νυκ̣[τόϲ.

Κρόνο̣ϲ	Κριῷ
Ζεύϲ	Ἰχθ̣[ύϲι
Ἄρηϲ	Ταύρ̣[ῳ
Ἀφροδ̣ίτη	Διδύμοι[ϲ
Ἑρμῆϲ̣	Διδύμο[ιϲ
ἥλιοϲ	Διδύμο[ιϲ
ϲελήνη	Ἰχθύϲ̣ι̣
ὡροϲκόποϲ	Αἰγ[όκερῳ

διευτυχε̣ῖ

1 ∟ʹ

4248. HOROSCOPE, A.D. 229

Year] 8, Pachon [22,
hour 4 about the end.
Sun, Mercury in Taurus.
Venus in Gemini.
Saturn in Pisces.
Jupiter in Capricorn.
supr. Pachon? 21 Pachon? 22
Mars, moon [in Leo.
Ascendant ⟦...⟧
Cancer 14°? [

4249. HOROSCOPE, A.D. 172 (?)

Year 12? of Commodus?...
Payni 8 [...
according to the Egyptians...
hour 3 of night.

Saturn in Aries.
Jupiter in Pisces.
Mars in Taurus.
Venus in Gemini.
Mercury in Gemini.
Sun in Gemini.
Moon in Pisces.
Ascendant in Capricorn.

Good luck!

4250. Horoscopes, A.D. 237 and undated

— — — — — — — — — — —

ὠ̣]ρο̣ς[κό]π̣[ος] Δ̣ιδύμο̣ις
γ]ένεσις Σαραπ̣ί̣ων
(ἔ̣τ̣ο̣υ̣ς̣) γ Μαξιμείνου [Ἐπ]ε̣ὶφ $\overline{ιγ}$
ὥ]ρ̣α̣ $\overline{ε}$ ἡμέρας πλ̣η̣ρωμένης
ὠ]ρ̣ο̣σκόπος Ζυγῷ
ἥ]λ̣ι̣ς Ἑρμῆς Καρκίνῳ
Ἀφρ]οδε[ί]τη [Λέ]ο̣ντι
]ῳ̣

— — — — — — — — — — —

3 (ἔ̣τ̣ο̣υ̣ς̣) γ : two diagonal strokes above the line; at the bottom of the second stroke, a trace of a horizontal stroke that may be part of γ ; *l.* Μαξιμίνου 4 *l.* ὥρας 7 *l.* Ἀφροδίτη

4251. Horoscopes, A.D. 244 and undated

γένεσις Κ̣ . . ρ̣ι̣[
α (ἔτους) Φιλ[ί]ππων κ[ατ' Αἰγυ-
π̣τίων Φαρ(μοῦθι) κδ ο . . [
Μ]ε̣χεὶρ ιζ ὥρ(α) γ νυκ̣[τός.
ὡρ(οσκόπος) Κ]ρόνος Παρθέ(νῳ)
. .] . α̣ Ζυγ̣ῷ
ἥλιος Ἑ]ρ̣μ̣ῆς Ὑδροχόῳ
Ζεύς Ἰ]χθύ(σι)
Ἀφροδίτη] Κ̣ριῷ
? Τ]α̣ύρ̣ῷ
Ἄρης Κα]ρκ̣ίνῳ

] . [.]μου
Τ]ῦ̣βι $\overline{κη}$
Γ̣
]
⇓

4250. Horoscopes, A.D. 237 and undated

Ascendant in Gemini

Nativity of Sarapion.
Year 3 of Maximinus, Epeiph 13,
Hour 5 of day completed.
Ascendant in Libra.
Sun, Mercury in Cancer.
Venus in Leo.
...

4251. Horoscopes, A.D. 244 and undated

Nativity of K...
Year 1 of the Philips, according to
the Egyptians Pharmuthi 24, which is?
Mecheir 17, hour 3 of night.

Ascendant], Saturn	in Virgo.
...	in Libra.
Sun], Mercury	in Aquarius.
Jupiter]	in Pisces.
Venus]	in Aries.
...]	in Taurus.
Mars]	in Cancer.

...
...] Tybi 28
...

⇓

⇑

] ̣ϲ [Ϲ]κορπίῳ
] Τοξ(ότῃ)
] Α̣ἰ̣γό̣κ̣(ερῳ)
Ὑ]δ̣ρ̣η̣χ̣(όῳ)
] Ἰ̣χ̣θύ(ϲι)
]
traces of ends of 5 lines

19 *l.* Ὑδροχόῳ

4252. Horoscopes, a.d. 248/249 and undated

— — — — — — — — — — —
ὥρᾳ α̅ ἡμ[έ]ραϲ [
Κρόνοϲ Ζυγῷ̣ [
Ἄρηϲ Λέ̣οντι [
Ἀφ̣ροδί(τῃ) Ἑρμῆϲ [
ϲελ̣ήνη Καρκ[ίνῳ
(m. 2) Ἀφρο[δ]ίτηϲ ε ̣ ̣ ̣α ̣ ̣[

(m. 1) γέν[ε]ϲιϲ Νεικ[
(ἔτουϲ) ϛ Φιλίππω̣[ν
ἡμ̣[έραϲ
Κρόνοϲ Ϲκορ̣[πίῳ
Ἄρη̣ϲ Ἰχθ̣ύϲ̣[ι
Ἀφ[ρ]οδε[ίτῃ
ϲε[λ]ή̣νῃ [
(m. 3) ̣[*c. 4 letters*]ο̣υφ ̣[
— — — — — — — — — — —

4 αφ[ρ]οδ̣ 11 ϊχ̣θυ̣ϲ̣[12 *l.* Ἀφροδίτῃ

⇑

...]	Scorpio
...]	Sagittarius
...] Capricorn	
...] Aquarius	
...] Pisces	

22-26 *traces of ends of 5 lines*

4252. Horoscopes, A.D. 248/249 and undated

— — — — — — — — — — —

hour 1 of day. [
Saturn in Libra. [
Mars in Leo. [
Venus, Mercury [
Moon in Cancer. [
(m. 2) Venus' ...

(m. 1) Nativity of Neik...
Year 6 of the Philips [
 of day. [
Saturn in Scorpio. [
Mars in Pisces. [
Venus [
Moon [
(m. 3) ...

— — — — — — — — — — —

4253. Horoscope, a.d. 249/250

γένεϲιϲ [
α (ἔτουϲ) Δεκκί̣[ου
ὥρᾳ η̅ ἡ̣μ̣[έραϲ
ὡ̣ροϲκόπο[ϲ
ϲελήνη [
Ζεὺϲ Ζυγ[ῷ
Κρόνο̣ϲ Ἄ̣ρ̣[ηϲ Τοξότῃ
ἥλιϲ Ἑρμῆ[ϲ
Ἀ̣φ̣[ρ]ω̣δ[ί]τῃ [

2 ∟ ; *l.* Δεκίου 8 *l.* ἥλιοϲ 9 *l.* Ἀφροδίτῃ

4254. Horoscope, a.d. 254

		Ἄρ̣η̣ϲ
	ἥλιϲ	Τ̣αύρ̣ῳ
	Ἑρμῆϲ	Κρόν̣οϲ
Ταῦροϲ	Ἀφροδίτη	Ἐγόκαιρῳ
	Σκορπίῳ	ὥρα Κριῷ
	[. . . . . .].	[Ζ]ε̣ὺϲ Ὑτρηχῷ

2 *l.* ἥλιοϲ 4 *l.* Αἰγόκερῳ 5 *l.* ὡροϲκόποϲ 6 *l.* Ὑδροχόῳ

4253. Horoscope, a.d. 249/250

Nativity [
Year 1 of Decius [
Hour 8 of day. [
Ascendant [
Moon [
Jupiter in Libra. [
Saturn, Mars [in Sagittarius.
Sun, Mercury [
Venus [

4254. Horoscope, a.d. 254

		Mars
	Sun	in Taurus.
	Mercury	Saturn
Taurus	Venus	in Capricorn.
	in Scorpio.	Ascendant in Aries.
	...	Jupiter in Aquarius.

4255. Horoscope, A.D. 256

ὡρ(οσκόπος) Τοξότης
Κ̣ρ̣ο̣ν̣[ο]ς̣ Ἡ̣δρυχόῳ
Ζ̣ε̣[ὺς] σελήνης̣ κ̣αὶ Ταύρῳ
Ἄρ[ης] Διδ̣ύμοις
ἥλ̣[ιος] Ἑρμ̣ῆς Παρθέν̣ης
Ἀφροδίτη Ζ̣υγῷ

1 ϕ ; *l.* Τοξότῃ   2 *l.* Ὑδροχόῳ   3 *l.* σελήνη   5 *l.* Παρθένῳ

4256. Horoscope, A.D. 269–275

(ἔτους) ̣]´ Αὐρηλιανο̣[ῦ
ὥρ(α̣) [*vac.?*] ε̣ πληρωμέν[η
κατ̣[ὰ δὲ] τοὺς ἀρχαίους̣ [
ὥρ(α) ἡ π̣ρ̣[ο]κειμένη
ἥλιος Κριῷ
οἰκοὶ Ἄρεως̣ [
σελ̣ή[νη] Παρθέν̣[ῳ
— — — — — — — — — — —

2 ϕ (so line 4)   6 *l.* οἴκῳ

4257. Horoscope, A.D. 279

Παῦνι ιγ εἰς $\overline{\text{ιδ}}$, ὥρα̣ $\overline{\text{γ}}$ νυκτός. ὥρα Ἐγόκερῳ.
μης(ουράνημα) Ταύρῳ. •— Ζεὺς Κριῷ.) Κρόνος Σκορπίῳ.
Ἄ̣ρ̣η̣ς̣ Διδύμοις. • Ἀφ[ρ]ο̣δ̣(ί̣τ̣η̣) Διδύμοις
Ἑρμῆς Ταύρῳ. • ὡ̣ρ̣ο̣σ̣κ̣ό̣πος Αἰγόκ̣αιρῳ. ⫯ σελήνη Ζυγῷ.
[(ἔτους)] δ Πρόβ̣[ου]

1 ὥρα: *l.* ὡροσκόπος ; *l.* Αἰγόκερῳ   2 μης : *l.* μεσουράνημα   3 αφ[ρ]οδ̣ ̣ ̣ (letters after gap very uncertain)   4 *l.* Αἰγόκερῳ

4255. Horoscope, a.d. 256

Ascendant in Sagittarius.
Saturn in Aquarius.
Jupiter, moon and(?) in Taurus.
Mars in Gemini.
Sun, Mercury in Virgo.
Venus in Libra.

4256. Horoscope, a.d. 269–275

Year x] of Aurelian [
hour 5 completed. [
according to the ancients [
the foregoing hour [
Sun in Aries,
house of Mars [
Moon in Virgo [
— — — — — — — — — —

4257. Horoscope, a.d. 279

Payni 13/14, hour 3 of night. Ascendant in Capricorn.
Midheaven in Taurus. — Jupiter in Aries.) Saturn in Scorpio.
Mars in Gemini. • Venus in Gemini.
Mercury in Taurus. • Ascendant in Capricorn. I Moon in Libra.
[Year] 4 of Probus.

4258. Horoscope, A.D. 283

γένεϲιϲ Ἀπίαϲ. (ἔτουϲ) α Κάρου καὶ Καρίνου
καὶ Νουμεριανοῦ, Παοῖνι κε εἰϲ κϛ,
νοικτὸϲ ὅρᾳ δ.

1 ∟ 2 *l.* Παῦνι 3 *l.* νυκτὸϲ ; *l.* ὥρᾳ

4259. Horoscope, A.D. 284

ἀγαθῇ τυχῇ.
γ]ένεϲιϲ Ἱέρακοϲ.
(ἔτουϲ)] α̣΄ Διοκλητιανοῦ, Χοίακ ιγ̣,
ὥ̣ρᾳ ϛ΄ ἡ̣μέραιϲ.
ὥ̣ρ̣α Ὑδρηχόῳ.
Ἄρηϲ Κριῷ.
ϲελήνη Διδύμοιϲ.
Ζεὺϲ Παρθέ̣ν̣ῳ.
ἥλιοϲ, Ἑ̣ρμῆϲ, Ἀφροδ̣[ίτη
Τοξότῃ.
Κ̣[ρ]ό̣[νοϲ] Αἰγόκ[ερῳ.

2 ϊεραkοϲ 4 *l.* ἡμέραϲ 5 *l.* ὡροϲκόποϲ ; ϋδρηχοω : *l.* Ὑδροχόῳ

4258. Horoscope, a.d. 283

Nativity of Apia. Year 1 of Carus and Carinus and Numerianus, Payni 25/26, hour 4 of night.

4259. Horoscope, a.d. 284

Good fortune.
Nativity of Hierax.
Year] 1 of Diocletian, Choeac 13,
hour 6 of day.

Ascendant	in Aquarius.
Mars	in Aries.
Moon	in Gemini.
Jupiter	in Virgo.
Sun, Mercury, Venus	
	in Sagittarius.
Saturn	in Capricorn.

4260. Horoscope, a.d. 296(?)

γ]ε̣νεϲιϲ Ϲαπρίων̣[οϲ
] ιγ´´ Διοκλ̣η̣τιαν̣ο̣ῦ̣
] Κωνϲταντ̣ί̣ο̣υ̣ κ̣α̣ὶ̣
Φα]ῶφι ϛ ὥρ(ᾳ) ιβ ̣
Κα]ρκίνῳ
Το]ξό̣τ̣η̣ϲ
]τη Παρθέ̣ν̣ῳ
]ϲ̣ Ζυ̣γῷ
] ̣ου Ἑρμῆϲ Τοξότῃ
]ρῶ
]
] ̣

4 ϕ

4261. Horoscope, a.d. 289

ὡρ(οϲκόποϲ) Λέοντι
Ἀφροδίτη Ἑρμῆϲ Ϲκορπίῳ
ἥλιϲ Ἄρηϲ Τοξότῃ
Ζεὺϲ Ἐγόκερῳ
Κρόνοϲ Ἰχθύϲι
ϕ κοϲποϲ Ταύρῳ
ϲελήνη Καρκίνῳ

2 ϕ 4 *l.* ἥλιοϲ 5 *l.* Αἰγόκερῳ 6 ϊχθυϲι 7 *l.* ὡροϲκόποϲ ? (but cf. line 1)

4260. Horoscope, a.d. 296(?)

] Nativity of Saprion [
] 13 of Diocletian
] Constantius and
] Phaophi 6, hour 12 ...
] in Cancer.
] Sagittarius.
]... in Virgo.
]... in Libra.
]... Mercury in Sagittarius.
]...
]
]...

4261. Horoscope, a.d. 289

Ascendant in Leo.
Venus, Mercury in Scorpio.
Sun, Mars in Sagittarius.
Jupiter in Capricorn.
Saturn in Pisces.
Ascendant[?] in Taurus.
Moon in Cancer.

4262. Horoscope, a.d. 292/293

θ (ἔτους) Διο̣[κλητιανοῦ
ὥρ(ᾳ) δ̄ ̣[
Κρόνος Ζ̣[εὺς Ταύρῳ
Ἄ̣ρ̣[ης
Ἀφροδίτ̣[η
ἥ̣λιος ̣[
— — — — —

1 ✓ 2 ϕ

4263. Horoscope, a.d. 299

ὡρ(οσκόπος) Παρθένων
Κρόνος Λέωντι
Ἀρέα̣ν̣ Ὑδρηχων̣
Ζεὺς Σκορπίῳ
σελήνη Κρίῳ
ἥλιον Ἑ̣ρ̣[μ]ὴν Ἀφρο-
δίτη Διδύ̣μ̣ο̣ι̣ς

1 ϕ ; *l.* Παρθένῳ 2 *l.* Λέοντι 3 *l.* Ἄρης ; *l.* Ὑδροχόῳ 6 *l.* ἥλιος ; *l.* Ἑρμῆς

4262. Horoscope, a.d. 292/293

Year 9 of Diocletian [
hour 4 ...[
Saturn, Jupiter [in Taurus.
Mars [
Venus [
Sun [
– – – – –

4263. Horoscope, a.d. 299

Ascendant in Virgo.
Saturn in Leo.
Mars in Aquarius.
Jupiter in Scorpio.
Moon in Aries.
Sun, Mercury, Venus
in Gemini.

4264. Horoscope, A.D. 300

ἀγαθῇ τύχῃ. γένεϲιϲ
ἔτουϲ (ἔτουϲ) ιϛ̅ Διοκλητιανοῦ
Φαρμοῦθι κγ̅ ὥρ(ᾳ) δ̅ ἡμέραϲ
ὡρ(οϲκόποϲ) Ἄρηϲ Διδύμοιϲ
Κρόνοϲ Λέοντι
ϲελήνη Πάρθενο̣ν
Ζεὺϲ Αἰγόκερῳ
ἥ̣λ̣ιοϲ Ἑ̣ρ̣[μῆϲ] Ἀ̣φροδ̣[εί]τη
Κρ̣ί̣ῳ δ̣ιευτ̣υ̣χ̣ε̣ῖ̣

2 ∟ 3 ῳ 4 ῳ 6 *l.* Παρθένῳ 8 *l.* Ἀφροδίτη

4265. Horoscopes, A.D. 313 and 348

ἀγαθῇ τύ̣χῃ. γ[ένεϲιϲ
(ἔτουϲ) κθ̅ Δ[ι]ο̣κλη[τιανοῦ Ἐπεὶφ] κζ̣ [
ἡμέραιϲ ὡρ(ᾳ) α̣ [
Ζεὺϲ Ὑδ̣ρ̣ηχόῳ [Κρόν]οϲ ϲελήν̣[η
Ἄρηϲ κ̣(αὶ?) Ἀ̣φροδί[τη] ἥ̣λ̣ιο̣ϲ̣ Κα̣ρ̣κ̣ί̣νῳ

ἀγαθῇ τύχ[η. γ]ένεϲιϲ λαμπρο̣τάτηϲ
ἔτουϲ [ξ]ε Δ̣ι̣οκλητιανοῦ Φαῶφι
κθ ὥρ(ᾳ) ̣[̣] ἡμέρα̣ϲ̣.
ὡρ(οϲκόποϲ) [̣ ̣ ̣] ̣ ̣ ̣ [ϲ]ε̣λήνη Διδύμοιϲ
Ἑρμῆ[ϲ *c. 8 letters*] *vac.*
ἥλιο̣ϲ [*c. 12 letters*] ̣[
ῳ ̣[

2 ∟ 3 *l.* ἡμέραϲ ; ῳ (so line 8) 4 *l.* Ὑδροχόῳ 5 κ̣(αὶ?) : κ broken on right, with no room for further letters 9 ῳ

4264. Horoscope, a.d. 300

Good fortune. Nativity.
Year 16 of Diocletian,
Pharmuthi 23, hour 4 of day.

Ascendant, Mars	in Gemini.
Saturn	in Leo.
Moon	in Virgo.
Jupiter	in Capricorn.
Sun, Mercury, Venus	
	in Aries. Good luck!

4265. Horoscopes, a.d. 313 and 348

Good fortune. Nativity [
Year 29 of Diocletian, [Epeiph] 27? [
hour of day 1? [
Jupiter in Aquarius. Saturn, moon [
Mars and? Venus, sun in Cancer.

Good fortune. Nativity of her ladyship?.
Year [6]5 of Diocletian, Phaophi
29, hour x[x] of day.
Ascendant … moon in Gemini.
Mercury …
Sun …
Ascendant? …

4266. Horoscopes, a.d. 312 and 314

Col. i

(ἔτους) λα
Ἁθὺρ κ̣γ
ἡμέραις ι̣̅
Κρόνος ὡρ(οσκόπος) Ὑδρηχώῳ
Ζεὺς Ἰχθύσι
Ἄρης αὐθημέρᾳ Ἰχθύσι
Ἀφροδίτη Ζυγῷ
ἥλιος Ἑρμῆς
αὐθημέρᾳ Σκορπίῳ
σελήνη Παρθένῳ
δευτυχεῖ
ὁ γράψας Διάπαλος

1 / 4 *l.* Ὑδροχόῳ

Col. ii

(ἔτους) κθ
Φαῶ[φι
ἡμέρ[ας ̣ ̣̄
Ἄρης Κρόνο[ς Ζεὺς ὡρ(οσκόπος) Αἰγόκερῳ
Ἑρμῆ[ς σελήνη
ἥλιος [Ζυγῷ
Ἀφρ[οδίτη Τοξότῃ
διευτ̣[υχεῖ

1 /

4266. Horoscopes, a.d. 312 and 314

Column i

Year 31
Hathyr 23
of day 10?
Saturn, ascendant in Aquarius
Jupiter in Pisces
Mars, that day in Pisces
Venus in Libra
sun Mercury
that day in Scorpio
moon in Virgo
Good luck!
The scribe, Diapalos.

Column ii

Year 29,
Phaophi [xx,
(hour of) day [xx.
Mars, Saturn, [Jupiter, ascendant in Capricorn.
Mercury, [Moon,
Sun [in Libra.
Venus [in Sagittarius.
Good luck!

4266a. HOROSCOPE, A.D. 313

γέ]ν̣εϲιϲ
θ (ἔτουϲ)] ζ (ἔτουϲ) ε̣ (ἔτουϲ), ὅ ἐϲτιν̣
κθ] (ἔ̣τ̣ο̣υ̣ϲ̣) Διοκλητιανοῦ,
Μεϲωρ]ὴ η̅, ὥρ(ᾳ) β̣´ ἡμέραϲ.
ὥ̣ρ̣(οϲκόποϲ), [ἥ]λ̣ι̣οϲ, Ἄρηϲ Λέοντι.
Ἑρμῆϲ Παρθένῳ
Κρόνοϲ Αἰγόκερῳ
Ζεὺϲ Ὑδρηχόῳ
ϲ]ε̣λ̣ήνη Ταύρῳ
— — — — — — — — — — — —

2 (ἔτουϲ)[1] : ʃ/ ; ε̣ (ἔτουϲ) : ϛ/ 3 (ἔ̣τ̣ο̣υ̣ϲ̣) : end of diagonal stroke at edge of papyrus 4 ϕ 5 descending stroke of ϕ, the rest broken off 8 ϋδρηχοω : *l.* Ὑδροχόῳ

4267. HOROSCOPE, A.D. 318

λε (ἔτουϲ), Ἀθὺρ ιβ´´, ὥρ(ᾳ) ἐ̣νάτη ἡμέραιϲ.
ὥρα Κριῷ. Ἀφροδίτη Τοξότῃ.
ϲελήνη Σκορπίῳ̣. Κ̣[ρόνοϲ
ἥλιοϲ Σκορ[πίῳ]. Ἰχθύϲ̣ι̣ν.
Ἑρμῆϲ̣ Σκορπίῳ.
Ζεὺ̣ϲ̣ Λέοντι.
Ἄρηϲ Διδύμοιϲ.

1 ʃ ; ϕ ; *l.* ἡμέραϲ 2 *l.* ὡροϲκόποϲ

4266a. Horoscope, A.D. 313

Nativity.
Year 9,] year 7, year 5, which is
year [29] of Diocletian,
Mesore 8, hour $2^{?}$ of day.
Ascendant, sun, Mars in Leo.
Mercury in Virgo.
Saturn in Capricorn.
Jupiter in Aquarius.
Moon in Taurus.

— — — — — — — — — — — — —

4267. Horoscope, A.D. 318

Year 35 (of Diocletian), Hathyr 12, ninth hour of day.
Ascendant in Aries. Venus in Sagittarius.
Moon in Scorpio. Saturn
Sun in Scorpio. in Pisces.
Mercury in Scorpio.
Jupiter in Leo.
Mars in Gemini.

4268. Horoscope, A.D. 318

ιγ (ἔτους) ια [(ἔτους)] γ (ἔτους), Θὼ[θ κς, ὥρᾳ] θ̣ νυκτός.
ὡρ(οσκόπος), Ζε̣ὺς Λέον̣[τ]ι̣.   [ἥλι]ο̣ς, Ἑρμῆς,
Ἀφροδίτη Ζυ[γῷ.   σε]λ̣ή̣νη Ὑ̣δ̣ρηχ̣ό̣ῳ.
κ̣ρ̣ό̣ν̣ο̣ς Ἰχ̣[θυσι.]   Ἄ̣ρης Διδύμ̣ο̣ις.
διευτ[υχεῖ.

1 ιγ𐅹ια 𐅹   2 ῳ   4 ιχ̣[

4269. Horoscope, A.D. 320

ἔτους λς Διοκλητιανοῦ,
Μεχεὶρ κβ, ὥρ(ᾳ) β ἡμέραις.
ὡρ(οσκόπος) Ἰχθύσι.
Κρόνος Κρίῳ.
Ζεὺς Λέοντι, αὐθη-
μερὶ Παρθένῳ.
ἥλιος Ὑδρηχόῳ.
Ἄρης Ὑδρηχόῳ.
Ἀφροδίτη Ὑδρηχόῳ.
Ἑρμῆς Ὑδρηχόῳ.
σελήνη Σκορπίῳ.
δ̣ιευτυχεῖ.

2 ῳ ; *l.* ἡμέρας   3 ῳ   7 *l.* Ὑδροχόῳ (so lines 8, 9, 10)

4268. Horoscope, A.D. 318

Year 13, [year] 11, year 3, Thoth [$26^{?}$, hour] 9 of night.
Ascendant, Jupiter in Leo. Sun, Mercury,
Venus in Libra. Moon in Aquarius.
Saturn in Pisces. Mars in Gemini.
Good luck!

4269. Horoscope, A.D. 320

Year 36 of Diocletian,
Mecheir 22, hour 2 of day.
Ascendant in Pisces.
Saturn in Aries.
Jupiter in Leo, on the same
day in Virgo.
Sun in Aquarius.
Mars in Aquarius.
Venus in Aquarius.
Mercury in Aquarius.
Moon in Scorpio.
Good luck!

4269a. Horoscope, a.d. 325

ἀγα]θ̣ῇ̣ τύχῃ.
γένεcιc.
βμ (ἔτουc) Διοκλητιανοῦ,
Ἁθὺρ κ̣ʹ, ὥρ(ᾳ) ι νυκτόc.
ὡρ(οcκόποc), Ἀφροδίτη̣ Ζ̣υ̣γ̣ῷ̣.
ἥλιοc, Ἑ̣ρ̣[μῆc Cκορπίῳ?.
Ἄρηc Εὐόκερῳ.
Ζεὺc Ὑδρυχόῳ.
Κρόνοc Καρκίνῳ.
cελήνη [Παρθένῳ.
— — — — — — — — — — —

3 βμ// 4 ϕ 5 ϕ 7 *l.* Αἰγόκερῳ 8 *l.* Ὑδροχόῳ

4270. Horoscope, a.d. 350

/ ἀγαθῇ τύχῃ. γένηcιc
Ἀμμωνᾶc, ἔτουc
ἀπὸ κρατήcεωc Διοκλητι-
ανοῦ ξϛ̄, Φαμενὼθ κγ
ὥρ(ᾳ) ζ νυκτὸc π̣λ̣ήραιc.
ὡρ(οcκόποc) Αἰγόκαιρῳ.
Ἄρηc Ὑτρηχόῳ.
ἥλιοc Ἑρμῆc.
Ἀφροδίτη, Ζεὺc
Ἰχθύcι.
Κρ̣ό̣νοc Ταύρῳ.
cελήνη Αἰγοκαίρῳ.
διευτυχεῖ.
(m. 2) κλῆροc γάμοc Cκορπίῳ.

1 *l.* γένεcιc 5 ϕ 6 ϕ ; *l.* Αἰγόκερῳ (so line 12) 7 *l.* Ὑδροχόῳ 14 *l.* γάμου

4269a. Horoscope, a.d. 325

Good fortune.
Nativity.
Year 42 of Diocletian,
Hathyr 20?, hour 10 of night.
Ascendant, Venus in Libra.
Sun, Mercury [in Scorpio?.
Mars in Capricorn.
Jupiter in Aquarius.
Saturn in Cancer.
Moon [in Virgo?.
– – – – – – – – – – – –

4270. Horoscope, a.d. 350

Good fortune. Geniture
of Ammonas, year
from the accession of Diocletian
66, Phamenoth 23,
7 hours of night completed.
Ascendant in Capricorn.
Mars in Aquarius.
Sun, Mercury,
Venus, Jupiter
in Pisces.
Saturn in Taurus.
Moon in Capricorn.
Good luck!
(m. 2) Lot of marriage in Scorpio.

4271. Horoscope, a.d. 354

ἀ̣[γ]αθῇ τύχῃ. γέ[νεcιc
ἔτουc (ἔτουc) οα΄΄ Διοκλητια̣[νοῦ,
Τῦβι ε ὥρ(ᾳ) ιβ ἡμέραιc.
[. . .] ε̣λ̣ . . . . . ε̣νη̣
(*stripped*)
ἥλιο]c̣ Αἰγ[ό]κε̣ρῳ.
Κρόνο]c̣ Διδύμοιc̣.
Ζεὺ]c̣ Λέοντι̣.
Ἄρη]c̣ Λέ̣οντι̣.
Ἀφρο]δίτη Τοξ[ό]τ̣ῃ̣

— — — — — — — — — — — — —

2 L 3 φ ; *l.* ἡμέρας

4272. Horoscopes, a.d. 370 and 375

$\overline{\pi\varsigma}$΄΄ ῑ \`Παῦνι΄ κβ [
ὡ̣ρ(οcκόποc), Ζ(εὺc) Cκορπίῳ̣.
Κρ(όνοc) Τοξότ(ῃ).
(ἥλιοc), (Ἑρμῆc) Διδύμ(οιc).
(Ἀφροδίτη) Κ̣αρκίνῳ.
(cελήνη), (Ἄρηc) Παρθένῳ.
Κλῆρ(οc) τύχ(ηc) Ὑδ̣ρο̣χ(όῳ).
$\overline{\text{ϙβ}}$ β̣΄΄ κ̣η̣ . . [
ὡ̣ρ(οcκόποc), Ζ(εὺc) Τα̣ύ̣ρ̣ῳ.
Κρ(όνοc) Ἰχθύc(ι).
[. .], (Ἑρμῆc) . . .
(Ἀφροδίτη) Ζυγῷ̣.
(*traces*)
(*traces*)

1 ῑ ; παυνι above the line 2 φ ; ♃ (so line 9) 3 ♄ (so line 10) ; τοξοτ 4 ☉ ; ☿ (so line 11); διδυ^μ 5 ♀ (so line 12) 6 ☾ ; (Ἄρηc): very faint symbol, apparently something like ♂ ; παρθε^ν 7 [symbol] ; υδρο^χ 9 apparently ταυρ^ω 10 ιχθυ^c

4271. Horoscope, a.d. 354

Good fortune. Nativity,
year 71 of Diocletian,
Tybi 5, hour 12 of day
...
(*stripped*)
Sun in Capricorn.
Saturn in Gemini.
Jupiter in Leo.
Mars in Leo.
Venus in Sagittarius.

– – – – – – – – – – – –

4272. Horoscopes, a.d. 370 and 375

(Year) 86 (of Diocletian), (month) 10 (=) Payni, (day) 22, [
Ascendant, Jupiter in Scorpio.
Saturn in Sagittarius.
Sun, Mercury in Gemini.
Venus in Cancer.
Moon, Mars in Virgo.
Lot of fortune in Aquarius.
(Year) 92, (month) $2^?$, (day) $28^?$, ...
Ascendant, Jupiter in Taurus.
Saturn in Pisces.
[...], Mercury ...
Venus in Libra.
(*traces*)
(*traces*)

4273. Horoscopes, undated and A.D. 399

— — — — — — — — —
ἥλιος Αἰγοκέ̣[ρῳ
σελήνη Σκορ[πίῳ
———
γένεσις Ἀφ ̣ [
(ἔτους) ριε Μεχεὶρ̣ [
ὥρα Κρόνος [Τοξότῃ
Ζεὺς Κρίῳ
Ἄρης Ἀφροδιτ̣[η Αἰγόκερῳ?
ἥλιος Ἑρμῆς [Ὑδροχόῳ
σελήν̣η̣ Κ̣αρ̣[κίνῳ
— — — — — — — — —

4274. Horoscopes, A.D. 480 and 503

Along the fibres:
— — — — — — — — — — — —
. . . [
ἔτ]ους ρϙζ̣ [
(ἥλιος)] Σ̣κορπίῳ̣ [
(σελήνη) Αἰγόκερῳ μο̣(ίρᾳ) [
Ζ(εὺς) Ἰχθύσι
(Κρόνος) Ζυγῷ
(Ἄρης) Λέοντι
(Ἀφροδιτη) Τ]ο̣ξότ̣ι̣ ζ(ῳδίῳ?) Ζυγῷ
(Ἑρμῆς)] Ζυγῷ
(ὡροσκόπος)] Αἰγόκ[ερ]ῳ̣ μο(ίρᾳ) ϛ̣

4 ☾ ; presumably μ̊ (the superscript is broken off) 5 ♃ ; ϊχθυcι 6 ♄ 7 ♂
8 ζ(ῳδίῳ?): ζ

4273. Horoscopes, undated and A.D. 399

- - - - - - - - - -

Sun in Capricorn.
Moon in Scorpio.

———

Geniture of Aph[
Year 115 (of Diocletian), Mecheir [
Ascendant, Saturn [in Sagittarius.
Jupiter in Aries.
Mars, Venus [in Capricorn?
Sun, Mercury [in Aquarius.
Moon in Cancer.

- - - - - - - - - -

4274. Horoscopes, A.D. 480 and 503

Along the fibres: - - - - - - - - - - - - -

. . .
Year 197 [
Sun] in Scorpio [
Moon in Capricorn [xx]°.
Jupiter in Pisces.
Saturn in Libra.
Mars in Leo.
(Venus)] in Sagittarius? in sign? Libra
(Mercury)] in Libra
(Ascendant)] in Capricorn 6°

Across the fibres:

cιθ´ Δι]οκλη[τια]νοῦ .
.[Π]α̣ῦνι κ̣̅ ἐπὶ κα̅, νυ(κτὸc)
ὥρ(ᾳ) [ ]. κατὰ (cελήνην) ϛ
γ̣ι̣[ ]. ἐπὶ (ἥλιον).
(ἥλιος) [Δι]δ̣ύμοιc μο(ίρᾳ) κα
(cελήνη) Λέ]οντι μο(ίρᾳ) κϛ
(Κρόνοc) Κ]α̣ρκίν̣[ῳ
(Ζ̣ε̣ὺ̣c̣) [Ὑ]δριχ̣[όῳ
— — — — — — — — — — — — —

2 ♐ 3 ☉ ; ☾ 4 ♂ (so line 5) 5 μ̊ (so line 6) 8 (Ζ̣ε̣ὺ̣c̣): speck on isolated fibres; *l.* Ὑδροχόῳ

4275. Horoscope, a.d. 508

☧ ἐπ' ἀ̣γαθοῖc Θεόδωροc.
Διοκλητιανοῦ cκδ Μεχεὶρ ϛ̅ εἰc ζ
ὥρ(ᾳ) νυκτερ(ινῇ) ζ, α̅ ἰνδικ(τιῶνοc)

(ἥλιοc) Ἱδρυχόου	μο(ίρᾳ) ιγ	λ(επτοῖc) ια
(cελήνη) Καρκίνου	μο(ίρᾳ) κϛ	λ(επτοῖc) δ
ἀφ(αιρετικὸc) (Κρόνοc) Παρθένου	μο(ίρᾳ) ιδ	λ(επτοῖc) ιθ
ἀφ(αιρετικὸc) Ζ(εὺc) Καρκίνου	μο(ίρᾳ) γ	λ(επτοῖc) β
πρ(οcθετικὸc) (Ἄρηc) Αἰγοκέρωτι	μο(ίρᾳ) ζ	λ(επτοῖc) θ
ἀ̣φ̣(αιρετικὸc) (Ἀ̣φ̣ρ̣ο̣δ̣ί̣τ̣η̣) Αἰγοκέρωτι	μ̣ο̣(ίρᾳ) κα̣	λ(επτοῖc) λϛ
ἀφ(αιρετικὸc) (Ἑρμῆc) Ὑδροχόου]	μο(ίρᾳ) ιε	λ(επτοῖc) ιδ
ὡρ(οcκόποc) ἐν Cκορπίῳ	μο(ίρᾳ)] κ̣ζ	λ(επτοῖc) λ.

ἐπ' ἀγαθο]ῖc

2 διοκλητιανο\ 3 ☉ ; νυκτερ (descender of ρ crossed); ινδικ (κ crossed at lower right) 4 ♂ ; ϊδρυχοо̄: *l.* Ὑδροχόου ; μ̊ (so in subsequent lines); λ´ (so in subsequent lines) 5 ☾ ; καρκινο̊ (so line 7) 6 αφ (φ crossed; so in subsequent lines); (Κρόνοc): ♄ ; παρθενο̊ 6 ♃ 7 ♀ ; (Ἄρηc): ♂ ; *l.* Αἰγόκερω (so line 8)

Across the fibres:

[(Year) 219] of Diocletian ...
Payni 20/21, hour of night
[x]x, according to the moon (day) 6,
... [the (day of) Saturn]/the (day of the) sun.
Sun in Gemini 21°.
Moon] in Leo 26°
(Saturn)] in Cancer [
(Jupiter) in Aquarius [

4275. Horoscope, a.d. 508

⳨ For the good, Theodoros
(Year) 224 of Diocletian, Mecheir 6/7
hour 7 of night, indiction 1.

Sun in Aquarius	13°	11'
Moon in Cancer	26°	4'
Subtractive Saturn in Virgo	14°	19'
Subtractive Jupiter in Cancer	3°	2'
Additive Mars in Capricorn	7°	9'
Subtractive Venus in Capricorn	21°	36'
Subtractive Mercury in Aquarius]	15°	14'
Ascendant in Scorpio]	27°	3x'

For the] good!

4276. Deluxe Horoscope

Col. i

ἐ]πὶ τοῦ βορί–

ψ]ηφιcθεὶc̣
τ]ροπ̣ικῷ
ο]ἴ̣κῳ Ἄρ[ε–]
]̣[̣ ̣]̣
]̣
ὁ δὲ τῆc γεν[έ]c̣εωc ὡροcκόποc ψηφιc̣–

1 *l.* βορείου 4 *l.* Ἄρεωc

Col. ii

θεὶc ἐκ τοῦ ἀν̣αφορικοῦ̣ ὡ̣c πρὸc πρώ–
τὴν ὥραν ἡμέραc Ἱππαρχικοῦ cυ̣[ν–
τάγματοc κα̣ὶ πρὸc τὸ κλίματοc τ[ῆc
Αἰγύπτου κα̣ὶ κατὰ cύμφωνον τοῦ
[γν]ορίμου τ̣όπου ψηφιcθεὶc ἀκρει–
[βῶ]c εὑρέ[θη] ἐν ζῳδίῳ ἀρρενικῷ τρο̣–
[πικῷ] Κριῷ [μο]ιρῶν $\overline{\kappa\eta}$ λεπτ[ῶ]ν̣ $\overline{\lambda\epsilon}$
]ρ̣η ̣[ ]υ [ὁ]ρείοιc Κρόνο[υ
] ̣ ̣[ ] ̣ ̣ ̣ ̣ιω (*vac.?*)

τὸ δὲ τῆc [γενέcεωc
τιν πρακ[
]̣[
]̣ β̣α̣[
14 (*traces*)
15–21 (*vac.*)
22 (*traces*)
23

6 *l.* γνωρίμου 6–7 dash used as line-filler at end of line 6: *l.* ἀκριβῶc 9 *l.* ὁρίοιc

4276. Deluxe Horoscope

Col. i

] in the northern

] computed
] tropical
] house of Mars
]...
]...

— — — — —

Col. ii

The ascendant of the nativity, computed
on the basis of the table of ascensions for the
first hour of day from the compilation of Hipparchus
and for the *klima* of
Egypt and compatible
places accurately computed
has been found in the masculine tropical sign
Aries 28 degrees 35 minutes
... terms of Saturn
...

The [...] of the nativity [...
...[
...
...

16–20 (*traces*)
21 (*vac.*)
22 (*traces*)

4277. Deluxe Horoscope

Fr. 1, col. i

Ἑρμῆϲ

Ϲτίλβων ὁ τοῦ Ἑρμο[ῦ] ἀϲτ[ήρ, ὁ] τὴ[ν] ζώνην τοῦ ἀέροϲ λελογχώϲ, εὑρέθη [τὴ]ν κατὰ μῆκοϲ κείνηϲιν ποιούμενοϲ ἐ[ν] ζῳδίῳ θηλυκῷ τροπικῷ χερϲ[α]ίῳ Αἰγόκερῳ μο(ιρῶν) ϛ̅ λε[π(τῶν)] τῆϲ ἑξηκοντάδοϲ κ̅β̅, ο[ἴ]κῳ Κ[ρό]νου, ὁρίοιϲ ἰδίοιϲ, μο[νο]μοιρ[ίᾳ] Ἀφροδείτηϲ, τριγώνῳ ϲελήν[η]ϲ [κ]ατὰ μέτοχον Κρό[νο]ῦ, ὑ[ψ]ώματι Ἄρεω[ϲ], ταπεινώματ[ι] Διόϲ. ... ἀπὸ τῆϲ μ[..]ρηϲ δωδεκατημορίου καταλήγ[ει] Ἰχθ[ύ]ϲι, ζῳδίῳ θ[η]λυκῷ διϲώμῳ μο(ιρῶν) ι̅δ̅ λεπ(τῶν) [τ]ῆϲ ἑξηκ[ο]ντά[δο]ϲ ι̅β̅. βαθμῷ ϛ̅ ἀνέμῳ Λ[ι]β[εί, τ]όπῳ ζ[ῳδια]κῷ Αἰγόκερῳ, ἐ[π'] ἀναποδιϲμοῦ ἀνωμαλίαϲ δευτ[έ]ρ[ῳ], δ[ε]καν[ῷ] α̅.

>———

ὡροϲκόποϲ

οἴαξ ὁ [κ]αλούμενοϲ ὡ[ροϲκό]ποϲ, .[......] κυβερ[ν]ήτηϲ κατὰ τ[..] [...]ει[...]αρ[...] ἀκρειβῶϲ ψηφε[ι]ϲ[θε]ὶ[ϲ εὑρέθη] τὴν κατὰ μῆκοϲ κ[εί]νη[ϲ]ιν πο[ιο]ύ[με]νοϲ [ἐν] ζῳδίῳ ϲ]τερεῷ ἀρρε[ν]ικῷ βαϲιλικ[ῷ] Λ[έο]ντι μο(ιρῶν) κ̅β̅ λ]επ(τῶν) τῆϲ ἑξηκοντάδ[οϲ] ι̅α̅, [ο]ἴκῳ ἡλίου, ὁρί[ο]ι[ϲ] Ἑρμοῦ, τριγώνῳ Δ[ιὸ]ϲ κατὰ μέτοχον Ἄ]ρεωϲ, ἐπεὶ τμήματοϲ ε̅, ἐπεὶ βαθμῷ .], τ[ό]πῳ ζῳδιακ[ῷ] Παρ[θ]ένῳ, ἐπεὶ μο(ίρᾳ) θηλυκῇ], δεκανῷ [γ̅.

4277. Deluxe Horoscope

Fr. 1, col. i

Mercury

Stilbon, the star of Mercury, which has been allotted the belt of the air, has been found making its longitudinal motion in the feminine, tropical, terrestrial sign Capricorn, at 6 degrees 22 minutes of the sixty, its own house and its own terms, single degree of Venus, triangle of the moon shared with Saturn, exaltation of Mars, depression of Jupiter, ... from the *dodekatemorion* fell in Pisces, the feminine bicorporal sign, 14 degrees 12 minutes of the sixty. Step 6, wind Lips, zodiacal place Sagittarius, at the second change of direction of anomaly, decan 1.

Ascendant

The rudder called 'ascendant' ... steersman ... accurately computed has been found making its longitudinal motion in the solid, masculine, royal sign Leo, at 22 degrees 11 minutes of the sixty, house of the sun, terms of Mercury, triangle of Jupiter shared with Mars, in segment 5, in step [x], zodiacal place Virgo, in a feminine degree, decan [3.

>———————————

δῦν̣ο̣ν̣

τ̣[ὸ] δὲ̣ δῦνο̣ν̣ [ε]ὑ̣[ρ]έ̣θ̣η̣ [τὴν κατὰ μῆ]κο̣ϲ̣ κ̣ε̣ί̣ν̣η̣ϲ̣ι̣ν̣ ποιούμε̣νο̣ν̣ ἐν ζῳ̣[δίῳ ϲτερεῷ] ἀ̣ρ̣ρ̣ενι̣κ]ῷ παρύ̣[γρ]ῳ̣ Ὑδρ̣[οχ]ό̣ῳ [μο(ιρῶν) κ̅β̅ λεπ(τῶν) τῆϲ ἐ]ξηκοντ[άδο]ϲ ι̅α̅, οἴκ[ῳ Κρόνοῦ] ὁρίοιϲ Ἄρεω̣ϲ̣, τριγώ̣ν̣ῳ Ἑρμο̣ῦ̣ κα[τὰ μέτοχο]ν̣ Κ̣ρό̣[ν]ο̣[υ], ἐ[πὶ] τ̣μ[ήμ]α̣τ̣ο̣ϲ̣ ̅ ̣, ἐπ[ὶ βαθμῷ] ϛ̅ κα̣[ὶ] μ̣ο̣(ίρᾳ) θ[ηλυ]κ̣ῇ [̣ ̣] δεκανῷ γ̣̅.

>———————————

μεϲουράνη[μα]

τὸ δὲ μεϲουράνημα ε̣[ὑρέθη τὴν] κατὰ μῆκοϲ κε̣ί̣νηϲιν ποιούμε̣ν[ον ἐν ζῳδίῳ ϲτ]ε̣ρ̣εῷ θηλυκῷ Ταύρῳ μο(ιρῶν) κ̅ζ̅ λ̣[επ(τῶν) τῆϲ ἐξηκ]οντά̣δοϲ η̅, οἴκῳ Ἀφροδείτη[ϲ], τρ̣[ιγώνῳ ϲελ]ή̣ν̣η̣ϲ̣ κατὰ μέτοχον Κρόνου, ὁρ̣[ί]ο̣ι̣[ϲ Ἄρεωϲ, μο]νο–

4 *l.* κίνηϲιν 5 μ̊ (so henceforth) 6 λε̄ (so henceforth) 7 ὁριοιϲ (with breathing; so henceforth) 7–8 *l.* Ἀφροδίτηϲ 11 ϊχθυϲι 19 *l.* ἀκριβῶϲ ; *l.* ψηφιϲθεὶϲ 20 *l.* κίνηϲιν 24 *l.* ἐπὶ (bis) 25 *l.* ἐπὶ 28–29 *l.* κίνηϲιν 37 *l.* κίνηϲιν

Fr. 1, col. ii

μοιρίᾳ Ἑρμοῦ, ἐπὶ τμήματοϲ θ̅, ἐπε̣ὶ̣ βαθμῷ ε̅, ἀνέμῳ βορρᾷ, δεκανῷ γ̣̅.

>———————————

ὑπὸ γῆ̣ν̣

τὸ δὲ ὑπὸ γῆν εὑ̣ρ̣έθη τ̣ὴ̣ν̣ κατὰ μ̣ῆκο̣[ϲ] κείνηϲιν ποιούμενον ἐν ζ̣ῳ̣δί̣ῳ θηλυκῷ ϲτερ[ε]ῷ Ϲ̣κορπίῳ ἐ̣ν̣ μ̣ο̣(ίραιϲ) κ̅ζ̅ λεπ(τοῖϲ) τῇ̣[ϲ] ἐξηκοντάδοϲ η̅, οἴκῳ Ἄ[ρ]εωϲ, τ̣ρ̣[ι]γώ̣νῳ Ἄρεωϲ κατὰ μέτοχο̣ν̣ Δ̣ι̣ὸϲ, ὁρίοιϲ̣ Κ̣ρ̣[ό]–

Setting point

The setting point has been found making its longitudinal motion in the solid, masculine, damp sign Aquarius, at 22 degree, 11 minutes of the sixty, house [of Saturn], terms of Mars, triangle of Mercury shared with Saturn, in segment x, in [step] 6, and (in) a feminine degree, [...] decan 3.

Midheaven

The midheaven has been found making its longitudinal motion in the solid, feminine sign Taurus, at 27 degrees 8 minutes of the sixty, house of Venus, triangle of the moon shared with Saturn, terms of Mars, single

Fr. 1, col. ii

degree of Mercury, in segment 9, in step 5, wind Boreas, decan 3.

Lower midheaven

The lower midheaven has been found making its longitudinal motion in the feminine, solid sign Scorpio, at 27 degrees, 8 minutes of the sixty, house of Mars, triangle of Mars shared with Jupiter, terms of Saturn,

νου, μονομοιρίᾳ Διός, ἐπεὶ τμήματ[ος] ια, ἐπεὶ βαθμῷ ι, ἀνεμῷ Λίβει, δεκανῷ γ.

κ[λ]ῆρος τύχης

ὁ πρῶτος κλῆρος ὁ καλούμενος τύχης εὑρέθη τὴν κατὰ μῆκος [κίνησιν] ποιούμενος ἐν ζῳδί[ῳ θ]η[λ]υκῷ ἀνθρωποειδεῖ Παρθένῳ [ἐν] μο(ίραις) [ζ?] λεπ(τοῖς) [τῆ]ς [ἐ]ξηκοντάδ[ος] ‾, [οἴκῳ Ἑρμοῦ, ὑψώματι αὐτοῦ [*c. 10 letters*] Ἀφροδείτης ταπεινώματι [*c. 6 letters*] ὁρίοις Ἀφροδείτης, μονομοιρίᾳ Ἑρ[μοῦ, ἐπεὶ τμήματος δ, ἐπεὶ βαθμῷ ‾, [ἀνεμῷ Λίβει, δεκανῷ α.

κ[λῆρος δαίμ]ονος

ὁ δεύτερος [κλῆρος ὁ καλούμενος δαίμων εὑρέθη τὴν [κατὰ μῆκος κίνησιν ποιούμεν[ος ἐν ζῳδίῳ στερέῳ βασιλικῷ Λέοντι ἐν [μο(ίραις) ‾ λεπ(τοῖς) τῆς ἐξηκοντάδος ις, οἴκῳ [ἡλίου, τριγώνῳ Διὸς κατὰ μέτοχον [Ἄρεως, ὁρίοις ?, μονομοιρίᾳ [?, ἐπὶ τμήματος ‾, ἐπεὶ βαθ[μῷ ‾, ἀνεμῷ ?, δεκανῷ α.

[κλῆρος ἔρωτος?]

ὁ τρί[τος κλῆρος ὁ ἔρωτος εὑρέθη τὴν [κατὰ μῆκος κίνησιν ποιούμενος

1 *l.* ἐπὶ 5 *l.* κίνησιν 9 *l.* ἐπὶ 10 *l.* ἐπὶ (*bis*) 21 *l.* ἐπὶ 31 *l.* ἐπὶ

single degree of Jupiter, in segment 11, in step 10, wind Lips, decan 3.

Lot of fortune

>————————————————

The first lot, called of fortune, has been found making its longitudinal [motion] in the feminine, anthropomorphic sign Virgo, at [7?] degrees xx minutes of the sixty, [house] of Mercury, exaltation of the same (Mercury) [...] depression of Venus [...] terms of Venus, single degree of Mercury?, in segment 4, in step x, wind Lips, decan 1.

>————————————————

Lot of daimon

The second lot, [called] daimon, has been found making [its longitudinal motion] [in the solid], royal [sign] Leo, at [xx degrees] 16 [minutes of the] sixty, house [of the sun, triangle of Jupiter] shared [with Mars, terms of ...,] single degree [of ..., in segment x,] in step [x, wind ...,] decan 1.

>————————————————

Lot [of eros?]

The third [lot, of eros?, has been found] [making] its [longitudinal motion]

Fr. 1, col. iii

ἐ̣ν̣ ζ̣ω[δί]φ θηλυ̣[κῷ ἀνθρωποειδεῖ
Π̣α̣ρ̣θ̣έ̣ν̣ῳ̣ (*traces*)
3–5 (*traces*)
6 ̣ ἀ̣ν̣ε̣μ̣ῷ̣ Λίβε̣[ι
>———————————

κ̣λ̣ῆρ̣ο[ϲ ἀνάγκηϲ

[δ] τ̣έ̣τ̣α̣[ρ]τ̣ο̣ϲ κ̣λ̣[ῆροϲ ὁ καλούμενοϲ
[ἀ]ν̣ά̣γ̣κ̣[ηϲ εὑρ]έ̣θ̣η τ̣ῆ̣[ν κατὰ μῆκοϲ
(*no further traces visible*)

Fr. 2

— — — — — — —
οἰκ]ῷ Διὸϲ κα̣[ὶ
] ̣η̣ ̣ ̣μ̣[
] ̣[̣]α̣γει ̣[
] ̣ ̣ μο(ίραιϲ) κε̣ [
] ̣[̣ ̣] ̣ ̣[
— — — — — — —

4278. Deluxe Horoscope

— — — — — — —
]η̣[
] ιθ [λεπ]τῶν ι[
] ιθ λεπτῶν [
] ια λεπτῶν ̣[
] ιϛ λεπτῶν ι ̣[
] ̣ ϛ λεπτῶν ι̣ ̣[
]ο̣υ̣ϲ οἰκοδεϲπό[ταϲ
] ̣ χρόνον διαιρε ̣[
] ϲ̣υνοικοδεϲποτη[
] ̣ τὴν ἡλικίαν β̣[
] ̣ μὲν τὸ κακῶϲ θα[
] ̣ουϲ. αὐτὸν δὲ το[
] ̣δ̣ ̣ε̣ϲ̣ ̣[
] ὁμοίωϲ μη ̣ ̣[
— — — — — — —

Fr. 1, col. iii

in the feminine, [anthropomorphic] sign
Virgo...
3–5 (*traces*)
6 ... wind Lips, [...

Lot [of necessity]

The fourth lot, called
of necessity, has been found [making] its [longitudinal]
(*no further traces visible*)

Fr. 2

– – – – – – – –
] house of Jupiter and [
]...[
]...[
]... 25 degrees [
]...[
– – – – – – – –

4278. Deluxe Horoscope

– – – – – – – –
...
] 19 [degrees] 1[x] minutes [
] 19 [degrees xx] minutes [
] 11 [degrees] x minutes [
] 16 [degrees] 1x minutes [
]... 6 [degrees] 1x minutes [
]... lord of the house [
]... time divide? [
] joint ruler of the house [
]... time of life ...[
]... bad death? [
]... the very...[
]...[
] likewise ...[
– – – – – – – –

4279. Deluxe Horoscope

— — — — — — — —
]. . [. .] ἡ̣μέρα[
οι̣[. .]δ̣ιω. . . δι. [
Ἑρ̣μ̣ῆ̣ς ἐν Ζυγῷ
μ̣ο̣ι̣ρ̣ῶ̣ν̣ $\overline{κβ}$ λ̣(επτῶν) $\overline{ι}$
οἴκῳ καὶ ὁρίοις Ἀφρο-
δείτης, δυτ̣ικὸς ἑῷ-
ος συνοδικός.
ὡροσκοπεῖ Τοξότης.
(m. 2) ὀχληθεὶς πέμψεις
μοι ἐπὶ μοίραις
δωσε. . ε̣ τ̣ῷ κομ̣ίζον-
τ̣ί̣ σ̣οι τὸ πιττάκιον.
(m. 3?) ὁ δαίμ(ων) Τοξό(τη) . . . . .

4 λ̣(επτῶν) : uncertain traces   5–6 *l.* Ἀφροδίτης   13 $δαι^{μ}$ $τοξ^{ο}$

4280. Deluxe Horoscope

— — — — — — — —
δε]κ̣α̣νῷ $\overline{α}$.

Ζεὺς Ζυγῷ μοιρῶν [
οἴκῳ Ἀφροδίτης ὁρί[οις
ἑσσπέριος δυτικός, δ̣[

4 *l.* ἑσπέριος

4279. Deluxe Horoscope

– – – – – – – –

]... day [
...[
Mercury in Libra
22 degrees 10 minutes,
house and terms of
Venus, setting in the
west, at conjunction.
Sagittarius ascendant.
(m. 2) Troubled, you will send
me concerning? degrees?,
You? give... to the person who brings
you the ticket.
(m. 3?) The daimon is? Sagittarius ...

4280. Deluxe Horoscope

– – – – – – – –

decan 1.

Jupiter in Libra, [xx] degrees [
house of Venus, terms [
setting in the west, ...[

4281. Deluxe Horoscope

— — — — — — —
ἐγλίψεωϲ̣ [

Κρόνοϲ ἐν Ὑδρηχόωι μοιρῶν ι̅ε̅ καὶ
λεπτῶν λ̅ϛ̅, οἴκωι ἰδίῳ καὶ ὁρ̣ί̣ο̣ι̣ϲ
Ἀφροδείτηϲ, ϲτηριγμῷ πρώ̣τ̣[ῳ.

1 *l.* ἐκλείψεωϲ 2 *l.* Ὑδροχόῳ 3 ϊδιω 4 *l.* Ἀφροδίτηϲ

4282. Deluxe Horoscope

ἡ προγενομένη
πανϲέληνοϲ
ἐγένετο τῷ
αὐτῷ μηνὶ
Παῦνι κ̅ε̅ εἰϲ
κ̅ϛ̅ ὥρ(ᾳ) β̅ νυκ(τόϲ)
Τοξότῃ μοι-
ρῶν κ̅θ̅ λεπτῶ(ν)
ι̅ζ̅, οἴκῳ Διὸϲ
ὁρίοιϲ Ἄρεωϲ.
>——————
ἡ δὲ π̣α̣ρ̣[̣] ̣ ̣ ̣
ϲύνοδοϲ [
 ̣ ̣ ̣ ̣ ̣] ̣[

6 ῳ ; νυκ 8 λεπτω̅

4281. Deluxe Horoscope

– – – – – – – –

of the eclipse [

Saturn in Aquarius, 15 degrees and 36 minutes, its own house and terms of Venus, at its first station.

4282. Deluxe Horoscope

The preceding full moon took place in the same month, Payni 25/26, hour 2 of night, at Sagittarius 29 degrees 17 minutes, house of Jupiter, terms of Mars.

———

The following[?] conjunction [

4283. Deluxe Horoscope

Fr. 1

– – – – – – – –
]. . .[
τριγ]ώνῳ̣ ἰδ[ί]ῳ̣, μ̣ονο̣[μοιρίᾳ̣
] δ̣εκανῷ π̣ρ̣ώ̣τῳ [
Πυρόειc ὁ τοῦ] Ἄρεοc [ἀ]cτὴρ ὁ τ[ὴν
ζώνην τοῦ π]όλου κλ[η]ρωθ̣ε̣ὶ̣c ψ[
]. ζ̣[ῳδίῳ] θ̣ηλυκῷ [
] Π̣[αρθέ]νῳ μο(ιρῶν) .[
] ὑψ[ώματι
τ]ριγώ̣[νῳ
– – – – – – – –

Fr. 2

– – – – – – – – – – – – – – – – – –
Cτίλ]β̣[ων, ὁ το]ῦ Ἑρμ[ο]ῦ̣ [ἀ]c̣τ̣[ὴ]ρ̣ [. .] .[
ὁ τὴ]ν ζώνην τοῦ π̣ό̣λ̣ο̣υ̣ κ̣[λ]ηρωθ[εὶc
εὑρ]έθη ἐν ζῳδίῳ ἀ̣[ρρ]ε̣ν̣ι̣[κῷ] δισώμ̣[ῳ
ἀνθ]ρ̣ωπωειδῆ μουc̣ι̣κ̣ῷ φωνα[cκῷ
Δι]δ̣ύμοιc μο(ιρῶν) κ̅θ̅ λεπτ̣ῶν ν̅, οἴκ[ῳ
ἰδί]ῳ̣, ὁρίοιc Κ̣ρόνου, ὑψώματι οὐθέ[νοc,
ταπε]ι̣ν̣ώματ̣ι̣ οὐ̣θ̣έ̣νοc, τριγώνῳ ἰ[δίῳ
] .ω̣η[
– – – – – – – – – – – – – – – – – –

4283. Deluxe Horoscope

Fr. 1

]...[
] triangle of the moon [
] first decan [
Pyroeis] the star of Mars which
has been allotted the [belt of the] heavens,
]... feminine sign [
] Virgo at x degrees [
] exaltation [
] triangle [

Fr. 2

Stilbon, the star of Mercury, ...
[which] has been allotted the belt of the heavens,
was found in the masculine, bicorporal,
anthropomorphic, musical, voice-training sign
Gemini, at 29 degrees 50 minutes, its own house,
terms of Saturn, exaltation of none,
depression of none, its own triangle
]...[

4284. Deluxe Horoscope

Παρ]θ̣έ̣ν̣[ῳ
ὁρ]ί̣ο̣ι̣ϲ̣ [. .]ο̣[*c. 6 letters*]ωμ̣[
ζῳ]δίῳ [. .] (*vac.*)
] . ικῷ . [. .]ω̣α̣[. .] . Λέοντ̣ι
λεπ]τ̣ῶ̣ν̣ ε̅, οἴκ̣ῳ ἡλί[ου, ὁ]ρίοιϲ Δ[ι]όϲ,
δεκαν]ῷ̣ πρώτῳ.
] Λ̣έοντι μ̣ο̣(ιρῶν) ε̅ λεπτῶν ι̅β̅, οἴκῳ
ἡλίου, ὁρίοιϲ] Δ[ιό]ϲ̣, δεκ̣[α]ν̣ῷ πρώτῳ.
] κλῆροϲ τῆϲ τ̣ύ̣χηϲ εὑρέθη ἐν ζῳ-
δίῳ] ϲτερέῳ [ἀρ]ρενικῷ β̣αϲιλικῷ Λέοντι μο(ιρῶν) κ̣̅,
οἴκῳ ἡλίου, ὁρίοιϲ] Ἑρμοῦ, δεκ̣α̣ν̣ῷ πρώτῳ.
κλῆρο]ϲ̣ δαίμονοϲ . . . ωϲ . .
λε]πτῶν δύο [ὁρίοι]ϲ̣ Διόϲ,
δεκανῷ πρώτ]ῳ̣———

7 μο¯ (so line 10)

4285. Deluxe Horoscope

i	ii
ζῳδι . [ ]	ἐτύγχανεν [
γρω φερ̣ . [ ] . [.] .	λ̣ῳ περ[ι]γραφ[
ν̣ου δ̣ι̣α̣[ ] ἐ̣π̣ὶ̣ ἀ-	κλῳ ἐν Αἰγόκε[ρῳ μοιρῶν
π̣η̣λ̣ιώ[τη, δε]κανοῦ	κ̅ λεπτῶν θ̅, [ζῳδίῳ θη-
δ̣ε̣υ̣[τέρου . .]ϲ̣τ̣ο̣	λ[υκῷ τ]ροπικῷ, οἴκῳ Κρό-
ρ̣ε̣υ̣[ ]	[νου, ὁρί]οιϲ Ἀφρ[οδίτηϲ,
κ̣η̣[ ]	ἐπ[ὶ δ]εκ̣ανοῦ [
[ ]	ταη̣[. .]ερατικῇ [
[ ]	τα . [. .] (*vac.*)

4284. Deluxe Horoscope

] Virgo [ ] terms ...[ ] sign ...]... Leo] 5 minutes, house of the sun, terms of Jupiter,] first decan.] Leo 5 degrees 2 minutes, house of the sun, terms] of Jupiter, first decan.] lot of fortune was found in the solid, masculine, royal sign Leo at 20? degrees, house of the sun, terms] of Mercury, first decan.] lot of daimon ...] two minutes, terms of Jupiter, first [decan].

4285. Deluxe Horoscope

i	ii
sign? ...	It happened to be [
...	... circuit [
... to?	... in Cancer, at 20 [degrees]
Apeliotes, second	9 minutes, feminine,
decan ...	tropical [sign], house of
...	Saturn, terms of Venus,
...	in decan [
[ ]	...
[ ]	...

4286. Deluxe Horoscope

]κ̣θυ̣γενν ̣ ̣ [
] ̣ ̣ ̣ ̣ κατα̣τ̣ ̣ο̣υ̣ ̣ ̣ ̣[

] ̣ κδ // λ̣ε̣π̣(τῶν) // [

] ια // λε̣π̣(τῶν) // ζ̣ // ̣[

] ̣ κ̣ζ // λε̣π(τῶν) // λς [

] ̣ ̣ λ̣ε̣π̣(τῶν) // λ ̣ // ̣ ̣[

] λ̣ε̣π(τῶν) // ̣ ̣ // ὑ̣ψώ̣[ματι

] ̣ // λ̣ε̣π̣(τῶν) // κ̣γ̣ // ὑψώ̣μ̣[ατι

] λ̣ε̣π̣(τῶν) // ε̣ // ὑψώμα[τι

] ̣α̣ι̣ ̣ (*illegible traces*) [

] κλυ̣ρωc τύχ̣[ηc
] κλυ̣ρωc ε̣κ̣[̣ ̣ ̣] ̣[
κ]λ̣υ̣ρ̣ω̣c̣ γαμωc Καρ̣[κίνῳ
] Τ̣αύρ̣[̣] ̣ ἀναφωρ̣α̣ ̣[

3 λεπ (so in subsequent lines) 11 *l.* κλῆροc (so lines 12, 13) 13 *l.* γάμου 14 *l.* ἀναφορα…

4286. Tabular Horoscope

]...[
]...[

]... 24 // minutes // [

] 11 // minutes // 7 // ...[

]... 27 // minutes // 36 [

] ... minutes // 3x // ...[

] minutes // ... // exaltation [

] ... // minutes // 23 // exaltation [

] minutes // 5 // exaltation [

]...[

] lot of fortune [
] lot of ...[
] lot of marriage in Cancer [
] Taurus ... [

4287. Horoscope, a.d. 221–234

] γέν̣[εϲιϲ
Ἀλε]ξ̣άνδρ̣[ου
]α̣ Ἀφροδε[ίτη
] ̣ω ἧλιϲ, Ἑρμῆϲ Ζ ̣[
Λέ]οντι. ϲελήνη Καρκίνῳ απ̣[
] ̣ ̣ ̣ ̣κρ̣[̣ ̣ ̣] ̣ ̣ιζ̣[
(*stripped*)
]ϲ̣ ὡροϲκ̣[όποϲ

3 *l.* Ἀφροδίτη 4 *l.* ἥλιοϲ

4288. Horoscope

— — — — —
Ἄ]ρηϲ [
ϲελήνη [
Ἀφροδί[τη
ἥ]λ̣ιοϲ Ἑ[ρμῆϲ?
— — — — —

4289. Horoscope

— — — — —
] ̣[̣]β̣[
Ἀφρο]δ̣είτη [
Δι]δ̣ύμοιϲ [
]υ̣ Ἑρμ[
]ι̣ ι̅η̅ ο ̣[
]νεα [
]ο̣ν κα[
]κ̣αι[̣] ̣υ[
— — — — —

2 *l.* Ἀφροδίτη

4287. Horoscope, a.d. 221–234

] Nativity [
] of Alexander [
]... Venus [
]... Sun, Mercury ...[
in Leo. Moon in Cancer ...[
]...[
(*stripped*)
]... ascendant

4288. Horoscope

— — — — —
Mars! [
Moon [
Venus [
Sun, Mercury [
— — — — —

4289. Horoscope

— — — — —
]...[
] Venus [
] in Gemini. [
]... Mercury [
]... 18 ...[
]...[
]...[
]...[
— — — — —

4290. Horoscope, a.d. 285

– – – – – – – – – – – – – – –
] ̣[
Διοκλη]τιανοῦ ⟨τὸ⟩ καὶ̣ Μ̣[αξιμιανοῦ?
] ὡρ(α) α̅ ἡμέρας̣ [
Ζ]εὺς Ἑρμῆς ὡρ(οσκόπος) Ζυ̣[γῷ
Ἄρ]η̣ς̣ σελήνη Π̣α̣[ρθένῳ
– – – – – – – – – – – – – – –

3 φ 4 φ

4291. Horoscope, a.d. 285–293

ἀγαθῇ τ]ύ̣χ̣η. γένεσις [
]δ̣ώ̣ρου
] ̣ Διοκλητιαν̣[οῦ
] Μαξιμινια̣[νοῦ
] η̅ ὥρ(α) α̣̅ [
ἡμέρ]ας.
Ἑρ]μῆς Σκορπί̣[ῳ
– – – – – – – – –

4 *l.* Μαξιμιανοῦ 5 φ

4292. Horoscope

Κρόνος καὶ [
Λέοντος [
Ἀφροδίτη [
Π̣αρθένῳ [
Ζ̣εὺς Ἰχθύσι [
ὥρα, σελήν̣η̣ [
Εὐόκερῳ [

5 ϊχθυσι 6 *l.* ὡροσκόπος 7 *l.* Αἰγόκερῳ

4290. Horoscope, a.d. 285

— — — — — — — — — — — — — — — —
]...[
] of Diocletian and Maximian [
] hour 1 of day. [
] Jupiter, Mercury, ascendant in Libra [
] Mars, moon in Virgo. [
— — — — — — — — — — — — — — — —

4291. Horoscope, a.d. 285–293

Good fortune. Nativity [
...]dorus.
]... of Diocletian [
] of Maximian [
] 8, hour 1 [
] of day.
] Mercury in Scorpio. [
— — — — — — — — — —

4292. Horoscope

Saturn and [
in Leo [
Venus [
in Virgo [
Jupiter in Pisces [
Ascendant, moon [
in Capricorn [

4293. Horoscope

γένεϲιϲ ἔτουϲ β [
νυκτὸϲ ὥρ(ᾳ) δ̄ [
Ἑρμῆϲ Αἰγ[όκερῳ
Κρόνοϲ Δ[ιδύμοιϲ
] ̣ ̣[
— — — — — — —

2 φ

4294. Horoscopes, later than a.d. 284

— — — — — — — — — —
] ̣ ̣ ̣ ̣[
Κ]α̣ρκίνῳ [
ἀγαθῇ τ]ύχη. γέν[εϲιϲ
Διοκ]λη̣τιανοῦ̣ [
] ̣ ὥρ(ᾳ) δ̄ νυκτό̣ϲ̣. [
]ϲ Ζυγῷ. Ἑ̣ρ̣μ̣[ῆϲ
] Α̣ἰ̣γ̣[όκ]ε̣ρῳ [
— — — — — — — — — —

5 φ

4295. Horoscope

— — — — —
ϲελήνη Καρκίνῳ̣.
κλῆροϲ τύχηϲ
Κριῷ.
κλῆροϲ δαίμονοϲ
Ϲκορπίῳ.
διευτυχεῖ.

4293. Horoscope

Nativity, year 2 [
hour 4 of night. [
Mercury in Capricorn. [
Saturn in Gemini. [
]...[
— — — — — — —

4294. Horoscopes, later than A.D. 284

— — — — — — — — — —
]...[
] Cancer [
] Good fortune. Nativity [
] of Diocletian [
]... hour 4 of night. [
]... in Libra. Mercury [
] in Capricorn [
— — — — — — — — — —

4295. Horoscope

— — — — —
Moon in Cancer..
Lot of fortune
in Aries.
Lot of daimon
in Scorpio.
Good luck!

4296. Horoscope

\- - - - - - - - - -

] ̣ ̄ ̣ νυκτόϲ.
Ὑδ]ρ̣ηχόφ̣. Κ̣ρόνοϲ Ἰχθύϲι.
Διδύ]μοιϲ. [Ἀφ]ρ̣ο̣δ̣ε̣ίτη̣

2 *l.* Ὑδροχόῳ   3 *l.* Ἀφροδίτη

4297. Horoscope (?)

\- - - - - - - - - - - - -

Θοώνιοϲ   Ἑρμῆϲ [
ὥ̣ρ̣(α) νυ(κτὸϲ) ι [
τονπλ̣ ̣ ̣[

\- - - - - - - - - - - - -

3 φ̣ νυϊ

4298. Horoscope

\- - - - -

]αιϲ [
]οϲ ζ[
]ϲ [
]ϲ̣ φ [
Ἀφρ]οδίτη̣ [

\- - - - -

4296. Horoscope

]... of night.
] in Aquarius. Saturn in Pisces.
] in Gemini. Venus

4297. Horoscope (?)

Of Thoonis   Mercury [
Hour of night 10 [
...[

4298. Horoscope

]...[
]...[
]...[
]... ascendant? [
] Venus [

4299. Horoscope

— — — — — — — — — — —

] Ζεὺϲ Καρκίνῳ
] ὡροϲκόποϲ Τοξότῃ
] (*vac.*)
ἥ]λ̣ι̣ο̣ϲ̣, Ἀφ̣ρ̣ο̣δ̣[ί]τ̣η̣
] ̣ Διδύμο̣ι̣ϲ

4300. Horoscope

— — —

] ̣[
] ̣ ἥλιοϲ̣ [
] ̣ ϲελήν̣[η
]ρ̣ω̣ ̣ ̣[

4300a. Horoscope

— — — — — —

] ἔ̣τ̣ο̣υ̣[ϲ
] Ϲκορ̣πίῳ̣.
] Ταύρ[ῳ
Ἀφροδί]τη Ζυγ[ῷ̣.
] Λέοντι.
ϲελήν]η Ὑδρηχόῳ.
] ̣ ̣[

— — — — — —

6 ϋδρηχοω ; *l.* Ὑδροχόῳ

4299. Horoscope

— — — — — — — — — — — — —
] Jupiter in Cancer.
] Ascendant in Sagittarius.
] (*vac.*)
] sun, Venus
]... in Gemini.

4300. Horoscope

— — —
]...[
]... sun [
]... moon [
]...[

4300a. Horoscope

— — — — — — — —
] year [
] in Scorpio.
] in Taurus.
] Venus in Libra.
] in Leo.
] moon in Aquarius.
]...[
— — — — — — — —

INDEXES

INDEXES

Square brackets indicate that a word is wholly or substantially restored by conjecture or from other sources, round brackets that it is expanded from an abbreviation or a symbol. The article is not indexed.

I. EMPERORS AND REGNAL YEARS

AUGUSTUS

no name (Years 36–42) **4160a** ii 1–6 (Years 17–24) **4199** recto ii 1, 14, 27, 40, verso i [13], [26], 39.

TIBERIUS

no name (Years 16–22) **4160** iii 2–7, ii 8 (Years 9–22) **4161** i [1–13].

GAIUS

Γάϊος (Years 1–4) **4160** iii 8–11 (Years 1–3) **4161** i 14–17.

CLAUDIUS

Κλαύδιος (Years 1–2) (**4160** iii 12–13).
Τιβέριος (Year 14) **4204** i 3.

NERO

no name (Years 4–13) **4160** iii 27–35.
Νέρων (Year 10) **4236** 2.

GALBA

Γάλβα (Year 1) **4160** iii 36.

VESPASIAN

Οὐεσπασιανός (Years 1–9) (**4160** iii 37–39, [40–45]).

TITUS

no name (Years 1–2) [**4160** iii 46–47].

DOMITIAN

no name (Years 2–12) **4161** vii 1–10.
Δομιτιανός (Year 4) **4237** 1.

TRAJAN

ἔτος? τῆς Τραϊανοῦ Καίσαρος ἡγεμονίας (Year 8) **4133** i 5–6.
no name (Years 10–19) **4157a** ii 2–8.

HADRIAN

no name (Years 2–7) **4157a** ii 9–12.
ʿΑδριανός Αἴλιος ὁ κύριος (Year 14) [**4239** 1–2].

ANTONINUS PIUS

no name (Years 6–9) **4186** i–iv 1.
Αἴλιος ʾΑντωνῖνος (Year 14) [**4240** 2].

MARCUS AURELIUS, LUCIUS VERUS

ʾΑντωνῖνος καὶ Λούκιος (Years 1–32, includes Commodus) [**4148** iii 1–33].
ʾΑντωνῖνος καὶ Οὐῆρος (Years 1–6) [**4149** back i 1–9].
ʾΑντωνῖνος καὶ Οὐῆρος οἱ κύριοι ϲεβαϲτοί (Years 2–5) [**4188** 1, iii 7, v 1].
Αὐρήλιος ʾΑντωνῖνος καῖϲαρ κύριος (Year 23) **4240** 11–13.
no name (Year 1) **4187** 1.

COMMODUS

Κόμμοδος (Years 28 and 3) [**4241** 1, 5] (Year 12?) **4249** [1].

SEPTIMIUS SEVERUS AND CARACALLA

Ϲεουῆρος καὶ ʾΑντωνῖνος (Years 1–25) (**4148** vii 1–26) (Year 20) **4242** 1.
Ϲεουῆρος (Year 24) [**4203** verso 1].
no name (Years 15–23) **4153** ii 1–25. (Years 12, 16) **4188a** → 8, ↓ [6].

ELAGABALUS

ʾΑντωνῖνος (Years 1–4) **4148** vii 27–31 (Years 1–4) [**4151** –II 3–9].
ὁ κόρυφος (Year 1) **4244** 3.

SEVERUS ALEXANDER

ʾΑλέξανδρος (Years 1–13) **4148** vii 32–34, [xi 1–11] (Years 1–8) **4151** 15, [–II 10–21] (Year 2) **4246** 1 (Year 3) **4247** 8 (Year lost) **4287** 2.
no name (Years 5–12) [**4159** O 1–4].

MAXIMINUS

Μαξιμῖνος (Years 1–3) **4148** xi 12–15 (Year 1) **4159** 5, [O 6] (Year 3) **4250** 3.

GORDIAN III

Γορδιανός (Year 1) **4148** xi 16–17 (Year 6) **4151** viii 3–5 (Years 5–6) **4190** recto 1, verso 1.

PHILIPPI

Φίλιπποι (Year 1) **4251** 2 (Year 6) **4252** 8.
no name (Years 1–6) **4151** viii 6–15.

DECIUS

Δέκιος (Year 1) (**4151** vii–viii 14, 16) **4253** 2.

GALLUS

Γάλλος (Years 1–3) [**4151** vii–viii 17–20].

VALERIAN AND GALLIENUS

Οὐαλεριανός (Year 7) **4163** verso text 18.
no name (year 1) [**4151** viii 21].

AURELIAN

Αὐρηλιανός (Year lost) **4256** 1.

PROBUS

Πρόβος (Year 4) **4257** 5.

CARUS, CARINUS, AND NUMERIANUS

Κάρος καὶ Καρῖνος καὶ Νουμεριανός (Year 1) **4258** 1–2.

DIOCLETIAN

Διοκλητιανός (Year 1) **4259** 3 (Year 9) **4262** 1 (Year 16) **4264** 2.

DIOCLETIAN AND MAXIMIAN

Διοκλητιανός καὶ Μαξιμιανός (Year lost) **4290** 2 **4291** 3–4.

DIOCLETIAN AND MAXIMIAN AUGUSTI, CONSTANTIUS AND GALERIUS CAESARS

Διοκλητιανός... Κωνστάντιος καὶ... (Years 13 and 12 and 5) **4260** 2–3.

MAXIMINUS, CONSTANTINE, AND LICINIUS

(Years 9 and 7 and 5) **4266a** 2.

CONSTANTINE, LICINIUS, AND THREE CAESARS

(Years 13 and 11 and 3) **4268** 1.

II. INDICTIONS

A.D. 508 $\overline{\alpha}$ ἰνδικ(τιῶνος) **4275** 3.

III. ERAS

CALLIPPIC PERIODS

κατὰ Κάλλιπ(πον) τῆς $\overline{ϛ}$ περιόδου ἔτος $\overline{ϛ}$ (**4137** 18–19).

ERA DIOCLETIAN

(ἔτος) Διοκλητιανοῦ (Year 29) [**4266a** 3] (Years 29 and 65) **4265** 2, [7] (Years 31 and 29) **4266** i 1, ii 1 (Year 35) **4267** 1 (Year 36) **4269** 1 (Year 42) **4269a** 3 (Year 71) **4271** 2 (Years 86 and 92) **4272** 1, 8 (Year 115) **4273** 4 (Years 197 and 219) **4274** → 2, ↓ [1] (Year 224) **4275** 2.
ἀπὸ κρατήσεως Διοκλητιανοῦ (Year 66) **4270** 3.

IV. MONTHS AND DAYS

(a) Months

Ἀθύρ (**4151** ix 4, 7, 18, 21) **4153** iii 22, 25 **4154** i 13 [**4156a** i 5] **4157a** iii 2, 7, 12 **4158** fr. 1 ii 7, 22 **4159** i 6 **4160** ii 11 [**4180** 35] [**4202** i 10, 23] **4206** fr. 1 verso ii 16 **4207** recto 6 **4211** recto 14 **4236** 2 **4266** i 2 **4267** 1 **4269a** 4.

ἐπαγόμεναι **4148** viii 12–31 [**4151** –I 8] (**4191** fr. 1 verso 27) (**4193** fol. A verso ii 18, fol. B recto i 11) (**4196** verso i 8) **4204** ii 15 (**4208** ii 11).

Ἐπείφ **4137** 21 (**4151** ix 5, 19) **4153** iii 24 [**4156a** i 4, 7] **4158** fr. 1 ii 13, 20 **4159** i 2 **4160** ii 7, 18 [**4160a** iii 3] [**4202** i 5, 18, 31] **4204** ii 5, 13, iii 10 **4205** verso i 1 **4206** fr. 2 recto ii 2, verso i 2 **4209** ii 3 [**4212** ii 2] **4250** 3.

Θαργηλιών **4137** 20.

Θώθ **4146** fr. 1 9 **4148** [viii 33–34], xii [1–11], 13–17 **4149** front i [1], 4, back ii 3 [**4151** –I 11, 14] **4153** iii 12, 15 **4154** i 3, 6 **4158** fr. 1 ii 21 **4160** ii 9, 20 **4160a** iii 5 **4175** 1 **4177** fr. 3 4 [**4202** i 8, 21] **4205a** → 3 **4206** fr. 1 recto ii 1, verso ii 2 [**4211** recto 0] **4212** ii 12 **4237** 1 **4241** 6 **4244** 3 **4268** 1.

Ἰανουαρίων (καλανδῶν) **4133** i 7.

Μάρτιος **4183** fr. 1 recto 1–3.

Μεσορή **4148** iv 1–26, [27–32], viii 2–11 [**4149** front i 8, 11] (**4151** [–I 5, 19], ix [8], 11, 15) (**4153** iii 18, 21) (**4154** i 9, 12) [**4156a** i 1] (**4157a** iii 5, 10) (**4158** fr. 1 ii 6, fr. 2 i 3) (**4159** i 3) (**4160** ii 8, 19) **4160a** iii 4 [**4175** 1] **4177** fr. 2 8 [**4202** i 6, 19, 32] **4204** ii 6, 14, iii 12 **4205** verso i 8 (**4205a** → 4) **4206** fr. 2 recto ii 9, verso i 9 **4208** ii 4 (**4212** ii 6) **4266a** 4.

Μεχείρ **4133** i 7 **4138** 13 **4149** front i 6, back ii 5, 8 (**4151** ix 17) **4153** iii 7, 26 **4154** i 1 **4156a** i 9, 12 (**4158** fr. 1 ii 2, 9, 17, 24, fr. 2 i 6) (**4160** ii 3, 14) **4202** i 13, 26 **4206** fr. 2 recto i 9, verso ii 9 **4209** i 3 [**4210** fr. 1 recto 8] **4251** 4 **4269** 2 **4273** 4 **4275** 2.

Παῦνι **4149** front i 5, back ii 7 [**4151** –I 16] **4153** iii 8 **4154** i 2 [**4156a** i 10] (**4157** iv 2) **4157a** iii 3, 8 (**4158** fr. 1 ii 5, fr. 2 i 2) (**4160** ii 17) [**4160a** iii 2] (**4177** fr. 1 xi 5) [**4202** i 4, 17, 30] **4204** ii 4, 12, iii 7 **4210** fr. 1 verso ii 11 **4240** 13 **4249** 2 **4257** 1 **4258** 2 **4272** 1 **4274** ↓ 2 **4282** 5.

Παχών **4149** front i 2, back ii 4 (**4151** [–I 9, 12], ix 16) **4153** iii 11, 14 **4154** i 5 [**4156a** i 13] (**4158** fr. 1 ii 12, 20, 26, fr. 2 i 1, 9) (**4160** ii 6) [**4160a** iii 1] (**4177** fr. 1 viii 5) [**4202** i 3, 16, 29] **4204** ii 3, 11, iii 6 **4205** ii 10 **4206** fr. 1 recto i 15 **4207** verso 6 **4210** fr. 1 verso ii 4 (**4212** i 10, iii 9) **4242** 2 **4248** 1, 7 *bis*.

Cεξτίλιος (**4175** v 15).

Cεπτέμβριος (**4175** xiii 11).

Τῦβι **4149** front i [3], back 2 [**4151** –I 10, 13] **4153** iii 10, 13 **4154** i 4, 7 **4157a** iii 4, 9 **4158** fr. 1 ii 1, 16, 23, fr. 2, 5 **4160** ii [2], 13 **4202** i 12, [25] **4206** fr. 2 recto i [2], verso ii 2 **4210** fr. 1 recto 1 **4239** 2 **4246** 1 **4251** 13 **4271** 3.

Φαμενώθ (**4151** ix [6], 20) (**4153** iii 20, 23) (**4154** i 14) [**4156a** i 3, 6] (**4158** fr. 1 ii 3, [10], 25, fr. 2 i 7) (**4160** ii 4, 15) (**4177** fr. 1 ii 5) **4178** 10a **4183** fr. 1 recto ii 1 (**4202** i [1], 14, [27]) **4204** ii 2, 8, iii 2 **4206** fr. 1 recto i 1, verso i 2 **4209** i 11 **4210** fr. 2 2 **4211** verso 1 [**4212** i 2] **4270** 4.

Φαρμοῦθι [**4149** front i 9] (**4151** [–I 6, 17, 20], ix [9], 12) (**4153** iii 17) (**4154** i 8, 11) (**4157a** iii 6, 11) (**4158** fr. 1 ii 4, [11], 18, fr. 2 i 8) (**4159** i 1) (**4160** ii 5, 16) (**4177** fr. 1 v 5) (**4202** i [2], 15, [28]) **4204** ii 10, iii 4 **4205** verso ii 3 **4206** fr. 1 recto i 8, verso i 9 **4210** fr. 2 9 (**4211** verso 8) (**4212** i 6, iii 2) (**4251** 3) **4264** 3.

Φαῶφι **4149** front i 7, back ii 6, 9 **4153** iii 9 [**4155** i 8, 11] (**4158** fr. 1 ii 14, fr. 2 i 4) (**4159** i 4) (**4160** ii 10, 21) **4160a** iii 6 [**4202** i 9, 22] **4205a** → 7 **4206** fr. 1 recto ii 8, fr. 2 verso ii 9 **4211** recto 7 **4240** 3 **4260** 4 **4265** 7 **4266** ii 2.

Χοίακ [**4149** front i 10] (**4151** [–I 4, 7, 18, 21], ix 10, 13?, 14) **4153** iii 16, 19 **4154** i 10 (**4156a** i 2) **4158** fr. 1 ii 8, 15 **4160** ii [1], 12 **4174** verso i 10 [**4180** verso 35] **4202** i 11, [24] **4205** recto 7 [**4209** iii 3] **4213** 4 [**4245** 1] **4247** 8 **4259** 3.

(b) Days

διχόμηνος [**4137** 19].
εἰδοί [**4175** xiii 23] **4179** i [15], xv 13 [**4180** recto i 21, verso –I 21] **4181** fr. 2 i 9.
ἥλιος (**4274** ↓ 4).
καλάνδαι **4133** i 7 [**4179** i 1, xv 1] [**4180** recto i 9, verso –I 9].
νεομηνία [**4146** fr. 1 9].
νῶναι (**4175** v [21], xiii 15) [**4179** i 7, xv 5] [**4180** recto i 12, verso –I 13] [**4181** fr. 2 i 1].

V. PERSONAL NAMES

Α[**4240** 15.
Ἀμμωνᾶς **4270** 2.
Ἀλι̣ ̣[**4244** 2.
Ἀπ[**4236** 1.
Ἀπία **4258** 1.
Ἀφ̣[**4273** 3.
Διάπαλος **4266** i 12.
Θεόδωρος **4275** 1.
Θοῶνις **4297** 1.
Ἱέραξ **4259** 2.
Κ̣ ̣ρι̣[**4251** 1.
Νεικ[**4252** 7.
Πτολεμαῖος **4242** 1.
Σαπρίων **4260** 1.
Σαραποῦς **4240** 10.
Σαραπίων **4250** 2.
]δωρος **4291** 2.

VI. GEOGRAPHICAL

Αἰγυπτιακός **4133** i 4.
Αἰγύπτιος (**4137** 21) **4249** 3 **4251** 2.
Αἴγυπτος **4276** ii 5.
Ἀλεξάνδρεια **4242** fr. 2 ii 11, 12, 15, (32), 39.
Ἀπάμεια **4168** fr. 6 back ii 3.
Γαλλία **4168** fr. 6 back ii 5.
Δαμασκός **4168** fr. 6 back iii 7.
Ἕλλην **4239** 2.
Κιλικία **4168** fr. 6 back ii 7.
Ὀρχῆνοι **4139** 8.
Ῥώμη **4142** fr. 2 ii 6, 7, 11, 12, 14, 22, 26, 29, 35.
Συρία **4168** fr. 6 back iii 5.
Φρυγία **4168** fr. 6 back ii 2.

VII. AUTHORITIES CITED

Ἱππαρχικὸν σύνταγμα **4276** ii 2–3.
Τιμοχάρις **4135** 36.

VIII. ASTRONOMICAL AND ASTROLOGICAL

(a) Zodiacal Signs

Αἰγόκερως **4147** fr. 1 ii 5 (**4149** front iii 4) (**4150** front ii 13) [**4151** i 11] (**4152** front i 10, back i 6, 9) (**4155** fr. 1 iii 3, 6, fr. 2 7, 10) [**4156a** iv 6, 9] (**4156b** iv 3) (**4156c** ii 5) (**4159a** iii 5) **4166** back iii–iv 28 (**4172** fr. 2 iii 8) [**4174** recto ii 1, 12] [**4175** xvi 20] **4176** [2], i 27 (**4177** fr. 1 ii 8, v 7, 19, viii 6, 18, xi [16], fr. 2 iii 13, fr. 3 ii 7) **4179** v [3], 31 (**4180** recto iv 16, verso ii [14]) (**4181** fr. 2 iv 16) (**4182** recto ii 3) (**4184a** ii 2) (**4199** verso iii 5, [6–12], [18]) (**4202** vii 22) (**4214** 6) (**4222** fr. 1 i 9) (**4226** fr. 1 i [11], iii 5, 17, fr. 2 ii 5, iv 5, fr. 3 i 3) **4236** 10 **4239** 3 *bis*, 4 **4240** 7 **4243** 5 **4246** 3 **4247** 3, 11 **4248** 6 **4249** 12 (**4251** 18) **4254** 4 **4257** 1, 4 **4259** 11 **4261** 4 **4264** 7 [**4266** ii 4] **4266a** 7 **4269a** 7 **4270** 6 **4270** 12 **4271** 7 **4273** 1, [7?] **4274** → 4, 10 **4275** 8, 9 **4277** fr. 1 i 5, 14 **4285** ii 3 **4292** 7 **4293** 3 **4294** 7.

Δίδυμοι **4138** 2 (**4149** front iii 9, 22) (**4150** front ii 5) [**4151** i 4, 17] (**4152** front i 2, 5, back i 1, 4) (**4155** fr. 1 iii 1, 20, 23, fr. 2 iii 5, [8]) (**4156a** iv 4) (**4156b** iv 4) (**4156c** ii 3) (**4159a** iii 8, [15]) **4162** recto i 20 (**4163** recto iii 8) **4166** back iii–iv 25 [**4167** fr. 1 viii 1] (**4172** fr. 2 iii 13) (**4174** recto ii 5) (**4175** viii 12) (**4176** i 11) (**4177** fr. 1 ii 3, iv 1, 2, 3, ii 13, v [12], viii 11, xi 9, fr. 2 iii 18) (**4177a** i 10) **4178** iv 10 (**4179** v 15, xiv 16a, ix 21a) (**4180** recto iv 27, verso ii [24], 36) (**4181** fr. 2 iv 1) (**4202** vii 4, 29) (**4203a** 23) **4204** ii (12), iii 5 **4222** fr. 1 i 2 (**4226** fr. 1 i [16], iii 10, fr. 2 ii 10, iv [10], fr. 5 4) **4232** (3) i 3 (**4236** 6) (**4236a** 1) **4238** 2 **4240** 8 (**4241** 7, 11) **4246** 6 **4248** 4 **4249** 8, 9, 10 **4250** 1 **4255** 4 **4257** 3 *bis* **4259** 7 **4263** 7 **4264** 4 **4265** 9 **4267** 7 **4268** 4 **4271** 8 (**4272** 4) **4274** ↓ 5 **4283** fr. 2 5 **4289** 3 **4293** 4 **4296** 3 **4299** 5.

Ζυγόν **4142** fr. 1 19, fr. 2 i 17 **4149** front iii 1, 26 **4150** front ii 9 [**4151** i 8] (**4152** front i 6, back i 2) **4155** fr. 1 iii 2, 21, fr. 2 iii 3, 6 (**4156a** iv 2, [5]) (**4156c** ii 1, 4) (**4159a** iii 11) **4172** fr. 2 iii 5 (**4174** recto ii 3) (**4175** 3, viii [21], xiii 3, xvi [13]) **4176** i 21 (**4177** fr. 1 ii 17, v 16, viii 15, xi 13, fr. 2 iii 10, 22, fr. 3 ii 2, 3) [**4177a** i 20] **4179** v 24 **4180** recto iv 10, verso ii [7], [34] (**4181** fr. 2 iv 9) (**4199** recto iii 18–37, [38]) (**4202** v 23–27, vii 11–12) **4203** verso 3–4, [5], 6–8 (**4203a** 15) **4204** ii 6 (**4214** 3) **4222** fr. 1 i 6 **4226** fr. 1 i 8, iii 2, 14, fr. 2 ii 2, iv 2, fr. 4 i (3) **4240** 9 (**4241** 10, 13) **4243** 6 [**4245** 22] **4250** 5 **4251** 6 **4252** 2 **4253** 6 **4255** 6 **4257** 4 **4260** 8 **4266** i 7, ii [6] **4268** 3 **4269a** 5 **4272** 12 **4274** → 6, 8, 9 **4279** 3 **4280** 2 **4290** 4 **4294** 6 **4300a** 4.

Ἰχθύες **4140** ii 3 **4149** front iii 6 (**4150** front ii 2, 15) [**4151** i 13] (**4152** front i 1) (**4155** fr. 1 iii [13], 16, fr. 2 iii 1) (**4159a** iii 13) (**4172** fr. 2 iii 10) (**4175** xvi 24) **4176** 1 (**4177** fr. 1 ii 10, v [9], viii 8, xi 6, 18, fr. 2 iii 15, fr. 3 ii [9]) [**4177a** i 3] **4178** iv 3 **4179** v (8), xii 6a **4180** recto iv [20], verso ii [18] (**4181** fr. 2 iv 21) (**4182** recto iii 3) (**4184a** ii 6) (**4199** verso ii [30], 31–36, [37], 38–44) (**4202** vii 24, 25) (**4203a** 6, 20) **4204** ii 9, iii 2 (**4205** verso ii 4) (**4214** 8) **4222** fr. 1 i 11 (**4226** fr. 1 i [13], iii 7, fr. 2 ii 7, iv 7, fr. 3 i 5) (**4232** (1) fr. 1 i 1–3, iii 1, 2) **4243** 3 **4246** 4 **4247** 4, 12 **4248** 5 **4249** 6, 11 (**4251** 8, 20) **4252** 11 **4261** 5 **4266** i 5, 6 **4267** 4 **4268** 4 **4269** 3 **4270** 10 (**4272** 10) **4274** → 5 **4277** fr. 1 i 11 **4292** 5 **4296** 2.

Καρκίνος (**4149** front iii 10, 23) (**4150** front ii 6) (**4151** i [5], 18) (**4155** fr. 1 iii 14, 17, fr. 2 iii 2) (**4156a** iv 1) (**4159a** iii 1, 9, 16) (**4162** recto iv 23) **4165** fr. 1 ii [1], fr. 2 i 1 [**4167** fr. 1 ii 1] **4168** fr. 1 back 3 **4171** back 1 (**4172** fr. 2 iii 2) (**4175** v 5, viii 14, xvi 8) (**4176** i 13) (**4177** fr. 1 ii 3, ii 14, ii [13], viii 12, xi 10, fr. 2 iii 19) (**4177a** i 12) **4178** iv 13 (**4179** v 18) (**4180** recto iv 30, verso ii [27]) (**4181** fr. 1 iv 28a, fr. 2 iv 3) **4183** fr. 2 6 (**4202** v [1], [2], vii 5, 6, 30) (**4203a** 10) **4204** ii 13, iii 6, (9), 11 (**4222** fr. 1 i 3, [17], iii 11) **4232** (3) i 4 **4238** 3 **4240** 5 **4242** 3 **4248** 10 **4250** 6 **4251** 11 **4252** 5 **4260** 5 **4261** 7 **4265** 5 **4269a** 9 **4272** 5 **4273** 9 **4274** ↓ 7 **4275** 5, 7 **4286** 13 **4287** 5 **4294** 2 **4295** 1 **4299** 1.

Κριός (**4149** front iii 7, 20) **4150** front ii 3, 16 [**4151** i 14] (**4152** back i 10) **4155** fr. 1 iii 10, fr. 2 iii 14 **4156a** iv [10], 13 (**4159a** iii 6) **4167** fr. 7 i 2 **4171** front ii 1 **4272** fr. 2 iii 11 (**4174** recto ii 7, 10) (**4175** viii 8) **4176** 3, i 7 (**4177** fr. 1 ii 1, ii 11, v [10], viii 9, xi 7, fr. 2 iii 16) **4177a** i 5 **4178** iv 5 **4179** v 10 [**4180** recto iv 23, verso ii 20] (**4181** fr. 1 i 2, iii [2]) **4184a** ii 9 (**4199** verso ii 45) (**4202** vii 26) **4204** i 7–13, ii 10, iii 3 (**4205** verso ii 5–7), **4226** fr. 1 i [14], iii 8, fr. 2 ii 8, iv [8], fr. 3 i 6, fr. 5 2 (**4232** (1) fr. 1 i 4–5, [6], iii 3–5) [**4232** (3) i 1] **4237** 5, 7 **4242** 7 **4246** 5 **4249** 5 **4251** 9 **4254** 5 **4256** 5 **4257** 2 **4259** 6 **4263** 5 **4264** 9 **4267** 2 **4269** 4 **4273** 6 **4276** ii 8 **4295** 3.

Λέων **4141** 4 **4147** fr. 1 ii 4 (**4149** front iii 24) (**4150** front ii 7) (**4151** i [6], 19) (**4152** back i 11) (**4155** fr. 1 iii 11, fr. 2 iii 12) [**4156a** iv 11] (**4159a** iii 2) (**4162** recto vii 25) **4165** fr. 1 iii 1 [**4167** fr. 1 vi 1] **4168** fr. 1 back 3 (**4172** front iii 3) (**4174** recto ii 8) (**4175** viii 5, [17], xvi [9]) **4176** i 16 (**4177** fr. 1 ii 4, 15, v 14, viii 13, xi 11, fr. 2 iii 20) **4177a** i 15 **4179** v 20 (**4180** recto iv [5], 32, verso ii [29]) (**4181** fr. 1 ii 2, fr. 2 iv 5) (**4199** recto iii 8–10) (**4202** iii [3], 4–8, v 3–9, [10–13], 14–16, vii 8, 31, 32) **4204** ii 14, iii 12 (**4205** verso i 2–5) **4205a** → 4, 5 (**4214** 1) **4222** fr. 1 i 4 **4226** fr. 1 i [18], iii 12, fr. 4 i 1 **4232** (3) i 5 [**4236** 8] **4236a** 4 **4238** 5, 6 **4239** [5], 5 (**4241** 3, 8) **4242** 6 **4243** 7 **4247** 6, 14 [**4248** 8] **4250** 7 **4252** 3 **4261** 1 **4263** 2 **4264** 5 **4266a** 5 **4267** 6 **4268** 2 **4269** 5 **4271** 10 **4274** → 7 **4277** fr. 1 i 21, 27 **4284** 4, 7, 10 **4287** 5 **4292** 2 **4300a** 5.

Παρθένος (**4149** front iii 25) (**4150** front ii 8) [**4151** i 7, 21] (**4152** front i 9, back i 5, 8) (**4155** fr. 1 iii 5, 8, 24, fr. 2 iii [9]) [**4156a** iv 8] (**4156b** iv 2, 5) (**4159a** iii 3, 10) [**4162** recto x 26] (**4172** fr. 2 iii 4) (**4174** recto ii 11) (**4175** 2, v 6, viii 6, [19], xiii 5, 6, xvi [11]) **4176** i 18 (**4177** fr. 1 iv 4, ii 16, v 15, viii 14, xi 12, fr. 2 iii 9, 21) [**4177a** i 17] (**4179** v 22) (**4180** recto iv 7, viii 2, verso ii [5], [32]) (**4181** fr. 2 iv 7) (**4199** recto iii 3–7, 11–17, 22–25) (**4202** iii 14–21, [22], 23–24, [25], 26–28, [29], 30–32, v 17–22, 28–31, vii 9–10, [32]) **4203** recto 3–9 **4204** ii 5, 15, iii 13, 14 (**4205** verso i 6–10) (**4214** 2) (**4222** fr. 1 i 5) (**4226** fr. 1 i 7, iii 1, 13, fr. 4 i 2) (**4236a** 3) (**4237** 3) **4245** 8 (**4251** 5) **4255** 5 **4256** 7 **4259** 8 **4260** 7 **4263** 1 **4264** 6 **4266** i 10 **4266a** 6 **4269** 6 [**4269a** 10] **4272** 6 **4275** 6 **4277** fr. 1 i 25, ii 16, iii 2 **4283** fr. 1 7 **4284** 1 **4290** 5 **4292** 4.

Cκόρπιος **4141** 6 **4147** fr. 3 5 (**4149** front iii 2, 27) (**4150** front ii 10, 11) [**4151** i 9] (**4152** front i 3) (**4155** fr. 1 iii 15, 18) (**4159a** iii 4) (**4172** fr. 2 iii 6) (**4174** recto ii 6) (**4175** 4 *bis*, xvi 16) **4176** [4], i 23 (**4177** fr. 1 ii 6, 18, v 17, viii 16, xi 14, fr. 2 iii 11, 23, fr. 3 ii 5) (**4179** v 26) [**4180** recto iv 12, 35, verso ii 9] (**4181** fr. 2 iv 12) (**4199** recto iii 31–35, 39–43) (**4202** vii 13–19) (**4203a** 2) (**4205** recto [1], 2–3) (**4214** 4) (**4221** fr. 1 ii 9) **4222** fr. 1 i 7, fr. 2 ii 3 (**4226** fr. 1 i 9, iii 3, 15, fr. 2 ii 3, iv [3], fr. 3 i 1, fr. 4 i 4) **4238** 7 **4239** 6 [**4240** 4] (**4241** 10, 13) **4245** 12, 25 **4246** 2 **4247** 13, 15 **4251** 16 **4252** 10 **4254** 5 **4257** 2 **4261** 2 **4263** 4 **4266** i 9 **4267** 3, 4, 5 **4269** 11 [**4269a** 6] **4270** 14 **4272** 2 **4273** 2 **4274** → 3 [**4275** 11] **4277** fr. 1 i 6 **4291** 7 **4292** 5 **4300a** 2.

Ταῦρος **4135** 38 [**4141** 2] (**4149** front iii 8, 21) (**4150** front ii 4) [**4151** i 16] (**4152** front i 8, 11, back i 7) **4155** fr. 1 iii 4, 7, fr. 2 iii 11 [**4156a** iv 7] (**4156c** ii 6) (**4159a** iii 7, 14) **4163** recto i 1 **4167** fr. 1 iv 1 **4172** fr. 2 iii 12 (**4174** recto ii 2) (**4175** viii 9) **4176** i 9 (**4177** fr. 1 ii 2, 12, v [11], viii [10], xi 8, fr. 2 iii 17) **4177a** i 7 **4178** iv 8 **4179** v 13, viii 15 **4180** recto iv 25, (36) [**4181** fr. 1 iv 2] (**4202** vii 27–28) (**4203a** 9) **4204** ii 11, iii 4 (**4205** verso ii 8–9) (**4221** fr. 2 i 3) **4222** fr. 1 i 1 (**4226** fr. 1 i [15], iii 9, fr. 2 ii 9, iv [9], fr. 5 3) **4232** (3) i 2 **4238** 4 **4240** 6 **4242** 4 **4243** 2 **4246** 5 **4248** 3 **4249** 7 [**4251** 10] **4254** 2, 4 **4255** 3 **4257** 2, 4 **4261** 6 [**4262** 3] **4266a** 9 **4270** 11 **4272** 9 **4277** fr. 1 i 38 **4286** 14 **4300a** 3.

Τοξότης (**413y** front iii 3, 28) (**4150** front ii 12) [**4151** i 10] (**4155** fr. 1 iii 9, 12, fr. 2 iii 13) [**4156a** iv 12] (**4159a** iii 12) [**4162** verso i 27] (**4172** fr. 2 iii 7) (**4174** recto ii 9) [**4175** xvi 18] (**4176** 4, [5], 5, i 25) (**4277** fr. 1 ii 7, 19, v 6, 18, viii 17, xi [15], fr. 2 iii 12, 24, fr. 3 ii 6) (**4179** v [1], 28, viii 30) (**4180** recto iv [14], verso ii [11], 35, ix 25a) (**4181** fr. 2 iv 14) [**4199** verso iii 1–5] (**4202** vii 19, 21) (**4203a** 3) (**4205** recto 4–8) (**4205a** ↓ i 1–4) (**4214** 5) **4222** fr. 1 i 8, fr. 2 ii 2 (**4226** fr. 1 i [10], iii 4, 16, fr. 2 ii 4, iv 4, fr. 3 i 2, fr. 4 i [5]) **4236** 4 (**4241** 11) **4242** 7 **4243** 4 **4244** 5 **4246** 3 **4247** [2], 10, 13 (**4251** 17) [**4253** 7] **4255** 1 **4259** 10 **4260** 6, 9 **4261** 3 [**4266** ii 7] **4267** 2 **4271** 11 (**4272** 3) [**4273** 5] **4274** → 8 (**4279** 13) **4272** 7 **4299** 2.

Ὑδροχόος (**4149** front iii 5) (**4150** front ii 1, 14) [**4151** i 12] (**4152** front i 4, 7, back i 3) (**4155** fr. 1 iii 19, 22, fr. 2 iii 4) (**4256a** iv 3) (**4156c** ii 2) **4166** back iii–iv 26 **4168** fr. 1

back ii 22 (**4172** fr. 2 iii 9) (**4174** recto ii 4) (**4175** xvi 22) (**4176** 2, i 29) (**4177** fr. 1 ii 9, v [8], viii 7, 19, xi [17], fr. 2 i 1, iii 14, fr. 3 ii 8) [**4177a** i 1] [**4178** iv 1] [**4179** v 6] (**4180** recto iv 18, vii 2, verso ii [16], viii 30a) (**4181** fr. 2 iv 18) (**4184a** ii 4) (**4199** verso ii [19–22], 23–25, [27–29]) (**4202** vii 23) (**4203a** 19) **4204** ii 8 (**4214** 7) (**4222** fr. 1 i 10) (**4226** fr. 1 i [12], iii 18, fr. 2 ii 6, iv 6, fr. 3 i 4) (**4236a** 2) **4239** 6 **4245** 16, 19 **4246** 4 **4247** 12 **4251** 7, 19 **4254** 6 **4255** 2 **4259** 5 **4263** 3 **4265** 4 **4266** i 4 **4266a** 8 **4268** 3 **4269** 7, 8, 9, 10 **4269a** 8 **4270** 7 **4272** 7 [**4273** 8] **4274** ↓ 8 **4275** 4, [10] **4277** fr. 1 i 30 **4281** 2 **4296** 2 **4300a** 6.

(*b*) Planets and Luminaries

Ἄρης **4181** fr. 1 ii 1 **4182** recto i 2 **4197** verso i 1 **4204** ii 1 **4206** fr. 1 verso ii 1 **4226** fr. 2 ii 1 **4237** 6, [8 *bis*] **4238** 4 **4239** 3 *bis* **4270** 7 (**4241** 7) **4242** 6 **4243** 1 **4244** 6 **4245** [13], 18, 26 **4246** 4 **4247** 5, 13 **4248** 8 **4249** 7 [**4251** 11] **4252** 3, 11 **4253** 7 **4254** 1 **4255** 4 **4256** 6 **4257** 3 **4259** 6 **4261** 3 **4262** 4 **4263** 3 **4264** 4 **4265** 5 **4266** i 6, ii 4 **4266a** 5 **4267** 7 **4268** 4 **4269** 8 **4269a** 7 **4270** 7 **4271** 10 (**4272** 6, [11]) **4273** 7 (**4274** → 7) (**4275** 8) **4276** i 4 **4277** fr. 1 i 9, 24, 31–32, [40], ii 7, 8, [29] **4282** 10 **4283** fr. 1 4 **4288** 1 **4290** 5.

Ἀφροδίτη [**4181** fr. 1 iii 1] (**4182** recto ii 2) **4197** verso ii 1 **4204** ii 7 (**4236** 7) **4237** 4 **4238** 5 **4239** 6 **4240** 4 **4241** 3 **4242** 5 (**4243** 3) **4244** 6 **4245** 21 **4246** 4 **4247** 2, 10, 15 **4248** 4 **4249** 8 **4250** 7 [**4251** 9] **4252** (4), 6, 12 **4253** 9 **4254** 4 **4255** 6 (**4257** 3) **4259** 9 **4261** 2 **4262** 5 **4263** 6 **4264** 8 **4265** 5 **4266** i 7, ii 7 **4267** 2 **4268** 3 **4269** 9 **4269a** 5 **4270** 9 **4271** 11 (**4272** 5, 12) **4273** 7 [**4274** → 8] (**4275** 9) **4277** fr. 1 i 7, 39, ii 19, 20 **4279** 5 **4280** 3 **4281** 4 **4285** ii 6 **4287** 3 **4288** 3 **4289** 2 **4292** 3 **4296** 3 **4298** 5 **4299** 4 **4300a** 4.

Ἑρμῆς [**4181** fr. 1 iv 1] (**4182** recto iii 2) **4204** iii 1 [**4236** 6] **4236a** 3 **4237** 3 **4238** 6 **4239** 4 **4240** 4 (**4241** 3, 8, 10, 13) **4242** 4 **4243** 1 **4245** 9 *bis*, 13, 17, 24 **4246** 3 **4247** [2], 10 **4248** 3 **4249** 9 **4250** 6 **4251** 7 **4252** 4 **4253** 8 **4254** 3 **4255** 5 **4257** 4 **4259** 9 **4260** 9 **4261** 2 **4263** 6 **4264** 8 **4265** 10 **4266** i 8, ii 5 **4266a** 6 **4267** 5 **4268** 2 **4269** 10 **4269a** 6 **4270** 8 (**4272** 4, 11) **4273** 8 [**4274** → 9] [**4275** 10] **4277** fr. 1 i 1, 2, 23, 32, ii 1, 17–18, 20 **4279** 3 **4283** fr. 2 1 **4284** 11 **4287** 4 **4288** 4 **4289** 4 **4290** 4 **4291** 7 **4293** 3 **4294** 6 **4297** 1.

Ζεύς **4180** recto viii 1 **4181** fr. 1 i 1 **4183** fr. 2 7 [**4204** i 10] [**4209** iii 2] **4226** fr. 1 iii 6 **4236** 5, 10 **4236a** 1 **4238** 3 **4239** 5 **4240** 6 (**4241** 7, 10, 13) **4242** 3 **4243** 5 **4244** 5 **4245** 6 *bis*, 15 **4246** 5 **4247** 6, 14 **4248** 6 **4249** 6 [**4251** 8] **4253** 6 **4254** 6 **4255** 3 **4257** 2 **4259** 8 **4261** 4 **4262** 3 **4263** 4 **4264** 7 **4265** 4 **4266** i 5, ii [4] **4266a** 8 **4267** 6 **4268** 2 **4269** 5 **4269a** 8 **4270** 9 **4271** 9 (**4272** 2, 9) **4273** 6 (**4274** → 5, ↓ [8]) (**4275** 7) **4277** fr. 1 i 10, 23, ii 8, 9, [28], fr. 2 1 **4280** 2 **4282** 9 **4284** 5, 8, 13 **4290** 4 **4292** 5 **4299** 1.

ἡλιακός [**4142** fr. 3 col. i 12]

ἥλιος (**4142** fr. 1 18) (**4168** fr. 1 front iv 1, vi 1) (**4171** front i 4) **4176** iii 11, 16 **4180** recto vi 1 **4236** [4], 9 **4237** 3 [**4239** 3] **4240** 4 **4242** 4 **4243** 2 **4245** 4 **4246** 3 **4247** [2], 10 **4248** 3 **4249** 10 **4250** 6 [**4251** 7] **4253** 8 **4254** 2 **4255** 5 **4256** 5 **4259** 9 **4261** 3 **4262** 6 **4263** 6 **4264** 8 **4265** 5, 11 **4266** i 8, ii 6 **4266a** 5 **4267** 4 **4268** 2 **4269** 7 **4269a** 6 **4270** 8 **4271** 7 (**4272** 4) **4273** 1, 8 (**4274** → [3], ↓ 4, 5) (**4275** 4) **4277** fr. 1 i 22, ii [28] **4284** 5, [8], [11] **4287** 4 **4288** 4 **4299** 4 **4300** 2.

Κρόνος **4168** fr. 2 front 1 **4280** recto vii 1 **4204** i 6 **4209** i 2, ii 2 **4226** fr. 1 i 6, fr. 5 1 [**4236** 8, 10] (**4236a** 2) **4237** 6, 7 **4238** 2 [**4239** 6] **4240** 5 (**4241** 11) **4242** 6 **4243** 6 **4245** 12, [16], 20, 26, 27 **4246** 2 **4247** 2, 11 **4248** 5 **4249** 5 **4251** 5 **4252** 2, 10 **4253** 7 **4254** 3 **4255** 2 **4257** 2 **4259** 11 **4261** 5 **4262** 3 **4263** 2 **4264** 5 **4265** 4 **4266** i 1, ii 4 **4266a** 7 **4267** 3 **4268** 4 **4269** 4 **4269a** 9 **4270** 11 **4271** 8 (**4272** 3, 10) **4273** 5 (**4274** → 6, ↓

[7]) (**4275** 6) **4276** ii 9 **4277** fr. 1 i 7, 9, [31], [32], 40, ii 8–9 **4281** 2 **4283** fr. 2 6 **4285** ii 5 **4292** 1 **4293** 4 **4296** 2.
Πυρόεις [**4283** fr. 1 4].
σελήνη **4133** i 1 [**4137** 11, 13] **4138a** ii 9 **4140** ii 3, 6, 9 **4146** fr. 1 11 **4147** fr. 2 4, ii 10 (**4168** fr. 1 front ii 1, iv 1) (**4182** verso iv 1) **4183** fr. 1 recto iii 3 [**4236** 6] **4237** 5 **4238** 1 **4239** 4 **4240** 8 (**4241** 10) **4242** 7 (**4243** 7) **4246** 5 **4247** 4, 12 **4248** 8 **4249** 11 **4252** 5, 13 **4253** 5 **4255** 3 **4256** 7 **4257** 4, 7 **4261** 7 **4263** 5 **4264** 6 **4265** 4, 9 **4266** i 10, ii [5] **4266a** 9 **4267** 3 **4268** 3 **4269** 11 **4269a** 10 **4270** 12 (**4272** 6) **4273** 2, 9 (**4274** → 4, ↓ 3, [6]) (**4275** 5) **4277** fr. 1 i 8, 39 **4287** 5 **4288** 2 **4290** 5 **4292** 6 **4295** 1 **4300** 2 **4300a** 6.
σεληνιακός **4133** i 12, 16, 20, ii [15], 19, 27.
Στίλβων **4277** fr. 1 ii 2 **4283** fr. 2 1.
Φαέθων **4133** ii 23.

(c) Stars and Constellations

Δίδυμοι **4137** 13.
κύων **4146** fr. 1 3.
ὁ δεξιὸς... τοῦ ἡγουμένου διδύμου **4137** 14–15.
ὁ ἐν τῷ ὀμφαλῷ τοῦ ἑπομένου διδύμου **4137** 16–17.
ὁ νότιος ὄνος **4133** i 8, 14, 19, ii 18?
ὁ νότιος τῶν περὶ τὸ νεφέλιον Φάτνην τετραπλεύρου προηγουμένων **4133** i 9–11.

(d) Winds and Directions

ἀνατολή **4133** i 15 (**4137** 6) **4140** 5?
ἀνατολικός **4142** fr. 2 ii 24, 25.
Ἀπηλιώτης **4285** i 3.
ἄρκτος **4133** i 11, 14.
Βορέας **4277** fr. 1 ii 2.
βόρειος **4133** i 2, 22 **4140** ii 4, 7 [**4141** 3, 5] (**4168** fr. 1 back iii 9) **4276** i 1.
δύσις **4137** 9.
δυσμή **4133** i 12.
δυσμικός **4142** fr. 2 ii 14, 17, 23.
δυτικός **4142** fr. 2 ii 23, 44.
Λίψ **4277** fr. 1 i 13, ii 10, 22, iii 6.
μεσημβρία **4137** [5], [7], (9), (13).
νότιος **4133** i 8, 9, 14, 19, ii 18 [**4141** 7] (**4168** fr. 1 back iii 7, 14, 26).
νότος **4137** 1.

(e) Astrological Objects

δεκανός **4245** 7, 11, 14, [17], [20], 23, 27 **4277** fr. 1 i 15, 26, 34, ii 2, 10–11, 22, 31–32 **4280** 1 **4283** fr. 1 3 **4284** 6, 8, 11, [14] **4285** i 4, ii 7.
δύνον **4277** fr. 1 i 27, 28.
κλῆρος ἀνάγκης **4277** fr. 1 iii 7, 8.
κλῆρος γάμου **4270** 14 **4286** 13.
κλῆρος δαίμονος **4277** fr. 1 ii 23, [24] **4284** 12 **4295** 4.
κλῆρος εκ[? **4286** 12.
κλῆρος ἔρωτος [**4277** fr. 1 ii 33, 34].
κλῆρος τύχης (**4272** 7) **4277** fr. 1 ii 12, 13 **4286** 11 **4295** 2 κλῆρος τῆς τύχης **4284** 9.
μεσουράνημα **4238** 8 (**4257** 2) **4277** fr. 1 i 35, 36.
μονομοιρία **4277** fr. 1 i 7, 40–ii 1, ii 9, 20, 30 **4283** fr. 1 2.
οἶκος **4237** 3, 5, 7 **4245** 6, 9, [13], 16, 20, 23, 26 **4256** 6 **4276** i 4 **4277** fr. 1 i 6, 22, 31, 39, ii 7, [17], 28, fr. 2 1 **4279** 5 **4280** 3 **4281** 3 **4282** 9 **4283** fr. 2 5 **4284** 5, 7, [11] **4285** ii 5.
ὅριον **4236** [5], [7], 9 **4237** 4, 6, [8] **4245** 6, 9, 13, 17, 20, 23, 26 **4276** ii 9 **4277** fr. 1 i 7, 23, 31, 40, ii 8, 20, 29 **4279** 5 **4280** 3 **4281** 3 **4282** 10 **4283** fr. 2 6 **4284** 2, 5, [8], [11], 13 **4285** ii 6.
ταπείνωμα **4277** fr. 1 i 9, ii 19 **4283** fr. 2 7.
τόπος ζῳδιακός **4277** fr. 1 i 13, 25.
τρίγωνον **4277** fr. 1 i 8, 23, 32, 39, ii 7–8, [28] **4283** fr. 1 2, 9, fr. 2 7.
ὕψωμα **4277** fr. 1 i 9, ii 18 **4282** fr. 1 8, fr. 2

6 **4286** 7, 8, 9.
ὥρα (= ὡροσκόπος) **4254** 5 **4257** 1 **4259** 5 **4267** 2 **4273** 5 **4292** 6.
ὡροσκοπεῖν **4240** 9 **4279** 8.
ὡροσκόπος **4238** 7 (**4241** 8) **4242** (3), 3 **4243** 4 **4245** 29 **4246** 2 (**4247** 6, 14) **4248** 9 **4249** 12 **4250** 1, 5 [**4251** 5] **4253** 4 (**4255** 1) **4257** 3 (**4161** 1, 6?) (**4263** 1) (**4264** 4) (**4265** 9) (**4266** i 4) (**4266a** 5) (**4268** 2) (**4269** 3) (**4269a** 5) (**4270** 6) (**4272** 2, 9) [**4274** → 10] [**4275** 11] **4276** ii 1 **4277** fr. 1 i 16, 17 **4287** 7 (**4290** 4) (**4298** 4?) **4299** 2.

IX. GENERAL INDEX OF WORDS

ἀγαθός **4236** 1 [**4239** 1] **4247** 7 **4259** 1 **4264** 1 **4265** 1, 6 **4269a** 1 **4270** 1 **4271** 1 **4275** 1, 12 [**4291** 1] [**4294** 3].
ἄγειν **4144** 2.
ἀήρ **4277** fr. 1 i 3.
αἰεί **4139** 6.
ἀκμή **4144** 10.
ἀκριβῶς **4276** ii 6–7 **4277** fr. 1 i 19.
ἀκρόνυχος (**4199** recto 21, 35, verso ii 25, 40, iii 11) (**4200** recto ii 8, verso i 6) (**4201** i 3).
ἄκρος **4248** 2.
ἄλλος **4135** 6, 22.
ἅμα **4133** i 11, 15.
ἄν **4142** fr. 2 ii 27 [**4163** verso text 15].
ἀναβαίνειν **4140** ii 7 **4141** 3.
ἀνάγκη **4277** fr. 1 iii [7], 9.
ἀναλύειν **4142** fr. 1 17, fr. 2 i 12, (19), ii (35).
ἀνάπαλιν **4142** fr. 2 ii 27, 3 i (9).
ἀναπληροῦν **4137** 8.
ἀναποδισμός **4277** fr. 1 i 14.
ἀνατέλλειν **4135** 32 **4204** i 8, 12, iii 10, 14.
ἀνατολή (*see also* VIII *d*) **4147** fr. 1 ii 9 **4167** fr. 7 ii 1.
ἀνατολικός (*see also* VIII *d*) (**4181** fr. 1 i 3) (**4236** 8) **4245** 14, 17.
ἀναφορά **4167** fr. 1 iii 2, vii [2], i 2, v [2], ix [2] **4171** front ii 2 **4286** 14?
ἀναφορικός **4276** ii 2.
ἄνεμος **4143** recto 10, 18 **4277** fr. 1 i 13, ii 2, 10, 21–22, [31], iii 6.
ἄνεσις **4144** 10.
ἀνθρωποειδής **4277** fr. 1 ii 16, iii [1] **4283** fr. 2 4.
ἀνομογενής **4143** recto 14.
ἀντί **4140** ii 5.
ἀντιστροφή **4134** fr. 1 12.
ἀνωμαλία **4277** fr. 1 i 14.
ἀνώμαλος **4142** fr. 3 i (4), 8, 10, 13, [16] **4144** 1?, 5, 11 (**4181** fr. 1 ii 3).
ἀνωμάλως **4144** 9.
ἅπαξ **4143** verso 9.
ἀπαρτίζειν **4134** fr. 1 23 **4145** back 6.
ἀπέχειν [**4142** fr. 2 ii 16) [**4143** recto 12].
ἀπό **4133** i 13 **4134** fr. 1 3, 7, 9, 19, 20, 24 **4136** 8, 15, 16 **4142** fr. 2 ii (22), (29), fr. 3 i 4, [11] **4143** recto 5 **4144** 5 **4147** fr. 1 ii 9 **4204** i 2 **4270** 3 **4277** fr. 1 i 10.
ἀπόδειξις **4147** fr. 1 ii 10.
ἀποκατάστασις **4139** 9?
ἀπολαμβάνειν **4138** 8.
ἀπόστημα (**4168** fr. 1 back iv 5, 24).
ἀποφράς **4180** recto iii 16, 17, 29.
ἀποχή (**4142** fr. 2 ii 16).
ἄρα **4133** i 1, 19 **4142** fr. 2 ii 6, [31].
ἀριθμός (**4142** fr. 1 7) **4143** recto 2, [6] **4147** fr. 1 ii 7 **4167** fr. 8 ii 1 **4168** fr. 1 front iv (2), v (1), fr. 2 front i 3 **4245** 10.
ἀρρενικός **4245** 5, 15, 18, 21 **4276** ii 7 **4277** fr. 1 i 21, 29 **4283** fr. 2 3 **4284** 10.
ἀρχαῖος **4256** 3.
ἄρχειν **4138a** ii 2 **4146** fr. 1 12.
ἀρχή **4135** 35 **4183** fr. 2 5.
ἀστήρ **4133** i 3 **4137** 17 **4140** ii 8 (**4170** front v 1) **4245** [21], 24 **4277** fr. 1 i 2 **4283** fr. 1 4, fr. 2 1.
αὐθημερί **4269** 5.
αὐθήμερος **4266** i 6, 9.
αὔξη **4134** fr. 1 13?
αὔξησις (**4136** 17).
αὐτός **4133** i 1 **4142** fr. 2 i 16 **4144** 4, 5 **4146** fr. 1 12 **4170** front v 1 **4277** fr. 1 ii 18 **4278** 12 **4282** 4.
ἀφαιρεῖν **4134** fr. 1 3 **4136** 8, (15), 17 **4138a** ii

6 **4142** fr. 2 ii 21, 27, fr. 3 i (10) **4143** recto 15 **4245** 10.
ἀφαίρεϲιϲ (**4136** 11) (**4168** fr. 2 front v 4, vi 4, back i 3).
ἀφαιρετικόϲ (**4275** 6, 7, 9, 10).

βαθμόϲ **4245** 11 **4277** fr. 1 i 13, 24, [33], ii 2, 10, 21, 31.
βαίνειν **4138** 1?
βαϲιλικόϲ **4277** fr. 1 i 21, ii 26–27 **4284** 10.
βιβάζειν **4138** 3.
βραχύϲ **4133** i 2, 21.

γάμοϲ **4270** 14 **4286** 13.
γάρ **4138** 7 **4138a** i 4 **4142** fr. 1 14, fr. 2 ii 16, 25 **4144** 2.
γένεϲιϲ **4140** ii 2 **4236** 1 **4239** 1 **4240** 1, 10, 15 **4242** 1 **4244** 2 **4245** 29 **4250** 2 **4251** 1 **4252** 7 **4253** 1 **4258** 1 **4259** 2 **4260** 1 **4264** 1 **4265** 1, 6 **4266a** 1 **4269a** 2 **4270** 1 **4271** 1 **4273** 3 **4276** ii 1, [11] **4287** 1 **4291** 1 **4293** 1 **4294** 3.
γῆ **4277** fr. 1 ii 3, 4.
γίγνεϲθαι **4134** fr. 1 10, 17 **4142** fr. 2 ii (10), 36, 39, fr. 3 ii 7 **4282** 3.
γίνεϲθαι **4134** fr. 1 2, 8, 21, 25, 30, fr. 2 7, 9 (**4135** 45) (**4136** 3, [9], 12, 21) **4139** 6? (**4142** fr. 1 7, 9, fr. 2 i 14, 4, 5, fr. 3 ii 4, 5) (**4145** front 2, 3, 4, 7, back 7) **4146** fr. 1 7 (**4163** verso text 20, 21, [22], 26) (**4166** front 5).
γνώριμοϲ **4276** ii 6.
γράφειν **4143** recto 3 **4266** i 12.

δαίμων **4277** fr. 1 ii 23–25 (**4279** 13) **4284** 12 **4295** 4.
δαψιλῶϲ **4133** ii 20.
δέ **4133** i 2, 3, 13, ii 24 **4134** fr. 1 12 **4135** 35 **4136** 8, 13 **4137** 1, [6], 10, 13, 18, 19, 21 **4138** 5, 9 **4139** 7 (**4142** fr. 2 ii 8, 12 *bis*) **4143** recto 14 **4144** 6 **4146** fr. 1 5, 7 **4147** fr. 2 6 **4163** verso text 20 **4245** 7, [12], [15], [18], 21, 24, 29 [**4256** 3] **4276** ii 1, 11 **4277** fr. 1 i 28, 36, ii 4 **4278** 12 **4282** 11.
δεῖν **4136** 1 **4143** verso 4.
δέκα **4163** verso text 20.
δεκατρεῖϲ **4145** back 5.
δεξιόϲ **4137** 14?
δεύτεροϲ **4277** fr. 1 i 15, ii 24 **4284** i 5.
δηλοῦν **4143** verso 22.
διά **4133** i 1, 8, 21 **4137** 18 (**4142** fr. 2 ii 11 *bis*, 12 *bis*, 13, 15, 26) **4243** recto 13, [17].
διαιρεῖν **4138** 8 **4278** 8?
διάμετροϲ **4133** i 13, 16, 17, 20, ii 15, [27] **4137** [2], 12 (**4142** fr. 2 i 8).
διάϲταϲιϲ **4143** recto 8, 16.
διάϲτημα **4138a** ii 12.
διαφέρειν **4142** fr. 3 i 1.
διαφορά **4142** fr. 2 ii (38), (45), fr. 3 i (7), 11 **4167** fr. 8 i 1 **4168** fr. 2 front vi 3, back i (2), iv (2).
διδόναι **4142** fr. 1 16, fr. 2 i 11, 19, ii 1, 6, 32, 35, 43, fr. 3 ii 1, 8 **4279** 11?
διεκβάλλειν **4163** verso text 14.
διεξαγωγή **4135** 33.
διευτυχεῖν **4249** 13 **4264** 9 **4266** i 11, ii 8 **4268** 5 **4269** 12 **4270** 13 **4295** 6.
διϊϲτάναι **4133** i 15.
δίϲωμοϲ (**4180** recto iii 7, 8, 9, [14], 15, [20], [21], [22], 27, 28) **4245** 5, 8 **4277** fr. 1 i 12 **4283** fr. 2 3.
διχοτομεῖν **4137** 11.
δύνειν (*see also* VIII *e*) **4135** 29 **4204** i 7, 11, iii 7.
δύναϲθαι **4142** fr. 3 i 15.
δύο **4142** fr. 2 ii 37, 45 **4284** 13.
δύϲιϲ (*see also* VIII *d*) **4147** fr. 1 ii 9 **4181** fr. 1 iv 18 **4245** 28.
δυτικόϲ (*see also* VIII *d*) **4279** 6 **4280** 4.
δωδεκατημόριον **4277** fr. 1 i 10–11.

ἐάν **4138** 7 **4142** fr. 1 9, fr. 3 i 4, 9 **4143** recto 9, 14 **4163** verso text 1, 23.
ἔγγιϲτα **4133** ii 28 **4142** fr. 2 ii 5, 21.
ἐγώ **4133** i 4 **4138** 11 **4279** 10, 12.
εἰ **4142** fr. 2 ii 25.
εἰδέναι **4136** 1.
εἶναι **4133** i 1, 2 **4135** 37 **4136** 4, 6, 13, 20 **4137** 3, 5, 13, 22 **4138** 6 **4138a** ii 10 **4140** ii 3, 9 **4142** fr. 2 i 17, ii 2, 7, 9, 12, 14, 16, 18, 24, 25, 41, fr. 3 ii 2 **4143** verso 4 **4146** fr. 1 14 **4147** fr. 1 ii 12 **4163** verso text 16, 23 **4245** 4, [8], 12, 15, 18, [21], [24] **4266a** 2.
εἰϲ **4133** i 8 **4134** fr. 1 6? **4135** 12, [21], 34 [**4137** 21] **4138** 8 **4142** fr. 1 17, fr. 2 i [12], 20, 36,

fr. 3 i 2, [17] **4147** fr. 1 ii 8 **4163** verso text 26 **4180** recto 35, 36, verso [36] **4245** 1 **4246** 1 **4247** 8 **4257** 1 **4258** 2 **4275** 2 **4282** 5.
εἰcφέρειν **4143** recto 2.
ἐκ **4133** i 2 **4134** fr. 2 8 [**4136** 5] **4142** fr. 1 8, 14, fr. 2 i 9, ii 13, 43 **4143** recto 3 **4276** ii 2.
ἕκαcτοc **4135** [3], 4, [5], 7, 8, [10], 14, 15, 17, 18, 22, 23, [25], [26], [27], [28], [31] **4147** fr. 1 ii 7, 11.
ἑκάτεροc **4143** recto 11.
ἐκεῖ [**4163** verso text 15].
ἐκεῖνοc (**4142** fr. 3 i 2).
ἔκκειcθαι **4136** 5.
ἔκκεντροc (**4168** fr. 2 front iii 3).
ἐκλειπτικόc **4138a** ii 1 (**4180** iii 29, 30).
ἐκλείπειν **4137** (5), 7, (10).
ἔκλειψιc **4137** (1), 3 **4138** 6 **4138a** ii 7 **4139** 5 **4281** 1.
ἐκλογίζεcθαι **4143** recto 4.
ἐκτιθέναι **4134** fr. 1 15.
ἐλάccων [**4133** i 19 **4137** 1 **4143** recto 15].
ἐλάχιcτοc **4134** fr. 1 27, fr. 2 12 **4136** 9 **4143** verso 3.
ἐμφανήc **4138a** ii 3.
ἐν **4135** [2], [4], [5], 6, [8], 9, [14], [15], 16, 21, [23], [24], 25, 27, 31 **4137** [13], 16 **4138a** i 3, ii 7, 10 **4140** ii 3 **4142** fr. 2 ii 6, 7, 18, 22, 23, 24, [29], 32, 39, 44, fr. 3 i 12 **4143** verso [2], 9, 15 **4147** fr. 1 ii 11 **4183** fr. 2 5 **4204** i 8, 13 **4236** 4, 6, [8], 10 **4236a** 4 **4237** 3, 5, 7 **4239** [3], 3, 4 *bis*, 5, [6], 6 **4245** [5], 8, 11, 12, [18], 21, [24], 28, [11] **4276** ii 7 **4277** fr. 1 i 4, [20], 29, [37], ii 5, 6, 15, [16], [26], 27, iii 1 **4279** 3 **4281** 2 **4283** fr. 2 3 **4284** 9 **4285** ii 3.
ἔνατοc **4136** 22 **4267** 1.
ἑξαμηνιαῖοc **4138a** ii 11.
ἑξηκοντάc **4277** fr. 1 i 6, 12, 22, 31, 38–39, ii 7, 17, 27–28.
ἑξηκοcτόc **4142** fr. 3 i 1, 7, (11) **4143** verso (1), 3, 5 (**4167** fr. 3 ii 3) (**4168** fr. 2 front v 3).
ἑξῆc **4134** fr. 1 13.
ἐπεί **4136** 13 **4138a** ii 4 **4142** fr. 2 ii 1, fr. 3 ii 1 **4143** verso 1.
ἐπειδήπερ (**4142** fr. 2 ii 23).
ἔπειτα **4135** [5], 6, [8], [9], 13, 16, 25, 27, 30 **4143** recto 4.
ἐπέχειν **4142** fr. 1 18, fr. 3 i (19).
ἐπί **4133** i 1 **4134** fr. 1 7, 10, 12, 14, 17 **4136** 2, [4], 7, 8 **4142** fr. 1 [5], fr. 2 ii (3), fr. 3 i 12, ii (3) **4143** verso 4 **4145** 2 **4147** fr. 2 6 **4163** verso text 13, 17, 21 **4183** fr. 2 2 **4245** [7], 11, 14, 17, [20], [23], 26, 28 **4274** ↓ 2, 4 **4275** 1, [12] **4276** i 1 **4277** fr. 1 i 24 *bis*, 25, 33 *bis*, ii 1 *bis*, 9, 10, 21 *bis*, [30], 31 **4279** 10 **4285** i 4, ii 7.
ἐπίκυκλοc (**4168** fr. 2 back ii 2).
ἐπιλαμβάνειν **4138a** ii 5.
ἐπίλειψιc **4136** 18.
ἐπιπροcθετικόc **4143** verso 2, 11.
ἐπιcυνάγειν **4136** [1], (7), (13).
ἐπιτολή **4146** fr. 1 10.
ἕπεcθαι **4133** ii 26 **4135** [21], 34 [**4137** 16].
ἐποχή **4142** fr. 3 i 2.
ἔρχεcθαι **4245** 28.
ἔρωc [**4277** fr. 1 ii 33, 34].
ἑcπέρα **4204** iii 7, 14.
ἑcπέριοc **4135** 10, 32 (**4175** [2], ix 6, xv 3, 6) (**4180** recto vii 3) **4280** 4.
ἔcχατοc **4137** 8.
ἔτι **4143** verso 7.
ἔτοc **4133** i 3 (**4137** 18, [19]) (**4149** back i 2) (**4160** iii 2–39) (**4160a** ii 1–5, [6]) **4163** verso text (18), 19 (**4184** fr. 1 iii 2) (**4186** i [1], ii 1, iii 1, iv [1]) (**4187** 1) (**4188** 1, iii 7) (**4190** recto 1, verso 1) (**4200** verso i 5) **4203** recto 1, 2, verso 2 [**4236** 2] (**4237** 1) (**4239** 1) (**4240** 2, 11, 16) (**4241** 1, 5) (**4242** 1) (**4246** 1) **4247** 8 (**4248** 1) (**4249** 1) (**4250** 3) (**4251** 2) (**4252** 8) (**4253** 2) [**4256** 1] [**4257** 5] (**4258** 1) [**4259** 3] (**4262** 1) (**4264** 2 *bis*) **4265** (2), 7 (**4266** i 1, ii 1) (**4266a** 2 *bis*) (**4267** 1) (**4268** 1 *bis*, [1]) **4269** 1 (**4269a** 3) **4270** 2 (**4271** 2 *bis*) (**4273** 4) **4274** → 2 **4293** 1 **4300a** 1.
εὐθύc **4133** i 11.
εὑρίcκειν **4139** 1? **4276** ii 7 **4277** fr. 1 i 3, [19], 28, 36, ii 4, 14, 25, [34], iii 9 **4283** fr. 3 3 **4284** 9.
ἐφεξῆc **4144** 5.
ἐφήμεροc **4209** 1.
ἐφιcτάναι **4134** fr. 1 12, 28, 31.
ἔχειν **4133** i 12 **4142** fr. 1 (11), fr. 2 i 22, fr. 3 i (8), (13) **4147** fr. 1 ii 11 **4245** 11.

ἐῷος **4135** 29 **4157a** iii 1 **4180** recto viii 3 **4182** recto ii 4, iii 4 **4204** iii 10 **4236** 8 **4245** 14, 17, 28 **4279** 6.

ζητεῖν **4163** verso text 24.
ζῳδιακός **4143** verso 15 **4277** fr. 1 i 13, 25.
ζῴδιον **4135** 13, 21, [34] **4147** fr. 1 ii 12 **4180** verso ii 1 **4182** verso iv 2 **4204** i 2 **4222** fr. 2 ii 1 [**4236** 4, 6, 9, 10] **4245** 5, 8, [12], 15, [18], 21, 24 (**4274** → 8) **4276** ii 7 **4277** fr. 1 i 4, 11, 20, 29, [37], ii 5, 15, [26], iii 1 **4283** fr. 1 6, fr. 2 3 **4284** 3, 9 **4285** i 1, ii [4].
ζώνη **4277** fr. 1 i 2–3 **4283** fr. 1 [5], 2 2.

ἤ **4133** i 2, 17 **4137** 2 **4138a** ii 3 **4139** 2 **4140** ii 4 **4142** fr. 1 6, 12, fr. 2 i 13.
ἡγεμονία **4133** i 6.
ἡγεῖcθαι **4137** 15.
ἤδη **4144** 8.
ἡλικία **4278** 10.
ἡμέρα **4133** i 4 **4135** 2, 3, [4], 4, 5, [6], 7, 8, [8], 9, [10], (11), [14], [15], 15, [16], 16, 17, (18), 19, [23], 23, [24], 24, [25], (25), [26], [27], (27), (28), 29, 31, 32, 33 **4163** verso text 25 **4176** iii 13, (18), 21, (23), (25), (27), (29) [**4177a** iv 17] (**4178** vii 5, 10) (**4179** vii [1], 3, 8, 10, 13, 15, 22, 24, 26, 28, viii 31a) (**4180** recto iv 16, 27, 32, verso ii 22, 24, 29, 32, 34, 35) **4236** 3 **4237** 2 **4239** 2 **4240** 14 **4242** 2 **4244** 4 **4250** 4 **4252** 1, 9 **4253** 3 **4259** 4 **4264** 3 **4265** 3, 8 **4266** i 3, ii 3 **4266a** 4 **4267** 1 **4269** 2 **4271** 3 **4276** ii 3 **4279** 1 **4290** 3 **4291** 6.
ἡμερινός **4142** fr. 1 16.
ἡμικύκλιος **4138a** ii 4, 9, 15.
ἥμιcυc **4133** i 13 **4147** fr. 2 7.
ἤτοι **4142** fr. 1 [5], 11.

θεός **4146** fr. 1 15 **4245** 7.
θηλυκός **4245** 8, 24 **4277** fr. 1 i 5, 11, 25–26, 34, 38, ii 5–6, 15, iii 1 **4283** fr. 1 6 **4285** iii 4.
θνήcκειν **4278** 11?

ἴδιος **4236** 9 **4245** 23 **4277** fr. 1 i 7 **4281** 3 **4283** fr. 1 2, fr. 2 6, 7.
ἰcημερινός **4138** 5 **4142** fr. 1 11, 17, fr. 2 i 10, 12, 14, 20, ii 5, (7), 9, 20, (22), (28), (29), 36 *bis*, 38, 40, 42, fr. 3 ii 6, 8 (**4168** fr. 1 back i 5).
ἶcoc **4133** i 3.

κάθετος **4144** 2.
καθόλου **4142** fr. 2 ii 43.
καί **4133** i 2, 4, 8, 12, 15, 17 **4134** fr. 1 22, 24, fr. 2 7 **4135** [3], [7], 10, [11], 13, [15], 19, [20], 21, 28, 32 **4136** [1], 3, [9], 10, 20 **4137** [2], [5], 9, [15] **4138a** ii 11 **4139** 3 **4140** ii 4 **4142** fr. 1 (7), fr. 2 i 19, 21, ii 4, (11), (13), (20), 28, 31, (33), (35), (36), (37), (39) *bis*, (40), (42), (43), fr. 3 i (1), (4), (8), (9), (13), ii [3] **4143** recto 9, verso 7, 8, 12, 17, 18 **4143** 3? **4144** 10 *bis* **4145** front 3, 6, back 7, 11 **4146** fr. 1 10 **4147** fr. 1 i 5 **4148** iv [1], viii (1) **4149** back 1 **4163** verso text 11, 15, (27) [**4165** fr. 1 ii 1] **4168** fr. 1 front iv 1 [**4236** 4] **4242** 1 **4245** 7 **4255** 3 **4258** 1, 2 **4260** 3 (**4265** 5?) **4276** ii 4, 5 **4277** fr. 1 i 33, fr. 2 1 **4279** 5 **4281** 2, 3 **4290** 2 **4292** 2.
καιρικός (**4142** fr. 1 16, fr. 2 i 19, 20, ii 1, 2, 6, 33, 35, 44).
κακός **4180** recto iii 8, 28.
κακῶc **4278** 11.
καλεῖν **4277** fr. 1 i 17, ii 13, [24], iii [8]
κανόνιον **4142** fr. 3 i 13.
κανών (**4142** fr. 3 i 18).
κατά **4133** ii 24 **4136** 1 [**4137** 9, 18] **4142** fr. 2 i (8), 24, [26], fr. 3 i (1), (9), 14 **4146** fr. 1 11? **4163** verso text 30 **4168** fr. 1 back 2 **4183** fr. 1 recto iii 1 **4239** 2 **4249** 3 **4251** 2 **4256** 3 **4274** ↓ 3 **4276** ii 5 **4277** fr. 1 i 3–4, 8, 18, 19, 23, [28], 32, 36, 40, ii 4, 8, 14, [25], 28–29, [35], iii [9].
καταβαίνειν **4140** ii [1] **4141** 5, 7.
καταλείπειν **4136** (6), 15, (18) **4138** 11 (**4142** fr. 2 ii 30).
καταλήγειν [**4163** verso text 15] **4277** fr. 1 i 11.
καταρχή **4180** verso i 2.
κινεῖν **4135** [2], 30 **4144** 9.
κίνηcιc **4134** fr. 2 [12] **4135** 12, (20) [**4136** 2] **4144** 4 **4277** fr. 1 i 4, 20, 28–29, 37, ii 5, [14], [25], [35].
κληροῦν **4283** fr. 1 5, fr. 2 2.
κλίμα **4142** fr. 1 20 **4167** fr. 4 vi 2, viii [2] **4276** ii 4.

κοινός **4167** fr. 3 ii 2, fr. 8 ii 2 **4168** fr. 2 front i 4 **4173a** fr. 3 front i 1.
κομίζειν **4279** 11.
κόρυφος **4244** 3.
κράτηcιc **4270** 3.
κυβερνήτηc **4277** fr. 1 i 18.
κύκλοc **4136** 19 **4138a** i 5.

λαμβάνειν **4142** fr. 2 ii 10, fr. 3 i 6, (10) **4143** recto 6.
λαμπρόc **4265** 6.
λαγχάνειν **4277** fr. 1 i 3.
λέγειν **4142** fr. 2 ii 16 **4143** recto 12 **4144** 6 **4145** back 8.
λείπειν **4134** fr. 1 4 **4136** 14.
λεπτόc **4138** 8 (**4182** recto 1) (**4200** recto ii [2], 3, 4, 5, 6, [7], 8, 9, verso i [1], 2, 3, 4, [6], 7, 8) **4239** [4], 4, 5, 6, 6 **4245** 6, 9, (13), [16], (19), (22), (25) (**4275** 4–11) **4276** ii 8 (**4277** fr. 1 i 6, 12, 22, [30], 38, ii 6, 17, [27]) **4278** 2, 3, 4, 5, 6 (**4279** 4) **4281** 3 (**4282** 8) **4283** 5 **4284** 5, 7, 13 **4285** ii 4 (**4286** 3–9).
λοιπόc **4134** fr. 2 10 **4139** 11 **4140** ii 8 **4143** verso 21 **4163** verso text [14], 20, 22, 25, 27 (**4166** front 3, 7) **4236a** 4.
λόξωcιc **4168** fr. 1 front vi 2.

μέγαc **4145** back 9.
μεθοδεύειν **4142** fr. 2 i 23.
μέλλειν **4138** 12.
μέν **4137** 4 (**4142** fr. 2 ii 12) **4143** recto 9 **4278** 11.
μερίζειν **4134** fr. 1 5, 20 **4147** fr. 1 ii 8.
μέροc **4133** i 17 **4137** 2.
μεcημβρία (*see also* VIII *d*) **4142** fr. 1 5, 6, 12 *bis*, fr. 2 i 12, [13], ii 2, 8 *bis*, 22, 30, (32), 34, 41, fr. 3 ii 2, 7.
μεcημβρινόc **4142** fr. 2 ii 14, 15, 17, (26).
μεcονύκτιοc **4138** 9, 12.
μέcoc **4133** i 1, 21 **4134** fr. 1 14 **4136** 2 **4138** 6 **4143** recto 5, 13, [17], verso 11, 27 **4180** iii recto 9, 10, 11, 24, 25, 27, 30, 31, 32, 33.
μετά **4133** i 3 **4142** fr. 1 6, 12, fr. 2 i 13, ii 9, 33, 42 **4183** fr. 1 verso i 2?
μεταβάλλειν **4142** fr. 3 i 16.
μεταξύ **4137** 5, 9, 14.
μέτοχοc **4277** fr. 1 i 8, 23, 32, 40, ii 8, 29.
μέχρι **4147** fr. 1 ii 9.
μῆκοc **4142** fr. 2 ii 10, 16, 18 **4143** recto 5 (**4168** fr. 1 back ii 5, 24) **4183** fr. 2 2 **4277** fr. 1 i 4, 20, 28, 36–27, ii 4, 14, [25], [35], iii [9].
μήν **4137** 18, 20 *bis* **4237** 1 **4282** 4.
μοῖρα **4134** fr. 1 2, 20, 28 (**4135** [3], 4, 5, [7], 8, 9, 11, 14, 15, [17], 17, [20], 22, 23, 24, 25, [28], 30, 31, [34]) (**4136** 9 *bis*, 14, 19) (**4141** 2, 4, 6) (**4142** fr. 1 19, fr. 2 i 17, ii 12, 13, 18, fr. 3 i 12, ii 4) **4171** front i 2 (**4180** recto [35], [36], verso iii 1, 35, 36) (**4200** recto ii 2, [3], 4, 5, 6, [7], 8, 9, verso i 1–8) (**4236** 4, 6, [8], 10) (**4237** 3, 5, 7) **4239** 3 *bis*, 5 *bis*, 6 *bis* **4245** 6, 9, 13, 16, [19], 22, 25 (**4274** → 4, 10, ↓ 5, 6) (**4275** 5–10, [11]) **4276** ii 8 (**4277** fr. 1 i 5, 12, 21, 25, [30], 33, 38, ii 6, 16, [27], fr. 2 4) **4279** 4, 10 **4280** 2 **4281** 2 **4282** 7 (**4283** fr. 1 7, fr. 2 5) (**4284** 7, 10) [**4285** ii 3].
μουcικόc **4283** fr. 2 4.

νεφέλιον **4133** i 9.
νυκτερινόc (**4275** 3).
νῦν **4146** fr. 1 8.
νύξ [**4138** 13] (**4176** iii 9) [**4177a** iv 14, 19] (**4178** vii 7, 12) (**4179** vii 5, 16a, 17, 18, 30) (**4180** recto iv 23, 25, 35, 36, verso ii 20, 27, [36]) **4240** 3 **4246** 2 **4247** [1], 9 **4249** 4 **4251** 4 **4257** 1 **4258** 3 **4268** 1 **4269a** 4 **4270** 5 (**4274** ↓ 2) (**4282** 6) **4293** 2 **4294** 5 **4296** 1 (**4297** 2).
νυχθήμεροc **4138** 7 **4142** fr. 3 i 3, (5), 17.

ὅθεν **4138a** ii 2.
οἴαξ **4277** fr. 1 i 17.
οἰκεῖοc **4142** fr. 1 20.
οἰκοδεcπότηc **4278** 7.
οἷοc **4163** verso text 16.
ὅλοc **4136** 21 **4138a** i 5, ii 6, 7.
ὁμαλόc **4134** fr. 2 8 **4142** fr. 3 i 3, 5, 8, (9), (11), 17, 20.
ὁμογενήc **4143** recto 10.
ὁμοίωc **4136** 22 **4145** front 2 **4278** 14.
ὁμόc **4134** fr. 1 11, 25, fr. 2 9.
ὀρθόc **4142** fr. 3 i 12.
ὅc **4142** fr. 3 i 1 **4143** verso 4 *bis* **4266a** 2.

ὅcoc **4138a** ii 10 **4142** fr. 1 9.
ὅcπερ **4163** verso text 15.
ὅcτιc **4142** fr. 2 ii 19, 41.
ὅτι **4138a** ii 4 **4142** fr. 2 ii 13, 43.
οὐθείc **4283** fr. 2 6, 7.
οὖν **4137** 4 **4142** fr. 2 i 17, ii 1, 18, 28, fr. 3 ii 1 **4143** verso 6.
οὗτοc **4134** fr. 1 6, 13, 15, 19, 20, 29, 32 **4136** 3, 5, 10, 14, 15, 18, 20 **4142** fr. 2 i 17, ii 4, 10, 13, 18, 21, 43 **4143** verso 1, 25? **4144** 6.
οὕτωc **4147** fr. 1 ii 10.
ὀχλεῖν **4279** 9.

πάλιν **4133** i 17, 22 **4136** 16.
πανcέληνοc (**4137** 22) **4163** verso text 23, 28 (**4179** viii 29) [**4180** recto 36, verso 36] **4282** 2.
παρά **4134** fr. 1 5, 29 **4136** 11 (**4142** fr. 1 9, fr. 2 i 15, 18, ii 5, 19, fr. 3 ii 5) **4163** verso text 20, [21].
παραβάλλειν (**4142** fr. 1 8, fr. 2 i 15, ii 4, 19).
παραγίνεcθαι **4138a** ii 2.
παράδειγμα (**4142** fr. fr. 1 [14], fr. 2 i 21).
παρακεῖcθαι **4138** 10 (**4142** fr. 2 i 16, fr. 3 i 11).
παραλλάccειν **4168** fr. 1 back 2.
παράλληλοc (**4142** fr. 2 ii 11).
παρέχειν **4142** fr. 2 i 11, (18), (19).
πάρυγροc **4277** fr. 1 i 30.
πᾶc **4135** [11], [19], 29 **4136** 7 [**4137** 3].
πέμπειν **4279** 9.
πεμπτημόριον **4142** fr. 2 ii 20, (29), 31, 33, 39, 40, (42).
πενταμηνιαῖοc **4138** ii 11.
πεντάμηνοc **4138a** ii 6.
πέντε **4209** 1?
περί **4133** i 9 **4137** 4 **4183** fr. 1 verso i 3? **4248** 2.
περιγραφή **4285** ii 2.
περιϊcτάναι **4137** 6.
περιοδικόc **4146** fr. 1 2.
περίοδοc **4137** 19 **4138a** ii 8, [11], 14?
περιφέρεια **4138a** ii 1 (**4168** fr. 1 back iv 6, 25).
πίπτειν **4183** fr. 2 4?
πιττάκιον **4279** 12.
πλανᾶν [**4209** 1].
πλάτοc **4143** recto 6, verso 13?, 14, 23 **4168** fr. 1 front ii 1, back iii (5), (24) **4245** 11.
πλεῖcτοc **4137** [6], 10.
πλῆθοc **4142** fr. 1 10.
πλήρηc **4163** verso text 19 **4164** vi 17 **4270** 5.
πληροῦν **4250** 4 **4256** 2.
ποιεῖν **4163** verso text 12 **4277** fr. 1 i 4, 20, 29, 37, ii 5, 15, 26, [35].
πόλοc **4283** fr. 1 5, fr. 2 2.
πολυπλαcιάζειν (**4136** 2, 4) **4142** fr. 1 4, fr. 2 ii 3?, fr. 3 ii 2 **4143** verso 5, 6, 7.
πολυπλαcιαcμόc **4142** fr. 1 8.
πορεύειν **4135** 35.
πραγματεία **4142** fr. 3 i 18.
πρό **4133** i 6 **4138** 11 **4142** fr. 1 5, 11, fr. 2 i [12], ii 2, 7, 31, fr. 3 ii (1), 6 **4176** iii 11, 16.
προαιρεῖν **4142** fr. 3 i 5.
προγίγνεcθαι **4282** 1.
προηγεῖcθαι **4133** i 10 [**4135** 13].
προκεῖcθαι **4256** 4.
προπορεύειν **4245** 27.
πρόc **4133** i 11, 12, 14, 15 **4136** 14, 19 [**4137** 7, 12] **4143** recto 7, 11, 13, [17], verso 14 **4276** ii 2, 4.
προcάγειν **4136** 21.
προcθαφαίρεcιc (**4168** fr. 2 front iii 4, back ii 3, iv 3).
πρόcθεcιc **4134** fr. 1 9, 11, 16, 26?, fr. 2 6 (**4136** 4, 11) **4167** fr. 8 i 2.
προcθετικόc (**4275** 8).
προcκεῖcθαι **4136** 22.
προcνεύειν [**4137** 10].
πρόcνευcιc (**4137** [4], 7).
προcτιθέναι **4136** 10, (18) **4142** fr. 2 ii 27, (37), 45, fr. 3 i (6) **4143** recto 10, 19 **4145** front 3, 6 **4163** verso text 7 **4167** fr. 3 v 24a (**4168** fr. 2 front v 31a).
πρότεροc **4142** fr. 2 ii 8, 24, 34, 41.
πρῶτοc [**4137** 4] **4144** 6?, 7 **4146** fr. 1 13 **4245** 7 **4276** ii 2 **4277** fr. 1 ii 13 **4281** 4 **4282** fr. 1 3 **4284** 6, 8, 11, 14.

cελαcφόροc (**4245** 7).
cελίδιον (**4136** 16, 19, 20, [23]).
cτερεόc (**4180** recto iii 5, 6, [12], [13], 18, 19, 25, [26], 32, 33, [34]) **4245** 12, 15, 19, 25 **4277** fr. 1 i 21, [29], 37–38, ii 6, [26] **4284** 10.
cτηριγμόc **4199** recto iii 20, 23, (34), (37), verso

(9), (14), (23), 28), (37), (42) (**4200** recto ii 6, verso i [3], 8) (**4201** i 2) (**4202** iii 12, 15, 29, v 15, 29, vii 14, 17) **4204** iii 9 **4281** 4.
cτηρίζειν **4135** 10 **4204** i 13.
cτίχοc (**4136** [1 *bis*], 3, 4, 5, 7, 8, [11], 12).
cτρέφειν **4135** [12], 20.
cύμφωνοc **4276** ii 5.
cυνάγειν (**4164** ii 2).
cυνάπτειν **4140** ii 6.
cύναψιc **4175** (4), i 5, vi (5), ix (5), vi (6), xiv (5).
cυνέρχεcθαι **4144** 8.
cυνιcτάναι **4142** fr. 3 i 19?
cυνοδικόc (**4179** xiv 12a) **4279** 7.
cύνοδοc **4157a** iii 1 **4163** verso text 22, 29 (**4179** viii 14) [**4180** recto 35, verso 35] **4282** 12.
cυνοικοδεcπότηc **4278** 9.
cύνταγμα **4276** ii 3–4.
cφαῖρα **4142** fr. 3 i 12.
cχέcιc **4143** 13.

τε **4134** fr. 1 31.
τέταρτοc **4277** fr. 1 iii 8.
τετράπλευροc **4133** i 10.
τηλικοῦτοc **4136** 13.
τηρεῖν **4133** i 5, ii 21.
τιc **4133** i 5?
τμῆμα **4137** 12 **4138** 4 **4277** fr. 1 i 24, 33, ii 1, 9, 21, [30].
τόποc (*see also* viii *e*) **4142** fr. 2 ii 24, 37, (45) **4276** ii 6.
τοcοῦτοc **4138a** ii 8 **4142** fr. 1 10, fr. 2 ii 31 **4143** recto 12, 17.
τουτέcτι **4142** fr. 2 ii 33, 38 **4145** back 10.
τρεῖc **4142** fr. 2 ii 20, [30], 39, 40.
τρίγωνον **4277** fr. 1 i 8, 23, 32, 39, ii 7–8, [28] **4283** fr. 1 2, 9, fr. 2 7.
τρίτοc **4245** 20, 26 **4277** fr. 1 ii 34.
τροπικόc (**4180** recto iii 10, 11, 16, 17, 23, 24, 31) **4245** 2 **4276** i 3, ii 7–8 **4277** fr. 1 i 5 **4285** ii 5.
τυγχάνειν [**4143** recto 9, 14] **4285** ii 1.
τύχη (*see also* viii *e* s.v. κλῆροc τύχηc) **4236** 1 [**4239** 1] **4247** 7 **4259** 1 **4264** 1 **4265** 1, 6 **4269a** 1 **4270** 1 **4271** 1 **4291** 1 **4294** 3.

ὑπεροχή **4136** 11, 21 **4142** fr. 2 ii 18.
ὑπό **4277** fr. 1 ii 3, 4.
ὑπόδειγμα (**4142** fr. 1 15).
ὑποκεῖcθαι **4142** fr. 1 14, fr. 2 i 11.
ὑπολείπειν **4133** i 18, ii 15.
ὕcτεροc **4139** 7.
ὑφαιρεῖν **4163** verso text 25.

φαίνειν **4142** fr. 3 i 3, 14.
φάναι **4139** 10 **4142** fr. 3 i 6?
φανερόc (**4142** fr. 2 ii 13, 43).
φάcιc **4180** recto vii 4 **4245** 28.
φαῦλοc **4180** recto iii 26.
φέρειν **4147** fr. 1 i 6?
φορά **4144** 7, 8, 10, 11.
φυλάccειν **4139** 4.
φωναcκόc **4283** fr. 2 4.

χάριc **4142** fr. 1 15, fr. 3 i 15.
χερcαῖοc **4277** fr. 1 i 5.
χρόνοc [**4137** 3] **4138** 7 **4139** 5, 6 **4142** fr. 2 i (8), 16, ii 1, fr. 3 i 2, (16) **4167** fr. 1 i [2], v [2], ix (2), iii [2], vii [2] **4171** back 3 **4278** 8.

ψηφίζειν **4276** i 2, ii 1–2, 6 **4277** fr. 1 i 19.
ψῆφοc **4135** 40.

ὥρα (*see also* viii *e*) (**4180** verso ii [32], [34], [35], [36], ix 25b, viii 30b) (**4181** fr. 2 vi 21) **4236** 2 **4237** 2 **4239** 2 (**4240** 3, 13) (**4241** 6) (**4242** 2) **4246** 1 (**4247** 1, 9) (**4248** 2) **4249** 4 **4250** 4 (**4251** 4) **4252** 1 **4253** 3 (**4256** 2, 4) **4257** 1 **4258** 3 **4259** 4 (**4260** 4) (**4262** 2) (**4264** 3) (**4265** 3, 8) (**4266a** 4) (**4267** 1) [**4268** 1] (**4269** 2) (**4269a** 4) (**4270** 3) (**4274** ↓ 3) (**4275** 3) **4276** ii 2 (**4282** 6) (**4290** 3) (**4291** 5) (**4293** 2) (**4294** 5) (**4297** 2).
ὡριαῖοc **4167** fr. 1 i [2], v 2, ix 2, iii [2], vii [2].
ὡc **4136** 1 [**4137** 7, 8, 12] **4142** fr. 2 ii 16, 21, fr. 3 i 6 **4276** ii 2.
ὡcαύτωc **4142** fr. 1 18 **4163** verso text 27.

X. Concordance of Publication and Inventory Numbers

1. By publication number.

4133	100/124(a)
4134	16 2B.47/B(c)
4135	115/92(b)
4136	12 1B.133/C(d), part
4137	38 3B.85/H(1)c
4138	68 6B.25/H(1-2)a
4138a	35 4B.101/G(1–3)c
4139	82/87(b)
4140	18 2B.64/F(3–4)b
4141	68 6B.19/G(3)e
4142	100/3-5
4143	96/69(b)
4144	100/103(d)
4145	68 6B.24/D(1–3)c
4146	72/29(b)
4147	103/40(c)
4148	103/49(a)
4149	93/Dec. 11/C.3
	+ 93/Dec. 11/E.7
4150	A387/5D
4151	64 6B.60/G(1)a
4152	51 4B.21/D(1-3)b
4153	68 6B.20/L(8)a, part
4154	68 6B.20/L(8)a, part
4155	66 6B.26/K(1–2)i
	+ 66 6B.29/L(1–4)d
	+ 68 6B.25/H(3)b, part
4156a	19 2B.78/B(1–3)c
	+ A188/3F
4156b	20 3B.31/C(4–7)b
4156c	38 3B.81/B(1–2)c, part
4157	93/Jan. 1/E.2
4157a	38 3B.81/B(3–4)c, part
4158	62 6B.82/H(2-4)b
4159	68 6B.20/D(4)a, part
4159a	19 2B.83/E(c), part
4160	(P. Berol. 16511 +) 30 4B.39/D(1-3)a
4160a	19 2B.81/C(s), part
4161	A145/6H
4162	32 4B.1/E(1-2)a
4163	85/2(d)
4164	64 6B.60/G(1)a
4164a	15 1B.201/F(b), part
4165	65 6B.34/G(1–2)f
	+ 68 6B.20/D(4)a, part
4165	68 6B.25/H(3)b, part
4165a	115/81
4166	100/52(c)
4167	65 6B.30/M(1)a, part
	+ 66 6B.27/G(5–6)b, part
	+ 67 6B.16/K(2–4)b, part
	+ 68 6B.20/L(8)a, part
	+ 68 6B.25/H(3)b, part
	+ 68 6B.25/H(4)b, part
4168	48 5B.107/D(1)
4169	66 6B.27/G(5–6)b, part
4170	93/Dec. 8/B1
4171	68 6B.22/P(1–2)a
4172	12 1B.142/D(b)
	+ 12 1B.142/E(b), part
4173	72/94(a)
4173a	65 6B.30/M(1)a, part
	+ 66 6B.26/D(1)c, part
	+ 68 6B.25/H(3)b, part
	+ 68 6B.25/H(4)b, part
4174	72/29(d)
4174a	9 1B.173/J(c), part
4175	93/Dec. 28/F.1
4176	50 4B.34/A(1-3)a
4177	118/11(b)
4177a	15 1B.201/F(b), part
4178	25 3B.58/L(a)
4179	67 6B.10/L(1)a
4180	96/69(a)

4181 ... 5 1B.42/C(c)
4182 ... 68 6B.24/D(1-3)d
4183 ... 93/Dec. 16/E.1, part
4184 ... 2 1B.100/K(c)
4184a ... 66 6B.26/D(1)c, part
+ 68 6B.20/L(8)a, part
4185 ... 5 1B.43/B(b)
4186 ... A20/1H
4187 ... 19 2B.78/D(11–13)
4188 ... 11 1B.151/C(d)
4188a ... 40 5B.95/J(1–4)c
4189 ... 105/147(c)
4190 ... 103/14(b)
4191 ... 118/9(a)
+ A22/C251
4192 ... 103/20(a)
4193 ... 51 4B.19/C(1)c
4194 ... 68 6B.20/L(8)a, part
4195 ... 65 6B.36/E(1–3)d
4196 ... 68 6B.20/L(8)a, part
4196a ... 65 6B.30/M(1)a, part
4197 ... 28 4B.62/E(1–5)a
4198 ... 106/80(a)
4199 ... 15 2B.31/A(b), part
+ 81 2B.85/17(a)
4200 ... 100/65(f)
4201 ... 19 2B.75/D(d)
4202 ... 8 1B.197/F(5)
4203 ... 104/153(b)
4203a ... 15 2B.31/A(b), part
4204 ... 103/142(a)
4205 ... 65 6B.30/C(1)b
4205a ... 66 6B.26/K(1–2)g
4206 ... 68 6B.25/H(3)d, part
4207 ... 103/70(b)
4208 ... 68 6B.25/H(3)d, part
4209 ... 68 6B.20/L(8)a, part
4210 ... 68 6B.20/L(14–16)a & b
4211 ... 68 6B.25/H(3)d, part
4212 ... 68 6B.25/H(1–2)i, part
+ 68 6B.25/H(3)d, part
4213 ... 68 6B.20/D(5–8)f
4214 ... 101/106(c)
4215 ... 68 6B.20/L(3)d
4216 ... 31 4B.9/H(1–4)b
4217 ... 39 5B.121/B(1–2)a
4217a ... 68 6B.20/D(4)a, part
4217b ... 65 6B.30/M(1)a, part
+ 66 6B.26/D(1)c, part
+ 68 6B.20/L(8)a, part
4217c ... 68 6B.20/L(8)a, part
4218 ... 69/10(h)
4219 ... 18 2B.70/C(1–3)c
4220 ... 68 6B.25/E(1–2)
4221 ... 68 6B.25/H(3)b, part
4222 ... 50 4B.30/J(1–2), part
4223 ... 20 3B.31/F(1–3)b
4224 ... 68 6B.20/D(4)a, part
4225 ... 68 6B.25/H(3)b, part
4226 ... 65 6B.30/M(1)a, part
+ 68 6B.20/L(8)a, part
+ 68 6B.25/H(3)b, part
+ 68 6B.25/H(4)b, part
4227 ... A22/C287
4228 ... 9 1B.172/D(e)
4229 ... 22 3B.21/B(c)
4230 ... 35 4B.101/G(1–3)c
4231 ... 65 6B.30/M(1)a, part
4231a ... 66 6B.26/D(1)c, part
4232 ... 68 6B.20/D(4)a, part
4233 ... 68 6B.20/L(8)a, part
4234 ... 68 6B.25/H(3)b, part
4235 ... 68 6B.25/H(4)b, part
4236 ... 21 3B.23/D(1-3)b
4236a ... 103/124(c)
4237 ... 75/82(b)
4238 ... B3.6/C320
4239 ... 47 5B.46/H(3-6)d
4240 ... 12 1B.133/C(d), part
4241 ... 12 1B.133/C(d), part
4242 ... 36 4B.110/H(1–3)a
4243 ... 71/84(e)
4244 ... A188/4E
4245 ... 22 3B.20/E(1–3)b

4246	10B4/C319
4247	31 4B.13/H(6–8)a, part
4248	20 3B.37/J(5)a
4249	19 2B.78/B(4–7)c
4250	15 2B.39/A(e), part
4251	100/127(b)
4252	49 5B.98/K(1-4)a
4253	22 3B.19/F(a)
4254	16 2B.45/D(b), part
4255	85/72(a)
4256	114/68(b)
4257	101/155(b)
4258	101/168(a)
4259	16 2B.45/F(b), part
4260	40 5B.116/H(1-5)c
4261	28 4B.60/B(4–5)a
4262	85/82(b)
4263	102/67(b)
4264	36 4B.110/H(4–6)b
4265	104/69(a)
4266	69/24(c)
4266a	40 5B.108/K(5–10)c, part
4267	50 4B.33/J(1–5)b
4268	A141/6A
4269	37 4B.104/A(1-3)a
4269a	A604/5A
4270	42 5B.72/B(9-11)a
4271	A46/2F
4272	58/A(6)

4273	104/171(a)
4274	93/Dec. 26/C.1
4275	68 6B.20/F(1-2)a
4276	15 2B.40/C(b)
4277	21 3B.27/H(8)a
4278	28 4B.62/E(6–8)b
4279	29 4B.42/J(4)a
4280	30 4B.37/F(1–3)b
4281	46 5B.53/G(1–5)d
4282	17 2B.63/G(c)
4283	105/117(d)
4284	119/71(b)
4285	46 5B.51/F(5–7)c
4286	106/84(a)
4287	20 3B.31/E(6–7)b
4288	67 6B.16/M(4–5)b
4289	68 6B.19/G(3)d
4290	102/49(c)
4291	102/58(b)
4292	102/150(a)
4293	102/167(b)
4294	18 2B.69/E(1)b
4295	18 2B.69/F(2–4)c
4296	125/25(j)
4297	6153/C10
4298	6153/C11
4299	A188/3F
4300	A31/10
4300a	A608/32A

2. *By inventory number.*
(marks items drawn from more than one inventory number.)*

2 1B.100/K(c)	**4184**
5 1B.42/C(c)	**4181**
5 1B.43/B(b)	**4185**
8 1B.197/F(5)	**4202**
9 1B.172/D(e)	**4228**
9 1B.173/J(c), part	**4174a**
11 1B.151/C(d)	**4188**
12 1B.133/C(d), part	**4136, 4240, 4241**
12 1B.142/D(b)	**4172***
12 1B.142/E(b), part	**4172***
15 1B.201/F(b), part	**4164a, 4177a**

15 2B.31/A(b), part	**4199*, 4203***
15 2B.39/A(e), part	**4250**
15 2B.40/C(b)	**4276**
16 2B.45/D(b), part	**4254**
16 2B.45/F(b), part	**4259**
16 2B.47/B(c)	**4134**
17 2B.63/G(c)	**4282**
18 2B.64/F(3–4)b	**4140**
18 2B.69/E(1)b	**4294**
18 2B.69/F(2–4)c	**4295**
18 2B.70/C(1–3)c	**4219**

19 2B.75/D(d) **4201**
19 2B.78/B(1–3)c **4156a***
19 2B.78/B(4–7)c **4249**
19 2B.78/D(11–13) **4187**
19 2B.81/C(s), part **4160a**
19 2B.83/E(c), part **4159a**
20 3B.31/C(4–7)b **4156b**
20 3B.31/E(6–7)b **4287**
20 3B.31/F(1–3)b **4223**
20 3B.37/J(5)a **4248**
21 3B.23/D(1-3)b **4236**
21 3B.27/H(8)a **4277**
22 3B.19/F(a) **4253**
22 3B.20/E(1–3)b **4245**
22 3B.21/B(c) **4229**
25 3B.58/L(a) **4178**
28 4B.60/B(4–5)a **4261**
28 4B.62/E(1–5)a **4197**
28 4B.62/E(6–8)b **4278**
29 4B.42/J(4)a **4279**
30 4B.37/F(1–3)b **4280**
30 4B.39/D(1-3)a **4160**
31 4B.9/H(1–4)b **4216**
31 4B.13/H(6–8)a, part **4247**
32 4B.1/E(1-2)a **4162**
35 4B.101/G(1–3)c **4138a, 4230**
36 4B.110/H(1–3)a **4242**
36 4B.110/H(4–6)b **4264**
37 4B.104/A(1-3)a **4269**
38 3B.81/B(1–2)c, part **4156c**
38 3B.81/B(3–4)c, part **4157a**
38 3B.85/H(1)c **4137**
39 5B.121/B(1–2)a **4217**
40 5B.95/J(1–4)c **4188a**
40 5B.108/K(5–10)c, part **4266a**
40 5B.116/H(1-5)c **4260**
42 5B.72/B(9-11)a **4270**
46 5B.51/F(5–7)c **4285**
46 5B.53/G(1–5)d **4281**
47 5B.46/H(3-6)d **4239**
48 5B.107/D(1) **4168**
49 5B.98/K(1-4)a **4252**
50 4B.30/J(1–2), part **4222**
50 4B.33/J(1–5)b **4267**
50 4B.34/A(1-3)a **4176**
51 4B.19/C(1)c **4193**
51 4B.21/D(1-3)b **4152**
58/A(6) **4272**
62 6B.82/H(2-4)b **4158**
64 6B.60/G(1)a **4151, 4164**
65 6B.30/C(1)b **4205**
65 6B.30/M(1)a, part **4167*, 4173a*, 4196a, 4217b*, 4226*, 4231**
65 6B.34/G(1–2)f **4165***
65 6B.36/E(1–3)d **4195**
66 6B.26/D(1)c, part **4173a*, 4184a*, 4217b*, 4231a**
66 6B.26/K(1–2)g **4205a**
66 6B.26/K(1–2)i **4155***
66 6B.27/G(5–6)b, part **4167*, 4169**
66 6B.29/L(1–4)d **4155***
67 6B.10/L(1)a **4179**
67 6B.16/K(2–4)b, part **4167***
67 6B.16/M(4–5)b **4288**
68 6B.19/G(3)d **4289**
68 6B.19/G(3)e **4141**
68 6B.20/D(4)a, part **4159, 4165*, 4217a, 4224, 4232**
68 6B.20/D(5–8)f **4213**
68 6B.20/F(1-2)a **4275**
68 6B.20/L(14–16)a & b **4210**
68 6B.20/L(3)d **4215**
68 6B.20/L(8)a, part **4153, 4154, 4167*, 4184a*, 4194, 4196, 4209, 4217b*, 4217c, 4226*, 4233**
68 6B.22/P(1–2)a **4171**
68 6B.24/D(1–3)c **4145**
68 6B.24/D(1-3)d **4182**
68 6B.25/E(1–2) **4220**
68 6B.25/H(1-2)a **4138**
68 6B.25/H(1–2)i, part **4212***
68 6B.25/H(3)b, part **4155*, 4165*, 4167*, 4173a*, 4221, 4225, 4226*, 4234**
68 6B.25/H(3)d, part **4206, 4208, 4211, 4212***
68 6B.25/H(4)b, part **4167*, 4173a*, 4226*, 4235**
69/10(h) **4218**
69/24(c) **4266**
71/84(e) **4243**
72/29(b) **4146**
72/29(d) **4174**
72/94(a) **4173**
75/82(b) **4237**

81 2B.85/17(a) **4199***
82/87(b) **4139**
85/2(d) **4163**
85/72(a) **4255**
85/82(b) **4262**
93/Dec. 8/B1 .... **4170**
93/Dec. 11/C.3 .... **4149***
93/Dec. 11/E.7 .... **4149***
93/Dec. 16/E.1, part **4183**
93/Dec. 26/C.1 .... **4274**
93/Dec. 28/F.1 .... **4175**
93/Jan. 1/E.2 .... **4157**
96/69(a) **4180**
96/69(b) **4143**
100/3-5 .... **4142**
100/52(c) **4166**
100/65(f) **4200**
100/103(d) **4144**
100/124(a) **4133**
100/127(b) **4251**
101/106(c) **4214**
101/155(b) **4257**
101/168(a) **4258**
102/49(c) **4290**
102/58(b) **4291**
102/67(b) **4263**
102/150(a) **4292**
102/167(b) **4293**
103/14(b) **4190**
103/20(a) **4192**
103/40(c) **4147**
103/49(a) **4148**
103/70(b) **4207**
103/124(c) **4236a**
103/142(a) **4204**
104/69(a) **4265**
104/153(b) **4203**
104/171(a) **4273**
105/117(d) **4283**
105/147(c) **4189**
106/80(a) **4198**
106/84(a) **4286**
114/68(b) **4256**
115/81 .... **4165a**
115/92(b) **4135**
118/9(a) **4191***
118/11(b) **4177**
119/71(b) **4284**
125/25(j) **4296**
A20/1H **4186**
A22/C251 .... **4191***
A22/C287 .... **4227**
A31/10 .... **4300**
A46/2F **4271**
A141/6A **4268**
A145/6H **4161**
A188/3F **4156a*, 4299**
A188/4E **4244**
A387/5D **4150**
A604/5A **4269a**
A608/32A **4300a**
6153/C10 .... **4297**
6153/C11 .... **4298**
10B4/C319 .... **4246**
B3.6/C320 .... **4238**

PLATE I

4137

4133

PLATE II

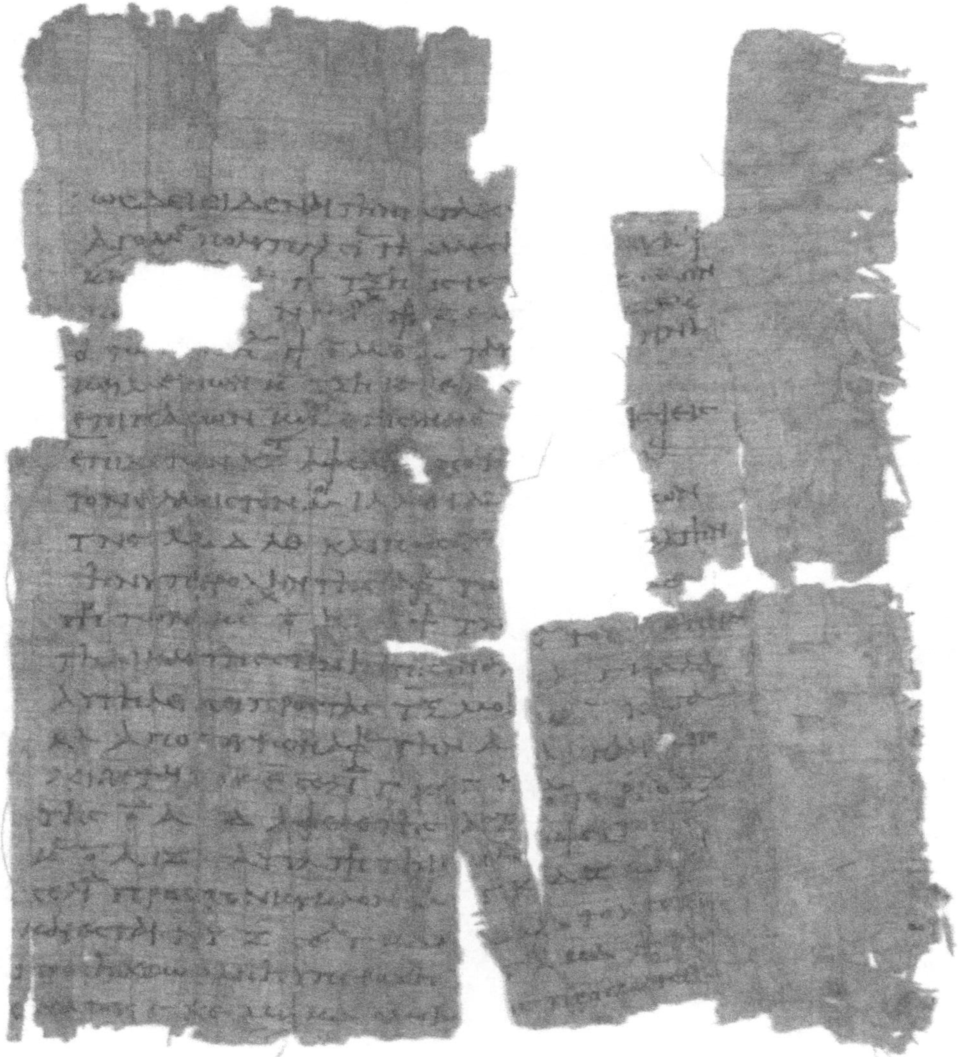

4136

4144

4139

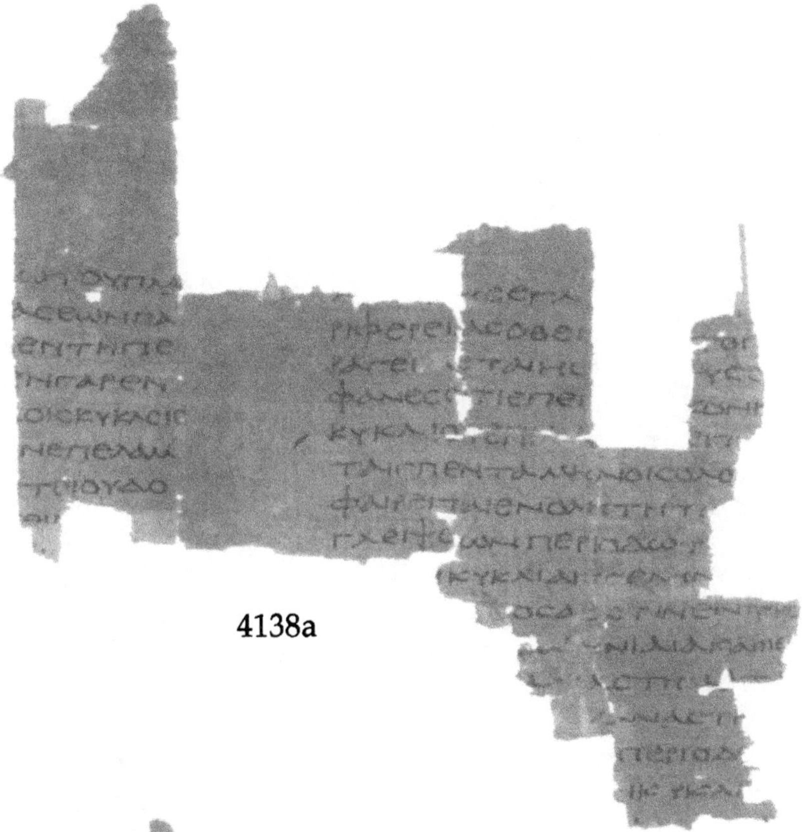

4138a

4152

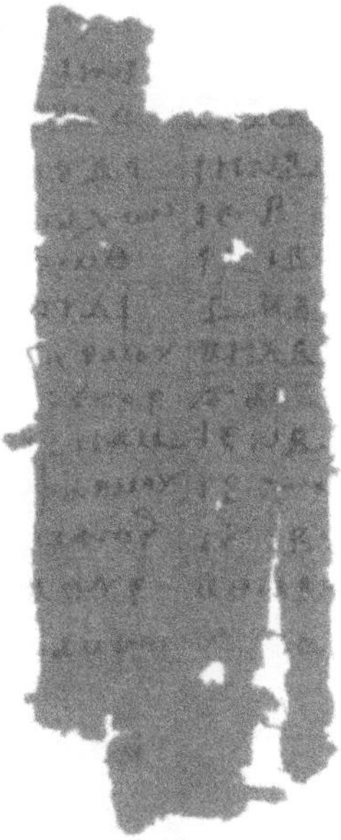

4154

4155

PLATE IV

4163 (verso)

4161

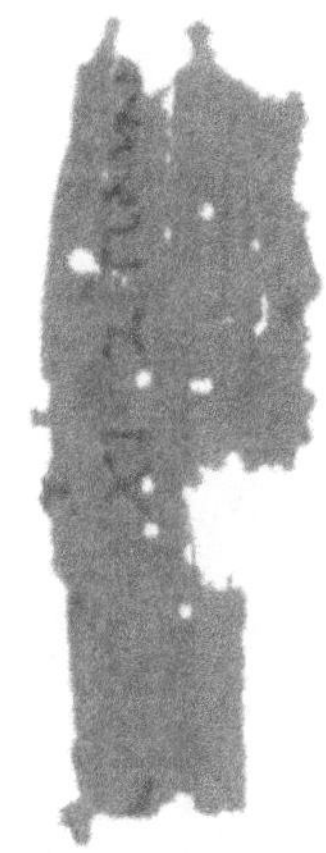

4157

4158 fr. 1

4205 (verso)

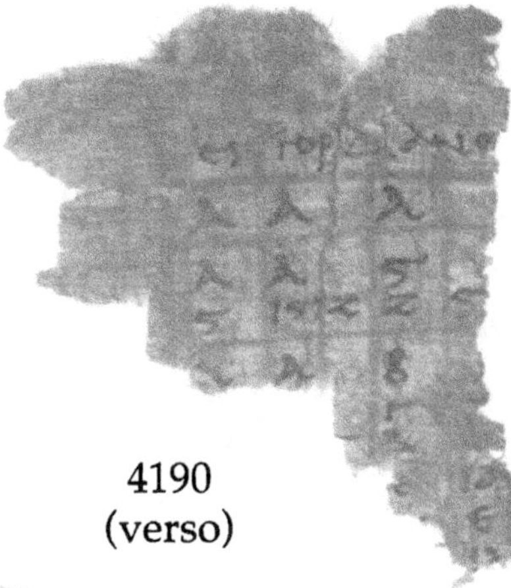

4190 (verso)

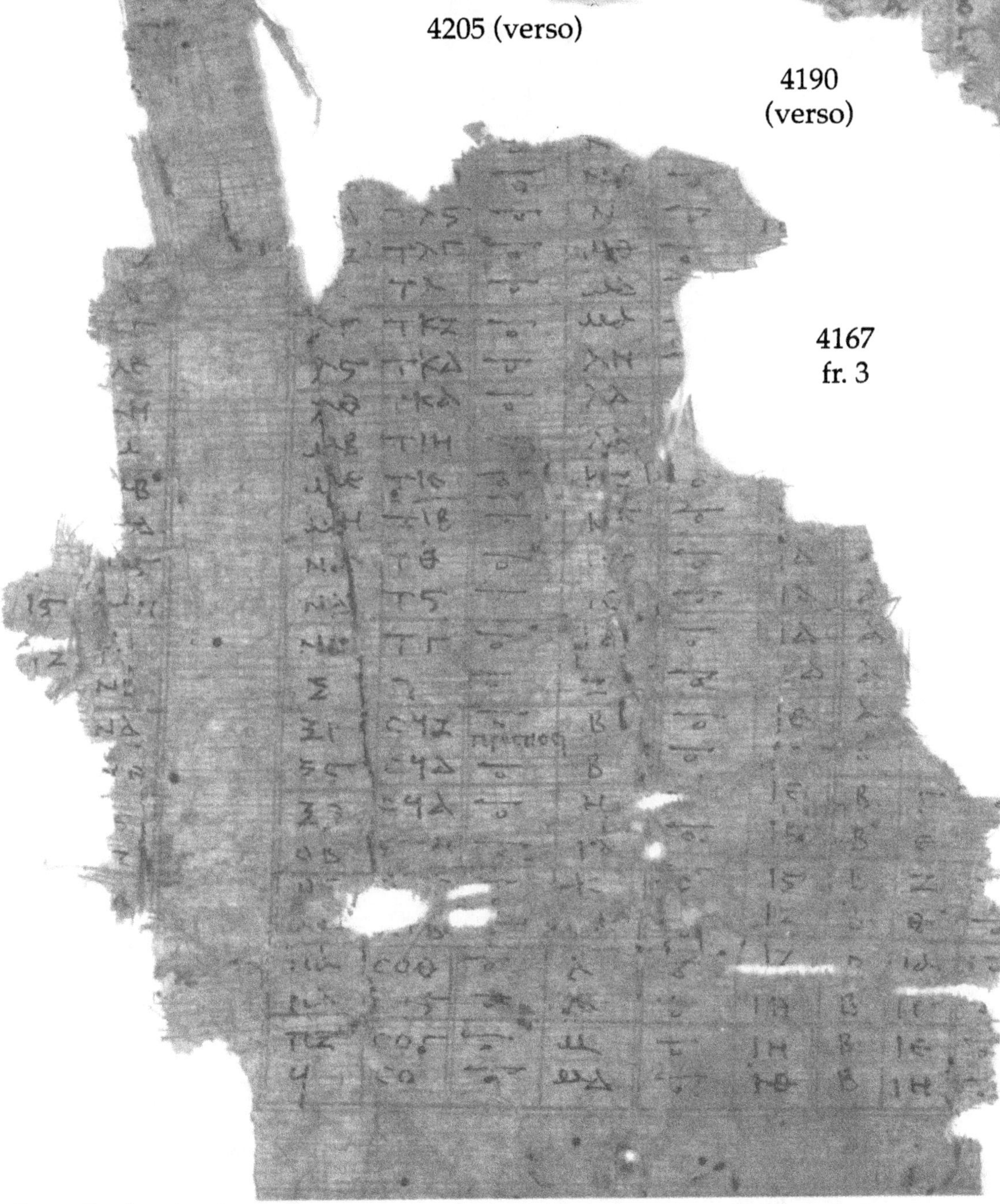

4167 fr. 3

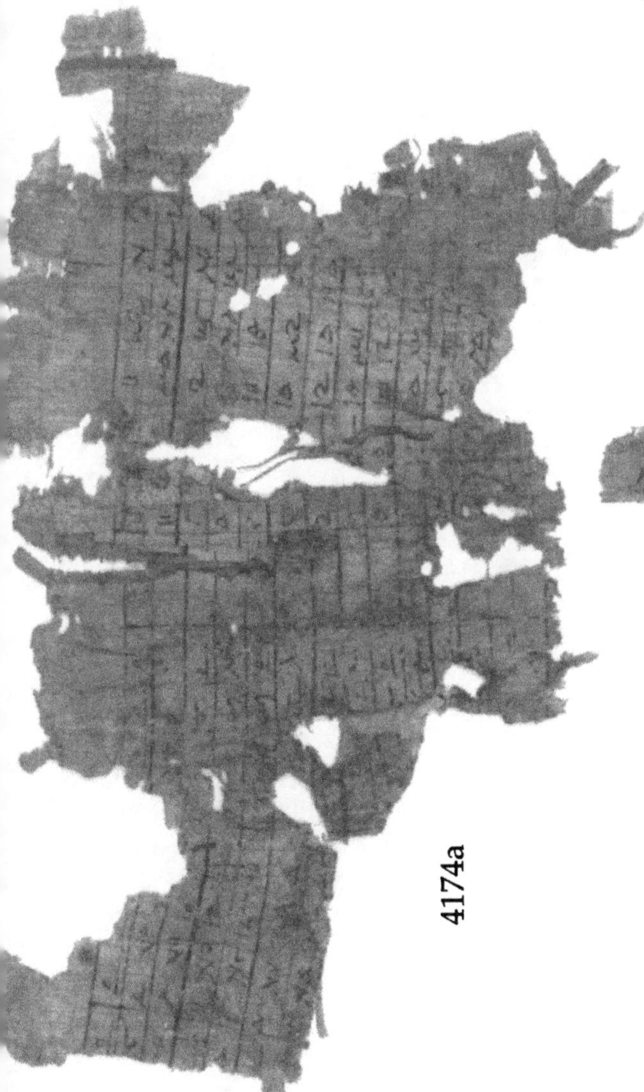

4174a

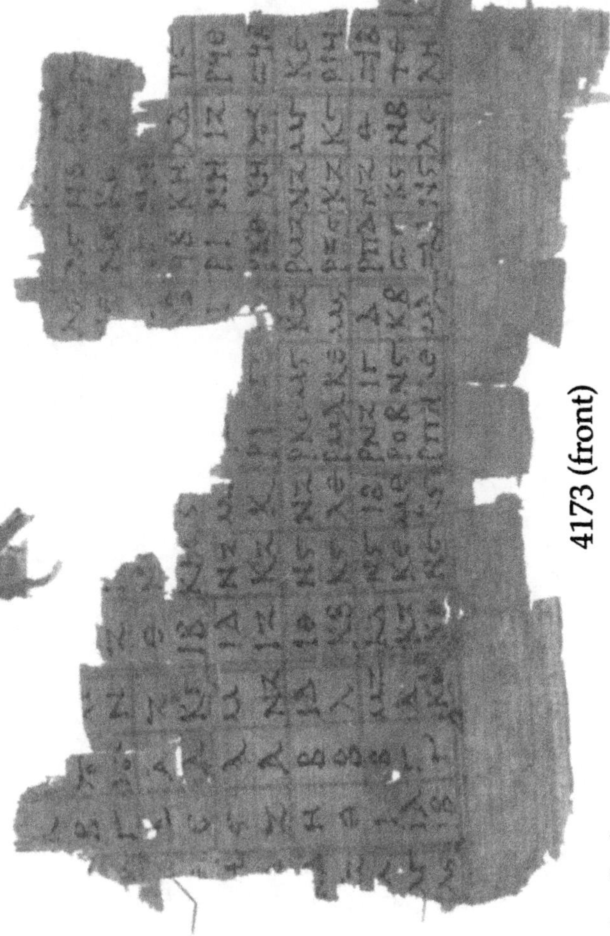

4173 (front)

4165 fr. 1

PLATE VII

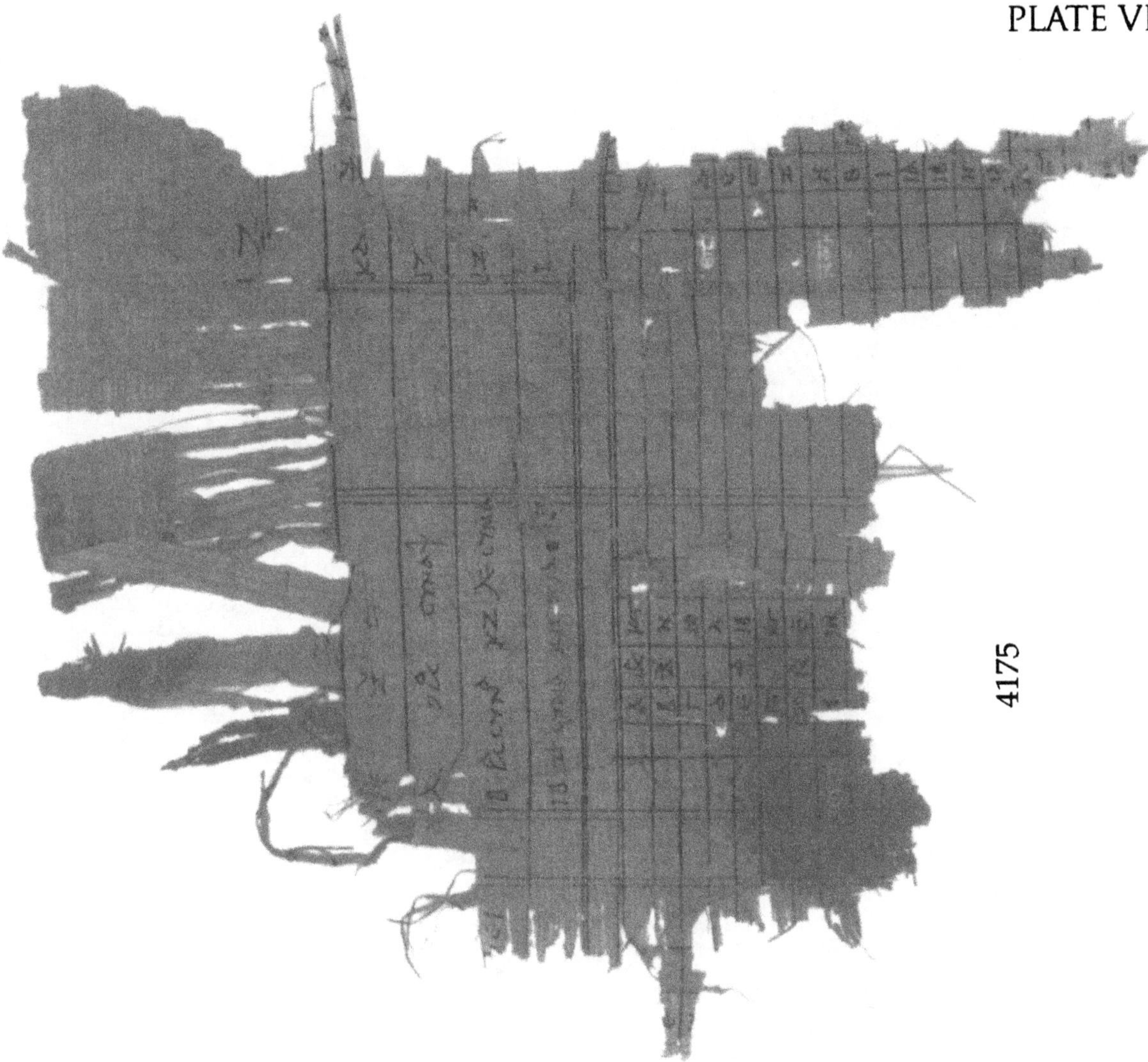

4175

4191 fr. 1
(recto)

4226 fr. 2

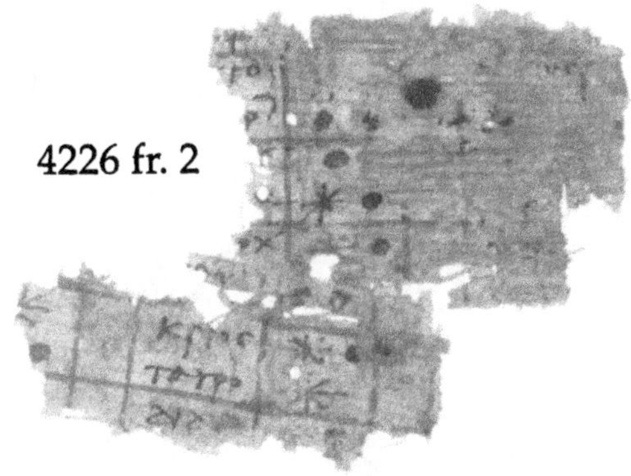

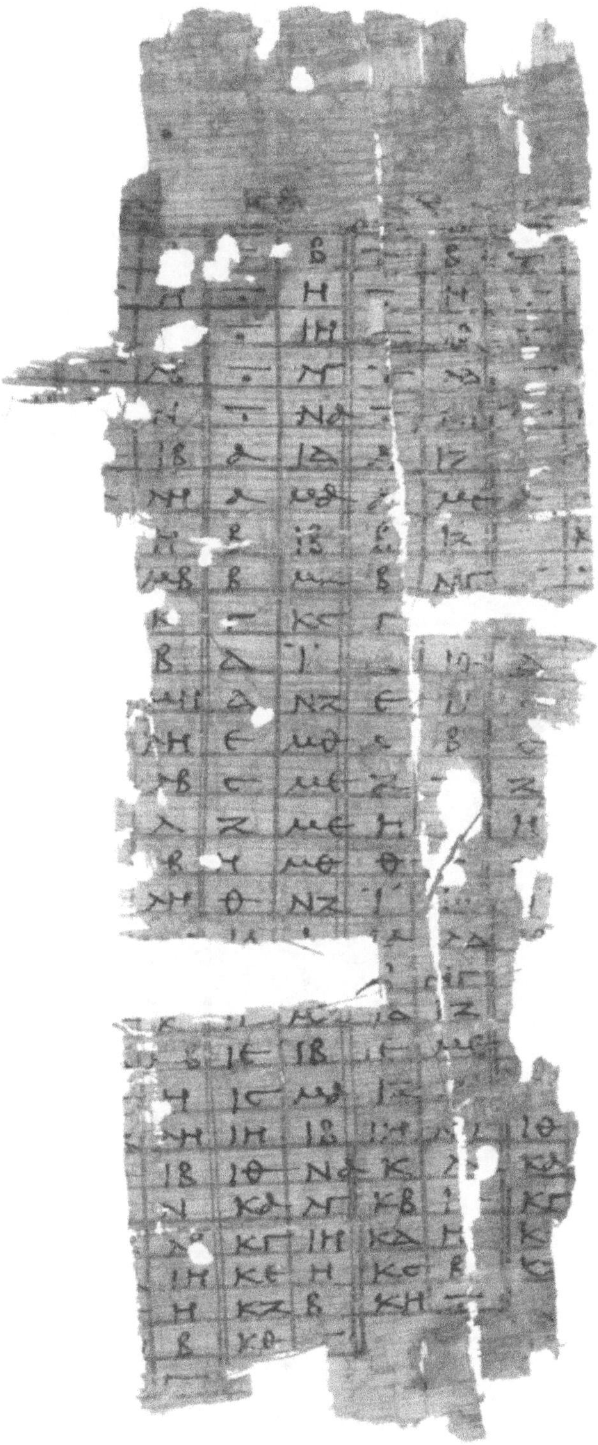

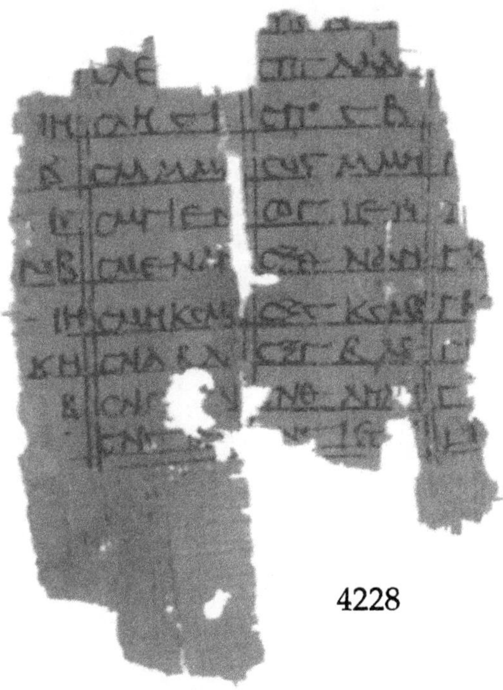

4228

4217

4140

4276

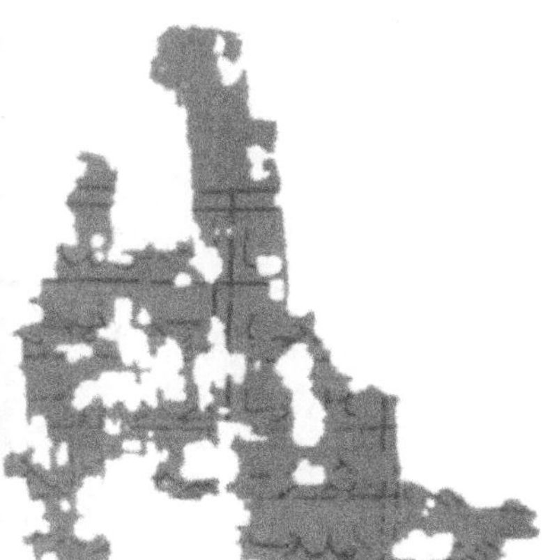

4160a

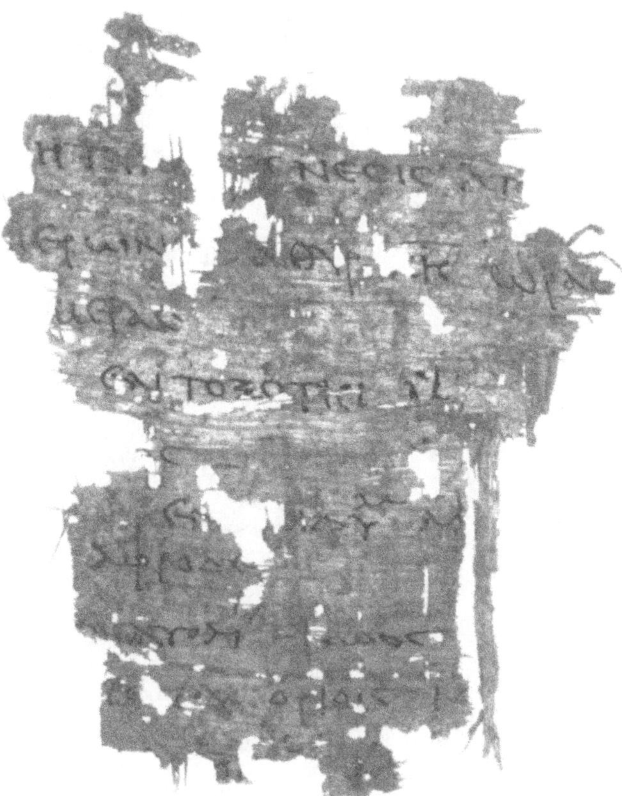

4236

PLATE X

4203a

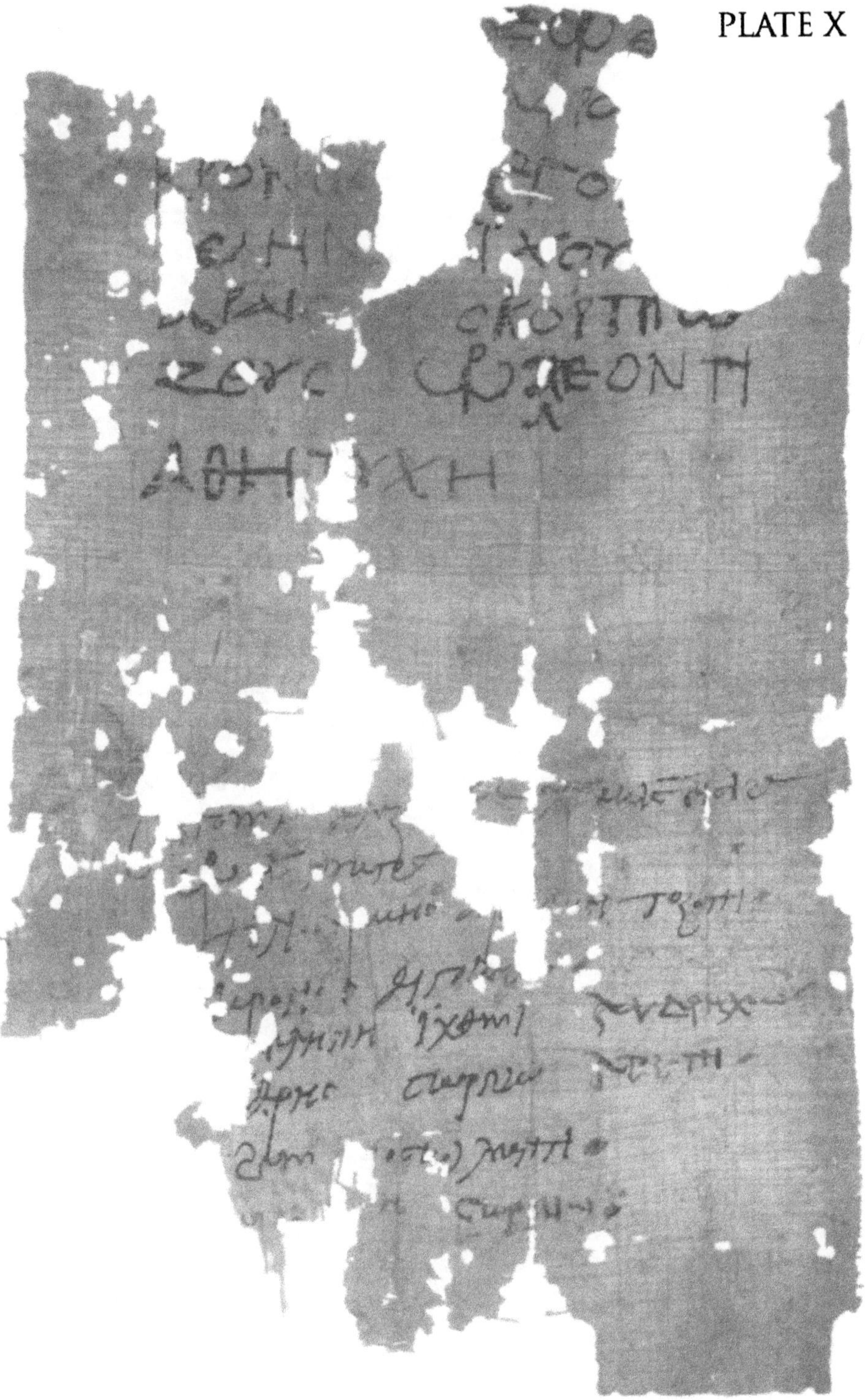

4247

PLATE XI

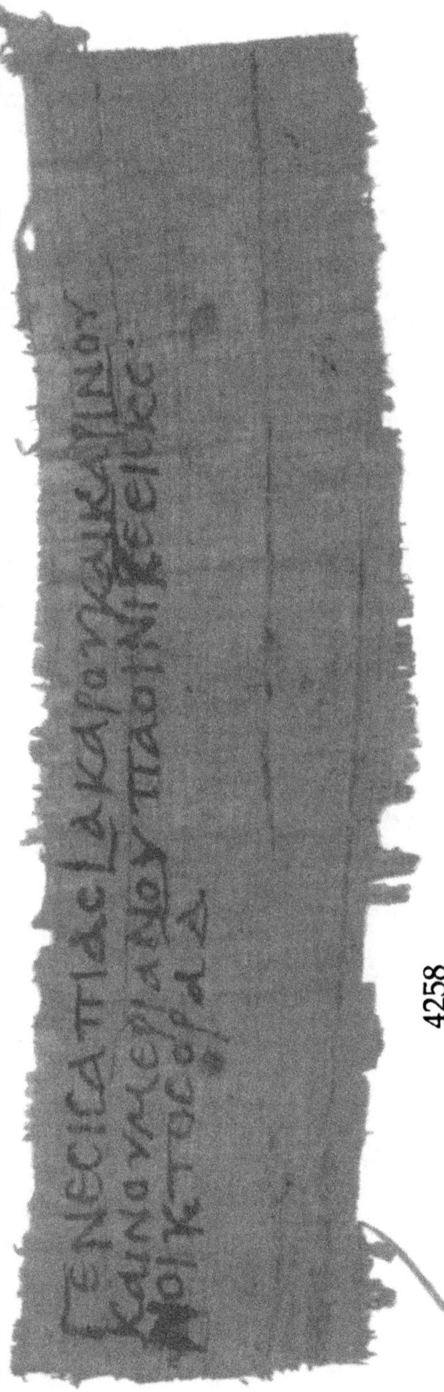

4258

4274 (→)

4138

4275

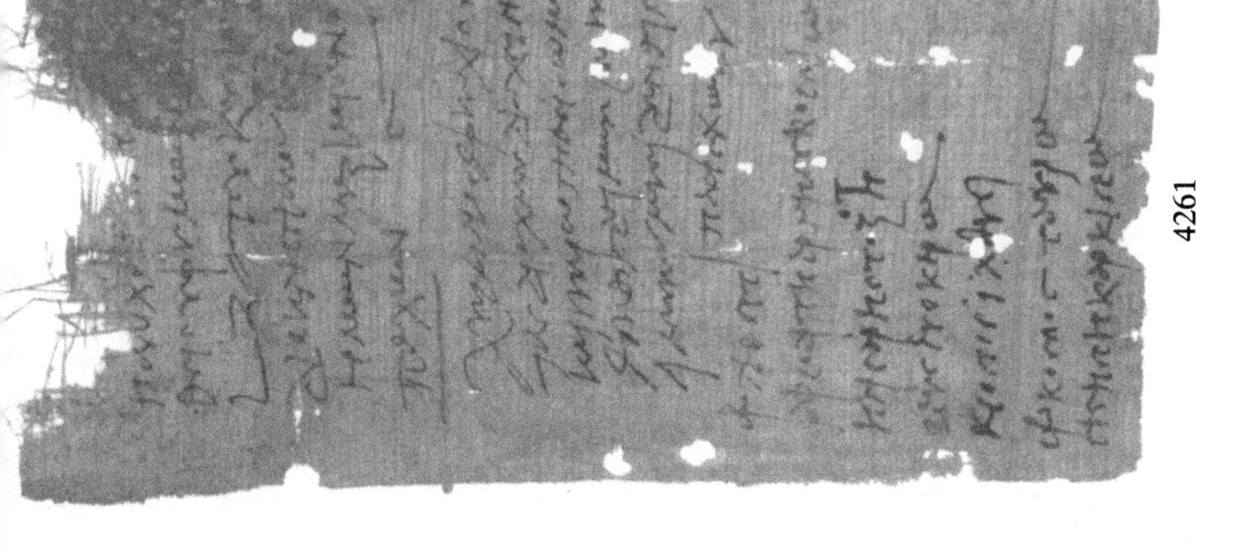

4261

www.ingramcontent.com/pod-product-compliance
Lightning Source LLC
LaVergne TN
LVHW081943200926
848484LV00009B/2504